Materials Modification by High-fluence Ion Beams

NATO ASI Series

Advanced Science Institutes Series

A Series presenting the results of activities sponsored by the NATO Science Committee, which aims at the dissemination of advanced scientific and technological knowledge, with a view to strengthening links between scientific communities.

The Series is published by an international board of publishers in conjunction with the NATO Scientific Affairs Division

A Life Sciences	Plenum Publishing Corporation
B Physics	London and New York
C Mathematical	Kluwer Academic Publishers
and Physical Sciences	Dordrecht, Boston and London
D Behavioural and Social Sciences	
E Applied Sciences	
F Computer and Systems Sciences	Springer-Verlag
G Ecological Sciences	Berlin, Heidelberg, New York, London,
H Cell Biology	Paris and Tokyo

Materials Modification by High-fluence Ion Beams

edited by

Roger Kelly

IBM T. J. Watson Research Center,
Yorktown Heights, NY, U.S.A.

and

M. Fernanda da Silva

Laboratório Nacional de Engenharia
e Tecnologia Industrial, Sacavem, Portugal

Kluwer Academic Publishers

Dordrecht / Boston / London

Published in cooperation with NATO Scientific Affairs Division

Proceedings of the NATO Advanced Study Institute on
Materials Modification by High-fluence Ion Beams
Viana do Castelo, Portugal
August 24 – September 4, 1987

Library of Congress Cataloging in Publication Data

Materials modification by high-fluence ion beams / edited by Roger
Kelly and M. Fernanda da Silva.
        p.   cm. -- (NATO ASI series. Series E, Applied sciences ; no.
155)
    Proceedings of an institute held in Viana do Castelo, Portugal,
Aug. 24-Sept. 4, 1987.
    "Published in cooperation with NATO Scientific Affairs Division."
    Includes index.
    1. Materials--Effect of radiation on--Congresses. 2. Ion
bombardment--Industrial applications--Congresses.   I. Kelly, Roger.
II. Silva, M. Fernanda da (Maria Fernanda), 1938-   . III. North
Atlantic Treaty Organization. Scientific Affairs Division.
IV. Series.
TA418.6.M377 1988
620.1'1228--dc19   88-31612
                                                                          CIP

ISBN-13: 978-94-010-7063-8   e-ISBN-13: 978-94-009-1267-0
DOI: 10.1007/978-94-009-1267-0

Published by Kluwer Academic Publishers,
P.O. Box 17, 3300 AA Dordrecht, The Netherlands.

Kluwer Academic Publishers incorporates the publishing programmes of
D. Reidel, Martinus Nijhoff, Dr W. Junk, and MTP Press.

Sold and distributed in the U.S.A. and Canada
by Kluwer Academic Publishers,
101 Philip Drive, Norwell, MA 02061, U.S.A.

In all other countries, sold and distributed
by Kluwer Academic Publishers Group,
P.O. Box 322, 3300 AH Dordrecht, The Netherlands.

TABLE OF CONTENTS

PREFACE

This volume is the proceedings of a NATO Advanced Study Institute held at the Hotel do Parque in Viana do Castelo (Portugal). The school was directed by Roger Kelly (U.S.A.) and M. Fernanda da Silva (Portugal), and the organizing committee consisted of Harry Bernas (France), George Carter (U.K.), Paolo Mazzoldi (Italy), and Edmund Taglauer (Germany). The dates were 24 August to 4 September, 1987.

The School was conceived to satisfy the needs of those who study particle-surface interactions at high fluence, i.e. under conditions leading to chemical, electronic, mechanical, and structural changes rather than just doping. This community of scientists is surprisingly large even though demonstrated applications are still relatively few. As such, the community is as much engaged in pure research as in applied work. It is very highly developed in Western Europe but less so in North America. This interesting state of balance has a number of causes, not the least of which is the way in which science is funded in North America: public relations plays a large role.

The idea of organizing the School took shape at a related School on "Erosion and Growth of Solids Stimulated by Atom and Ion Beams" held in Crete in September, 1985. Portugal was chosen by virtue of being a NATO country with still embryonic science but in a state of vigorous self-help. Two Portuguese students, Eduardo Alves and Rui da Silva suggested that Fernanda da Silva of the national laboratory at Sacavem be the Co-director, while Fernanda got unofficial help from Jose Soares and Manuel da Silva of the Universidade de Lisboa. The School secretary was Ofelia Ferreira from the national laboratory at Sacavem.

Viana do Castelo, in northern Portugal near the Spanish border, was chosen over better known locations to avoid the atmosphere which prevails when the beaches are covered with tourists, especially topless tourists. In addition it had one of the very few hotels in northern Portugal able to host a School such as ours: the Hotel do Parque. The owner of the hotel (Mr. Crispim) helped us to plan the logistics of the School and organized meetings with the Bürgermeister (Mr. Araujo) and the head of tourism for the province of Alto Minho (Dr. Sampaio). An exceptional contribution was made by Jose and Olimpia de Souza, who helped us establish our four School outings and in general made both the organizational visit and the School itself agreeable experiences. Jose is the Manager of the Hotel do Parque, while Olimpia teaches biology at the local secondary school.

The main part of the financing of the School came, of course, from the Scientific Affairs Division of NATO. We were completely overwhelmed, however, by the extent to which Portuguese institutions helped us. The list, with somewhat simplified spelling, is as follows: (1) Fundacao Calouste Gulbenkian, (2) Lab. Nac. de Engenharia e Tecnol. Industrial, (3) Junta Nac. de Investigacao Cientifica e Tecnol., (4) TAP Air Portugal, (5) Inst. Nac. de Investigacao Cientifica, (6) Hotel do Parque (Viana do Castelo), (7) Camara Municipal de Viana do Castelo, (8) Governo Civil de Viana do Castelo, (9) Camara Municipal de Ponte de Lima, (10) Regiao de Turismo do Alto Minho (Costa Verde), (11) Inst. Politecnico de Viana do Castelo, (12) Adega Cooperativa de Ponte de Lima, (13) Banco Portugues do Atlantico, (14) Tecnodidactica, (15) Rank Xerox, (16) FOC Escolar, (17) Canon-Copicanola.

Other funding, in some cases unexpected, was as follows: (1) Alcan Research and Development Center (Canada), (2) Natural Sciences and Engineering Research Council (Canada), (3) Danfysik (Denmark), (4) Leybold and Siemens (F.R.G.), (5) MPI für Plasmaphysik (F.R.G.), (6) IBM (U.S.A.), (7) Ionwerks (U.S.A.), (8) National Science Foundation (U.S.A.), (9) Office of Naval Research, London Branch (U.S.A.; U.K.), (10) Oxford Instruments (U.K.).

Roger Kelly
IBM Research Center
Yorktown Heights, NY 10598, U.S.A.

M. Fernanda da Silva
Laboratorio Nacional de Engenharia
    e Tecnologia Industrial (LNETI)
2685 Sacavem, Portugal

June 1988

DEDICATION TO BILL GRANT

This volume is dedicated to the memory of Bill Grant, who in the original plan of the School was to have been one of the lecturers. The following notes were written by George Carter (Salford Univ.) and transmitted to us by Robin Procter (Univ. of Manchester).

Bill Grant died at the tragically early age of 46 on 23 October 1987 after a year of long and debilitating illness. He bore this with fortitude and courage, and maintained a lively and committed interest in his work right until the end.

Bill obtained his first degree and then his PhD in Electrical Engineering at Liverpool University, where he first became involved in research in Atomic Collision Processes in Solids. After a brief lectureship appointment in Liverpool he moved to Salford University in 1968, where he started developing and using medium and high-energy ion accelerators for property modification and for analysis of solids. Under his guidance, Salford was the first U.K. University to acquire a, then, state-of-the-art implanter. Initially his interests were associated with semiconductor substrates but then he made a major move and concentrated on metal systems. He developed collaborative contacts and lasting friendships with Robin Procter and Vic Ashworth at the Univ. of Manchester Institute of Science and Technology, and collectively they undertook early definitive work on the use of implantation in both understanding and controlling aqueous corrosion. It was also through this contact that the International Conference series on Surface Modification of Metals by Ion Beams began. His other major interest was in phase changes induced by ion implantation and again he was involved in the early definitive work on crystalline-amorphous transitions induced by metalloid implantation into metals.

In addition to his own research work, Bill participated fully in the professional and institutional activities related to his work. This included being chairman of the Atomic Collisions in Solids group for the Institute of Physics and being the deputy editor for the journal Vacuum. His publication list was extensive and included a popular text on Ion Implantation in Semiconductors written with George Carter. He was the Director of the Thin Film and Surface Research Center at Salford.

In recognition of his research and international standing he was appointed Reader and then, in 1984, Professor of Ion Beam Engineering at Salford. Sadly, his ambitions to further develop and support research have been frustrated by his untimely death.

Bill not only had a wide circle of national and international colleagues and friends but also a happy and fulfilling family life. In addition to his widow, Mary, he leaves three teenage children, all with bright and hopeful futures because of Bill's dedication and care, and who, like all his colleagues, will miss him badly.

AUTHOR INDEX

OVERVIEW

HIGH—FLUENCE ION IRRADIATION, AN OVERVIEW

G CARTER, I V KATARDJIEV AND M J NOBES
University of Salford, Dept. of Electronic and Electrical Engineering,
Salford M5 4WT, U.K.

1. INTRODUCTION

Historically the effects of high—fluence ion beams upon materials modification were observed, and partly recognised as such processes, long
before lower—fluence investigations were attempted or fashionable. The
disintegration of cathodes in X-ray tubes observed by Grove (1) over
130 years ago and a nearly contemporary observation of gas disappearance
from electrical discharge tubes by Plücker (2) were the earliest recorded
instances of high fluence sputtering and of ion implantation respectively.
Since such devices were, for many years, most common sources of intense
ion fluxes it is hardly surprising that much of the early understanding,
and in many cases misunderstanding, of the effects of ion bombardment on
materials and their properties derived from studies in such systems.
Three major factors changed this situation. Firstly the fundamental
interest of Niels Bohr (3) in the passage of swift particles through
matter required the elimination of at least one variable from consideration
in physical description of the processes. This variable was incident
particle fluence (and, by association, flux density) since the collision
kinematics of individual incident particles only were considered and the
cooperative effects of many incident particles were assessed by linear
superposition of individual particle effects, fully recognising the
statistical nature of these events. This dilute medium approach tradition
was followed with great success, by Lindhard (4) and his colleagues and
many others (5)(6), in predicting such important parameters as ion slowing
down ranges in solids, energy deposition (and related defect production
rate) profiles, ion reflection and surface atomic ejection or sputtering.
Equally importantly experimental techniques had developed sufficiently
well that both highly controllable ion accelerators and sensitive methods
for measurement of low concentrations of impurities in solids and atomic
ejection from solids were also available. The measure of agreement between
theory and experiment was, and continues to be, quite excellent and, since
the early 1960's such approaches have been complemented by increasingly
more complex computer simulation studies. Interestingly the first
important clue to the reality of atomic lattice steered particle motion
(channelling) derived from simulations (7).
Thirdly, the late 1960's heralded the beginning of very large scale
integration concepts, and later devices, for semiconductor electronic
applications and these demanded doping levels which were not only quantitatively and geometrically precise and predictable but, in atomic concentration terms, were low (<< 1%). Low fluence ion implantation was quickly,
but not without initial problems, found to fulfil such needs admirably
and supporting theory, experiment and simulation have continued much in
demand.
This is not to say that higher fluence processes have been ignored and,
in such areas as ion pumping, high rate sputtering sources for film growth,

3

R. Kelly and M. Fernanda da Silva (eds.), Materials Modification by High-fluence Ion Beams, 3–27.
© *1989 by Kluwer Academic Publishers.*

impurity depth profiling of semiconductors and other materials, lithographic patterning or etching of semiconductor and other devices and first wall fusion reactor technology, interest and application has grown rapidly. High fluence ion irradiation has also continued to play a significant role in simulation of other radiation situations such as those occurring in fission reactor materials, in solar wind and other extra-terrestial phenomena and in analysis of organic and biomolecular materials. More recently, increasing attention has been paid to the use of ion, and indeed other directed energy beams such as photon beams (lasers) and electron beams, for modifying the properties, either by direct implantation or by beam mixing of the atomic constituents, of solids, including semiconductor substrates. These areas include processes in which the solid boundary is virtually static during irradiation, or recedes by etching or grows by deposition during irradiation and so surface effects are equally important to bulk effects. In such areas property changes occur as a result of $>1\%$ modification of composition and/or structure and therefore require high fluence irradiation conditions.

The underlying theory and detailed physical mechanisms involved in all of the above situations is only just emerging and discussions of these and a significant sample of the current and potential applications of high fluence ion irradiation are presented in these Proceedings. It is hardly surprising that problems in the high-fluence regime are more difficult to treat than low fluence, essentially individual particle, events since cooperative phenomena, by their very nature, are more intractable. Nevertheless progress is being made, as evidenced by the accompanying papers, each of which analyses specific topics in great detail, rendering it pointless and ineffective to rehearse or repeat the arguments in this overview. Instead we will confine our attention to attempting to answer the simple but fundamental question of what is meant by high fluence.

Ion, and indeed many other forms of irradiation, modify both the structure and the composition of the substrate upon which they are incident. The former occurs generally, but not exclusively through the generation and transport of atomic recoils and defects (which may be charged or uncharged), electrons and holes, phonons and photons. Incorporation of atomic species, other than native to the substrate, by external implantation or internal transmutation can also modify structure although the primary effect will be one of composition modification. However, in the case of implantation each individual ion can introduce at most only one additional moiety to the substrate (fragmented molecular ions introduce more than one atom of course) whereas, in general, each ion can introduce orders of magnitude more recoils and defects. It is immediately transparent that high ion fluence for spatially extensive restructuring of a substrate may be substantially less than high ion fluence for composition modification and this distinction will be explored in the third and fourth sections of this communication. In the restructuring area the incident irradiation may be regarded as a catalyst which promotes a reaction. In the composition modification area the irradiation must be regarded as a participant in the reaction. Under both circumstances irradiation tends to perturb the substrate system from its local thermodynamic equilibrium and such perturbation can be impeded, or enhanced, by thermodynamic influences and driving energies and their gradients (forces) and the nature and extent of these processes for a variety of systems will also be explored in the fourth section. However, the time and space evolution of these effects will be intimately determined by whether the

substrate surface is static or moving during irradiation and the speed
of surface motion will control the rate and integrated extent of structure
and composition modifications and their dependence on fluence and flux
density. These problems will be addressed in the third section.

Most importantly perhaps is that each individual ion may be regarded,
over a limited spatial volume mediated by its passage through the substrate as, effectively, a low or a high-fluence process dependent upon
the local density of events it creates. The next section will therefore
devote, firstly, attention to individual ion impact, and then explore the
requirements that the sequential effects of many impacts may be, to first
order, regarded as linearly additive. In this section, as in later
discussion, no attempt will be made to give detailed quantitative analysis,
since this appears elsewhere in these proceedings, but concepts will be
developed with simplified quantitative arguments. Exemplification of these
concepts will be made via reference to other presentations here and in
the literature.

2. LINEARITY AND SUPERPOSITION

2.1 Individual Ion Impacts

An individual ion slows down, during penetration of a substrate, via
a sequence of atomic collisions, transferring ballistic energy to complete
atoms and excitation energy to the electronic subsystems. These energy
loss and transfer mechanisms are usually considered, as a good first
order approximation, to be separable and uncoupled. The dominant energy
transfer mechanism depends upon the reduced energy, ε, (4) in a single
collision and is ballistic (elastic) for $\varepsilon \leq 0.1$. ε depends upon the
actual kinetic energy E, and the atomic numbers and masses of the colli-
sion partners and increases with E but decreases with increasing values
of the other atomic parameters. For relatively low energy, reasonably
heavy ions, most of the energy transfer initially results in enhanced
atomic motion which couples slowly into the electronic system of the
substrate. For electron and photon irradiation, however, first energy
transfer is into the electronic system which then couples into the atomic
system.

Models and analysis (3-6) of projectile slowing down and energy transfer to substrate subsystems which employ this sequential collision
approach, even when energy loss is considered to be quasi continuous
rather than discrete, are based upon a linear superposition of sequential
events in that the progeny of any one collision event do not influence
time sequential later collisions. This amounts to considering, always,
collisions only between pairs of moving and stationary atoms which renders
mathematical analysis tractable. The more general case of collisions
between individual moving atoms can be given formal analytic description
but solution is only feasible using numerical computational methods,
and even then with considerable cost. An alternative to employing
numerical computation of a theoretical analysis of moving, interacting
atomic systems is to employ, **ab** initio, computational simulation of such
systems using Molecular Dynamics Codes (8,9) which again, for many-
atom systems are expensive. Sequential collision or binary codes are,
often, good approximations to such simulations (10). As pointed out by
Williams (11) the linearisation approximation is equivalent to a dilute
collisional system in which the density of moving atoms, n, is a small
fraction of the local atomic density N, i.e.

$$\frac{n}{N} \ll 1. \tag{1}$$

When this level of approximation is employed it turns out that the production rate of atomic recoils, of excited electrons and of sputtered surface atoms are all linear functions of energy deposition density in the bulk or the surface and many experiments reveal the validity of those predictions (12,13,14). When, however, local energy deposition densities are large, experiments reveal (12-14) that such parameters no longer scale linearly with the appropriate local energy deposition density and a 'collision spike' is advocated. Whatever the fine details of such a spike, and opinions and descriptions vary (15,16,17,18), they must involve, at some stage of their evolution, momentum or atomic mass transport as well as energy transport. Such spikes are therefore better described, fundamentally, by hydrodynamic concepts than by thermodynamic approaches, although, during later time stages of spike evolution where energy is shared by increasing numbers of atoms, a thermal diffusion model driven by energy gradients may be a good approximation (19,20). Such an approach is implicit in spike mediated atomic and defect migration processes which lead to local phase change phenomena in models such as those of Johnson (21), Peak and Averback (22) and Hobbs(23).The correlation between low collision density $\frac{n}{N}$ and energy deposition density is readily established by using the results of the linearised (cascade) theory. This theory indicates (24,25) that the number of recoils generated by a projectile of energy E is given by $\frac{kf(E).E}{E_d}$, where k is a constant of order unity, f(E) is the fraction of the energy E dissipated in atomic recoil motion and E_d is an average 'adiabatic' displacement energy which a recoil must receive in order to travel sufficiently far in the substrate to be stable against instantaneous recombination (strain relief) with its associated vacancy. These recoils are generated over a spatially limited atomic volume, Ω, which is also predictable from linearised theory. The collision density $\frac{n}{N}$ is therefore equivalent to $\frac{kf(E).E}{E_d\Omega N}$ and the linearity criterion becomes

$$\frac{kf(E).E}{E_d\Omega N} \ll 1.$$
(2)

Writing $\frac{kf(E).E}{\Omega N}$ as the mean energy per atom, θ, in the collision volume leads to the criterion

$$\theta \ll E_d$$
(3)

E_d is an adiabatic energy, usually substantially larger than any isothermal equivalent energy for atomic motion (interstitial-vacancy formation and migration, heat of melting per atom, heat of sublimation per atom) and so an acceptable equivalent to equation (3) is that

$$\theta \lesssim U$$
(4)

where U is the appropriate isothermal energy parameter. This is the form of criterion developed by Sigmund (26), and given analytic form through calculation of Ω etc., to describe the boundary between linear cascade and spike events. It must be regarded as guiding rather than exact since it employs parameters only easily calculated from linearised theory.

It is therefore evident that, even in individual impact events, large collision densities may ensue which may, or may not, be equivalent to the time integrated effect of a number of lower collision density events occupying the same spatial region and generated by a higher fluence of individual impacts. Distinction between the individual event and time integrated events seems to be of particular importance in describing certain phase-change phenomena in different substrate systems such as local crystal-amorphous phase transitions in a range of semiconductor substrates (27) and some alloys (28-30). In some of these systems dense collision cascades appear to lead directly to local collapse to amorphousness whilst much lower density events (stimulated by electron or light ion irradiation) even when fluence accumulated lead to dissimilar structural transformation. In other systems, which may be composition dependent, dense and weak (or even no) cascades, when accumulated, lead to similar transitions. It is worth noting here that, in the linear approximation, the local defect density is directly proportional (through $\frac{1}{E_d}$) to the collision density and so a collision density criterion becomes equivalent to a defect concentration criterion which may be the more appropriate requirement for structural transformation, not only in semiconductors (31) but in elemental metals also (32-34).

Because of these clear experimental differences in system behaviour as a function of local collision density it is now important to consider the effects of successive impacts and the necessary conditions for linear superposition of events.

2.2 Multiple Ion Impacts

Multiple ion impacts onto a surface, effected by either a continuous or pulsed ion beam, must properly be regarded as a stochastic process with random impact of individual ions in space and time over the bombarded area. Except for very low ion flux densities this level of generality of description is usually unnecessary and it is acceptable to time average the flux density, J, at a given instant but to retain the random spatial distribution of impacts. If each ion permanently modifies the state of a zone, volume Ω, extending over a mean depth Z from the surface, then the time rate of change of the fraction, f_v, of the solid volume from the surface to depth Z which is modified, is simply,

$$\frac{df_v}{dt} = \frac{J\Omega}{Z} (1 - f_v) \tag{5}$$

with the straightforward solution,

$$f_v = 1 - \exp. \left(- \frac{J\Omega.t}{Z}\right) \tag{6}$$

This is the well known result (35) which describes the evolution of total amorphousness in ion implanted semiconductors in which each (heavy) ion directly and permanently transforms a local zone to this state. The parameter $\frac{J\Omega}{Z}$ is an inverse time constant for the process and is indicative of the time for which modification is approximately linear in time or the time at which individual volumes begin to intersect spatially. Equation (6) also describes the area fraction, f_A, modified by successive impacts and if the approximation is made that $\frac{\Omega}{Z} \approx \sigma_o$, where σ_o is a cross-sectional area for modification per ion impact then

$$f_A = 1 - \exp (- J\sigma_o t) \tag{7}$$

In a linear (low fluence) approximation this expression reduces to

$$f_A \simeq J\sigma_o t = \Phi \sigma_o \tag{8}$$

where Φ = Jt = total fluence

Equation (8) indicates that successive ion impacts, which lead to permanent local changes over an area σ_o, interfere spatially after a fluence

$$\Phi = \frac{1}{\sigma_o} \tag{9}$$

If only one impact is required to effect permanent local modification then interfering subsequent impacts create no further change and so equation (9) indicates a fluence level above which linearity with ion fluence (or in other words linear superposition) of extensive modification becomes invalid.

This, however, is a very special case since it was assumed that local modifications were both instantaneous and permanent. Any individual ion will develop a time evolving zone of ballistically energised atoms and electrons, which subsequently relaxes by both athermal and thermally activated processes. Such processes may extend or diminish the spatial volume over which modification has occurred by, for example, defect transport, accumulation and annihilation. A very simplistic description of the zone modification process is to assume, as above, initial quasi instantaneous zone generation over a cross-section σ_o and subsequent zone quenching which extends or reduces the zone area to a maximum σ_L or minimum σ_S value via an exponential growth or decay process with unique time constant τ. In a time interval dt at t after commencing irradiation, the area fraction of modified substrate then changes by introduction of further local zones and expansion or contraction of zones formed at t' in the interval $0 < t' < t$. In the low fluence limit, described above, zone interference may be ignored, and the defining equation for f_A becomes,

$$\frac{df}{dt} = J \left\{ \sigma_o + \frac{(\sigma_L - \sigma_o)}{\tau} \int_o^t dt'. \exp\left(- \frac{t-t'}{\tau}\right) \right\} \tag{10a}$$

for area expansion, and

$$\frac{df}{dt} = J \left\{ \sigma_o - \frac{(\sigma_o - \sigma_S)}{\tau} \int_o^t dt'. \exp\left(- \frac{t-t'}{\tau}\right) \right\} \tag{10b}$$

for area contraction.

These expressions are readily integrated to yield

$$f(\text{expansion}) \equiv f_e = J\sigma_L \left\{ t - \tau(1-e^{-t/\tau}) \right\} + J\sigma_o \tau(1-e^{-t/\tau}) \tag{11a}$$

and

$$f(\text{contraction}) \equiv f_c = J\sigma_S t + J(\sigma_o - \sigma_S)\tau(1-e^{-t/\tau}) \tag{11b}$$

For irradiation times $t \gg \tau$, equations (11a) and (11b) relax, respectively, to

$$f_e = \Phi\sigma_L - J(\sigma_L - \sigma_o)\tau \tag{12a}$$

and

$$f_e = \Phi\sigma_S + J(\sigma_o - \sigma_S)\tau \tag{12b}$$

Although simplistic these equations reveal that the stage at which individual event interference or superposition begins to occur ($f \rightarrow 1$) is no longer simply dependent on total fluence ϕ as in the case of instantaneous, permanent zone modification. For zone expansion the critical fluence, ϕ_c, increases with increasing flux density, zone areal expansion and expansion time constant whereas for zone contraction the critical fluence decreases with increase in these parameters. Consequently the linear superposition requirement is no longer only fluence dependent but depends on these other parameters also. If, for example, the effect of an individual ion becomes more spatially delocalised by long range defect transport, the overall defected volume and defect density will exhibit fluence and flux density dependence. In fact the parameter dependences will be more complex than outlined above and depend intimately upon the nature of the delocalisation process as indicated by Johnson (21) in these proceedings and by others (22, 23, 36, 37) but the simple analysis serves to illustrate the importance of the contributory parameters.

A specific case of the contraction mode is when the localised modified zone area contracts to zero ($\sigma_s = 0$) and can be thought to be analogous to the generation and athermal reorganisation phases of an impact event with no further atomic transport. Thus the near prompt, ballistic generation and athermal reordering processes occur over an average area σ_o and with a development and quenching time constant τ. The factor f then takes on the meaning of a probability for individual event interference whilst quasi prompt processes are still active. In this case equation (12b) relaxes to

$$f = J\dot{\sigma}_o \tau \tag{13}$$

The criterion for linear superposition of prompt processes from sequential events thus becomes

$$J \ll \frac{1}{\sigma_o \tau} \tag{14}$$

Typical cross-sections for ballistic transport in an ion bombardment event are in the range $10^{-14} - 10^{-16}$ cm^2 and generation and reordering time constants of the order 10^{-12} s, which leads to the requirement that the ion flux density should be less than about $10^4 - 10^7$ A.cm^{-2}. These are, of course, quite enormous current densities and much higher than achieved in conventional continuous ion beams (even microbeams) but may be approachable in pulsed 'ion cannon' type systems. In general therefore linear superposition of ballistic processes is quite an acceptable approximation to successive ion impact events for all fluences and flux densities and allows diffusion-like approximations to ion irradiation induced ballistic mixing processes (38). Non-ballistic processes, on the other hand, such as thermal transport and non radiative electronic de-excitation, may possess much larger relaxation time constants and the critical current densities for linear superposition of events to be valid can be reduced substantially (e.g. to 1 mA. cm^{-2} for $\tau = 1$ μs). As long as this criterion is fulfilled and each individual event quenches to a different final state than before the event, then the spatially extensive modified fraction of the solid increases with fluence ϕ, according to equations (7) - (9) and the critical fluence for linear variation is $\frac{1}{\sigma_o}$. This criterion is also valid when the appropriate modification is not achieved in a single event but is the result of successive overlap or cooperative phenomena which mutate the local state incrementally. However the criterion only describes the fluence limit up to which the unmutated extensive modification is a linear function of this fluence. The fluence dependence of mutated fraction initially follows an mth power law where m is the number of mutations (35,39-41) in reaching the observed state. It is this type of approach which has been used to describe the fluence behaviour of amorphous fraction in light ion irradiated semiconductors (35,39-42) which is assumed to result from the local accumulation of point and complex defects, generated by successive ion impacts, up to a critical level where lattice instability against phase transformation occurs and collapse to local amorphousness ensues. This model cannot be entirely valid since studies with very light ions (H and He) and electrons appear not to induce amorphousness and suggests that the local spatial and time evolution of ballistic, athermal and thermal transport processes in individual cascades must be important in critical phenomena such as the crystal-to-amorphous phase change. The need for specification of effective high fluence or high flux density processes in individual events is reinforced. It is interesting to note in this context that if equation (13) is considered descriptive of an individual event, the flux density, J is equivalent to n , where n is the moving atom (collision density) and ν is the atom mean speed. σ_o is equivalent to $\frac{1}{\lambda N}$ where λ is the mean collision free path and τ, the mean collision free time is equivalent to $\frac{\lambda}{\nu}$. Consequently equation (13) becomes $f = \frac{n}{N}$, which, as argued earlier must be much less than unity to ensure linear superposition of discrete events.

3. ION COLLECTION AND DEFECT PRODUCTION. QUASI-STATIONARY AND MOVING
   SURFACES

3.1   Quasi-Stationary Surfaces - Ion Collision

Penetrating ions slow down via the elastic and inelastic energy loss mechanisms already discussed and, once they have been retarded to a sufficiently low energy that they can no longer surmount the potential energy barrier of neighbouring atoms, they stop ballistic motion. These ballistic stopping positions are statistically distributed spatially because of the statistical nature of individual collision histories. These spatial distributions can be predicted theoretically (4) and shown to agree very well with experimental results (43, 44) and computational simulations (44) for low fluence irradiation conditions where accumulation of implanted ions does not perturb the slowing down processes. The only area where major uncertainties exist in low-fluence heavy ion range profile prediction is when correlated atomic steering (channelling) processes assume importance. Upon ballistic stopping, however, the projectile will still retain some tens of eV energy which must be dissipated to local substrate atoms in their enhanced thermal motion (a local thermal spike is created) which may result in further thermal migration of the implanted species. This thermal migration may occur anyway if the migration activation energy of the implant species in the substrate is low. This is a well documented phenomena in He irradiation of many substrates (45). Even if further migration does not occur the implant species must either dissolve into, or be rejected back out of the solid. Dissolution may occur via interstitial occupancy in the host, by occupancy of vacant lattice sites and extended defects either native to or irradiation induced in the substrate or by substitution of substrate atoms. This latter process probably occurs in heavy inert gas ion implanted solids to relieve local strain and may be enhanced by chemical affinity of implant and substrate.

For example oxygen and nitrogen irradiation of Si produces silicon oxides
and nitrides (46). As impurity accumulates in the substrate, with continuing and congruent defect production and migration, new material phases
can develop in order to reduce the total free energy of the system, i.e.
thermodynamic influences assume importance although it should be recognised
that the system is very different from a binary (or more complex) system
produced by slow thermal or transient thermal (laser and electron beam
irradiation) processing. This is because the defect distributions and
concentrations in the ion irradiated systems are far different to thermal
equilibrium systems. The contributions of Johnson (21) and Kelly (47) in
these proceedings explore these differences. Specific examples of the
influence of this restructuring is in the production of gas filled bubbles
(48) in (insoluble) inert gas implanted solids and the production of two
dimensional defect (dislocation) structures and three dimensional (void)
structures (49) in many implant situations. The production of these new
phases is usually strongly substrate temperature dependent illustrating
the importance of defect migration and interaction as well as implant atom
interactions with the substrate.

Whatever the emerging nature of the implanted-substrate system one
fact is incontrovertible. Unless constrained the substrate must change
in its physical dimensions since new material has been incorporated. This
change will generally be an expansion unless the new phases formed are much
denser than the initial substrate. This expansion can result in dramatic
effects such as local surface rupture and exfoliation (50).

In attempting to give analytic form to this behaviour it is easiest,
initially to consider the case of self-ion implantation into an equal
atomic mass substrate of atomic density N and in which the stopping probability of projectiles, in a depth interval dz' at z', relative to the
position of the instantaneous surface and in the incident ($0z$) ion flux
direction, is given by $p(z')dz'$ where $p(z')$ is a stopping probability per
unit depth.

In the absence of any substrate expansion as a result of implant
incorporation the rate of change of the implant concentration, C, at
depth z', would be simply given by

$$\frac{\partial C}{\partial t} . dz' = J.p(z')dz' \qquad (15)$$

with the local solution

$$C(z')dz' = \Phi p(z')dz' \text{ cm}^{-2} \qquad (16a)$$

and the extensive, depth integrated, solution, i.e. the total implanted
concentration

$$C_T = \Phi \int_o^\infty dz' p(z') \qquad (16b)$$

For relatively energetic ions $\int_o^\infty dz' p(z') = \eta$ (the collection coefficient)
$\to 1$, and

$$C_T \to \Phi \qquad (16c)$$

Expansion, however, causes the surface to advance in the $-0z$ direction
at a speed, ν_a, given by

$\frac{J}{N} \int_o^\infty dz' p(z')$, which, again for energetic ions tends to $\frac{J}{N}$.

The defining equation for implant accumulation at a depth z' from the instantaneous surface position ($z' = 0$) now becomes

$$\left\{\frac{\partial C}{\partial t} + \nu_a \frac{\partial C}{\partial z'}\right\} dz' = Jp(z')dz' \tag{17}$$

where the second term in the bracket reflects the effect of the moving boundary (advancing surface).

The effect of surface advance is to modify the operator $\frac{\partial}{\partial t}$ in equation (15) to $\frac{\partial}{\partial t} + \nu_a \frac{\partial}{\partial z'}$ in equation (17) and equations of this latter type are readily solved (51) by, for example, Laplace transform methods or by a simple substitution $z^* = z' - \nu_a t$, provided that the parameters ν_a and $p(z')$ are time (or fluence) independent. These requirements will be most valid for self ion-substrate systems where phase and density changes do not occur and the local solution of equation (17) for depth z below the initial surface, becomes:

$$C(z)dz = N \int_z^{z + \nu_a t = z + \Phi/N} dz'p(z').dz \tag{18a}$$

and for depth z' below the instantaneous surface (at $z = -\Phi/N$)

$$C(z')dz' = N \int_o^{z'} dz' \, p(z').dz' \tag{18b}$$

For large fluences equation (18a) relaxes to

$$C(z)dz = N\,dz \int_z^{\infty} dz'p(z') = Ndz\left\{\eta - \int_o^z dz'p(z')\right\} \tag{19a}$$

whilst, also for large fluence and large z' (but less than Φ/N) equation (18b) relaxes to

$$C(z')dz' = N \int_o^{\infty} dz'p(z').dz' = N\eta dz' \tag{19b}$$

The physical meaning of equation (19a) is that for any depth $z > 0$, at high fluence Φ, the implant concentration density is Nx (forward integral of the stopping probability from z to $z = \infty$) whilst equation (19b) indicates that for any depth $o < z' < \Phi/N$, but again for high fluence Φ, the implant concentration is simply $N \times$ (depth integrated stopping probability). These results were already pointed out by Carter et al. (52) for conditions where the surface was considered to advance, not by implant accumulation, but by externally imposed growth processes. A qualitative description of this evolution is shown in Figure 1 which shows how, at a sufficiently late stage of surface advance, the concentration at depth z is related to the integrated stopping probability, and, for this condition, how this concentration saturates over a large depth interval.

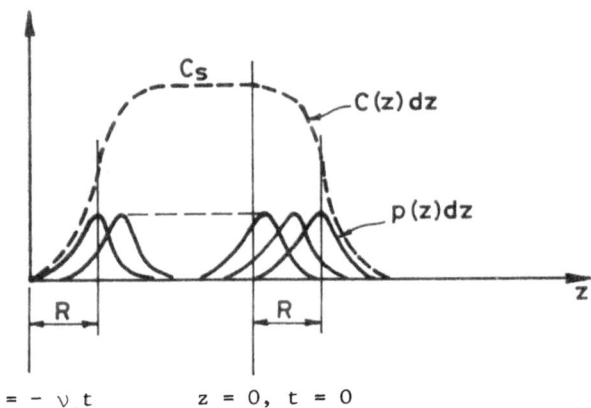

$z = -\, \nu_a t$ $z = 0,\ t = 0$

Fig. 1 A qualitative description of the stopping probability profile p(z)dz and how this becomes depth integrated to give the implant concentration profile C(z)dz for a substrate expanding due to projectile accumulation.

The magnitude of this saturated concentration C_s is given by $N\eta$ ($\to N$ as $\eta \to 1$) whilst the positions of points of inflection in the C(z) depth function are located (53) at depths R below the initial and instantaneous surfaces, where R is the most probable stopping range. It is to be noted that C_s extends outwards from the initial surface through the expanded region. The total, depth integrated, quantity of accumulated implant C_T is given by

$$C_T = N \int_o^{\frac{\Phi\eta}{N}} dz' \int_o^{\infty} dz\, p(z) = \Phi\eta \qquad (20)$$

This result is identical to that for a non-expanding system (equation (16c)) and so, as far as total implant concentration is concerned there is no high fluence limit. The same conclusion for the concentration at any depth below either initial or instantaneous/surface is not valid, however, since this local concentration varies continuously with irradiation time and fluence until a saturation is reached. Definition of any fluence limit is therefore arbitrary but might, for example, be related to the maximum fluence Φ_C which yields an expansion

$$\frac{\Phi_C \eta}{N}$$

equivalent to the depth resolution of whatever method is used to measure concentration depth profiles.

The results given above are very simple approximations and only given as guidelines. More sophisticated treatments can be developed to include, for example (a) implant diffusion effects where a further operator

$$D \frac{\partial}{\partial z^2}$$

is involved (54) (b) collisional relocation of implant atoms both into and out from a given depth interval (55,56) which involves writing the righthand side of equation (17) as an integral over all input and output events leading to accumulation in dz ; and (c) unequal mass conditions which continuously modifies stopping power of the changing substrate-implant complex and renders p(z) fluence dependent (57). All of these can assume considerable importance for specific systems, and in various applications and the literature cited above should be consulted for the more complete analyses.

It is notable that, even with implant collection alone considered, the
surface is only quasi-stationary since expansion occurs.
Real zero surface motion or further modified motion can be forced, however,
situations to which attention is now turned.

3.1.2 Moving Surfaces - Ion Collection

The preceding discussion ignored the fact that for incident ion energies
greater than a few tens of eV at most, the sputtering process occurs in
which atoms are ejected from the surface. This constitutes a congruent
erosion process to expansion and so the net speed of surface motion is

$$\nu_a = \frac{J\eta}{N} - \frac{JY}{N}$$

where Y is the sputtering yield in atoms ion^{-1}. Y depends upon a wide
variety of incident ion and substrate parameters, many of which have been
documented experimentally (12) and elegant theory (6) for prediction of
Y is available. One of the most important of the variables is incident
ion energy E and both theory predicts and experiment confirms that $Y(E)$
rises from very low values at some threshold energy E_{TS}, reaches a maximum
$Y_M(E)$ at an energy E_m, and then declines. The absolute magnitudes of
E_{TS}, $Y(E)$, $Y_m(E)$ and E_m depend upon projectile and substrate parameters
of mass and incident angle, surface orientation and temperature and atomic
binding energy (decreasing more or less inversely with increase of this
final parameter).

The behaviour of η, the collection coefficient, is nowhere near as well
known as that of Y but measurements and simple theory suggests (45) that
$\eta(E)$ depends upon similar projectile-substrate parameters to **those** of $Y(E)$,
increasing from low values at a threshold energy E_{TC} and then increasing,
rapidly at first and then more slowly to values near unity at higher
energies (generally greater than a few keV). For many projectile-substrate
systems the difference $\eta(E) - Y(E)$ will be positive, particularly for very
low energy ions and for high surface atomic binding energy substrates, and
so net growth of the substrate occurs. For equal mass conditions the
analysis presented earlier applies with η replaced by $\eta(E) - Y(E)$ for
given incident energy conditions.

If the projectile energy is such that $\eta(E) = Y(E)$ then no net surface
motion occurs and a true stationary boundary condition exists so that the
results given in equation (16) apply. In this condition continued increase
in implant is balanced by loss of substrate and implant atoms. The situation is of course more complex if non-equal mass conditions are used and
detailed balancing of input and output of different atomic species must be
considered (58). This process may be further complicated by atomic supply
from the bulk and the operation of both normal thermal diffusion and
radiation diffusion processes, possibly mediated by thermodynamic influences
such as Gibbsian segregation, as discussed in these proceedings by Kelly
(47).

The approximate validity of the simple first order treatment may be
perceived in the equal mass case however by studies of the growth of Si and
Ge substrates using low energy self ion implantation (59) and of diamond
(low sputtering yield) implanted with rather higher energy C ions (60).
This approximate treatment can also guide expectations for situations where
the difference $\eta(E) - Y(E)$ is negative in which net erosion occurs and
$\nu_a \rightarrow \nu_e$. The analysis of implant collection in this condition is identical to that of equation (17) et seq. except that the sign of ν is reversed
and accumulation of implant only occurs within a constantly inward moving
surface. The resulting expresssion for $C(z)dz$ beyond the instantaneous

surface becomes (52):

$$C(z)dz \ = \ \frac{N\eta(E)dz}{Y(E)-\eta(E)} \ \int_{z}^{z+\nu_e t} dz'p(z')$$

(21)

i.e. the concentration at depth z is the backwards integral of the stopping profile from z to the instantaneous surface. For large times this behaviour saturates with the type of depth profile indicated in Figure 2.

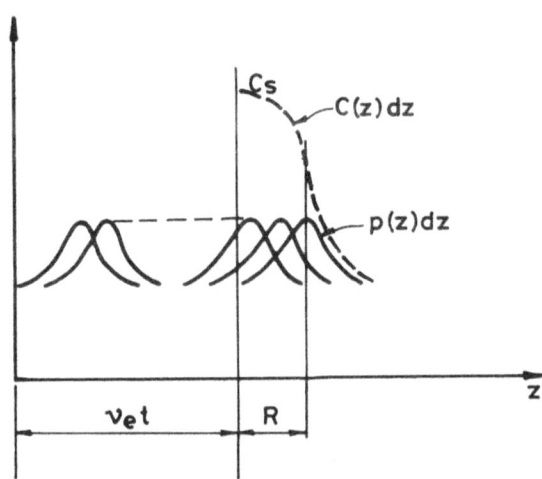

Initial Surface (t=0)   Surface at time t

Fig. 2 A qualitative description of the stopping probability profile p(z)dz and how this becomes depth integrated to give the implant concentration profile C(z)dz for a substrate undergoing net erosion by sputtering. The magnitude of this, near surface, saturated concentration C_s is, at t → ∞ given by $C_s \ = \ \frac{N\eta(E)}{Y(E)-\eta(E)}$, which for high energies where $\eta(E) \to 1$ and Y(E) can be much larger than unity, tends to $C_s \to \frac{N}{Y(E)}$.

The point of inflection in the C(z) depth function is located, after large time or high fluence (53), at a depth R below the instantaneous surface, whilst the total, depth integrated concentration C_T is given by (53)

$$C_T \ = \ \Phi \ \{\eta \ (E) \ - \ \int_{o}^{\frac{\Phi}{N}(Y(E)-\eta(E))} dz'p(z') \ \}$$

(22)

Equations (21) and (22) indicate that, like the case for substrate expansion unperturbed by sputtering erosion, the concentration is never a linear function of ion fluence, Φ, but, unlike this case, neither is the collected quantity a linear function of the ion fluence. Only for fluences for which little erosion of the stopping depth profile (through a distance

$\frac{\Phi}{N}$ (Y(E)-η(E)) has occurred is the accumulated quantity approximately equivalent to $\Phi\eta(E)$. The fluence at which this occurs will clearly depend intimately upon the detailed form of the stopping profile p(z')dz' and the value of Y(E)-η(E). In power law approximations to interatomic potentials the spatial extent of the stopping profile will scale (5) with E^{2m} whilst the sputtering yield (6) will scale with E^{1-2m} where m increases from zero to unity as energy E increases. Consequently the critical fluence for non-linear fluence dependence of implant collection will scale, approximately, with E^{4m-1}, i.e for $m > 1/4$ it will increase with energy. Actual magnitudes of this critical fluence value can be determined from range (5) and sputtering (6) theory or from experimental data but, for a sputtering yield of, for example, 10 atoms ion^{-1} and a depth profile with a mean plus one standard deviation of 500 atomic layers, the critical fluence would be of the order of 3.10^{16} ions.cm^{-2}. Measurements (45,53,57) of fluences required to observe departure from linearity of collected quantity as a function of incident fluence tend to be in the neighbourhood of this value indicating the approximate validity of the analysis.

An interesting result of the above analysis is that the steady state implant surface concentration should be approximately proportional to the inverse sputtering yield. This result is often used as an indication of maximum concentration achievable during energetic implantation and indeed, presents a limitation in direct ion implantation alloying or compound formation processes discussed elsewhere in these proceedings (61-63). One way of increasing this limit is to increase ion energies to very large values where $Y(E)$ tends towards lower values whilst $\eta(E)$ remains large (the quotient η/Y is the important parameter) or another method is to coat the substrate with a low sputtering yield material. This latter method suffers more disadvantage in that coating atomic species can be recoil implanted into the substrate.

Coating, however, can lead to a further interesting behaviour, particularly if coating, or atomic deposition, is conducted simultaneously with implantation. Again, if for simplicity, an equal mass situation is considered in which, during implantation, a flux J_a of like atoms also strikes the substrate surface, with a sticking (condensation) coefficient, S, the net speed of surface advance is given by

$$\nu_a = \frac{1}{N} \{ J_a S + J\eta - JY \} \qquad (23)$$

Analysis of implant accumulation proceeds exactly as before, with results for $C(z)$ and $C(z')$ quite analogous to the expansion-only situation but in which η is replaced by $\eta^1 = \frac{J\eta}{J_a S + J\eta - JY}$ since the numerator represents implant collection rate and the denominator represents surface advance speed. The general behaviour in space and time (fluence) is quite identical to the expansion-only case but all local concentrations are moderated by η^1. The depth integrated concentration, C_T, however remains at $\Phi\eta$ since the everywhere reduced local concentrations extends over a greater thickness (the coating plus net expansion thickness). The conclusions with regard to linearity with incident fluence are therefore similar to the expansion alone case in that total collected implant quantity always maintains linearity with fluence but the loss of linearity of collected implant at any depth occurs at lower fluences since the growth process, if very fast ($J_a S \gg J\eta - JY$), rapidly changes the stopping probability as a function of a fixed space location. Again definition of a critical fluence depends upon comparison of stopping depth profile changes with a minimum observable depth concentration distance.

Since the ratio $J_a:J$ can be varied by independent control, implant incorporation concentrations can be substantially modified, a result used to advantage in ion bombardment assisted atomic deposition techniques (64) for preparation of variable stoichiometry films. As in the other cases considered however, treatment must be more detailed for non equal mass, and atomic relocation by ballistic and radiation enhanced diffusion and thermodynamically driven cases.

3.2 Defect (Recoil) Production

In the first two sections of this overview it was stressed how energy and momentum transfer between projectile atoms (and electrons) and substrate species and between substrate species can lead to Frankel pair production of an energetic interstitial recoil and a vacancy. Just like the implant atoms discussed in the previous subsection the interstitials can, when thermalised, continue to migrate in the substrate whilst the vacancies may also migrate. These migrations will be conditioned by concentration gradients, thermal stress and chemical driving gradients as discussed by Johnson (21) and may couple with implant and substrate atoms to effect their relocation as discussed by Wiedersich (65). Local accumulation of defects may give rise to extended defect structures.

Ballistic collisions represent, therefore, only source terms for a wide range of defect generation and transport processes which are considered more fully in other contributions to these proceedings. Nevertheless they are fundamental to all the other processes (excepting normally active thermal processes) and consideration of their own spatio-temporal dependence is initially helpful. Defect production is, of course, not confined to incident ion impact alone and other, more penetrating radiation such as electrons, neutrons (and photons in some cases) produce recoils. More penetrating radiation generally produces simple defects, with large separation between events, more or less uniformly throughout the thickness of a substrate so that the recoil generation rate per atom of the solid is given by:

$$r = \frac{J.q}{N} \qquad (24)$$

where J is the radiation flux density and q is the defect generation rate per unit depth of the substrate per unit radiation flux density. In the absence of competing effects, total defect production increases linearly with time and the displacements per atom of the substrate become

$$D = \frac{\phi q}{N} \qquad (25)$$

Most sources of irradiation result in little dimensional change to a substrate (except for ionic crystals where electron and photon sputtering is well known and in solids where neutron irradiation generates trans-mutation products which can cause swelling) and so equation (25) for total recoil production is analogous to the non-expanding substrate approximation for implant collection in equation (16). Ion bombardment does, as fully illustrated in Section 3.1., lead to dimensional change of a substrate and the same conditions of stationary, quasi-stationary and moving surfaces apply to, and modify defect or recoil generation, as to implant collection. The only major difference is in the magnitude of recoil generation rates per unit depth and integrated overall depths, as compared to collection coefficients which cannot exceed unity. Without specificity of recoil generation mechanisms one could write a recoil (38) generation rate function Jq(z)dz analogous to a stopping distribution function. If the mechanism was of the linear cascade type then q(z) is subject to the Kinchin-Pease (24), modified by Sigmund (25), normalisation:

$$Q(E) = \int_0^\infty dz q(z) = \frac{0.42f(E).E}{E_d} \qquad (26)$$

with $f(E)$ and E_d defined as in section 2 .

q(z) can also be determined, for linear cascade conditions, by theoretical analysis (5). Generally speaking E_d is of the order of some tens of eV so that for incident ion energies greater than a few hundred eV, the total depth integrated recoil production rate is measured in many tens to many thousands per incident ion, whereas $\eta \leq 1.0$. Nevertheless exactly the same formalisms employed to deduce $C(z)dz$, the time integrated local implant concentration and C_T, the time and depth integrated concentration can be used to evaluate the local displacements per atom $D(z)dz$ and the depth integrated displacements per atom D_T. As before it is easier to consider the equal mass case (ion $\equiv$ deposit atom $\equiv$ substrate atom) and to assume, as discussed in section 2, linear superposition of events so that total recoil densities are equivalent to time integrated rates.

In the truly stationary surface condition ($J_a S + J\eta - JY = 0$) the local displacements per atom in the depth interval dz is given by

$$D(z) = \frac{Jq(z)t}{N} = \frac{\Phi q(z)}{N} \tag{27a}$$

and the depth integrated displacements per atom is given by:

$$D_T = \frac{\Phi . Q(E)}{N\int dz} \tag{27b}$$

wherethe integral in the denominator extends over the full depth of recoil production. In the situation, without deposition and $J_a = 0$, but with net erosion, $Y(E) - \eta(E) > 0$, the spatial distribution of the displacements per atom is equivalent to that shown in Figure 2, i.e. after high fluence irradiation, a maximum at the surface and decreasing with increasing depth into the substrate. This maximum number of displacements per atom reaches a time (fluence) independent value, at the surface, of

$$D_S = \frac{Q(E)}{Y(E) - \eta(E)} \tag{28}$$

and decreases inwards into the substrate according to a multiplication of equation (28) by a factor $\frac{1}{Q(E)} \int_z^\infty dz\, q(z)$.

The total, depth integrated displacements per atom is readily determined, for all fluences and, as in the case of implant collection, remains linear with ion fluence only as long as sputter erosion of the surface does not etch a significant depth of the recoil generation depth profile. Since stopping and recoil generation profiles are not too dissimilar the fluence limit derived earlier for linear fluence dependence of implant collection is not a bad guide.

In the net surface advance condition, where $J_a S + J\eta(E) - JY(E) > 0$, the analysis for ion collection under surface advance is applicable so that, under high-fluence irradiation (with accompanying thick film growth) the displacements per atom are approximately constant throughout the film and that the total, depth integrated displacements per atom, through most of the film thickness reaches this equilibrium value. The steady state displacements per atom throughout most of the film is given by

$$D(z) = \frac{JQ(E)}{J_a S + J\eta(E) - JY(E)} \tag{29}$$

but this is reduced near the film surface and below the initial substrate surface.

The importance of these conclusions can be judged through three technical applications. In ion mixing studies of bilayer, multilayers and buried
marker geometries high ion energies are used so that recoil production
distributions are, approximately, uniform throughout the depth of the
system and sputtering yields tend to be low. In this condition it is not
a bad first approximation to consider the surface as stationary and employ,
as is often done, equation (27a) to estimate the displacements per atom
at a given depth. This is much less acceptable for lower energy ions
where both the distribution $q(z)$ is closer to the surface and sputtering
yields are, for not too low energies, higher. In depth profiling of
impurities by sputter sectioning, where rather low energy ions are usually
employed, equation (28) indicates that after sputtering a layer thickness
of the order of $R + 4\sigma$ (where σ is the standard deviation of the recoil
production profile) the displacements per atom of surface atoms has saturated at a level of approximately $\dfrac{Q(E)}{Y(E)}$ and subsequently remains at this
value. Consequently atoms of the substrate (and impurities) are continuously being relocated and a broadening of the impurity profile will
result. Andersen (66) and others (67,68) have employed equation (28),
or a similar version, to deduce the number of relocation steps suffered by
an atom and, using a diffusion approximation (superposition of successive
steps), have derived a profile broadening parameter which scales with
$\dfrac{Q(E)}{Y(E)}$. This figure of merit increases with E, slowly at first, then more
rapidly and it is concluded that profile distortion will be minimised the
lower the projectile energy which can be usefully used.

At the other extreme much interest is now being shown in ion bombardment
assisted film deposition (64) and it is clear from equation (29) that,
during film growth, the displacements per atom scale approximately with
$\dfrac{JQ(E)}{J_a S}$, provided that the condition $J_a S > J(Y(E) - \eta(E))$ is satisfied.

If, during growth, it is required to obtain the highest possible atomic
relocation and intermixing, the criterion $Q(E) > \dfrac{J_a S}{J} > Y(E)$ must

be satisfied. This condition, derived already by Carter and Armour(69),
implies relatively slow net growth rates and the use of high energy
ions. If, on the other hand, recoil generation is to be avoided, for
example to minimise the density of built in defect structures, rapid
deposition rates and low-energy ion beams are required.

3.3 Saturation

In the preceding analyses of both ion collection and recoil generation
it was explicitly assumed that individual events were linearly additive
in time (fluence) and without modification of the event probabilities
with fluence. This assumption can be relaxed by allowing either $p(z)$
or $q(z)$ to be prescribed functions of fluence, or of C or D or both, and
some limited further analysis of the problems can be undertaken (70).
A simple, and physically meaningful, approximation is to assume saturation
of processes at each depth in the substrate other than via surface
motion constraints. In the case of ion collection this could represent
the production of a stable new phase (e.g. O implanted into Si generates
SiO_2) beyond which no further implant collection is possible. In the
case of recoil production it could represent, not a change in recoil rate,
but in the production of stable disordered or amorphous regions which,
although recoil production within them occurs, suffer no further phase
change. Both these processes can be represented by a limiting behaviour
of a general local quantity $F(z)$ (equivalent to $C(z)$ or $D(z)$) such that

the time rate of change of this variable is given by $u(z) \left\{ 1 - \dfrac{F(z)}{F_S} \right\}$ dz
where $u(z)$ is equivalent to $p(z)$ or $q(z)$. The second term in the brackets
indicates the declining probability of change as the process increases
in density towards a local, but depth independent, saturation level F_S .
    Insertion of this relation into a moving surface system results in
a modified defining   equation of the form (51)

$$\left\{ \frac{\partial F}{\partial t} - \nu \frac{\partial F}{\partial z'} \right\} \, dz' = J(1 - \frac{F}{F_S}) \, u(z')dz' \tag{30}$$

where ν represents either $- \nu_a$ or ν_e

    This equation is no more difficult to solve than the earlier, non satur-
able form of equation (17), using the substitution of $z^* = z' + \nu t$ with
the general result (71):

$$Fdz = F_S \left\{ 1 - \exp \, - [\frac{J}{\nu F_S} \int_z^{z+\nu t} dz^* \, u(z^*)] \right\} \, dz \tag{31}$$

    In the limit of no surface motion ($\nu \to 0$) this equation relaxes to

$$Fdz = F_S \left\{ 1 - \exp (- \frac{\Phi u(z)}{F_S}) \right\} \, dz \tag{32}$$

which is analogous to equation (7) for the formation of amorphousness by
direct individual ion impact. In the limit of no saturation processes
($F_S \to \infty$) equation (31) relaxes to

$$Fdz = \frac{J}{\nu} \int_z^{z+\nu t} dz^* \, u(z^*) \tag{33}$$

which is analogous to equation (21) for ion collection in the presence of
net erosion without saturation.
    In the limit of low fluence, for all conditions of ν, equation (31)
relaxes to

$$Fdz \approx \Phi u(z)dz \tag{34}$$

which indicates that neither surface motion nor saturation processes
moderate the low fluence behaviour of linearity of F with this parameter.
    Other, fluence (time) dependent and (high) fluence independent values
of F are readily deduced (51) by appropriate expansion of equation (31)
for assumed values of the process (i.e. the nature of ν and of F_S) and the
quite general result of including a saturation process is to further limit
the density of the process, for all fluences, by comparison with the moving
boundary imposed limitations discussed earlier. For example in the net
erosion case F reaches a limiting value, at the surface, of F(0)

$$F(0) = F_S \left\{ 1 - \exp \, (\frac{-J}{\nu_e F_S} . U(E)) \right\} \tag{35a}$$

where $U(E) = \int_o^\infty dz \, u(z)$, which is to be compared with the result, in
the absence of saturation processes, of

$$F(0) = \frac{J}{\nu_e} . U(E) \tag{35b}$$

Only if $\dfrac{J}{\nu_e F_S} U(E) \equiv \dfrac{N}{F_S} \dfrac{U(E)}{Y(E) - \eta(E)}$ is small are these results approximately equivalent which implies low ion energy situations, but in general

$F(0)$ will tend towards F_S rather than the limitations set by surface motion. The limiting fluence, Φ_c, for which linear behaviour with fluence will occur, will therefore be of order $\Phi_c \approx \dfrac{F_S}{u(0)}$.

The critical fluence for linear/ity will be less, in the saturation limitation case compared to no saturation by a factor $\dfrac{F_S}{N} \cdot \dfrac{Y(E)-\eta(E)}{U(E)}$

which, as already outlined will be considerably less than unity. The effect of the operation of saturation processes is, therefore, to impose even more severe constraints on the limits of linear behaviour with fluence.

4. LOCAL AND EXTENSIVE PHENOMENA

All of the desirable, and sometimes undesirable, property changes which result from high-fluence ion irradiation of substrates, and discussed in detail by a number of authors in these proceedings, are related to the composition and structural changes which result from the incorporation of external species into the substrate and/or the change in the local energy and defect state of the substrate. In direct ion implantation processes all three parameters are important whilst in ion mixing applications, unless the irradiating species is a desired end component of the system, only the latter two parameters are considered influential. It is true that even in this latter case the nature and energy of the projectile will influence local energy and defect density conditions, and indeed it is this point which we first wish to reinforce. In section 2 a qualitative description of the evolution of a collision event was given and this is now amplified slightly. In any individual projectile generated event there will be a prompt series of processes followed by longer time or delayed processes. The prompt series of processes is described in time sequence as follows.

1. The production of a family of moving atoms and associated vacancies which occupies a time scale of $10^{-14} - 10^{-13}$ s. and in which the number of moving atoms first increases and then decreases with time as the interstitials slow down to rest whereas the energy per moving atom always decreases with time. In this phase the volume occupied by the event increases with time until, at the end of the event, there will be a spatial distribution of vacancies which is a maximum at the core of the event and decreases radially outwards and a distribution of interstitials which is a minimum at the core of the event, increases radially outwards, and then decays beyond the boundary of the vacancy distribution. This behaviour results from the net outward flow of momentum in an event. The temperature in this zone will also be modified, spatially and temporally, by sub-displacement threshold collisions, particularly as the interstitial recoils slows down mainly at the boundaries of the event.

2. This phase will result in a net rarefied central zone and a net densified peripheral region with, in addition, atoms resited via replacement events. Athermal relaxation of neighbouring defects will modify the defect and atomic concentration distributions and, via stored energy release, give rise to further temperature excursion. This phase may occupy $10^{-13} - 10^{-12}$ s, equivalent to the ratio of the atomic spacing to the accoustic wave velocity in the solid.

3. Overlapping with the late time stages of recoil generation, athermal readjustment occurs and, continuing for about another 10^{-11} s, the excess energy of the recoils will be quenched to the eV region and quasi thermal migration of these defects, vacancies and substrate atoms will occur

within the event volume. This quenching will continue as the mean particle energy density (temperature) falls by conduction out of the event volume, a process which will occupy about 10^{-9} s. to reach ambient conditions. In the first two stages processes are most aptly described as 'quasi-hydrodynamic' whilst in the second and third stages there is an increasing tendency towards a 'quasi-thermodynamic' description.

In all of these stages local atomic concentration profiles will change by ballistic, quasi-thermal and true thermal processes and the diffusion equations which describe the relocations must, for full generality, have full regard for all of the spatio-temporally evolving distributions of atoms, defects and 'effective' temperature. Phase changes can occur during each and all of these stages but again any description of these must acknowledge the spatial defect concentration profiles as well as the atomic and thermal profiles, i.e. any thermodynamic quantities employed (heats of mixing, energies of cohesion etc) must be those appropriate to the defective state not the isothermal, equivalent phase state. Johnson addresses these considerations in greater detail (21). Essentially these arguments can be summarised by indicating that phase stability is conditioned by topological (associated with entropy and system disorder) as well as compositional (associated largely with free energy) considerations.

Any model to describe phase transitions in irradiated substrates must therefore include all the above elements and must be capable of describing differences between differently ordered and bonded atomic systems. No such comprehensive model yet exists but progress is being made by comparison of effects between, for example, semiconductors, elemental metals, alloys and compounds in which ballistic processes should be similar but where thermodynamic quantities are very different. Model simplifications may be achieved by, for example, considering that either or both compositional and/or defect profiles are 'frozen in' during the quasi-thermal quench phase if not in the athermal phase, because of the limited number of activated jumps which can occur in the time scale of the former.

It should be noted here that, because of the very many recoils generated locally the formation of new phases may proceed much more rapidly with ion fluence in interfacial mixing studies or in precipitate dissolution or segregation in alloys where very high atomic fractions of each of the atomic constituents participate in an event which intersects an interface than in direct implantation where, at most, only one foreign atom participates in the local event. Of course in this latter area, as fluence increases, local impurity concentrations will increase, limited by sputtering processes discussed in section 3, and critical concentrations for phase change may ensue.

As indicated above the first regime of relaxation will be under high pressure (concentration) gradient conditions and will be essentially adiabatic. Further relaxation may then occur as the quasi-thermal energy of the event redistributes in space and time both within the event and into the surrounding substrate. An approximate idea of the possibility of relaxation processes may be obtained by considering the total probability of atomic, defect (and their complexes) migration and relaxation over a defined time scale. For a thermally activated process, requiring an activation energy Q to proceed, the frequency of relaxation processes for a given entity is given by

$$f = f_o \exp\left(-\frac{Q}{KT}\right) \tag{36}$$

where f_o is generally of order 10^{13} s^{-1}. From this relationship it is readily shown that the relaxation only proceeds rapidly at and above a temperature

$$T_c \text{ given by } kT_c \simeq Q/\log_e f_o \simeq \frac{Q}{30} \tag{37}$$

During the quasi-thermal quench of the event the temperature, at any spatial point, will fall rapidly and the total relaxation probability will be given by $\int_0^t f dt$, where T, and therefore f, are time dependent. This integrated probability can be approximated by $f_i \tau(f)$ where f_i is the initial (highest temperature) rate and $\tau(f)$ is a relaxation time constant. This relaxation time constant $\tau(f)$ may be approximated (71, 72) by

$$\tau(f) = f_i \cdot \left(\frac{df}{dt}\right)^{-1} = \frac{kT_i}{Q} \cdot T_i \left(\frac{dT}{dt}\right)^{-1} = \frac{kT_i}{Q} \tau(T_i) \tag{38}$$

where $\tau(T_i)$ is a temperature quenching time constant. The total relaxation probability then becomes

$$f_i \cdot \frac{kT_i}{Q} \tau(T_i) \tag{39}$$

If the initial temperature T_i is much greater than the critical temperature T_c for the process (i.e. $\frac{Q}{kT_i} << 30$) then $f_i \simeq f_o$ and the temperature quench time constant, which depends upon the thermal properties of the solid and the initial conditions but very weakly upon the background substrate temperature, is of order $10^{-11} - 10^{-9}$ s. Consequently during the prompt thermal quench regime of the event, the number of attempts for relaxation for an entity will be of order $10^2 - 10^4$, which indicates a high probability of atomic and defect relaxation over the whole event volume. This relaxation from the quenched-in athermally defined state will drive the system towards the local minimum free-energy configuration which will be conditioned by thermodynamic considerations mediated by the non-thermal defect population. Thermodynamic quantities and their relationships in the form of phase diagrams may therefore by used as first order guides to predict which phase transformation may result.

However the prompt quench regime does not necessarily represent the end of local restructuring since if the ambient temperature is greater than T_c, further relaxation in the event volume can occur with a total probability per entity of $f_c \tau$, where f_c is the relaxation frequency at the ambient temperature T_c and τ is the time between spatially overlapping events. This time, for a non-time varying event volume is given, from equation (13) by $\tau \simeq (J\sigma_o)^{-1}$. For ion flux densities in the range 1 μA cm^{-2} to 1 mA cm^{-2} and an event area of order 10^{-12} cm^2 this leads to overlap time constants in the range $10^{-4} - 10^{-1}$ s, which are much larger than the prompt quench relaxation times. It is therefore not surprising that substrate ambient temperature can play a very significant role in determining the phase formation if the delayed or long-time process of relaxation becomes significant (higher ambient temperatures).

Following the prompt (ballistic and thermal) stages of an event, delayed thermal and other relaxation processes can occur with defect and atomic migration and energy transport into the surrounding substrate. Composition and phase change process can therefore occur more extensively in both space and time than the localised prompt phases of the event. If all migration processes are frozen after the end of the prompt regimes then the conditions for event integration discussed in section (2) must be examined to adduce possible flux density and fluence behaviour. Most atomic mixing events in

the prompt regime allow linear integration with fluence. If migration
occurs, however, and this will be dependent on atomic and defect concentrations, the thermal evolution of the decaying prompt event and the
substrate temperature, then radiation enhanced diffusion processes occur,
the details of which are considered by Johnson (21), Wiedersich (65) and
Kelly (47) in the contexts of further phase and composition evolution.
Where the projectile acts only as a catalyst to drive the reaction then
structural modifications of the substrate will generally emerge from
point-defect aggregation, and the formation of high densities of dislocations and voids are well documented phenomena in ion irradiated materials.
The existence of such local modifications will change atomic binding
energies and the spatial development of recoil cascades in crystalline
substrates and may exert profound influences on properties of substrates
including, specifically, the details of sputtering. The contribution
by Wilson (73) demonstrates the role of such local phase modifications on
the evolution of sputter erosion generated surface topography. Indeed it
should be noted in this context that the presence of a free surface near
the points of volumetric defect generation will usually act as a sink
for such defects and any considerations of the evolution of defect
structures via interactions between defects migrating from and between
individual collisional events must, properly, include this loss centre.
This has not, unfortunately, been included explicitly in most analyses,
although internal sinks figure prominently. The formation of precipitates, either by implantation or by mixing, is an important outcome of
many experimental studies and these processes may initiate in either the
prompt or delayed regimes and their existence and stability will depend
upon atomic and defects fluxes into and out from them, originating from
both (enhanced) migration and ballistic events all play contributory
roles. Strain relief at high concentrations of insoluble species by
relaxation to more compact forms such as bubbles can also be decisive.
In addition even if local events do not lead to phase transition then
accumulation of implant, under low sputtering yield conditions, such
as achieved by higher/ion energy implantation, and extensive defect production,
migration and coalescence, may lead to sufficient local strains to
effect phase transformations (74, 75). Indeed processes occurring after
the prompt and delayed regimes cannot be ignored and the propagation of
dislocation structures (loop punching) from radiation induced structures
is a particular example of this phenomenon.

Finally it should be noted again that the atomic complexes, defect
complexes and atom-defect structures formed in all regimes will depend
intimately upon the details of the generation and quench process and may
lead to production of nuclei from which new phases may grow. It is
therefore not surprising that different phases may evolve in high collision
density spikes, in superposed low density events and in thermal processing.
Examples of these differences lie in amorphousness production in semiconductor and some alloys by heavy ion impact but not via cumulative
light ion or electron irradiation, but amorphization in other systems
with relative independence of individual event density. Differences in
final phase formation in an ion impact event and via rapid cooling
techniques may also be associated with the existence of different initial
defect and other nucleation centre configurations.

There can be few doubts that many complex and interacting processes
occur during individual and cumulative ion impact events and that detailed
understanding of materials modification will only result from inclusion
of many cooperative and competing effects. In this overview many simplifying approaches have been used in order to guide understanding of some

of these effects and indeed it may be suspected that some of the more
complex systems may not admit detailed theoretical analyses and that the
computational approaches discussed by Møller (10) may be vital, provided
that all effects are properly incorporated into the modelling and that,
in particular, the spatio-temporal effects of both individual and cumulative impacts are fully described.

5. CONCLUSIONS

This overview has not attempted to provide a comprehensive but superficial synopsis of the many and varied contributions to these proceedings
but to address the more fundamental concepts involved in high-fluence
and flux-density effects in ion bombardment substrates. This was first
accomplished by an examination of processes involved in individual ion
impacts and the conditions necessary to be able to cumulatively and linearly
sum the effects of many successive impacts. Ion bombardment introduces
both extraneous atoms to and generates atomic recoils within substrates
and the spatial and temporal evolution and integration of these processes
was examined in various situations of stationary and moving boundaries –
the latter mode itself arising from the sputtering effects of bombardment.
It was shown how this boundary condition alone can totally modify impurity
incorporation and atomic redistribution effects with resulting impact on
the magnitude and extent of possible substrate property changes.

Finally a further examination of prompt and delayed processes within
and between successive ion impact events outlined the necessity to consider
the influence of thermodynamic influences with full incorporation of
transient atomic, defect and local energy concentrations and gradients
in these constraints. It was argued that such complete incorporation of
all these processes may lead to analytic intractability, that universal
models may be difficult to erect, and that, on the one hand, simplified
models may guide qualitative understanding but, on the other hand,
complex and interacting subsystem behaviours may only be modelled by
comprehensive computational simulations which properly account for the
evolutionary nature of effects which occur as the result of multiple
impact phenomena.

6. REFERENCES

1. Grove W R. Trans Roy Soc, London 142, 87 (1852).
2. Plucker J. Pogg Ann 105, 84 (1858).
3. Bohr N. Kgl Danske Vid Selsk Mat Fys Medd, 18, No. 8 (1948).
4. Lindhard J, Scharff M and Schiøtt H E. Kgl Danske Vid Selsk Mat Fys
   Medd, 33, No. 14 (1963).
5. Winterbon K B, Sigmund P and Sanders J B. Kgl Danske Vid Selsk Mat
   Fys Medd, 37, No. 14 (1970).
6. Sigmund P. Phys Rev 184, 383 (1969).
7. Robinson M T and Oen O S. Phys Rev, 132, 2385 (1963).
8. Gibson J B, Goland A N, Milgram M and Vineyard G H. Phys Rev, 120,
   1229 (1960).
9. Harrison D E. Rad Effects 70, 1 (1983).
10. Møller W. These Proceedings.
11. Williams M M R. Rad Effects 37, 131 (1978).
12. Andersen H H and Bay H L. In Topics in Applied Physics Vol 47,
    Sputtering by Particle Bombardment 1. (Ed. R Behrisch, Springer-
    Verlag, Berlin) Chapter 4, 1981.
13. Thompson D A. Rad Effects 56, 105 (1981).
14. Carter G. Nucl Instrum and Meth 209/210, 1 (1983).
15. Brinkman J A. J Appl Phys, 25, 961 (1954).

16. Seeger A. Proc 2nd Int. Conf. on Peaceful Uses of Atomic Energy, 6, 250 (1958).
17. Sigmund P and Claussen C. J Appl Phys, 52, 990 (1981).
18. Carter G. Rad Effects Letters, 50, 105 (1980).
19. Guinan M W and Kinney J H. J Nucl Mat, 103/104, 1319 (1981).
20. King W E and Benedek R. J Nucl Mat, 117, 26 (1983).
21. Johnson W L. These Proceedings.
22. Peak D and Averback R S. Nucl Instrum and Meth, B7/8, 561 (1985).
23. Hobbs J E. PhD Thesis. University of Salford, 1985.
24. Kinchin G H and Pease R S. Rep Prog Phys, 18, 1 (1955).
25. Sigmund P. Appl Phys Letters, 14, 114 (1969).
26. Sigmund P. Appl Phys Letters, 25, 169 (1974).
27. Carter G and Grant W A. Nucl Instrum and Meth, 199, 17 (1982).
28. Jaouen C, Delafond J and Riviere J P. J Phys F Met Phys, 17, 335 (1987).
29. Ossi P M. To be published 1987.
30. Pedraza D F and Mansur L K. Nucl Instrum and Meth in Physics Res, B16, 203 (1986).
31. Swanson M L, Parsons J R and Hoelke C W. Rad Effects 9, 249 (1971).
32. English C A, Eyre B L and Summers K L. Phil Mag 34, 603 (1976).
33. Eyre B L. J Phys F Met Phys, 3, 22 (1973).
34. Holz M, Ziemann P and Buckee W. Phys Rev Lett, 51, 1584 (1983).
35. Gibbons J P. Proc IEEE, 60, 1062 (1972).
36. Myers S M. J Appl Phys, 47, 1812 (1976).
37. Myers S M. Nucl Instrum and Meth, 168, 265 (1980).
38. Littmark U and Gras-Marti A. These proceedings.
39. Denmis J R and Hale E B. J Appl. Phys, 49, 1119 (1978).
40. Carter G and Webb R. Rad Effects Letters 43, 19 (1979).
41. Webb R and Carter G. Rad Effects, 42, 159 (1979).
42. Thompson D A, Golanski A, Haugen H K, Stevanovic D V, Carter G and Christodoulides C E. Rad Effects, 52, 69 (1980).
43. Oetzmann H and Kalbitzer S. Rad Effects, 47, 57 (1980).
44. Fichtner P F P, Behar M, Olivieri C A, Livi R P, de Souza J P, Zawislak F C, Fink D and Biersack J P. Nucl Instrum and Meth, B15, 58 (1986).
45. Carter G, Armour D G, Donnelly S E, Ingram D C and Webb R P. Rad Effects, 53, 143 (1980).
46. Reeson K J. Nucl Instrum and Meth, B19/20, 269 (1987).
47. Kelly R. These Proceedings.
48. Donnelly S E and Roussouw C. Nucl Instrum and Meth B13, 485 (1986).
49. Krishan K. Rad Effects, 66, 121 (1982).
50. Scherzer B M U. In 'Erosion and Growth of Solids Stimulated by Atom and Ion Beams.' Proc NATO ASI Series E112, (Eds. G Kiriakidis, G Carter and J L Whitton, Martinus Nijhoff Publ, Dordrecht) p222 (1986).
51. Carter G, Webb R and Collins R. Rad Effects, 37, 21 (1978).
52. Carter G, Colligon J S and Leck J H. Proc Phys Soc, 79, 299 (1962).
53. Carter G, Baruah J N and Grant W A. Rad Effects 16, 107 (1972).
54. Collins R and Carter G. Rad Effects, 26, 181 (1975).
55. Sigmund P and Gras-Marti A. Nucl Instrum and Meth, 168, 389 (1980).
56. Littmark U and Hofer W O. Nucl Instrum and Meth, 168, 329 (1980).
57. Carter G, Armour D G, Kostic S, Jimenez-Rodriguez J J, Karpuzov D S and Nobes M J. Nucl Instrum and Meth, B19/20, 758 (1987).
58. Betz G and Wehner G K. Topics in Applied Physics, Vol 52, Sputtering by Particle Bombardment II (Ed. R Behrisch, Springer-Verlag, Berlin) Ch. 2, (1983).

59. Appleton B R, Pennycook S J, Zuhr R A, Herbots N and Noggle T S. Nucl Instrum and Meth, B19/B20, 975 (1987).
60. Rusbridge K and Nelson R S. Communicated at Proc 2nd Int Conf on Ion Beam Modification of Materials. Albany, New York, July 1980.
61. Bernas H. These Proceedings.
62. Mazzoldi P. These Proceedings.
63. Dran J C. These Proceedings.
64. Harper J M E, Cuomo J J and Kaufman H R. Ann Rev Mat Sci, 13, 413 (1983).
65. Wiedersich H. These Proceedings.
66. Andersen H H. Appl Phys 18, 131 (1979).
67. Schwarz S A and Helms C R. J Vac Sci Technol, 16, 781 (1979).
68. Haff P K and Switkowski Z E. J Appl Phys 48, 3383 (1977).
69. Carter G and Armour D G. Vacuum 36, 337 (1987).
70. Cerofolini G F, Bresolini C, Meda L and Volpones C. To be published (1987).
71. Carter G, Armour D G, Donnelly S E and Webb R. Rad Effects,36, 1 (1978).
72. Carter G and Cruz S A. Rad Effects Letters, 58, 125 (1981).
73. Wilson I H. These Proceedings.
74. Linker G. Nucl Instrum and Meth, B19/B20, 526 (1987).
75. Egami T and Waseda Y J. Non Cryst Solids, 64, 113 (1984).

SPUTTERING

HISTORICAL OVERVIEW ON THE FUNDAMENTALS OF SPUTTERING

A.OLIVA

Dipartimento di Fisica - Università della Calabria
I-87036 Rende (CS) - Italy
and
Unità GNSM - CISM di Cosenza

1.INTRODUCTION

Since the first appearence of the word "spluttering" (the original word used by Sir J.J.Thomson) in the scientific literature [1], many papers have been published on the subject and the widespread applications of the techniques and concepts involved in sputtering studies are now of everyday use. The uses extend to fields far away from that in which sputtering phenomena were first studied and developed (e.g. I found an application of sputtering to blood cells, to remove surface layers from red blood cells! [2]).

Today within the classical definition of sputtering as the erosion of solid surfaces under particle bombardment we must imagine the variety of sputtering experiments in which various targets (metals, liquids, ices, organic compounds) have been combined with several bombarding beams (ions, electrons, photons, charged molecules, dust particles). In any case the erosion is quantitatively expressed by the sputtering yield Y which is the mean number of emitted target particles per incident beam particle (see ref.[3] for an historical survey and as source of basic bibliography).

In its monograph [1] Sir Thomson introduced the word spluttering indicating the sputtering process as "analogous to that which occurs when a marble falls upon the surface of water, a crater is formed under the marble but the rim of the crater moves upwards and escapes, if the marble has fallen from a considerable height, from the surface of the liquid in drops." However, he also prefigured an emission picture which is close to the modern cascade theory saying that "it is not the atom struck by the positive ray which is torn from the metal, but a ring of atoms in its neighbourhood".

In fig.1 a photograph recording the results of laser irradiation on an Al target [4] is reported together with a picture illustrating Sir

R. Kelly and M. Fernanda da Silva (eds.), Materials Modification by High-fluence Ion Beams, 31–81.
© *1989 by Kluwer Academic Publishers.*

Thomson example. A splash similar to that caused by laser pulses (2 J/cm^2 is the energy used, the reflectivity of Al being 92%) can be generated on a water surface with a ball of 1cm radius and ~10g mass, falling from a height of roughly 50cm above the water surface: assuming that all the available energy goes into the splash creation one can estimate that the ball hits the surface with roughly 0.15 J/cm^2.

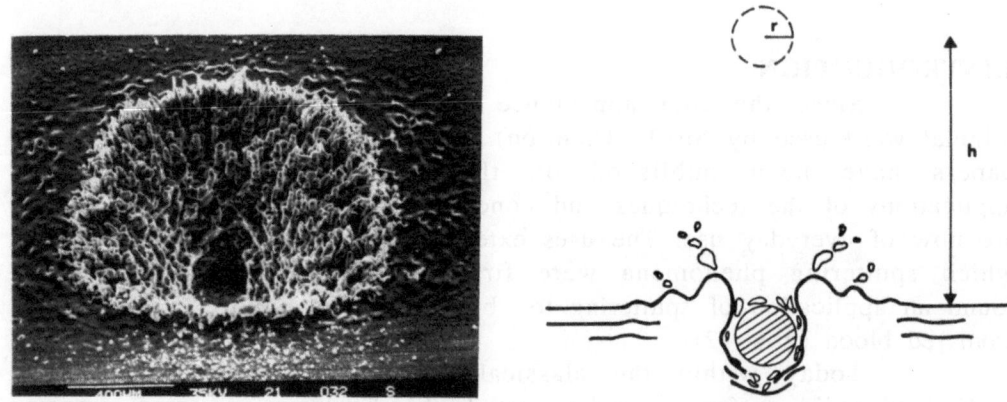

Fig.1. Schematic representation of a ball falling on a water surface (right) and a picture of an Al target irradiated with a laser beam (left).

If we look closer to the scientific production about the subject sputtering in the last twenty years , we obtain a total number of papers around 11,000 as classified in Physics Abstracts [5], where for total number is intended the number of works containing in their abstracts the word "sputtering" or "sputtered" or "sputter" regardless of the context in which such words are used. If one extends the search including abstracts different from those of Physics Abstracts , this number is even larger (~16,000). Confining ourselves to Physics Abstracts we can try to split up this total number of papers into three categories. The first category contains the "theoretical" works, where we include papers in which some calculation or explanation of mechanisms are involved, computer simulations, simple models or review of models and of course genuine theoretical works; the number of works pertaining to this category is ~735. The second category contains the works concerning experimental data and measurements oriented towards the explanation of a model or supporting a theory, including those papers where experimental systems are described or data are collected. The number here is ~1856. In the last category we place the works which have something to do with sputtering

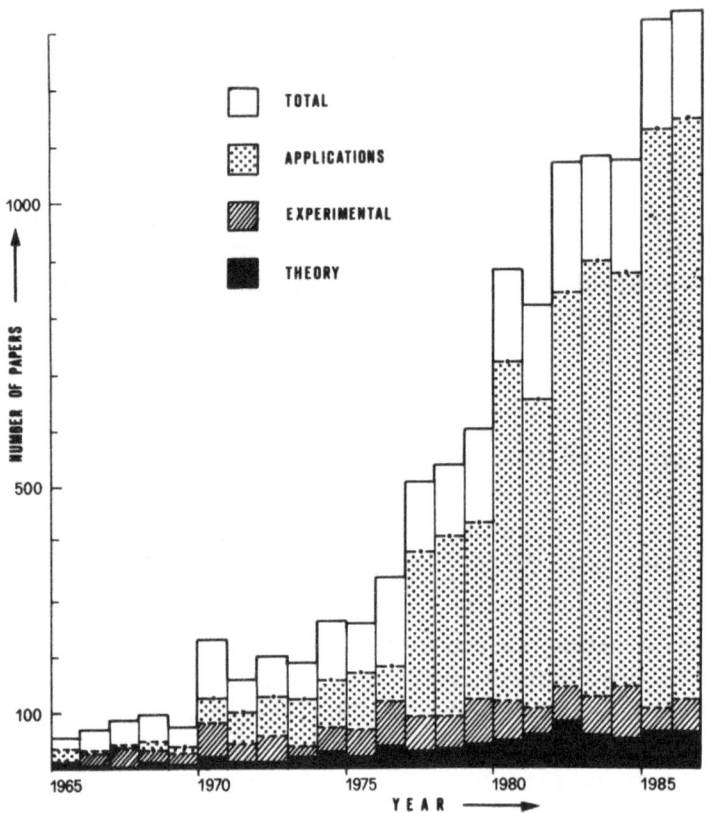

Fig.2. Statistics of scientific publications in the field of sputtering covering roughly the last 20 years; a distinction is made between theoretical, experimental and application oriented works.

in the sense that the sputtering technique is used as a general application. This group then contains all the works using ion beam etching techniques or coating and deposition techniques, with a total number of ~8634 published papers.

The details for each year covering the period 1965-1986 are reported in fig.2, where we can see the explosion in the practical applications of the sputtering technique, while the community active in basic research seems to be in a stationary state with a production in the last 5 years of roughly 200 papers per year both in experiments and theories. The field is still active and future trends on developments of both theories and applications will easily receive input from the frontiers of research and development in new materials [6], the same need of knowledge about materials in fact having previously pushed the development of research in radiation damage in solids.

In this lecture we restrict our contribution to the background on the essentials of sputtering theory and then to extended considerations about the use of the recoil density concept in sputtering . Applications will be shown for the case of sputtering from monoatomic targets, by light ion bombardment and from compoud targets.

2. BACKGROUND

Let an ion of initial energy E, mass M_1 and atomic number Z_1, impinge on a target of of randomly distributed atoms (M_2, Z_2). It is useful to distinguish three regimes concerning the target atoms set in motion by the incident primary ion [3], using as leading parameters the incident ion energy and the mass ratio M_2/M_1. In the case (see fig.3) of low incident energy and/or low incident mass $(M_1 << M_2)$ the recoils do not receive enough energy to generate cascades and they get sputtered only if the received energy is sufficient to overcome surface binding forces; this is the so called single-knock-on regime (fig.3b). In the linear cascade regime (fig.3c), i.e. in the case of moderate incident energy and/or similar target and primary ion masses $(M_1 \cong M_2)$ the recoils receive enough energy to generate a cascade , but the density of moving atoms is dilute enough to disregard collisions between moving atoms and multiple collisions. In fig.3d the spike regime is sketched, where the spatial density of moving atoms is large so that within the spike volume the majority of atoms is moving, this being the proper regime for high incident energy and/or large incident mass $(M_1 >> M_2)$. A second classification can be outlined following the time evolution of sputtering phenomena [7],i.e. taking as t=0

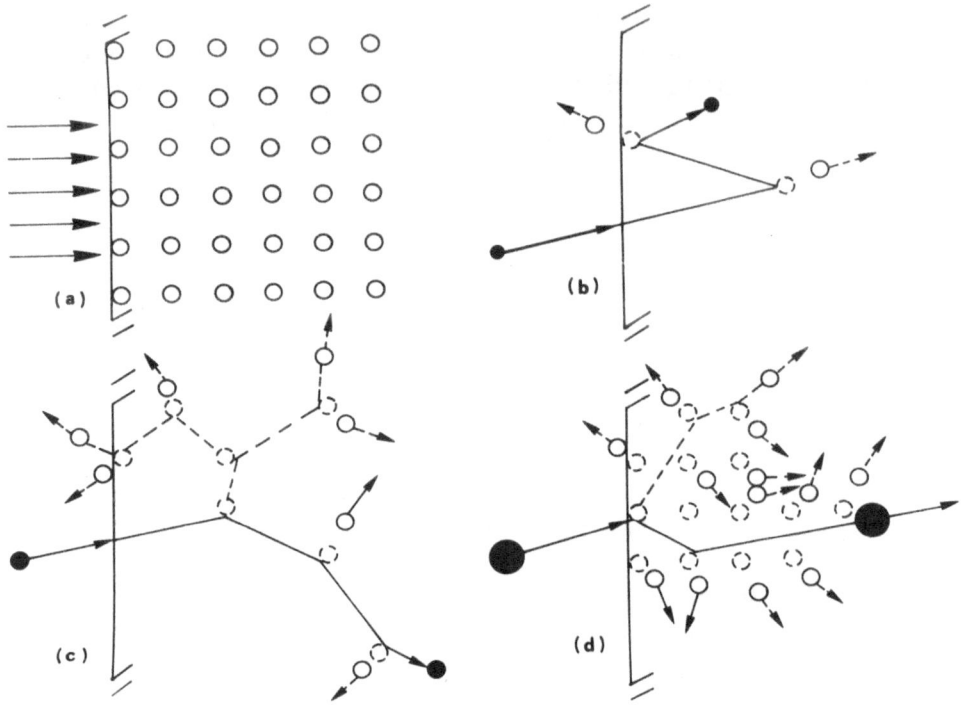

Fig.3. Schematic representation of three regimes in sputtering phenomena: (a) beam of particles impinging on a target; (b) single knock–on regime; (c) linear cascade regime; (d) spike regime.

in a time scale the instant where an incident particle hits the surface. The prompt collisional processes take place for $10^{-15} \leq t \leq 10^{-14}$ s and involve direct interaction of the incident particle with the target atoms. Then the slow collisional processes follow for $10^{-14} \leq t < 10^{-12}$ s caused by cascade development. Prompt thermal processes, the origin of which is still under debate to be a thermal spike or vaporization, which are in principle distinct processes, occur for $10^{-13} < t < 10^{-10}$ s. Finally, slow thermal processes, i.e. the vaporization of metal from halides or oxides by electronic sputtering, occur for t >> 10^{-11} s. Alternatively one can group in the <u>electronic sputtering</u> all processes based on electronic transitions occurring for t ≥ 10^{-13} s. This classification scheme projects over the development of new theories and peculiar experiments.

What might be called the refence formulation of sputtering theory is that of Sigmund which deals only with the linear cascade regime and is contained in the 1969 paper [8].The preexisting approaches to the problem turned out to be unsuccessful for a number of reasons, so it is interesting infact to briefly overview some of these approaches and their salient points in order to see why the Danish school achieved such a success in this field. The discussion of the sputtering models proposed by Keywell,Harrison and Joyes are representative, in my opinion, of a research situation where the need of knowledge about collision cross sections constituted a limitation for the model; this limitation is absent in Sigmund's theory, which is based on the background supplied by Bohr and Lindhard.

In the formulation of Keywell (1955) an expression of the sputtering yield has been derived in terms of the lethargy concept [9]. Lethargy is defined as the logarithmic energy variable:

$$u = \ln E/E_0 \tag{1}$$

where E is the initial energy of a particle undergoing a series of collisions in a moderator and cooling to the energy E_0 by energy loss during the process. This has been a pioneeering concept in elementary neutron transport theory [10]. If ξ is the mean lethargy change after a collision, then

$$n_c = \ln(E/E_0)/ \xi \tag{2}$$

is the mean number of collisions suffered by the particle cooling down to the energy E_0. Assuming for E_0 a value similar to the atomic displacement energy E_d or to the sputtering threshold E_{th}, Keywell recognized that eq.(2) is close the number of sputtered atoms per incident ion in the case of silver bombarded with argon. But since ξ decreases with decreasing incident ion mass, so that n_c increases while the trend of the sputtering yield is inverse, Keywell stated that eq.(2) is not enough to explain the sputtering mechanism. It is a collision mechanism involving also the atoms set in motion by the recoiling partners in each collision. He proceeded to modify eq. (2) in terms of radiation damage theory. By means of Seitz's formula [10], the number of displaced atoms $n_{s,i}$ in the i-th collision can be calculated as a function of E, E_d, M_1 and M_2; then assuming that, of the $n_{s,i}$ displaced atoms at depth x below the surface, only a fraction n_e will be able to escape from the surface with an escaping probability decreasing exponentially, i.e.:

$$n_e = n_{s,i} \, \exp(-\beta \, x_i)$$ (3)

where x_i is related to the number of collisions suffered by the primary ion (e.g. assuming a random-walk-like model $x_i \propto i^{1/2}$). The sputtering yield is obtained by summing over the total number of collisions n_c (given by a relation similar to eq.(2)), or

$$Y = \sum_{i=1}^{n_c} n_{s,i} \, \exp(-\beta \, x_i)$$ (4)

The basic point here is, that the parameter β is treated as a fitting parameter not related to the various mechanisms involved in the loss of energy by atoms travelling towards the surface.

In Harrison's formulation (1956) the sputtering process is described in terms of four distribution functions [12]. The first distribution function concerns the particles of the beam which have not suffered a collision, while the second distribution accounts for those beam particles which, having made one or more collisions, are still not in thermal equilibrium with the target. The third distribution includes target particles which are in motion but have not "cooled" to thermal equilibrium with the surroundings. Lastly, the fourth distribution contains all particles in thermal equilibrium with the target atoms. Then a transport equation is written for the collision frequency per unit of volume , assuming that no collisions between moving particles take place. The final result for the sputtering yield is expressed as:

$$Y(u) = \sum_{n=1}^{3} S_n [\exp(r_n u) - 1] \tag{5}$$

where $u = \ln(E/E_{th})$ is the lethargy corresponding to E_{th}, the sum being extended to the only three terms which are not zero. The key point here is the dependence of S_n and r_n upon some physical parameters characterizing the system, although their algebraic expressions are very complicated[12]. These parameters are: the mass ratio M_1/M_2, the mean free path ratio λ_1/λ_2 and the trapping parameter h which is defined as

$$h(u) = \sigma_s(u)/[\sigma_s(u) + \sigma_t(u)] \tag{6}$$

where σ_s is the cross-section for scattering and σ_t is the cross-section for trapping in the lattice. The latter corresponds to what is meant by capture cross-section in neutron transport theory. Essentially S_n and r_n depend on the

Tab.1. Values of threshold energies as proposed in ref.[12] and compared with those of ref.[13], for various ion-target combinations.

Combination	M_2/M_1	E_{th} (eV)	E_{th} ref.[13] (eV)	U_0 (eV)
He - Pb	51.81	336	40	2.03
Ar - Pb	5.18	46	45	
He - Ag	26.95	275	45	2.97
Ne - Ag	5.35	72	45	
Ar - Ag	2.70	48	23	
Kr - Ag	1.29	39	85	
Ar - Cu	1.60	57	34	3.52
Kr - Cu	0.76	56	56	

mean-free-path ratio and the trapping parameter. Since at that time the knowledge about these parameters was not available, in Harrison's formulation they are treated as fitting parameters some values of which are reported in table 1 [12]. In conclusion we have again a need for cross-sections, i.e. a more detailed knowledge of collision mechanisms.

The use of a cross-section model for atomic collision appears in the sputtering theory formulated by Joyes (1968) [14]. In this formulation the sputtering yield is expressed as:

$$Y = \int d^2\Omega \int du \; \lambda(u) \; \phi(z=0,u,\Omega) \tag{7}$$

where the function $\phi \; dz \; du \; d^2\Omega$ is the number of particles in the interval $dz \; du \; d^2\Omega$ around the direction Ω, at a distance z inside the solid target, u being still the lethargy unit and $\lambda(u)$ the mean free path related to the cross-section for total collision. A transport equation can be written for the function ϕ with the following assumptions:

(a) the incident beam reaches an isotropic distribution after few collisions;

(b) all particles, which are set in motion at $t = t_0$, have the same energy after a certain time interval Δt;

(c) target atoms set in motion have an isotropic distribution.

The main problem is again the use of an interatomic potential for the determination of the mean free path λ and the cross-section describing binary collisions. In the model of Joyes the choice is a cut-off potential whose constants are fixed in relation to the Bohr potential. In practice the cut-off is active for $r = r_0$, r being the distance between the colliding partners and r_0 a parameter entering the interatomic potential. The determination of the mean free path is made through a total cross-section equal to $\pi r_0 R$, where R is the closest approach distance (i.e., a cross-section typical for hard sphere potentials) and r_0 is calculated matching the interatomic potential used in this model to the Bohr potential. On the other hand, the resulting analytical form of the cross-section for binary collisions is somewhat similar to the Lindhard power cross-section (see section below) characterized by the parameter m. The connection is given by

$$m+1 = 2(1+2a)/(1+a)^2, \tag{8}$$

a being the characteristic parameter of the cross-section calculated by Joyes. A further comment on this point is found in the following sections.

What can be pointed out here is that this model is formulated to apply for crystal structures, which is in contrast with the assumption (a) made in writing the pertinent transport equation, and the calculated sputtering yield for copper and silver are underestimated as compared with the experimental data. In fact using this theory in the case of 8 keV Ar^+ bombardment, a yield of 1.2 and 0.4 has been evaluated for Cu and Al targets respectively [14], while experimental results give a yield of 5 to 7 for Cu and 3 to 4 for Al [15].

In order to go further on in our historical analysis we need some detailed concepts in atomic collision theory and in general in the penetration of atomic particles in solids; these concepts are summarized in the next section.

3. ESSENTIAL BACKGROUND

In general a collision between two atomic systems consisting of nuclei and electrons is a compicated many body problem. But a distinction on how an atom penetrating matter loses its energy, can greatly simplify the problem. This distinction dates back to the beginning of this century[16] and was found to be useful in the development of atomic collision theory. The underlying assumption is that we can distinguish between the energy spent by the incoming ion in elastic collisions with the target atoms (we will term these as nuclear collisions) and the energy spent in inelastic collisions with the electrons resulting in excitation or ionization processes (termed electronic collisions). This separation in the energy spent by the incoming particle seems artificial since we are dealing with a non-equilibrium system during the collisions, but the transfer of the energy, acquired by electrons, back to atoms is generally a very slow process and quantitatively small so that it can be considered as a one-way process; while direct transfer of energy to electrons at moderate energy is inefficient since target electrons move much faster than the incoming ions or the collision time is large compared with the orbiting period of electrons (see discussion on the adiabaticity argument in refs.[17,18]). On the other hand, since the mass of the nucleus is large compared with that of the electron, under the same transferred momentum, the energy drain by electrons is more efficient since energy is the square of momentum divided by twice the mass, so that the two processes are competitive and in many problems in fact, the two contributions can be separated by physical considerations allowing for a simplified treatment in analytical theories.

The statistics of a collisional event is conveniently described in terms of cross sections. Considering a particle which penetrates a layer with thickness ΔR of a target with N randomly distributed atoms/cm^3, the quantity

$$N \, \Delta R \, \sigma \qquad\qquad\qquad (9)$$

represents the average number of occurrences for the event characterized by the cross section σ. Since $N \Delta R$ can be interpreted as the number of target atoms/cm^2 seen by the incoming particle and σ, in the classical sense, as the effective area for the occurrence of the specified event, so that the number of trials giving rise to the event is practically unlimited while the probability of occurrence is very small; then the statistics of the event can be described by means of the Poisson law with the mean number given by eq.(9). Moreover in the case of $N \Delta R \sigma \ll 1$, eq.(9)

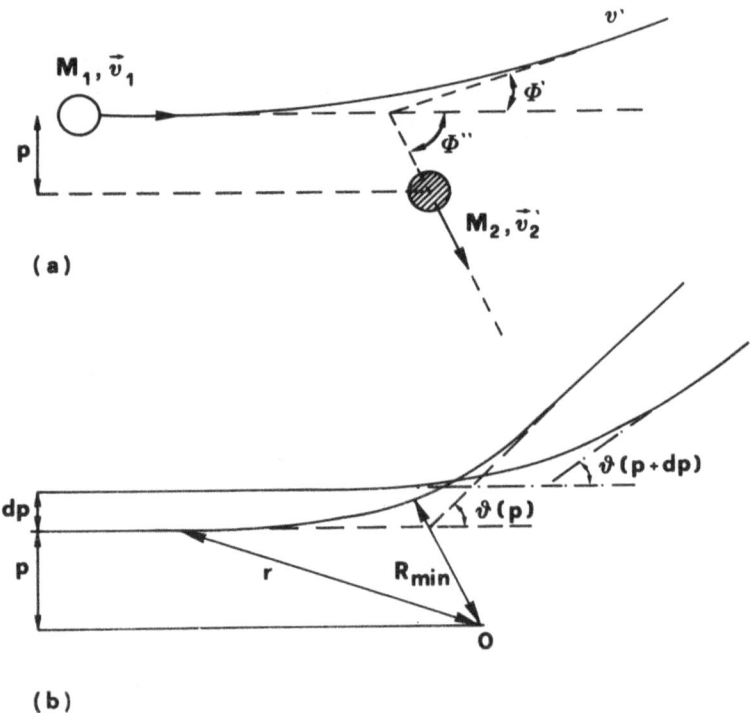

Fig.4. Parameters involved in a single collision event: (a) laboratory system; (b) center of mass system.

represents the probability for the incoming particle to undergo the event specified by the pertinent cross section, or in differential form we can write for this probability

$$dP = N\Delta R \, d\sigma \qquad (9a)$$

$d\sigma$ being the differential cross section.

We are now interested in quantifying the energy lost by the incident particle in elastic collisions with target atoms, so that the event is characterized by the initial energy E of the incident particle and the energy T transferred to the recoiling particle (see fig.4). In terms of cross sections this process is described by the energy loss cross section:

$$d\sigma = d\sigma(E,T) = \frac{d\sigma}{dT} \, dT \qquad (10)$$

which is an essential quantity in solution of basic integral equations as we will see below. Cross sections are discussed in detail elsewhere in this school [19]. What we need to recall here is the following. Assuming the classical definition of the differential cross section [21,22,23], i.e. $d\sigma=d(\pi p^2)$, p being the impact parameter, we write:

$$d\sigma = -2\pi \, p \, \frac{dp}{d\theta} \, d\theta = -\frac{p}{\sin\theta} \frac{dp}{d\theta} \, d\Omega \qquad (11)$$

where θ is the scattering angle in the center of mass system (c.m.s.) and $d\Omega = 2\pi \sin\theta \, d\theta$. Observing conservation of energy and momentum this angle can be related to the interatomic potential V(r) by:

$$\theta = \pi - 2p \int_{R_{min}}^{\infty} \frac{dr/r^2}{(1-V(r)/E_r - p^2/r^2)^{1/2}} \qquad (12)$$

r being the distance from the scattering center, R_{min} the closest approach distance, and $E_r = M_2 E/(M_1+M_2)$ the c.m.s. energy (the relative energy). In other words eq.(12) means that when speaking of a cross section we are speaking about the interatomic potential from which the cross section has been obtained applying eqs.(12,11) by inversion of eq.(12) in order to obtain p(θ) and calculate dp/dθ.

Unfortunately from eq.(12) is not often possible to obtain exact analytical solutions, except in the well known cases of Coulomb interaction ($V(r) \propto r^{-1}$) and "inverse-square" interaction ($V(r) \propto r^{-2}$). The latter doesn't correspond to any real potential. A reference tabulation of scattering integrals of type eq.(12) is that published by Robinson [24] in which data concerning R_{min}, θ, the transferred energy T, the time integral and the differential scattering cross section are dislplayed for the following interatomic potentials: 1. Born-Mayer (BM), 2. Bohr (B), 3. Thomas-Fermi (TF) with Molière and Sommerfeld approximations.

The philosophy underlying the classical scattering calculations is to assume an ion-atom potential of the form:

$$V(r) = \frac{Z_1 Z_2 e^2}{r} \, u(x) \tag{13}$$

where u(x) is the so called "screening function" which tends to 1 for $r \to 0$ and vanishes for $r \to \infty$. In a simplified Thomas-Fermi potential $u \equiv u(r/a)$, i.e. independent of Z_1 and Z_2 which implies a similarity for all ion-atom combinations, a being a characteristic screening length. Lindhard et al.[25] (extending the investigation already initiated by Bohr [16]) proposed a power form for the screening function:

$$u(r/a) = \frac{k_s}{s} \, (a/r)^{s-1} \tag{14}$$

from which for several integer values of s, exact scattering formulae are obtained. In fact, using eq.(12) in the form given by the momentum approximation (which means essentially $V(r)/E_r \ll 1$ [22,23]), one obtains:

$$\theta(p) = \frac{\gamma_s k_s}{\varepsilon} \, (a/p)^{s-1} \tag{15}$$

Here γ_s and k_s are well defined constants [20] and

$$\varepsilon = \frac{a\,E_r}{Z_1 Z_2\,e^2} \qquad\qquad (16)$$

is the dimensionless unit for the c.m.s. energy. By inversion of eq.(15) and insertion in eq.(11), we obtain:

$$d\sigma(\theta) = \text{const}\ a^2\,\varepsilon^{-2m}\,\theta^{-1-2m}\,d\theta \qquad\qquad (17)$$

(m=1/s), or in terms of recoil energy T

$$d\sigma(E,T) = C_m\,E^{-m}\,T^{-1-m}\,dT \qquad\qquad (18)$$

assuming, when replacing θ in eq.(17), that $T/T_m \ll 1$ [25,23], i.e. eq.(18) is derived for the case of soft collisions ($T_m = \gamma E = 4M_1 M_2 (M_1+M_2)^{-2}\,E$ is the maximum energy transfer and C_m a constant which depends on the ion-target combination [25,23,3]). Eq.(18) is a powerful tool in the theory of atomic collisions. The parameter m varies slowly with energy (see the discussion in ref.[26]), ranging from m=1 at high energy (Rutherford scattering) to m=1/2 for $\varepsilon \le 2$, m=1/3 for $\varepsilon \le 0.2$ down to m=0 at very low energies, even if the basis for deriving eq.(18) were the study of moderate deviation from Rutherford scattering, i.e. m≥0.2. Then we can further comment on the value suggested by Joyes for the parameter a in eq.(8). In the energy range 1-10 keV (which corresponds to $0.016 \le \varepsilon \le 0.16$ for the combination $Ar^+ \rightarrow Al$ and to $0.009 \le \varepsilon \le 0.09$ for $Ar^+ \rightarrow Cu$) the suggested range for the parameter a is 1 to 0.25 which gives a corresponding range for m of 0.5 to 0.9. These values are too high compared with the recommended m-values in the corresponding energy ranges when using power-law scattering.

　　We are now able to quantify the energy lost by the incoming particle when crossing a thickness ΔR of target material. The average energy loss, for collisions which transfer an energy T to the collision partner and which have an occurrence probability given by eq.(9a), is then

$$< E > = \int T \, dP = N \, \Delta R \int T \, d\sigma \tag{19}$$

or introducing the nuclear stopping cross section

$$S_n(E) = \int T \, d\sigma \tag{20}$$

we can write, in the limit of ΔR small, the well-known expression for the nuclear stopping power:

$$\left(\frac{dE}{dR}\right)_n = -NS_n(E) \tag{21}$$

the minus sign being introduced to indicate that energy is obviously lost by the incident particle.

The picture for the energy spent in inelastic collisions with electrons is more involved and we refer to these proceedings for details [19]. What we need here is to keep in mind that the drain of energy by electrons can even be the dominant process, especially at high incident energy where theories like that of Bethe [27] are valid. For the energy range pertinent for sputtering, the problem can be formally treated following the same outline as in the case of nuclear stopping power, in the sense that one can introduce the electronic stopping power as [28,16,23]:

$$\left(\frac{dE}{dR}\right)_e = -NS_e(E) \tag{22}$$

where in analogy with eq.(20), S_e is the electronic stopping cross section, which takes into account the energy transferred to electrons in a single collision. In the energy range interesting to sputtering, the commonly used expressions for S_e are those of Firsov[29] and Lindhard et al.[30], both proportional to the incident ion velocity, even if the experimental results, as supplied by the rapidly improving techniques, indicate deviations from the dependence on both Z_1 and velocity[31]. Stopping cross sections are conveniently expressed in terms of dimensionless quantities like ε (eq.(16)) and ρ, the dimensionless measure of the path length defined as:

$$\rho = \pi a^2 \gamma NR \tag{23}$$

In ρ-$\mathcal{E}$ units Lindhard's expression for the electronic stopping power, reads:

$$s_e(\mathcal{E}) = k \, \mathcal{E}^{1/2} \tag{24}$$

with

$$k = \xi_e \frac{0.0793 \, Z_1^{1/2} \, Z_2^{1/2} \, (A_1 + A_2)^{3/2}}{(Z_1^{2/3} + Z_2^{2/3})^{3/4} \, A_1^{3/2} \, A_2^{1/2}} ; \quad \xi_e \cong Z_1^{1/6} \tag{25}$$

A being the atomic weight, and the total energy loss will be (in reduced

Tab.2. Values of the reduced energy parameter for various ion-target combinations. The values of the constant k entering the electronic stopping cross section are also given.

Combination	$\mathcal{E}$ corresponding to 1 keV	k
He - Cu	.15884	.96757
Ne - Cu	.02269	.29045
Ar - Cu	.00935	.29045
Cu - Cu	.00445	.15793
Kr - Cu	.00294	.14070
Xe - Cu	.00137	.12791
Ar - C	.02139	.09939
Ar - Al	.01568	.12218
Ar - Ar	.01355	.14493
Ar - Fe	.01015	.17168
Ar - W	.00404	.40330
Ar - U	.00321	.50403

notation)

$$\frac{d\mathcal{E}}{d\rho} = -(s_n(\mathcal{E}) + s_e(\mathcal{E})). \tag{26}$$

Table 2 shows values of $\mathcal{E}$ corresponding to 1 keV and values of k [32] for various ion-target combinations. Fig.5 shows a comparison between s_n and s_e as a function of $\mathcal{E}$. We use the plot versus $\mathcal{E}$ instead of the traditional plot versus $\mathcal{E}^{1/2}$ in order to show that at very low energy the drain of energy by electrons can be competitive with the drain by elastic collisions. The potential of Wilson et al.[33] has been included for what concerns

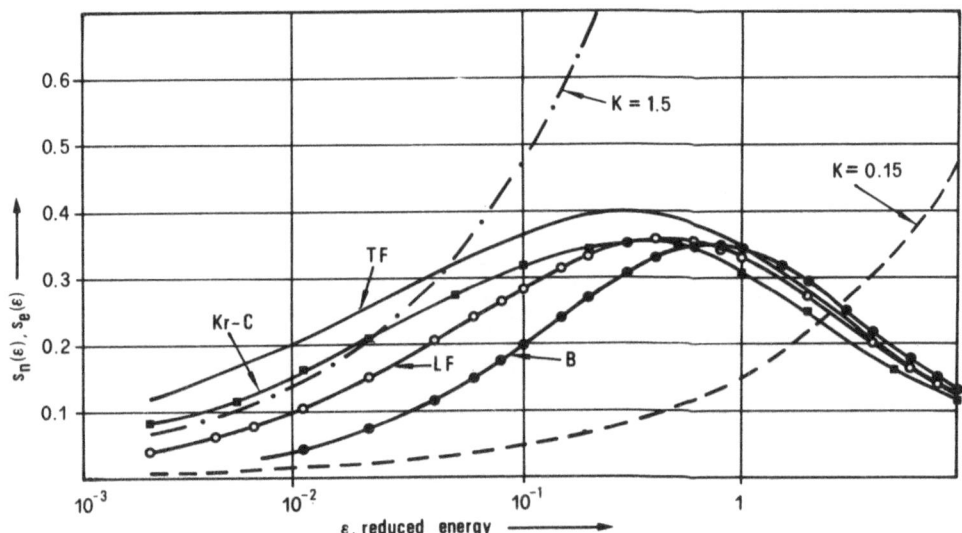

Fig.5. Reduced nuclear (solid lines) and electronic (dashed lines) stopping cross sections versus reduced energy: Thomas-Fermi potential (TF), Lenz-Jensen (LJ), Bohr (B) and Wilson et al. potential (Kr-C).

nuclear-stopping cross-section calculations (see also the curves labeled 'universal reduced nuclear stopping' in fig.2-18 of ref.[34]), since this potential shows a better agreement with experimental sputtering yield toward the low energy region[35]. It can be seen that for low ε, electronic stopping is of the same order of magnitude or greater than nuclear stopping, expecially for low energy light ion bombardment on heavy substances (i.e. for values of k near or greater than 1).

From the specific energy loss, eq.(26), or from the corresponding equation $dE/dR = -NS(E)$, $S(E)$ being the total stopping cross section, we can define by integration the basic range concept[16,20,3]:

$$R(E) = \int_0^E \frac{dE'}{N \, S(E')} \tag{27}$$

which is the mean penetrated path length of a particle with initial energy E. Eq.(27) gives a straightforward estimate of R(E) when the energy is

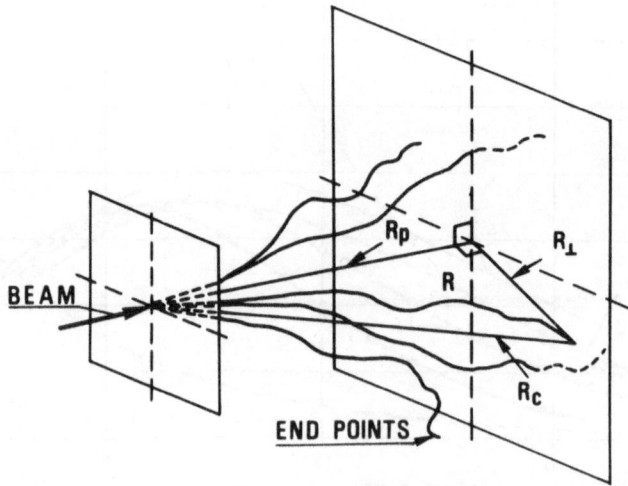

Fig.6. Pertinent penetration parameters: path length (R), projected range (R_p), chord range (R_c), transverse range ($R_\perp$).

spent in elastic collisions only, inserting eqs.(18,20) in (27). Another important penetration parameter is the mean projected range $R_p(E)$, see fig.6. The projected range correction $(R-R_p)/R_p$ gives an idea about the scattering suffered by the projectile during the penetration. This quantity is qualitatively shown in fig.7 as a function of M_2/M_1. In fact a simple approximation valid up to $M_2/M_1 \cong 1$ is given by[20]:

$$\frac{R-R_p}{R_p} \cong \frac{M_2}{M_1} \frac{1}{4m(2-m)} \qquad (28)$$

where electronic stopping is neglected and m is the usual power law parameter. Fig.7 shows that the projected range correction is <<1 (or

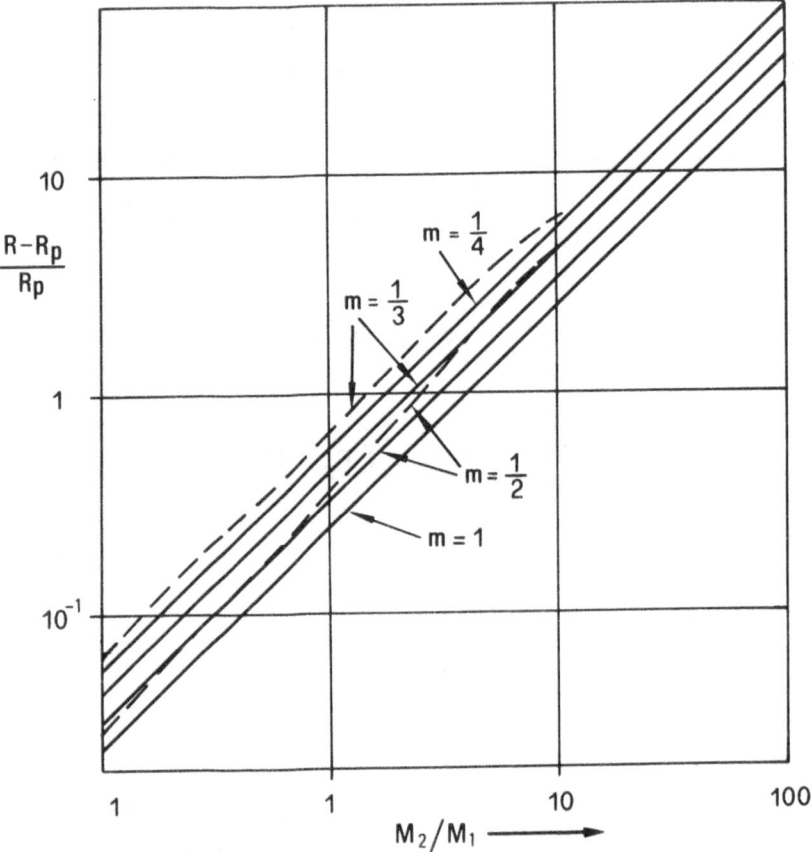

Fig.7. Projected range correction as function of the mass ratio; solid lines: eq.(28), dashed lines: Winterbon et al. calculations [35].

$R_p/R \cong 1$) for $M_2 << M_1$ and $>>1$ (or $R_p/R << 1$) for $M_2 >> M_1$. For comparison in the same figure is shown the projected range correction as calculated in ref.[36]. It can be seen, as already stated, that the approximation given by eq.(28) is rather good up to $M_2/M_1 \cong 1$, especially for $m=1/2$.

A further comprehension of sputtering theory now requires introduction to the basic integral equations which is the subject of the next section.

4. BASIC INTEGRAL EQUATIONS

Let us now look for a physical quantity concerning radiation damage (e.g. the number of recoils generated by an incoming ion, the number of Frenkel pairs, the number of vacancies, the distribution in space of the implanted particles etc.etc.). We indicate this quantity by the generic symbol φ [37], while $\bar{\varphi}(E)$ will be its mean value after irradiation with a particle of energy E. Consider first the equal mass case in which $Z_1=Z_2$ and $M_1=M_2$. It is important to state that our physical quantity $\bar{\varphi}(E)$

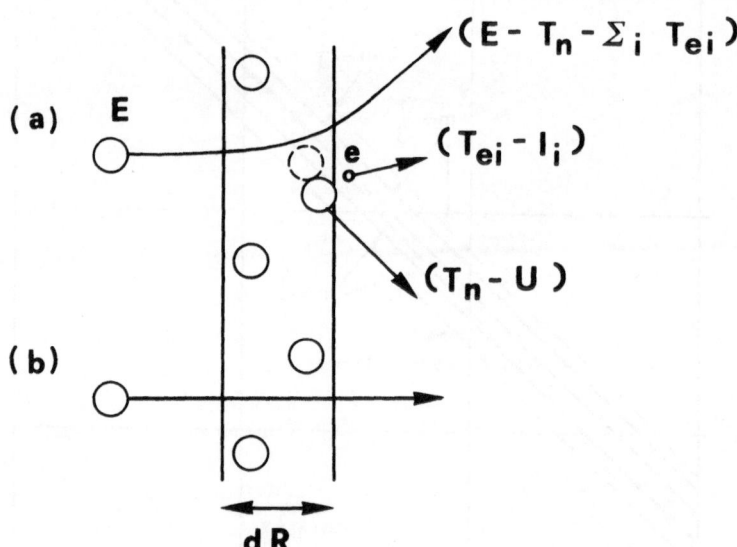

Fig.8. Schematic representation of situations concerned with a particle of energy E crossing a small path length dR: (a) collision with a target particle and related events (the label "e" indicates an electron); (b) non−occurrence of collisions.

is additive, or in other words that the damage effects under consideration are additive, i.e. each independent slowing down event suffered by the incoming particle will contribute additively to the quantity $\bar{\varphi}$. From a collisional point of view, looking at the simplified picture of fig.8, a particle with energy E travelling a path length dR (small enough to avoid multiple collisions) in a medium with N particles per unit of volume, will be concerned with the following situations:

(a) Denoting by $d\sigma_{n,e}$ the differential cross section for collisions which transfer an energy T_n to the nucleus of the colliding partner and an energy T_{ei} to an electron (produced during the collision and labeled by the suffix 'i'), there is a probability for such a collision to occur given by N dR $d\sigma_{n,e}$. As a consequence of this collision the incoming particle has changed its energy contributing then to the quantity $\bar{\varphi}$ with a value given by $\bar{\varphi}(E - T_n - \sum_i T_{ei})$. On the other hand the struck particle will contribute to the same quantity with a value given by $\bar{\varphi}(T_n - U)$, where U is the energy lost by the target atom in disrupting bonds ("bulk binding energy"). In the same way the electrons set in motion after the collision will contribute to $\bar{\varphi}$ with a value given by $\sum_i \bar{\varphi}_e(T_{ei} - I_i)$, summed over all produced electrons, I_i being the associated ionization potential. Note that another symbol, $\bar{\varphi}_e$, has been used to indicate the contribution of electrons to the physical quantity, i.e. the $\bar{\varphi}$-function for electrons is described by a different function.

(b) There is a probability (1- N dR $\int d\sigma_{n,e}$) that no collision occur, or that the contribution to the quantity $\bar{\varphi}(E)$ remains unchanged at its original value $\bar{\varphi}(E)$.

Considering the events described in (a) and (b) and recalling the stated additivity for contributions to the quantity $\bar{\varphi}$, integrating over all possible collisions, we get the following equation:

$$\bar{\varphi}(E) = N \, dR \int d\sigma_{n,e} [\bar{\varphi}(E - T_n - \Sigma_i \, T_{ei}) + \bar{\varphi}(T_n - U) + \Sigma_i \, \bar{\varphi}_e (T_{ei} - I_i)] +$$

$$+ (1 - N \, dR \int d\sigma_{n,e}) \bar{\varphi}(E) \qquad (29)$$

or

$$\int d\sigma_{n,e} [\bar{\varphi}(E) - \bar{\varphi}(E - T_n - \Sigma_i \, T_{ei}) - \bar{\varphi}(T_n - U) - \Sigma_i \, \bar{\varphi}_e (T_{ei} - I_i)] = 0. \qquad (30)$$

Eq.(30) is the basic equation in atomic collision theory and can be interpreted as follows: the $\bar{\varphi}$-value of the particle before the collision is equal to the sum of $\bar{\varphi}$-values carried by the particle itself, the struck atom and the produced electrons after the collision, averaged over all possible collisions.

The key quantity in eq.(30) is the cross section $d\sigma_{n,e}$, and its knowledge or some actual hypothesis on it is crucial for the solution of eq.(30). We will see below that this problem can be simplified using reasonable approximations. Two comments must be added here. First there is no necessity that the total cross section ($\int d\sigma_{n,e}$) be finite, in fact cross sections of practical interest, like those of power-law scattering, are divergent; but since the actual quantities entering the equations are integrals of the cross section times quantities tending to zero, the problem is normally immaterial. Second, when cross sections larger than the atomic size are important it is no more possible to treat events as single collision events, so that the integral equations must be restated.

In order to solve eq.(30), the knowledge of the function $\bar{\varphi}_e$ is needed and, apart from those cases where this function is known a priori, an additional equation is necessary to describe $\bar{\varphi}_e$. Denoting by $d\sigma^*_{n,e}$ the differential cross section which describes the energy transfer T_n to the nucleus and T_{ei} to an electron in collisions initiated by a primary electron of initial energy E, the same argument used to obtain eq.(30), leads to

$$\int d\sigma^*_{n,e} [\bar{\varphi}_e(E) - \bar{\varphi}_e(E - T_n - \Sigma_i \, T_{ei}) - \bar{\varphi}(T_n - U) - \Sigma_i \, \bar{\varphi}_e (T_{ei} - I_i)] = 0 \qquad (31)$$

which in combination with eq.(30) gives the solution for $\bar{\varphi}$ and $\bar{\varphi}_e$.

We briefly discuss now the approximations often used in solving equations of the type discussed above. We will not enter into the details concerning solution techniques, which are out of the purpose of this contribution, even if they are of great practical importance so we just mention here that a look at these techniques is very useful (e.g., refs.[23,20,36,37]). The following five approximations have been discussed by Lindhard et al. in the "Notes on Atomic Collisions III" [37]:

(i.) The first approximation consists in neglecting the $\bar{\varphi}$-term in eq.(31), i.e. the electron energy is not enough to produce recoil atoms. This approximation is valid only until the disruption of atomic bindings occurs as a consequence of electron excitation or until the electron energy is so high to dislodge atoms directly. Within this approximation eq.(31) gives a separate solution for $\bar{\varphi}_e$, and transforming $d\sigma^*_{n,e}$ in $d\sigma^*_e$, the differential cross section for energy transfer in collisions between electrons, a simpler equation can be obtained

(ii) The second approximation concerns the atomic binding energy U. If the energy transferred in elastic collisions is high, then the recoil term in eq.(30), $\bar{\varphi}(T_n-U)$, can be replaced by $\bar{\varphi}(T_n)$. This approximation is not correct when the incident particle energy is low, so that the binding energy U plays a fundamental role in the energy balance. In this case, since the electronic stopping is also small (except in near threshold problems where this mechanism of energy loss can be competitive, as we have shown in fig.5., one obtains a simplified version of eq.(30):

$$\int d\sigma_n [\bar{\varphi}(E) - \bar{\varphi}(E-T_n) - \bar{\varphi}(T_n-U)] = 0 \qquad (32)$$

in which $d\sigma_{n,e}$ reduces to $d\sigma_n$, the differential cross section for elastic ion-atom collisions. Eq.(32) has also been used by other authors (e.g. [38]).

(iii) The third approximation assumes $T_e \ll E-T_n$, which still holds if the particle energy is not too low and then, together with approximation (ii), by a series expansion of the second term under the integrand of eq.(30) the pertinent equation can be obtained.

(iv) Since when elastic collisions are important (small impact parameter, see the discussion on this argument in ref.[39]), electronic excitation is quantitatively small, which conversely take place at large impact parameter, a separation between electronic and nuclear collisions can be assumed. This approximation allows us to split up the cross section into separate contribution for elastic nuclear collisions and for electronic stopping.

(v) The last approximation is the assumption that T_n is small as compared with E. Even if the validity of this approximation is questionable, since e.g. for equal target and ion masses we can have T_{max} =E, its use enables the evaluation of approximate solutions.

We procede now in the illustration of two examples where the equations just discussed are applied.

Let P(E,R) dR be the probability that an ion with initial energy E comes at rest after travelling a path length (R,dR) in a random medium of density N, i.e. R represents the total distance traversed by the particle and measured along its path length. An integral equation for P can be derived following the steps leading to eq.(29), having in mind that in this case we follow only the primary ion, disregarding then the fate of both the recoiling particle and the electrons. The equation reads:

$$P(E,R) = N \, \delta R \int d\sigma_{n,e} \; P(E-T_n-\sum_i T_{ei}) +$$

$$+ (1-N \, \delta R \int d\sigma_{n,e}) \; P(E,R-\delta R). \tag{33}$$

Note, in eq.(33), terms like P(E,R- δR) have the meaning that the particle, in order to come to rest in the interval (R,δR), needs only to travel a path length (R- δR) after having travelled the portion δR. Also note that in the second term on the r.h.s. of eq.(33), energy loss to electrons is included in the cross section $d\sigma_{n,e}$, but an alternative formulation, in which electronic energy loss is explicitly taken into account, is also possible[23]. Taking first the limit $\delta R \to 0$ and including then approximations (iii) and (iv), i.e. small energy transfer to electrons and separation between electronic and nuclear collisions, we obtain

$$- \frac{\partial P(E,R)}{\partial R} = N \int d\sigma_n \, [\, P(E,R) - P(E-T_n,R) \,] + N \, S_e(E) \, \frac{\partial}{\partial E} P(E,R) \tag{34}$$

which is the now famous equation for the path length distribution. Eq.(34) can be solved by moments technique, reconstructing, in principle at least, the distribution from its moments. In fact the following normalization holds for P(E,R):

$$\int_0^\infty dR\, P(E,R) = 1 \qquad (35)$$

and adding the definition of spatial moments,

$$R^n(E) = \int_0^\infty dR\, R^n\, P(E,R) ; \quad n = 0,1,2........ \qquad (36)$$

it is possible to write the pertinent integral equations for the average path length (n=1) and for the average projected range [20,23,40]. Since generally a finite number of moments is calculated, the reconstruction of the distribution follows from more or less good approximation methods; the uniqueness of the distribution is, however, determined by an infinite set of moments, so the problem is not trivial [36,41,42]. Reasonable zero-order approximations seem to be gaussian distribution functions [23,43,44] and reference tabulations are those of Brice [45] and Winterbon [46].

   The second example is concerned with the distribution of deposited energy which answers the question (with sufficient information in many experimental situations), "where is the energy deposited in a target spatially located after the end of a slowing down process initiated by a primary ion [36]?" Again an integral equation can be derived following the arguments stated in Notes III [37]. Let $F_D(E,\Omega,r)\, d^3r$ be the average amount of energy deposited in the volume element (r, d^3r) in a collisional process initiated by an ion (of mass and atomic number same as the target atoms) starting at $r=0$ with energy E and direction Ω. The deposited energy is determined after termination of all slowing down processes, those concerning the primary ion and those concerning the recoil atoms. Considering the case of a random infinite medium of density N and confining the problem to the case of elastic collisions only and negligible binding energies, with the steps leading to eq.(29) as a guide, the integral equation for F_D reads [36,42,47]:

$$F_D(E,\Omega,r) = N \, |\delta R| \int d\sigma_n \, [\, F_D(E-T_n,\Omega\,',r) + F_D(T_n,\Omega\,'',r) \,] +$$

$$+ \, (1-N \, |\delta R| \int d\sigma_n \,) \, F_D(E,\Omega\,,r-\delta R) \tag{37}$$

where δR is a small vector distance travelled by the incoming particle. An analogous equation has been derived by Sanders [48] on the same basis. From eq.(37) by a series expansion in the final term on the r.h.s., one obtains:

$$- \, \Omega\cdot\nabla F_D(E,\Omega,r) = N\int d\sigma_n \, [\, F_D(E,\Omega\,,r) - F_D(E-T_n,\Omega\,',r) - F_D(T_n,\Omega\,'',r) \,] \tag{38}$$

which allows the determination of the function F_D through its spatial moments [23,36], with the same problems as mentioned above. The function F_D will obey the condition:

$$\int d^3r \, F_D(E,\Omega\,,r) = v(E) \tag{39}$$

which holds from energy conservation, if [37]

$$v(E) = E - \eta(E) \tag{40}$$

is the average energy spent in elastic collisions, while $\eta(E)$ is the average energy going in electronic stopping. The efforts in the characterization of F_D were directed, both experimentally (e.g. [49]) and theoretically (e.g. [41]), toward its depth profile defined as

$$F_D(E,\vartheta,x) = \int_{-\infty}^{\infty} dy \int_{-\infty}^{\infty} dz \, F_D(E,\Omega,r) \tag{41}$$

the x-axis (inward oriented) having the same direction as the surface normal, and ϑ being the angle between Ω and this axis.

To conclude this section, since we are showing only the iceberg tip concerning the research in this field, a special mention is customary, as style example to imitate, about Lindhard's care in checking the accuracy of adopted approximations or the validity of calculated solutions, and two further remarks.

Eq.(29) and similar equations have been written for the so-called equal masses case, i.e. the incoming particle has the same mass and atomic number as the target particle, but the actual case is the opposite: in general, in fact, one deals with experiments where primary beams contain particles which are different from those of the target. The problem is still solvable but a parallel set of integro-differential equations needs to be considered. To briefly illustrate the problem take as example the equation just discussed for F_D . In the case of a projectile with mass M_1 and atomic number Z_1 bombarding a target (M_2, Z_2 and density N), in addition to the function F_D, we need to determine the function $F_D^1(E, \Omega, r)$ which describes the spatial distribution of energy deposited by the projectile in its slowing down in the target, the ion-target atomic collisions being characterized by a cross section $d\sigma_n^1$. The usual arguments lead to

$$- \Omega \cdot \nabla F_D^1(E, \Omega, r) = N \int d\sigma_n^1 \ [\ F_D^1(E, \Omega, r) - F_D^1(E - T_n, \Omega', r) - F_D(T_n, \Omega'', r) \ ] \qquad (42)$$

with the condition

$$\int d^3r \ F_D^1(E, \Omega, r) = v(E) \qquad (43)$$

Note that eq.(42) is non-homogeneous and if, as already mentioned in the case of the path length distribution (eq.(33)), we need to follow primary particles only as in the case of range distributions [40,50], the inhomogeneous term (F_D in eq.(42)) disappears [37,8,36].

The second remark concerns the interesting discussion about the apparently crowded family of equations inherent to transport phenomena [51,52,39]. The type of equations just discussed belong to the backward or adjoint forms of the Boltzmann equation. The equivalence between this type of equation and the so-called forward form has been demonstrated by Lindhard et al.[51] and the backward one seems to be the most useful form in ion beam transport problems [39]. The equivalence was well known in neutron transport field [53] and the adjoint form (backward) was found to be useful for very interesting applications in this field [54,55] by the introduction of the importance concept.

5. SIGMUND'S SPUTTERING THEORY

From the previous sections we extract now the stopping power and deposited energy concepts. Consider a one-dimensional geometry,

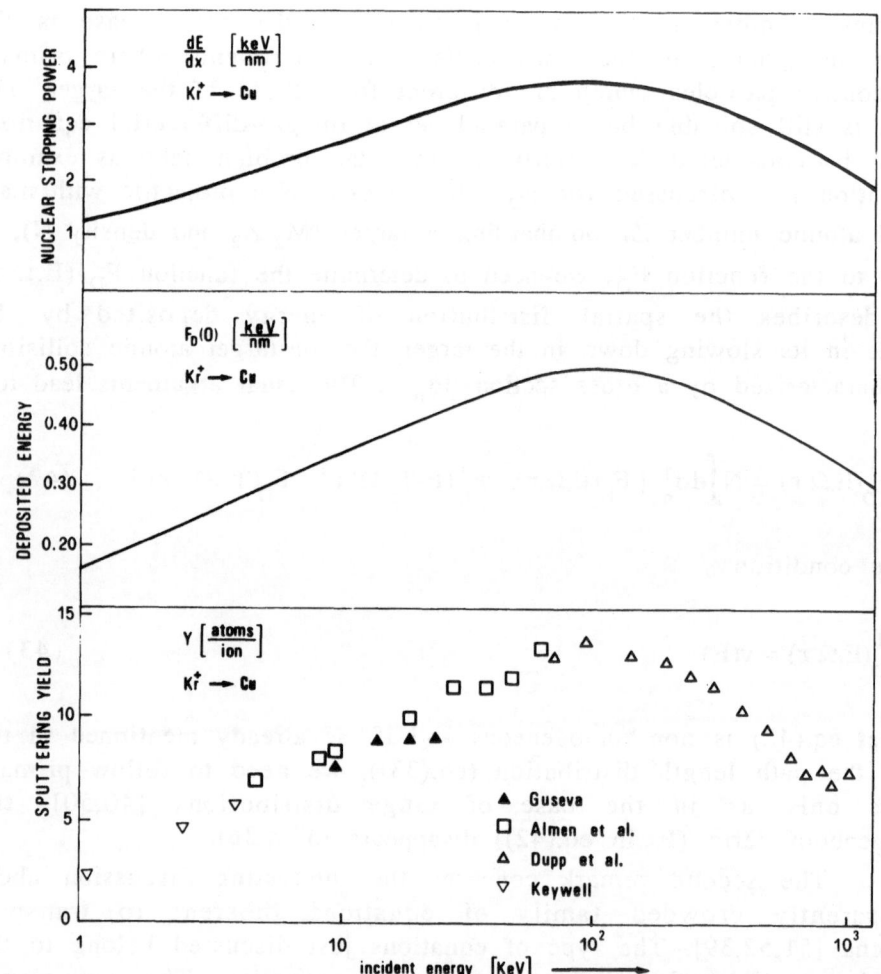

Fig.9. Nuclear stopping power, deposited energy near the surface, and sputtering yield in the case of Kr bombarding a Cu target.

assuming an x-axis perpendicular to the target surface and take as reference the classical example of Kr^+ ions bombarding a Cu target

[25,56,57]. Since for the combination Kr-Cu an incident energy of, e.g., 100 keV corresponds to roughly $\mathcal{E}=0.3$ (see tab.2) we can disregard electronic stopping and consider only nuclear stopping. In fig.9 a comparison between nuclear stopping power $(dE/dx)_n$, deposited energy near the surface $(F_D(x=0)$, calculated from [46]) and sputtering yield measurements is reported as function of beam incident energy (experimental data from refs. [58,59,60,9]). The shape of the curves are encouragingly similar so that one can argue that the sputtering yield in the elastic region is proportional to the energy deposited near the surface or, with a suitable coefficient, to the nuclear stopping power. This is a very simple conclusion, but is of course a posteriori and based on roughly 40 years of research in the atomic collisions field. Now the question is: can we build-up a theory in which the sputtering yield can be formally expressed as

$$Y = \Lambda \, F_D(E,\vartheta,x=0) \tag{44}$$

where all pertinent parameters are well defined according to first principles and to compatible physical models? A positive answer to this question is contained in the sputtering theory formulated by Sigmund in 1969 [8].

We will briefly outline this theory in its original formulation, switching then to a more recent presentation of the teoretical concepts involving sputtering [3], since the latter in our opinion is more instructive from a tutorial point of view.

Define a function G in a slab geometry with the x-axis inward oriented so that $G(x,v_0,v,t) \, d^3v_0 \, dx$ is the average number of atoms moving at time t with velocity (v_0,d^3v_0) in the layer located at (x,dx) as consequence of the motion initiated by an atom starting with velocity v in the plane x=0 at time t=0. Then, if v_{0x} is the x-component of v_0, $G \, |v_{0x}| \, d^3v_0 \, dt$ is the number of atoms crossing the plane x with velocity (v_0,d^3v_0) in a time interval dt, and the sputtering yield (backward sputtering) through a plane surface located at x=0 can be written as:

$$Y = \int d^3v_0 \; |v_{0x}| \int_0^\infty dt \; G(x=0, \, v_0,v,t) \tag{45}$$

Integration limits on d^3v_0 are extended to all negative x-components of v_0 and must be compatible with surface binding forces. An analogous formulation can be given for transmission sputtering through an arbitrary surface[8]. An equation for G can be written by linearization of the Boltzmann (non–linear) equation [3,8], but within the framework traced in this contribution, it is more consistent to follow the approach containing the steps outlined in the previous section. In fact after an interval of time δt the standard arguments gives:

$$G(x,v_0,v,t) = Nv \; \delta t \int d\sigma_{n,e} \; [\; G(x,v_0,v',t) + G(x,v_0,v'',t) \;]$$

$$(1-Nv \; \delta t \int d\sigma_{n,e}) \; G(x-\cos\vartheta \; v \; \delta t,v_0,v,t-\delta t) \qquad (46)$$

where $\cos\vartheta = v_x/v$ with obvious meaning of other symbols and expansion to the first order in δt yields

$$-\frac{\partial}{v\partial t} G(x,v_0,v,t) - \cos\vartheta \; \frac{\partial}{\partial x} G(x,v_0,v,t) = N\int d\sigma_{n,e} \; [\; G(x,v_0,v,t)$$

$$- G(x,v_0,v',t) - G(x,v_0,v'',t)] \qquad (47)$$

Eq.(47) is derived for single collisions between atoms and the cross section $d\sigma_{n,e}$ contains implicitly the electronic energy loss. Extension to the non-equal mass case has been outlined above, so we do not repeat the argument here although we stress that this extension is essential for a complete solution of the problem. To solve eq.(47) in view of eq.(45), we define the following auxiliary functions:

$$H(x,v) = \int d^3v_0 \; |v_{0x}| \; F(x,v_0,v) = \int d^3v_0 \; |v_{0x}| \int_0^\infty dt \; G(x,v_0,v,t) \qquad (48)$$

so that

$$Y(E,\vartheta) = H(E,\vartheta,x=0) \qquad (49)$$

for backward sputtering in terms of energy variables. To illustrate the compatibility of eqs.(49) and (44) let us recall the main steps followed in solving the integral equations describing function H, obtained by using eq.(47) with the definition (48) (see ref.[8] for details and boundary conditions). The procedure for analytic solution adopts the following basic steps:

1. Expansion of function H in terms of Legendre polynomials for the angular dependence.

2. Use of the moment technique.

3. Separation between elastic and inelastic collisions with subsequent use of Lindhard's expressions for electronic energy loss and power cross section for nuclear (elastic) energy loss (see sections above).

4. Reconstruction of function H by an Edgeworth expansion with gaussian weight function [61,62].

5. Neglect of displacement,threshold and bulk binding energies and inclusion of surface binding forces.

6. Proof, by physical and mathematical arguments [8], that the leading term in the asymptotic expansion used in the calculations [63,64] has the same form at all energies $E \gg U_0$ (U_0 being the height of a planar potential barrier which accounts for surface binding forces) and is proportional to E or, if electronic stopping is taken into account by Lindhard's model [37], the leading term is proportional to $v(E)$, so that the function H can be linked to the deposited energy function F_D [8], i.e.

$$H(E,\vartheta,x) = \Lambda \, F_D(E,\vartheta,x) \tag{50}$$

with

$$\Lambda = \frac{3}{4\,\pi^2\,N\,C_0\,U_0} = \frac{0.0420 \ \text{\AA}^{-2}}{N\,U_0} \tag{51}$$

C_0 being the constant entering the power cross section for $m=0$ (cfr. eq.(18)). It can be also shown that in the elastic collision region the sputtering yield is proportional to the nuclear stopping power or

$$Y(E) = \Lambda \, \alpha \, N \, S_n(E) \tag{52}$$

where α is a factor which depends on the power-law parameter and on the mass ratio M_2/M_1. Comparison of eq.(44) with experimental data has now roughly 20 years of history and the reference work is that of Andersen and Bay [15]. It is worthy and healthy to recall the basic assumptions under which eqs.(44,52) have been derived [3]:

1. Assumption of linear regime in the cascade, i.e. within the volume of the cascade only a small fraction of target atoms is set in motion (see the discussion on fig.3 above).

2. The cascade is developed enough to justify the asymptotic expansion used in recovering the solution of the transport equation and the isotropy of the recoiling atoms set in motion (see discussion about the recoil density below).

3. The presence of a surface does not influence the development of the cascade, e.g. this is not valid in the case of light ion bombardment since the ion can cross the surface several times by backscattering.

4. The elastic and inelastic energy loss models which are used bring implicitly their limitations in the theory.

5. Apart from the inclusion of surface binding forces, other type of binding energies have been disregarded.

6. No crystalline structures or preferential directions enter the theory (e.g. channeling or blocking effects need further considerations).

The results of the theory have of course been used also outside the range of the discussed assumptions, mainly to recover suitable interpretations in more compicated physical situations, with more or less encouraging results.

The alternative derivation of the sputtering formula is more qualitative but enlightening from a physical point of view and clearer for what concerns the basic assumptions in the theory. This derivation is based on the recoil density concept [64], which originated as an answer to classical problems in radiation damage [65]. Suppose that a nearly monochromatic beam of particles impinges on a surface (fig.10) of a solid and look at what happens inside the solid, i.e. the primary quantity of interest is how many atoms are set in motion, since the answer to this question allows us to determine other connected and important quantities [23]. Let $F(E,E_0)$ dE_0 be the average number of atoms set in motion with initial energy (E_0, dE_0) as consequence of a collision cascade initiated by a primary atom starting with energy E. $F(E,E_0)$ is the recoil density and obeys the condition

$$F(E,E_0) = 0, \quad \text{for } E < E_0 .$$

(53)

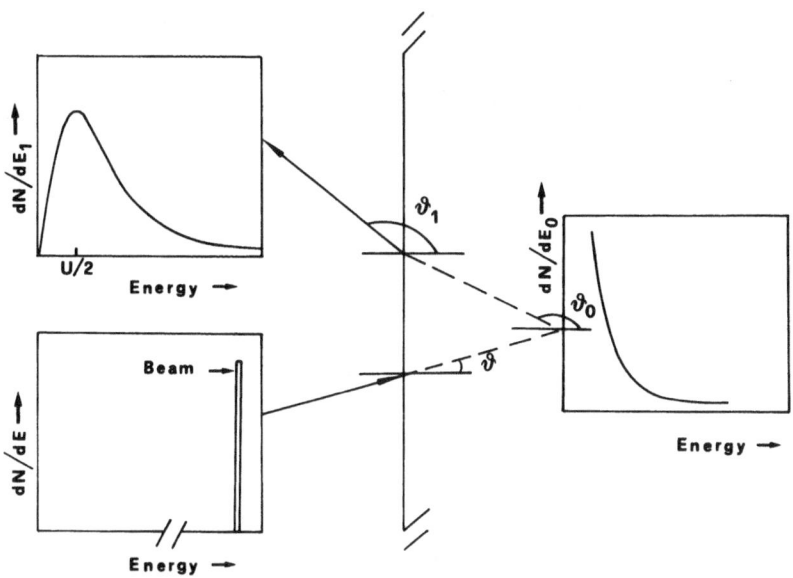

Fig.10. Steps involved in sputtering calculations: a beam impinges on a target and first the energy spectrum of moving atoms inside the target is needed, then the spectrum of sputtered particles outside the target is obtained.

The integral equation governing F can be derived following the usual arguments [64,23], the only difference being that in crossing a path length Δx, the probability that a collision with recoil energy (E_0, dE_0) occurs, must be taken into account. The equation, for the equal mass case, reads:

$$F(E, E_0) = N\Delta x \, \frac{d\sigma_n(E, E_0)}{dE_0} + N\Delta x \int d\sigma_n \, [\, F(E-T_n, E_0) + F(T_n, E_0) \,] +$$

$$+ (1 - N\Delta x \int d\sigma_n) \, F(E - N\Delta x \, S_e, E_0) \qquad (54)$$

where the electronic energy loss is now explicitly taken into account. Expansion to first order in Δx gives:

$$\int d\sigma_n \, [\, F(E, E_0) - F(E - T_n, E_0) - F(T_n, E_0) \,] + S_e(E) \, \frac{\partial}{\partial E} \, F(E, E_0) =$$

$$= \frac{d\sigma_n(E, E_0)}{dE_0} \, . \tag{55}$$

The most famous solution [23] of eq.(55) is that where electronic stopping is assumed to negligible and the power cross section is inserted. The details of calculations are very instructive and make use of Laplace transform evaluating the inversion by an asymptotic expansion for $E \gg E_0$. The well known solution (daily used I would add) is:

$$F(E, E_0) \cong \Gamma_m \, \frac{E}{E_0^2} \tag{56}$$

with Γ_m a well defined function of the power–law parameter m [23,3], but we stress that eq.(56) is derived for elastic collisions only and in the energy region $E \gg E_0$ ($\gamma E \gg E_0$ for the non-equal mass case). An analogous result can be derived for compound targets [66] on the basis of an extension of similar integral equations. At this point it is possible to include in eq.(56) the effect of electronic stopping (marginal, resulting in a limitation for the energy available for elastic collisions), as well as the angular and spatial dependence by physical considerations. Actually the amount of energy available for generating recoils is not E but the quantity specified in eq.(40), so we can replace E in eq.(56) by $v(E)$. Since in a completely developed cascade the recoils of higher generations dominate, recoils lose memory about the initial direction of primary atoms, so that the angular distribution of recoils can be assumed isotropic. The last two sentences included in eq.(56) give:

$$F(E, E_0, \Omega_0) \, dE_0 \, d^2\Omega_0 \cong \Gamma_m \, \frac{v(E)}{E_0^2} \, dE_0 \, \frac{d^2\Omega_0}{4\pi} \, . \tag{57}$$

The number of recoils generated in the layer (x,dx) is proportional to the energy available in this layer for elastic collisions, i.e. the spatial distribution of recoils can be assumed to be given by the spatial distribution of the deposited energy function F_D and taking into account the condition (39), we can write:

$$F(E, \Omega; E_0, \Omega_0, r) \cong \Gamma_m \frac{F_D(E,\Omega,r)}{E_0^2} dE_0 \frac{d^2\Omega_0}{4\pi} d^3r \qquad (58)$$

or, if the definition (41) is also included

$$F(E, \vartheta; E_0, \Omega_0, x) \cong \Gamma_m \frac{F_D(E,\vartheta,x)}{E_0^2} dE_0 \frac{d^2\Omega_0}{4\pi} dx \qquad (59)$$

which indicates that the spatial distribution of the cascade is determined by the terms belonging to high energy particles, while the detailed structure belongs to the behaviour of low-energy atoms.

Once one knows how atoms are set in motion inside the solid, at least in the case where the various assumptions leading to eq.(58) are valid, the next question to be answered is on the existence of some simple treatment which permits the determination of the sputtering yield outside the surface. In other words (see fig.10) we ask if it is possible to write the differential sputtering yield as [67,68]

$$Y(E_1,\Omega_1) = \int_0^\infty dx \int_0^\infty dE_0 \int_{4\pi} d^2\Omega_0 \, F(E,\Omega;E_0,\Omega_0,x) \, P(E_1,\Omega_1;E_0,\Omega_0,x) \qquad (60)$$

where F is the recoil density and P dE_1 $d^2\Omega_1$ is the probability for an atom, set in motion in the layer (x,dx) with energy E_0 and direction Ω_0, to be ejected outside the surface with energy (E_1, dE_1) and direction $(\Omega_1, d^2\Omega_1)$. The form is desirable also in view of further applications [69], but obviously difficulties are inherent in the determination of appropriate F and P. In the case of linear cascade theory, assuming that the recoils slow down continously along a straight line [67], the results obtained from transport theory can be reproduced. In this case in fact the expression for P reads [68]:

$$P(E_1,\Omega_1;E_0,\Omega_0,x) = \delta[E_1 + U_0 - f(E_0,\Omega_0,x)] \; \delta(\chi_1 - \chi_0) \times$$

$$\times \delta\{\cos \vartheta_1 - [(1 + U_0/E_1) \cos^2 \vartheta_0 - U_0/E_1]^{1/2} \} \qquad (61)$$

where $f(E_0,\Omega_0,x)$ describes the energy of a recoil atom (E_0,Ω_0) after having travelled a path length x [67]; a planar surface barrier of height U_0 has been assumed with refraction at the surface [70], ϑ_1 and ϑ_0, χ_1 and χ_0 being respectively polar angles and azimuths associated with the directions Ω_1 and Ω_0; δ is the Dirac delta function. If one assumes expression (58) for F and

$$f(E_0,\Omega_0,x) = E_0 \left(1 - \frac{x}{R(E_0) \,|\cos \vartheta_0|} \right)^{1/2m} \qquad (62)$$

where $R(E_0) = E_0^{2m}/2m$ A (with A constant [8] once the appropriate value of m has been choosen) is the range of the recoils, integration of eq.(60) over E_0 and Ω_0 yields:

$$Y(E_1,\Omega_1) = \frac{\Gamma_m \cos \vartheta_1 \, E_1}{2 \, (E_1+U_0)^{5/2} \, (E_1 \cos^2 \vartheta_1 + U_0)^{1/2}} \times$$

$$\times \int_0^\infty dx \, F_D(E,\vartheta,x) \left[1 + \frac{x \, (E_1+U_0)^{1/2}}{R(E_1+U_0) \, (E_1 \cos^2 \vartheta_1 + U_0)^{1/2}} \right]^{-1-1/2m} \qquad (63)$$

which for m=0 reduces to

$$Y(E_1,\Omega_1) = \frac{\Gamma_0 \cos \vartheta_1 E_1}{2 (E_1+U_0)^{5/2} (E_1 \cos^2 \vartheta_1 +U_0)^{1/2}} \quad x$$

$$x \int_0^\infty dx \; F_D(E,\vartheta,x) \; \exp\left(- \frac{x A_0 (E_1+U_0)^{1/2}}{(E_1 \cos^2 \vartheta_1 + U_0)^{1/2}} \right) \tag{64}$$

Integration of eq.(63) or (64) over x assuming $F_D(x) \cong F_D(x=0)$ [67] (we remark here that this assumption is not necessary in the transport theory discussed above where the sputtering yield is inherently proportional to $F_D(x=0)$) yields the already mentioned results; in particular, e.g. for m=0, one reproduces the well known energy spectrum of sputtered particles ($\propto$ $E_1/(E_1+U_0)^3$) and by further integration over E_1 and Ω_1 the sputtering formula, eq.(44), is obtained. Information on depth of origin of sputtered atoms [67] can be obtained by integration of eqs.(63,64) over ϑ_1 and E_1; consider for sake of simplicity, the case m=0 (eq.(64)), the mentioned integration gives:

$$Y = \frac{\Gamma_0}{8 \; U_0} \int_0^\infty dx \; F_D(E,\vartheta,x) \; 4 \; E_4(x \; A_0) \tag{65}$$

$E_n(z)$ being the exponential integral [71]. Since the following approximation can be introduced [67]:

$$4 \; E_4(x \; A_0) \cong \frac{4}{3} \; \exp(- \frac{4}{3} x \; A_0) \tag{66}$$

it is clear that the characteristic depth of origin of sputtered atoms is

$$\Delta x = \frac{3}{4} \frac{1}{A_0} \tag{67}$$

in complete agreement with the results of transport theory [8].

6. APPLICATION

In this paragraph we discuss two applications where the concept of recoil density are applied together with the simplified model and implications assumed in eq.(60).

The first example discusses the sputtering of atoms by light-ion bombardment in the keV region. This problem has been treated by evaluating the energy distribution of sputtered atoms from the spectrum of backscattered ions [72]. In this model it is shown that the contribution to sputtering from cascade recoils is negligible and then only primary recoils (recoils directly generated by incoming particles) can be assumed as candidate for sputtering. Since the backscattering model can be in principle included in transport theory [73,74], the idea is to use the model expressed in eq.(60) and to modify the function F, i.e. including instead a function which accounts for the distribution of primary recoils only. Defining $F_P(E,E_0)dE_0$ as the average number of primary recoils (Z_2,M_2) created in the energy interval (E_0,dE_0) by a slowing down process initiated by an ion (Z_1,M_1) of initial energy E. It is clear that this distribution is close to the definition of the recoil density, so that F_P satisfies the integral equation [23]

$$\int d\sigma_n \left[F_P(E,E_0) - F_P(E-T_n,E_0) \right] + S_e(E) \frac{\partial}{\partial E} F_P(E,E_0) = \frac{d\sigma_n(E,E_0)}{dE_0} \tag{68}$$

which holds from the usual arguments and differs from the corresponding one related to the recoil density (eq.(55)) only for the omission of the recoil term, since recoil atoms cannot generate primary recoils. A straightforward solution of eq.(68) can be found if an expansion of the second term in the l.h.s. for $T_n \ll E$ is used and the approximation $S_e \gg S_n$ is included, both approximations being justified by the physical situation $M_1 \ll M_2$ and by the energy range under investigation (keV range). The solution reads [23,75]:

$$F_P(E,E_0) = \int_{E_0/\gamma}^{E} \frac{dE'}{S_e(E')} \frac{d\sigma_n(E',E_0)}{dE_0} \tag{69}$$

with the already mentioned meaning of symbols. If we assume the

Lindhard [30] electronic stopping cross-section ($S_e = K E^{1/2}$) and a power law with m=1/2 for $d\sigma_n$, eq.(69) becomes:

$$F_P(E,\Omega; E_0,\Omega_0,x) \cong \frac{C_{1/2}}{4\pi K X_{max} E_0^{3/2}} \ln(T_m/E_0) \qquad (70)$$

where isotropy for the primary recoils distribution is assumed together with a constant spatial dependence over the maximum damage depth $X_{max} = R_p(E) \cos\vartheta$, R_p being the mean projected range and ϑ the angle of incidence with respect to the surface normal. Insertion of (70) in eq.(60)

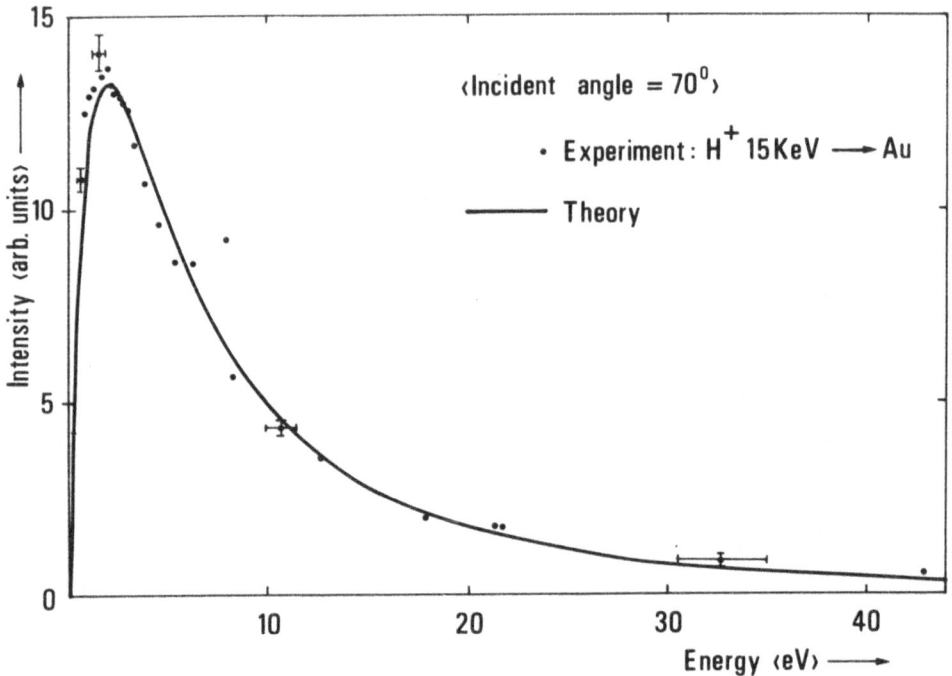

Fig.11. Energy spectrum of sputtered gold atoms compared with experimental results.

and repetition of integration steps with function P as given in eq.(61), leads to the following double differential sputtering yield [75]:

$$Y(E_1,\Omega_1) \propto \frac{E_1 \cos \vartheta_1}{(E_1+U_0)^{5/2}} \; \{ \ln(T_m/(E_1+U_0)) + 2[(E_1+U_0)/T_m]^{1/2} - 2\}. \qquad (71)$$

Figs.11,12 show comparison of eq.(71) with experimental data (fig.11) and the prediction of shift toward lower energies (fig.12, theory) showed up in measured velocity distributions from light-ion bombardment [76,77,78], despite the assumption of isotropy assumed in eq.(70). This prediction is also contained in the work of Roosendal and Sanders [79] with inclusion of an anisotropic term in the energy distribution of sputtered atoms. The

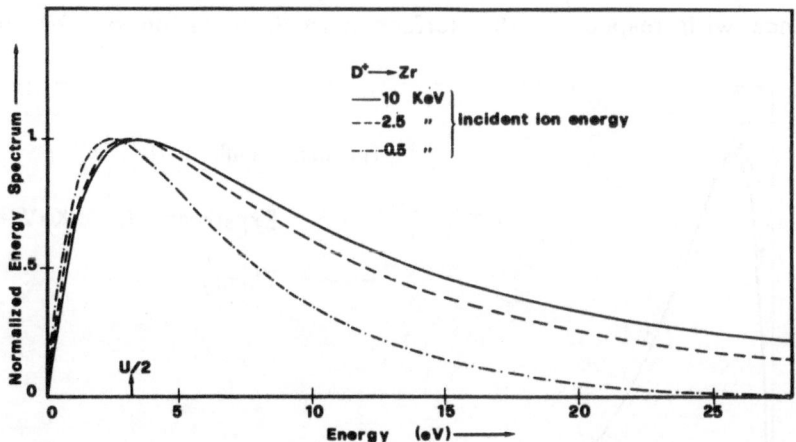

Fig.12. Normalized energy spectra (theory) of zirconium atoms sputtered at different incident energies showing the peak shift toward lower energy.

sputtering yield has also been obtained from eq.(71) after integration, including a surface correction [80] which takes into account the overestimation of deposited energy near the surface in case of light-ion bombardment, and reads [81]:

$$Y = \frac{C_{1/2}}{\gamma^{1/2} \, C_0 \, N \, S_e(E) \, R(E) \, \cos \vartheta} \; y(T_m/U_0) \qquad (72)$$

with $y(x)$ a well defined function of the argument [81]. Fig.13 shows a comparison of eq.(72) with experimental data. The calculations using eq.(72) have also been extended to the upper keV region (25-100 keV) and in tab.3 is reported a sample of such calculation [82] compared with experimental [83] and simulation [84] results. The overall conclusion is that a better agreement with experiments goes through the inclusion in eq.(70) of a correct angular and spatial dependence of the primary recoils, which supply the main contribution to sputtering by light-ion bombardment.

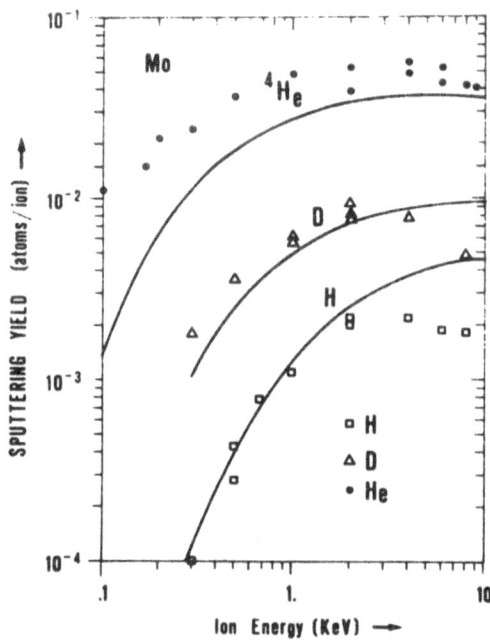

Fig.13. Sputtering yield from a molybdenum target by light-ion bombardment: solid lines are obtained by using eq.(72), experimental data from ref.[13b].

The second example concerns sputtering from compound targets. Due to the kind of measurements available in this sputtering process [85,86], the most appropriate theoretical description should be the determination of the percent concentration $\alpha_i(x,\phi)$ for each component "i" as function of the depth x inside the target and the incident ion fluence ϕ (ions/cm^2). A balance equation for this quantity has been determined by Sigmund et al.[68,87] in which only relocation by collisional processes is taken into account and segregation phenomena have been deliberately excluded; the equation is similar to that derived by Littmark and Hofer [88].We do not enter the details of such derivations, which would require a background slightly different of that contained in this contribution, but we direct the attention to the results of these theories which are expressed in fig.14 where the stationary concentration profile is shown as function of depth [87].

Tab.3. Comparison of calculated yields (eq.72) with data from experiments and simulation.

Target	Ion energy [keV]	Ion	Angle of incidence	Measured total yield	Calculated total yield (isotropic)	Calculated total yield (TRIM)
Ti	50	He^+	0	3.7×10^{-2}	3.86×10^{-2}	
	100	He^+	0	1.26×10^{-2}	3.19×10^{-2}	
	100	He^+	40	1.75×10^{-2}	4.18×10^{-2}	
	100	He^+	70	6.00×10^{-2}	9.30×10^{-2}	
Ta	25	H^+	0	2.32×10^{-3}	2.93×10^{-3}	1.95×10^{-3}
	25	H^+	30	5.32×10^{-3}	3.40×10^{-3}	3.63×10^{-3}
	25	H^+	50	9.94×10^{-3}	4.65×10^{-3}	7.48×10^{-3}
	25	H^+	70	1.66×10^{-2}	8.89×10^{-3}	1.93×10^{-2}
	50	He^+	0	2.78×10^{-2}	2.19×10^{-2}	
	100	He^+	0	2.01×10^{-2}	1.94×10^{-2}	
	100	He^+	70	1.36×10^{-1}	5.64×10^{-2}	
Ni	25	H^+	0	3.67×10^{-3}	8.31×10^{-3}	
	50	H^+	0	4.91×10^{-3}	7.47×10^{-3}	2.78×10^{-3}
	50	H^+	20		7.97×10^{-3}	2.55×10^{-3}
	50	H^+	30	6.60×10^{-3}	8.65×10^{-3}	
	50	H^+	40		9.84×10^{-3}	4.09×10^{-3}
	50	H^+	60	1.45×10^{-2}	1.52×10^{-2}	
	50	H^+	60			9.68×10^{-3}
	50	H^+	80	4.52×10^{-2}	4.51×10^{-2}	4.42×10^{-2}
	50	H^+	85		8.47×15^{-2}	8.97×10^{-2}
	100	He^+	0	4.58×10^{-2}	3.54×10^{-2}	
	100	He^+	0	6.41×10^{-2}		2.32×10^{-2}
	100	He^+	30	7.06×10^{-2}	4.08×10^{-2}	3.28×10^{-2}
	100	He^+	60	1.48×10^{-1}	7.12×10^{-2}	8.44×10^{-2}
	100	He^+	80	2.96×10^{-1}	2.08×10^{-1}	3.27×10^{-1}
	50	He^+	0	5.92×10^{-2}	4.27×10^{-2}	
W	50	D^+	0	4.34×10^{-3}	5.55×10^{-3}	
	100	D^+	0	2.7×10^{-3}	5.12×10^{-3}	
Mo	25	H^+	0	1.1×10^{-3}	4.68×10^{-3}	
	50	H^+	75	1.34×10^{-2}	1.79×10^{-2}	1.91×10^{-2}

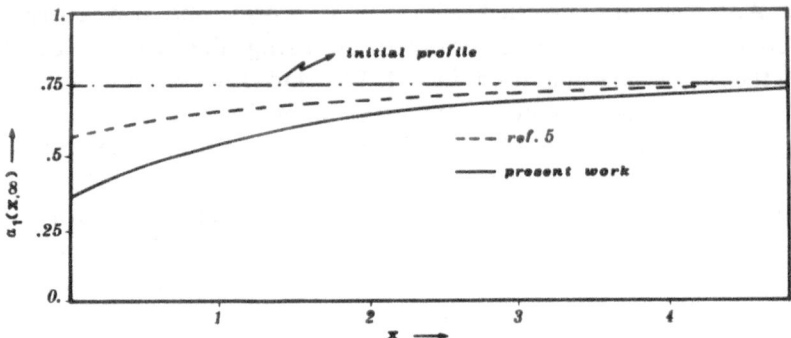

Fig.14. Calculated stationary profile from a balance equation [68] where segregation phenomena are not included.

Experimental results [89,90,91] show, of course, the presence of a segregated layer on the surface, which now is detected by using both Ion Scattering Spectroscopy and Auger Electron Spectroscopy. Fig.15 shows the stationary concentration profile of a Cu_3 Ni sample bombarded

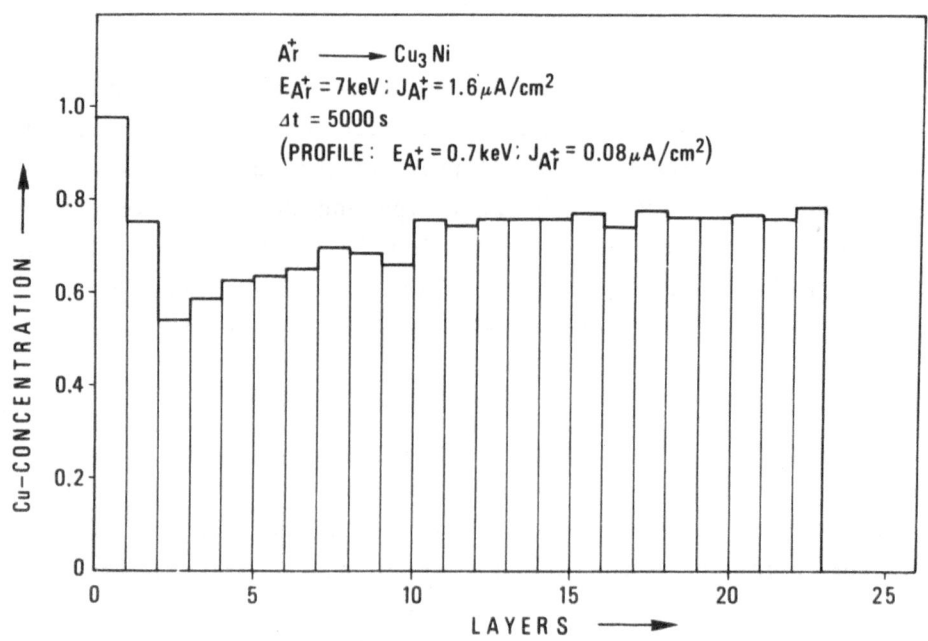

Fig.15. Measured stationary depth profile of a segregating alloy by means of Auger spectroscopy.

with 7 keV Ar^+ ions and profiled with a low energy and low current beam [92]. The concentrations were determined by using the method of Pons et al.[93], i.e. the same method used in ref.[91] and low energy Auger peaks. So segregated systems provide a peculiar experimental situation, where there is a natural calibration (more or less pronounced, depending on the segregation ratio, i.e. on the amount of material which segregates on the surface) for the depth of origin sputtered atoms and this information is contained in measurements involving sputtered flux ratios or angular distributions of sputtered atoms. Another possibility is the experiments carried out by Jørgensen et al.[94] (see these proceedings) where the system of 1-2 monolayers of Cu on a Ru(0001) single crystal has been chosen to measure sputtered fluxes. This system is known to have a monolayer—by—monolayer growth on the substrate for the first two monolayers [95]. In other words the system supplies an experimental situation where Cu is emitted only from the first, or the first two layers.

The question is now how to correlate experimental information with the available theories; this can be done going back to eq.(60) and applying this equation, within the already mentioned assumptions, to compound targets [68]. Consider a compound target with the generic component labeled "i". The problem reduces to the substitution in eq.(60) of F with a function F_i (defined so that $F_i(E,E_0)dE_0$ is the mean number of i-atoms recoiling in the energy interval (E_0,dE_0) per incident ion) and to the substitution of P with a function P_i describing the probability for an i-atom to escape from the target. The expression for P_i is reported in ref.[68] and is the same as eq.(61) provided the substitutions of U_0 with U_i and f with f_i, i.e. surface binding energy and description of energy loss by the atoms of type "i". An expression for F_i has been calculated in ref.[66], on the basis of integral equations similar to that describing the recoil density for monoatomic targets (eq.(55)), and reads:

$$F_i(E,E_0) = \frac{E}{E_0^2} \, \alpha_i(x) \, K_i \qquad \text{for } E_0 << E \qquad (73)$$

K_i being the displacement efficiency which depends on the concentration [66]. The detailed results obtained applying eq.(60) are reported in ref.(68). The point is that the final result for the sputtering yield is an expression which is differential in energy and angle of emission as well as in depth from which particles originate. What we need here is an expression of the yield still differential in x , after integration over energy and angle so that, taking again the case m=0, one obtains [66,96]:

$$Y_i(x) \propto \frac{\alpha_i(x)}{U_i} E_4(A_0 x) \qquad (74)$$

consistent with eq.(65), where E_n is the exponential integral. From (74), assuming that $\alpha_i(x)$ is reasonably constant, we can define the fractional emission from beyond a certain depth x_0 as

$$f_0(x_0) = \frac{\int\limits_{x_0}^{\infty} dx\ Y_i(x)}{\int\limits_{0}^{\infty} dx\ Y_i(x)} = 4E_5(A_0 x_0) \equiv 4E_5(x_0/L_0) \qquad (75)$$

so that if $x_0 = \lambda$, the mean atomic spacing, $f_0(\lambda)$ will be the fractional emission from beyond the first atomic layer. Inversion of eq.(75) gives an estimate of L_0, the characteristic depth of origin of sputtered atoms. Fig.16 shows the quantity x_0/L_0 as function of x_0 with data as supplied by computer simulation (note that the majority of information on depth of origin comes easily from simulation, see works quoted in tab.2 of ref.[96]). Fitting the reported data an estimate of $L_0/\lambda = 0.95$ or $L_0/\lambda = 0.80$ is obtained.

The experimental results of ref.[94] which gives a ratio of 3 between Cu and Ru in the sputtered flux can be interpreted as the ratio

$$[1 - f_0(\lambda)]/f_0(\lambda) \qquad (76)$$

i.e. the ratio between the fractional emission from within atom layer <u>one</u> and that from beyond atom layer <u>two</u>, or

$$[1 - f_0(\lambda)]/[f_0(\lambda) - f_0(2\lambda)] \qquad (77)$$

where the denominator gives the fractional emission from <u>within atom</u> <u>layer two</u>. Eq.(76) gives $L_0/\lambda = 0.91$, while eq.(77) $L_0/\lambda = 1.2$. These results have to be compared with the theoretical prediction of eq.(67) [8] which gives $\Delta x/\lambda \cong 3$.

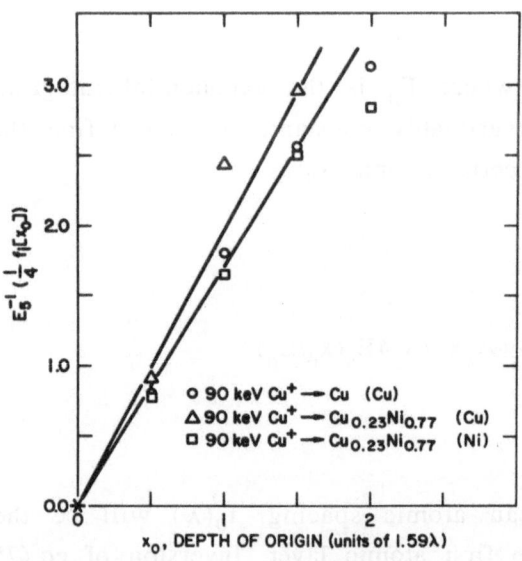

Fig.16. Information on depth of origin of sputtered atoms as obtained from computer simulation (points) and from eq.(75).

7. SUMMARY AND CONCLUSIONS

The research field on sputtering is still active with new proposals of investigation in both sputtering from single component and from multicomponent target, the last one extending also to mixing and related problems.

The brief historical remarks showed the importance of the knowledge about atomic collisions in solids when one is concerned with sputtering problems and the difficulties entering theories where a lack of this knowledge is evident.

The Danish school, with the fundamental research in atomic collisions in solids carried out by Bohr and Lindhard, supplied some helpful tools in this field; we recall just the useful analytical expression of the power-law scattering, the concept of separability between elastic and inelastic energy loss in collision between two atomic systems, and the basic integral equations.

These tools all entered Sigmund's theory with the successful determination of his sputtering formula of which we would like to point out the simplicity: the yield is expressed as the product of two factors, one related to the properties of the material constituting the target, the other one related to the energy deposition of the incoming ion.

Another important concept entering the field is that of the recoil density. This concept was found to be fruitful as we showed in the case of sputtering from single component targets, sputtering by light-ion bombardment, and sputtering from compounds; its application is even larger and will be found to be useful also in the future.

The problem of depth of origin of sputtered atoms is still open and its solution seems to generate new research efforts both experimentally and theoretically, because of consequences, e.g., in the sputtering yield formula where it enters through the target material constant. In this constant also enters the surface binding energy. Until now such a quantity has been taken as the heat of vaporization in the case of metals, but some investigations indicate that probably it is a more complex quantity[97]. Segregation phenomena seem to provide some unique experimental situations to supply information for these problems.

Finally, mention must be made of the research involving computer simulation of sputtering problems, a field which requires a special attention which is out of this contribution, but a look to the enormous production and accumulation of data is of sure interest.

REFERENCES

[1] J.J.Thomson, in "Rays of positive electricity and their application to chemical analyses", Longmans Green and Co., London (1921) 2nd edition.
[2] J.Vac.Soc.Jpn. 24(1981)456 (in japanese).
[3] P.Sigmund in Sputtering by Particle Bombardment I, R.Behrisch ed., Top.Appl.Phys. 47(Springer, 1981).
[4] R.Kelly, private communication.
[5] Physics Abstracts, Subject Index 1965-68.
       "   "   "   "   1969-72.
       "   "   "   "   1973-76.
       "   "   "   "   1977-80.
       "   "   "   "   1981-84.
       "   "   "   "   1985.
       "   "   "   "   1986.
[6] Scientific American   vol.37 n.220(December 1986).

[7] R.Kelly,Radiat.Eff. **80**(1984)273.

[8] P.Sigmund, Phys.Rev. **184**(1969)383; **187**(1969)768.

[9] F.Keywell, Phys.Rev. **87**(1952)160;**97**(1955)1611.

[10] E.Fermi, in **Nuclear Physics**, University of Chicago Press, Chicago 1950, p.181; E.Persico, in **Lezioni sulla Fisica del Reattore**, 2nd ed., ENEA-Ufficio Pubblicazioni, Roma 1963.

[11] F.Seitz, Discussion Faraday Soc. **5**(1949)271.

[12] Don E.Harrison Jr., Phys.Rev. **102** (1956) 1473; Phys.Rev. **105** (1957) 1202.

[13] N.Matsunami, Y.Yamamura, Y.Itikawa, N.Itoh, Y.Kazumata,S.Miyagawa, Atomic Data and Nuclear Data Tables **31**(1984)1; J.Roth, J.Bohdasky,W.Ottenberger, report IPP-9/26, Garching(1979).

[14] P.Joyes, J. de Physique **29**(1968)774.

[15] H.H.Andersen and H.L.Bay in ref.3, p.145.

[16] N.Bohr, Mat.Fys.Medd.Dan.Vid.Selsk. **18** n.8 (1948).

[17] N.Bohr, Phil.Mag. **25**(1913)10.

[18] P.Sigmund, in "**Radiation Damage Processes in Materials**", C.H.S.Dupuy ed., Noordhoff, Leiden (1975) p.3.

[19] P.Sigmund, these proceedings.

[20] J.Lindhard, Mat.Fys.Medd.Dan.Vid.Selsk. **34** n.14(1965), "Notes on Atomic Collisions II".

[21] H.Goldstein, "**Classical Mechanics**", Addison-Wesley(1953). L.D.Landau, E.M.Lifshitz, "**Mecanique**", MIR Publishers Moscou(1969).

[22] C.Lehmann, "**Interaction of Radiation with Solids and Elementary Defect Production**", North-Holland Publ. Co., Amsterdam(1977).

[23] P.Sigmund, Rev.Roum.Phys. **17**(1972)823,969,1079;

[24] M.T.Robinson, ORNL-4556, Oak Ridge Nat.Lab. report(1970).

[25] J.Lindhard, V.Nielsen and M.Scharff, Mat.Fys.Medd.Dan.Vid.Selsk. **36** n.10(1968), "Notes on Atomic Collisions I".

[26] P.Sigmund, Radiat.Eff. **1**(1969)15.

[27] H.A.Bethe, Ann.Phys.(Leipzig) **5**(1930)325.

[28] J.J.Thomson, Phil.Mag. **23**(1912)449.

[29] O.B.Firsov, Sov.Phys. JETP **9**(1959)1076.

[30] J.Lindhard, M.Scharff, Phys.Rev.**124**(1961)128.

[31] L.Erikson,J.A.Davies and P.Jespersgaard, Phys.Rev.**161**(1963)219; Ya.A.Teplova,V.S.Nkolaev,I.S.Dmtriev and L.N.Fateeva, Sov.Phys. JETP **15**(1962)31; B.Fastrup,P.Hvelplund and C.A.Sautter, Mat. Fys. Medd. Dan.Vid.Selsk. **35** n.10(1966); P.Hvelplund, B.Fastrup, Phys.Rev. **165**(1968)1046.

[32] K.Winterbon, report AECL-3194(1968).

[33] W.D.Wilson,L.G.Haggmark,J.P.Biersack, Phys.Rev. **B15**(1977)2458.

[34] J.F.Ziegler,J.P.Biersack,U.Littmark, in **"The Stopping and Range of Ions in Solids"**, vol.1, Pergamon Press, New York(1985).

[35] P.Sigmund, Nucl.Instr.Methods **B**(1987).

[36] K.B.Winterbon,P.Sigmund,J.B.Sanders, Mat.Fys.Medd.Dan.Vid.Selsk. **37** n.14(1970).

[37] J.Lindhard,V.Nielsen,M.Scharff and P.V.Thomsen, Mat.Fys.Medd.Dan.Vid.Selsk. **33** n.10(1963), "Notes on Atomic Collisions III".

[38] W.S.Snyder and J.Neufeld, Phys.Rev. **97**(1955)1636; Phys.Rev. **103**(1956)862.

[39] P.Sigmund, Phys.Scripta **28**(1983)257.

[40] H.E.Schiøtt, Mat.Fys.Medd.Dan.Vid.Selsk. **35** n.9(1966).

[41] K.B.Winterbon, Phys.Lett. **32A**(1970)265;Radiat.Eff. **13**(1972)215.

[42] U.Littmark, Thesis H.C.Ørsted Institute, Copenhagen(DK) 1974 (Italian transl. by M.Salvatore).

[43] J.F.Gibbons,W.S.Johnson and S.W.Mylroie, **Projected Range Statistics,Semiconductors and Related Materials**,2nd ed. Dowden, Hutchinson and Ross, Stroudsberg PA U.S.A.(1975).

[44] R.Kelly, **Ion Bombardment Modification of Surfaces**, Ch.2, O.Auciello and R.Kelly eds., Elsevier Science Publs., The Netherlands(1984).

[45] D.K.Brice, **Ion Implantation Range and Energy Deposition Distributions**, vol.1, Plenum Press, New York and London(1975).

[46] K.B.Winterbon, **Ion Implantation Range and Energy Deposition Distributions**,vol.2, Plenum Press, New York and London(1975).

[47] P.Sigmund,J.B.Sanders, Proc. Int. Conf. on Application of Ion Beam to Semiconductors Technology, Editions Ophrys (1967) p.215.

[48] J.B.Sanders, Thesis University of Leiden, The Nederlands(1968).

[49] E.Bøgh, P.Høggild, I.Stensgaard, Radiat.Eff. **7**(1971)115.

[50] J.B.Sanders, Can.J.Phys. **46**(1968)455.

[51] J.Lindhard and V.Nielsen, Mat.Fys.Medd.Dan.Vid.Selsk. **38** n.9(1971).

[52] W.Huang, H.M.Urbassek and P.Sigmund, Phil.Mag. **52**(1985)763.

[53] B.Davison, **Neutron Transport Theory**, Clarendon Press, Oxford(1957).

[54] L.N.Usachev, **"Equation for the Importance of Neutrons, Reactor Kinetics and Theory of Perturbation"**, Proc.1 ICPUAE-UN- Geneva(1955).

[55] A.Gandini, Nucl.Sci.Eng. **35**(1969)141; in **"Elementi di Fisica e**

Calcolo dei Reattori Veloci", RT/FI(72)47, ENEA Roma(1972).

[56] P.Sigmund in "Inelastic Ion-Surface Collisions", N.H.Tolk ed., Academic Press Inc., New York(1977) p.121.

[57] J.C.Pivin, J.Mat.Sci. **18**(1983)1267.

[58] M.I.Guseva, Soviet Phys.-Solis State **1**(1960)1410.

[59] O.Almen and G.Bruce, Nucl.Instr.Methods **11**(1961)257.

[60] G.Dupp and A.Sharmann, Z.Physik **192**(1966)284.

[61] H.Cramer, "Mathematical Methods of Statistics", Princeton University Press, Princeton(1946).

[62] E.M.Barody, J.Appl.Phys. **36**(1965)3565.

[63] M.T.Robinson, Phil.Mag. **12**(1965)741;**17**(1968)639.

[64] P.Sigmund, Appl.Phys.Lett. **14**(1969)114.

[65] G.H.Kinchin and R.S.Pease, Rep.Progr.Phys. **18**(1955)1.

[66] N.Andersen and P.Sigmund, Mat.Fys.Medd.Dan.Vid.Selsk. **39** n.3 (1974).

[67] G.Falcone and P.Sigmund, Appl.Phys. **25**(1981)377.

[68] P.Sigmund, A.Oliva and G.Falcone, Nucl.Instr.Methods **194**(1982)541.

[69] P.Sigmund, in "Secondary Ion Mass Spectrometry", SIMS IV. A.Benninghoven, J.Okano, R.Shimizu and H.W.Werner, eds., Springer-Verlag(1984).

[70] M.W.Thompson, Phil.Mag. **18**(1968)377; Phys.Rep. **69**(1981)335.

[71] M.Abramowitz, I.Stegun, in "Handbook of Mathematical Functions", Dover, New York(1964).

[72] U.Littmark, S.Fedder, Nucl.Instrum. Methods **194**(1982)607.

[73] R.Weissmann, P.Sigmund, Radiat.Eff. **19**(1973)69.

[74] R.Weissmann, R.Behrish, Radiat.Eff. **25**(1981)307.

[75] G.Falcone, A.Oliva, Appl.Phys. **A32**(1983)201.

[76] H.L.Bay, W.Berres, E.Hintz, Nucl.Instrum.Methods **194**(1982)555.

[77] H.L.Bay, B.Schweer, E.Hintz, J.Nucl.Mater. **111/112**(1982)732.

[78] H.L.Bay, W.Berres, Nucl.Instrum.Methods **B2**(1984)606.

[79] H.Roosendaal, J.B.Sanders, Radiat.Eff. **52**(1980)137.

[80] J.Bohdansky, Nucl.Instrum.Methods **B2**(1984)587.

[81] G.Falcone, A.Oliva Radiat.Eff.Lett. **86**(1984)57.

[82] A.Oliva,G.Falcone, unpublished results.

[83] J.Bohdansky, G.L.Chen,W.Eckstein, J.Roth, B.M.U.Sherzer and R.Behrish, J.Nucl.Mater. **111/112**(1982)717.

[84] J.P.Biersack, L.G.Haggmark, Nucl.Instrum.Methods **174**(1980)257.

[85] H.H.Andersen, in "The Physics of Ionized Gases SPIG'80", M.Matic ed., Boris Kidric Inst. of Nucl. Sci., Beograd(1980) p.42.

[86] G.Betz,G.K.Wehner, in "Sputtering by Particle Bombardment II",

R.Behrisch ed., Top.Appl.Phys. **52**(Springer, 1983) p.11.

[87] G.Falcone, A.Oliva, Appl.Phys. **A33**(1984)175.

[88] U.Littmark, W.O.Hofer, Nucl.Instrum.Methods **168**(1980)329;
U.Littmark, Nucl.Instrum.Methods **B7/8**(1985)684.

[89] D.G.Swarzfager, S.B.Ziemeki, M.J.Kelley, J.Vac.Sci.Technol.**19**(1981)185.

[90] R.S.Li, T.Koshikawa, Surface Sci. **151**(1985)459.

[91] J.Bertella,H.Oechsner, Surface Sci. **126**(1983)581.

[92] S.L.Jiang, A.Oliva, A.Amoddeo, to be published.

[93] F.Pons, J.Le Héricy,J.P.Langeron , Surface Sci. **69**(1977)547;
69(1977)565.

[94] B.Jørgensen, M.J.Pellin, C.E.young, W.F.Calaway, E.L.Schweitzer,
D.M.Gruen, J.W.Burnett, J.T.Yates, these proceedings.

[95] J.E.Houston,C.H.F.Peden,D.S.Blair,D.W.Goodman, Surface Sci.
167(1986)427.

[96] R.Kelly,A.Oliva, Nucl.Instrum.Methods **B13**(1986)283.

[97] R.Kelly, Nucl.Instrum.Methods **B18**(1987)388.

DEPTH OF ORIGIN OF SPUTTERED ATOMS*

B. JØRGENSEN**, M. J. PELLIN, C. E. YOUNG, W. F. CALAWAY, E. L. SCHWEITZER, and D. M. GRUEN
MATERIALS SCIENCE AND CHEMISTRY DIVISIONS, ARGONNE NATIONAL LABORATORY, ARGONNE, ILLINOIS 60439

J. W. BURNETT AND J. T. YATES
SURFACE SCIENCE CENTER, UNIVERSITY OF PITTSBURGH, PITTSBURGH, PENNSYLVANIA 15260

1. INTRODUCTION

The depth of origin of sputtered atoms is currently a very important quantity in surface physics. The importance is best observed by examining the literature on computer simulations and theoretical calculations. Despite this interest, it is very hard to find experimental measurements of the depth of origin of sputtered atoms, which is probably due to the difficulty of such experiments. A good determination of the depth of origin should be done on a well characterized system and without damaging it. The results reported here are obtained from 1-2 monolayers of Cu on a Ru(0001) single crystal surface. The first 2 monolayers of Cu evaporated onto Ru(0001) is known to exhibit a layer by layer growth[1]. The sputtering is studied in the Surface Analysis by Resonance Ionization of Sputtered Atoms(SARISA) apparatus[2], where it is possible to do the determination before the overlayer is damaged by the ion beam. One monolayer of Cu on Ru is found to give a ratio of 3.2 between Cu and Ru in the sputtered flux for 4 keV normal incidence Ar ions.

2. EXPERIMENTAL

The experiments are performed in an ultra high vacuum system with a base pressure of $3*10^{-8}$ Pa. The pressure will increase a little during the sputtering experiments, but it is always below 10^{-7} Pa. The Cu is evaporated from a Cu wire wrapped around a tungsten filament. The system is equipped with a Cylindrical Mirror Analyzer (CMA) with a coaxial electron gun. This setup is used for Auger Electron Spectroscopy (AES), which is used to monitor the cleanliness of the sample and to calibrate the Cu coverage. The sputtering results are obtained with the SARISA facility. The 5 keV Ar ion pulse is 500 ns long and the beam current is 20 nA. The ionization is done nonresonantly with an excimer laser running on XeCl giving a photon energy of 4.03 eV.

3. RESULTS AND DISCUSSION
3.1. Calibration of Cu coverage

The calibration of Cu coverage is done with AES. Plotting the Cu(MNN) peak-to-peak height as a function of the Ru(MNN) peak height (see Fig. 1) will produce a curve with breaks at the full monolayer points. These breaks are due to the layer by layer growth, and the fact that the inelastic mean free path of 60 eV and 274 eV electrons in Cu are different. The data are obtained by deposition of 2-3 monolayers of Cu on Ru. After deposition we measure the concentrations with AES,

R. Kelly and M. Fernanda da Silva (eds.), Materials Modification by High-fluence Ion Beams, 83–86.

and prepare a new surface with lower Cu coverage by heating the sample to desorb some Cu. Continuing this process until all the Cu is desorbed gives the datapoints. The solid curve is a least square fit to the experimental data. The resulting fitting parameters are:

Inelastic mean free path of a   60 eV electron in Cu   2.3±0.3Å
Inelastic mean free path of a 274 eV electron in Cu   7.0±0.8Å
Peak-to-peak height of Cu from bulk Cu   146±10
Peak-to-peak height of Ru from bulk Ru   458±10

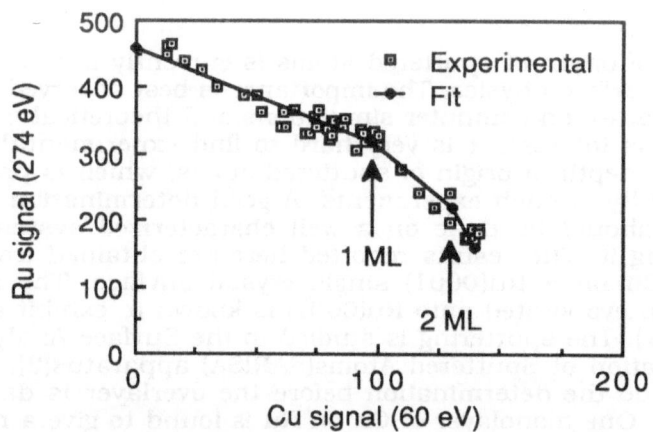

FIGURE 1.   Calibration of the Cu coverage by
                 Auger Electron Spectroscopy.

3.2. Sputtering experiment

The first problem to address is damage to the target during the experiment. A 500 ns pulse of Ar ions with a beam current of 20 nA contains $6.2*10^4$ ions. 1000 ion pulses is generally enough to obtain the data, giving a total ion dose of $6.2*10^7$ ions. The spot size of the ion beam is 0.04 mm², resulting in a total flux of $1.6*10^9$ ions/mm². Since 1 mm² of a solid normally has $\sim10^{13}$ surface atoms, this corresponds to one ion impact for every $6.5*10^3$ surface atoms. The corresponding damage is sufficiently low not to influence the measurement. Normally the SARISA method involves resonant ionization, but in this case we want to see 2 different species at the same time, so we are doing the measurement without a resonant photon. This means that we will have to calibrate the ionization efficiency with respect to laser power. A calibration curve is shown in Fig. 2, with data obtained by sputtering of bulk samples of Cu and Ru. The signal per sputtered atom is calculated from the known sputter yield of Cu (4.5) and Ru (2.7)[3]. At 6 MW/cm² (1 mJ/pulse in an area of 1 by 3 mm²) the useful yield (defined as particles detected/particles sputtered) is approximately 10^{-5}, giving a few counts per ion pulse, which is quite adequate for the experiment.

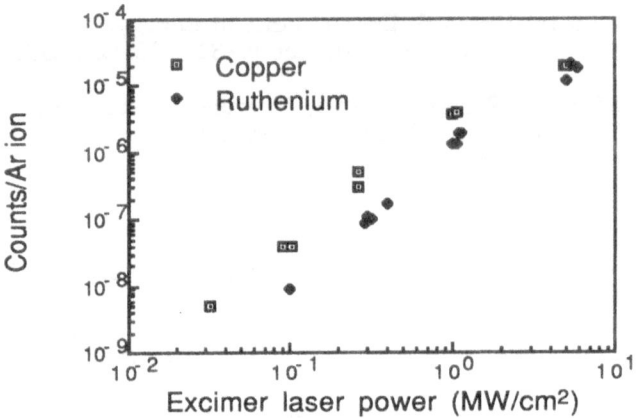

FIGURE 2. Laser power study of the ionization of Cu and Ru sputtered from the bulk samples.

3.3 Sputtering results and analysis

The sputtering results for 4 keV normal incidence Ar ions are shown in Table 1. The ratio at 1 monolayer of Cu on Ru, gives 75% Cu in the sputtered flux, meaning that 75% is originating in the first layer. If we assume an exponential decay of the emitted particles with respect to depth, this will give a characteristic depth of origin of 2 Å. If we take into account the difference in sputter yields for the bulk materials (Cu 4.5 and Ru 2.7 [3]), we will get a ratio of 1.9, indicating 66% from the first layer. This last approach is an overcorrection, but it is giving a lower limit on the first layer contribution.

Table 1.   Sputtering results for 4 keV normal incidence Ar ions on monolayers of Cu on Ru(0001)

Copper coverage (monolayers)	Cu/Ru sputter yield ratio
0.69±0.07	2.1±0.4
1.0±0.1	3.2±0.6
1.4±0.1	5.8±1

4. CONCLUSION

Sputtering of one monolayer of Cu adsorbed on a Ru(0001) single crystal surface is used to obtain the depth of origin of sputtered atoms. The measurements are performed in the SARISA apparatus to minimize the damage to the sample. The contribution to the sputtered flux from the first layer is determined to be between 66% and 75%, which is suggesting a characteristic depth of about 2Å.

*   Work supported by the U.S. Department of energy,
        BES-Material Science, under contract W-31-109-Eng-38.

* *   Permanent address: Fysisk Institut, Odense Universitet,
        Campusvej 55, DK-5230 Odense M, Denmark

References

1.   J. C. Vickermann and K. Christmann; Surf. Sci. 120 (1982) 1

        J. C. Vickermann, K. Christmann, G. Ertl, P. Heimann,
        F.J. Himpsel, and D.E. Eastman; Surf. Sci. 167 (1983) 367

        J. E. Houston, C. H. F. Peden, D. S. Blair, and D. W. Goodman;
        Surf. Sci. 167 (1986) 427

        J. T. Yates, C. H. F. Peden, and D. W. Goodman;
        J. Catal. 94 (1985) 576

2.   M. J. Pellin, C. E. Young, W.F. Calaway, J. W. Burnett,
        B. Jørgensen, E. L. Schweitzer, and D. M. Gruen;
        Nucl. Instr. methods B18 (1987) 446

3.   I. Matsunami, et. al.,
        Atomic Data and Nucl. Data Tables 31 (1984) 1

MAGNETRON SPUTTERING - PHYSICS AND DESIGN

J. B. ALMEIDA

University of Minho, Physics Lab., P-4719 BRAGA Codex, PORTUGAL

1. INTRODUCTION

Magnetron sputtering has proven to be a very useful deposition and etching technique in many spheres for several reasons, among which are the low heating effect and minimal damage caused to delicate substrates, the ability to cope with almost any metal or alloy, and indeed many insulating materials when reactive or radio frequency (R.F.) sputtering are used, the high rate of coating compared to ordinary diode sputtering, the versatility and adaptability to different shapes and geometries and, last but not the least, the polution-free nature of the technique.

Magnetron sputtering is a technique for industrial applications, rather than for physicists. In fact, the objective is to obtain high rate etching and deposition, by means of a large flux of bombarding particles into large areas. There is virtually no concern about the characterization of the bombarding flux in terms of speed distribution, thus rendering the technique useless for the investigation of the sputtering mechanisms.

On the other hand, where the emphasis is on the magnitude of the sputtered flux, this technique has few competitors.

2. BASIC PHYISICS

The physics of sputtering in this technique is no different from what it is in all other cases of sputtering and will not be dealt with here; what is characteristic of magnetron sputtering is the mechanism of creation of the bombarding particles and to this we will now dedicate our attention.

2.1. Glow discharge

This is a method of creating a low pressure plasma by means of an electric field established between a positive electrode (anode) and a negative electrode (cathode). The gas pressure for setting up a glow discharge is usually of the order of 1 Pa and the degree of ionization is of the order of 1% or less.

The discharge is initiated by the collisions of the few free electrons existing in the gas, by virtue of gamma ray radiation, when they are accelerated by the electric field. After the steady state is reached, which happens in a very short time, the gas begins to glow and a current starts flowing between the two electrodes.

The electric field between anode and cathode is not uniform and, in fact, most of the voltage drop appears adjacent to the cathode in a dark region called the "cathode dark space" with a thickness **d** given by (S.I. units) (9):

R. Kelly and M. Fernanda da Silva (eds.), Materials Modification by High-fluence Ion Beams, 87–92.

$$d^2 = \frac{4\varepsilon_o}{9j}\left(2\,\frac{e}{m}\right)^{1/2} V^{3/2}$$

where **V** is the cathode voltage, **j** is the ionic current density, **m** is the ionic mass, **e** is the electric charge and ε_o is the permitivity of vacuum.

It follows from the expression above, that the cathode dark space thickness grows with the cathode voltage and decreases with the pressure, through the ionic current **j**.

The fact that most of the voltage drop appears across the cathode dark space means that virtually only the ions that exist in this region will be accelerated towards the cathode and used for sputtering; all the others will be lost in the gas.

Next to the cathode dark space, in the direction of the anode, there is a region of high ionic density, called the negative glow, where the secondary electrons, generated by collisions of the ions with the cathode and accelerated by the electric field in the cathode dark space, make ionizing collisions with the atoms of the gas.

Following towards the anode we now find a region where the electrons have lost most of their kinetic energy and proceed slowly to the anode.

The thickness of the cathode dark space is a lower limit for the cathode-anode distance for a stable discharge.

2.2. Diode sputtering

In ordinary diode sputtering the cathode is the target material and the anode is the substrate. In this system the working pressure is usually between 1 and 10 Pa and the applied voltage is of the order of a few kilovolts. The inefficiency of the system derives from the compromise that must be achieved between pressure and anode-cathode distance.

The gas pressure must be sufficiently high for the electrons to have a probability of ionizing the gas; as we have seen, increasing the pressure reduces the cathode dark space. On the other hand, a high gas pressure also reduces the mean free path of the sputtered material, suggesting that the anode should be brought closer to the cathode.

2.3. Magnetron effect

The idea of using a magnetic field for incresing the ionisation of the plasma was first suggested and put into practice by Penning in 1940, but the full potential of the idea was not realized then and only in the seventies was magnetron sputtering developed as a technique in its own right.

The magnetic field adds one degree of freedom to the system, making it a great deal more flexible. The effect of the magnetic field is to increase the distance the electrons have to travel, by bending their trajectories into helices, thus increasing the probability of collisions.

With a suitably shaped magnetic field, the trajectories can be so elongated as to break the compromise between pressure and anode-cathode distance. As a consequence, very high degrees of ionization are possible with relatively low pressures, as is required for high efficiency sputtering. The positioning of the anode becomes of little importance and it can generally be placed wherever is suitable.

In order to explain the mechanisms of the magnetron we refer to figure
1, which depicts an idealized cross section of the discharge in an inwards
sputtering cylindrical magnetron. This special geometry has been chosen
because the representation is easier than with more common geometries.
Later, when we discuss the design of the magnetrons, we will transpose all
the conclusions to the more general case.

Although the plasma inside the magnetron is known to be rich in
collective behaviour, i.e. the particles interact with each other, a
single particle picture of the discharge provides some useful
understanding of the phenomena taking place (4).

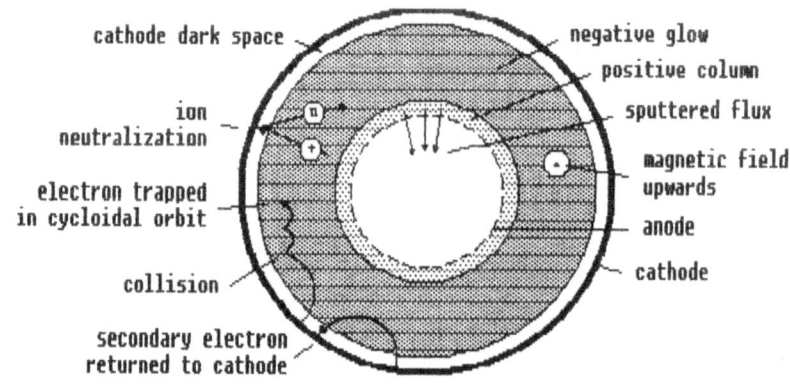

FIGURE 1 - Cylindrical magnetron

In figure 1 the cathode is represented by the outermost circumference
and the anode by the dashed inner circumference. The dashed line means
that the anode is not solid and allows the sputtered material through it.

The system is completely filled with argon at a pressure around 0.1 Pa,
and a voltage of a few hundred volts is applied between anode and cathode.
When this voltage is applied, any positive ions within the electric field
are accelerated towards the cathode and some will have an energy of 1 eV
per volt of the applied voltage.

The trajectories of the ions are mostly insensitive to the magnetic
field, between 0.02 T and 0.05 T, and remain essentially straight. If the
ions are energetic enough, they release secondary electrons from the
cathode, which contribute to further ionize the gas. Once the secondary
electrons are released, there is no real distinction between them and
those that are generated within the plasma and are known as primary
electrons; they all work together to guarantee the ionization of the
plasma.

When the steady state is reached, the three main regions of the
discharge, "cathode dark space", "negative glow" and "positive column" are
established (4,6) and the main voltage drop appears across the cathode
dark space, while the main ionization takes place within the negative glow.

An electron leaving the cathode is accelerated by the strong electric
field of the cathode dark space and enters the negative glow; as it
crosses the vertical magnetic field lines, its trajectory is bent back
towards the cathode and, unless it makes one collision, it is decelerated

by the electric field and ends up on the cathode surface with the same energy as it started with. One such electron has been depicted in figure 1.

If the electron makes one collision on its path, namely one ionizing collision, some of its energy will be lost and the electron will be trapped in a cycloidal orbit. In fact it will drift slowly towards the anode. The path to the anode is thus increased greatly and so is the likelihood of collisions and the degree of ionization. The net effect is the substantial reduction of the cathode dark space thickness, as if the pressure were higher.

If the electron has a component of the speed along the magnetic field lines, it will be lost at the ends if no precautions are taken. In order to avoid this, the magnetic field lines are made to emerge and immerge from the cathode, so that the electron that moves along them will merely be reflected back and forth.

In a uniform magnetic field, and in the absence of an electric field, the electrons orbit around the field lines in such a way that the magnetic force balances the centrifugal force:

$$ev_\perp B = m_e \, v_\perp^2/r_g$$

where e is the charge of the electron, m_e is its mass, $v_\perp$ is the component of the velocity at right angles to the field, B is the the magnitude of the magnetic field and r_g is the radius of the orbit, called gyro or Larmor radius.

In the presence of an electric field, the drift velocity of the electrons is the sum of their parallel velocity $v_{||}$ with a velocity v_E of magnitude,

$$v_E = E_\perp/B$$

parallel to $\vec{E} \times \vec{B}$ (4). $E_\perp$ is the component of the electric field normal to the magnetic field.

The magnetron effectively removes most of the electrons kinetic energy, allowing for high degrees of ionization. As a result, the thickness of the cathode dark space is reduced and the following consequences can be observed:

- The working pressure can be reduced, because there is a higher likelihood of collisions.
- The voltage can be reduced to a few hundred volts, because fewer secondary electrons are needed.
- The sputtered flux is increased by virtue of the reduction in the working pressure.

3. DESIGN CONSIDERATIONS

One important point of magnetron sputtering is the ease of adaptation to different geometries and sizes, namely the popular planar geometry in either circular or rectangular shape. The author has been involved in the design of several magnetrons (10) leading him to the following conclusions:

3.1. Magnetic field

The magnetic field lines must all emerge from the target (cathode) and plunge back into it, in order to reflect the electrons drifting along them. Besides that, the tunnel formed by the arches of magnetic field over the target must close on itself, so that the drift trajectory due to the combination of electric and magnetic field is closed and electrons are not lost.

The erosion of the target depends on the magnitude of the magnetic field parallel to it at any given point. The optimum magnetic field design would then leave the target at right angles, bend 90 degrees to form a section parallel to the target and plunge into it, again at right angles. The best aproximations to this give the best results in target erosion uniformity. The magnitude above the target must be greater than 0.02 T and there seems to be little point in going above 0.05 T.

Any stray magnetic field that doesn't abide the rules above, should be avoided; one can prevent erosion in places where stray magnetic field exists, by placing an anode in the cathode dark space area, but this isn't always easy.

3.2. Cathode and target

The cathode is at a negative potential and so must be electrically insulated. Any attempts to ground the cathode and have a live anode will usually result in unwanted erosion of all grounded metal parts. If possible the cathode should be made part of the vacuum chamber walls, so that the magnets and water cooling are left outside the vacuum; this will reduce the problems with the vacuum.

The erosion rate of the target is directly related to the current density and this is limited just by the capability of the cooling system. The cathode must be water cooled and the target must be in close contact, electrically and thermally, with it. Some care must also be exercised in order to prevent electric current from flowing through the water.

3.3. Power supply

Virtually any DC power supply capable of suppling a regulated voltage between 300 V and 700 V will be usable; the author has worked with a variac rectified by a full wave silicon rectifier. In order to get good results, though, the power supply should be current controlled rather than voltage controlled and should have a current capability in relation to the amount of cooling that is available; current densities up to 60 mA/cm^2 have been reported.

The placement of the anode inside the chamber is of minor importance; the fact that the electrons have high mobility along the magnetic field lines means that these are equipotentials and so one convenient place to locate the anode is usually near the place where the field lines emerge from the cathode.

3.4. RF operation

All magnetrons are in principle suitable for R.F. operation, as is necessary for sputtering from insulating targets; the author, though, has no hands-on experience of this subject and will not give any details on it.

92

REFERENCES

1. Weston GF: Cold Cathode Glow Discharge Tubes. Iliffe Books Ltd., London, 1968.
2. Maissel L, Glang R (ed): Handbook of Thin Film Technology. Mc-Graw Hill, N.Y., 1970.
3. Wehner GK, Anderson GS: The Nature of Physical Sputtering. in "Handbook of Thin Film Technology".
4. Thornton JA: J.Vac.Sci.Technol. - 15 (2), 171 and 188 (1978).
5. Waits RK: J.Vac.Sci.Technol. - 15 (2), 179 (1978).
6. von Engel A: Ionized Gases. Oxford, London, 1965.
7. Behrish R (ed): Sputtering by Particle Bombardment. Springer-Verlag, Berlin, 1981.
8. Sigmund P: in "Sputtering by Particle Bombardment" pages 9 to 71.
9. Bessot JJ: Dépôts par pulverization cathodique. Techniques de L´ingénieur, M 1 657, 1985.
10. Almeida JB, Ferreira MIC, Santos MAP, Ramos MD: Nucl. Instr. Meth. - B 18, 651 (1987).

ON THE FORMATION AND CHARACTERIZATION OF MICROCRYSTALLINE Si:H PREPARED BY RF MAGNETRON SPUTTERING

G. KIRIAKIDIS AND S. LOGOTHETIDES*

RESEARCH CENTER OF CRETE AND PHYSICS DEPARTMENT, UNIVERSITY OF CRETE, CRETE, GREECE.
*FIRST PHYSICS LAB., UNIV. OF THESSALONIKI, THESSALONIKI, GREECE

1. INTRODUCTION

Over the last few years, microcrystalline silicon (μc–Si) has attracted increasing attention as a new potential material for thin film device applications such as contact interlayers in solar cells [1-4]. Preparation of thin films of μc–Si can be achieved via chemical transport in a hydrogen plasma [1], deposition from a glow-discharge (GD) of silane diluted in hydrogen or inert gas at a higher power level [5], or by reactive-sputtering, [6-8] a technique applied in the present report.

Microcrystalline material is polycrystalline material with crystallite size ranging between ~20 and 300 Å [1]. Recent research focusing on the structural analysis of these films has shown that while GD produced μc–Si exhibits the presence of columnar structure in the range of submicrons [9], μc–Si produced by chemical transport shows a network composed only of microcrystallites [10]. Furthermore, for GD produced μc–Si it has been observed that an increase in the volume fraction of microcrystallites is accompanied by a decrease in bonded hydrogen present in the amorphous tissue, leading to hydrogen rich regions acting as microvoids. It has also been argued that an increase in the average grain size and volume fraction of microcrystallite results in the reduction of electron density fluctuations in the matrix.

In this paper we present a wide spectrum of characterization techniques (optical, electrical, structural) providing clear evidence towards distinguishing between crystalline, amorphous and microcrystalline structures of the developed films. In addition, for μc–Si films prepared by sputtering there is information of columnar structure development.

2. EXPERIMENTAL

We prepare our samples in a standard RF/magnetron system described previously [8]. We vary the hydrogen partial pressure ratio ($P_{H_2}/[P_{H_2} + P_{Ar}]$) from 0.1 to 0.5 under constant total pressure (Ar + H$_2$) of either 10 or 20 mTorr, while the substrate temperature varies from $150 \leqslant T \leqslant 250$ °C, the power density of the target varies from 0.9 to 1.2 W/cm^2, and deposition rate between 0.8 and 3.3 Å/sec.

Optical, transport and structural characterization of the samples has also been described previously [8]. Details of the rotating analyser ellipsometer [11] used in this work can also be found elsewhere [12]. In the present study, ellipsometry measurements have been made in the energy range of 1.66 – 5.6 eV. Organic or inorganic contamination effects were minimized in our samples by cleaning the surfaces in a windowless cell with methanol and distilled water and then maintaining the samples in dry N$_2$ during measurements. Etching of

R. Kelly and M. Fernanda da Silva (eds.), Materials Modification by High-fluence Ion Beams, 93–100.
© 1989 by Kluwer Academic Publishers.

the surfaces was avoided due to danger of creating additional surface damage or even destroying the sample.

3. RESULTS/DISCUSSION

A systematic study of twenty different sets of samples prepared not only by conventional but also by magnetron sputtering has been made [8]. This has lead to the establishment of the specific conditions required to trip-on microcrystalline formation in the same chamber where a-Si is produced. The two most critical conditions have been identified to be the substrate temperature (T_s) and the ratio of hydrogen partial pressure to total pressure (P_{H_2}/P_T). Films produced at $200 < T_s < 250$ ^0C and with $P_{H_2}/P_T > 0.4$ systematically exhibit microcrystalline structure, otherwise amorphous-like behaviour is detected.

One way of monitoring the change in the film structure from amorphous to microcrystalline is by plotting the dependence of the Absorption Coefficient, (a), upon energy (measured by an optical spectrophotometer) and observe the decrease in its slope as the film becomes more crystalline-like. Fig.(1) shows the a(hv) function at high (a) values in the 10^4 region for a typical crystalline sample, two amorphous and a typical microcrystalline produced by magnetron sputtering. The above behaviour of the optical absorption may be explained in terms of unsaturated bonds. At the lowest hydrogen partial pressure (< 0.4) numerous bonds are unsaturated. As hydrogen pressure increases, the bonds saturate at about $P_{H_2}/P_T = 0.4$. This slope is proportional to an optical transition matrix element which is smaller for crystalline than amorphous silicon because of the momentum selection rule for the crystalline material [13].

Another way of detecting the internal changes in the film structure is by applying transport measurements. In particular, co-planar dark conductivity (σ_d) measurement of the individual films as a function of temperature can provide useful information about the film bulk structure changes. Fig.(2) shows how room temperature conductivity increases by about five orders of magnitude (from 10^{-9} to 10^{-4}) once the film structure becomes microcrystalline. In the meantime, the activation energy is drastically reduced from ~0.8 eV at $\sigma_d = 10^{-9}$ Ohm^{-1} cm^{-1} to ~0.3 eV at $\sigma_d = 10^{-4}$ Ohm^{-1} cm^{-1}. These changes are attributed to the fact that the film structure becomes more crystalline-like, with islands of crystallites embedded in a sea of amorphous material, and conductivity is then preferentially taking place through the crystallites leading to the enhancement of its value [14].

A third way of detecting the film structure changes is, as shown in Fig.(3), spectroscopic ellipsometry spectra (for typical c-Si, a-Si, a-Si:H and μc-Si:H films) of the imaginary part of the pseudo-dielectric function in the region between 3 to 4.5 eV. This region represents the smooth and superposed ε_1 and ε_2 edges of the c-Si. It should be noted that the ε_1 structure in the c-Si, corresponds to transitions from the highest valence band to the lowest conduction band along the <111> direction, while the ε_2 transitions correspond to electronic states not well localized in k space which include or are close in energy to the lowest gap at the X point. Calculations of complex band structures by means of the Green's Function technique have shown that the band stucture is less affected by disorder in the <111> direction than in the <100> direction. Consequently the structure around the ε_2 edge in c-Si is then expected to be smoothed out by disorder much more than the structure in the ε_1 edge. Therefore, by monitoring the shape of the ε_2 function, information can be extracted regarding the film structure.

Finally, traditional transmission and scanning microscopy can provide additional information not only on the film structure (i.e. amorphous,

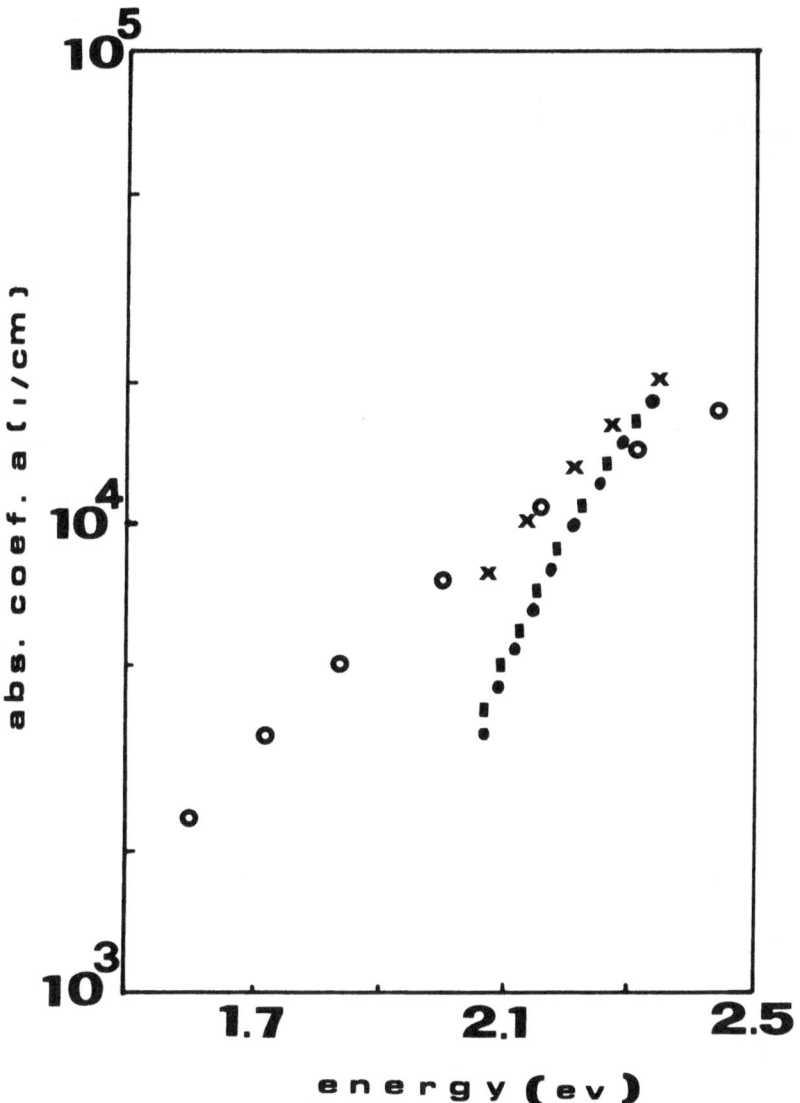

FIGURE 1. Absorption Coefficient (a) vs. Energy
(a) O, typical c–Si, (b) ■, ●, typical a–Si:H, (c) x, typical μc–Si:H

microcrystalline, size and volume function of the crystallites) but also on the film structure determining whether the conditions applied favour or not a homogeneous film development. Fig.(4a) and (4b) present typical micrographs of TEM analysis of μc–Si:H (with the dark areas being the microcrystals) and its

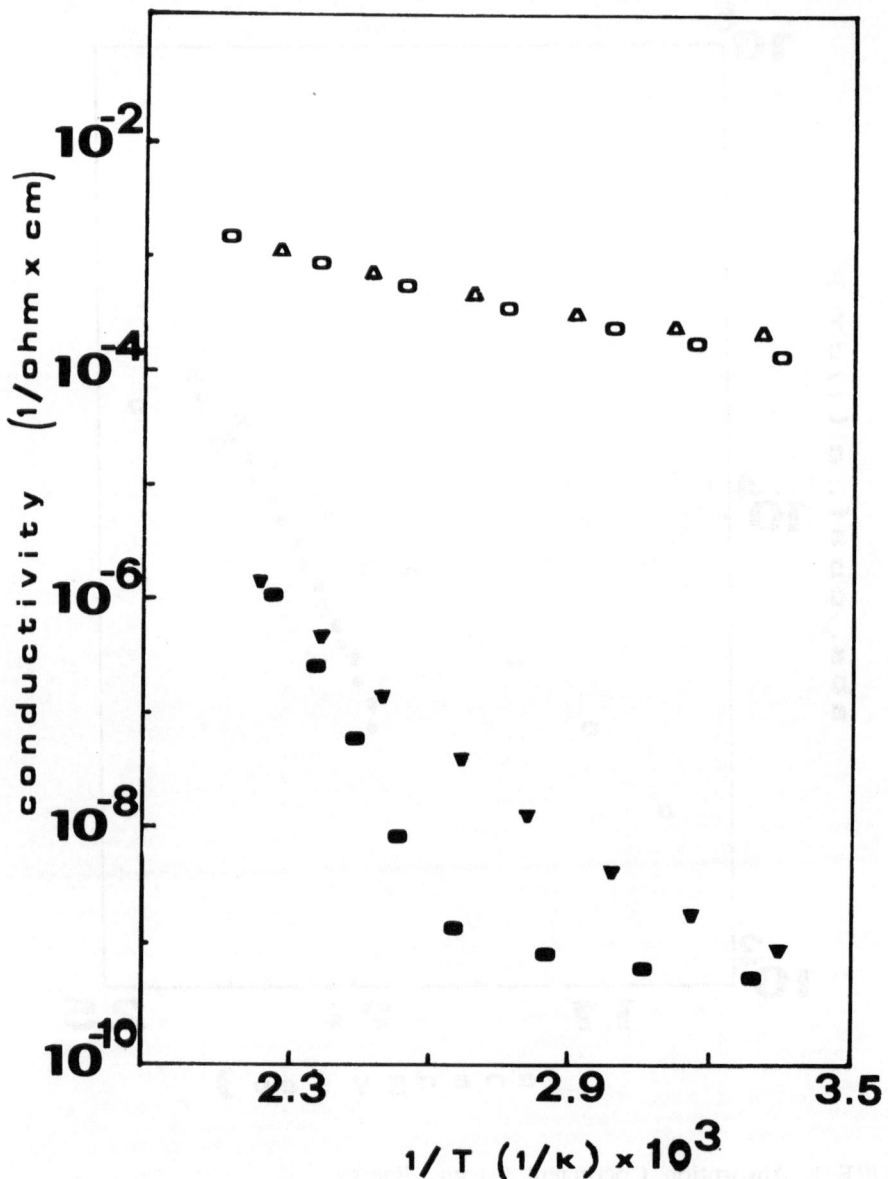

FIGURE 2.   Conductivity vs. 1/T : (a) △, 0, typical μc-Si:H,
(b) ●, ▼, typical a-Si:H.

corresponding electron diffraction pattern with rather diffused rings, confirming the film structure to be mixed phase (amorphous with crystallites). SEM

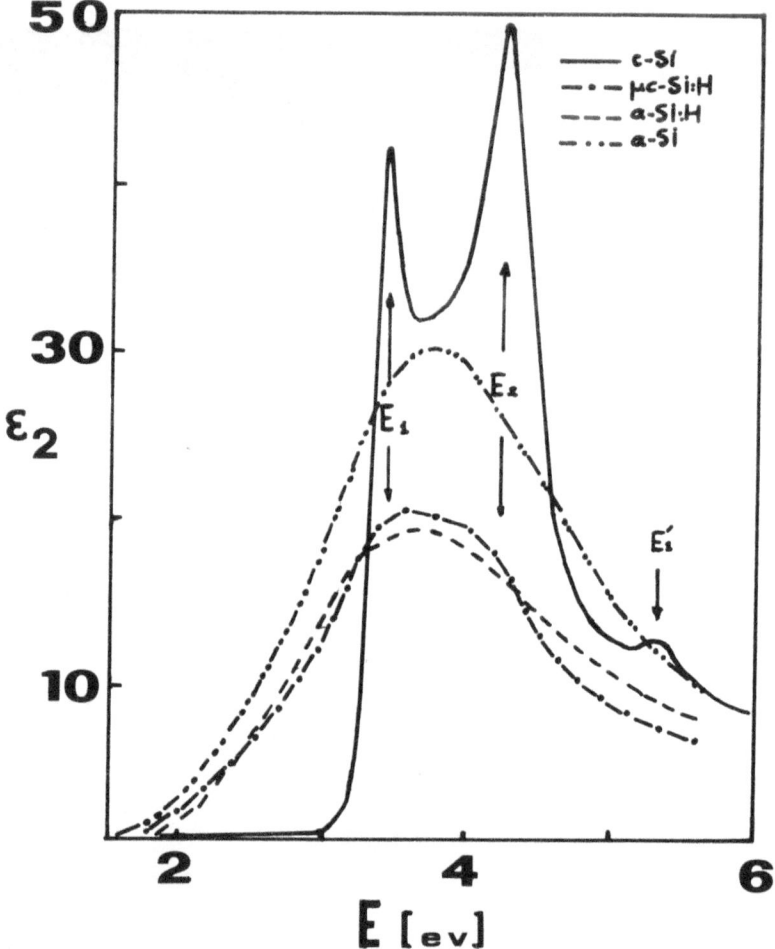

FIGURE 3. Dependence of ε_2 on Energy from spectroscopic ellipsometry measurements

analysis micrograph shown in Fig.(5a) indicates that the final surface is far from being smooth. Viewing thin sections of the μc-Si:H films under cross-section TEM (XTEM) it was observed in Fig. 5b that the reason for the rough surface is the columnar structure of the bulk. Similar bulk structure has also been reported by Das et al. [15] under deposition conditions similar to ours, and attributed by them to increased atom mobility.

4. CONCLUSIONS
There is still wide speculation as to whether the mechanism of microcrystallite formation is: a) H-plasma etching of the growing film attacking

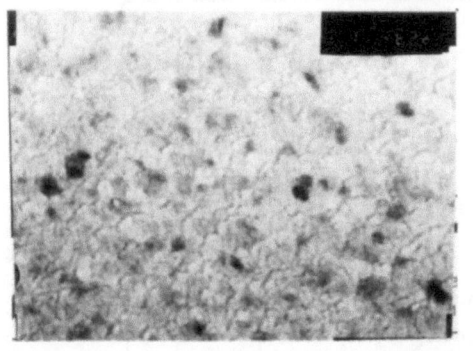 

FIGURE 4a.   TEM 120 KV - x210 K   FIGURE 4b.   EDX, 74

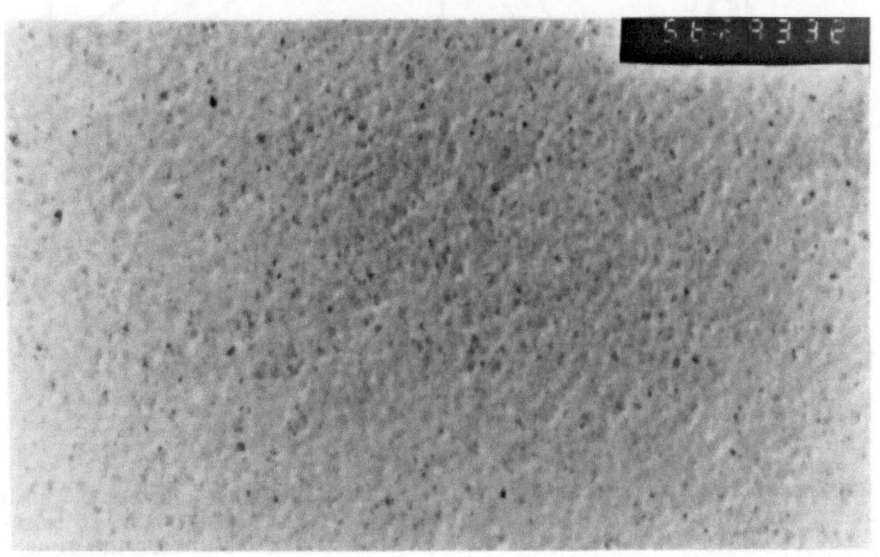

FIGURE 5a.   SEM, x56 K

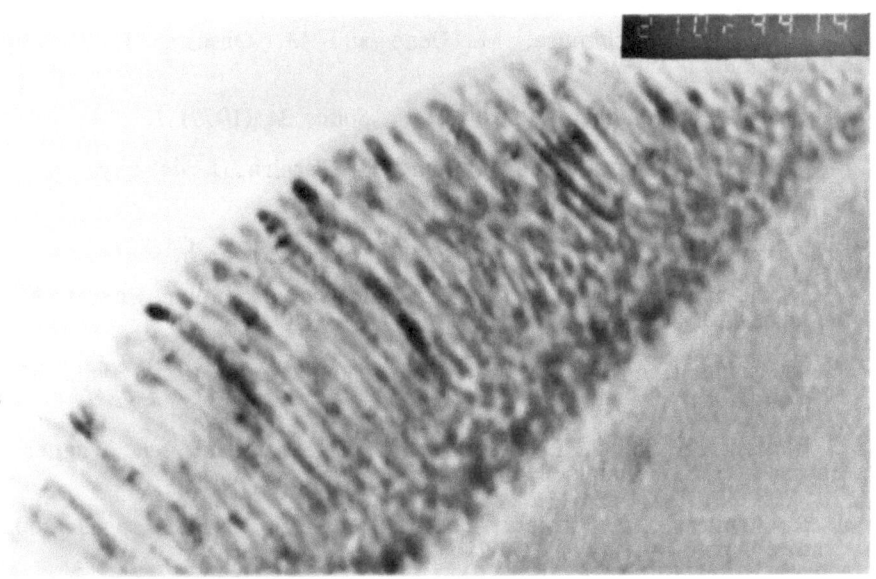

FIGURE 5b.   XTEM, x210 K

the weaker Si-Si bonds and thus leading to stable crystalline nuclei [10], and b) high electron bombardment of the growing film as the result of higher secondary electron emission yields of silicon hydride compounds formed on the target, which can lead to the crystallization of the substrate through local heating [14]. We have shown that:

a) μc-Si:H can be produced in an RF magnetron system provided the substrate temperature is between 200 and 250 ^{0}C and $P_{H_2}/P_T > 0.4$ in the plasma.

b) Evidence for microcrystallite formation is:
- the slope of the Absorption Coefficient, (a).
- the jump of $> 10^4$ orders of the room temperature conductivity.
- the shape of the dielectric function ε_2 and the appearance of two peaks at ~3.5 and ~4.2 eV.
- TEM and in particular EDX micrographs.

c) Evidence of columnar structure film development is provided from cross-section XTEM analysis.

5. REFERENCES

1.   S. Veprek and V. Maracek, Solid State Electron. 11 (1968) 683.

2.   A. Matsuda, T. Yoshida, S. Yamasaki and K. Tanaka, Jpn. J. Appl. Phys. 20 (1981), L439.

3.   W.E. Spear, G. Willeke, P.G. LeComber and A.G. Fitzerald, J. de Phys. 42 (1981), Suppl. C4-257.

4. Y. Ushida, T. Ichimura, M. Ueno and M. Ohsawa, J. de Phys. 42 (1981), Suppl. C4-265.

5. S. Usui and M. Kikuchi, J. Non-Cryst. Solids 34 (1979) 1.

6. A. Hiraki, T. Imura, K. Mogi and M. Tashiro, J. de Phys. 42 (1981), Suppl. C4-277.

7. H. Richter and L. Ley, J. de Phys. 42 (1981), Suppl. C4-261.

8. G. Kiriakidis, M. Marder and Z. Hatzopoulos, Solar Energy Mat., to be published.

9. Y. Mishina, S. Miyazaki, M. Hirose and Y. Osoka, Phil. Mag. 46 (1982), 1.

10. S. Veprek, Z. Iqbal, H.R. Oswald, F.A. Saratt and J.J. Wagner, J. de Phys. 10 (1981), Suppl. C4-251.

11. D.E. Aspnes, Opt. Commun. 8 (1973) 232; D.E. Aspnes and A.A. Studya, Appl. Opt. 44 (1975) 220.

12. L. Vina, S. Logothetides and M. Cardona, Phys. Rev. B30 (1984) 1979.

13. S. Abe and Y. Toyozawa, JARECT, Amorphous Semiconductors Technology and Devices, Y. Hamakawa ed., Vol. 6 (1983) 22.

14. T.D. Moustakas, Solar Energy Mat. 13 (1986) 373.

15. S.R. Das, D.F. Williams and J.B. Webb, J. Appl. Phys. 54 (1983) 3101.

DEPTH PROFILING OF Ta$_2$O$_5$ THIN LAYER ON Ta FOIL BY ION SCATTERING SPECTROMETRY AND ION SPUTTERING

F. RUMMENS, P. BERTRAND and Y. DE PUYDT

Université Catholique de Louvain
PCPM, Place Croix du Sud, 1
B-1348 Louvain-la-Neuve, Belgique

1. INTRODUCTION

Ion sputter profiling in conjunction with surface analysis is becoming nowadays the most important method to obtain composition depth profiles. Its application concerns all fields of Materials Science: microelectronics, thin films technology, catalysis, adhesion, corrosion, ... However, from the time dependence of the surface signal, the correct derivation of the true composition profile is not obvious. To achieve this goal the knowledge of the depth resolution achieved by ion beam sputtering is crucial. A lot of experimental and theoretical work was performed on these lines to determine the different factors contributing in this depth resolution (1,2).

A reference material for depth profiling with a step interface is available (3). It consists of an amorphous Ta$_2$O$_5$ layer of certified thickness grown on a Ta foil. This interface has been studied by different surface analytical techniques (AES, XPS, NMS)(4-6).

In this work, we have obtained depth profiles on this material by ISS for various experimental conditions: Ar$^+$, Ne$^+$, and He$^+$ bombardment with different energies and current densities. Owing to its monolayer surface sensitivity, ISS seems very appropriate for this purpose (7). However, obtaining good interface resolution requires a constant ion current density on the analysed area in order that the crater walls do not contribute to the analysis (2). This can be done by rastering the ion beam on the surface and by considering in the analysis only the signal coming from the central part of the crater by means of an electronic gate. This requirement is more easily fullfilled in AES, where a much thinner beam is used for the analysis than in ISS. (An electron beam is used in the first case, an ion beam in the second.)

Moreover, for a given experimental system, if the rastered area is increased too much, the ion current density may prove to be insufficient to maintain a clean surface. Contamination of the surface from the residual gases then induces inhomogeneities in the sputtering yield which produces in turn a deterioration of the interface resolution (8). There will be a compromise between the "crater wall" effects and surface contamination in the interface resolution.

A model is proposed to take correctly into account the ion current density variation during the rastering of the beam. With its help, satisfactory simulations of the experimental profiles are achieved taking also into account the initial ion impact cleaning of the surface, preferential sputtering of O in Ta$_2$O$_5$, and atomic mixing at the interface.

2. EXPERIMENTAL

The ISS spectra are measured with a Kratos spectrometer (Wg 541-515). This consists of a cylindrical mirror energy analyzer (CMA) with an integrated coaxial ion source. The samples are bombarded at normal incidence and the ions, backscattered in a annular solid angle of 0.18 sr centered at 138° with respect to the beam direction, are energy analyzed and detected. A programmable multichannel analyzer (Lecroy MCA 3500) is used for the spectra acquisition, for the spectrometer control and to perform automatic sequences of bombardment and analysis. The MCA is interfaced with a microcomputer (Digital Rainbow) for spectra

101

R. Kelly and M. Fernanda da Silva (eds.), Materials Modification by High-fluence Ion Beams, 101–108.
© *1989 by Kluwer Academic Publishers.*

processing and output facilities. The ISS energy spectra are converted into mass spectra by means of the kinematic factor (7). The ISS intensities are obtained by evaluating the peak areas after background subtraction . The ion current intensity is measured by a movable Faraday cup with a 1 mm black hole, and the beam full width at half maximum, measured by rastering the beam along a step metallic edge, is 450 μm for a primary beam energy E_0 of 2 keV. In typical working conditions, an ion beam current of about 35 nA is rastered on the sample to form a square of 1.2 mm side, with a frequency of 2.6 kHz in one direction and one hundred times lower in the perpendicular direction. This corresponds to an ion current density of 2.5 μA/cm^2. To prevent the formation of doubly charged Ar ions in the beam, the electron excitation energy in the ion source was kept as low as 50 eV. In these conditions the Ar^{++} contribution to the beam was shown to be negligible.

Without special bakeout the base pressure in the analysis chamber is $4*10^{-7}$ Pa of residual gases. It is obtained by means of a turbomolecular pump (Balzers TPU 330) and during the bombardment it rises to $4*10^{-5}$ Pa of noble gas.

The sample consists of a sputter profiling reference material produced by the National Physical Laboratory (NPL), Teddington. A Ta_2O_5 amorphous films of certified thickness (28.4 ± 1.7 nm) is grown by anodic oxidation of pure polycrystalline Ta foil. An interface resolution of 1.53 nm has been measured by AES and Ar$^+$ sputter profiling and this is interpreted as corresponding to a real step concentration profile (4). No bakeout of the samples is done before the measurements.

3. RESULTS

The samples are sputter profiled through the Ta_2O_5–Ta interface with Ar$^+$ and Ne$^+$ ion beams. He$^+$ ions are only used for ISS analysis since its sputter yield is too low for profiling. Typical ISS spectra obtained before and after the interface for a primary beam energy E_0 of 2 keV are presented on Figs. 1 to 3 respectively for Ar$^+$ ions, He$^+$ ions during Ar$^+$ profiling, and a mixing of Ne$^+$ and He$^+$ ions. The peaks are shifted to lower values than the theorical $E/E0$ values indicating inelastic energy losses. Their absolute evaluation requires however an accurate spectrometer calibration (9).

For Ar$^+$ bombardment (Fig 1), it is observed that the Ta peak is broader and less symmetric in the oxide and that the low energy background is higher. This may be due to inelastic effects and multiple collisions with O and Ta in the oxide. In both cases the high energy shoulder indicates the contribution of multiple collisions with Ta atoms.

Figure 1: Ar$^+$ (2 keV) ISS spectra obtained
in the same experimental conditions
a) on Ta_2O_5
b) on Ta after the interface
profiling

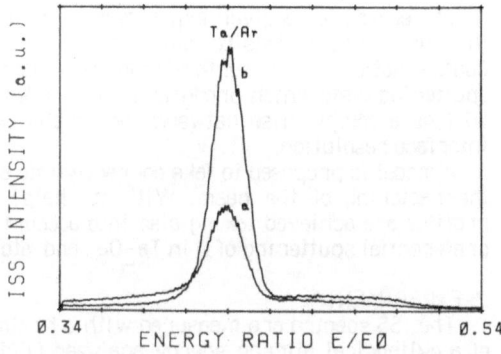

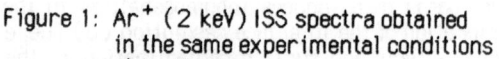

The He$^+$ spectra (Fig. 2,a and b) show that oxygen disappears after the interface. This indicates that clean surfaces may be achieved during the Ar$^+$ bombardment. An important low energy tail is associated with the Ta(He$^+$) peak, which origin has been recently discussed (9).

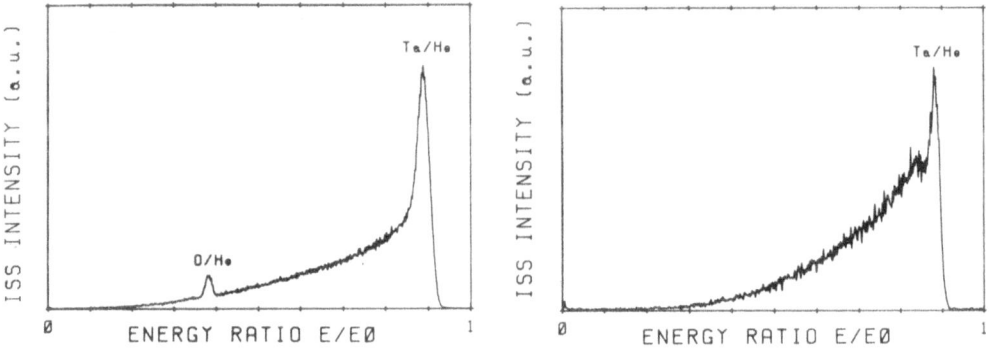

Figure 2: He$^+$ (2 keV) ISS spectra obtained

a) on Ta$_2$O$_5$ b) on Ta after the interface profiling
                                        with Ar$^+$ (2 keV) ions

For Ne$^+$ and He$^+$ bombardment (Fig. 3a and b), the spectra obtained before and after the interface are very similar in shape and intensity. The detection of oxygen without overlapping peaks is possible by mixing He$^+$ ions in the Ne$^+$ beam. Oxygen is shown to be still present after the interface. The low energy peak before the O (He$^+$) peak is attributed to a Ne$^+$ multiple collision with Ta and O atoms.

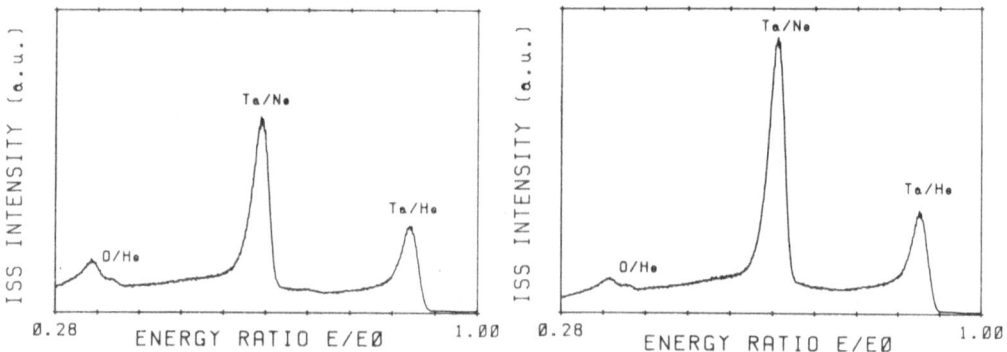

Figure 3: Ne$^+$ (2 keV) + He$^+$ (2 keV) ISS spectra obtained in the same experimental conditions
a) on Ta$_2$O$_5$ b) on Ta after the interface profiling

The variation of the Ta intensity with the bombardment time produces the depth profile. Figures 4 to 6 show some results obtained for 2 keV Ar$^+$ and Ne$^+$ bombardments (the experimental conditions are specified in the figure captions). The measured interface width ΔZ_m, is obtained by taking the difference from the abscissa values corresponding to 16% and 84% of the signal intensity variations. (This definition corresponds to twice the variance σ of an error-type function). Values of ΔZ_m varying from 3.6 nm to 9 nm are measured.

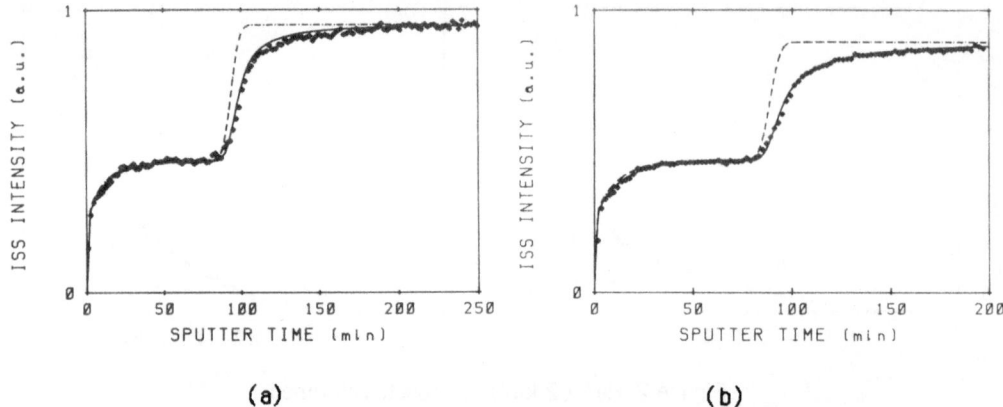

(a)   (b)

Figure 4: ISS depth profiles obtained with Ar$^+$ (2 keV) ions, J = 2.8 μA/cm^2, raster size
D = 1.2 mm, and (a): gate G = 33 %, ΔZ_m = 4.2 nm (measured interface width)
(b): gate G = 50 %, ΔZ_m = 6.2 nm.
Also reconstructed profiles $I_m(Z)$ (full line) and $I(Z)$ (broken line) with the
simulation parameters: σ_{Beam} = 0.2 mm, Z_D = 0.25 nm, Z_S = 4 nm, c = 0.43
and ΔZ_i = 2.7 nm. (The definition of the different parameters is given farther in
the text.)

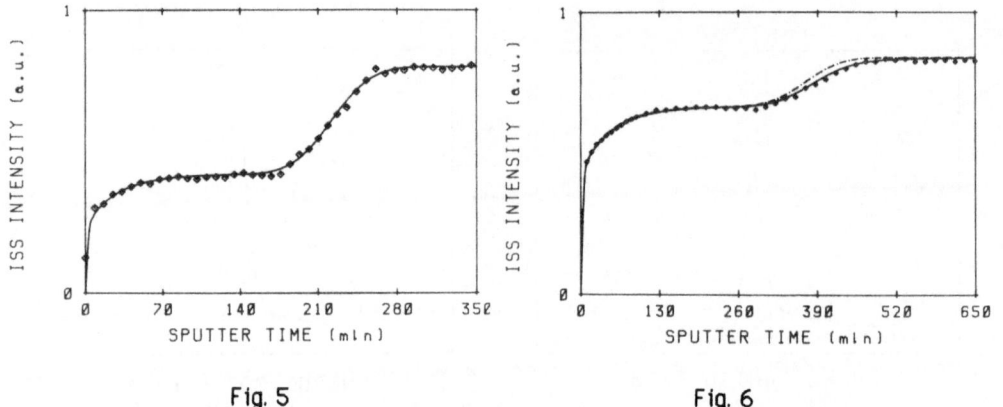

Fig. 5   Fig. 6

Figure 5: ISS depth profile obtained with Ar$^+$ (2 keV) ions, J = 1 μA/cm^2, raster size
D = 2.55 mm, and gate G = 30 %, ΔZ_m = 7.5 nm.
Also reconstructed profiles $I_m(Z)$ (full line) and $I(Z)$ (broken line merged with
$I_m(Z)$) with the simulation parameters: σ_{Beam} = 0.2 mm, Z_D = 0.25 nm, Z_S = 4
nm, c = 0.36 and ΔZ_i = 7.5 nm

Figure 6: ISS depth profile obtained with Ne$^+$ (2 keV) + He$^+$ (2 keV) ions, J (Ne$^+$) = 1.1
μA/cm^2, J (He$^+$) = 0.7 μA/cm^2, raster size D = 1.35 mm, and gate G = 30 %,
ΔZ_m = 8.7 nm. Also reconstructed profiles $I_m(Z)$ (full line) and $I(Z)$ (broken
line) with the simulation parameters: σ_{Beam} = 0.2 mm, Z_D = 0.25 nm, Z_S = 4
nm, c = 0.35 and ΔZ_i = 7.5 nm

They depend on the primary energy E_O and the ion type but also on the residual gas pressure, P_r, the ion current density J, and the bombarded and accepted area dimensions. These values of ΔZ_m are higher than those measured by AES at the NPL (ΔZ_m = 1.9 nm)(4). This shows that measured interface broadening in our ISS measurements results in large part from experimental factors. Moreover, small beam shifts in position ($\sim$ 1mm) during the profiling as produced by instabilities in the power supplies, lead to dramatic increases of ΔZ_m.

From the time needed to reach the interface situated at a certified depth the sputter rate $\dot{Z}$ and hence the sputter yield Y may be deduced (4): Y= 20.89 $\dot{Z}$/j where $\dot{Z}$ (nm/min) and $j(\mu A/cm^2)$ for a Ta_2O_5 mass density of 8200 kg/m³. A mean value of Y = 2.14 is found for Ar^+(2 keV) bombardment and is in agreement with the ASTM42 round-robin results (Y= 2.07 ± 0.21)(10).

4. DISCUSSION

To separate the contribution of the crater walls in ΔZ_m we evaluate the spatial dependence of the ion current density J(x,y) during the scanning of the ion beam over a surface rectangle with dimensions D_X and D_Y. For a beam with a Gaussian shape, the normalized current density found at a point (X, Y) for the spot centered at (x,y) may be written

$$ j(x-X, y-Y) = \frac{1}{2\pi\sigma_x\sigma_y} \exp\left(-\frac{(x-X)^2}{2\sigma_x^2}\right) \exp\left(-\frac{(y-Y)^2}{2\sigma_y^2}\right), $$

where σ_x and σ_y are the variances in both surface directions. Then the total current density J(X,Y) deposited at the point (X,Y) during the scanning is

$$ J(X,Y) = \frac{1}{D_X D_Y} \int_{-\frac{D_X}{2}}^{+\frac{D_X}{2}} \int_{-\frac{D_Y}{2}}^{+\frac{D_Y}{2}} j(x-X, y-Y)\, dx\, dy $$

and

$$ \int_{-\infty}^{+\infty} \int_{-\infty}^{+\infty} J(X,Y)\, dX\, dY = 1. $$

With the electronic gate system, the signal is only accepted when the beam top is in the centered rectangle ($D_X.G_X$, $D_Y.G_Y$) where G_X and G_Y represent the X and Y gate sizes (from 0 to 100%). Since the contribution of the point XY to the signal S(X,Y) is proportional to the total current density while the beam spot is in the gate rectangle:

$$ S(X,Y) = \frac{1}{D_X D_Y} \int_{-\frac{D_X G_X}{2}}^{+\frac{D_X G_X}{2}} \int_{-\frac{D_Y G_Y}{2}}^{+\frac{D_Y G_Y}{2}} j(x-X, y-Y)\, dx\, dy $$

$$\text{and} \int_{-\infty}^{+\infty} \int_{-\infty}^{+\infty} S(X,Y) \, dX \, dY = G_X G_Y$$

the gate reduces the total signal intensity.

During the profiling, if the sputtering yield is homogeneous the crater shape will follow the spatial distribution of the total current density $J(X,Y)$. After a bombardment time t the center of the crater, where the ion density $J(0,0)$ is a maximum, has reached a maximum depth $Z_m = Z(0,0)$. Around this point, since the ion density is lower, only the depth $Z(X,Y)$ will be reached:

$$Z(X,Y) = Z_m \, J(X,Y) / J(0,0)$$

Supposing an ideal intensity profile $I(Z_m)$ without crater wall effect, the measured intensity $I_m(Z_m)$ will become

$$I_m(Z_m) = \int_{-\infty}^{+\infty} \int_{-\infty}^{+\infty} S(X,Y) \cdot I(Z(X,Y)) \, dX \, dY .$$

We proceed by discretisation to evaluate $J(X,Y)$ and $S(X,Y)$ with $\sigma_x = \sigma_y$ and $D_x = D_y$. Figure 7 (a and b) represents in one dimension the crater shapes or total current distributions $J(X,Y)$ and the accepted parts of the signal $S(X,Y)$ for two experimental conditions corresponding respectively to the profiles shown on figures 4 (a) and 5. For $\sigma = 0.2$ mm, with $D = 1.2$ mm and $G = 30\%$, it is evident that the crater may not be considered as flat in the accepted area whereas this is realized for $D = 2.55$ mm and $G = 30\%$.

The ideal intensity profile $I(Z_m)$ has to contain all the interface broadening sources except that originated from the crater walls. With a step interface in composition, the atomic mixing produced by the bombardment will contribute to the measured interface width. For 2 keV Ar^+ bombardment, the linear cascade regime will apply for the mixing (5,11). The first order model of isotropic diffusion (12) will then convert the step composition profile into an error-type function symmetric with respect to the interface. (The loss of symmetry due to the sputtering is neglected). This function seems more appropriate than only exponential tail as proposed in (4).

The ion-impact cleaning of the surface at the beginning causes the first increase of the intensity observed in the measured profile. An exponential increase with a characteristic length Z_D, will account for this effect in $I(Z)$. A second exponential increase must count for the preferential sputtering of O in Ta_2O_5 (13,14). The length Z_s will characterize the distance over which the preferential sputtering is established. The ideal intensity profile is then written :

$$I(Z) = \left(1 - \exp\left(-\frac{Z}{Z_D}\right)\right) \cdot \left(1 - c \cdot \exp\left(-\frac{Z}{Z_s}\right)\right) \cdot I_{ox}^{\infty} \cdot (1 - b \cdot R(Z))$$

where

$$R(Z) = \frac{1}{\sigma_i \sqrt{2\pi}} \int_{-\infty}^{Z} \exp\left(-\frac{(z - Z_i)^2}{2\sigma_i^2}\right) dz$$

is the normalized error-type function. Z_i is the interface depth (30 nm) and σ_i is the variance with the interface width $\Delta Z_i = 2\sigma_i$. In addition we have

$b = (I_{met}^{\infty} / I_{ox}^{\infty} - 1)$, where I_{met}^{∞} and I_{ox}^{∞} are the saturation Ta intensities in the metal and in the oxide,

$c = (1 - I_{ox}^{0} / I_{ox}^{\infty})$, where I_{ox}^{0} is the Ta intensity before the setting up of the preferential sputtering. At the equilibrium for Ar^+ (2 keV), it is known that the surface is modified to give a TaO composition rather than Ta_2O_5 (13, 14). The theoretical value for $I_{ox}^{0} / I_{ox}^{\infty}$ is then Ta atomic concentrations ratio 2/7 : 1/2 = 0.57. This value must hold if there is no oxygen contamination during the sputtering.

Z_S is related to the ion range. A value of 1.33 nm has been calculated for Ar^+ (2 keV) range in Ta_2O_5 (15) and a value of Z_S = 2 nm is deduced in (4). However, we use the arbitrary value Z_S = 4 nm to fit the data. This is compatible with a distance of the order of 20 monolayers as suggested by E. Taglauer for 1 keV Ar^+ (16) and about 3 times the damage range as concluded in (14).

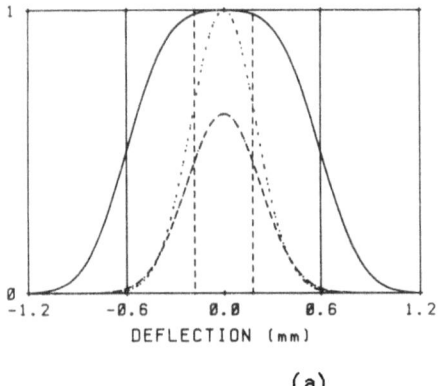 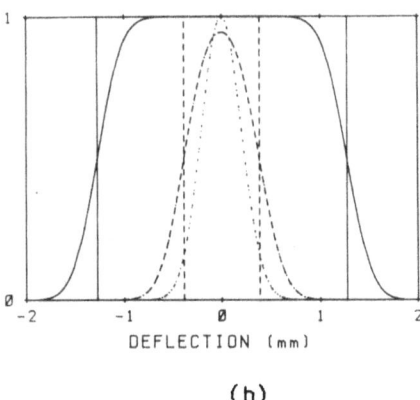

(a)　　　　　　　　　　　　　　(b)

Figure 7: One dimensional representation of the calculated crater shapes or total current density distributions J (X,Y) (full line) and the accepted parts of the signal S (X,Y)(broken line) for Gaussian beam shapes j (x,y) (dots). We assume $\sigma = 0.2$ mm and (a) a raster size D = 1.20 mm with a gate G = 30 %
(b) a raster size D = 2.55 mm with a gate G = 30 %.
The raster and gate dimensions are represented by the full and broken vertical lines respectively.

For Ar^+ bombardment, $I_{met}^{\infty} / I_{ox}^{\infty}$ is found very reproducible for clean surfaces and its value decreases with the primary energy E_0 (4.1 for 1 keV, 2.4 for 2 keV and 1.25 for 4 keV). In the simulations, the characteristic length for ion-impact cleaning is taken of the order of one monolayer (Z_D = 0.25 nm). I(Z) and I_m(Z) are calculated and represented on the different profiles shown in figures 4 to 6. Only the value of ΔZ_i is adjusted to obtain the best fit of the experimental profiles. The two profiles shown in figure 4 (a and b) differ only by the accepted area. A same value of ΔZ_i = 2.7 nm is found and the difference in ΔZ_m is correctly explained by the crater wall effect. For the atomic mixing contribution to the profile broadening, Hunt and Seah fitted an exponential decay with a characteristic length of 0.9 nm for Ar^+ (1 keV) at 45° incidence (4).
By extrapolating this value to our experimental conditions with the help of the Andersen model (12)(ΔZ_i is proportional to $\sqrt{(E_0/Y)}$) and applying the correspondence between the error and exponential function, ΔZ_m = 2.4 nm is expected. The difference with the 2.7 nm fitted value may be due to a possible contamination of the surface.
In the profile shown figure 5, there is no more crater wall contribution and I(Z) =I_m(Z) but the ion current density is not enough to prevent a partial reoxidation of the surface during the profiling. A value of $I_{ox}^{0} / I_{ox}^{\infty}$ = 0.64 higher than the predicted one has to be

used in the simulation. An interface width $\Delta Z_m = \Delta Z_i = 7.5$ nm is observed. A dominant contribution ΔZ_c due to the surface contamination is suspected (8), where $\Delta Z_c \equiv (\Delta Z_m^2 - 2.4^2)^{1/2} = 7.1$ nm.

For the Ne^+ bombardment (see Fig. 6) both surface contamination and crater effect are present and the value of $\Delta Z_i = 7.5$ nm is needed for the fitting. During the Ne^+ and He^+ profiling the $O(He^+)$ and $Ta(He^+)$ peaks do not exhibit any detectable intensity variation though the interface, indicating an important Ta reoxidation.

5. CONCLUSION

The depth resolution achieved with our experimental system is shown to be a compromise between the contribution of the crater walls and surface contamination. The proposed model allows to take into account the former effect. An estimation of the atomic mixing contribution to the interface width may be done which is in agreement with the other published values.

ACKNOWLEDGEMENTS

The authors are indebted to J-M. Beuken for his help in the automation of the data acquisition and treatment. The ISS equipment was acquired thanks to the support of FRFC Belgium and Y. De Puydt acknowledges a grant from PREST (Belgium).

REFERENCES

(1) e.g. Thin Film and Depth profile Analysis, Ed. H. Oechsner (Springer-Verlag, Berlin Heidelberg, 1984)
(2) S. Hofmann, Surface Interface Anal. 2 (1980) 148
(3) Certified Reference Material NPL n° S7B83, BCR n° 261
(4) C.P. Hunt and M.P. Seah, Surface Interface Anal. 5 (1983) 199
(5) H.J. Mathieu, in: Thin Film and Depth Profile Analysis, Ed. H. Oechsner (Springer-Verlag, Berlin Heidelberg, 1984) p. 39
(6) H. Oechsner, in : Thin Film and Depth Profile Analysis, Ed. H. Oechsner (Springer-Verlag, Berlin Heidelberg, 1984) p. 63
(7) e.g. T. Buck, in: Methods of Surface Analysis, Ed. A.W. Czanderna (Elsevier, Amsterdam, 1975)
(8) H.J. Mathieu and D. Landolt, J. Microsc. Electron. 3 (1978) 113
(9) P. Bertrand, E. Pierson and J.M. Beuken, Proc. ICACS12, to be publ. in Nucl. Instrum. Meth. Phys. Res. B (1988)
(10) J.A Bevolo, Surface Interface Anal. 3 (1981) 240
(11) P. Sigmund, in: Sputtering by Particle Bombardment I, Ed. R. Behrisch Springer-Verlag, Berlin Heidelberg New York, 1981) p. 9
(12) H.H. Andersen, Appl. Phys. 18 (1979) 131
(13) E. Taglauer and W. Heiland, Appl. Phys. Lett. 33 (1978) 950
(14) D.K. Murti, R. Kelly, Z.L. Liau and J.M. Poate, Surf. Sci. 81 (1979) 571
(15) P.H. Holloway and G.C. Nelson, J. Vac. Sci. Technol. 16 (1979) 793
(16) E. Taglauer, Appl. Surf. Sci. 13 (1982) 80

BOMBARDMENT OF ALKALI AND ALKALI-EARTH HALIDES BY IONS AND ELECTRONS

P. WURZ, G. BETZ, W. HUSINSKY, K. MADER, B. STREHL AND E. WOLFRUM

Institut für Allg. Physik, Technische Universität Wien, A-1040 Wien, Wiedner Hauptstr. 8-10, Austria

1. INTRODUCTION

For alkali and alkali earth halides ion as well as electron bombardment are efficient processes leading to continuous surface erosion [1,2]. While sputtering of metals under ion bombardment is well understood in terms of a collision cascade emission mechanism [3], for electron bombardment no such process can account for any particle emission due to the negligible momentum transfer. Electronic processes have been proposed to explain electron or photon induced desorption [4,5]. The key steps are the generation of an exciton, which is trapped at a halogen site. Then the lattice ion which is raised to an excited state, thermally relaxes to form a molecular-like state with a neighbouring halogen ion. In this new system the excited electron can make a non-radiative transition to the new ground state, and the energy is then released to the ions as kinetic energy directed along the <110> crystal direction. Thus a replacement sequence along the line of halogen ions takes place, and if the sequence reaches the surface, emission of a halogen atom is caused. Therefore the expelled halogen atoms feature directional emission and hyperthermal energies [6]. The replacement sequence and the diffusion of defects to the surface lead to an alkali or alkali-earth enrichment on the surface. These excess metal atoms prevent further halogen emission or, if the vapour pressure is high enough may evaporate in random directions [6].

2. ENERGY DISTRIBUTION OF SPUTTERED PARTICLES

To gain the velocity (energy) distribution of the sputtered particles we used Laser Induced Fluorescence (LIF) and Doppler Shift Laser Fluo-

R. Kelly and M. Fernanda da Silva (eds.), Materials Modification by High-fluence Ion Beams, 109–115.

rescence Spectroscopy (DSLFS) [8,17] for neutral ground-state atoms and high resolution optical spectroscopy [9,11] for excited atoms and ions.

The first measurements using LIF and DSLFS were performed to study the emission of ground-state Na atoms from a NaCl target (pressed powder target) under ion bombardment [7]. These measurements were redone under better vacuum conditions and with a single crystal target. We compared these measurements with our results obtained under electron bombardment.

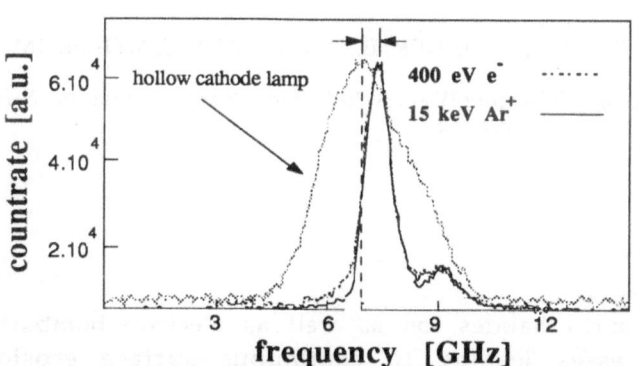

Fig.1: DSLFS spectra from emitted ground-state Na atoms from a NaCl single crystal under ion and electron bombardment. The maximum of the spectrum from a Na - hollow cathode lamp determines the zero point of the velocity spectrum. A shift of the signal towards higher frequencies indicates Na atoms with increasingly higher velocities. In addition, the spectra feature the hyperfine splitting of the Na (588.9nm) line. The indicated shift of the maxima agrees with a thermal Maxwell-Boltzmann distribution of 300 K target temperature.

For electron bombardment on NaCl the energy distribution is purely thermal for all target temperatures (300 to 700 K). This distribution can always be fitted with a Maxwell-Boltzmann distribution at target temperature. Figure 1 shows clearly that the energy distribution of sputtered Na atoms under ion bombardment is the same as for electron bombardment. We therefore conclude that, within the accuracy of our measurements, there is no collisional component in the energy distribution of ion–beam sputtered Na atoms from NaCl. The hyperfine splitting of the Na ground-state makes it difficult to extract this information from the measured spectra under ion bombardment, but can be clearly seen from direct comparison of the two spectra. However, if the target is bombarded in an oxygen atmosphere, the magnitude of the thermal component decreases and in addition a cascade contribution appears (see figure 2). This can be understood in terms of a lower vapour pressure for Na_2O than Na. Thus the difference to the results by Husinsky et al. [7] can be attributed to the high oxide content of their pressed powder samples. Szymonsky et al. [15] also report ToF (Time of Flight) measurements of halides, where they fit the flux of emitted alkali atoms with three contributions. Be-

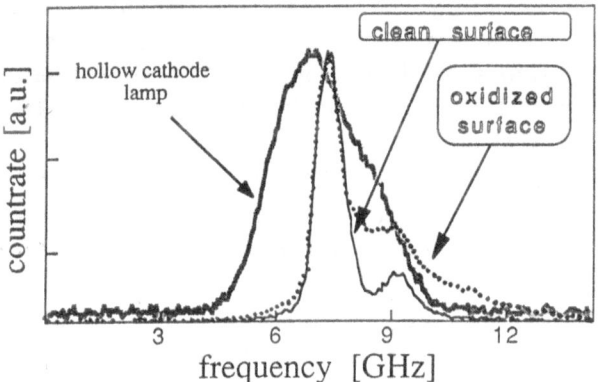

Fig. 2: DSLFS spectra from emitted ground-state Na atoms as in fig.1 at a target temperature of 300 K from a clean surface (base pressure 10^{-7} Pa) and a surface exposed to an oxygen partial pressure of 5.10^{-4} Pa. The signal from the clean surface was higher by a factor of 5.

sides a collision cascade and a thermal evaporation they introduce a thermal spike component with higher temperatures than target temperature. In our DSLFS measurements we did not detect a spike contribution. The strong cascade contribution in the ToF results is probably also caused by the use of a pressed powder sample (see figure 2).

The excited sputtered atoms, studied with high resolution optical spectroscopy, behave differently. For electron bombardment the energy distribution of Na I (588.9nm) again is thermal, but for ion bombardment we derive an energy distribution which has its maximum at about 2 eV and can be fitted very well with the Thompson formula [3]. These results are in good agreement with the experiments done by Walkup et al. [10], also performed on NaCl.

At room temperature Ca ground-state atoms sputtered from a CaF_2 target under ion bombardment exhibit an energy distribution which peaks at 0.7 eV. Collision cascade theory [3], which is also applicable for binary targets with mass ratios less than 2 [15], gives a good fit and leads to a surface binding energy of 1.4 eV, which is in perfect agreement with previous measurements that we have performed

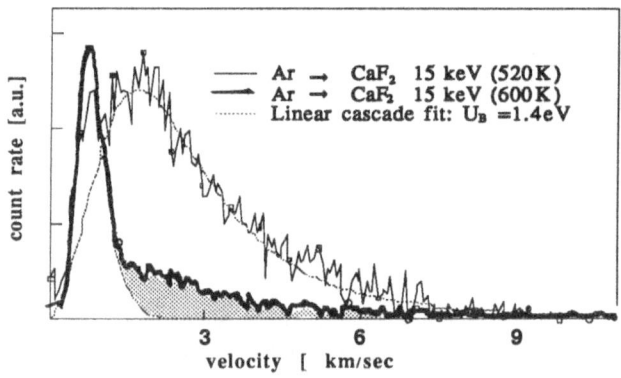

Fig. 3: Normalized velocity spectra of sputtered ground-state Ca atoms from a single crystal CaF_2 target. At 520 K the distribution is dominantly collisional (thin curve) and can be fitted by linear cascade theory assuming a surface binding energy of 1.4eV (shaded thin curve). At 600 K the thermal component is already dominant and the difference to a pure thermal distribution (Maxwell-Boltzmann) at target temperature is indicated by the shaded area. The 600K spectra was higher by a factor of 7.

on metallic Ca [11]. Above 500 K an additional thermal component in the energy distribution of sputtered Ca atoms appears and becomes dominant at higher temperatures (see figure 3). Under electron bombardment Ca atoms are only desorbed above 500 K, where ion beam induced evaporation starts and in addition the distribution is purely thermal.

No desorption of excited Ca or F particles from a clean CaF_2 under electron bombardment was observed over the investigated temperature range. Only if the surface was contaminated, desorption of excited particles was observed to a minor extent. For ion bombardment of CaF_2 the excited Ca levels are populated in the same manner as for a metallic target. A detailed investigation of the Ca I (422.7nm) line with high resolution optical spectroscopy leads to a maximum in the energy distribution of this state at 6eV. This is the same result that we derived for excited and also for metastable atoms sputtered from a metallic Ca target [12].

3. TIME DEPENDENCE OF THE DESORBED FLUX FROM CALCIUM FLUORIDE.

In order to get more information about the composition of the desorbed particles, we studied the time dependence of the LIF signal when CaF_2 was exposed to a chopped electron or ion beam.

3.1 <u>Under ion bombardment:</u> (15 keV Ar ions, 5 $\mu A/cm^2$). As can be seen in figure 4, one can distinguish three contributions to the sputtered flux at temperatures above 500 K. i) *A prompt response* of the LIF signal by almost an order of magnitude was observed immediately after the ion beam was switched on or off.

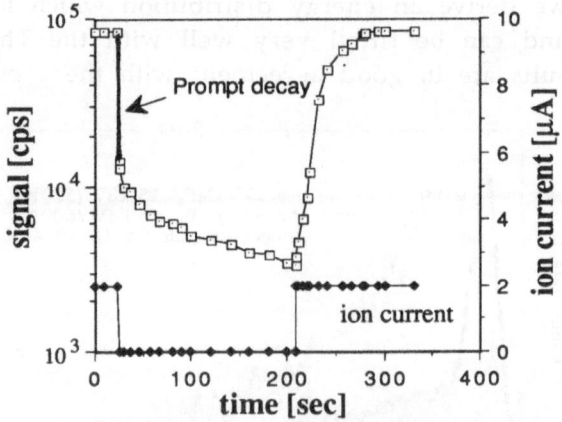

Fig. 4: Time dependence of the yield of Ca ground-state atoms sputtered from a CaF_2 target at 600 K during an interrupt of the ion beam. The bold drawn part of the curve, which is labelled "prompt decay", corresponds to a signal decrease within less than 0.1 sec.

Prompt in this context means faster than 0.1 s due to our limited time resolution. We assume that the prompt response consists in reality of two contributions. A collision cascade contribution at non-thermal energies, which should not last longer than 10^{-12} -

10^{-11}s [13], and a defect-induced component due to H-centers with time constants of 10^{-3} to 10^{-5}s [5]. As outlined before the decay of a H-center is accompanied by the emission of a halogen atom and the formation of a surface metal layer. If the vapour pressure of the metal at the selected target temperature is high enough, the metal will evapourate in addition to purely collisional sputtering. We know from our DSLFS measurements that the sputtered flux above 550 K is dominated by the thermal component. As the life time of these defects is in the order of ms, they will also contribute to this "prompt response" [5]. Such delay times in the order of ms have first been observed by Townsend et al. [6] under ion and by Overeijnder et al. [14] under electron bombardment. ii) *A thermal contribution* exhibiting a decay constant in the order of 30 s, which we assume to be caused by the migration of these defects to the surface and in turn cause Ca emission similar to the direct emission by decay of self trapped excitons. Recently such long life times have been reported for photon stimulated desorption of Li from LiF [2]. iii) *A purely thermal contribution* which is very weak and is not caused by ion bombardment. This is the steady state value after a long period without sputtering (after 30 min in fig. 4), representing only the pure evaporation rate at target temperature for an unbombarded target.

3.2 <u>Under Electron Bombardment:</u> Under electron bombardment (400 eV, 200 μA/cm^2) CaF$_2$ showed no emission of Ca ground-state atoms up to 500 K, the same temperature for which under ion bombardment the thermal contribution appeared. The desorbed flux has always a thermal velocity distribution at target temperature as our DSLFS measurements show and was four times larger than for purely thermal evaporation at 350 K. The prompt response, which we have observed for the chopped ion beam measurement, did not occur for the measurements with a chopped electron beam. The time constants for the decay are about 3 min, that is considerably longer than for ion bombardment. This may be due to the larger penetration depth in the solid of electrons compared to ions. In addition, at a high electron fluence we observed a saturation of the target with defects, which are responsible for the desorption. Then switching the electron beam on or off has no effect on the desorbed flux. Only after a long period of annealing the target recovers to its original behaviour.

4. TEMPERATURE DEPENDENCE OF SPUTTERED NEUTRAL AND EXCITED PARTICLES

The temperature dependence of the neutral ground state atoms and

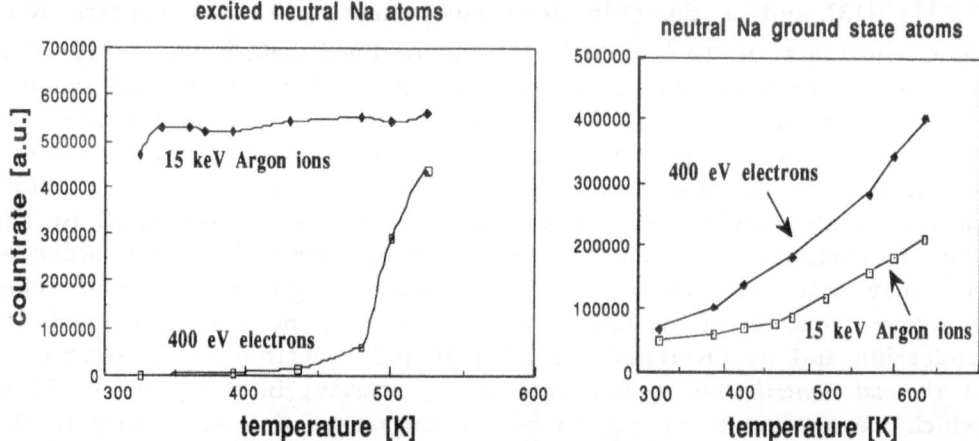

Fig. 5: Temperature dependence of the yield of Na ground- and excited-state (588.9 nm) atoms under 15 keV Ar$^+$ or 400 eV electron bombardment of NaCl measured by LIF or BLE, respectively.

excited atoms has been studied with LIF or BLE (Bombardment induced Light Emission), respectively, to gain additional information if excited atoms or ions are created either due to a surface intrinsic effect as proposed by Tolk et al. [16] or due to excitation or ionisation by secondary effects (secondary electrons, primary beam), which is favoured by Szymonsky et al. [1] and Walkup et al. [10]. Figure 5 shows the temperature dependence of neutral ground-state and excited state (588.9nm) Na atoms sputtered from a NaCl target. For electron bombardment the yield of excited Na atoms is very similar to the yield of ground-state Na atoms, and both increase exponentially with temperature. On the other hand, under ion bombardment only the ground state Na yield exhibits this exponential dependence on temperature and the excited atoms remain nearly unaffected by temperature. The observed increase in the ground-state yield is a consequence of the thermal evaporation of the Na layer on the surface.

As mentioned before, for CaF$_2$ under electron bombardment, no emission of excited atoms was observed over the investigated temperature range. Ca ground state atoms behave similar to Na ground state atoms, but the exponential increase of the yield starts at about 500K, which is the onset for ion-beam induced thermal evaporation (see paragraph 2 and ref. [11]). Below 500K the yield of sputtered Ca atoms from CaF$_2$ is constant under ion bombardment and zero under electron bombardment. While results on NaCl do not contradict a creation mechanism of excited particles due to excitation in the gas phase, the absence of excited Ca atoms desorbed from CaF_2 contradicts such a model and favours a surface intrinsic mechanism.

REFERENCES

1. Z. Postawa, J. Rutkowsky, A. Poradzisz, P. Czuba and M. Szymonski, Nucl. Instr. Meth. B18 (1987) 574
2. R.F. Haglund Jr., N.H. Tolk, G.M. Loubriel and R.A. Rosenberg, Nucl. Instr. Meth. B18 (1987) 549
3. P. Sigmund, Phys. Rev. 184 (1969) 383
4. N. Itoh, Nucl. Instr. Meth. 198 (1987) 155
5. P.D. Townsend, Nucl. Instr. Meth. 198 (1982) 9
6. P.D. Townsend, R. Browning, G.J. Garlant, J.C. Kelly, A. Mahjoobi, A.J. Michael and M.Saidoh, Rad. Effects 30 (1976) 55
7. W. Husinsky and R. Bruckmüller, Surface Science 80 (1979) 637
8. H.L. Bay, Nucl. Instr. Meth. B18 (1987) 430
9. G. Betz, Nucl. Instr. Meth. B27 (1987) 104
10. R.E. Walkup, Ph. Avouris and A.P. Gosh, Phys. Rev. Lett. 57 (1986) 2227
11. W. Husinsky, I. Girgis and G. Betz, J. Vac. Sci. Techn. B3 (1985) 1543
12. W. Husinsky and G. Betz, Scanning Microscopy (1987) in print
13. M. Szymonsky, Rad. Effects 52 (1980) 9
14. H. Overeijnder, R.R. Tol, A.E. de Vries, Surface Science 90 (1979) 265
15. M. Szymonsky, H. Overeijnder and A.E. de Vries, Rad. Effects 36 (1978) 189
16. N.H. Tolk, P. Bucksbaum, N. Gershenfeld, J.S. Kraus, R.J. Morris, E. Murrick and J.C. Tully, Nucl. Instr. Meth. B2 (1984) 457
17. W. Husinsky, J. Vac. Sci. Techn. B3 (1985) 1546

EFFECTS OF Ar+ ANGLE-OF-INCIDENCE ON THE ETCHING OF Si WITH Cl₂ AND LOW-ENERGY Ar+ IONS.

J. VAN ZWOL, A.W. KOLFSCHOTEN, J. VAN LAAR, and J. DIELEMAN

PHILIPS RESEARCH LABORATORIES, P.O. BOX 80000,
5600 JA EINDHOVEN, THE NETHERLANDS.

1. INTRODUCTION

To study the fundamental aspects of processes involved in plasma etching, for some years beam experiments have been performed, in which a low energy (<5 keV) ion beam and a reactive gas beam are directed at a solid surface and the properties of the products leaving the surface are measured. This model system of plasma etching has been studied for various gas-ion-solid combinations[1]. In some of these systems both spontaneous etching (i.e. by the gas alone) and ion-influenced etching is observed. In several cases ion bombardment enhances the etch rate, sometimes to a level higher than the sum of the spontaneous etch rate and the pure sputter rate. (However, in other cases ion bombardment causes a rate reduction)[2].

In order to explain the role of the ions, three models have been proposed. The first suggests that surface damage produced by the ion bombardment enhances the reaction with reactive gases. The second model states that the ions stimulate the thermal reaction path by promoting the formation of the volatile product which is followed by thermal desorption of the products, possibly at elevated temperatures that occur during ion impact[2,3].

According to the third model, as soon as ion bombardment is involved, we are mainly dealing with physical sputtering. Also, ion beam mixing causes the reactive gas to enter the top atomic layers of the substrate, where they form products (not necessarily volatile) that are subsequently sputtered from the substrate. Thermal desorption is not ruled out but is not the main effect. In this model the etch rate enhancement/reduction is caused by a change in the sputter yield due to the chemical and structural modification of the top atomic layers[4].

Experimentally one can distinguish between these models by determining the mass, energy and angular distributions of the reaction products leaving the substrate. Mass spectroscopy combined with time-of-flight (T.O.F.) measurements (MS-TOF) gives direct information about these distributions. In conventional mass spectroscopy one has the problem that the fragmentation patterns of the product molecules to be detected are usually unknown. If, however, T.O.F. distributions are measured as well, it is often possible to separate the contribution of different product molecules to a particular mass signal in the mass spectrometer. Also, because the T.O.F. distribution is measured, one can correct for the velocity dependent ionization probability in the mass spectrometer. For these reasons we have chosen MS-TOF for our study of model systems of plasma etching.

Here we will focus on some aspects of the etching of Si with Cl₂ gas and an Ar+ beam (to be described as Si(Cl₂,Ar+)). For this system experiments show that the major part (>80 %) of the product molecules (found to be mainly Si, Cl, SiCl and SiCl₂) leave the surface with an energy distribution typical of physical sputtering, viz. a collision cascade distribution $\phi(E) = E/(E + U_0)^3$ where U_0 is the binding energy of the partic-

R. Kelly and M. Fernanda da Silva (eds.), Materials Modification by High-fluence Ion Beams, 117–121.
© *1989 by Kluwer Academic Publishers.*

ular sputtered product. The low U_0 of the products SiCl and SiCl$_2$, found in these experiments, leads to the conclusion that these product molecules are not adsorbed at the surface, from which they would have evaporated, but are trapped in the bulk. A small fraction of the products (<20 %) leaves the surface with a Maxwell-Boltzmann distribution at target temperature[4] . So from these experiments one can deduce that for the case of Si(Cl$_2$, Ar$^+$) the "chemically enhanced physical sputtering" model seems appropriate.

Here we will discuss some further investigations on the Si(Cl$_2$,Ar$^+$) system, viz. the effect of the Cl$_2$ to Ar$^+$ flux ratio and the Ar$^+$ angle-of-incidence on the T.O.F. distributions of the product SiCl.

2. EXPERIMENT

A schematic representation of the beam apparatus in which the experiments are performed is given in fig.1. It consists of a main vacuum chamber in the center of which a target can be placed. The target can be rotated about a vertical axis. We direct an ion beam and a thermal reactive gas beam at this target. For the neutral product detection a differentially pumped mass spectrometer with a T.O.F. measurement capability is used, which can be rotated about the same (vertical) axis as the target. Thus we can vary the angle-of-incidence of the ion beam on the target independently from the detection angle. The angle between the ion beam and the reactive gas beam is fixed. The Ar$^+$-ion beam is electrostatically chopped with a pseudorandom code to distinguish ion-induced processes from spontaneous ones. The pseudorandom code also reduces measuring time when compared to "single pulse" chopping.

The base pressure of the main chamber is 3×10^{-7} Pa (without bake-out). At present we use Ar$^+$ ions of energy up to 3 keV. The maximum beam current is 2 μA corresponding to a flux of about 10^{14} ions cm^{-2} s^{-1} . The reactive gas is Cl$_2$ with a direct flux up to 10^{17} molecules cm^{-2} s^{-1} . This gas flow gives rise to a background pressure in the main chamber of up to 10^{-4} Pa , which means that the Cl$_2$ background contributes < 1% to the total Cl$_2$ flux at the target.

3. RESULTS AND DISCUSSION

First, we investigated the effect of variation of the Cl$_2$ flux at constant Ar$^+$ flux and angle-of-incidence $\theta_{inc} = 50°$ from the surface normal on the T.O.F. distribution of the product SiCl. The detection angle was kept fixed at 0° from the surface normal. The Ar$^+$ beam current was 1 μA, the beam spot area was about 0.1 cm^2, which leads to an Ar$^+$ flux of about 5×10^{13} ions cm^{-2} s^{-1} . The Cl$_2$ flux was varied between 1×10^{15} and 5×10^{16} molecules cm^{-2} s^{-1} , and thus the Cl$_2$ to Ar$^+$ flux ratio was varied between 20 and 1000. The substrate temperature was kept at 300 K.

The T.O.F. distributions obtained for SiCl can be fitted quite well with a collision cascade distribution. Under all flux ratio's and angles-of-incidence fewer than 20 % of the SiCl molecules leave the target with a Maxwell-Boltzmann energy distribution at substrate temperature. No indications are found that the $m/e = 63$ signal in the mass spectrometer contains a contribution from other products than SiCl. (This should have been visible as an extra peak or a broadening of the peak in the T.O.F. distribution). An example of a measured distribution and the fitted curves is given in fig. 2 for flux ratio 40. The signal has been corrected for the $1/v$ dependence of the ionization probability in the mass spectrometer. (v is the particle velocity). The U_0 values obtained from these fits are plotted in fig. 3. The increase in U_0 with decreasing Cl$_2$ to Ar$^+$ flux ratio has already been mentioned qualitatively in ref. 4, but no data were given. This increase in U_0 is interpreted as a decrease of the extent to which the topmost layers of the substrate are modified: as less Cl$_2$ is mixed into these layers, they will be more similar to the strongly bound original solid, whereas more Cl$_2$ mixing

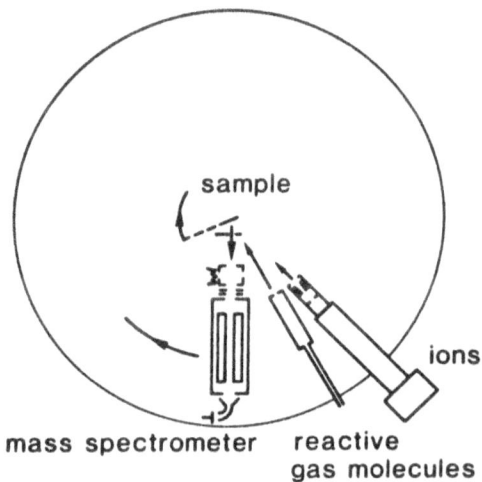

Fig. 1 - Schematic drawing of the crossed beams reactor

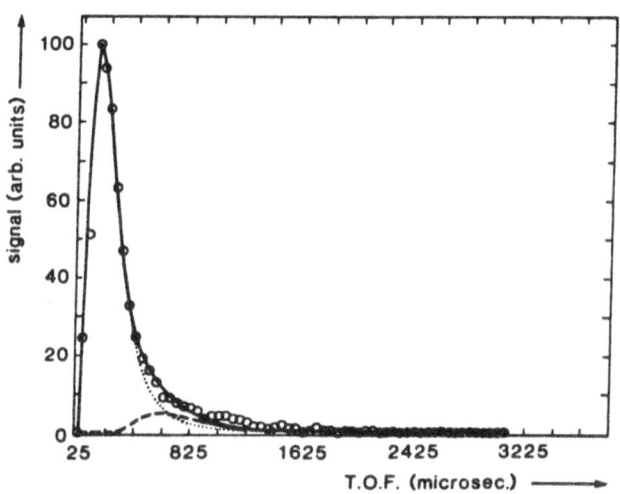

Fig. 2 - Time-of-flight distribution of the product SiCl for $\phi_{Cl_2}/\phi_{Ar^+} = 40$, $\theta_{inc} = 50°$, $\theta_{det} = 0°$. As demonstrated, a combination of 92 % of a collision cascade distribution and of 8 % of a Maxwell-Boltzmann distribution at T = 300 K fits the data quite well.

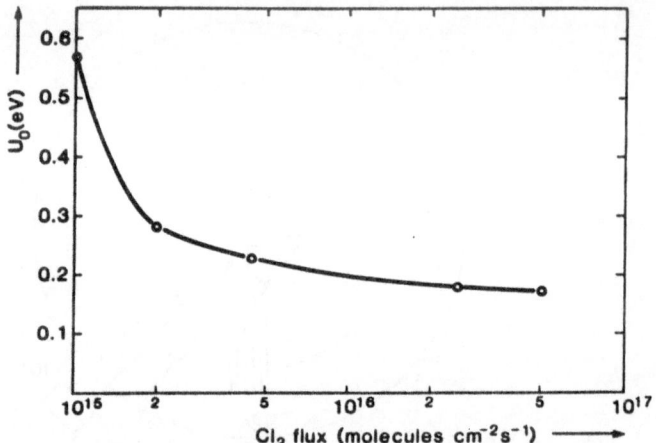

Fig. 3 - Variation of U_0 as a function of ϕ_{Cl_2}. The line is a guide to the eye.

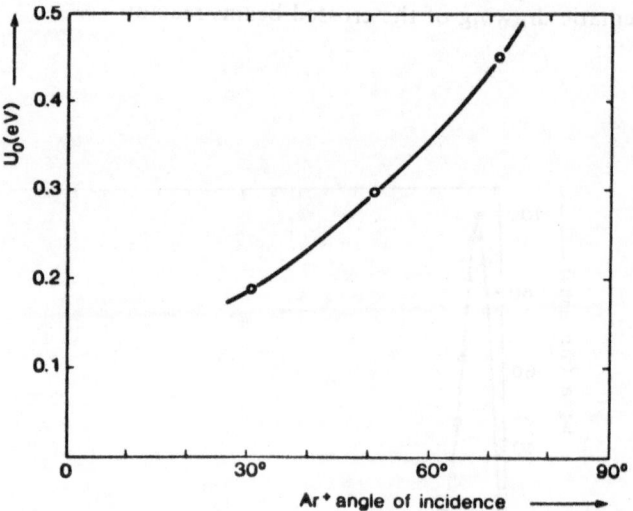

Fig. 4 - Variation of U_0 as a function of the Ar^+ angle-of-incidence. The line is a guide to the eye.

leads to a weakened layer with low binding energies. (A possible model that explains how such low binding energies can be found is put forward in ref. 4).

At the highest Cl_2 flux (Cl_2 to Ar^+ flux ratio about 1600) we observe a saturation effect in U_0. This might be connected with a Cl coverage of a full monolayer, but also with a saturation effect in the altered layer. At the lower Cl_2 fluxes U_0 increases.

Secondly, we varied the ion angle-of-incidence θ_{inc} at an intermediate (2×10^{15} molecules $cm^{-2} s^{-1}$) Cl_2 flux, Cl_2 to Ar^+ flux ratio about 40. We took T.O.F. distributions at $\theta_{inc} = 28°, 50°$ and $70°$ from the surface normal. Again $\theta_{det} = 0°$. The U_0 values obtained from the fits are plotted in fig. 4. We observe a significant increase in the value of U_0 with increasing ion angle-of-incidence.

When these findings are related to the results of Barish et al.[5], it is clear that the two shifts in U_0 probably have the same origin. Their experiments show that the Cl removal cross section by Ne^+ ions increases by a factor of about 40 when the Ne^+ angle-of-incidence is increased from $0°$ to about $70°$ from the surface normal. If a similar effect is also present with Ar^+ ions, the Cl coverage will depend strongly on the Ar^+ angle-of-incidence, and the trends in the U_0 variations with Cl_2 to Ar^+ flux ratio and with ion angle-of-incidence are consistent. Since U_0 is determined by the amount of Cl mixed into the substrate, this supports the notion that ion beam mixing is important in the studied process.

4. CONCLUSION

It has been observed that the binding energy U_0 of the product SiCl in the etching of Si with Cl_2 and Ar^+ ions increases with decreasing Cl_2 to Ar^+ flux ratio and with increasing Ar^+ angle-of-incidence. These observations give further support for the "chemically enhanced physical sputtering" model for the etching of Si with Cl_2 and Ar^+ ions.

5. REFERENCES

[1]P. C. Zalm, Vacuum 36, 787 (1986).
[2]H. F. Winters, J. W. Coburn and T. J. Chuang, J. Vac. Sci. Technol. B1, 469 (1983).
[3]J. W. Coburn and H. F. Winters, Appl. Surface Sci. 22/23, 63 (1985).
[4]J. Dieleman, F. H. Sanders, A. W. Kolfschoten, P. C. Zalm, A. E. de Vries and A. Haring, J. Vac. Sci. Technol. B3, 1384 (1985).
[5]E. L. Barish, D. J. Vitkavage and T. M. Mayer, J. Appl. Phys. 57, 1336 (1985).

IRRADIATION EFFECTS IN ICES BY ENERGETIC IONS

J. BENIT, J-P. BIBRING and F. ROCARD
Laboratoire Rene Bernas du CSNSM, Bat. 108, 91406 Orsay, France

1. INTRODUCTION

It has been shown that the irradiation of ices leads to "giant" erosion yields, in some specific conditions (1,2). In order to understand the physical processes involved, we have performed direct in-situ measurements of the erosion during irradiation, by means of infrared spectrometry of the ices. In this paper, we summarize the main results we obtained, during the irradiation of a number of ices (pure H_2O or CH_4; mixtures: $H_2O + CO_2$; $H_2O + NH_3$; $CO_2 + NH_3$) with ions in a wide range of energy (a few keV/u to a few MeV/u) and mass. We have performed experiments both with chemically reactive ions (C, N, H, H_2) and unreactive ones (He, Ne, Ar, Kr). In addition to the erosion process, we discuss the synthesis of molecular species, induced by the irradiation. Furthermore, we have analyzed the ions desorbed from the targets during the irradiation, by time of flight mass spectrometry. The comparison between the yields of erosion and of ion desorption gives information on the ion/neutral ratio of the ejected material, as well as its dependence with mass and energy of the incident particles. We conclude this paper with some astrophysical implications of our results.

2. EXPERIMENTAL PROCEDURE

Irradiations have been made both with keV and MeV ions. An ion implanter, located in our lab, provides ions of a wide variety of masses, with energy in the range 10 - 50 keV and ion beams of a few $\mu A \cdot cm^{-2}$ (3). We have performed irradiations with the following ions: $^1H^+$, $^2H^+$, $^4He^+$, $^{12}C^+$, $^{13}C^+$, $^{14}N^+$, $^{20}Ne^+$, with energies from 10 to 50 keV. Ne^{8+}, Ar^{12+} and Kr^{20+} ions of energy 1.16 MeV/amu have been obtained at Orsay using the LINAC heavy ion facility (4). Finally, irradiations with 400 MeV ^{84}Kr ions were obtained, using the "ALICE" accelerator at Orsay. Thin aluminium foils allowed to decrease this energy down to 12 MeV total. Typical beam intensities on the targets range from a few tens of $nA \cdot cm^{-2}$ for MeV ions up to tens of $\mu A \cdot cm^{-2}$ for keV ions.

The same target chamber has been used for all irradiations. It allows production of icy films by vacuum deposition of vapor phases onto KBr substrates maintained at low temperature. Using liquid nitrogen as coolant, we have prepared, at 77 K, icy films of H_2O, D_2O, CO_2, NH_3 and mixtures. The abundance ratios of these composite ices are controlled by pressure measurements in the gas phase prior to condensation, and measured after condensation by infrared spectrometry of the condensed films. We have also used a closed-cycle He cryostat which allows to condense organic gases at 35 K, in particular CH_4. We have to take into account that in these conditions, species like O_2 and N_2, present within the residual vacuum, do also condense onto the KBr cold substrate. In order to minimize the concentration of contaminants, as H_2O, we have improved the thermal screens surrounding the target. Typical film thicknesses were in the range 100 - 2000 nm.

R. Kelly and M. Fernanda da Silva (eds.), Materials Modification by High-fluence Ion Beams, 123–138.
© 1989 by Kluwer Academic Publishers.

Two distinct means of analysis of the irradiated ices have been used. We have monitored the modifications of the irradiated targets by infrared spectrometry, and analyzed the ionized desorbed species by mass spectrometry.

To allow the infrared analyses to be performed on line during the irradiation, the infrared beam was perpendicular to the ion beam, and the irradiations were made at an incidence of 45°. We used a Nicolet MXS Fourier Transform Infrared Spectrometer, operating in the spectral range 4300 - 400 cm^{-1}. This spectrometer has been modified first to enable purging with dry nitrogen, and second by coupling to a LeCroy 3500 microcomputer. It allows the repetitive scanning of spectra, thus increasing the signal/noise ratio, as well as an efficient reduction of the data. A more detailed description of the irradiation chamber and experimental procedure can be found in reference (5).

The mass analysis of the desorbed ions was obtained using a time of flight mass spectrometer developed at Orsay (6). The detector system consists of microchannel plates, located at a distance of ∼ 15 cm from the targets. The targets consist of films of ice deposited from vapor onto a polished copper disk in thermal contact with the cryostat cooled to liquid nitrogen temperature. The mass resolution $M/\Delta M$ was ∼ 300. The high voltage ($\pm$ 6 kV) allows acceleration of either the positive or the negative ions produced. The system is optimized when the average number of primary ions on the target is of the order of one thousand per second.

3. EROSION

The erosion rates are measured by taking IR spectra of the icy films before and during the irradiation. Each spectrum exhibits bands corresponding to the main vibration modes of the molecules. For example, H_2O ice bands are located at the following wavenumbers, in decreasing order of intensity: 3250 cm^{-1}, 830 cm^{-1}, 1620 cm^{-1} and 2250 cm^{-1}, approximately. The intensities of the bands are directly related to the integrated number of molecules in the IR beam. Given the oscillator strength of each band, we derive the integrated number of molecules per unit area. Assuming the film homogeneous, and for a given density, this number can be expressed as an "equivalent thickness" of the H_2O film, hereafter called "thickness" for sake of simplicity. The decrease of the band intensities during the irradiation leads to the number of H_2O molecules per incident ion which disappears. Although, as we demonstrate below, this disappearance does not predominantly come from the direct surface ejection of H_2O molecules, we call it "erosion".

3.1. Erosion of H_2O ice by keV ions

The erosion yields have been studied as a function of the beam intensity, mass and energy of the incident ions, and film thickness. At energies in the 10 - 50 keV range, we have not observed any significant change in the yields as a function of beam current for beam currents up to 10 μA.cm^{-2}. Consequently, we have restricted ourselves to currents below this value. The main results concerning the variation with mass and energy are plotted in figure 1. We observe a linear dependence of the yield upon the energy of the incident ions, H^+, He^+ and Ne^+. Furthermore, the slopes are similar. The main difference concerns the yields at a given energy, much higher for the Ne irradiation than for the H or He ones. Our interpretation is as follows: the erosion proceeds through two distinct processes, a surface process and a "track-correlated" one. The surface process is due to the collision cascades initiated by the particles close to the surface, its efficiency being directly related to the nuclear stopping power at the incident energy. This sputtering process occurs predominantly for Ne, with an intensity remaining nearly constant in the experimental energy range. However, the deposition of energy all along the track of the particles, mainly by electronic interaction, is also responsible for the erosion of the ice. The increase of the yield with energy, in the case of He and Ne, would result from the increase of the range. In the case of H_2 bombardment, an additional collective effect might result from the occurrence of two tracks, initiated by each of the two H atoms, very close to each other (7).

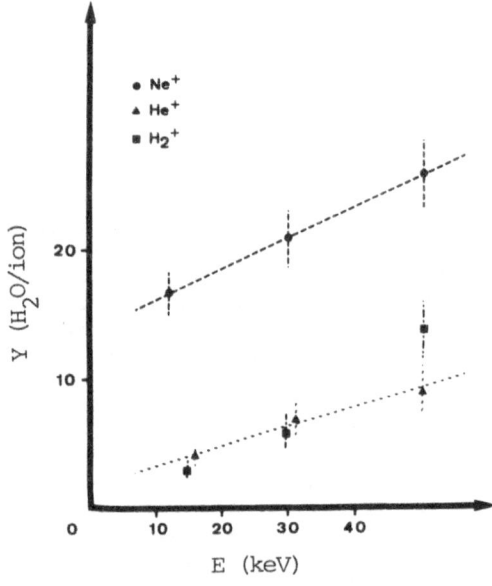

FIGURE 1. Erosion yields (Y), measured by infrared spectrometry expressed as the number of H_2O molecules disappearing per incident ion, for irradiations with Ne^+, He^+ and H_2^+ at energy 10 to 50 keV.

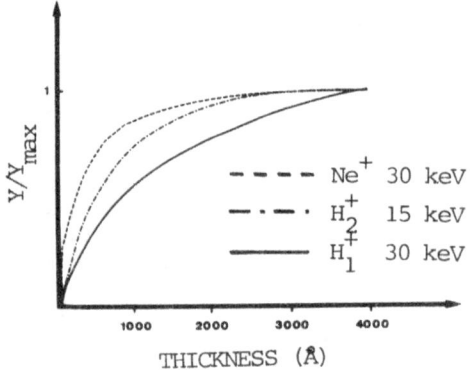

FIGURE 2. Dependence of the yield on film thickness for irradiation by Ne^+ at 30 keV, H_2^+ at 15 keV and H_1^+ at 30 keV. The yields are normalized to the plateau value obtained for thick films (cf. Fig. 3).

In order to test the validity of our interpretation, and specifically to measure the effect of the range on the yield, we have studied the dependence of the yields on film thickness, for a given energy of the incident particle. The results are plotted on figure 2. We have measured the erosion yields during the irradiation as a function of the remaining thickness of the irradiated films. We observe that, starting with films of high thicknesses, the yields remain constant (plateau value) down to a critical value depending upon the mass and energy of the incident particle. This value is well fitted by the calculated mean projected range within H_2O ice, for Ne, He and H irradiation (3). The decrease of the yield for film thicknesses below this critical value would then be due to the fact that when part of the range of the ions takes place within the KBr substrate, the incident energy cannot be entirely converted into erosion of the ice (Fig. 3).

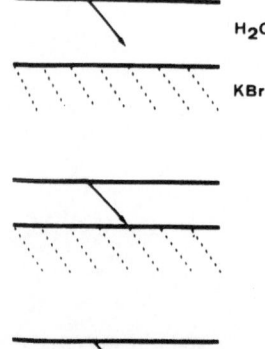

H₂O

KBr

FIGURE 3. Schematic view of the ion penetration at a given energy (constant range) in icy films of different thicknesses. In the two upper cases, the energy is deposited entirely in the icy film and thus totally contributes to the erosion process. In the third case, corresponding to a very thin film, the ions come to rest within the KBr substrate. Consequently, only part of the incident energy contributes to the erosion of the ice.

It seems very unlikely that the erosion of the films along the tracks occurs by release of H_2O molecules from the target. On the other hand, we do not observe, by IR spectrometry, the appearance of new bands that would correspond to the synthesis of molecules. Consequently, the disappearance of H_2O molecules is correlated with the synthesis of molecules either transparent in the IR (no permanent dipole) or volatile, that rapidly diffuse out of the films. Our interpretation is that the erosion of the H_2O film proceeds through the dissociation of H_2O followed by the synthesis of H_2 and O_2.

Not every dissociated H_2O molecule contributes to the erosion yield: a large fraction of the dissociation products does indeed recombine into H_2O. We have tentatively evaluated the efficiency of this recombination process by comparing the erosion yields obtained in similar irradiation conditions of two different ices, one made of pure H_2O, the second made of a mixture of H_2O and CO_2 (see section 4.1). In the first case the yield is approximately 7 H_2O per incident ion. In the case of the irradiation of a mixture of ($2H_2O + CO_2$) ice, we observe the synthesis of a rich variety of new molecules containing both H and C atoms, with a total yield of approximately 30 new molecules per incident He ion. It indicates that bombardment of H_2O leads to a number of dissociated H_2O molecules much in excess of 7 per incident ion. The presence of radicals resulting from the dissociation of CO_2 in the vicinity of the protons enables new reactions to take place with the products of the H_2O dissociation, preventing their recombination into H_2O (8).

In summary, the irradiation leads both to the surface sputtering of the ice and the dissociation of molecules within the films, all along the track of the incident ions. This latter process results from the synthesis of new molecules within the irradiated ice.

3.2. Erosion of H_2O ice by MeV ions

We have irradiated H_2O films with Ne, Ar and Kr ions at energies close to the maximum of the electronic energy losses. The variation of the yields with film thickness leads to the data shown in figure 4. In each case, we obtain a linear dependence, which can be accounted for in the framework described above. These irradiations correspond to the case of film thicknesses much lower than the range of the ions. The continuous decrease of the yield can be understood if erosion proceeds all along the track of the incident particles, with no dominant surface effect.

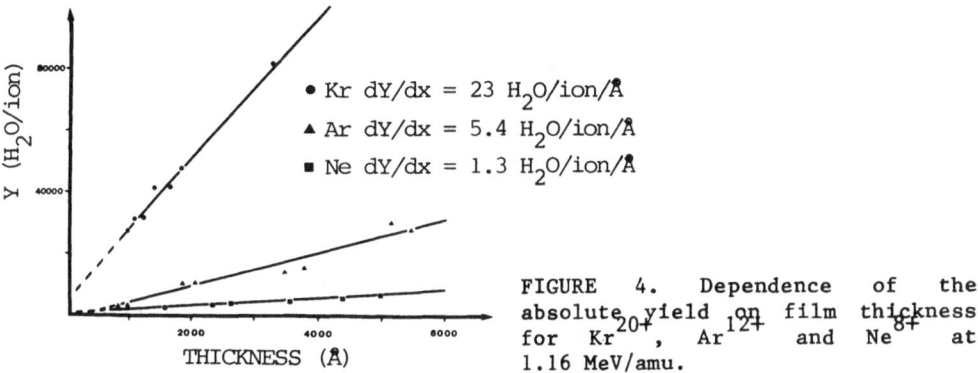

FIGURE 4. Dependence of the absolute yield on film thickness for Kr^{20+}, Ar^{12+} and Ne^{8+} at 1.16 MeV/amu.

The dependence of the yields on film thickness has been obtained in two ways, leading to the same results: for one set of experiments, we have irradiated successively different films, each with a different thickness. In an other experiment, we have measured the erosion yield of a thick film, all along its irradiation, which led to its progressive erosion.

An additional experiment has been performed, which confirms the occurrence of the erosion all along the path of the incident ions. We have produced a two-layered film by the deposition of 2500 Å of D_2O ice followed by 3300 Å of H_2O ice. We have monitored the erosion of this film by measuring the disappearance of both the H_2O and the D_2O molecules. The results (Table 1) show that the relative thicknesses of the H_2O and the

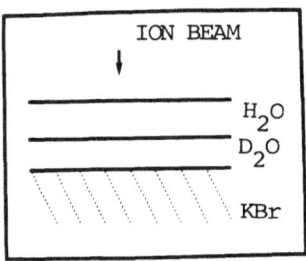

FIGURE 5. Schematic view of a two-layered film obtained by deposition of 2500 Å of D_2O ice followed by 3300 Å of H_2O ice. The results concerning the erosion of this film are given in Table 1.

TABLE 1. Erosion of a double layered film, consisting of 2470 Å of D_2O ice covered by 3300 Å of H_2O ice (see Fig. 5), by a beam of 1.16 MeV/amu Ar^{12+} ions. During the irradiation the thickness of each of the layers, measured by infrared spectrometry, is reduced (columns 1 and 3). Columns 2 and 4 indicate, for every measurement, the fraction of the total remaining film consisting of H_2O and D_2O respectively.

H_2O		D_2O	
Thickness (Å)	%	Thickness (Å)	%
3300	57	2470	43
1860	50.1	1850	49.9
1545	52.5	1400	47.5
1395	51.4	1320	48.6
1100	50.6	1075	49.4
790	50	790	50

D_2O layers decrease at similar rates: the film, constituted of 57% H_2O ice condensed onto 43% D_2O ice before the irradiation ends up with 50% of each of the layers after the erosion of most of it. It indicates that the incident ions are equally efficient in eroding away the H_2O layer and the underlying D_2O one.

The number of H_2O molecules eroded per incident ion and per Angstrom (23, 5.4 and 1.3 for Kr, Ar and Ne respectively) corresponds to a cylinder of diameter 30, 15 and 7.5 Å respectively. As we are in the limit of "low" fluences (10^{12} cm^{-2}, which corresponds, in the case of Kr irradiation, to "cylinders" covering less than 10 % of the volume), we would not observe the linear dependence if the lattice did not undergo long-range relaxation.

The thicknesses of the films being much lower than the mean range of the incident ions, the ions have almost constant energy and stopping power all across the targets. We have calculated the stopping powers according to the tables of Andersen and Ziegler (9). The yields per unit thickness, derived from figure 4, are then observed to vary as $(dE/dx)^2$.

4. MOLECULAR SYNTHESIS

The erosion of the ices results from the chemical synthesis of volatiles species induced between the dissociation products of the major constituent of the ices. Upon irradiation with chemically reactive ions, like C^+, we have observed the synthesis of molecules, associating the incident ion to atoms constituting the targets. We have thus studied the C irradiation chemistry in H_2O (10). We present here the results we have obtained first with the He^+ irradiation of different mixtures of ices, next with the irradiation of CH_4 by H_2^+ beams.

4.1. Irradiation of $(H_2O + CO_2)$

We have irradiated two mixtures of $(H_2O + CO_2)$ ice, at 77 K, with abundance ratios $H_2O/CO_2 = 2$ and 0.2 respectively. The primary beam consisted of 1.5×10^{16} cm^{-2} 30 keV He^+ ions. Table 2 summarizes the dominant species detected by their absorption features. The identification of the species is based on the comparison with known spectra obtained either in gaseous form or isolated in rare gas matrixes. The second column gives a rough estimate of the validity of our identification. Case (a) corresponds to the unambiguous identification: the various major bands of each compounds appear, with their relative intensities, at their expected spectral position ($\pm$ 10 cm^{-1}). Case (b) corresponds to species for which each major band is present, but with relative intensity differing from the gaseous ones. Case (c) corresponds to molecules for which some expected bands have not been detected. Figure 6 illustrates the modification of the ice induced by the irradiation. Newly synthesized species appear by their IR bands. For example, the CO stretching mode, at 2139 cm^{-1}, is clearly present. It appears that most of the synthesized molecules are oxides and not hydrides. A rough estimate of the overall synthesis yields have been obtained, using an oscillator strength of 10^{-17} cm.molecule^{-1} for the dominant mode of each species. The irradiation of the H_2O rich mixture leads to the synthesis of $\sim$ 30 molecules per incident ion, whereas in the case of the CO_2 rich one, the yield is $\sim$ 60 molecules per ion.

TABLE 2. Icy mixtures (H_2O + CO_2), at 77 K, irradiated with He$^+$ for 2 abundance ratios: H_2O/CO_2 = 2 and 0.2, labelled A and B respectively. The proposed identifications are divided into 3 categories (a, b, c) which indicate the degree of confidence (see text). The given intensities correspond to the integral of the absorption bands. (*) means that the band is identified after a subtraction with a spectrum corresponding to an unirradiated ice. (sh) shoulder; (-) non detected.

Identification	Category	Wavenumber (cm^{-1})	Bandwidth (cm^{-1})	Intensity A	Intensity B
OH	c	3470*	400	14.2	-
(?)		3345*sh			
(?)		3212*	400	-	8.6
(?)		3120*sh			
CH_2CO	b	3070*	180	2.1	-
CH_4	c	3000 sh			
H_2CO	a	2840	45-100	0.6	< 0.01
H_2CO	a	2780 sh			
CO	a	2139	50	0.14	1.85
O_3	c	2090	15	< 0.01	0.02
C_3	b	2044*	15	0.01	0.27
(?)		1946*	20	-	0.04
(?)		1919	15	-	0.01
HCO	c	1878	30	-	0.18
H_2CO	a	1714	140	1.81	3.16
(?)		1617	20	-	0.03
H_2CO	a	1494	110	0.1	1.24
CH_2CO	b	1391-1378	40	0.04	0.17
CH_4	c	1300	100	0.45	3.65
H_2CO	a	1230	30	< 0.01	-
H_2CO	a	1173	25	-	0.04
CH_2CO	b	1107	15-25	0.02	0.01
CO_3	b	1071	15	0.01	0.01
O_3	c	1010-1030	35	0.09	0.37
CH_2CO, CO_3	b,b	976	20	< 0.01	0.04
(?)		812-820	40	-	0.17
CH_2CO	b	501	55	0.16	-
CH_2CO	b	480 sh			

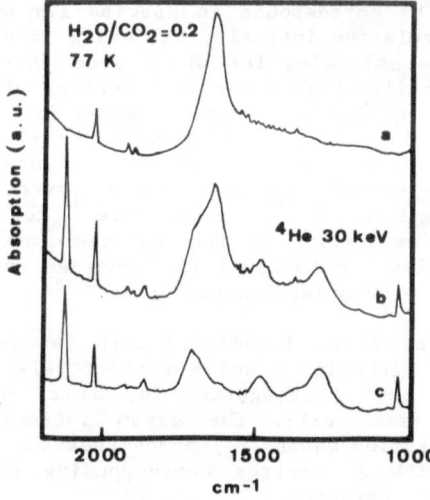

FIGURE 6. 2200 - 1000 cm^{-1} spectra of a (H_2O + CO_2) ice mixture (H_2O/CO_2 = 0.2) at 77 K; (a) before irradiation; (b) after irradiation with 1.5 10^{16} He$^+$.cm^{-2} at 30 keV; (c) subtraction of the spectra (b) and (a). The position of the bands are given in table 2.

4.2. Irradiation of (H_2O + NH_3)

The ice consists of the mixture of (H_2O + NH_3) at 77 K, with a ratio H_2O/NH_3 = 5. The irradiation was performed with 30 keV He$^+$ ions, at fluences of 3 10^{16} cm^{-2}. The spectrum after irradiation does not exhibit intense IR features. Some very small bands have been identified, and reported in Table 3. The irradiation does not synthesize detectable amounts of NO, NO_2, N_2O, N_2H_2, N_2H_4 and NH_4^+; only NH_2 and N_3 seem present in low concentration. Our interpretation is that N_2, which is not detectable in the IR, is the major N-containing synthesized species.

TABLE 3. Mixture (H_2O + NH_3), at 77 K, irradiated with 3 10^{16} He$^+$.cm^{-2} at 30 keV (H_2O/NH_3 = 5). See also Table 2.

Identification	Cat.	Wavenumber (cm^{-1})	Bandwidth (cm^{-1})	Intensity
(?)		2165	50	0.04
N_3	c	2110–2000	110	0.12
NH_2	c	1480	?	?
(?)		860	10	0.01
(?)		824	10	< 0.01
(?)		754	15	0.01
(?)		730	150	0.69
(?)		510	100 ?	0.40 ?
		497 sh		

4.3. Irradiation of (CO_2 + NH_3)

The CO_2/NH_3 abundance ratio of the mixture we condensed was 5, and its temperature 77 K. We used 1.5 10^{16} He$^+$.cm^{-2}, at 30 keV, as primary beam. The irradiation led to a rich variety of new compounds (Table 4). However, their identification is difficult, because large regions of the spectrum contain the intense signatures of CO_2, NH_3 and H_2O. The latter constituent, present in small amounts before the irradiation due to the condensation of some residual vapor, is efficiently synthesized by the irradiation. H_2O and CO are unambiguously identified. NO_2 and N_2O seem also present. The problem for their identification comes from the presence of a very broad feature, from 1300 to 1800 cm^{-1}, superimposed on their major peaks (Fig. 7). This broad band might correspond to a more complex molecule, similar to that observed in the case of irradiated CH_4 (section 4.4).

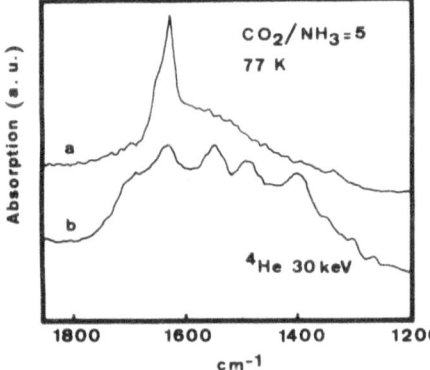

FIGURE 7. 1850 – 1200 cm^{-1} spectra of a $(CO_2 + NH_3)$ ice mixture $(CO_2/NH_3 = 5)$ at 77 K; (a) before irradiation; (b) after irradiation with 1.5 10^{16} $He^+.cm^{-2}$ at 30 keV. The positions of the bands are given in table 4.

TABLE 4. Mixture $(CO_2 + NH_3)$, at 77 K, irradiated with 1.5 10^{16} $He^+.cm^{-2}$ at 30 keV $(CO_2/NH_3 = 5)$. See also Table 2.

Identifications	Cat.	Wavenumber (cm^{-1})	Bandwidth (cm^{-1})	Intensity
N_2O	b	2227	20	< 0.01
(?)		2164	70	0.38
CO,CH_2CO	a,b	2140 sh		
$HCO,CONH_2$	c,c	1853	30	0.03
$CO(NH_2)_2$	c	1700 sh		
NO_2	b	1635	90	0.60
(?)		1548	75	0.52
$CO(NH_2)_2$	c	1490	60	0.42
NH_4	c	1399	80	0.63
(?)		1345	30	0.03
N_2H_2,N_2O	c,b	1300	50	0.09
N_2H_2	c	1264*	30	0.05
(?)		902*	65	0.19
(?)		827*	20	0.03
(?)		515*	35	0.10

4.4. Irradiation of CH_4

CH_4 films, at 35 K, of various thicknesses have been irradiated with 40 keV H_2^+. When we start with a thickness of the order of the range of the ions (800 nm to 1 μm), we observe after a fluence of 10^{17} $H.cm^{-2}$ the presence of a variety of new absorption features in the 4300 – 400 cm^{-1} IR spectrum (Table 5). They correspond to the synthesis of species containing C, H and O atoms. Two main groups of C-containing molecules are present, one in the 3000 – 2800 cm^{-1} region (Fig. 8), the other between

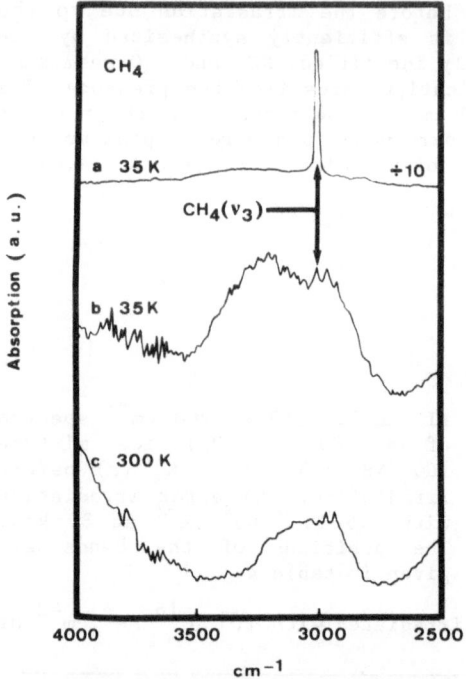

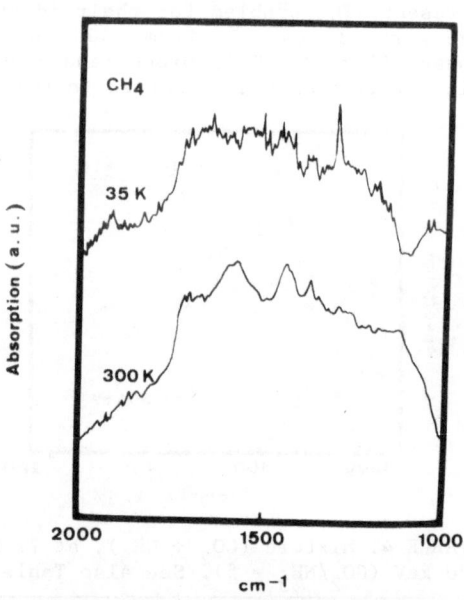

FIGURE 8. 4000 - 2500 cm^{-1} spectra of a CH$_4$ ice film: (a) at 35 K, before irradiation; (b) at 35 K, after irradiation with 10^{17} H.cm^{-2} at 30 keV; (c) after warming up at room temperature. The ordinate expansion in the spectrum (a) is 10 time smaller than in the spectra (b) and (c). The position of the bands are given in table 5.

FIGURE 9. 2000 - 1000 cm^{-1} spectra of a CH$_4$ ice film. The upper and lower curves correspond respectively to the curves (b) and (c) of the figure 8. The position of the bands are given in table 5.

1700 and 1300 cm^{-1} (Fig. 9). We have tentatively attributed the different peaks to compounds by comparison with library spectra: between 2960 and 2850 cm^{-1}, the bands correspond to the vibrational stretching of C-H bonds in the form of CH$_3$- and CH$_2$- functional groups, whereas in the spectral region in excess of 3000 cm^{-1}, the bands correspond more likely to unsaturated hydrocarbons, containing preferentially multiple bondings. In the 1700 - 1300 cm^{-1} region, the bending of C-H and the vibration of C-C dominate.

The irradiation is also responsible for the synthesis of H$_2$O, which appears as a band centered at 3287 cm^{-1}. H$_2$O likely results from the dissociation of CH$_4$ and O$_2$ (see section 2.).

Upon thermal annealing, by warming up the target to room temperature, the IR spectrum is modified: the volatile species (H$_2$O ice at 3287 cm^{-1}, CH$_4$ ice at 1300 and 3011 cm^{-1}, synthesized CO at 2135 cm^{-1}) are removed. Some new features appear, either corresponding to species synthesized during the annealing or rendered detectable after the removal of the volatile species. In the high frequency region, the spectrum of the residue is dominated by features between 3200 and 2800 cm^{-1} (Fig. 8c). Two small peaks, at 2975 and 2930 cm^{-1}, are slightly visible superimposed on a broad band. Optical observation of the residue indicates a yellow-brown stuff. We note that these peaks have positions similar to those observed in astrophysical spectra, either of the Halley inner coma (11) or IR sources (12).

We have observed that under subsequent irradiation, the residue is progressively transformed. Its IR spectra exhibits less and less features. After irradiation with an additional fluence of 10^{17} $H.cm^{-2}$, we observe a dark residue with very tiny spectral structures, in agreement with results obtained previously (13).

TABLE 5. Position and width of the bands detected in the irradiated CH_4 ice film at 35 K (right) and 300 K (left). The central column gives tentative identifications with some functional groups. The letters beside the position values refer to the intensity: (vs) very strong; (s) strong; (m) medium; (w) weak; (vw) very weak; (sh) shoulder.

300 K		Functional	35 K	
Wavenumber (cm^{-1})	Bandwidth (cm^{-1})	Group	Wavenumber (cm^{-1})	Bandwidth (cm^{-1})
4147 4044)s	250			
3370w	50	OH		
		H_2O	3287vs	
		CH_4	3011	
3020 2930)s	550	unsaturated [CH] saturated [CH]	2972-2961-2930)s 2907-2876-2853	240
2285vw	40		2260vw	10
		CO	2135s)	20
2104w	55		2111s	70
1720sh			1650vs	300
1590s	150	C=C	1590vw	20
1441m	70	bend CH_3 or $(CH_3)_n$-C	1440s	70
1374w	25	bend CH_3 or $(CH_3)_3$=C CH_4	1374m	40
			1300m	30
1295 1262)mw				
1130m	70	$(CH_3)_2$-C		
1063vw	50			
1030vw	40			
962m	50			
872s	100			
620s	130			

4.5. Comparison with previous in-situ detections

The only experiments dealing with the in-situ detection of species synthesized by ion irradiation in icy mixtures have been reported by Moore et al. (14). Their main IR results may be summarized as follows: the irradiation of $(H_2O + N_2 + CO)$ leads to the synthesis of CO_2 and of some CH_4; $(H_2O + N_2 + CO_2)$ leads to CO, some NO and CH_4; $(H_2O + NH_3 + CH_4)$ leads to CO_2, C_2H_6, CO and N_3. No compounds containing C and N have been detected. Their experiments differ from ours by the energy of the incident ions (1 MeV protons). However, in both cases, the electronic stopping power dominates, and has similar values.

Concerning the C-rich molecules, a main difference concerns the synthesis of H_2CO, which constitutes a major product of the irradiation of $(H_2O + CO_2)$ in our experiments. In the case of N-containing mixtures, the comparison of the data presented by Moore et al. with ours seems to indicate that the synthesis of compounds with both C and N is favored if one starts with a mixture containing CO_2 instead of CH_4.

Finally, it appears that the ion irradiation in both the keV and MeV regions is responsible for the synthesis of a rich variety of species, from small molecules to much more complex structures. The relative abundances of synthesized products are very sensitive to the composition of the initial ice as well as to the fluence of the irradiating beam. In particular, we have observed a progressive depletion of H in the products as a function of the irradiation fluence, leading to stable refractory residue highly enriched in C, exhibiting very weak IR features.

4.6. Molecular synthesis and astrophysical implications

There exists numerous astrophysical sites were icy grains or planetary surfaces are subjected to irradiation by energetic particles: solar and stellar wind ions (1-10 keV), cosmic-ray particles (1 MeV/u - 1 GeV/u, typically). They include solar system objects, like cometary nuclei, satellites and rings surrounding giant planets, as well as molecular clouds and circumstellar shells. Consequently, the experiments we perform in the lab provide a relevant simulation of the actual irradiation in space. In particular, we consider that our results concerning molecular synthesis under irradiation may account for observational data dealing with cosmic chemistry. We indicate two possible examples.

One of the most exciting results obtained during the VEGA and GIOTTO fly-bys of Comet Halley concerns the discovery of a carbonaceous phase, responsible for the very low albedo of the nucleus (< 4%) and its high surface temperature (> 350 K) (11), detected in the form of grains by mass spectrometry (15), and through their infrared features (11), centered around 3.4 μm, similar to those we obtain in the lab. The model we thus propose for the formation of comets includes an intense irradiation of icy grains within the solar nebula, prior to their accretion into cometary bodies, during an early stage of the Sun: young stars, during their "T-Tauri" phase, are known to suffer huge mass losses in the form of a stellar wind up to 10^8 times more intense than the present solar wind. The icy surfaces of planetary satellites undergo maturation effects in the form of a darkening with time. Young regions are systematically brighter than older ones. A plausible interpretation would be as follows: these surfaces, embedded in the magnetosphere of the planets, are irradiated by high fluxes of MeV ions. One of the results would be the synthesis of dark C-rich material, plausibly in the form of refractory polymer chains.

5. ANALYSIS OF THE DESORBED SPECIES
5.1. Masses and relative abundances of the desorbed ions

On figure 10 is reported the time of flight mass spectrum of the positive ions desorbed from H_2O ice when irradiated with 120 MeV Kr ions. Table 6 gives the mass and relative abundances of the ions. It appears that apart from H^+ ions (and to a much lesser extent H_2^+), the bombardment produces clusters with formula $H(H_2O)_n^+$. The abundance of the detected ions decrease as n^{-2} over this range.

The results concerning the negative species are reported in figure 11 and Table 7. As for the positive species, clusters are emitted together with single ions $(O^-$ and $OH^-)$. Two classes of clusters are produced, with formulae $O(H_2O)_n^-$ and $OH(H_2O)_n^-$, and rather similar yields. The large uncertainties are due to the low counting rate for negatively charged species. Nevertheless, it clearly appears that the decrease of the yield with increasing mass of the cluster is less pronounced than for positive ions, and varies as $n^{-1/3}$.

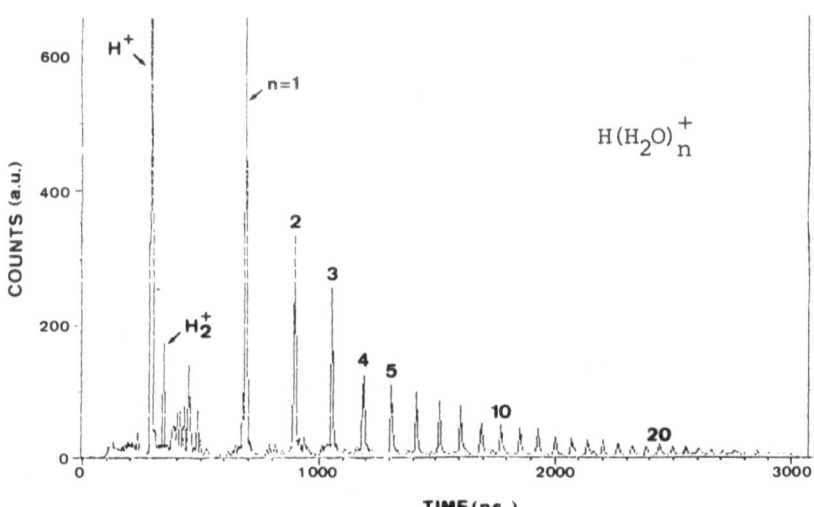

FIGURE 10. Mass spectrum of the positive ions desorbed after the irradiation by 120 MeV Kr ions of an H_2O ice film. The spectrum exhibits the presence of H_1^+, H_2^+ and the clusters $H(H_2O)_n^+$, up to masses exceeding 400.

TABLE 6. Masses and rough relative abundances of the positive ions.

ion	H^+	H_2^+	$H(H_2O)_n^+$							
n			1	2	3	4	5	6	7	8
relative yield (%) (H^+ = 100)	100	9	33	14	6	3	2	1	0.9	0.7

TABLE 7. Masses and rough relative abundances of negative ions.

ions	O^-	OH^-	$O(H_2O)_n^-$					$OH(H_2O)_n^-$				
n			1	2	3	4	5	1	2	3	4	5
relative yield (%) (OH^- = 100)	63	100	25	20	29	9	18	34	34	21	22	32

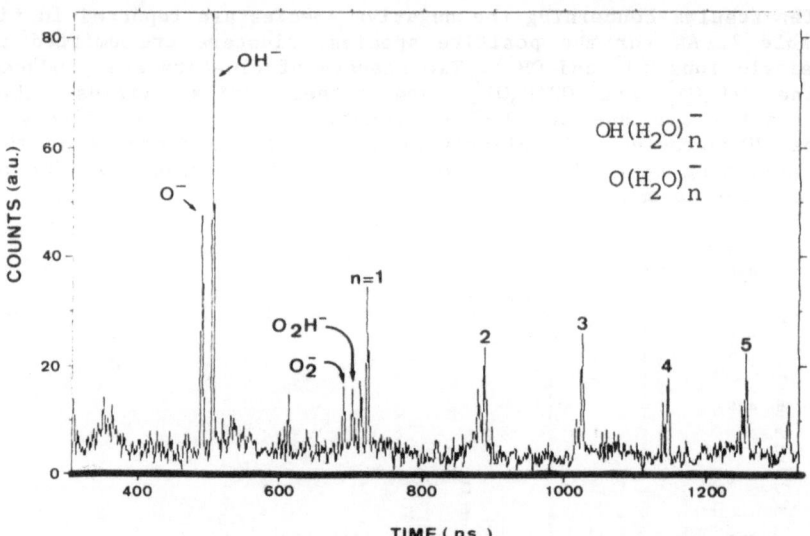

FIGURE 11. Same as figure 10, for negatively charged desorbed species. Clusters $O(H_2O)_n^-$ and $OH(H_2O)_n^-$ are well resolved.

5.2. Variation of the yields with the velocity of primary ions

The data are represented on figures 12 to 14, respectively for the positive ions, and the two families of negative clusters. The experiments have been performed with the beam at maximum energy (394 MeV) and with 4

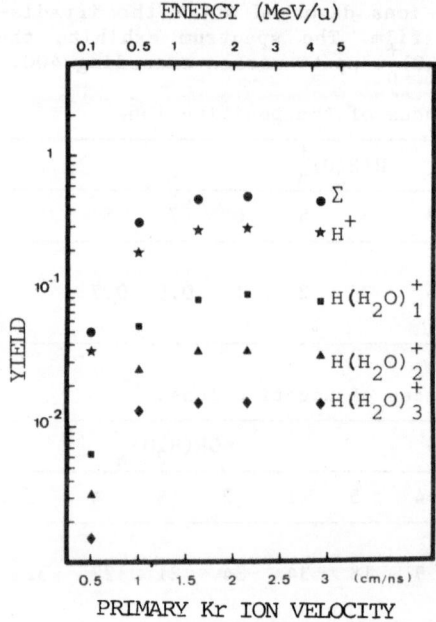

FIGURE 12. Variations with the velocity of the Kr incident ions of the desorption yields of H^+ and the more abundant clusters, together with that of the sum of all ions (see text).

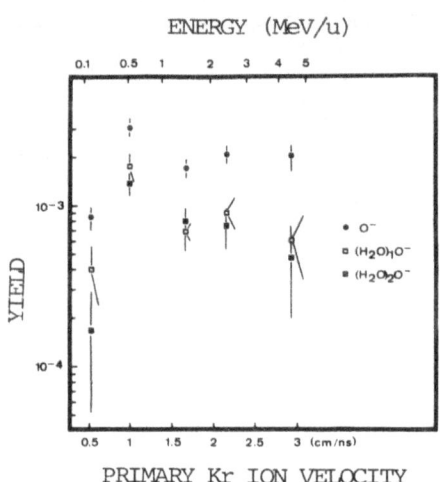

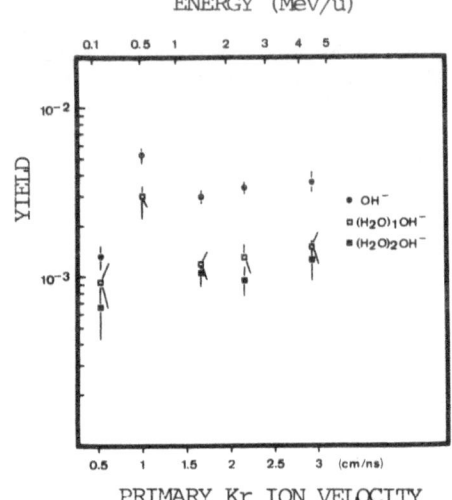

FIGURE 13. Same as figure 12, for O^- and the more abundant clusters with formula $O(H_2O)_n^-$.

FIGURE 14. Same as figure 12, for OH^- and the more abundant clusters with formula $OH(H_2O)_n^-$.

different aluminium degrading foils, leading to the following energies: 200 MeV, 120 MeV, 43 MeV and 12 MeV, measured by time of flight. The uncertainties on the yields are given by the vertical bars, when indicated, and smaller than the plotted data points otherwise.

We observe that all ions with same polarity (positive or negative) exhibit a similar variation of the desorption yield with beam velocity. The yields of desorption increase with energy up to a plateau value, starting at an energy E_p. The value of E_p is ~ 60 MeV.

It is interesting to compare this behaviour with that of the electronic stopping power of the ions. This stopping power, calculated with the Andersen and Ziegler tables (9), exhibit a maximum for an energy close to E_p. For $E > E_p$, the stopping power remains about constant according to (9). Consequently, there seems to be a good agreement between the dependence with energy of both the desorption yield and the stopping power (dE/dx).

5.3. Absolute yields of desorption and astrophysical implications

One of the most striking results is the high rate of desorption of ionized species under high energy ion bombardment. Taking into account the geometric factors entering our experiment, we calculate that the ratio of the desorbed ions, detected on the channel plates, to the incoming Kr ions is close to unity. Due to the efficiency of the detection, this number is underestimated by at least a factor of two. If we compare our data with those concerning the erosion rate of neutral species in similar conditions (16), where the maximum of the erosion rate lies between 10^3 and 10^4 H_2O/ion, we obtain an overall ionization rate of a few 10^{-4}.

Such a high rate might account for the production of ionized species in the gas phase within interstellar cloud, responsible for most of the cosmochemistry in molecular clouds. In such clouds, the UV stellar photons do not penetrate deep enough to provide an important source of energy, which allows molecules to survive without being photodissociated. The major process by which molecules disappear is their sticking onto the grains that they encounter after a mean time, short in comparison with the mean age of the cloud. Whatever their composition, the grains are then likely to be entirely covered with an "icy" mantle of frozen molecules. In that case, one needs an efficient mechanism to release the molecules from the grains (with the exception of H_2 molecules, which remain volatile even at the very low grain temperatures). The high energy cosmic rays constitute probably the only energetic source available in these dense clouds. Their interaction with the mantles would be responsible for the release into the gas phase of molecules and ions, with composition modified as a result of the irradiation itself. In particular, the desorbed ions might enter sequences of complex ion-molecule reactions leading to larger species as discovered by means of radioastronomy along the past ten years.

REFERENCES

1. W.L. Brown, L.J. Lanzerotti and W.M. Augustyniak: Phys. Rev. Lett., 40, 1027, 1978.
2. B.H. Cooper and T.A. Tombrello: Rad. Effects, 80, 203, 1984.
3. J. Camplan, R. Meunier and C. Fatu: Proc. 8th Int. EMIS Conference, Skovde, eds., G. Anderson and G. Holman, 186, 1973.
4. S. Della-Negra, O. Becker, R. Cotter, Y. Le Beyec, B. Monart, K. Standing and K. Wien: IPNO-DRE, 86.09, 1986
5. F. Rocard, J. Benit, J-P. Bibring, D. Ledu and R. Meunier: Rad. Effects, 99, 97, 1986.
6. S. Della Negra, D. Jacquet, I. Lorthiois, Y. Le Beyec, O. Becker and K. Wien: Intern. J. Mass Spectrom. and Ion Proc., 53, 215, 1983.
7. J-P. Thomas, P.E. Filpus-Luyckx, M. Fallavier and E.A. Schweikert: Phys. Rev. Lett., 55, 103, 1985.
8. F. Rocard: These de Doctorat, Universite Paris Sud, Orsay, 1986.
9. H.H. Andersen and J.F. Ziegler: The stopping and ranges in all elements. Vol. 3, New York: Pergamon Press, 1980.
10. J-P. Bibring and F. Rocard: Adv. Space. Res., 4, No 12, 103, 1984
11. M. Combes, V.I. Moroz, J-F. Criffo, J-M. Lamarre, J. Charra, N.F. Sanko, A. Soufflot, J-P. Bibring, S. Cazes, N. Coron, J. Crovisier, C. Emerich, T. Encrenaz, R. Gispert, A.V. Grigoryev, G. Guyot, V.A. Krasnopolsky, Y.V. Nikolsky and F. Rocard: Nature 321, 266, 1986.
12. S.P. Willner, R.C. Puetter, R.W. Russell and B.T. Soifer: Astroph. Space Scien., 65, 95, 1979.
13. L. Calcagno, G. Foti and L. Torrisi: Icarus, 63, 31, 1985
14. M.H. Moore, B. Donn, R. Khanna and M.F. A'Hearn: Icarus, 54, 388, 1983.
15. Y. Langevin, J. Kissel, J-L. Bertaux and E. Chassefiere: Astron. Astroph., in press, november 1987.
16. L.J. Lanzerotti, W.L. Brown and R.E. Johnson: Proc. Nato Adv. Res. Workshop: Ices in the Solar System. Nice, 1984.

ION FORMATION BY VERY HIGH ENERGETIC ION IMPACT ON SOLIDS[*]

K. WIEN, P. KOCZON AND M. WEBER

Institut für Kernphysik, Technische Hochschule Darmstadt, FRG

1. INTRODUCTION

A survey is presented of experimental investigations of secondary ion emission from solids performed at the heavy ion accelerator of the GSI in Darmstadt/FRG. The beams used by the authors had energies up to 1.4 MeV/n, their stopping power was close to the maximum value a heavy ion can ever have in solid material: the specific energy loss can reach 5 keV/A. Most of this energy is fed primarily into electronic excitation. The effects of this enormous energy deposit to the atomic structure are quite different, since they depend on the properties of the irradiated material: in a solid of high electrical conductivity like a metal any defects of the atomic system are hardly detectable - for example, a gold surface emits about 1 atom per impact, when it is irradiated by a 80 MeV I beam (1). On the other hand, an insulator shows a zone of vigorous damage close to the nuclear track (2) - an europium oxide film emits about 800 Eu atoms under bombardment with a 80 MeV Kr beam (3).

The number of secondary ions released per impact is about 2-3 orders of magnitude smaller than the total sputter yield. The question arises, if and how this small fraction of all desorbed particles can be utilized as a measure of the desorption process. The secondary ions are much easier to handle experimentally than the neutral particles. Particularly, refined time-of-flight (=TOF) techniques have been worked out to measure not only yields of certain ions but also their energy distribution (4) or metastable decay (5). The instrumental methods applied in the following investigations are described elsewhere (6).

The following section will be focussed on recent experimental observations and some qualitative conclusions. Several studies have already shown that desorption from clean metals is caused by atomic collision cascades; a fair description also of high energy sputtering in metals is given by linear sputter theory (7). Ionisation is, however, an open problem. For desorption from insulators in the 10-100 MeV regime several models have been developed during the recent years (8-12); most are based on a "hot core" formed in the vicinity of the nuclear track, from where the desorption process is initiated. Again, ionisation is usually not considered.

2. ION EMISSION FROM METALS

Various metal foils were cleaned by ion etching under UHV conditions (6) and then irradiated by U, Xe and Ni beams. The beam

---

[*]supported by Bundesminister für Forschung und Technologie, FRG

R. Kelly and M. Fernanda da Silva (eds.), Materials Modification by High-fluence Ion Beams, 139–147.
© 1989 by Kluwer Academic Publishers.

energy was varied by means of degrader foils. A typical example of a TOF spectrum is presented in Fig.1. It consists of the
metal ions Al^+, Al^{++} and Al^{+++} and a few impurity ions (namely
H^+, H_2^+, K^+, see also reference 6) having a needle-like line
shape. The two kinds of ions represent two different desorption
mechanism - the metal ions are related to atomic collision sputtering, the impurity ions to electronic sputtering. Arguments
for this statement are obtained from the energy dependence and
the line shape of the ions (4,6).

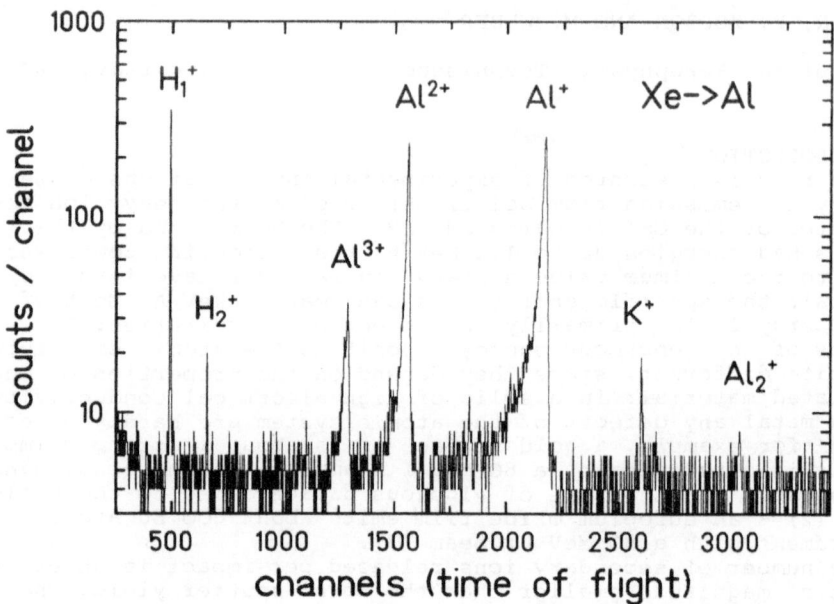

Fig.1: TOF spectrum of secondary ions ejected from Al by irradiation with 1.1 MeV/n Xe.

Contrary to impurity ions, the yields of the metal ions decrease with rising projectile velocity as illustrated in Fig.2.
The figure contains also curves of theoretical sputter yields
normalized to the low velocity part of the experimental data.The
essential ingredient of sputtering theory (7,13) is the nuclear
stopping power, which was taken from the work of Lindhard et al.
(14). The calculated stopping power was corrected for the fraction
of recoil energy lost to electronic excitation (15). The theoretical yields are total yields and therefore much higher than the
ion yields, but even the slopes of the predicted and measured curves are in disagreement, i.e. the ion yields decrease less with
projectile velocity than the theoretical total yields. That means
the fraction of ionized Al atoms seems to increase with rising
projectile velocity. A possible explanation could be that the
increasing electronic excitation in the neighborhood of each
collision cascade affects the ionization probability of the
ejected atoms.

We measured also the mean energy of the ejected ions as function of projectile velocity and recognized a rather small increase of this energy, when the projectile velocity decreases.
That means that the enhanced ionization probability is not accompanied by an increase of secondary ion energy.

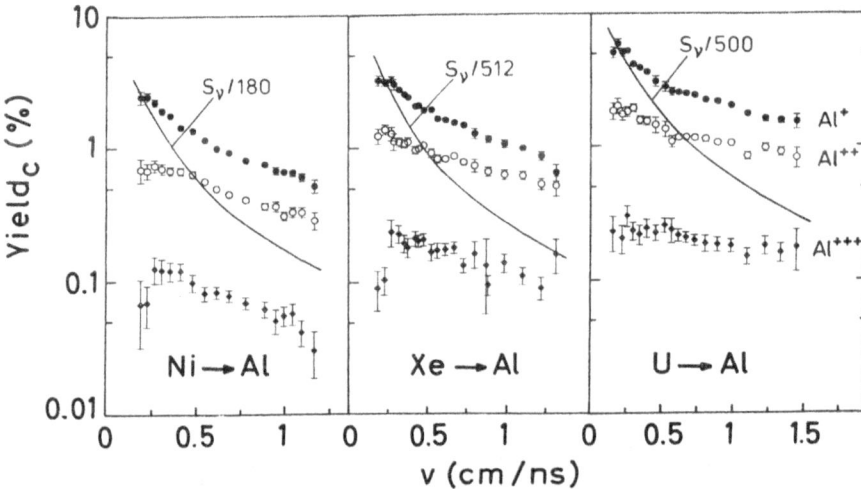

Fig.2: Yields of Al ions per 100 impacts as function of projectile
velocity compared with theoretical sputter yields. The yields are
corrected for ion detection efficiency.

The time distribution of each mass line in Fig.1 can be transformed into an energy distribution $\Delta N/\Delta E_x$, the x-direction being
parallel to the surface normal. Such a transformation suffers from
the difficulty to determine the correct $E_x=0$ point in the measured
time disttribution of a mass line as discussed in reference 4. In
Fig.3 a set of results is compared with energy distributions calculated by means of

$$\Delta N(E_x) / \Delta E_x = \int d\Omega_o \ Q_I(E_o,\Theta_o) \ S(\vec{v},E_o,\Theta_o)$$

This expression corresponds to a formalism developed by Sigmund
(16) in the framework of linear collision cascade sputter theory.
$S(\vec{v},E_o,\Theta_o)$ is the flux of recoil atoms of energy E_o per incident
ion having the velocity $\vec{v}$. The flux is directed into the solid
angle $d\Omega_o$ at an angle Θ_o. For the ionization probability $Q_I(E_o,\Theta_o)$
we choose the ansatz usually proposed in literature (17)

$$Q_I \sim \exp(-u_\lambda/u_x) = \exp(-\sqrt{E_\lambda/E_x})$$

Here, u_x is the x-component of the recoil velocity. With $E_o = E_x/\cos^2\Theta_o$ and the formalism of Sigmund (16) the energy distribution turns out to be

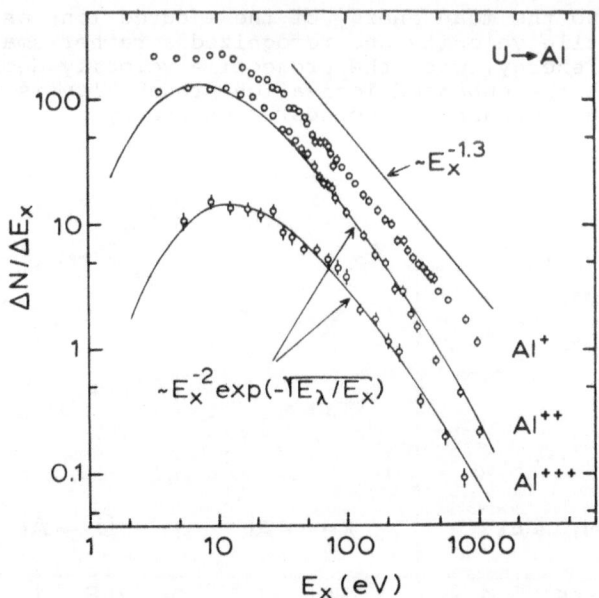

Fig.3: Energy distributions of Al ions ejected by 1.1 MeV/n U from a Al foil. To the data a function derived from sputter theory has been fitted.

$$\Delta N(E_x)/\Delta E_x \approx E_x^{-2} \exp(-\sqrt{E_\lambda/E_x})$$

Only in case of Al^{++} and Al^{+++} could this function be fitted reasonably to the experimental curves. The maxima of the curves at E_{xmax} are utilized to evaluate E_λ (=$16E_{xmax}$), the mean energy the recoils need to overcome neutralization, when they pass the surface. We obtained

$$E_\lambda(Al^{++}) = 120 \text{ eV}$$

$$E_\lambda(Al^{+++}) = 180 \text{ eV}$$

One has to be aware that E_{xmax} is very sensitive to the position of the $E_x=0$ point within a mass line. More precise measurements are planned in order to reduce this source of systematic errors. Certain fractions of the Al^{++} and Al^+ ions are probably generated by electron capture of Al^{+++} and Al^{++}, respectively. That might be the reason that the measured Al^+ curve is not describle by the proposed function, whereas in case of Al^{+++} the agreement is quite satisfactory.

3. ION EMISSION FROM SEMICONDUCTORS

Following the advice of Z. Sroubek (18) we started a program with semiconductors, in order to study the influence of electrical conductivity and band gap on ion formation. So far, only a pure GaAs crystal was investigated. The polished surface of

the crystal was cleaned by 4 keV-Ar etching. Sections of two
TOF-mass spectra of GaAs irradiated by 1.1 MeV/n ^{139}La are presented in Fig.4. Negative Ga and As ions were not found. After
cleaning the yield ratio of the Ga ions (^{69}Ga + ^{71}Ga) and the
As ions was determined as 11 $\pm$ 1.

The broadening of the line shapes in direction of higher ion
energies indicates that again atomic collisions are responsible for desorption from clean GaAs. Further evidence was deduced from the energy dependence of the Ga ions. We bombarded the crystal by 1.0 MeV/n and 0.09 MeV/n ^{74}Ge ions and found that the yield of Ga$^+$ did not decrease with energy, whereas the intensity of certain impurity ions went down by a factor of 3.5. We failed to observe the expected increase of the Ga$^+$ yield - probably because an unknown fraction of the rather diffuse low energy beam missed the sample. That caused problems with normalization.

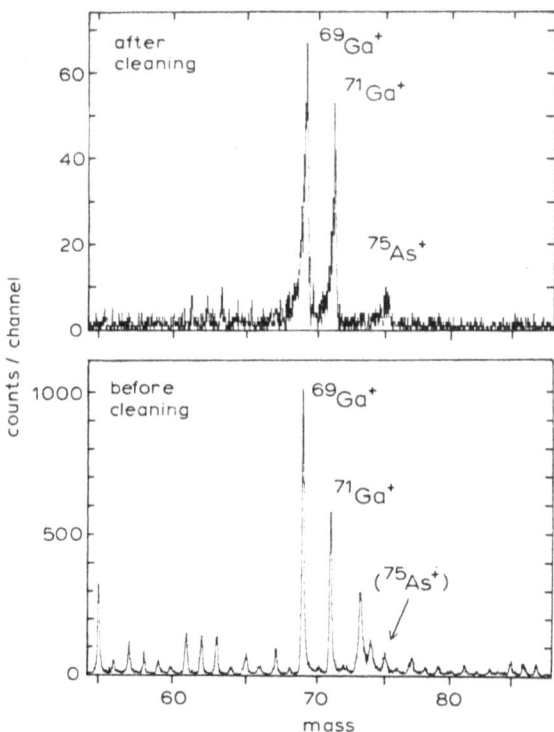

These first experiments supported the atomic collision mechanism, but further studies - in particular with semiinsulating material - are neccessary to reveal the general behaviour of semiconductors. The emission of Ga$^+$ ions is also observed before the ion etching procedure. In earlier studies (6) the M$^+$ metal ions were usually not detectable before ion etching. The mass lines have a narrow shape not extended

Fig.4: TOF spectra of ions ejected
from a GaAs crystal irradiated by
1.07 MeV/n La beams.

to shorter flight times. Also the yield of Ga$^+$ is larger by a factor
of 18 relative to the yield after ion etching. The yield enhancement
could be due to contaminants at the surface like oxygen; the line
shape, however, is typical for an electronic sputter mechanism as it
has been observed with small amounts of impurities on a metal surface (6).

4. ION EMISSION FROM CLEAVED FLUORIDE CRYSTALS

In order to investigate ion emission from clean and defined
surfaces of insulators, CaF$_2$, MgF$_2$ and NaF crystals were cut in

the vacuum chamber and the fresh planar surface irradiated by various heavy ion beams. The beams entered the vacuum chamber through a 0.5 mg/cm^2 Cu window.

Two TOF spectra of negative secondary ions taken with 1.1 MeV/n uranium are presented in Fig.5. Surprisingly in case of CaF_2 we observed a pronounced pattern of atomic, molecular and cluster ions, whereas in case of MgF_2 the total ion intensity was much lower and no peculiar mass lines are resolved. Both spectra were measured within the first 10 minutes after cleaving. The crystals were mounted on the same target holder, the rates of primary ion impacts were almost equal (40 % difference). So far, we explain this difference in ion yields by local surface charging due to the numerous secondary electrons (~10^3) ejected per impact. But why is the charging present at the MgF_2 surface (< 011> plane) during the emission process, but not at the CaF_2 surface (<111>plane)? In case of positive ion emission the charging effect is less pronounced: The CaF_2 crystal emits a spectrum of positive ions similar to the negative ion spectrum, while the spectrum of the

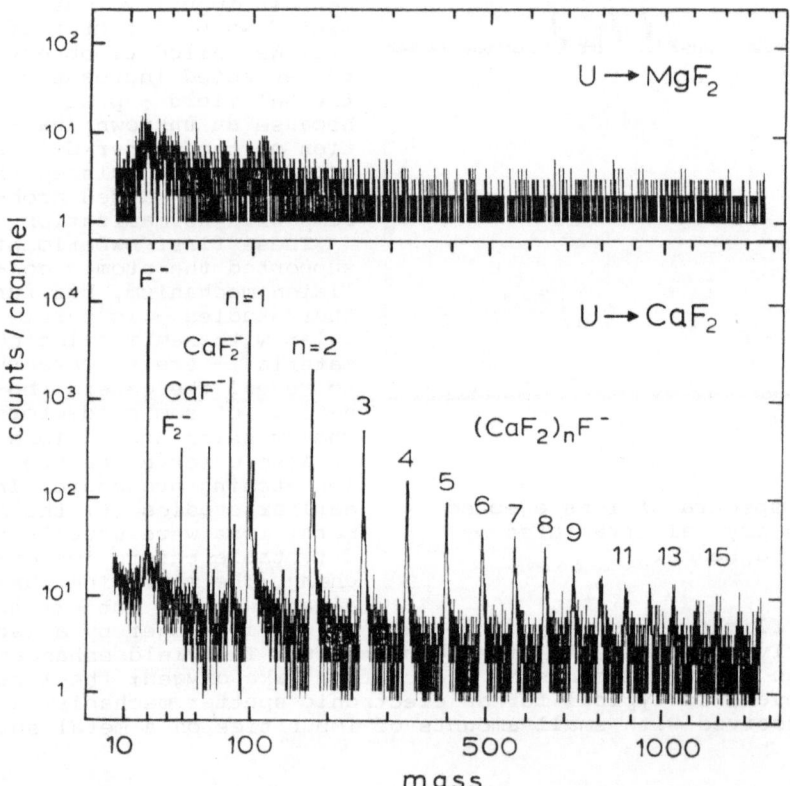

Fig.5: TOF spectra of CaF_2 and MgF_2 crystals irradiated by 1.1 MeV/n U. In both cases spectra of negative ions are shown.

MgF$_2$ surface shows intensive mass lines of Mg-F compounds including a cluster pattern (MgF$_2$)$_n$MgF$^+$. The mass lines in the positive MgF$_2$ spectrum are, however, considerably broadened - probably due to a positive charging of the crystal surface.

The TOF spectra taken with cleaved NaF crystals presented a
behaviour similar to that of MgF$_2$. Charging seemed to be even
higher, because also the positive ion spectrum consists of only
weak and hardly resolvable mass lines.

The second surprise was that about 15 hours after cleaving the
negative ion spectrum of MgF$_2$ exhibited a higher ion yield with
well resolved - but still broadened - mass lines including a
(MgF$_2$)$_n$F$^-$ cluster pattern. Contamination via the rest gas in the
vacuum chamber (10^{-7}mbar) had probably increased the surface conductivity.

Regarding the line shape and the energy dependence of the ion
yields, we conclude that MeV-ion induced desorption from the considered fluoride crystals is correlated with the electronic energy loss. As already observed with organic samples (19) two kinds
of secondary ions can be distinguished: atomic or small molecular
ions represented, for instance, by F$^+$, F$^-$ and Ca$^+$ and large molecular or cluster ions represented by (CaF$_2$)$_n$F$^-$ and (CaF$_2$)$_n$CaF$^+$.
Whereas F$^+$ and F$^-$ ions are emitted at all beam energies or energy
loss values available with our experimental facilities, cluster
ions vanish completely, when the sample is irradiated by a beam
having an energy loss of only 6.5 MeV/mg/cm^2. In Fig.6 the yield
ratio of the n=1 cluster ion and the F$^+$ or F$^-$ ion is plotted versus the specific energy loss
of the beam in the CaF$_2$ crystal.
It is clearly seen that the
formation of cluster ions requires a minimum energy deposit of about 9 MeV/mg/cm^2. It
would be of interest to compare this energy threshold
with that of latent nuclear
tracks in CaF$_2$ being etchable
by chemical solvents (20).

The emission of cluster
ions from epitaxal surfaces
is evidently not supported by
preformation of aggregates
being adsorbed at the sample
surface. The high energy impact breaks up the compound
of the solid and it is the
matrix material itself that
is blown into vacuum.

One can imagine two alternative ways of cluster formation: 1. clusters produced
directly as fragments of the
original matrix and 2. clusters
resulting from gas-phase collisions of certain fragmentation
residues very close to the sur-

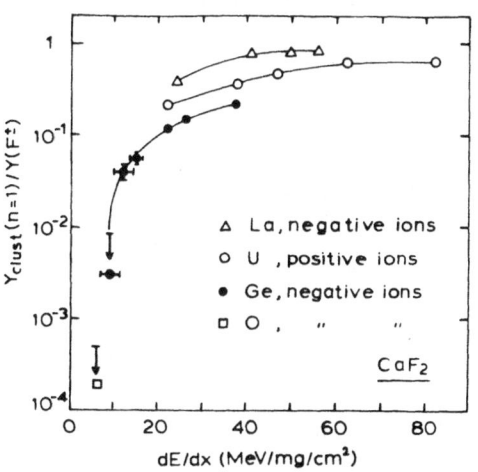

Fig.6: The yield ratio of (CaF$_2$)F$^-$
and F$^-$ ions or (CaF$_2$)CaF$^+$ and F$^+$
ions, respectivly, as function of
the energy loss of various beams
in the CaF$_2$ crystal. The arrows
indicate an upper experimental
limit of the cluster yields.

face (21). The second case requires high desorption yields: up to 10^4 molecules per impact have been observed at a dE/dx value of 50 MeV/mg/cm^2 (3). To a certain extent, however, cluster formation was found over the whole dE/dx range. Low total yields would favour the first formation process - for instance, organic molecular dimer and trimer ions have been observed below 7 MeV/mg/cm^2 (22), where the total yields are low. At high total yields collisions in the cloud of expanding particles might lead to larger aggregates. As an example, recently a $(UO_2)_n^+$ cluster series desorbed from uranylacetate has been reported. The authors found cluster numbers n larger than 50 (23). The fact that the acetate group is not included in the cluster series indicates that fast chemical reactions occur during the desorption process. So far in case of CaF$_2$ the situation of cluster formation is not clear and needs further investigations.

The studies with cleaved crystals are the first step of a program which is focused on secondary ion emission from defined surfaces under the bombardment with fast heavy ions. Our aim is to perform experiments on ion yields as function of the angle of incidence and of the initial charge state of the projectile. We think that this is a way to prove the various models (8-12) developed recently in the field of very high energy electronic sputtering. The numerous experiments on organic samples having a more or less "natural" surface have not be mentioned here. They revealed interesting results concerning, for instance, ionization of big biomolecules or contributions to desorption from layers deep underneath the surface (19). A detailed analysis, however, suffered often from undefined surface conditions.

REFERENCES

1. Nickel F: Thesis at the Justus-Liebig Universität Gießen FB 13, 1976
2. Albrecht P, Armbruster P, Spohr R, Roth M, Schaupert K and Stuhrmann H: Appl. Phys. A37 (1985) 37
3. Guthier W: Springer Proceedings in Physics 9 (1986) 17
4. Becker O and Wien K: Nucl. Instr. and Meth. B16 (1986) 456
5. Della Negra S and LeBeyec Y: Analyt. Chem. accepted for publ.
6. Becker O, Knippelberg W and Wien K: Physica Scripta T6 (1983) 117
7. Behrisch R(ed): Sputtering by Particle Bombardment I, Topics in Appl. Phys. 47, Springer Verlag, 1981
8. Watson CC and Tombrello TA: Rad. Effects 89 (1985) 263
9. Hedin H, Håkansson P, Sundqvist B and Johnson RE: Phys. Rev. B31 (1985) 1980
10. Lucchese RR: accepted for publication in J. Chem. Phys.
11. Bitensky IS and Parilis ES: Nucl. Instr. and Meth. B21 (1987) 26
12. Kammer HF and Hilf ER: Solid State Comm. 58, 7 (1986) 465
13. Sigmund P: Phys. Rev. 184,2 (1969) 383
14. Lindhard J, Nielsen V and Scharff M: Mat. Fys. Medd., Dan. Vid. Selsk. 36 10 (1968)
15. Thomsen PV: unpublished graphs., Institut of Physics University of Aarhus, Denmark
16. Sigmund P: in "Inelastic Ion-Surface Collisions", ed. Tolk NH, Tully JC, Heiland W and White CW, Acad. Press, N.Y. (1971) 121

17. Veksler VI: Rad. Effects 51 (1980) 129
18. Śroubek Z, Ceskoslovenska Akad. VED, Inst. of Radiotechniky and Elektroniky, Prag: he provided the GaAs crystal and supported the work by fruitful discussions.
19. Wien K, Becker O, Guthier W, Della Negra S, LeBeyec Y, Monart B, Standing K, Maynard G and Deutsch C: Int.J. of Mass Spect. and Ion Proc. 78 (1987) 273
20. Fleischer RL, Price PB and Walker RM: Nuclear Tracks in Solids, Principles and Applications, Univ. Calif. Press 1975
21. Kelly R: Private Communication, and Kelly R and Dreyfus RW: to be published in Nucl. Instr. and Meth. B (proc. REI 1987)
22. Guthier W, Becker O, Della Negra S, Knippelberg W, Le Beyec Y, Weikert U, Wien K, Wieser P and Wurster R: Int. J. of Mass Spec. and Ion Phys. 53 (1983) 185
23. Jungclas H and Schmidt L: Proceedings of INFOS IV, Proceedings in Physics Springer 1988

SIMULATION

COMPUTER SIMULATION OF STOPPING AND SPUTTERING

Wolfhard MÖLLER

Max-Planck-Institut für Plasmaphysik, EURATOM Association
D-8046 Garching/München, Fed. Rep. of Germany

1. INTRODUCTION

It is now about nearly 30 years ago that the first computer simulations of the type we will deal with have appeared in the literature. Interestingly enough, the basic principles of both main categories, i.e. molecular dynamics (MD) and binary collision approximation (BCA) calculations, were established nearly simultaneously. The first BCA simulations actually treated sputtering already /1/, whereas the initial purpose of MD calculations was the description of radiation damage in metals /2/. First BCA range calculations followed soon /3/. Since then, simultaneously with the increasing availability of fast and large computers, more than 400 papers on the computer simulation of atomic collisions in solids have been published.

We note that computer simulations of ranges and sputtering were performed before useful analytical descriptions in these fields were available /4,5/. This might indicate that computer simulations are able to produce theoretical predictions in a way which is easy to imagine, even for an experimentalist! This statement, however, shall not initialize here the controversy between analytical theory and computer simulation. Certainly, analytical theories often demand simplifications, which can be omitted in computer simulation - nevertheless, in many cases it remains questionable, if the simulations are really 'realistic'. Analytical calculations have the advantage of predicting physical dependencies, thereby encouraging the experimentalist to perform parameter studies which often contribute to the physical understanding. Corresponding parameter variations by computer simulation are often only meaningful on the basis of analytical concepts - in general, the simulation does not define, and does not need to define, the relevant quantities. (As an example, take the dependence of the sputtering yield on the deposited energy). In addition, simulated systematic variations are timeconsuming and subject to statistical uncertainties. On the other hand, simulations may be used to check simplifying assumptions made in analytical theory, and they may cover situations where analytical theory does not apply, e.g. effects of target crystallinity. Therefore, let us understand analytical theory and computer simulation to be complementary rather than representing subjects of controversy.

In spite of all legitimate enthusiasm which is common to all computer simulators, it is necessary to point out that today's simulations are by no means free of adjustable parameters. Therefore, any computer simulation which does not relate to experimental information must be interpreted with great care. In particular, this applies to cases where computer simulation represents the only way to obtain data on ion-surface interaction which are needed in other fields, and which are not accessible by experiments (see, e.g., ref. /6/).

R. Kelly and M. Fernanda da Silva (eds.), Materials Modification by High-fluence Ion Beams, 151–184.
© *1989 by Kluwer Academic Publishers.*

The present lecture cannot give a comprehensive review on the field of computer simulation of atomic collisions (earlier and recent reviews may be found in refs. /7-13/), nor can it be its purpose to join the mutual criticsm between analytical theory and computer simulation, or that between the MD and BCA branches of computer simulation. We shall rather try to describe the techniques employed in computer simulation of stopping and sputtering to some detail. The models and recipes seem to be simple enough that the reader should be able to set up his own computer simulation from the information given in this lecture! Powerful tools in the hand of non-experts might be dangerous, and we must therefore add a list of thorough remarks on the advantages and the applicabilites of certain computer models, as well as on their restrictions. A corresponding series of examples will be given finally, without any claim to be complete, but trying to emphasize areas of application where a specific type of computer simulation is especially useful.

2. SIMULATION PROCEDURES

As mentioned above, computer simulations of ranges and sputtering may roughly be subdivided into molecular dynamics (MD) and binary collision approximation (BCA) approaches. Physically, this classification is not strictly necessary: In fact, BCA simulations 'may be regarded as an extreme version of the quasistable (molecular) dynamical technique' (Robinson /9/). Nevertheless, the principles and the fields of application, which are characteristic for both classes, suggest a separate treatment in our course on the methods.

2.1 Molecular Dynamics (MD) Simulations

The most obvious simulation of the elastic effects of atomic collisions in solids is a complete description of the interaction of a projectile with the target atoms and the target atoms with each other. In MD simulations, a certain number of atoms is cut out from the target material, this cluster being large enough to prevent the moving atoms under consideration from escaping. Most MD calculations have been performed with crystalline clusters, but metastable models (see below) are also able to treat amorphous substances /14/. The cluster sizes range from $N = 500 \dots$ 1000 atoms in the earliest studies on radiation damage /2, 15/ to $N = 60\ 000$ /16/. As the cluster size determines computing costs (being proportional to $N \log N$ in recent sputtering simulations /14/), one should always use a minimum size which contains the collision cascades under investigation. As an example, recent sputtering studies with ~ 100 eV O impinging on Cu /17/ were performed with a $N = 13 \times 4 \times 9 = 256$ Cu cluster.

The MD simulation then consists of the simultaneous solution of the classical equations of motion for the cluster plus an incident projectile (if the atom set in motion initially is not one of the matrix). Pairwise central interaction is assumed for the ensemble of all atoms, so that each atom is subject to the force,

$$\vec{F}_j = - \sum_{\substack{i=1 \\ i \neq j}}^{N+1} \frac{\partial V(|\vec{r}_i - \vec{r}_j|)}{\partial(\vec{r}_i - \vec{r}_j)} \tag{1}$$

where V denotes the interatomic potential and $\vec{r}_i$ the position of an atom at the time t. With given target-target and projectile-target interaction potentials, the equations of motion

$$m_i \vec{\dot{v}}_i = \vec{F}_i \qquad (i = 1, \ldots, N+1) \qquad (2)$$

(m_i and $\vec{v}_i$ denoting the atom masses and velocities) can then be integrated, e.g., by means of a finite difference scheme employing sufficiently small time steps, Δt, according to:

$$\vec{v}_i(t + \frac{\Delta t}{2}) = \vec{v}_i(t - \frac{\Delta t}{2}) + \frac{\vec{F}_i(t)}{m_i} \Delta t \qquad (3)$$

$$\vec{r}_i(t + \Delta t) = \vec{r}_i(t) + \vec{v}_i(t + \frac{\Delta t}{2}) \Delta t$$

This choice neglects terms of order $(\Delta t)^3$ and higher. (Detailed descriptions of different numerical techniques are given in refs. /2, 10,18/).In any dynamic simulation, the choice of Δt and the integration procedure should be carefully checked, e.g. by requiring the conservation of the total (kinetic plus potential) energies of the system (in the absence of inelastic effects - see below).

It would waste computer time if the summation in eq. (1) would really extend over all atoms. In practice, it is sufficient to take into account close neighbours. This situation is depicted schematically in Fig. 1, where first- and second-nearest neighbours have been included.

2.1.1 Stable, Metastable, and Quasistable Models. In a stable dynamic simulation, which, for example, is necessary to study point defect configurations created in collision cascades /2,15,19/, the real static and dynamic properties of the substance have to be reproduced by the choice of the potential. Further, proper boundary conditions have to be established. Already the earliest studies /2,14/ employ a combination of a constant force, a spring force, and a velocity-proportional 'viscous' one as indicated in Fig. 1 allowing also for energy flow out from the cluster. Also, simple rigid /20/ or cyclic boundaries /21/ have been employed. The choice of the interaction potential determines also the stability of a given lattice structure. In this respect, it is sufficient to take into account

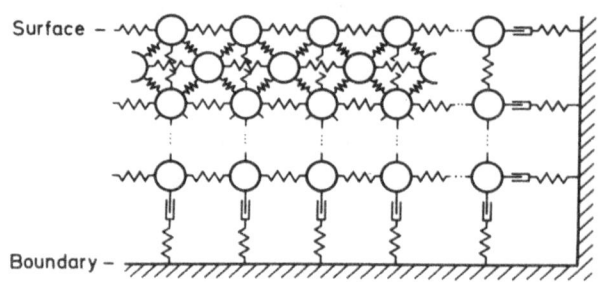

Surface —

Boundary —

FIGURE 1: Schematic representation of a cluster in a stable MD simulation. The boundaries are given by elastic and viscous forces.

nearest neighbour interactions for an fcc lattice, whereas for a bcc lattice second-nearest neighbours have to be included /15/.

For many applications, stable dynamical calculations are too time-consuming to produce results of satisfactory statistical quality with reasonable effort, so that approximations have been developed. Harrison and coworkers /11,14,22/ abandoned the requirement of stability, which is justified by the following consideration: The lifetime of a complete collision cascade is in the order of 0.1 to 1 ps. which is comparable to the vibration frequencies of individual atoms. This means that a model solid, which is not inherently stable, will loose its stability with a time constant considerably longer than the propagation time of energetic cascade atoms. Therefore, the thermal and elastic response of the lattice can be disregarded when treating the high-energy motion in a cascade only. Consequently, these 'metastable' dynamical models can be applied to phenomena associated with cascade atoms of sufficiently high energy, say E > 1 eV. Two simplifications result: As no 'realistic' boundary is needed sufficiently far away from the zone of interest, the size of the numerical crystallite can be kept small. Furthermore, only 'moving' atoms have to be considered /14/ instead of extending the calculation towards the boundaries of the whole cluster.

Still, dynamical simulations by means of the above models are limited to very small ranges of projectiles or recoil atoms. Pursuing the above concept, so-called 'quasistable' models have been set up. Torrens /23/ introduced a model where new clusters are generated along the path of moving atoms, when needed in order to trace the collision cascade. The equations of motion are solved only for atoms with a total energy above a cutoff energy of ~ 1.5 eV.

2.1.2 <u>Interatomic Potentials</u> Molecular dynamics simulations intend to realistically describe multiple pairwise interactions in the solid, and therefore essentially rely on the choice of proper interatomic potentials. In addition to the repulsive potentials investigated in the physics of nuclear stopping /4, 24/, an attractive part is needed at larger distances in order to reproduce the equilibrium nearest-neighbour distances in a solid, at least for unstable models. The idea to combine theoretical or semiempirical repulsive potentials with an attractive one of the Morse type, dates back to the first simulations of radiation damage /2,15/. Figure 2 shows a complete interaction potential employed for a metastable model /11/, which is given by a Born-Mayer potential at low separation

$$V(r) = V_{BM}e^{-r/a_{BM}}$$ (4)

with the so-called 'Gibson 2' parameters. Between 1.5 and 2 $\mathring{A}$, a cubic fit yields a smooth transition to the Morse function given by

$$V(r) = V_M\exp(-\frac{r-r_o}{a_M})(\exp(-\frac{r-r_o}{a_M})-2)$$ (5)

The potential is truncated just below the third-nearest neighbor distance. For comparison, the repulsive Molière approximation to the Thomas-Fermi potential is shown, according to

$$V(r) = \frac{Z_1Z_2e^2}{4\pi\varepsilon_o r}\phi(\frac{r}{a})$$ (6)

with the screening function

$$\phi(x) = 0.35\, e^{-0.3x} + 0.55\, e^{-1.2x} + 0.1\, e^{-6x}$$ (7)

The screening length a is given by Lindhard and Scharff according to

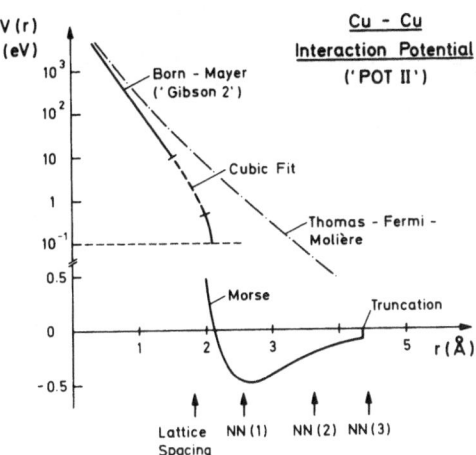

V(r) (eV)

Cu - Cu
Interaction Potential
('POT II')

Born - Mayer
('Gibson 2')

Cubic Fit

Thomas - Fermi -
Molière

Morse

Truncation

r(Å)

Lattice   NN(1)   NN(2) NN(3)
Spacing

FIGURE 2: Interatomic potential in a metastable MD simulation according to Harrison /11/. NN (1...3) denote the nearest-neighbour spacings.

$$a_{LS} = \frac{0.8853 \, a_0}{(z_1^{2/3} + z_2^{2/3})^{1/2}} \qquad (8)$$

a_0 denotes the Bohr radius, and Z_1 and Z_2 the atomic numbers of the collision partners. Alternatively, one might choose the Firsov screening length

$$a_F = \frac{0.8853 \, a_0}{(z_1^{1/2} + z_2^{1/2})^{2/3}} \qquad (9)$$

2.1.3 Inelastic Effects. Electronic energy losses have mostly been disregarded in MD simulations (however, see ref. /14/). This is justified in those cases where collisional events at sufficiently low energy are considered, where the nuclear energy loss exceeds the electronic one for $Z_1 \approx Z_2$ /25/. We will therefore postpone the treatment of inelastic losses and discuss them in connection with BCA simulations.

2.1.4 Thermal Vibrations. The treatment of thermal vibrations in MD calculations requires considerable additional computional effort and has therefore been included in a few instances only (e.g., ref. /26/).

2.1.5 Capabilities of Dynamic Models. There are some unique features with MD simulations in comparison with analytical theory and BCA calculations. Important for sputtering calculations is the fact that they should not only be valid in the linear cascade regime, as they allow for interactions between moving atoms, which is excluded by the linear Boltzmann equation which underlies the analytical treatment of sputtering /5/. Unfortunately, this principal advantage has never been made use of for a clear study of nonlinear effects, due to the shortcomings of the models with respect to limited capacity and the usually resulting poor statistics.

With BCA models and analytical treatments, a lot of discussions arose concerning the role of bulk and surface binding energies, as well as the choice of surface binding models /9,12,27-30/. In MD simulations, this is inherently included and needs no further modelling. We will readdress this topic in view of an example given in sect. 3.

The 'time-step' structure of the MD solution tracks automatically the full time evolution of the cascade. This, however, is also possible in BCA simulations, but requires additional storage. The real advantage of MD simulations is their ability to trace the time sequence of <u>overlapping</u> cascades, and the transition to the 'thermal' regime.

The main shortcoming of the MD simulations is their need for extensively long computing times, even in the study of low-energy events. The calculation of a single cascade, initiated by a ~ 100 eV atom, takes in the order of several minutes even on the fastest computers. From eq. (3), one might feel that MD codes written for vector processors might gain considerably higher speeds. However, the main computer time is needed for neighbour-search procedures which do not conveniently vectorize. In practice, the computing time can be reduced by a factor of about 3 due to vector-processing.

2.2 Binary Collision Approximation (BCA) Simulations

Two main shortcomings of MD calculations were evident in the preceding subchapter: Due to the time-consuming computation procedure, it is often difficult to obtain results with reasonable statistics, e.g., in the field of sputtering, and the limited cluster size excludes the study of large range phenomena, i.e. the treatment of high-energy or light-ion ranges. Metastable and quasistable MD models have been developed to partly overcome these difficulties. An extreme limit of a quasistable model is the situation where, along a particle track, collisions with individual atoms only are generated, rather than with clusters of atoms. This is the binary collision approximation.

Often, BCA simulations are termed 'Monte Carlo' calculations, addressing the random choice of the variables determining the trajectories of moving particles. However, e.g. in simulations for a single crystal at fixed energy and incident direction of an impinging projectile, the number of random variables reduces to the locus of incidence, from which the projectile trajectory and even the paths of atoms generated in a collision cascade are uniquely determined. Thus, there is a variety of models between complete 'random' and 'lattice simulation' codes applying the BCA model, the term 'Monte Carlo calculation' being more or less appropriate.

2.2.1 <u>Particle Trajectories</u>. Let us first assume a random medium. The trajectory of a moving particle is then approximated (see Fig. 3) by a sequence of straight lines, which represent the asymptotic flight paths between the collisions. The 'state' of a moving atom is at any time given by its locus $r = (x, y, z)$, its energy E, and its directional angles ϕ and θ with respect to a fixed coordinate system. Each collision is determined by the free pathlength s, the polar deflection angle θ and the azimuthal one, ϕ. These quantities are random variables, the actual values of which are determined in the computer by means of a random number r which is uniformly distributed in the interval $[0,1]$, i.e. with the distribution function

$$f(r) = 1 \qquad\qquad (10)$$

A random variable ρ, which is distributed according to the distribution function $f(\rho)$, must lie in the interval $[\rho, \rho + d\rho]$ with equal probability as its corresponding random number in $[r, r + dr]$:

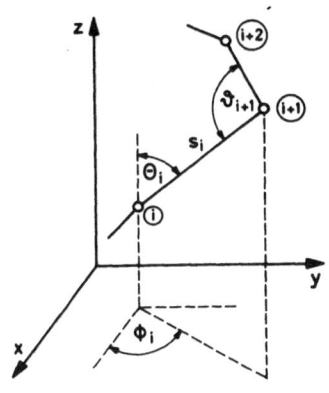

FIGURE 3: Part of an atom trajectory in a BCA simulation.

$$f(\rho)d\rho = f(r)dr = dr \qquad (11)$$

The integration shows the random number to be equal to the integral probability for the occurrence of ρ:

$$r = \int_{\rho_{min}}^{\rho} f(\rho)d\rho \qquad (12)$$

Applying eq. (12) to the free path s, being distributed according to

$$f(s) = \lambda^{-1} e^{-s/\lambda} \quad , \qquad (13)$$

we find

$$s = -\lambda \log(1-r), \qquad (14)$$

where λ denotes the mean free path length given by

$$\lambda = \frac{1}{\pi n \, p_{max}^2} \qquad (15)$$

n being the atomic density of the substance and p_{max} the maximum impact parameter to be discussed below.

Instead of a random choice of the polar deflection angle ϑ, most simulations for non-crystalline media vary the impact parameter p (see Fig. 4) with

$$f(p) = \frac{2p}{p_{max}^2} \qquad (16)$$

yielding for a given random number r

$$p = p_{max} \sqrt{r} \qquad (17)$$

From the impact parameter, the center-of-mass system scattering angle ϑ_{cm} is derived by the scattering integral

158

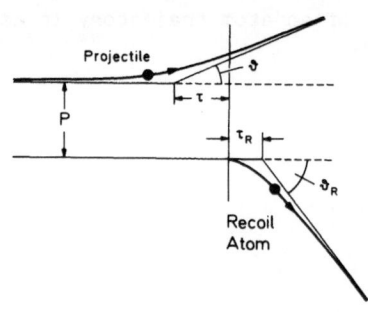

FIGURE 4: Scattering geometry of a binary collision in the laboratory system.

$$\vartheta_{cm} = \pi - 2p \int_{r_{min}}^{\infty} \frac{r^{-2} dr}{(1 - \frac{p^2}{r^2} - \frac{V(r)}{E_{cm}})^{1/2}} \qquad (18)$$

E_{cm} denotes the energy of the moving atom in the center-of-mass system, given by

$$E_{cm} = \frac{m_2}{m_1 + m_2} E \qquad (19)$$

(m_1 and m_2 masses of moving atom and target, respectively, E laboratory energy). The 'apsis' of collision, r_{min}, is given by

$$1 - \frac{p^2}{r^2_{min}} - \frac{V(r_{min})}{E_{cm}} = 0 . \qquad (20)$$

The center-of-mass angle is transformed into the laboratory one according to

$$\tan \vartheta = \frac{\sin \vartheta_{CM}}{\frac{m_1}{m_2} + \cos \vartheta_{CM}} \qquad (21)$$

It should be noted that equations (19) and (21) are only valid for target atoms being initially at rest.

Finally, the azimuthal deflection angle is uniformly distributed and calculated from its random number according to

$$\varphi = 2 \pi r \qquad (22)$$

In each collision, the particle **loses** an amount of energy

$$T = \frac{4 m_1 m_2 E}{(m_1 + m_2)^2} \sin^2 \frac{\vartheta_{cm}}{2} \qquad (23)$$

which is transferred to the target atom.

When taking into account inelastic energy losses properly (see sect. 2.2.5), the complete trajectory of any particle can now be constructed and traced down to a given cutoff energy, E_{co}. As seen from Fig. 4, the asymptotic deflection point differs from the original position of the target atom by a distance τ, as does the starting point of the recoiling target by τ_R. Some BCA programs (e.g. ref. /31/) offer an exact treatment of τ and τ_R, whereas others use the hard sphere approximation /32/

$$\tau = p \tan \frac{\vartheta_{cm}}{2} \tag{24}$$

and

$$\tau_R = 0, \tag{25}$$

or set even

$$\tau = 0 \tag{26}$$

The influence of τ on sputtering simulation results is small /37/. As stated above, the choice of s from a random number corresponds to a structureless medium without any neighbour correlation, similar to the treatment in analytical calculations. For an amorphous substance, one might approximate the structure by a constant free path **lengths**. This is given in, e.g., TRIM /38/ by

$$s = n^{-1/3}. \tag{27}$$

Compared to random medium simulations, the other extreme are BCA calculations for monocrystals. In those, the trajectory of any particle is uniquely determined by its initial conditions. Only for the point of projectile entrance into the crystal, random numbers are needed. In the simulation, each next collision partner is then chosen within a cluster of, e.g., nearest and second-nearest neighbours in the lattice (see Fig. 5). With $\vec{\lambda}_o$ denoting the current direction of the projectile, and $\Delta\vec{x}_i$ the distance to one of the neighbouring atoms, this atom is considered as collision partner provided /31/

$$p_{max} > p_i = (|\Delta\vec{x}_i|^2 - (\vec{\lambda}_o \cdot \Delta\vec{x}_i)^2)^{1/2} \tag{28}$$

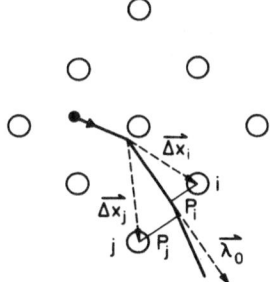

FIGURE 5: Search for the next collision partner and definition of simultaneous collision in a BCA simulation for crystalline media.

A special consideration is needed when more than one collision partner is found by this procedure, i.e. for the case of 'simultaneous' collision. According to ref. /39/, it is appropriate to make use of momentum conservation in order to calculate the final momentum of the projectile from the momenta of the collision partners, after these have been calculated from individual binary kinematics. The energies are scaled from the momentum requiring energy conservation.

Codes derived for crystalline simulations can also be applied to polycrystalline or amorphous targets by changing randomly the incident lattice direction for each projectile or rotating randomly the crystal after each collision, respectively /9,36,39/.

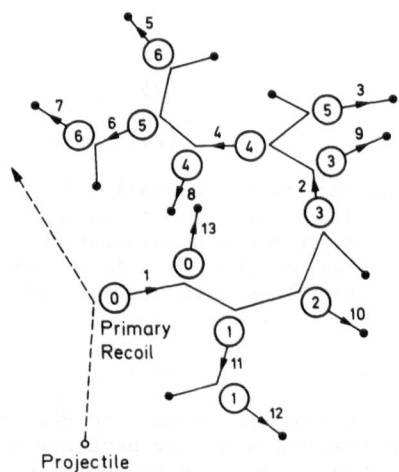

FIGURE 6: Cascade generation and event-store logics in a BCA simulation. The recoil atoms are denoted by their storage address, whereas the trajectory numbers indicate the sequence of their completion in the computer model.

Within the above frame, complete non-interacting cascades can be set up. This requires certain 'event-store' logics. Figure 6 shows an example for the TRIM.SP /32/ program: If the energy transfer to any target atoms is sufficiently high, recoil atoms are generated with a start energy

$$E = T - E_b \qquad (29)$$

where E_b denotes a given bulk binding energy. Their trajectories are simulated down to the cutoff energy E_{co}; next-generation recoils created on their way are stored with respect to their position, their energy, their direction, their atomic species and other properties (e.g. depth of origin) when of interest. After the slowing down of a recoil, the stored atoms are picked up in descending manner, and are allowed to generate new recoils with ascending indices in the storage vectors.

For reflection or sputtering studies, the surface requires an addtional treatment. First, atoms leaving a solid are exposed to an attractive potential. This is commonly simulated by a planar surface threshold with the surface binding energy E_s. The emerging energy is then

$$E_1 = E_0 - E_S, \qquad (30)$$

where E_0 denotes the energy just below the surface. From the conservation of the velocity component parallel to the surface, one obtains (see Fig.7)

$$\cos^2 \alpha_1 = \frac{E_0 \cos^2 \alpha_0 - E_s}{E_0 - U_s} \qquad (31)$$

Furthermore, Fig. 7 shows that one has to take into account the interaction of a leaving atom with the surface. This is performed, e.g., in TRIM /32/ up to a distance of

$$h_{max} = 2\, p_{max} \cdot \qquad (32)$$

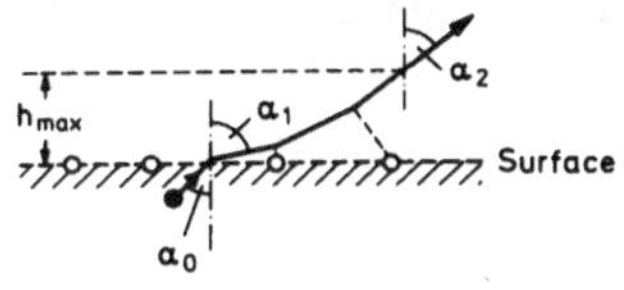

FIGURE 7: Surface treatment in a BCA simulation

2.2.2 <u>Maximum Impact Parameter</u>. The choice of the maximum impact parameter p_{max} in BCA simulations is not obvious. Generally, it is necessary to test the influence of its choice on the results of a specific simulation; this influence might be different for different interaction potentials. In the literature, both the choice of a fixed p_{max} /32, 34, 36, 38/ and of a fixed ϑ_{min} /33, 35/, corresponding to an energy-dependent p_{max}, are preferred for different codes. The original version of TRIM /38/ assigns one collision to each atomic volume, i.e.

$$\pi p_{max}^2 s = n^{-1}, \qquad (33)$$

and according to eq. (27)

$$p_{max} = \frac{n^{-1/3}}{\sqrt{\pi}} \qquad (34)$$

As, for low energies, the elastic energy losses and the angular deflections resulting from more distant collisions are found to be significant, TRIM.SP /32/ includes normally two additional distant collisions with

$$p_{max}^{soft} = p_{max}\,\sqrt{n} \quad ; \; n = 2,3. \qquad (35)$$

In the latest version of TRIM, Biersack /24, 40/ uses an analytic function, which determines p_{max} from the energy and the atomic species of the collision partners, thus, through eq. (33), shortening the free pathlength at low energies. The influence on, e.g., sputtering yields has to be tested in detail. In contrast, at high energies it is desirable to reduce p_{max} in order to save computer time by longer free pathlengths. TRIM /24,38/ determines p_{max} at high energy from a preset mean deflection angle along s, provided the electronic energy loss along s does not exceed 5 % of the initial energy (see sect. 2.2.5).

2.2.3 <u>Interatomic Potentials</u>. The interatomic potential to be inserted into the scattering integral, eq. (18), is commonly written as a screened Coulomb potential, eq. (6). For the screening function, earlier codes used the universal Lenz-Jensen /33/ or Thomas-Fermi /34/ functions. Often, the Molière function (eq. (7)) is applied /31, 35/. Both the Lindhard-Scharff (eq. (8)) and the Firsov (eq. (9)) screening radius have been taken; Robinson /9/ suggests a 'universal' value of $7.5 \cdot 10^{-10}$ cm for target-target interactions from a matching of the Molière and Born-Mayer potentials at the nearest-neighbour distance.

In principle, the interatomic potential can be computed individually for each combination of collision partners using simplified quantum-mechanical free electron gas models on the basic of Hartree-Fock isolated atom radial distributions /24, 41, 42/. 'Universal' screening functions derived from the average over such results for different projectile-target combinations have been applied especially in TRIM /24,32,38,40/ simulations. One of these is the so-called 'Kr-C' potential /42/, which can be written in the Molière form

$$\phi_{KrC}(x) = 0.191\ e^{-0.279\ x} + 0.474\ e^{-0.637\ x} + 0.335\ e^{-1.919\ x} \qquad (36)$$

with the Firsov screening parameter, eq. (9). Recent systematic calculations /24/ yielded the 'universal' potential with

$$\phi_U(x) = 0.182\ e^{-3.2x} + 0.51\ e^{-0.942\ x} + 0.28\ e^{-0.403\ x}$$
$$+ 0.0282\ e^{-0.202\ x} \qquad (37)$$

with the 'universal' screening parameter

$$a_U = \frac{0.8853\ a_o}{Z_1^{0.23} + Z_2^{0.23}} \qquad (38)$$

Figure 8 shows the band of individual screening functions derived from the quantum-mechanical calculations using a_U, in comparison to the other 'classical' screening functions.

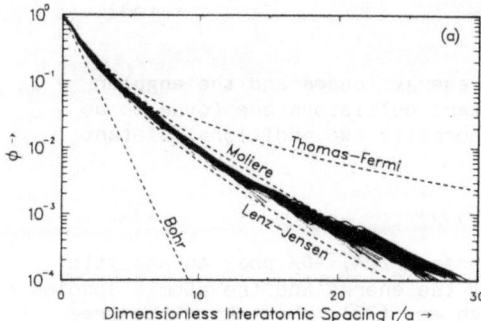

FIGURE 8(a): Different screening functions for the interatomic potential, with the 'universal' screening distance. The 'universal' screening function is an average within the band of individual potentials, see text.

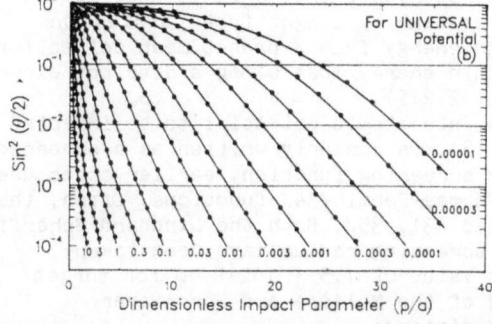

(b): CMS deflection angle as function of reduced energy (parameter on each curve) and impact parameter, calculated by the 'magic formula' and from an exact integration of the scattering integral. (From ref. /40/).

It should be noted that eqs. (37) and (38) do not necessarily represent the 'best' interatomic potentials for BCA simulations. For special

questions, it is sometimes advisable to use individual potentials /24,40/.
In some applications, seemingly similar potential function yield significantly different results. For example, this is observed in recent
simulations of single crystal sputtering, where the choice of the Kr-C or
the universal potential influences the directional emission of sputtered
particles drastically /43/.

2.2.4 <u>Calculations of the Scattering Integral</u>. The calculation of the
scattering integral, eq. (18), from the projectile energy and the potential is the most time-consuming step in a BCA simulation. It can be
evaluated at each collision /31/, but many codes have sought a possibility
to speed up the computation. One way is to set up an interpolation table
at the beginning of a specific run, which is then looked up during the
simulation procedure /34-36/. The TRIM program uses the so-called 'magic
formula' /24,38/, which is derived in an approximate way from the scattering geometry. Figure 8(b) shows an excellent agreement for the exact
integration and the 'magic formula' approximation of the scattering angle
as a function of the reduced impact parameter.

2.2.5 <u>Inelastic energy loss</u>. The interaction of moving atoms with the
target electrons is normally taken into account in two different ways. As
one possibility, the projectile is assumed to loose energy 'nonlocally' to
a homogenous electron distribution along the straight fractional paths:

$$\Delta E_{nl} = n S_e (s - \tau) \tag{39}$$

where S_e denotes the electronic stopping cross section. Equation (39)
illustrates a slight inconsistency in the BCA simulations: Before a new
collision, τ has to be determined at an energy, which is not yet known.
As simplification, e.g., TRIM evaluates the scattering integral after an
electronic energy loss along s, and corrects for the inelastic energy loss
along τ after the collision.

For sufficiently low energies, S_e can be chosen according to the
Lindhard and Scharff /44/ formula

$$S_L = \frac{8\pi e^2 a_o}{4\pi\varepsilon_o} \frac{z_1^{1/2} z_2}{(z_1^{2/3} + z_2^{2/3})^{3/2}} \frac{v}{v_o} \tag{40}$$

where v denotes the projectile velocity and v_o that of the first Bohr
orbit, being valid for $v < v_o z_1^{2/3}$. At high energy ($v \gg v_o$), the Bethe
formula applies:

$$S_B = \frac{4\pi e^4}{(4\pi\varepsilon_o)^2} \frac{z_1^2 z_2}{m_e v^2} \left[\log \frac{2m_e v^2}{I} - \log(1 - \frac{v^2}{c^2}) - \frac{v^2}{c^2} - \frac{C_s}{Z_2} \right] \tag{41}$$

with the electron mass m_e, the mean ionization potential I, the velocity
of light c, and the so-called 'shell correction' C_s. The intermediate
energy range can be described by /45/

$$S_e^{-1} = S_L^{-1} + S_B^{-1} \tag{42}$$

It is advisable, however, to make use of semiempirical tabulations,
where they are available /46/, which give the energy dependence of S_e in
the form of simple fitted curves.

Alternatively, a 'local' inelastic energy loss can be included, which depends on the actual impact parameter. (In MAR1OWE /31/, this local energy loss is even taken into account in a quasielastic formulation of the scattering kinematics). For the nonlocal energy loss, the Firsov formula /47/

$$\Delta E_F = \frac{0.35(Z_1+Z_2)^{5/3} \hbar v}{a_o \left[1 + 0.16 \ (Z_1+Z_2)^{1/3} \frac{p}{a_o} \right]^5} \tag{43}$$

can be used /36/. More often, an exponential impact-parameter dependence has been employed according to Oen and Robinson /48/:

$$E_{OR} = \frac{0.045}{\pi \ a^2} \ e^{-0.3 \ r_{min}/a} \cdot S_L \tag{44}$$

In many applications, combinations of the local and nonlocal losses have been taken, e.g.

$$\Delta E = \frac{1}{2} \ (\Delta E_{nl} + \Delta E_{OR}). \tag{45}$$

It is not evident which choice of the inelastic energy-loss - local, nonlocal or combinations of both - is most realistic. This must be regarded as an adjustable parameter. In BCA sputtering simulations, the sputtering yield is found to vary by a factor of up to 2 between pure local and pure nonlocal losses /32,49/. As a recommendation, one might choose nonlocal electronic energy losses for light ions, and an equipartition of local and nonlocal ones (eq. (45)) for the target-target interactions. With this choice reasonable agreement between simulation and experiment has been achieved.

At very high energy, above about 1 MeV/amu, the range fluctuation of projectiles starts to be influenced by the inelastic energy loss straggling, and is entirely dominated by it above app. 10 MeV/amu. This can be taken into account by a random selection of the nonlocal energy loss according to a Gaussian distribution with the variance /50/

$$\Omega^2 = \Omega_B^2 \cdot \begin{cases} (1/2)L(x) & , \quad x \leq 3 \\ 1 & , \quad x > 3 \end{cases} \tag{46}$$

with
$$x = (1/Z_2)(v/v_o)^2 \tag{47}$$

$$L(x) = 1.36 \ x^{1/2} - 0.016 \ x^{3/2} \tag{48}$$

and
$$\Omega_B^2 = \frac{4\pi e^4}{(4\pi\epsilon_o)^2} \ Z_1^2 \ Z_2 \ \frac{c^2-1/2 \ v^2}{c^2-v^2} \tag{49}$$

denoting the Bohr /51/ straggling, which is independent of energy in the nonrelativistic limit. TRIM employs an approximation to L(x) /24/. For a given random number, the actual deviation from the mean energy loss, ΔE_{nl}, is calculated according to

$$\delta E_{el} = \sqrt{2} \ \Omega \ erf^{-1} \ (2r-1) \ , \tag{50}$$

where erf^{-1} denotes the inverse of the error function.

2.2.6 <u>Thermal Vibrations</u>. It is possible to take into account thermal vibrations in BCA simulation codes for crystalline media. As the thermal vibrations are slow compared to the scattering process, one can randomly choose a static displacement for each target atom on a trajectory /36,52/ according to a Gaussian distribution (in analogy to eq. (50)). The standard deviation can be calculated from the Debye theory of thermal vibrations.

2.2.7 <u>Cut-off and Binding energies</u>. In a BCA simulation, an individual particle trajectory is terminated below a cutoff energy, E_{co}. In order to save computer time, one will take E_{co} as high as possible, but low enough so that no influence of E_{co} on the results can be detected. In sputtering simulations, it is apparently necessary to define E_{co} below the surface binding energy, E_S.

Much more intricate is the definition of the bulk binding energy, E_b, which determines the generation of a new free recoil in a collision cascade, and its initial energy. It is certainly inadequate to apply the Snyder-Neufeld model which sets E_b equal to the mean displacement threshold, E_d, which is typically 30 ... 80 eV in metals. This saves a lot of computer time, but must yield wrong results on, e.g., sputtering yields /53-56/. In the limit of very small defect concentrations, E_d is the minimum energy transfer in order create a stable Frenkel pair (otherwise the knocked-on atom will fall back into its own vacancy). However, stable displacements are not required at all, as the transport of kinetic energy in collision cascades is possible without mass transfer.

Searching for a proper bulk binding energy, one might try to find it from the energy stored by a Frenkel pair, E_{FP}^f. This energy accounts for the elastic and electronic rearrangements when forming a vacancy and an interstitial atom, and is thus at least needed to remove an atom from its lattice site. The bulk binding energy should even be a little larger, as kinetic energy is lost to neighbouring atoms. The typical value of E_{FP}^f is about 5 eV in fcc metals and 8 eV in bcc metals which is in rough agreement with MD results for the energy loss in the first collision within a collision sequence /15/. However, for sputtering problems part of this energy (the interstitial fraction E_i^f) will be paid back at the surface, so that one is left in this picture with the vacancy formation energy, E_V^f, only, amounting to app. 1 eV in fcc metals and 3 eV in bcc ones. Concluding this intricate subject, the bulk binding energy appears to describe a multiple-interaction phenomenon, which is beyond the scope of the BCA. In recent sputtering calculations, it has either **been set to** 0 /32/ in the lack of better knowledge or set equal to small values of 0.2 ... 0.5 eV /9, 49/.

Similarly complicated, and possibly even correlated to the bulk binding as stated above, is the definition of the surface binding energy, E_s. For simplicity and without a better knowledge, most evaluations of analytical theory and BCA computer simulations used the heat of atomization or sublimation energy,

$$E_s = \Delta H^a \qquad (51)$$

(Values are given in ref. /57/). More exactly, ΔH^a represents the energy to remove an atom from a surface step (a 'half space' atom /28/), whereas sputtering normally is believed to occur from a flat surface element, yielding a binding energy of

$$E_s \approx 1.35 \cdot \Delta H^a \qquad . \qquad (52)$$

But are sputtered atoms really released from an undisturbed surface, in view of high interstitial atom fluxes to the surface under irradiation and of the surface micro-topography which is caused by sputtering?
From the above considerations, one would propose for metals a bulk binding energy of $\sim E_{FP}^i$ and a surface binding energy of

$$E_s \approx 1.35 \cdot \Delta H^a - E_i^f \tag{53}$$

This subject has never been systematically investigated. The correlation between bulk and surface binding energies in BCA simulation was confirmed by a TRIM.SP run, which showed no variation of the sputter yield for Ne bombardment of Ni when varying E_b and E_s at constant $(E_b + E_s)$ /32/.
Additional considerations are required for polyatomic media /28/. For example, in the case of diatomic compound with the constituents A and B, one might define

$$E_{s,A} = -\frac{1}{2} \Delta H_{AB}^f + \Delta H_A^a$$
$$E_{s,B} = -\frac{1}{2} \Delta H_{AB}^f + \Delta H_B^a \tag{54}$$

where ΔH_{AB}^f denotes the heat of formation of the compound.
   2.2.8 <u>Capabilities of BCA simulations</u>. The main advantage of BCA simulations is the largely reduced computer time and storage requirement compared to MD models, allowing to study high energy, large penetration phenomena, as well as complete sputtering simulations, including energy spectra and angular distributions of emitted particles with good statistics.
   As a disadvantage, simultaneously overlapping cascades cannot be treated by BCA simulations; they cannot be used to investigate so-called 'non-linear' phenomena. In this respect, they are comparable to analytical treatments. However, they are able to treat details (e.g., crystalline structure) or systems where the solution of the Boltzmann equation is difficult /13/ (e.g., for light ions with high reflection coefficients).
   BCA simulations have to introduce additional parameters, the choice of which is not trivial (p_{max}, E_b, E_s). It is relatively convenient to treat electronic stopping and straggling, but again the choice of the model is uncertain. The influence of all these parameters has to be checked again and again.
   The employment of vector computers offers no big advantage for BCA simulation as the codes do not easy vectorize due to many conditional instructions. The corresponding reduction of computer time for a static sputtering run is by about a factor of 3.
   Finally, due to the relatively efficient operation, BCA codes may be used to set up simulations, which take into account the collisional target modification due to ion bombardement. This extension is treated in the subsequent chapter.

2.3 Dynamic Target Composition
   Up to now, we have dealt only with 'static' simulations in the sense, that subsequent events in a simulation run do not take into account what happened before. Therefore, we covered the low-fluence limit only. Dynamic substance modifications can arise from defects, sputtering, implanted atoms, or atomic relocation. Of these, the treatment of defects would be rather academic and only valid for very low temperatures ($\leq$ 30 K in metals when no interstitial diffusion takes place). Nevertheless, already early studies with MARLOWE took into account the mutual interaction of

interstitials and vacancies /31/. In the following, we will treat dynamic
composition changes in multicomponent substances, which arise from
implantation, inhomogeneous relocation by collision cascades, and
preferential sputtering. Several codes have been described in the
literature /53-56, 58-63/. We will follow here the layout of TRIDYN
/60,61/.

2.3.1 Compositional Change and Relaxation. Dynamic composition changes
depend on the number of incident projectiles per unit area. As no computer
will be able to treat, e.g., 10^{16} projectile histories corresponding to
reality, we have to identify each incident particle in the computer
('pseudoprojectile') with an increment of fluence

$$\Delta\phi = \frac{\phi_{tot}}{N_H} \tag{55}$$

where ϕ_{tot} is the total fluence simulated by N_H pseudoprojectiles. The
target substance is subdivided into slabs of initially equidistant
thickness Δx^0 (see Fig. 9). Each atom of type j being relocated into (or
removed from) a layer i corresponds to a change of the fractional areal
density, ν_{ij}, in that layer by an amount given by eq. (55) per
pseudoprojectile. After termination of each pseudoprojectile history, the
fractional compositions in each layer are recalculated according to

$$q_{ij} = \frac{\nu_{ij}}{\sum\limits_{k} \nu_{ik}} \tag{56}$$

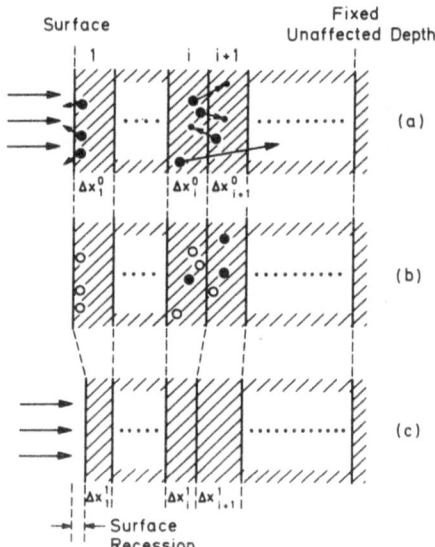

FIGURE 9: Principle of the dynamic
composition alteration in BCA
simulation codes. Relocation
caused by bombardment (a) leaves
vacancies and additional atoms
(b), which are allowed to relax
(c).

In order to account for the resulting excess or reduced densities, the
total atomic density, n_i, is allowed to relax (see Fig. 9 c) according to

$$n_i^{-1} = \sum_j q_{ij} \, n_{jo}^{-1} \qquad\qquad (57)$$

where n_{jo} denote the atomic densities of the pure components. However, n_{jo} might also be chosen to fit the real density of a given compound. Equation (57) corresponds to an altered thickness of the layer

$$\Delta x_i = \sum_j \nu_{ij} \, n_{jo}^{-1} \quad . \qquad\qquad (58)$$

Generally, the intervals turn nonequidistant, and provision is made to keep the interval width within certain limits, e.g. $0.5\Delta x^o < \Delta x_i < 1.5\Delta x^o$, by interval splitting or combination.

Δx^o has to be chosen small enough in order not to influence the results; $\Delta x^o \approx 5\ \text{\AA}$ has proved to be practical. The number of pseudo-projectiles chosen for a given total fluence determines the statistical quality of the simulation; normally it is chosen small enough so that the relative change of areal density per pseudoprojectile does not exceed 10 % in each interval.

2.3.2 Surface Binding and Relocation Threshold. In the simple model described above, new questions arise again with respect to energy parameters. For the surface binding, one has to be aware of the fact that the surface composition may change due to collisional effects, and thereby the surface binding energy may also be altered. One may setup schemes /64/ where the surface binding depends on the surface composition. An alternative simple choice is for a two-component material /60,65/

$$E_{s,A} = \Delta H_A^a$$
$$\qquad\qquad\qquad\qquad (59)$$
$$E_{s,B} = \Delta H_A^a + \Delta H_{AB}^f$$

for the case where the component A is enriched at the surface. Systematic variations of the surface binding energy input have not been performed up to date.

Of considerable influence on the results is the choice of a relocation threshold, E_r. At the first glance, one might identify it with the displacement threshold, E_d. However, we must recall that E_d is valid in the limit of a perfect crystal, whereas we deal with an already disturbed substance at high damage level. Thereby, an interstitial atom might find a new position very close, and be relocated at an initial energy which is considerably smaller than E_d. Therefore, E_r should be chosen somewhere between E_b and E_d. Comparisons to experimental findings /61,65,66/ were successful at $E_r = 8$ eV. However, this can certainly not be regarded as a 'universal' parameter.

2.3.3 Capabilities of Dynamic Composition BCA codes. BCA simulations are now capable of simulating high-fluence ion implantation, atomic mixing, and preferential sputtering provided these phenomena are dominated by collisional effects. In this way, the simulations may be used to check whether a system is dominated by collisions, and trace phenomena of ion-induced decomposition which are difficult to imagine /65,66/.

However, the computing time in present codes is often rather long. TRIDYN calculations on the sputtering of Ta_2O_5 by 1 keV He bombardment towards the stationary state require about 1 h on a CRAY-XMP in a non-vectorized version. The treatment of other problems (e.g. heavy-ion bombardment in the 100 keV range), is often prohibited by an exceedingly long computer time.

3. APPLICATIONS

In recent years, computer simulations have been employed to produce a vast amount of information and data. We shall try to pick out such examples where specific computer simulations have addressed special topics demonstrating their capability, e.g. in comparison to analytical theory or to a different type of simulation. Further, the listener should learn how powerful computer simulations are for the study of complicated systems, and how they might help to understand detailed mechanisms. However, he should also realize that the simulations are really meaningful only in connection to experiments. Any belief in the predictions themselves must be questioned in view of the many free parameters involved.

3.1 Ion Slowing Down and Deposition

We have seen that the requirements of MD simulation prevent to apply them for the slowing down of ions with higher energies, say $\geq$ 10 eV/amu. Therefore, we find mainly BCA calculations in this field.

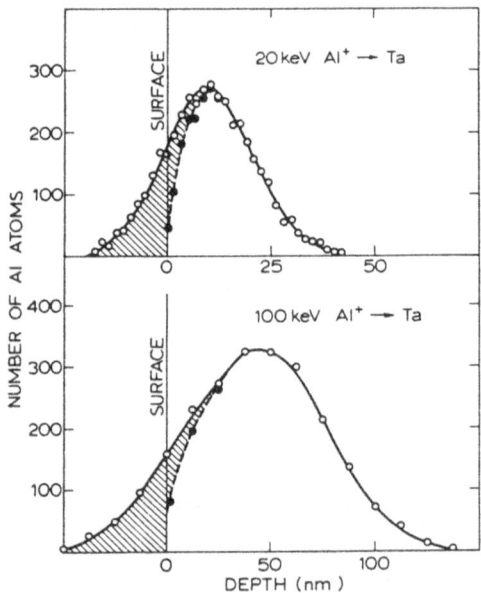

FIGURE 10: Surface correction to a range profile calculated for an infinite medium.
(From ref. /67/).

3.1.1 Ranges. Our first example (Fig.10) describes a situation where the analytical treatment of an ion range distribution is difficult. The analytical solution requires in general an infinite medium, so that projectiles being reflected through the surface plane might reenter. Figure 10 /67/ shows BCA simulations of Al ranges in Ta for a semi-infinite and an infinite medium, which both are easy to define in the simulation. One finds a significant 'surface correction'; the infinite medium results are in good agreement with the analytical prediction. Correspondingly, many BCA simulations have been applied to light ion ranges, and to light ion reflection /68/.

Analytical solutions of the Boltzmann equation are also difficult at very high energy, where the distributions become asymmetric and are largely influenced by electronic straggling. Here, a high-energy BCA model

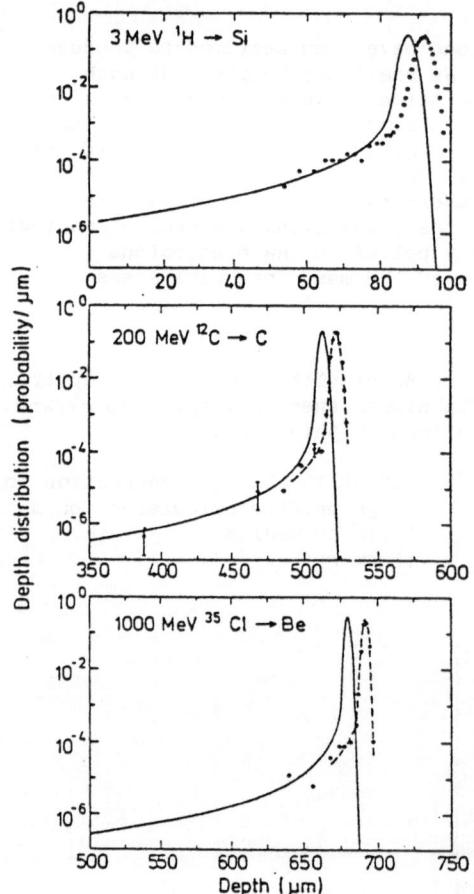

FIGURE 11: Range distributions of high-energy heavy ions generated by computer simulation (points) and an analytical single collision model (line). (From ref. /69/)

can be applied. In Fig. 11, its results are in reasonable agreement with an analytical single-collision model /69/ prediction.

Next, it is our obligation to recall one of the most spectacular findings with computer simulations: the discovery of channeling /70/. Based on the possibility to treat single crystals in the BCA simulation, pronounced differences of the Bragg attenuation profiles are seen in Fig. 12 for different crystallographic directions. Figure 13 /71/ shows a further example how crystalline effects may influence range distributions: Earlier investigations of deuteron ranges in Ni had found the width of the measured range distributions in significant disagreement with theoretical predictions /72/. Detailed investigations with MARLOWE then showed a broadening due to crystalline effects. The double peak structure found for the monocrystalline and polycrystalline substance could be removed by introducing Frenkel pairs into the model crystal, thus simulating radiation damage. The calculated distributions for the damaged, polycrystalline medium are then in excellent agreement with the experimental findings.

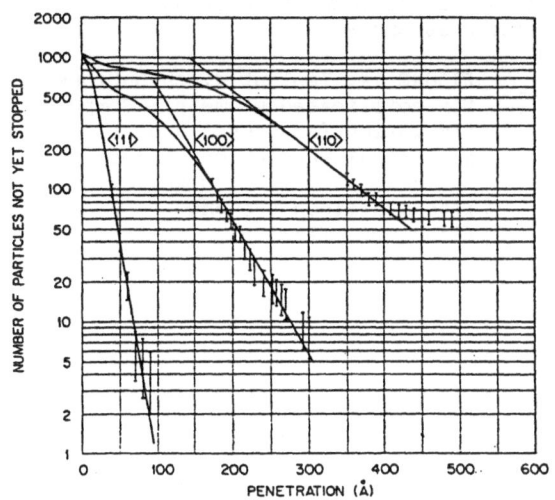

FIGURE 12: Integral range plots for 5 keV Cu ions incident along different directions in a Cu single crystal: The discovery of channeling.
(From ref. /70/)

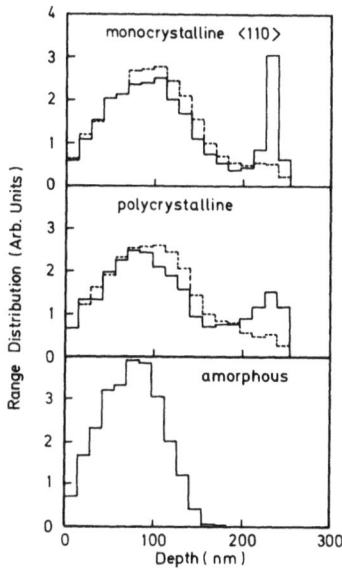

FIGURE 13: Crystalline effects on the range distribution of 10 keV deuterons in Ni. The distributions have been calculated for a target tilt 10° off the normal direction; the result for the monocrystal represents an average over different azimuthal directions of incidence.
(From ref. /71/).

Figure 14 shows an example of high technological relevance, where computer simulation is indispensable due to the complicated geometry /73/. The result of the computer simulation is in rough agreement with an analytical prediction for the surface perpendicular to the direction of incidence, but the simulated distributions are slightly more asymmetric. It is, however, impossible to predict analytically the deposition along

172

the inner walls of the trench, which is due to backreflected particles, and might influence electrical properties strongly.

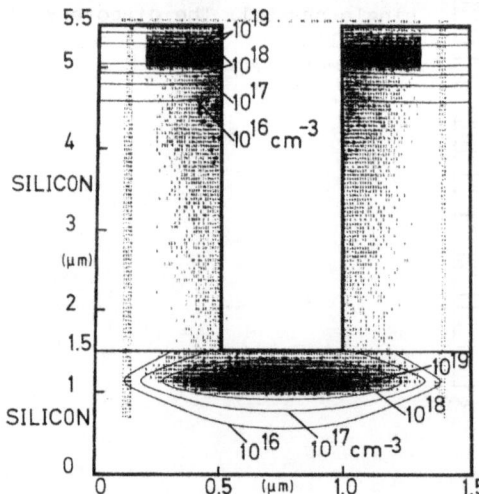

FIGURE 14: Implant distributions for complicated geometry: Computer simulation (shadowed) and analytical prediction (lines) for 100 keV, 10^{15}cm^{-2} B implanted into a silicon trench. (From ref. /73/).

FIGURE 15: Distribution of ranges and energy depositions generated in a BCA simulation

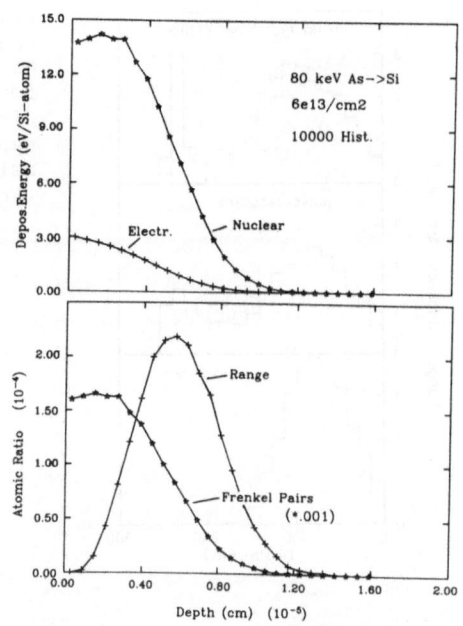

3.1.2 Energy Deposition It is a further advantage of computer simulation compared to analytical theory, that one run can provide not only information on ranges, but also on different kinds of energy deposition. They allow, for example, to compare the amount of energy dissipated into electronic and nuclear collisions. At sufficiently high ion energy, this is especially simple. Then the recoil ranges can be neglected compared to the ion ranges, so that only the primary distributions are important and the tracing of recoil atoms can be dispensed. In addition, it is then possible to calculate the number of generated Frenkel pairs by a primary recoil of energy transfer T, e.g., according to the modified Kinchin-Pease model:

$$
N_F(T) = \begin{cases} 0 & T \leq E_d \\ 1 & E_d \leq T \leq 2.5 \ E_d \quad (60) \\ \dfrac{0.8 \ \nu(\tau)}{2 \ E_d} & T \geq 2.5 \ E_d \end{cases}
$$

where $\nu(T) \approx 0.9 \ T$ denotes the fraction of energy dissipated into nuclear motion within the cascade. Figure 15 shows an example of a TRIM simulation for As ions in Si, calculated in less than 5 min /74/.

3.2 Sputtering

Within our field of interest, MD simulations have mainly been used for sputtering problems. However, the application of BCA calculations overwhelmed that of MD during recent years, due to their comparatively quick performance, and, arising from this, the possibility to study detailed mechanisms and differential distributions. Nevertheless, we should recall that MD simulations are a priori preferable to BCA ones for sputtering problems, as the energy parameters - bulk and surface binding, relocation threshold - are inherently defined in the model and do not have to be preset with sometimes ill-defined values. In principle, one should be able to derive the BCA energy parameters from MD simulations. However, detailed studies with this purpose are still missing, and may be questionable due to the uncertainty of the interatomic potential.

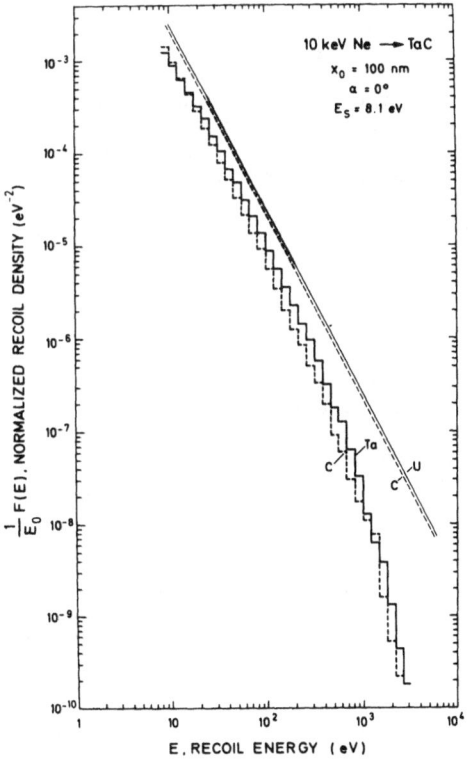

FIGURE 16: Normalized recoil density vs. recoil energy, in a cascade formed by a Ne atom in TaC, starting at a depth of x_0 = 100 nm with an initial energy of E_0 = 10 keV. BCA simulation results (histograms) are compared to analytical predictions /76/ (lines) for a similar compound system. (Thin lines: extrapolation)
(From ref. /75/)

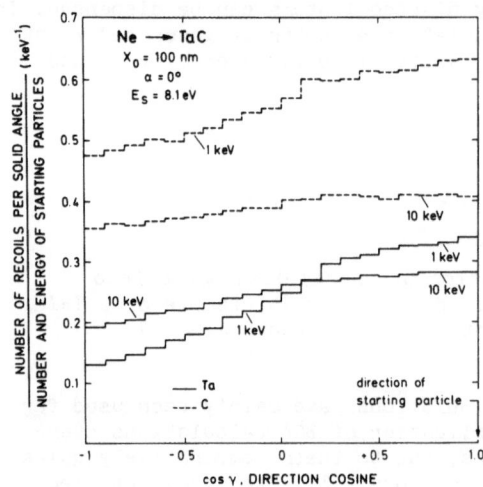

FIGURE 17: Angular distribution in a cascade started by 1 keV and 10 keV Ne atoms at a depth of 100 nm. The cut-off energy was chosen equal to the surface binding energy, E_s. (From ref. /75/)

3.2.1 <u>Cascade Characteristics</u>. Similarly, computer simulation might help analytical theory by checking its underlying assumptions. Figures 16 and 17 show examples for cascades generated in TaC by the BCA code TRSP2C /75/. For Figure 16, 10 keV Ne atoms were assumed to start in the bulk and to **initiate** a collision cascade. The recoil density, which represents the distribution of recoil atoms over their start energy E, is predicted by analytical theory /5,76/ to be proportional to E^{-2}. This is well-confirmed by the computer simulation in the energy range which is important for sputtering, $E \lesssim 100$ eV. However, the simulated slopes for the individual components are slightly different, and the absolute values differ by a factor 2-3 between analytical prediction and simulation.

Figure 17 raises some doubts on the isotropy of the collision cascade, which is assumed in analytical theory. Especially for lower primary energy, the dissipation of the cascade is insufficient for a complete momentum relaxation. It should be noted that the anisotropy is still maintained for the low-energy recoils, as these represent the overwhelming fraction of the cascade atoms (see Fig. 16).

3.2.2 <u>Total Sputtering Yield</u>. In BCA simulations, it is nowadays possible to reproduce experimental sputtering yields for widely varying ion-target combinations within about $\pm$ 50 %. We should, however, reformulate our reservations: BCA simulations are performed with sort of an 'agreed' input set of the energy parameters. Generally, the heat of sublimation appears to be a good choice for the surface binding energy in some contradiction to the arguments given in sect. 2.2.7. However, the fault introduced by an imprecise value of the surface binding energy might be balanced by other quantities, being imprecise as well, e.g. the interatomic potential, or the electronic energy loss.

For some materials, deviations from the above agreement have been found. E.g. in the case of copper, BCA simulations yield a result which is low by a factor of 2. Biersack /40/ explains this by a poor validity of the universal potential and the universal electronic stopping cross section. Using appropriate individual data, one can again reproduce the experimental finding. Another interpretation arises from MD calculations. Jakas and Harrison /77/ calculated Cu sputtering yields under Ar bombardment by

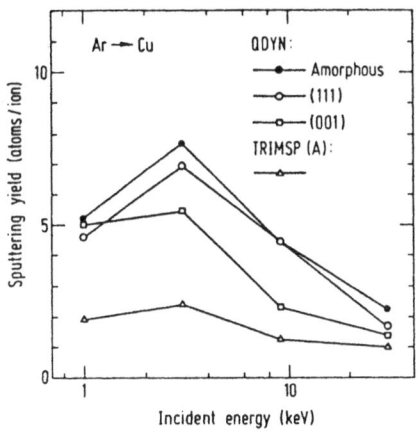

FIGURE 18: Sputtering yield for Ar
incident on Cu from MD (QDYN) and BCA
(TRIMSP(A)) simulations. The BCA code
models an amorphous substance.
(From ref. /77/)

MD and BCA using identical repulsive potentials (Fig. 18). The MD result
is significantly higher, which they attribute to the interaction of moving
atoms, which might be termed a 'nonlinear' effect. However, a lower
surface binding energy in the BCA simulation might remove the above
discrepancy, and yield a result in accordance with the experimental one!

3.2.3 Differential Sputtering Yield. Surprisingly detailed information
on the angular and energy-distributions of sputtered atoms can be extracted from BCA simulations. Figure 19 /32/ shows an example for a
polar/azimuthal angular distribution, which is in good agreement with
experimental data /78/. The angular distribution shows a pronounced ridge
which is due to a simple double collision mechanism, which has been
distinguished from larger-cascade sputtering in the simulation.

The angular distributions of the individual constituents in diatomic
compounds may be largely different, as shown in Fig. 20 /75/. The figure
compares the results for different bulk concentrations and to those
obtained with a dynamical code (see also sect. 3.3).

3.2.4 Molecular Ejection. Due to the possibility to treat multiple
interactions, MD simulations may be used to address special topics, where
the BCA method fails. An example is the ejection of sputtered molecules
/17,79/. Within the model, the criterion for dimer formation is that the
sum of the relative kinetic energy and the potential energy of the
constituents is negative. A recent example /17/ is shown in Fig. 21 for
Cu_2 and CuO formation under low energy oxygen bombardment. A rather
high yield of sputtered molecules is found, which, however, is difficult
to compare to experiments: During the emission electronic effects might
play a role, and, for example, lead to autodissociation. Obviously, this
cannot be taken into account in the present type of simulations.

3.2.5 Crystalline Effects. As for range simulations, we want to
mention an example of the influence of crystalline effects on sputtering.
A long-standing discussion concerns the explanation of the 'Wehner spots'
/80/, which represent a strongly anisotropic angular distribution of atoms
from single crystals due to the ejection along crystallographic axes. Two
contradictory models have been exposed in the literature: Silsbee /81/
explained the spots by long-range collision sequences, whereas Lehmann and
Sigmund /82/ proposed that a random cascade interacts with the periodic
surface structure. MARLOWE BCA simulations were employed to investigate

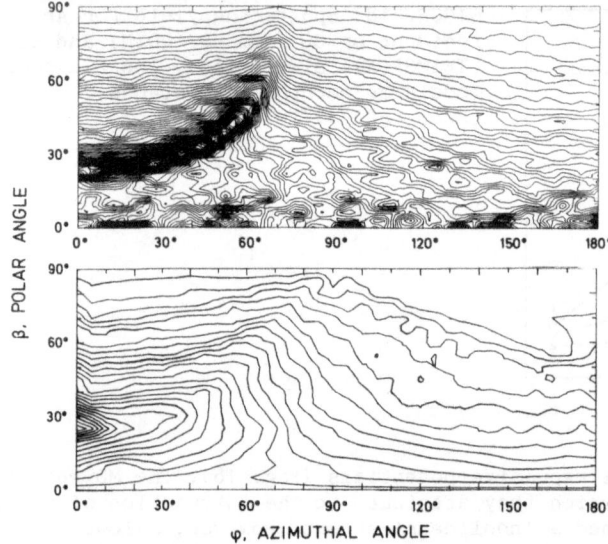

FIGURE 19: Angular distribution of sputtered particles by deuterons incident on Ni at an angle of 80° with respect to the surface normal, from computer simulation (top) and experiment /78/. The polar angle is defined with respect to the surface normal. The azimuthal one is $\varphi = 0$ for the forward direction.
(From ref. /32/)

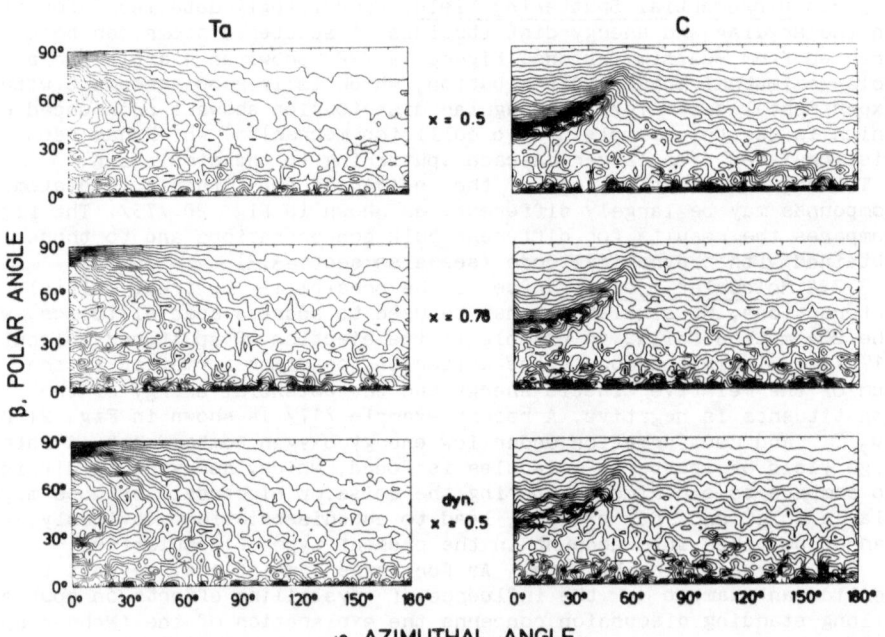

FIGURE 20: Angular distribution of sputtered atoms from TaC under 1 keV He bombardment at 70° with respect to the surface normal from BCA simulations: static with bulk composition (top), static with adjusted composition (middle), and dynamic (bottom). x denotes the Ta composition. For definition of angles, see Fig. 19. (From ref. /75/)

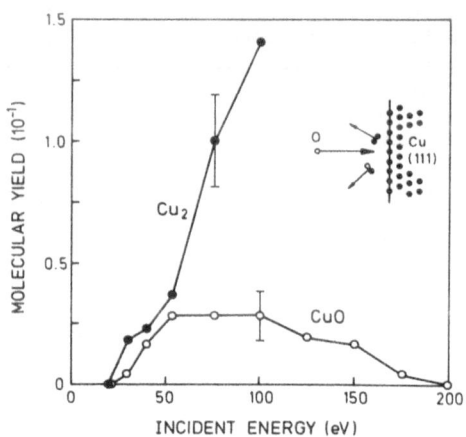

FIGURE 21: Molecular emission yield for oxygen incident on Cu, calculated by MD simulation. (Plotted from ref. /17/)

the phenomenon /83/. Figure 22 shows spots corresponding to the <100>, <110>, and <112> directions for Ar incident on Cu; due to the refraction within the planar surface barrier, they are shifted towards larger polar angles. Replacement sequences only are not sufficient to explain the anisotropy; their influence increases with larger incident energy. As found from the simulations, the Silsbee and Lehmann-Sigmund mechanisms cannot be clearly distinguished, both originating from momentum anisotropies in the collision cascades.

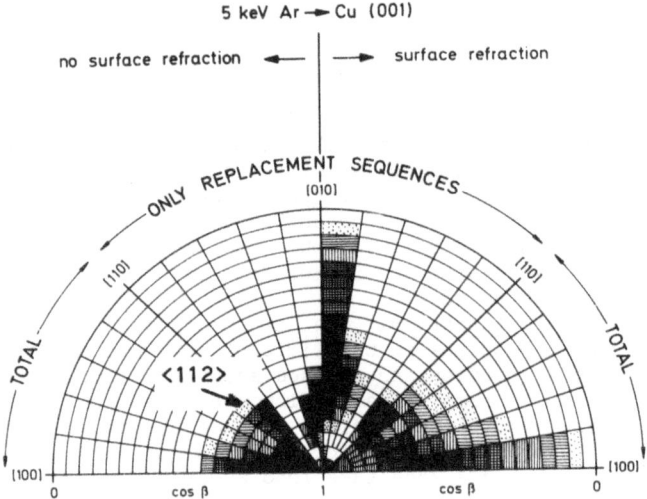

FIGURE 22: Angular distribution of sputtered particles for 5 keV Ar at normal incidence onto a Cu single crystal, as obtained from a BCA simulation. The four quarters of the plot correspond to all mechanisms/replacement sequences only, and surface refraction taken into account/neglected.
(From ref. /83/).

3.2.6 <u>Sputter Yield Fluctuation</u>. A field of recent interest for computer simulation are the statistics on the number of sputtered particles /17, 84/, where essentially no theoretical information is available until now. Figure 23 shows an example generated by means of a TRIM simulation /84/. The probability to sputter a certain number of atoms per ion exposes a rather broad distribution, much broader than the Poisson one around the mean yield. The detailed understanding is still missing. It should be pointed out again in view of this example that computer simulation may be used to identify problems, and promote an increased understanding of the detailed mechanisms.

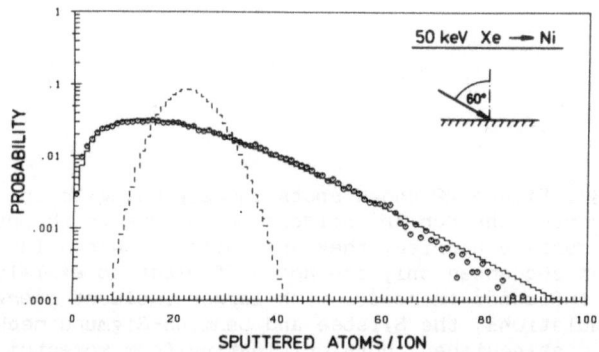

FIGURE 23: Probability to sputter N atoms for one Xe ion incident on Ni, as found by means of a BCA simulation. The distribution is compared to the Poisson one around the mean yield of 22.5 (dashed line), and an analytical fit (solid line).
(Ref. /84/)

3.3 High Fluence Phenomena

As some recent BCA codes are available which include the dynamic changes of composition during bombardment, we will finally address high fluence phenomena being simulated by means of the computer. It should be noted that nearly any experimental study or application of sputtering is performed with high ion fluences. Thereby, the implanted ions may take part in the energy dissipation and influence, e.g. the sputtering yield. This has often been neglected.

3.3.1 <u>High Fluence Implantation</u>. The stationary amount of implanted ions, and their depth profiles in the limit of high fluence are governed by a balance of ion deposition and surface erosion by sputtering (provided that no additional mechanisms of saturation are effective). Simple models (see e.g. ref. /85/) predict the implant surface concentration to be given by the inverse of the sputtering yield, when ion reflection is negligible. More detailed considerations /86/ take into account the 'range shortening' due to the implanted ions, possibly leading to a transient maximum of the collected amount of implants. Figure 24 /74/ shows preliminary results for Xe incident on Si. Indeed, a range-shortening peak is found in the simulation, which is clearly confirmed by the fluence-dependence of the depth profiles.

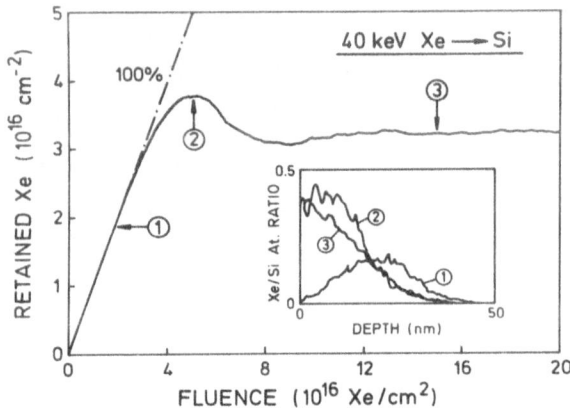

FIGURE 24: Amount of retained Xe during implantation in Si from a dynamic-composition BCA simulation. The insert shows the implantation profiles at different fluences denoted by the arrows pointing at the retention curve. (Ref. /74/)

The stationary surface concentrations of about 0.4 is significantly larger than the inverse sputtering yield of (3.5) derived from the simulation. (It should be noted, that the simulated sputter yield is in excellent agreement with the experimental one). Furthermore, the simulation also reproduces a sputter yield enhancement by about 1 atom/ion between the start of bombardment and the stationary state, which has also been observed experimentally /87/. Now, once the effects have been established in computer simulation, one can start to investigate their physical origin by studying the detailed mechanisms by means of the simulation. Unfortunately, computing times are still rather long prohibiting a fast progress in this area.

3.3.2 Preferential Sputtering. It is straightforward to apply dynamic-composition codes to problems of ion-induced decomposition. Fig. 25 shows the stationary (i.e., high-fluence) total sputtering yield of TaC as a function of the He energy. The TRIDYN results /75/ are in good agreement with experimental points /88/. The insert shows nicely the convergence of the partial sputtering yields towards a ratio of 1:1, i.e. the bulk composition ratio, as required by mass conservation.

There are now several examples where TRIDYN simulations fit experimental findings on surface concentrations as a function of fluence, and the depth profiles rather well /60, 65, 75/. For He bombardment of Ta_2O_5 the simulation was able to reproduce the surprisingly strong angle-of-incidence dependence of the fluence needed to reach the stationary state, and to clearify the detailed collisional mechanisms being responsible for it.

Sometimes it is not only possible to simulate the implantation phenomena, but also the effect of the experimental procedure: Figure 26 /89/ shows a compositional profile of Ta_2O_5 after He bombardment, as measured by AES and ISS in connection with Ar sputtering. Two simulated curves are shown, one for the depth profile after He bombardment only, and one for the surface composition during subsequent erosion by Ar. Although the conversion of the Ar fluence to the depth scale is subject to an error of ~ 50 %, we find a significant difference both with respect to the characteristic depth and the value at large depth. This confirms that the Ar beam itself induces a significant change of the compositional profile. The present result indicates that results obtained from sputter depth-profiling have to be interpreted with great care in general.

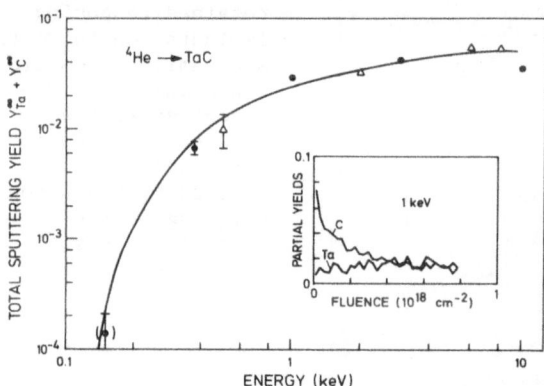

FIGURE 25: Total sputtering yield at large fluence for ^{4}He incident on TaC, from experiment /88/ (open triangles), computer simulation /75/ (full points), and an analytical fit /88/ (line). The insert shows the partial yields as obtained from computer simulation.

Unfortunately, the application of BCA simulation to preferential sputtering studies is often hampered by excessively long computer times (e.g., ~ 2h on a CRAY-XMP for the example given in Fig. 26). However, for certain problems the dynamic calculation can be approximated by a static one. Mass conservation requires that the ratio of the individual sputtered fluxes in a multiatomic compound is equal to ratio of the bulk atomic fractions of the constituents. Accordingly, one might vary the (homogeneous) composition in a static simulation until the correct ratios of sputtered fluxes are achieved. For He bombardment of TaC, the results of the dynamic simulation and the adjusted static one are in good agreement both for absolute sputter yields and the angular distributions of the individual components /75/. The latter is illustrated in Fig. 20.

The computer simulations mentioned before confirm a strong preferential sputtering of O and C from Ta_2O_5 and TaC, respectively, as the ratios of the stationary sputtered fluxes (or of the bulk compositons) is widely different from that of the surface compositions. A further important finding is that the sputtering and the mass transport during He bombardment in these systems is mainly governed by collisional effects. No different process in addition to collisional transport is needed to supply the component, which is preferentially sputtered, to the surface, in contrast to recent suggestions /13,90/.

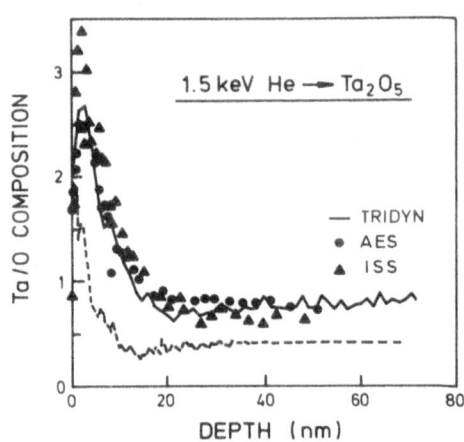

FIGURE 26: Surface composition of Ta_2O_5 after bombardment with 1.5 keV He at normal incidence to the stationary state. The experimental data have been obtained by Ar sputtering (1 keV, 60° with respect to the surface normal) in combination with Auger electron spectroscopy and ion scattering spectrometry. BCA calculations simulated the composition profile before Ar sputtering (dashed line) and the surface composition during Ar sputtering (full line). (Ref. /89/)

4. CONCLUSION

The present lecture was meant to emphasize some points concerning the computer simulations of slowing down and sputtering phenomena:

(i)   Their setup and programming is relatively easy
(ii)   They are able to describe complicated problems and to elucidate detailed mechanisms
(iii)   They represent by no means a 'theory from first principles'
(iv)   The input parameters to them have carefully to be selected and their influence has to be checked.
(v)   They must always be checked with experimental findings.

Thus, the enthusiasm in the field of computer simulation of atomic collisions seems justified, and one might invite more people to take part in it, by improving codes and working on interesting and relevant applications. However, the magical attraction exerted by the computer should never prevent you from asking the right questions, and doing the right experiments, which then you might interpret by means of computer simulation. Computer simulations are as good as the physics you investigate with their help.

The author is indebted to W. Eckstein and M. Hou for stimulating discussion on the present subject. They also made available some of the above material prior to publication.

REFERENCES

1. D.T. Goldmann, D.E. Harrison,Jr., and R.R. Coveyeau, ORNL 2729 (1959)
2. J.B. Gibson, A.N. Goland, M. Milgram, and G.H. Vineyard, Phys. Rev.
    120 (1960) 1229
3. M. Yoshida, J.Phys.Soc.Jpn. 16 (1961) 44
4. J. Lindhard, M. Scharff, and H.E. Schiott, Mat.-Fys.Medd. Kgl.Dan.
    Vid.Selsk. 33 (1963) No. 14
5. P. Sigmund, Phys. Rev. 184 (1969) 383
6. R.A. Langley, J. Bohdansky, W. Eckstein, P. Mioduszewski, J. Roth,
    E. Taglauer, E.W. Thomas, H. Verbeek, and K.L. Wilson: Data Compendium
    for Plasma-Surface Interactions, Nucl.Fus., Special Issue (1984)
7. D.P. Jackson and D.V. Morgan, Contemp. Phys. 14 (1974) 25
8. D.P. Jackson, in: Atomic Collisions in Solids, eds. S. Datz,
    B.R. Appleton, and C.D. Moak (Plenum, New York 1975) p. 185
9. M.T. Robinson, in: Sputtering by Particle Bombardment I,
    ed. R. Behrisch (Springer, Berlin-Heidelberg-New York 1981), p. 74
10. J.R. Beeler, Computer Simulations of Radiation Effects in Metals
    (North-Holland, Amsterdam 1983)
11. D.E. Harrison, Radiat. Eff. 70 (1983) 1
12. H.H. Andersen, Nucl.Instrum.Meth. B18 (1987) 321
13. P. Sigmund, Nucl. Instrum. Meth. B 27 (1987) 1
14. D.E. Harrison,Jr., and M.M. Jakas, Radiat. Eff. 99 (1986) 153
15. C. Erginsoy, G.H. Vineyard, and A. Englert, Phys.Rev. A 133 (1964) 595
16. M.W. Guinan and J.H. Kinney, J.Nucl.Mat. 103/104 (1981) 1319
17. D.E. Harrison,Jr., Ph. Avouris, and R. Walkup, Nucl.Instrum.Meth.
    B 18 (1987) 349
18. D.E. Harrison,Jr., W.L. Gay, and H.M. Effron, J.Math.Phys. 10 (1969)
    1179
19. A. Scholz and C. Lehmann, Phys.Rev. B6 (1972) 813
20. D.P. Jackson, Can.J.Phys. 53 (1975) 1513
21. D. Cherns, M.W. Finnis, and M.D. Matthews, Philos.Mag. 35 (1977) 693
22. D.E. Harrison,Jr., W.L. Moore,Jr., and H.T. Holcombe,
    Radiat. Eff. 17 (1973) 167
23. I.M. Torrens, J. Phys. F3 (1973) 1771
24. J.F. Ziegler, J.P. Biersack, and U. Littmark, Stopping Power and
    Ranges of Ions in Matter, Vol. 1, ed. J.F. Ziegler (Pergamon,
    New York 1985)
25. H.E. Schiøtt, Mat.-Fys.Medd. Kgl.Dan.Vid.Selsk. 35 (1966) No.9
26. J.R. Beeler, Jr., and M.F. Beeler in: Fundamental Aspects of Radiation
    Damage in Metals, Vol. 1, eds. M.T. Robinson and F.W. Young, Jr.,
    (U.S.E.R.D.A. Report CONF-751006, 1975) p. 21.
27. D.P. Jackson, Can.J.Phys. 53 (1975) 1513
28. R. Kelly, Nucl.Instrum.Meth. B18 (1987) 388
29. G. Falcone, Surf.Sci. 179 (1987) 498
30. B.J. Garrison, N. Winograd, D. Lo, T. Tombrello, M.H. Shapiro, and
    D.E. Harrison,Jr., Surf.Sci. 180 (1987) L 129
31. M.T. Robinson and I.M. Torrens, Phys.Rev. B 19 (1974) 5008
32. J.P. Biersack and W. Eckstein, Appl.Phys. A 34 (1984) 73
33. T. Ishitani, R. Shimizu, and K. Murata, Jap.J.Appl.Phys. 11 (1972) 125
34. W. Möller, G. Pospiech, and G. Schrieder, Nucl.Instrum.Meth. 130
    (1975) 265
35. D.P. Jackson, J.Nucl.Mat. 93&94 (1980) 507
36. M. Hautala, Phys.Rev. B 30 (1984) 5010

37. W. Eckstein, private communication
38. J.P. Biersack and L.G. Haggmark, Nucl. Instrum. Meth. 174 (1980) 257
39. M. Hou and M.T. Robinson, Nucl.Instrum.Meth. 132 (1976) 641
40. J.P. Biersack, Nucl. Instr. Meth. B27 (1987) 21
41. P. Gombás, Die Statistische Theorie des Atoms und ihre Anwendungen, Springer-Verlag, Austria (1949)
42. W.D. Wilson, L.G. Haggmark, and J.P. Biersack, Phys.Rev. B 15 (1977) 2458
43. W. Eckstein and M. Hou, private communication
44. J. Lindhard and M. Scharff, Phys.Rev. 124 (1961) 128
45. C. Varelas and J.P. Biersack, Nucl.Instrum.Meth. 79 (1970) 213
46. J.F. Ziegler (ed.): The Stopping and Range of Ions in Matter, Vols. 3-6 (Pergamon Press, New York 1978-1985)
47. O.B. Firsov, Sov.Phys. JETP 9 (1959) 1076
48. O.S. Oen and M.T. Robinson, Nucl.Instrum.Meth. 132 (1976) 647
49. M.T. Robinson, J.Appl.Phys. 54 (1983) 2650
50. J. Lindhard and M. Scharff, Mat.Fys.Medd.Kgl.Dan.Vid.Selsk. 27 (1953) No. 15
51. N. Bohr, Mat.Fys.Medd. Kgl.Dan.Vid.Selsk. 18 (1948) No. 8
52. J.H. Barrett, Phys.Rev. B3 (1971) 1527
53. S.T. Kang, R. Shimizu, and T. Okutani, Jap.J.Appl.Phys. 18 (1979) 1717
54. M.L. Roush, T.D. Andreadis, and O.F. Goktepe, Radiat. Eff. 55 (1981) 119
55. M.L. Roush, F. Davarya, O.F. Goktepe, and T.D. Andreadis, Nucl.Instrum.Meth. 209/210 (1983) 67
56. O.F. Goktepe, T.D. Andreadis, M. Rosen, G.P. Mueller, and M.L. Roush, Nucl.Instrum.Meth. B13 (1986) 434
57. N. Matsunami, Y. Yamamura, Y. Itikawa, N. Itoh, Y. Kazumata, S. Miyagawa, K. Morita, and R. Shimizu, At. Data Nucl. Data Tables 31 (1984) 1
58. H.J. Kang, E. Kawatoh, and R. Shimizu, Jap.J.Appl.Phys. 24 (1985) 1404
59. M. Hautala, Radiat. Eff. 51 (1980) 35
60. W. Möller and W. Eckstein, Nucl.Instr.Meth. B 2 (1984) 814
61. W. Möller and W. Eckstein, Nucl.Instr.Meth. B 7/8 (1985) 645
62. P.S. Chou and N.M. Ghoniem, J.Nucl.Mat. 141-143 (1986) 216
63. A.M. Mazzone, Appl.Phys. A42 (1987) 193
64. W. Eckstein and J.P. Biersack, Appl.Phys. A 37 (1985) 95
65. B. Baretzky, W. Möller, and E. Taglauer, Nucl.Instr.Meth. B 18 (1987) 496
66. W. Möller, Nucl.Instrum.Meth. B 15 (1986) 688
67. A. Anttila, M. Bister, A. Fontell, and K.B. Winterbon, Radiat. Eff. 33 (1977) 13
68. W. Eckstein, H. Verbeek, and J.P. Biersack, J.Appl.Phys. 51 (1980) 1194
69. N.E.B. Cowern and J.P. Biersack, Nucl.Instrum.Meth. 205 (1983) 347
70. M.T. Robinson and O.S. Oen, Appl.Phys.Lett. 2 (1963) 30
71. W. Möller and W. Eckstein, Nucl.Instr.Meth. 194 (1982) 121
72. P. Børgesen, J. Bøttiger and W. Möller, J.Appl.Phys. 49 (1978) 4401
73. H. Ryssel, J. Lorenz and W. Krüger, Nucl.Instrum.Meth. B 19/20 (1987) 45
74. W. Möller, unpublished results
75. W. Eckstein and W. Möller, Nucl.Instrum.Meth. B 7/8 (1985) 727
76. N. Andersen and P. Sigmund, Mat.-Fys.Medd.Kgl.Dan.Vid.Selsk. 39 (1974) No. 3
77. M.M. Jakas and D.E. Harrison,Jr., Nucl.Instrum.Meth. B14 (1986) 535

78. R. Becerra-Acevedo, J. Roth, W. Eckstein, and J. Bohdansky,
    Nucl.Instrum.Meth. B2 (1984) 631
79. D.E. Harrison,Jr., and C.B. Delaplain, J.Appl.Phys. 47 (1976) 2252
80. G.K. Wehner, J.Appl.Phys. 26 (1955) 1056
81. R.H. Silsbee, J.Appl.Phys. 28 (1957) 1246
82. C. Lehmann and P. Sigmund, Phys.Stat.Sol. 16 (1966) 507
83. M. Hou and W. Eckstein, Nucl.Instrum.Meth. B13 (1986) 324
84. W. Eckstein, unpublished results
85. F. Schulz and K. Wittmaack, Radiat. Eff. 29 (1976) 31
86. P. Blank, K. Wittmaack, and F. Schulz, Nucl.Instrum.Meth. 132 (1976)
    387
87. K. Wittmaack, Nucl.Instrum.Meth. B 2 (1984) 569
88. J. Roth, J. Bohdansky, and A.P. Martinelli,
    Radiat. Eff. 48 (1980) 213
89. B. Baretzky, W. Möller, and E. Taglauer, unpublished results
90. H.H. Andersen in: Ion Implantation and Beam Processing,
    eds. J.S. Williams and J.M. Poate (Academic Press, Australia 1984).

COMPUTER SIMULATION OF ION-BEAM MIXING OF COBALT ON SILICON

Ivan CHAKAROV and D.S. KARPUZOV

Institute of Electronics, Bulg. Acad. Sci., Blvd. Lenin 72, Sofia 1784, Bulgaria

1. INTRODUCTION

Many theoretical works deal with recoil mixing of thin or thick layers on a substrate [1,2,3,4]. These studies were carried out within the assumption of isotropic angular distribution of the moving low-energy cascade atoms. This is true if $E \gg E_o$, where E is the initial energy of the primary ions and E_o amounts to between 10 and 100 eV [5]. In this case one can expect a maximum recoil mixing efficiency at the depth of maximum energy deposition by the primary ions, i.e. where $F_E (E,\vec{\Omega},\vec{r})$ has its maximum [3,5]. There are also many experimental works which assess this suggestion [6,7,8].

Any deviations from the isotropic distribution of the cascade atoms affect the recoil mixing [9,10,11,12,18], which is altered by consideration of the deposited momentum distribution function $\vec{F_p} (E,\vec{\Omega},\vec{r})$. This function may substantially change the mixing efficiency, especially near the surface where its sign alters, so that sputtering of the surface layer instead of recoil mixing may occur. Such a situation is related for example to dynamic recoil mixing (DRM), a process where a thin layer of constant thickness is bombarded by an energetic ion beam to obtain recoil mixing at the interface between the layer and the substrate [17].

The necessity of accounting for two distribution functions, F_E and $\vec{F_p}$, as well as the low energies and small thicknesses typical for DRM require a sophisticated computer simulation method, which would allow the full kinetics of the process to be taken into consideration.

2. CALCULATIONAL METHOD

For the simulation of recoil mixing we have chosen the binary collision computer program MARLOWE [13,14]. The main assumptions made during the modeling are : a) taking into account the process kinematics only, i.e. a system without chemical driving forces, thermal spikes and radiation-enhanced diffusion, b) the mixing is considered as a linear superposition of results for cascades originated by different primary ions, and c) no composition or structural changes of both the film and the substrate are assumed during the bombardment.

The main advantage of computer simulations of cascade development is that the motion of each target atom is followed if the latter receives, in any collision, a kinetic energy greater than some threshold energy, E_m. This is important in the investigation of phenomena such as cascade mixing, recoil mixing from a thick source, or sputtering. Events such as focusons and simultaneous collisions

R. Kelly and M. Fernanda da Silva (eds.), Materials Modification by High-fluence Ion Beams, 185–190.
© *1989 by Kluwer Academic Publishers.*

which may occur in a crystalline target are also taken into account. The advantages of using the MARLOWE computer code at low energies was recently reviewed by Robinson [14].

The recoil mixing was simulated for a Co film on a Si substrate. A typical run consists of following the motion of 500 incident 10 keV Ar ions and all cascade atoms originated in the target. The information available for the transmission sputtered atoms, interstitials and vacancies created in each cascade is stored and later analyzed. Three basic structures were investigated. First, a thin cobalt film of depth between 5 and 60 Å was bombarded to determine the transmission sputtering coefficients corresponding to these thicknesses. Second, a semi-infinite Co target was irradiated with 10 keV Ar to determine the energy deposition profile. Third, a two-layer structure consisting of a thin Co film on a Si substrate with Co layer thicknesses between 10 and 60 Å was used to examine the recoil mixing efficiency. In all cases the Co film was amorphous and the Si substrate was polycrystalline. A polycrystalline target is modelled by rotating the basic crystal structure at two random angles before the first collision of the ion with the surface. The initiated atomic cascade is then developed as if taking place in a crystalline medium. Similar rotations before each collision are used to simulate amorphous targets [14,19]. Briefly, at every collision, the crystal structure is rotated about a lattice site through a rotation matrix with randomly selected elements. As commented in [19], this preserves the density of the crystal but destroys the directional correlations. Thus, the obtained medium is liquid-like. The model chosen here differs from those of Snyder and Neufeld which assume $E_b=E_c=E_m$ and the model used by Kinchin and Pease where $E_b=0$, $E_c=E_m$ [13 and ref. therein], E_c - cut-off energy, E_b - binding energy. The minimum energy E_m above which the target atom is incorporated in the cascade is 25 eV for Co and 15 eV for Si. The local binding energy, E_b, was always chosen to be 0.5 eV. The surface binding energy $U_{sb}=4.4$ eV was used. The same value was used for the cut-off energy, E_c. The latter determines the minimum energy of the atoms followed in the cascade. An atom or ion with lower energy is considered as stopped in a replacement or interstitial position. If a particle is in an act of escaping then it is followed below the cut-off energy. The dependence of the reflection sputtering yield on the cut-off energy, E_c, used in the computer code is demonstrated in Fig. 1a.

3. RESULTS

The transmission sputtering yield as a function of the Co film thickness is presented in Fig. 1b. Fig. 2 illustrates some distributions derived from the cascade development in Co films.

In Fig. 3a we summarize the results of the computer simulation of recoil mixing of a Co layer of various thickness on a silicon substrate. In addition, the deposited energy profile in a thick Co target is plotted in Fig. 3b. The comparision of the results illustrated in Figs. 1,2 and 3 shows that the recoil mixing efficiency correlates well with the transmission sputtering yield, the deposited momentum distribution, $\vec{F}_p(x)$, and the radiation damage profile or the deposited energy function $F_E(x)$.

Concentration profiles of recoil – transmitted Co atoms through the interface into the silicon substrate and of Si atoms into the Co film

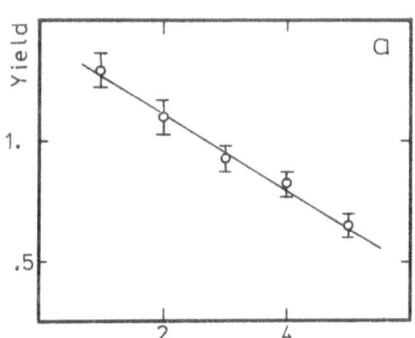 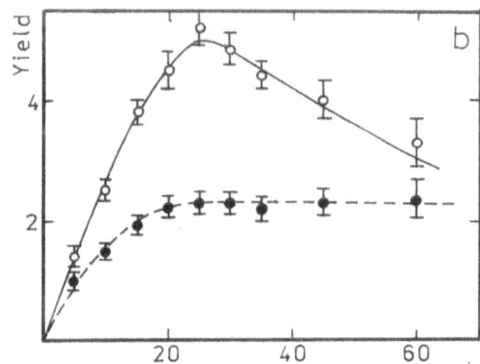

FIGURE 1. a) Reflection sputtering yield of a Co polycrystalline target bombarded with 1 keV Ar ions as a function of the cascade cut-off energy, E_c. The surface binding energy is 4.4 eV. b) Transmission and reflection sputtering yields of a thin Co film bombarded with 10 keV Ar ions versus the thickness of the Co film. $E_c = U_{sb} = 4.4$ eV. $\circ$ = transmission sputtering yield, $\bullet$ = reflection sputtering yield.

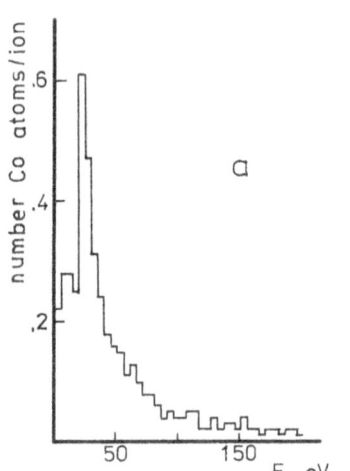 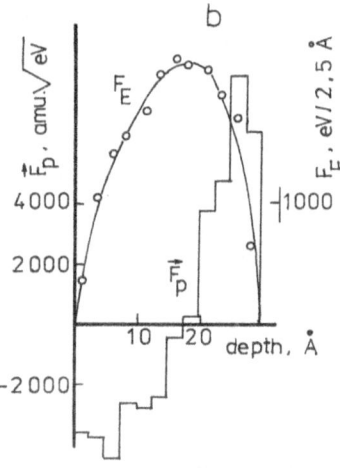

FIGURE 2. a) Energy spectrum of the transmission sputtered Co atoms from 25 Å thick Co film bombarded with 10 keV Ar ions. $E_c = U_{sb} = 4.4$ eV. b) Energy distribution function F_E and momentum distribution function $\vec{F}_p$ in a 30 Å Co film.

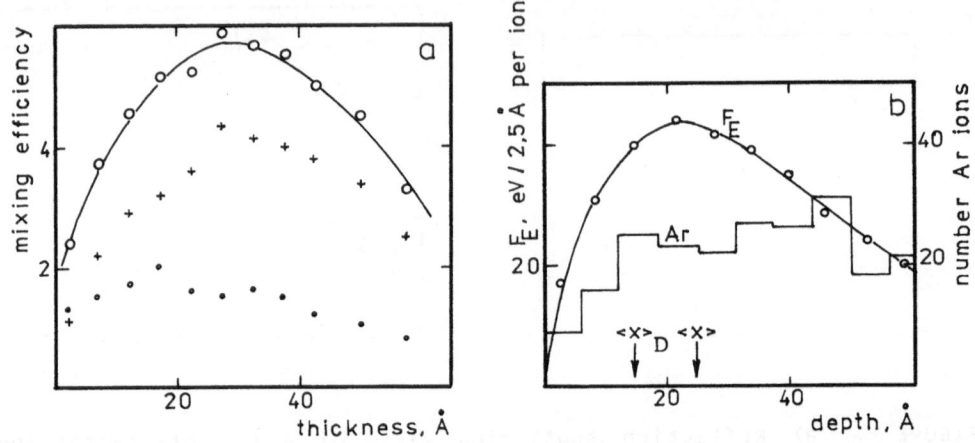

FIGURE 3. a) Mixing efficiency of a Co/Si structure bombarded with 10 keV Ar ions versus the thickness of the Co layer. The mixing efficiency is defined as the total number of atoms transmitted through the interface per ion, ○ (Co atoms into the Si substrate, +, plus Si atoms from the substrate into the Co layer, •). b) 10 keV Ar bombardment of a semi-infinite Co target. The full-drawn line is the deposited energy, $F_E(x)$, and the histogram is for stopped Ar ions. The projected range $<x>$ and damage range $<x>_D$ are taken from [15].

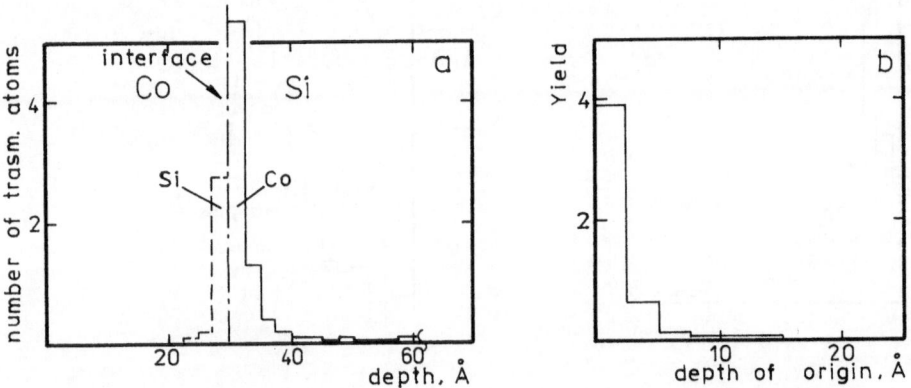

FIGURE 4. a) Depth profiles of recoil transmitted Co and Si atoms. The interface between the Co layer and Si substrate is also shown. b) Depth-of-origin of transmission sputtered Co atoms at the thickness of maximum transmission sputtering yield. The surface binding energy and cut-off energy are both equal to 4.4 eV.

are shown in Fig. 4a. The thickness of the cobalt layer corresponds to the maximum mixing efficiency. Short- and long-range recoil mixing can be distinguished for Co atoms in Si substrate. The FWHM is 5 Å. The cascade-mixed area of Si atoms in the Co film is narrower. A comparision of the recoil mixing with the depth-of-origin of the transmission sputtered Co atoms for this case shows that the layers closest to the interface contribute to the mixing. Our results are not very different from those obtained by Robinson [14] who found that in the case of sputtering of Xe from monocrystalline Au at 700 eV, 99% of the sputtered atoms originated within two atomic layers of the surface and the results of Rosen et al. [20] whose estimation is 88% for 90 keV Cu on a polycrystalline Cu target. It is interesting to point out that yield contributions for different layers (see Table 1) are in very good agreement with the general estimation of Kelly and Oliva [21] that the fraction of reflection sputtered atoms for the first layer is between 0.70 and 0.89. For instance, in our case the first four layers yield 97% of the transmission sputtered cobalt atoms, 80% of which should be attributed to the first layer, 13% to the second and 3% to the third. A presence of a potential barrier at the interface different from that of the binding energies of the host matrices used in these calculations would affect the concentration profiles of recoil-transmitted atoms [4,6], since most of the atoms ejected from the first layer have low energies (their distribution was found to be close to E^{-2}). Generally, a reduction of the barrier increases the transmission sputtering yield and the mixing rate [4,6]. At high fluences one also has to take into account the modification of the target composition near the interface.

TABLE 1. Depth-of-origin of sputtered Co atoms from 32 Å thick polycrystalline Co film bombarded with 10 keV Ar ions. The distance between the atomic layers is 2.5 Å.

Atom layer	1	2	3	all others
Refl. yield (Co/Ar)	2.0	0.25	0.03	0.02
Fraction of emissions	0.87	0.11	0.01	0.01
Transm. yield (Co/Ar)	3.88	0.64	0.17	0.15
Fraction of emissions	0.80	0.13	0.03	0.03

4. CONCLUSIONS

The present calculations illustrate the application of a BC computer model to atomic mixing at low energies and small layer thicknesses. In these circumstances the results are difficult to analyze using theoretical descriptions of F_E and $\vec{F}_p$.

REFERENCES

1. U. Littmark, Nucl. Instr. Meth. B7/8 (1985) 684
2. S. Dzioba and R. Kelly, J. Nucl. Mat. 76&77 (1978) 175

3. P. Sigmund, Phys. Rev. 184 (1969) 383
4. G. Fischer, G. Carter and R. Webb, Rad. Eff. 18 (1978) 41
5. P. Sigmund, Rev. Roum. Phys. 17 (1972) 823/969/1079
6. A. Grob, J.J. Grob, N. Nesli, D. Salles and P. Siffert, Nucl. Instr. Meth. 182/183 (1981) 85
7. R. Grotzschel, R. Klabes, U. Kreisig and A. Schmidt, Rad. Eff. 36 (1978) 129
8. M. Bruel, M. Floccari and J.P. Gailliard, Nucl. Instr. Meth.182/183 (1981) 93
9. U. Littmark and P. Sigmund, J. Phys. D: Appl. Phys. 8, (1975) 241
10. H.E. Roosendaal, U. Littmark and J.B. Sanders, Phys. Rev. B26 (1982) 5261
11. P. Sigmund in: Sputtering by Particle Bombardment I, ed. R. Behrish (Springer, New York, 1981)
12. U. Littmark and W.O. Hofer, Nucl. Instr. Meth. 168 (1980) 329
13. M.T. Robinson and I.M. Torrens, Phys. Rev. B9 (1974) 5008
14. M.T. Robinson in: Sputtering by Particle Bombardment I, ed. R. Behrisch (Springer, New York, 1981)
15. K.B. Winterbon, Ion Implantation Range and Energy Deposition Distributions, Vol. 2, IFI/Plenum, N.Y. 1975
16. Cui Fu-Zhai and Li Heng-De, Nucl. Instr. Meth. B7/8 (1985) 650
17. P. Argyrokastritis, D.S. Karpuzov, J.S. Colligon, A.E. Hill and H. Kheyrandish, Phil. Mag. A49 (1984) 547
18. H.J. Whitlow and M. Hautala, Nucl. Instr. Meth. B18 (1987) 370
19. M. Hou, M.T. Robinson, Nucl. Instr. Meth. 132 (1976) 641
20. M. Rosen, G.P. Mueller and W.A. Fraser, Nucl. Instr. Meth. 209/210 (1983) 63
21. R. Kelly and A. Oliva, Nucl. Instr. Meth. B13 (1986) 283

ON THE FRACTAL NATURE OF COLLISION CASCADES

YANG-TSE CHENG

Physical Chemistry Department, General Motors Research Laboratories, Warren, Michigan 48090-9055, U.S.A.

1. INTRODUCTION

Collision cascades induced by energetic ions are the origin of a variety of radiation effects in solids, such as those discussed in this Summer School. In this paper, a novel fractal geometry viewpoint of collision cascades is presented. The discussion is focused on the evolution of cascades to spikes.

Over the past several years, the fractal geometry, introduced by Mandelbrot,[1] has had a strong impact on a series of problems ranging from coastline geometry to disordered materials, to galaxy formation.[1-4] Recently, the fractal geometry has also found its application in collision cascades,[5-8] although it was recognized several years ago that fractals might be of interest to atomic collisions in solids.[9]

In section 2, the basics of fractal geometry are introduced. In section 3, the link between the fractal geometry and collision cascades is established by an "idealized cascade" model and the Winterbon-Sigmund-Sanders theory of cascades.[10] In section 4, the spike condition is formulated based on the concept of "space-filling" fractals.[1] In addition, the influence of the atomic number (Z) on spike formation and the spike energy density is discussed. Finally, the effect of many-body collisions on spike formation is shown. Section 5 summarizes the major results of the paper.

2. FRACTAL GEOMETRY
2.1. Self-similar exact fractals

A variety of shapes contains a self-similarity. Examples of shapes with self-similarity include intervals in a line, rectangles in a rectangular plane, and the like. In each case, the "whole" may be "paved" by N "parts." Each part is identical to the whole when all the length scales of the whole are reduced by a factor γ, $0 < \gamma < 1$. This ratio, γ, is called the similarity ratio. If a line segment is taken as a whole, it can be paved by N smaller line segments, each of which is obtained from the whole line segment by a similarity ratio $\gamma(N) = 1/N^1$, because the Euclidean dimension of a straight line is 1. Likewise, since the Euclidean dimension of a plane is 2, the "whole" rectangle $0 \leq x < X; 0 \leq y < Y$ can be "paved" exactly by $N = b^2$ parts (b is an integer). These parts are rectangles defined by $(k - 1)X/b \leq x < kX/b$ and $(h - 1)Y/b \leq y < hY/b$, where k and h equal 1, 2,..., b. Here, each part can now be deduced from the whole by a similarity ratio $\gamma(N) = 1/N^{1/2}$.

In general, for spaces with Euclidean dimension d, a "whole" polyhedron of dimension $D \leq d$ may be "paved" by N "parts." Each part is obtained by a similarity ratio $\gamma(N) = 1/N^{1/D}$, or equivalently,

$$D = - \ln N / \ln \gamma(N). \tag{1}$$

The exponent D in Eq. (1) is the self-similarity dimension of the corresponding self-similar geometric object. D is not restricted to integers. In order for the self-similarity

R. Kelly and M. Fernanda da Silva (eds.), Materials Modification by High-fluence Ion Beams, 191–203.
© *1989 by Kluwer Academic Publishers.*

dimension to be meaningful, the requirement is that the shape be self-similar, i.e., that the whole may be split up into N parts and that each part is obtainable from the whole by a similarity of ratio γ. The word fractal was coined by Mandelbrot as a generic name for objects which possess fractional self-similar dimensions, or fractal dimensions. As an example, we consider a triadic Koch curve (see Fig. 1). We assume that initially we have a straight line of length 1, to be called the *initiator* (Fig. 1a). We divide it into 3 pieces each of which is 1/3 of the original length, and we replace the middle third by an equilateral triangle. The resulting curve is a line formed of four intervals of equal lengths, to be called the *generator* (Fig. 1b). We can further replace each of the generator's four intervals by a generator whose size is reduced by a factor of one-third. The resulting curve is shown in Fig. 1c. Proceeding in this fashion, we see that each straight line can be decomposed into reduced-size pieces. We have $N = 4$ and $\gamma = 1/3$, resulting in an increasingly broken curve of a self-similarity dimension, $D = \ln N / \ln(1/\gamma) \sim 1.2618$. Fractals such as the Koch curve are exact fractals, because they are uniquely determined once the initiator and the generator are specified.

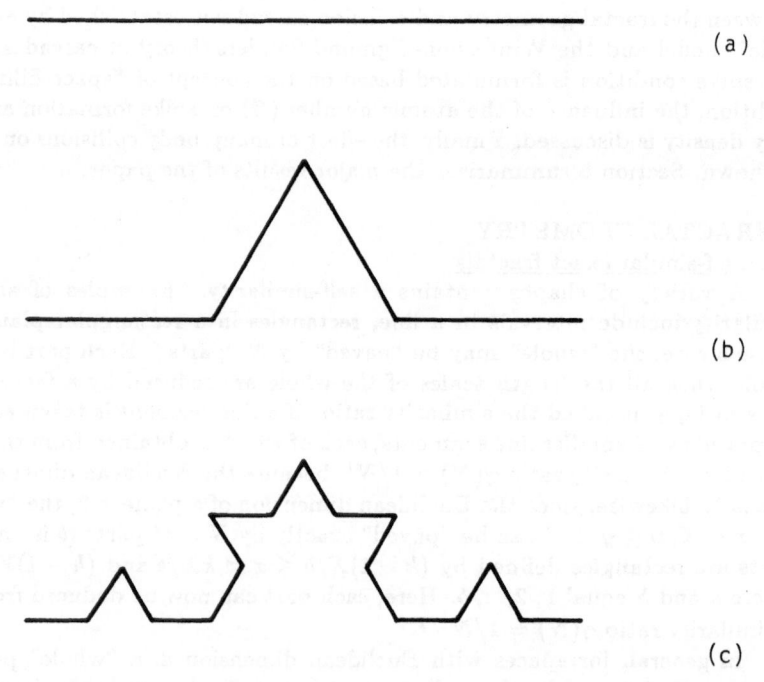

(a)

(b)

(c)

Fig. 1 The first three steps in constructing the triadic Koch curve ($D = \ln 4 / \ln 3 \cong 1.26$).

Exact fractals that are most relevant to collision cascades are perhaps the fractal trees illustrated in Mandelbrot's book.[1] Examples of fractal trees are shown in Fig. 2. The fractal trees have infinitely thin stems and the same angle θ between the branches at every branching point. Each tree is a self-similar geometric object because it is constructed in such a way that the length ratio $0 < \gamma < 1$ between the successive branches is fixed throughout. Since $N = 2$, the fractal dimension D is equal to $\ln 2 / \ln(1/\gamma)$, according to Eq. (1). In Fig. 2, D has values between 1 and 2.

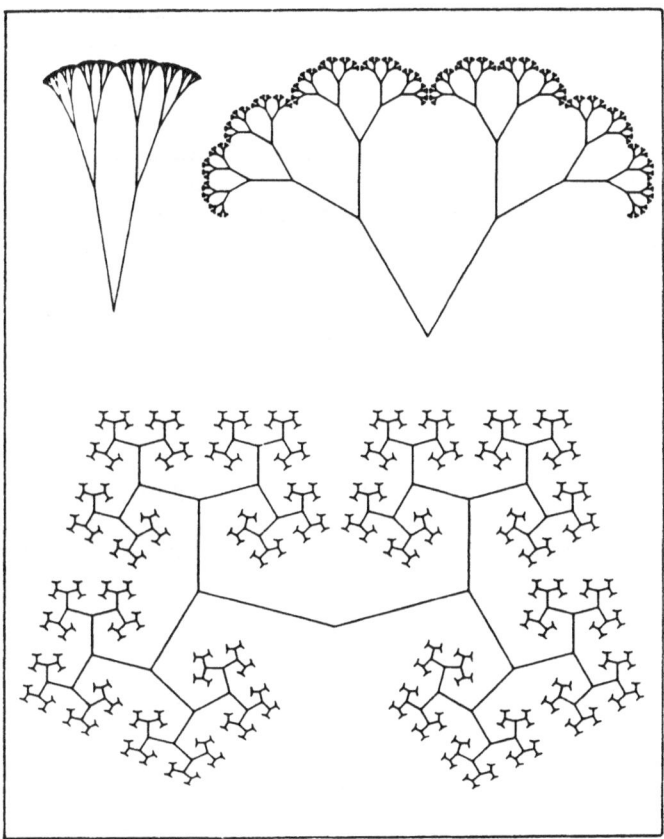

Fig. 2 Examples of trees from Mandelbrot's book, *The Fractal Geometry of Nature.* [1] The fractal dimensionality, measured by the present author, is about 1.06, 1.17, and 1.63, respectively, in the clockwise direction beginning from the upper left corner (Adapted from Ref. (1)).

2.2. Random fractals

Certain average quantities of random geometric patterns contain self-similarity as well. They are the random fractals. Examples are coastlines,[1] distribution of galaxies,[1] and vascular networks in the heart.[4] For a random fractal, it is convenient to think of the fractal dimensionality in terms of the scaling relationship between the number of objects, N, contained within a distance R, which is measured from an origin on the fractal

$$N \propto R^D. \tag{2}$$

The value of N can be measured, for example, by the mass or the number of occupied sites. For a random fractal, D is obtained from the dependence of N on R averaged over a large number of ensembles. For statistically self-similar fractals, D, obtained from Eq. (2), is equivalent to that obtained from the "all purpose" fractal dimensionality definition Eq. (1).[3,11]

2.3. "Space-filling" fractals

In many cases, the fractal dimension is smaller than the dimension (d) of the Euclidean space in which the fractal is embedded. Fractals of this type occupy a negligible volume. However, in some cases the fractal dimension is equal to the dimension of the embedding space $(D = d)$ and the fractal occupies a non-negligible fraction of the space. A fractal becomes a "space-filling" fractal when $D = d$. The effect of D approaching d is illustrated in Fig. 2 by these fractal trees. Mandelbrot described these trees as follows,[1] "For D barely larger than 1 (top left), the result is whisk-like, then broomlike. As D increases, the branches open up, and the outline or 'canopy' extends into folds hidden from the sunshine...." For larger D (bottom), "A Frenchman is reminded of the fortifications by Vauban. The values $D = 2$ and $\theta = \pi$ yields a plane-filling tree...." (Ref. 1, p. 154).

The space-filling condition may also be shown by using Eq. (2). Let N_D and N_d be the number of objects included in a radius R that are on the fractal, and that are in the embedding space, respectively. We consider the ratio $\eta = N_D/N_d$ as R approaches infinity

$$\eta = \lim_{R \to \infty} R^{D-d} = \begin{cases} \infty, & \text{if } D > d; \\ 0, & \text{if } D < d. \end{cases} \tag{3}$$

Thus, $D = d$ is the critical dimension for a space-filling fractal.

3. FRACTAL DIMENSION OF COLLISION CASCADES

3.1. Idealized cascade approach

In the idealized cascade approach a connection between a fractal tree and a collision cascade is established, if (1) each branching point of a tree represents a binary collision event and (2) each branch corresponds to a mean free path of a displaced atom. To construct an idealized collision cascade, we begin with the hard-sphere collision approximation in conjunction with an inverse-power potential $V(r) \propto r^{-1/m}$, where r is the internuclear distance and $0 < m \leq 1$. The inverse-power potential has been extensively used in studying collisional problems involving ion energies from several hundred keV to a fraction of an eV.[12,13,14] For a system of identical particles the radius of the particles is given as half of the distance of closest possible approach.

Using the differential and total cross section for hard-sphere collisions, we obtain the mean recoil energy $\overline{E_n}$ at the nth generation of collisions, which is given by[13]

$$\overline{E_n} = \frac{1}{2}\overline{E_{n-1}}. \tag{4}$$

and the average scattering angles become $\theta_1 = \theta_2 = \frac{\pi}{4}$.

The mean free path of the nth generation is given by $\lambda_n = 1(/N\sigma_n)$, where N is the atomic density. For inverse-power potentials, $V(r) = Gr^{-1/m}$, with G a constant which depends on m, the total effective cross-section σ_n equals $\pi(2G/\overline{E_n})^{2m}$. We obtain

$$\lambda_n = (\frac{1}{2})^{2m}\lambda_{n-1}. \tag{5}$$

The mean free path ratio, $\gamma = \lambda_n/\lambda_{n-1} = 2^{-2m}$, is independent of n, the generation of the cascade. The collision cascade, generated by $\gamma = 2^{-2m}$, is therefore self-similar and has a fractal dimension

$$D = \frac{\ln 2}{\ln 1/\gamma} = \frac{1}{2m}. \tag{6}$$

The fractal dimensionality of a cascade depends *explicitly* on the form of the interatomic potential. Idealized collision cascades with $D = 1$ and $D = 2$ ("plane-filling") are shown in Fig. 3. The overlapping of cascades is inevitable for $D = 2$ case (Fig. 3b).

3.2. WSS theory approach

The fractal dimensionality of collision cascades can also be obtained from the Winter-bon-Sigmund-Sanders (WSS) theory.[10] From the WSS theory, which is formulated in 3-dimensional physical space, the characteristic length scale, e.g., ion range, R, of a collision cascade scales with the incident ion energy E_0 and is given by[10]

$$R = \frac{E_0^{2m}}{NC_m}, \tag{7}$$

where C_m is a constant for fixed m. The number of displaced atoms N_c is obtained from a Kinchin-Pease type of argument[15,16,17]

$$N_c = \xi\frac{E_0}{E_d}, \tag{8}$$

where E_d is the threshold displacement energy and ξ is a constant between 0.40 and 0.65, which depends on the details of the models. From these two equations, we obtain

$$N_c \propto R^{\frac{1}{2m}}. \tag{9}$$

Comparing Eq. (9) with Eq. (2), we conclude that Eq. (9) is another manifestation of the fractal nature of the underlying collision cascade with the same fractal dimension $D = 1/(2m)$ obtained from the idealized cascade approach. This second derivation of Eq. (6) does not rely on the hard-sphere collision approximation.

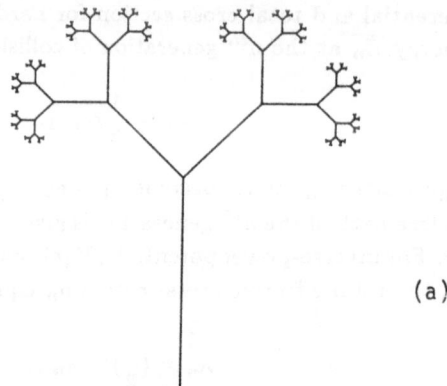

(a)

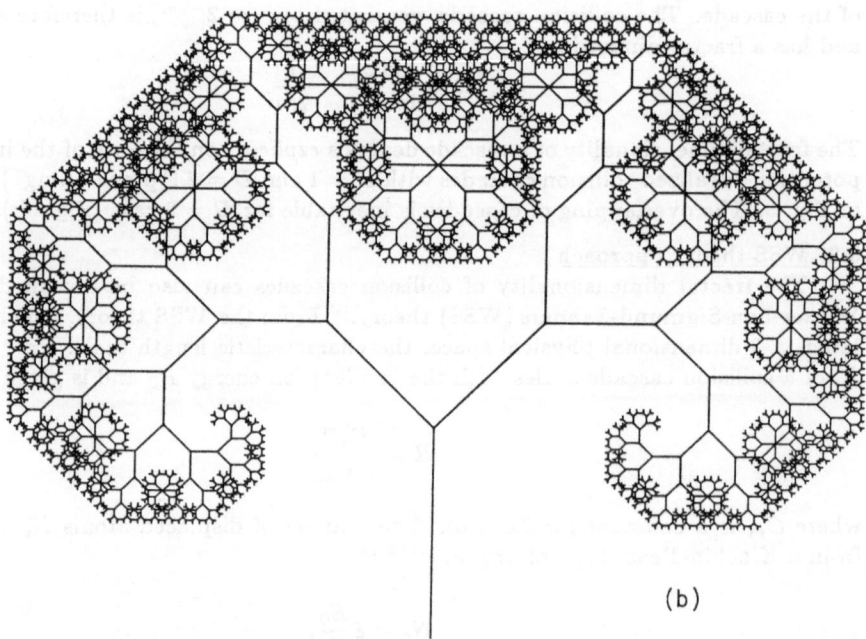

(b)

Fig. 3 Idealized collision cascades with (a) $\gamma = \frac{1}{2}$ so that $D = 1$, which corresponds to $V(r) \propto r^{-2}$, and (b) $\gamma = (\frac{1}{2})^{1/2}$ so that $D = 2$, which corresponds to $V(r) \propto r^{-4}$. The cascades must be terminated in the two cases when the mean free path approaches the same predetermined shortest length scale.

3.3. Cascades and non-uniform fractals

As an actual collision cascade evolves in time, the characteristic kinetic energy of atoms or ions changes from the primary incident ion energy typically exceeding 1 keV to the eV energy range. At the same time, inverse-power potentials $V(r) \propto r^{-1/m}$ with decreasing values of m must be used.[12,13,14] For example, it is known that over a major portion of the keV range and for medium to heavy-mass ions and atoms, the inverse-square potential (i.e., $m = 1/2$) is a fair approximation, while in the lower-keV and the upper-eV region, $m = 1/3$ should be adequate. In the eV region, m may be taken close to zero. An actual cascade is, therefore, not a fractal in the sense that has been given so far, but is a *nonuniform* fractal[1] with a variable dimensionality that increases from its initial dimensionality 1 (i.e., $m = 1/2$). *In a collision cascade, the fractal dimensionality increases as the cascade evolves because of the change of the interaction potential.* Non-uniform fractal trees with increasing and decreasing fractal dimensionalities are compared in Fig. 4.

4. SPIKE CONDITIONS

4.1. Space-filling fractals - a necessary condition

The "spike" is, as has been frequently stated, "a *limited volume* with the *majority* of atoms temporarily in motion."[18] This spike condition is far from quantitative because the terms "limited volume" and "majority" cannot be uniquely defined for a tree-like structure (see Fig. 2). Thus, the density of moving particles cannot be defined without further assumptions. To circumvent these difficulties, we propose[5] an alternative approach to the spike condition by applying the concept of "space-filling" which was introduced in the previous section: A fractal is "space-filling" when its fractal dimension D equals the physical dimension d of the space in which it is embedded.

For cascades in 3-dimensional space, a space-filling cascade, or a spike, can only occur when the fractal dimension of a collision cascade D exceeds 3. This implies, according to Eq. (6), *a spike can only occur when the interaction potential is characterized by an inverse-power potential with $0 < m \leq 1/6$*. In reality, such a potential corresponds to a particle kinetic energy in the hundreds eV ("upper-eV") to eV range.[12,14] Thus, when a spike occurs, the characteristic kinetic energy per particle must be in the "upper-eV" to eV range (further discussions in section 4.3).

4.2. Local spike formation

From the discussions in the previous sections, it is clear that a critical kinetic energy E_c at which the inverse-power potential takes the value $m = 1/6$ plays an important role in spike formation – *a particle will initiate a space-filling cascade, or a spike, only when its kinetic energy is smaller than E_c*. The critical energy E_c depends on the atomic species and can be estimated, in principle, from any one of the more realistic interaction potentials (to be further discussed). Furthermore, E_c serves to illustrate the following conceptual picture of spike formation.

As the initial kinetic energy of the incident ion E_0, which usually exceeds 1 keV and much greater than E_c, is shared by the displaced particles, each displaced particle will generate a "space-filling" sub-cascade, or a local spike, if the displaced particle has a kinetic energy just below E_c. A global spike will form if local spikes overlap. It is important to note that the formation of a local spike is determined by E_c, not

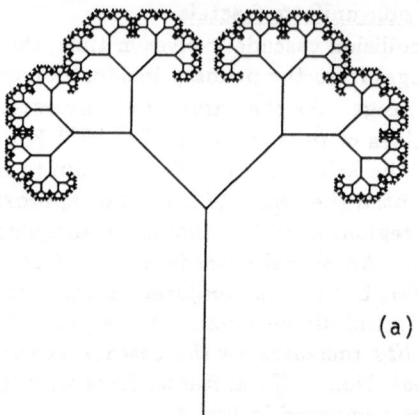

(a)

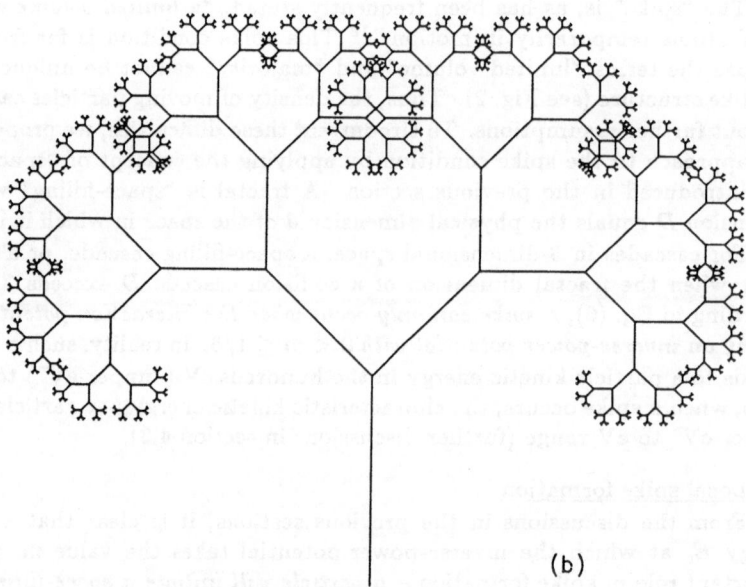

(b)

Fig. 4. Non-uniform fractal trees. (a) D increases linearly with generation from $D = 1$ to 2, and (b) D decreases linearly with generation from $D = 2$ to 1.

the incident ion energy E_0. Increasing E_0, however, can increase the number of local spikes, since the number of local spikes is roughly E_0/E_c.

The energy density of a local spike is considered next. A popular reference to the spike energy density calculation has been that of Sigmund.[18] His approach is based on the well established WSS theory for the energy-deposition profiles. In particular, he calculated the "effective maximum energy density $\theta_o(E_0)$ (energy per atom),"

$$\theta_o(E_0) = \frac{E_0}{(2\pi)^{3/2} N \alpha \beta^2}. \tag{10}$$

The projectile moves initially in the x-direction with an energy E_0. N is the number of target atoms per unit volume, α^2 is $\delta_x < \Delta x^2 >$, and β^2 is $\delta_y < y^2 >$. The factors δ_x and δ_y account for the difference between the cumulative energy profile of many cascades and a single typical one. By assuming elastic scattering with an inverse-power potential, Sigmund obtained[18]

$$\theta_o(E_0) = \begin{cases} G(1/2)N^2/E_0^2 & \text{for } m = 1/2, \\ G(1/3)N^2/E_0 & \text{for } m = 1/3. \end{cases} \tag{11}$$

The values of $G(1/2)$ and $G(1/3)$ are strongly dependent on the masses of projectile and target. Furthermore, Sigmund's analysis shows that θ_o decreases with increasing ion energy E_0.

From Eq. (10) we see that Sigmund's calculation of θ_o is in effect based on the assumption that the initial energy E_0 is shared through collision cascades by every atom in the cascade volume defined by α and β. But for non-space-filling cascades, as is the case of $m = 1/2$ or $1/3$, this assumption is invalid. However, Eq. (10) should apply to estimate the energy density of space-filling cascades. Using results from the WSS theory,[10] it is easily shown that for a space-filling cascade or sub-cascade initiated by a particle with energy $E \leq E_c$ ($0 < m \leq 1/6$),

$$\theta_o(E) = G(m)N^2 E^{1-6m}, \tag{12}$$

where $G(m)$ depends on m and the masses of projectile and target atoms. We see that the effective energy density θ_o for space-filling cascades ($0 < m \leq 1/6$) increases with increasing E. This is another manifestation of a space-filling cascade.

The mass dependence of $G(m)$ is considered next. Using the Thomas-Fermi potential, the mass dependence, for the case of equal masses of projectile and target, can be expressed as,

$$G(m) \propto Z^{2(7m-1)}. \tag{13}$$

For example, the exponent equals 5, 8/3, 1/3, -2 for $m = 1/2, 1/3, 1/6$, and 0, respectively. Sigmund's spike criteria show, a strong dependence of θ_o on the mass (or Z), because an inverse-power potential with $m = 1/2$ or $1/3$ was used.[18] The present analysis demonstrates a very weak mass-dependence of θ_o, when the cascade dimensionality reaches the critical dimension $D = 3$ or $m = 1/6$. Furthermore, the present analysis predicts an anomaly – local spike energy density can be *higher* for low Z materials than for the high Z materials, when $m \to 0$. For example, θ_o is proportional to Z^{-2} when $m = 0$.

4.3. Numerical examples of E_c

We want to calculate E_c the critical energy at which the inverse-power potential takes the exponent $m = 1/6$. The question of how to define an appropriate value of the exponent m in a given situation has been dealt with by a number of authors. Bohr[19] determined power potentials by matching in value and slope to a given screened Coulomb potential, thus defining an exponent that varied continuously as a function of interatomic distance. Marwick and Sigmund,[20] Vukanic and Sigmund[21] determined m by matching in value and slope the cross sections for the inverse-power potential and that for the Thomas-Fermi potential. Recently, Ziegler, Biersack, and Littmark[22] have extended the latter matching procedure to their "Universal Interatomic Potentials" and they obtained an interpolation expression for m:

$$m(x) = 1 - \exp\left[-\exp\sum_{i=1}^{6} a_i (0.1 \ln x/x_1)^i\right], \tag{14}$$

where $a_i = -2.432, -0.1509, 2.646, -2.742, 1.215, -0.1665$, $x_1 = 10^{-9}$, and $x = t^{1/2}$. Here t, according to Lindhard, Nielsen, and Scharff,[12] is proportional to the energy transferred T_n in the nth generation collision and to the particle energy E_n which initiates the nth generation collision cascade:

$$t = T_n E_n \left(\frac{M_2}{M_1}\right)\left(\frac{\alpha}{2Z_1 Z_2 e^2}\right)^2, \tag{15}$$

where $\alpha = 0.8854\alpha_o/(Z_1^{0.23} + Z_2^{0.23})$ is the "Universal Screening Length"[22] and α_o is the Bohr radius.

We determine the condition for $m = 1/6$ by solving Eq. (14) numerically and obtain $x_c = 0.903 \times 10^{-3}$. We then determine E_c for the case $M_1 = M_2$ by using Eq. (15) and the fact that on the average $T_n = (1/2)E_n$ (see Eq. (4)). We obtain,

$$E_c = \frac{2^{3/2}e^2 x_c}{\alpha} Z^2 = 3.923 \times 10^{-2} Z^{2.23} (eV). \tag{16}$$

The dependence of E_c on the atomic number of the elements Z is illustrated in Fig. 5. Note $E_c = 14, 31, 240, 670$ eV for $Z = 14, 20, 50, 80$, respectively.

We observe that E_c decreases as Z decreases. For light elements, E_c calculated according to Eq. (16) has a value smaller than the typical displacement energy E_d. In such cases cascades stop before reaching the condition $m = 1/6$. According to the necessary condition of spike formation discussed earlier, spikes cannot form in those light-element matrices. Since E_d is typically about 25 to 35 eV, we see *spikes cannot form in a matrix consisting of elements with $Z < 20$*. For example, spikes are not expected for cascades in C or Si matrix. In addition, E_c sets an upper-bound for the spike energy density, since the average energy per atom in a spike cannot exceeds E_c, which is, according to Eq. (16), in the "upper-eV" to eV range.

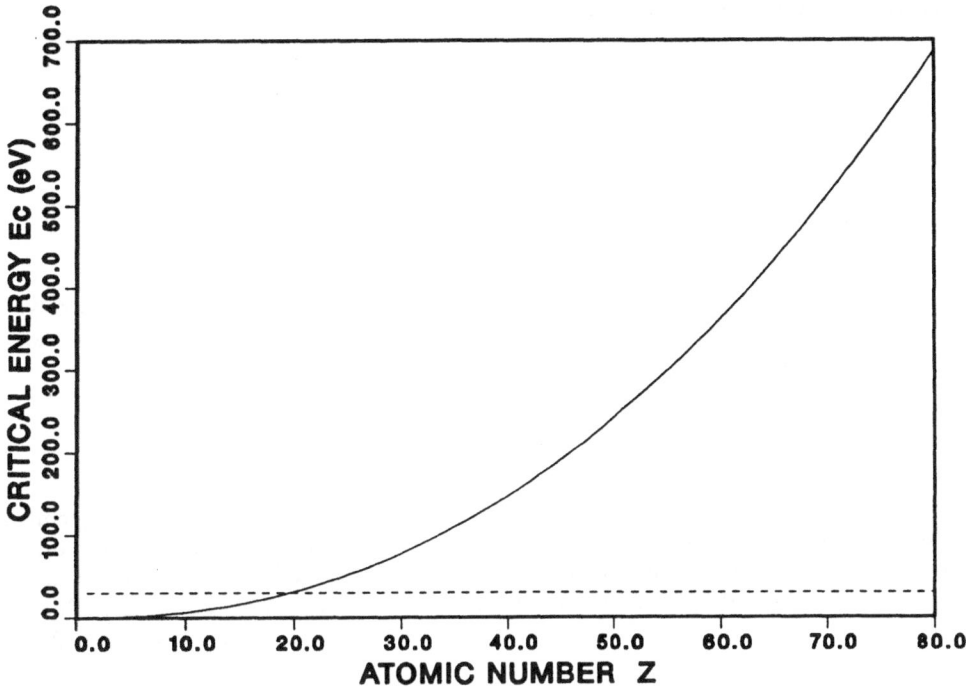

Fig. 5. The critical energy E_c as a function of the atomic number of elements Z (solid line). Spikes are not expected if E_c is lower than the typical displacement energy $E_d = 30$ eV (dash line).

4.4. Universality hypothesis and the effect of many-body collisions on spikes[7]

From the previous discussions, we observe that the "idealized cascade" and the stochastic cascade have the same fractal dimensionality $D = 1/(2m)$ despite their obvious differences. Therefore, we say that they belong to the same "universality class." Members of the same universality class have the same fractal dimensionality. We now consider M-body ($M \geq 2$) collision cascades. Does an idealized M-body collision cascade have the same fractal dimensionality as a stochastic M-body collision cascade? We know that they have the same fractal dimensionality for $M = 2$. Now we make a "universality hypothesis" that they have the same fractal dimensionality for all $M \geq 2$.

To study the effect of many-body collisions on spike formation, we construct a fractal tree which has M branches at each branch point. We consider M to be 3 or 4, because the probability of M-body collisions with M greater than 4 is very small. At each branching point, we have a M-body collision event and the mean energy transfer, using a symmetry argument, is simply

$$\overline{E_n} = \frac{1}{M}\overline{E_{n-1}}.$$ (17)

The iteration relation for the mean free path, Eq. (5), becomes[23]

$$\lambda_n = (\frac{1}{M})^{2m}\lambda_{n-1}.$$ (18)

In particular, Eqs. (17) and (18) become Eqs. (4) and (5) for $M = 2$, respectively. The mean free path ratio, $\gamma = M^{-2m}$, is independent of n. The fractal dimensionality of this idealized cascade is therefore

$$D = \frac{\ln M}{\ln(1/\gamma)} = \frac{1}{2m}.$$ (19)

Equation (19) is the same as the binary collision case (see Eq. (6)); many-body collisions do not have any effect on the fractal dimensionality in the idealized cascade model. By using the "universality hypothesis," we conclude that M-body collisions do not have any effect on the fractal dimensionality of stochastic collision cascades either. Therefore, the condition of spike formation, $D = 3$ or $m = 1/6$, remains valid when M-body collisions are included.

5. SUMMARY

In summary, we have shown that a collision cascade governed by the inverse-power potential $V(r) \propto r^{-1/m}$ $(0 < m \leq 1)$ is a fractal with a fractal dimension $D = 1/(2m)$. In an actual collision cascade, the fractal dimensionality increases as the cascade evolves because of the decrease in m as the kinetic energy of the atoms decreases. In 3-dimensional space, a "space-filling" collision cascade, or a spike, only occurs when $0 < m \leq 1/6$. These results are independent of the "binary collision" approximation.

The incident ion energy E_0 does not have a direct influence on the local spike formation, when E_0 is much greater than the critical energy E_c. Spikes cannot occur if $E_c < E_d$, as is the case for matrices consisting of single elements with $Z < 20$. The spike energy density increases as E_c increases. The spike energy density has only a weak mass dependence. It should not be a surprise that the energy density of a local spike is higher in some low Z materials than in high Z materials.

ACKNOWLEDGMENTS

I am grateful for the support of Professors W. L. Johnson and M-A. Nicolet of the California Institute of Technology, for the helpful discussions with Dr. M. Van Rossum of the Interuniversitair Micro-Elektronica Centrum (Belgium) and Dr. F. J. Rossi of the Center for Materials Science, Los Alamos National Laboratory, for the comments and suggestions from Drs. P. Sigmund, A. Gras-Marti, J. Biersack, U. Littmark, R. Kelly, and T. D. de la Rubia at the 1987 NATO Advanced Study Institute on Materials Modification by High-Fluence Ion Beams, for the discussions with Drs. F. W. Saris and

J. B. Sanders during my short visit to the FOM-Institute for Atomic and Molecular Physics (The Netherlands), and for the careful reading of the manuscript by Dr. D. N. Belton, A. A. Dow, Dr. G. B. Fisher, N. L. Lee, Dr. R. Richter, and Dr. Kathleen C. Taylor of the General Motors Research Laboratories.

REFERENCES

1. B. B. Mandelbrot, *The Fractal Geometry of Nature* (W. H. Freeman and Company, New York, 1983).
2. S. H. Liu, Solid State Physics, Vol. 39, 207 (1986).
3. *On Growth and Form*, edited by H. E. Stanley and N. Ostrowsky (Martinus Nijhoff, Boston, 1986).
4. B. J. West and A. L. Goldberger, American Scientist **75**, 354 (1987).
5. Y.-T. Cheng, M-A. Nicolet, and W. L. Johnson, Phys. Rev. Lett. **58**, 2083 (1987).
6. K. B. Winterbon, H. M. Urbassek, P. Sigmund, and A. Gras-Marti, Phys. Scripta (in press).
7. Y.-T. Cheng, Mat. Res. Soc. Conf. Proc. 1987, Boston (in press).
8. F. Rossi, M. Nastassi, and D. M. Parkin, Mat. Res. Soc. Conf. Proc. 1987, Boston (in press).
9. See, e.g., K. B. Winterbon, Rad. Effects, **60**, 199 (1982).
10. K. B. Winterbon, P. Sigmund, and J. B. Sanders, Mat. Fys. Medd. Dan. Vid. Selsk. **37**, 1 (1970).
11. B. B. Mandelbrot, in *Fractals in Physics*, Eds. L. Pietronero and E. Tosatti (North Holland, Amsterdam, 1986).
12. J. Lindhard, V. Nielsen, and M. Scharff, Kgl. Danske Videnskab. Selskab Mat.-Fys. Medd. **36**, 10 (1968).
13. M. W. Thompson, *Defects and Radiation Damage in Metals* (Cambridge University Press, London, New York, 1969).
14. P. Sigmund, Phys. Rev. **184**, 383 (1969).
15. G. H. Kinchin and R. S. Pease, Rep. Prog. Phys. **18**, 1 (1955).
16. P. Sigmund, Appl. Phys. Lett. **14**, 114 (1969).
17. M. T. Robinson, Phil. Mag. **12**, 741 (1965); **17**, 639 (1968).
18. P. Sigmund, Appl. Phys. Lett. **25**, 169 (1974); **27**, 52 (1975).
19. N. Bohr, Kgl. Danske Videnskab. Selskab Mat.-Fys. Medd. **18**, 1 (1948).
20. A. D. Marwick and P. Sigmund, Nucl. Instr. Meth. **126**, 317 (1975).
21. J. Vukanic and P. Sigmund, Appl. Phys. **11**, 265 (1976).
22. J. F. Ziegler, J. P. Biersack, and U. Littmark, *The Stopping and Range of Ions in Solids* Vol. 1 (Pergamon Press, New York, 1985).
23. This is true provided that, at the closest approach ρ_n of the M-hard-spheres, the total potential energy can be written as $f(M)\frac{1}{\rho_n^{1/m}}$, where $f(M)$ is some function of M.

SIMPLE STATISTICAL MODELS FOR EROSION AND GROWTH

R. SMITH, A. OSBALDESTIN,* G. CARTER, I.V. KATARDJIEV & M.J. NOBES†

* Loughborough University of Technology, Loughborough, Leics., LE11 3TU, U.K.
† Salford University, Salford, M5 4WT, U.K.

1. INTRODUCTION

The etching and shaping of materials by chemical, ion or plasma methods has an increasing number of technological applications [1]. Pattern delineation can involve the use of masks and deposition processes, in addition to etching.

Because of the importance of the technologies and the associated fundamental physics, a number of computer simulation programmes have been developed over recent years in order to model the changes in surface shape which can occur [2–8]. These programmes are almost entirely based on a description of the erosion or deposition process in terms of differential equation models and these are useful in predicting many of the resulting macroscopic surface structures. The programmes are generally based on first order erosion theory as primarily developed by Carter and co-workers [9]. The essence of these methods is a description of an eroding or growing surface as a non-linear wave. Many of the features predicted by the non-linear theory such as the development of cones, pyramids and edges from initially smooth surfaces, have been well-documented and are well understood. This theory was explained in some detail at a previous ASI in Crete [9–11].

The theory and computer simulations based on the theory have found widespread application, but since these are continuum models, they cannot fully incorporate all the physical features associated with the removal or deposition of discrete particles, on a surface, at the atomistic level. However, features with scales much larger than a few atoms have been observed to develop from initially flat surfaces subject to energetic inert gas ion bombardment [12]. The continuum models would predict that a flat plane of a pure material, subject to erosion by a monoenergetic ion beam of a single species, would continue to erode as a flat plane, and therefore cannot explain these phenomena. Examples of these features have been given for example by Whitton [12], who has shown that micron size features can develop on, for example, certain crystallographic planes of 99.999% pure Cu subject to inert ion bombardment from a monochromatic source.

The purpose of this paper is two-fold. First, to examine two-dimensional models of the development of surface shape based on the short-range interactions of individual surface atoms with their neighbours and, secondly, to examine if such short-range interactions can result in surfaces with large-scale developed features. The models are Monte Carlo probabilistic models where the probability of removal of a surface atom depends on the relationship of that atom with its physical neighbours. This idea has been used before [13,14] to describe chemical etching and evaporation/deposition processes.

R. Kelly and M. Fernanda da Silva (eds.), Materials Modification by High-fluence Ion Beams, 205–214.

2. SIMULATION MODELS

The material to be etched consists of an array of atoms, usually assumed to be squares in two-dimensions. A surface atom is described as one with at least one of its neighbours missing and only surface atoms are chosen for removal in this model, although it would be a relatively straightforward process to extend the model to remove subsurface atoms also. Figure 1 illustrates two different configurations which are described in the paper.

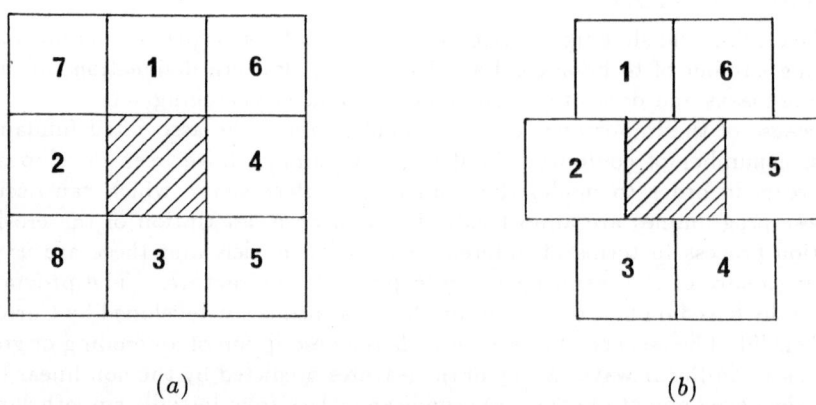

$$(a) \qquad\qquad\qquad (b)$$

Figure 1. The lattice used in the computer simulations. The atoms are represented by square boxes
(a) A regular square lattice.
(b) An offset, also termed "brick wall" or hexagonal lattice.

Figure 1a considers an atom to have four neighbours and is a surface atom if one of the squares labelled 1–4 is missing. Figure 1b represents a staggered square or hexagonal lattice where each atom is assumed to have six neighbours. It would also be possible to compute the next nearest neighbour models where, for example, the removal probability would depend on atoms 1–8 in figure 1a. However, next nearest neighbour models have not been computed in this paper. The removal probability of a surface atom is chosen to be dependent on its immediate configuration with its neighbours. Two basic rules have been used in the simulations presented here. The first is where the removal probability is chosen to be dependent only on the number and configuration of the missing neighbours, the second is where a direction is associated with an incoming beam and only those atoms which are in the line of the beam are removed. In this case the removal probability can be chosen to be dependent on the distance of separation of the next removeable atoms. These cases can be considered as simple models of, first, a chemical erosion process where exposed atoms can be attacked from all sides and, secondly, a physical erosion process where only those atoms in direct line of fire can be removed. At normal incidence on a square lattice, the second process can also be considered as a special case of the first, where atoms without a missing neighbour, along the side perpendicular to the beam direction are removed with probability zero. If the removal probability of a surface atom depends only on the number of missing neighbours then this represents an isotropic etch. However, material crystallography may be such as to mean that certain crystallographic planes are preferentially eroded compared to others, resulting in an anisotropic etch.

Both isotropic and anisotropic erosion will be considered here. The aim is therefore to specify some simple removal rules for surface atoms and to examine the effect of these rules on the macroscopic topography generated. In doing so it will be necessary to distinguish between those models which generate surface roughness as the erosion proceeds, which stabilise after the removal of sufficient numbers of atoms, and those models which generate structures which do not. For this latter case, small structures may grow and die but may eventually combine to produce topography of a scale which becomes large compared to the dimensions of an individual atom.

3. RESULTS

Figure 2a shows a simulation of the erosion of an initial string of 175 flat surface atoms, on a square lattice as depicted in figure 1a.

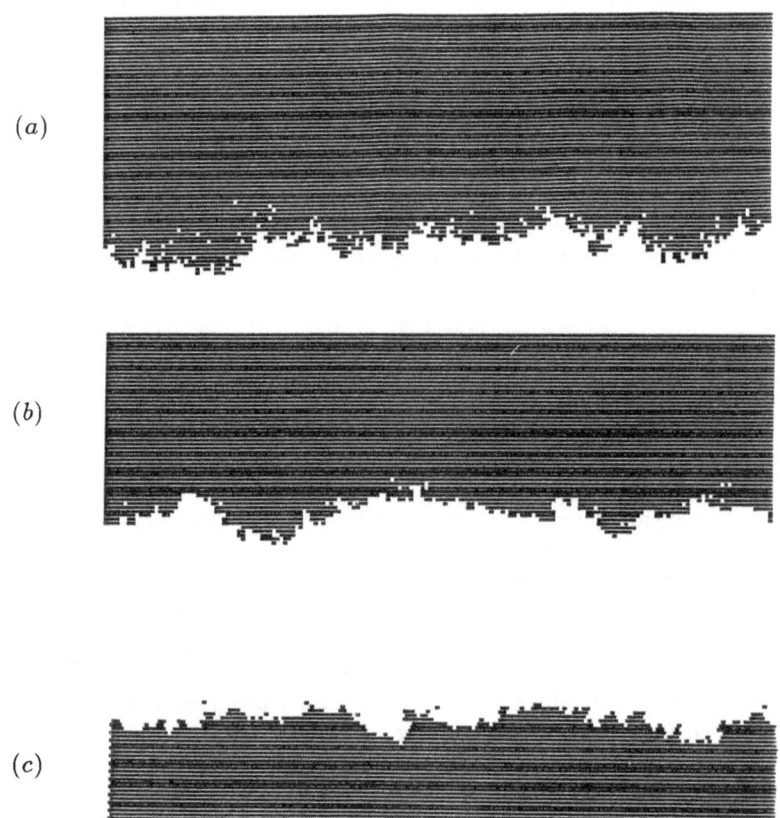

(a)

(b)

(c)

Figure 2. The random removal of atoms from an initially flat surface. In this and subsequent simulations the atomic string is 175 atoms, with periodic boundary conditions. That is, the left-hand neighbour of atom 1 is assumed to be atom 175, and the right-hand neighbour of 175, atom 1.

(a) The regular square lattice with $n = 0$.

(b) The regular square lattice with $n = 1$.

(c) The offset (hexagonal) lattice with $n = 1$. In (a) and (b) the shaded region corresponds to the depth of the etch. All square lattice simulations will depict the etched region as shaded, whereas the hexagonal simulations will depict the substrate as shaded. The quantity n is defined below.

A surface atom is defined to be one where a neighbour in direction 1, 2, 3 or 4 is missing. In this case a surface atom is chosen at random and removed with probability one. Note that isolated atoms exist, detached from the surface and these are considered to be part of the surface. This is because the erosion process is not specified. In a chemical dissolution process, it may be possible for isolated groups of atoms to exist detached from the surface for a short time. This is not possible in sputtering and it would be easy to modify the program to remove these atoms if required. A fairly rough surface is generated, although the RMS surface roughness stabilises after the removal of sufficient numbers of atoms. Blonder [13] has proposed a power law relationship for chemical etching where the probability p of the removal of an atom is given by

$$p = \left(\frac{b}{4}\right)^n \qquad n = 0, 1, 2, \ldots \tag{1}$$

where b is the number of missing neighbours of an atom. Thus the simulation presented in figure 1a corresponds to $n = 0$. Figure 2b corresponding to $n = 1$ produces a weaker etch and a smoother surface, and figure 2c corresponds to $n = 1$ but with the offset (hexagonal) lattice structure of figure 1b. In this case

$$p = \frac{b}{6} \qquad b = 1, 2, \ldots, 6. \tag{2}$$

Again a much smoother etched surface results. The figures show that the hexagonal and square lattice models do produce characteristic morphologies of different shapes and roughnesses. One way to classify roughness is by measuring the fractal dimensions of the etched surfaces [13,14]. The etched surfaces are not true fractals and their fractal dimension changes with time. However this classification of surface roughness dimension can be useful.

These isotropic etch models generate surface roughness from an initially flat surface but can also be used to show how erosion of initial surface topography can occur. Figures 3b, c illustrate the erosion of a square wave profile, fifty atoms in width. Figure 3a illustrates the erosion of the same profile using the non-linear wave model. The non-linear wave model shows that flat planes progress as flat planes and initially re-entrant corners expand into circular arcs [15]. Figure 3b, c show the same general profile but with "statistical noise" superimposed.

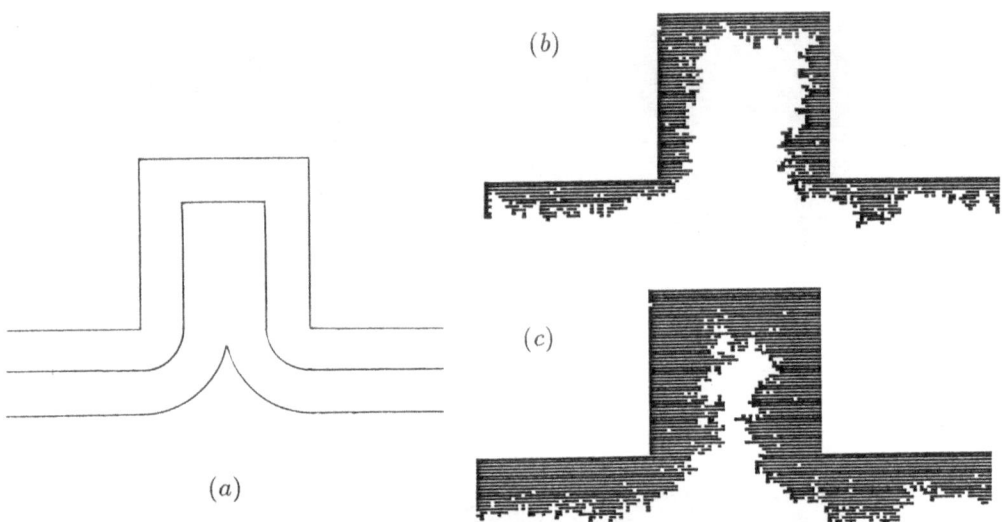

Figure 3. The isotropic erosion of an initial square wave profile.
(a) Using the differential equation of Huyghen's wavefront model.
(b) A simulation of the evolution of a square wave profile 50 atoms wide with 200
atoms in the surface string, after the removal of 2000 atoms.
(c) Same as (b) but with 4000 atoms removed.

For larger structures of the order of magnitude used in device fabrication this
noise would not be an important consideration, but for structures of the size of tens
of atoms, statistical effects can play an important role in determining the ultimate
etched shape. Figure 4a depicts an attempt to model a directional erosion process
where only atoms which have a missing neighbour in direction 1 are eroded. This
is the only one of the models considered which leads itself to some simple statistical
analysis. The model sets up a surface string of N atoms, and an atom is chosen
randomly from this string and removed. Its probability of removal is $1/N$ and the
probability, after n events, that r removals have taken place from the same position
in the string is

$$P(r) = {}^nC_r \left(\frac{1}{N}\right)^r \left(\frac{N-1}{N}\right)^{n-r}.$$

This is the binomial distribution.

Letting $n = Nm$ and letting the number of atoms in the string tend to infinity,
$N \to \infty$, gives

$$P(r, m) = e^{-m} m^r / r!,$$

the Poisson distribution. This has mean m and standard deviation $m^{1/2}$. Thus after
removing an average of m atoms from a given position in the surface string, the RMS
roughness of the surface is $m^{1/2}$. Although this model yields some simple statistics
it is unfortunately unphysical in practice because, as can be seen from figure 4a, it
would predict adjacent peaks and troughs in the surface of one atom in dimension,
but many atoms in height or depth. Diffusion or other physical effects would prevent
this occurring in practice. This is essentially the sequential layer sputtering model due
to Benninghoven [16].

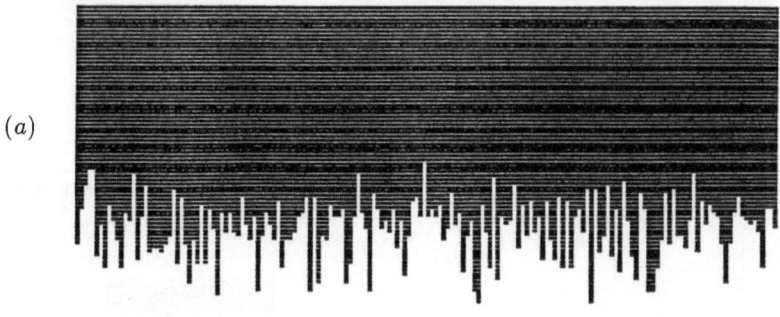

Figure 4. The random removal of atoms with a missing neighbour in direction 1.
Surface atoms with a neighbour in direction 1 are not removed.
(a) Random removal with no bond weighting.
(b) Random removal with bond weighting (see text).

Figure 4b illustrates the smoothing of this by bond strength weighting of removable atoms. Atoms are considered removable if they have a missing neighbour in
direction 1. If 2 and 3 are also missing they are removed with probability 1, if 2 and
3 are present they are removed with probability 1/4, otherwise with probability 1/2.
The effect is to produce a more slowly growing topography, but one where features
are still narrow in width.

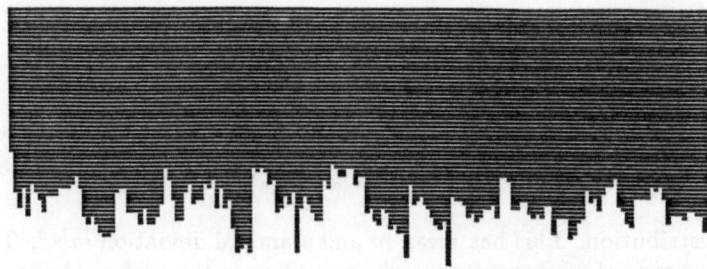

Figure 5. Erosion models with differential erosion rates for atoms situated on left-
hand sloping (upwards from left to right) and right-hand sloping planes.
(a) Where the probability of removal of atoms on right-hand sloping planes is $p =$
0.001, on left-hand sloping planes $p = 0.75$ (see text).

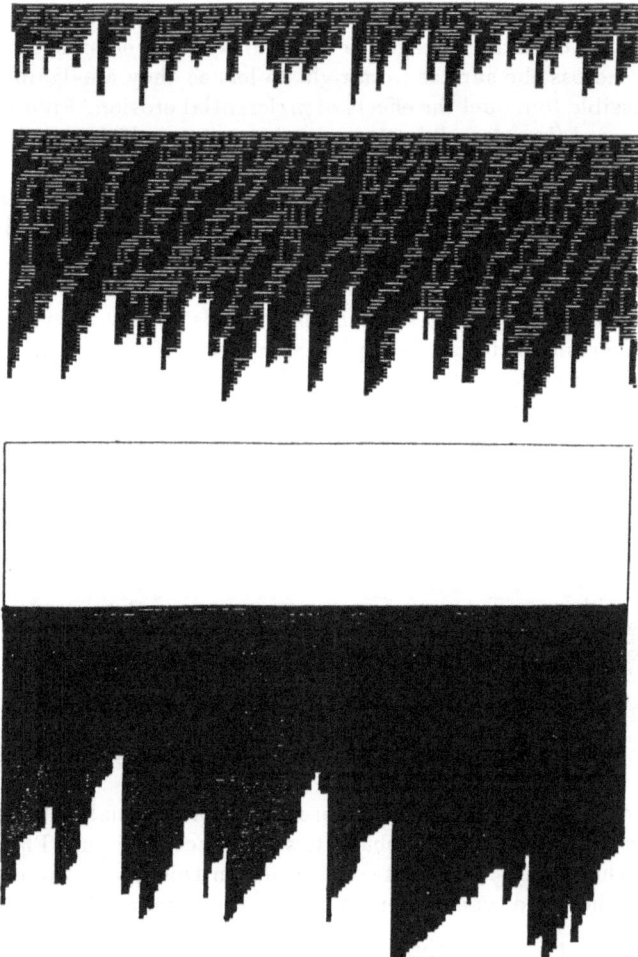

(b) A deterministic model where atoms from left-hand sloping planes are not removed, whereas points on right-hand sloping planes are removed to the level of their right-hand neighbour.

Figure 5a illustrates an erosion process where atoms on left-hand sloping planes are removed preferentially compared to those on right-hand sloping planes. The probability weightings were in the ratio $p = 0.001$ to $p = 0.75$. Individual atoms with two missing neighbours in positions 2 and 3 intact were removed with probability $p = 0.5$. The result of this rule is to produce a rough surface topography. A semi-deterministic rule of a similar nature is shown in figure 5b, and this produces a simulated etched surface with rapidly growing surface topography.

Atoms with a missing neighbour in direction 1 are selected at random and removed if neighbours 2 and 3 are present. However, atoms which sit on left sloping planes are not removed, whereas atoms with sit on right-handed planes are removed to the level of their right-handed neighbour. If an atom is selected which has neighbours 2 and 3 missing then this is removed to the level of the maximum of atoms 2 and 3.

The result of this rule is to produce wedge-like structures which grow as a result of the amalgamation of smaller wedge-like structures as the erosions proceeds. These appear to travel across the surface from right to left as they amalgamate and grow.

It is also possible to model the effects of preferential erosion. Figure 6 depicts the erosion of a staggered (hexagonal) lattice crystal structure which is as defined in figure 1b, subject to the rule given in equation (2) with the exception that the alternative layers have probabilities of removal which differ by a factor of 10. The surface becomes rich in the atoms of the less easily removed species and this considerably inhibits the erosion. It is possible, similarly, to model the effect of a chemical reaction between the substrate and the etchant, which could either speed up or inhibit the erosion. It is intended to use these models to see if ion beam assisted etching which results in a rapidly enhanced erosion rate can be explained in terms of the combination of effects on bond strengths of surface atoms. A possible mechanism would be the induction of subsurface damage by the ion beam which, besides sputtering, induces vacancies and defects which reduce the bond strengths of the surface atoms which are then more easily etched chemically. Such a statistical model would be relatively easy to construct.

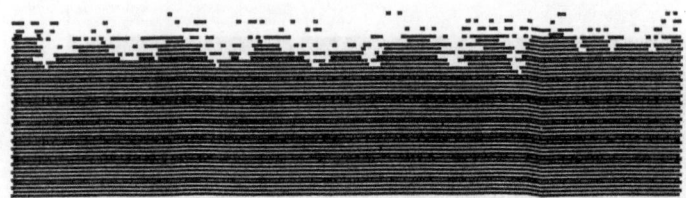

Figure 6. The erosion of a layered structure for the hexagonal lattice, where alternate layers differ in their removal probabilities by a factor of ten. The surface layer becomes rich on the less easily eroded component. In this model the removal probability of atoms of the same type is proportional to the number of missing neighbours.

Finally, it is worth noting that these ideas have been used to model ballistic deposition on surfaces [14]. An example of such deposition is described for a normally incident beam in figure 7a and a simulation in figure 7b. The growth produces a porous structure and if off-axis ballistic deposition at an angle α occurs, the structures have been observed to grow outwards from the surface at an angle β. The empirical observation was

$$\tan \beta = \frac{1}{2} \tan \alpha,$$

the so-called tangent rule. Computer simulation has shown that this relationship is only approximate and has no theoretical basis.

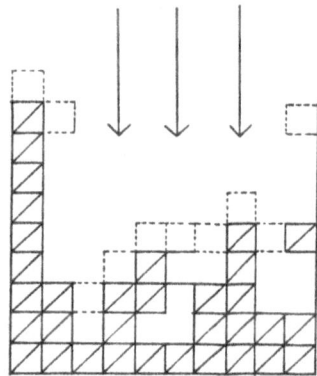

Figure 7a. A schematic diagram illustrating the ballistic deposition on a square lattice, at normal incidence. Sites occupied by either the original surface or the growing deposit are shaded, and sites where the deposit can grow are shown with a dotted line. The rule for the growth of deposit is that this occurs at the point of the suface nearest the incoming beam. This then shadows lower portions of the surface from the growth. Cyclical boundary conditions are assumed.

Figure 7b. A computer simulation based on the model shown in figure 7a after the deposition of 10,000 particles.

4. CONCLUSION

The simple models discussed in this paper are at an elementary stage. No attempt has been made to model subsurface damage, the ejection of subsurface atoms, and the destruction of the underlying crystal structure which would occur in ion beam etching. The bond strengths of the surface atoms have been defined with no chemical and little physical justification, except that the fewer the neighbours, the less well bound the surface atom. In addition three-dimensional effects have not been considered, nor has diffusion or any attempt to model the underlying dynamics of the process been incorporated.

The results and simulations presented here should therefore be seen as a precursor to more detailed calculations which would involved probably a binary collision program with a surface boundary condition. Unfortunately the amount of computing time required to model a meaningful development of surface shape is prohibitive. What is required is to use these computer programmes to try to define general rules which could be incorporated into erosion algorithms as described above.

Alternatively, the algorithms described here can be themselves used in order to examine possible etching mechanisms, by direct comparison of the simulated and practically eroded surfaces. It is in this sense that the simulations presented here have been useful, since distinct characteristic morphologies are associated with different etching rules.

REFERENCES

[1]   Auciello O: In Erosion and growth of solids stimulated by atom and ion beams (Proceedings of NATO Advanced Study Institute, Heraklion, Crete, Greece, 16–27 September 1985). p394, Holland: Martinus Nijhoff, 1986.

[2]   Ting CH and Neureuther AR: Solid State Technol **25**, 115, 1982.

[3]   Rossnagel SM and Robinson RS: J Vac Sci Technol A **1**, 426, 1983.

[4]   Lee RB:J Vac Sci Technol, **16**, 164, 1979.

[5]   Younger DW and Haynes CM:J Vac Sci Technol **21**, 677, 1975.

[6]   Ducommun JP, Cantagrel M and Moulin M: J Mater Sci **16**, 52, 1975.

[7]   Smith R and Walls JM: Philos Mag A **42**, 235, 1980.

[8]   Smith R, Carter G and Nobes MJ: Proc Roy Soc London, Ser A, **407**, 405, 1986.

[9]   Carter G: In erosion and growth of solids stimulated by atom and ion beams (Proceedings of NATO Advanced Study Institute, Heraklion, Crete, Greece, 16–27 September 1985), p70, Holland: Martinus Nijhoff, 1986.

[10]   Nobes MJ:Ibid, 103.

[11]   Smith R and Tagg MA: Ibid, 121.

[12]   Whitton JL:Ibid, 151.

[13]   Blonder GE: Phys Rev B, **33**, 9, 6157, 1986.

[14]   Meakin P, Ramanlal P, Sander LM and Ball RC: Phys Rev A, **34**, 6, 5091, 1986.

[15]   Smith R, Wilde SJ, Carter G, Katardjiev IV and Nobes MJ: J Vac Sci Technol B, **5** (2), 579, 1987.

[16]   Benninghoven A: Z Phys, **230**, 403, 1971.

DEFECTS AND DISORDER

DEFECTS AND DEFECT PROCESSES

A. M. STONEHAM

Theoretical Physics Division, B424.4,
Harwell Laboratory,
Didcot, Oxon OX11 ORA, U.K.

ABSTRACT
    The mechanisms of materials modification by ion beams exploit defects
and defect processes. This paper surveys the principal point defects in
ionic crystals and semiconductors and at their surfaces, and summarises
their properties in relation to such mechanisms.

1. INTRODUCTION
    In the modification of material properties by ion beams, defects and
defect processes are important components. They are involved at several
stages. First there are defects – some transient – which are created by
the incident beam. Once the particle energies have fallen to thermal
energies, there are the several defect processes which may lead to
defect annihilation or aggregation, or to processes like dislocation climb.
There may be more complex phenomena, like amorphisation, which may occur by
nucleation and growth in some cases; moreover, the effects of the
irradiation may sensitise the material to subsequent processes, as in
fission track formation, where etching becomes easier. At any stage the
defects present, or the impurities which have been implanted, can affect
observed behaviour. The electrical resistivity, the optical behaviour
(e.g. refractive index) and even mechanical properties will be altered.

    In the present paper, I shall discuss some of the main defects and
defect processes in the context of ion-beam processing. However, I shall
exclude the description of the initial collision events and the classical
collision dynamics, these being covered elsewhere at this meeting. The
defect phenomena to be described are representative (for a more complete
survey, see ref. 1) rather than comprehensive. Nevertheless, it is
important to draw several distinctions. Ion beam methods involve different
timescales, energy densities, and especially differences in distribution of
energy between electronic and nuclear systems (ref. 2). When electron
excitation occurs, it can have very different consequences depending on
whether the core electrons or the valence electrons are excited. The
mechanisms which involve ion beam removal of surface species also depend on
the chemical nature of the species, e.g. oxygen or caesium; there may be
dependence too on what happens just outside the surface, for example charge
exchange between the surface and an ion after its removal. Thus we must
consider a very complex, time-dependent system, and try to draw from this
complexity some general ideas of the key processes and their mode of
operation.

2. DEFECT ASPECTS OF RADIATION DAMAGE
2.1 <u>Ionic crystals, semiconductors and metals</u>
    It is helpful to begin by contrasting the behaviour of ionic crystals,
like NaCl or oxides, with semiconductors like silicon, and with metals like

R. Kelly and M. Fernanda da Silva (eds.), Materials Modification by High-fluence Ion Beams, 217–230.
© 1989 by Kluwer Academic Publishers.

Cu.   There are four main differences:
(1)   In polar crystals, there are two different chemical species.   This means that there are at least two types of vacancy (e.g. Ga vacancy or P vacancy in GaP) and also that the displacement energy will be different, either because the two types of ion have different radii, or for some more subtle reason.   In MgO, both Mg and O have similar displacement energies (about 53 eV), but the energy transferred from an incident particle will depend on the target mass.   Whereas 330 keV electrons will displace O, it needs electrons of at least 480 keV to displace the heavier Mg.
(2)   There can be charged defects in non-metals.   The anion vacancy in KCl can be empty (net charge +e), contain one electron (net charge 0) or trap two electrons (net charge −e).   In silicon, the vacancy can occur in charge states −e, 0, +e, with the +e vacancies disproportionating to populations of 0 and +2e vacancies (the so-called negative U behaviour).   Molecular ions like Cl_2^- (see sections 3.1.4, 3.1.5 below) may occur too, showing both the possibility of different charge states and that chemical bonds can occur in a way which conduction electron screening precludes in metals. Charge compensation and charge conservation are important rules because of the long range of Coulomb forces, with consequences such as space–charge formation near surfaces.   One should not forget that some species − especially transition metal ions − can occur stably in several different charge states, so that quite substantial electric fields can be built up by the redistribution of charge associated with the ion beam (which is, of course, directed rather than isotropic).   Such electric fields will tend to lead to motion of charged species, and may affect segregation to surfaces (ref. 3).

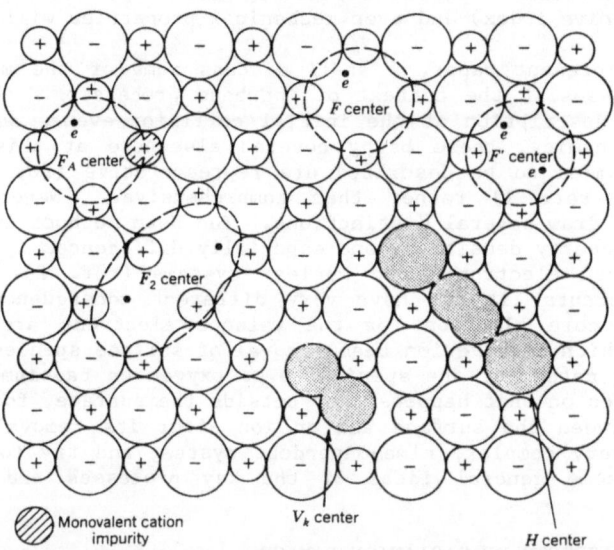

Figure 1   Common color centers in alkali halides *MX*. These include electron-excess centers, notably the *F* center and related defects (the *F'* center has an extra electron; the F_A center is an *F* center with an adjacent cation impurity; the F_2 center is two adjacent *F* centers), and two centers involving X_2^- molecular ions, namely, the self-trapped hole (V_K center) and neutral anion interstitial (*H* center).

(3) Damage initiated on one sublattice tends to stay on that sublattice. Thus if anions are displaced initially, the final damage state tends to involve only anion defects. This includes both thermal processes and dynamic processes like focussed collision sequences. There are several reasons, especially when there are strong Madelung terms in the energy, yet there are important exceptions too. The first exception is for alkali halides, where perfect dislocation loops are seen, which must involve disorder on both the anion and cation sublattices. Since only anions are involved at first, there must be a secondary step, and this is probably the generation of cation-cation

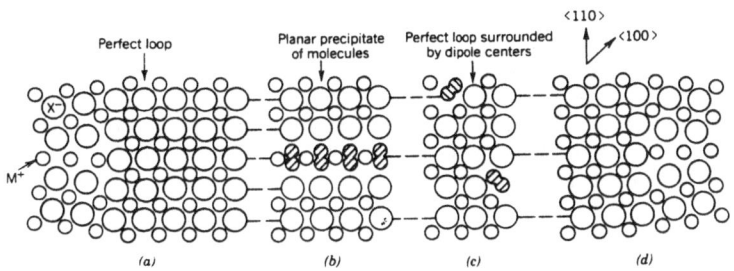

Figure 2   Perfect interstitial dislocation loop in alkali halides (see also Figure 3.22). (a) and (d) represent the edges of the perfect loop. (b) and (c) indicate two possible ways of incorporating halogen molecules into the dislocation loop. In (b), halogen molecules are precipitated at the loop; in (c) the dipolar halogen molecular centers surround the perfect loop. Alkali ions M^+ are small circles, halogen ions X^- large circles [after Hobbs et al. (1973)].

divacancies containing halogen molecules. Another exception is in III-Vs, where excess Ga interstitials can cause dislocations to climb, a process needing As interstitials too in GaAs (ref. 4). This may involve quite complex processes (ref. 4; ref. 1 p. 175). However, recent studies of models for the so-called EL2 centre led to calculations (ref.5) of the relative energies of Ga interstitials, As interstitials and As antisite defects, i.e. As on a Ga site, a well-established class of defects for III-Vs. This suggests there could be processes in which a Ga interstitial might displace an As antisite defect to give an As interstitial, i.e. disorder on the second sublattice:

$$Ga_i + As_{Ga} \rightarrow Ga_{Ga} + As_i$$

This may be contrasted with more complicated processes previously proposed (see e.g. figure 3.27 of ref. 1).

(4) Excited states of defects may occur during radiation phenomena in non-metals. In metals, de-excitation occurs rapidly, comparably with the reciprocal of the plasma frequency. But in non-metals, excited states may be quite long-lived, and the subsequent behaviour depends critically on which excited state is involved. The role of self-trapped excitons in the, production of anion vacancies and interstitials in alkali halides, and in photon-induced desorption is well-known (ref. 6). Other systems which appear to show an ionisation component of damage include silica and alumina where again there are descriptions in terms of self-trapped excitons (ref. 7; ref. 8; ref. 9).

2.2 Ionisation-related mechanisms

As noted in section 2.1, electronic excitation can lead to defect production and other defect behaviour. It is helpful to bring together some examples and to give a preliminary classification.

Firstly, one possible effect is simply heating, with the standard consequences of either evaporation or melting. Apart from noting that evaporation can be selective – e.g. removing O from some oxide, leaving metal excess – this aspect need not concern us. Secondly, if damage is to be produced, or if there is to be selective defect production, there must be energy localisation somehow. This is the reason self-trapped excitons (whatever their precise configuration) are effective in damage production. Thirdly, the local excess of energy must have a route to give rise to defects. This brings us close to studies of non-radiative transitions in solids, where excited systems return to their ground states other than by luminescence (ref. 10). It proves useful to distinguish between three main classes of such transition:

(a) Local excitation processes, where the excitation leads to an excited state which has an energy surface with a low (or even zero) barrier to defect formation. The important feature is that the energy has been given to the electronic degrees of freedom. This case includes the Itoh–Saidoh model of F-centre production (ref. 11; see also refs. 6 and 12) and the Menzel–Gomer–Redhead model of desorption (ref. 13 and section 4.3).

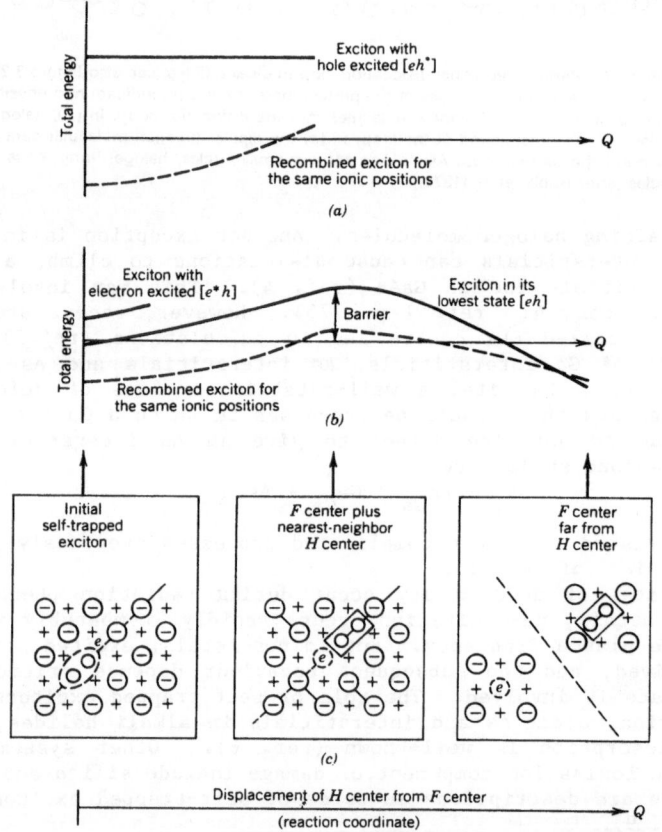

Figure 3 Photochemical damage associated with exciton decay, in alkali halides like KCl. The diagram compares energies for hole excited {case (a), [eh*]} and electron excited {case (h), [e*h] and also [eh]} as the reaction (c) proceeds [after itoh et al. (1977)]. The important point is that there is no barrier to F and H center production (Q increasing) when the hole is excited [as in (a)], whereas there is a large barrier [as in (b)] when the hole is in its lowest state. For reference, the broken line shows the energy for the same geometry when the exciton has recombined.

(b) Local heating processes, in which the excitation energy is given to vibrational degrees of freedom. Many models of recombination-enhanced diffusion fit this category. In principle, local heating effects can occur in metals, in that spikes cause strong local vibrations, and indeed the behaviour correlates with the temperature dependence of the thermal conductivity (refs. 14, 15).

(c) Cases where a change of charge state is involved (the distinction between this and (a) is not always clear, but is useful). This is found in the Bourgoin-Corbett mechanism of recombination-enhanced diffusion:

$$X_A^{N+} + e^- \rightarrow X_B^{(N-1)+}$$

$$X_B^{(N-1)+} + h^+ \rightarrow X_A^{N+}$$

and in the Knotek-Feibelman mechanism of desorption (ref. 16 and section 4.3). Thus in maximum valency oxides, core excitation can lead to Auger processes which leave O as a positive ion on a site which strongly favours negative ions, with a subsequent displacement event in suitable cases.

Sometimes the effects are indirect. An interesting example (ref. 13) occurs in GaP, where the implantation of hydrogen is used to produce a highly-insulating layer for isolation. The striking observation is that deuterium is far more effective than either protons or tritium. The explanation seems to be that the local mode frequency for D is almost exactly an integral fraction of the band gap of GaP, so GaP:D can cause efficient non-radiative recombination of electrons and holes. The less-efficient cases of GaP:H and GaP:T allow the electrons and holes to survive long enough to encourage recombination-enhanced recombination of the standard damage centres (e.g. vacancies and interstitials) which are actually responsible for the resistivity.

2.3 Phase changes and defect reactions

Whilst my present concern is with point defects, their subsequent reactions can lead to major changes. One such change is the appearance of a new phase (ref. 18). In some cases these are nucleated locally and grow; in other cases it is presumed that a radiation spike leaves a devastated region of finite extent, and that amorphisation simply corresponds to the overlap of these regions. Which is correct is still not clear, though it would seem that the straightforward overlap model is too simple in some respects. Specifically, one might assume that all measures of the fraction of amorphous component would have the same value for all types of measurement if amorphous zones simply accumulate. Yet the two-level systems in silicon (see e.g. ref. 1, p. 27; these are defects which affect the low-temperature thermal properties in a way consistent with systems which can tunnel between two configurations) do not seem to go directly with the amorphous fraction observed structurally. One possible explanation is that the two-level systems are in fact the nuclei from which larger amorphous regions develop (ref. 19).

So far as defect reactions are concerned following initial radiation damage, the fullest studies are those for alkali halides (see ref. 1, p.303, for a summary). The initial damage process yields anion vacancies and interstitials. Even at the lowest temperatures, some close-pair recombination can occur, but it is only when the neutral halogen interstitial (H centre, see fig. 1) becomes mobile at just below liquid-nitrogen temperature that more complex reactions begin to occur. There is, of course, more recombination, and also some trapping of H centres by impurities. In addition, various V centres form; these are not all identified, but some involve aggregates of H centres. Analogously, the anion vacancies aggregate to form their own clusters, the M, R and N centres. However, one of the key processes between 200K and room temperature is a sequence of reactions (ref. 20) which yields halogen molecules in cation-anion divacancies. It is these vacancy defects (to

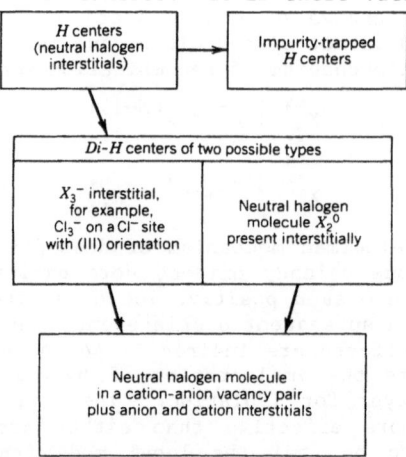

Figure 4   Reactions of neutral anion interstitials (H centers) produced in radiation damage in alkali halides to give defects on both anion and cation sublattices.

which correspond anion and cation interstitials), involving disorder on both sublattices, which lead to the perfect interstitial dislocation loops observed. At still higher temperatures, when the anion vacancies (F centres) become mobile, aggregation leads to metal colloids. The halogen molecules formed along with the dislocation loops may aggregate to form gas bubbles. Colloids are readily formed at halide surfaces too, though in that case the emission of halogen to the gas phase occurs if the temperature is high enough.

3.   SOME EXAMPLES OF SOLID STATE DEFECTS

In this section I describe the key properties of a small number of the characteristic defects formed in a wide range of non-metallic solids. It is not possible to be comprehensive, but the main features can be identified.

3.1   Intrinsic Defects

We begin with those defects which do not involve impurities.

3.1.1. Anion vacancies. In halides, these give the F centre (the neutral centre with a single electron trapped); the F centre may lose its electron, giving the (empty) vacancy known from ionic transport; the F centre may also gain a second electron, to give the F' centre. In oxides the same features occur, though the nomenclature has become confused. I shall refer to the one-electron centre in, say, MgO, as the F+ centre, and the two-electron centre as the F' centre.

The centres with trapped electrons are colour centres, for there is optical absorption to an excited state which usually lies within the visible spectrum. The energy level structure of the halide F centres is similar to that of the hydrogen atom, i.e. a 1s to 2p transition dominates; whilst this is simplistic, the analogy is often helpful.

In silica, the oxygen vacancy centre is also seen, but it shows one striking difference. Whereas the F+ centres in oxides (i.e. the centres obtained by removing O-) contain an electron whose charge is concentrated in the vacancy, with equal amplitude on the six neighbouring Mg ions in MgO, in quartz (and other silicas) the unpaired spin is localised on an Si sp3 hybrid, asymmetrically on one of the two neighbours (the E' centre).

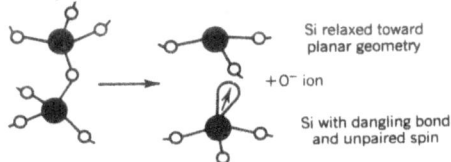

Si relaxed toward planar geometry

$+O^-$ ion

Si with dangling bond and unpaired spin

Figure 5   Possible formation mechanism for the E' center in irradiated a-SiO$_2$. Note that, while O$^-$ is removed in the sequence shown, removal of O$^+$ would be equally consistent with experiment. (Solid circles, Si; open circles, O.)

3.1.2. <u>Cation vacancies</u>. Cation vacancies trap holes, rather than electrons, and the holes tend to be localised on one or two of the neighbouring anions. Thus the Mg vacancy in MgO can trap one hole, which resides on a single oxygen neighbour, or two holes, these residing on opposite oxygen neighbours in the ground state. The important feature here is that there are two competing effects: the kinetic energy can be lowered by spreading the carrier over many sites (just like the formation of bands in perfect crystals) but the polarisation energy is lowered when the charge is concentrated on a small number of sites (see ref. 1), typically one or two. A second important feature is that there are two types of excited state: the "crystal field" transitions which involve only the excitation of the oxygen ion in which the hole is localised, and the "charge transfer" transitions, in which the hole passes to other oxygens neighbouring the vacancy. It is these charge transfer transitions which prove the most important. Cation vacancies of the type described are observed in many oxides and in ZnSe and other II-VI compounds. In alkali halides, the centre is hard to see, but involves a self-trapped hole localised on two neighbouring halogens. In transition metal oxides, like NiO, the hole may be localised on a cation.

3.1.3. <u>Vacancies in valence crystals</u>. Two cases have been studied in detail, namely the vacancies in diamond and in silicon. In both cases the simplest picture considers the four dangling bonds created when an atom is removed. There are four electrons in these bonds in the neutral vacancy, and the important electronic states can be inferred from the ways in which the electrons might be distributed over the molecular orbitals formed from the four dangling bonds combined with explicit calculations. Thus one is led to expect a spin singlet for the neutral vacancy, but an orbitally degenerate state of E symmetry; this is indeed found. In diamond, the optical studies have given a wealth of detail, for the neutral vacancy (which gives the GR1 zero phonon line and associated broad band) can be examined under stress to reveal accurate quantitative data on the Jahn-Teller effect. The negative vacancy is also seen. In silicon, the studies have been by spin resonance, and cover the 2-, -, o, + and 2+ states. Some of these are, of course, spin singlets and not directly visible, but their properties can be inferred from processes which change the charge state. Again, one finds strong Jahn-Teller effects asssociated with the orbital degeneracy. Indeed, these are so substantial that a population of vacancies in the + state is unstable against disproportionation into O and 2+ states. Many complexes of vacancies and impurities are observed in silicon. These are important in the radiation damage behaviour, where different defect complexes emerge or disappear as the temperature is raised (see e.g. ref. 1, p. 322).

3.1.4. <u>Interstitials</u>. It is a general rule that interstitials are more mobile (i.e. have a lower activation energy) than vacancies. It is also the case that very many interstitials (the exceptions being the closed shell ion interstitials, e.g. F^- in fluorite) have asymmetric structures, such as the "crowdion" or "split interstitial" structures. This seems to be equally true for the covalent crystals as for the ionic ones, though the precise geometries are not always well established. The best-understood interstitials are probably the H-centres, the neutral halogen interstitials in alkali halides, where a halogen molecular ion (Cl_2^-) lying along a close packed (110) direction replaces a Cl^- host ion. There are detailed studies of its motion, reorientation, optical and spin-resonance spectra and its interactions with impurities.

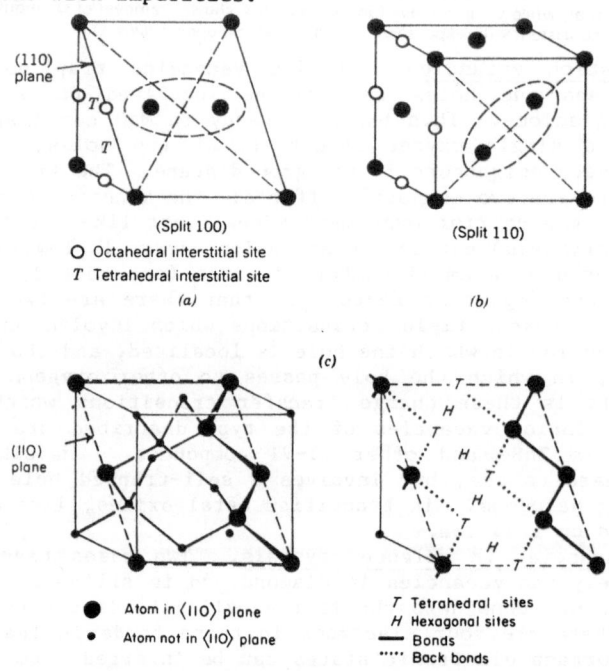

(110) plane

T

T

(Split 100)

O Octahedral interstitial site
T Tetrahedral interstitial site

(a)

(Split 110)

(b)

(c)

(110) plane

● Atom in ⟨110⟩ plane
• Atom not in ⟨110⟩ plane

T Tetrahedral sites
H Hexagonal sites
— Bonds
····· Back bonds

Figure 6 Two examples of split interstitial configurations, with (a) a ⟨100⟩ axis in a bcc host, corresponding to the cation sublattice in the cesium halide structure and (b) a ⟨110⟩ axis in an fcc host, corresponding to the cation sublattice in the NaCl structure. Split interstitials occur in the diamond structure also; (c) shows the tetrahedral (T) and hexagonal (H) interstitial sites for diamond but does not show the split form explicitly.

3.1.5. <u>Self-trapped holes and excitons</u>. The same (Cl_2^-) ion is observed as the self-trapped hole in alkali halides. For this centre, as for the H centre, the "molecule in a crystal" description works well, i.e. the host provides a minor perturbation of the molecular ion. The reason that self-trapping occurs is (cf. the discussion of cation vacancies above) that the polarisation and distortion energy gain from localisation exceeds the kinetic energy gain from delocalisation into bands like those in silicon. The self-trapped exciton is a key component of the radiation damage process (fig. 3). It can be regarded as an electron trapped at a self-trapped hole, so that there are excited states which involve the electron, or the hole, or both. One important issue at present is whether the exciton moves axially to form an asymmetric geometry. Self-trapped excitons have been proposed in oxides too, though there has been some speculation as to their precise form and importance (refs. 7,8,9). The underlying problem is that oxide ions have double charges, so a single hole on two O^{2-} ions still leaves substantial repulsion, whereas a single hole on two Cl^- gives a species which is stable even in isolation. Moreover, the Madelung energy in oxides is high, so there is a large energy penalty in putting a hole at an oxide ion site without some compensating effect. My own feeling is that the self-trapped excitons here can be created by removing an oxygen atom from one site (leaving the neutral F' centre) to join another to give a $(O_2)^{2-}$ molecular ion:

$$O^{2-} + O^{2-} \rightarrow [2e] + (O_2)^{2-}$$

in which [2e] is the F' centre, i.e. two electrons at an oxygen vacancy. Self-trapped holes in oxides are a feature of transition metal oxides, and are the small polarons of conductivity studies; in NiO, for instance, one would envisage charge localised on Ni^{2+} to form either Ni^+ (the electron carrier) or Ni^{3+} (the hole carrier).

   3.1.6. <u>Antisite defects.</u> Clearly the energy to put an Mg^{2+} at an O^{2-} site in MgO is very large indeed. Clearly too the energy to swap Si and Ge in an Si/Ge alloy is rather modest, for indeed (unlike SiC) there is little ordering, if any. For the III-V's and a few other special cases (the neutral halogen molecule in an anion-cation divacancy, mentioned earlier, is one such) the antisite defects are of major importance as a way of disordering. Consider GaAs. One knows that Si on a Ga site is a donor, like the other species from the Group IV column. As on a Ga site is thus a double donor; in the same way Ga on an As site would be a double acceptor. The As antisite gives a cluster of As_5 (the central antisite plus four As neighbours) and the Ga antisite a Ga_5 cluster. Such antisite defects are of importance as thermal defects, produced during the processing of semiconductors, as well as by radiation damage. They can exist in several charge states and, in some cases (with associated defects like interstitials) in metastable forms, like the EL2 defect (ref. 21), which is an As_{Ga} - As_i pair.

3.2 Impurity-related centres

   The range of possible impurities is enormous, and can only be touched on here. One important general point is that one should not assume there are no impurities present because this was said by the supplier of the sample. It is virtually impossible to eliminate some species like water (or its components, H and O) or carbon, and the introduction of transition metals from containers or during mechanical treatment is only too easy.

   3.2.1. <u>Centres with trapped electrons.</u> Donor centres are often well-behaved and have spectra which can be related to the hydrogen atomic spectrum after scaling. The case of GaP:O (i.e. O on a P site) is unusual, for there are two charge states (the neutral centre known as the one-electron centre and the negative two-electron centre); the two electron centre appears to exist in both a virtually unrelaxed ("centred") metastable form and a strongly-relaxed form; lattice relaxation cannot be ignored even for donors (ref. 22).

   3.2.2. <u>Centres with trapped holes.</u> The valence bands of many compounds are formed from anion p states, so there are the additional complexities of the electronic degeneracy and associated spin-orbit coupling. However, the scaled hydrogen atom picture, suitably generalised, still works well.

   Note that acceptors, like donors, can trap excitons in semiconductors; also, there are important cases of donor-acceptor recombination, in which the donor's electron recombines with the acceptor's hole.

3.2.3. <u>Isovalent impurities</u>. In semiconductors, these are often very shallow, like $GaP:N_p$, and hence of interest because they can assist luminescence with high energy. In ionic crystals, there are similar situations: $CsI:Na_{cs}$ is used as an efficient X-ray phosphor, where the high atomic number components lead to efficient absorption, and the isovalent Na traps an exciton which has a blue emission. CsI:Na also shows that isovalent impurities can trap both electrons (because of the difference in electronegativity) and holes (through an elastic interaction).

3.2.4. <u>Centres associated with hydrogen</u>. In halides, the various U centres (substitutional at anion sites, or interstitial as an ion or an atom) have been long studied and understood. In ionic oxides, the hydroxyl species is the commonest, though sometimes (as in MgO) this appears only stable when associated with an impurity. In Si and Ge hydrogen can passivate shallow impurities, an issue of some interest; the hydrogen appears to be stable in molecular form in the crystal. The case of H in GaP and its role in electrical isolation has been mentioned in section 2.2. In quartz, hydrogen occurs in many different forms, even in the best cases (it is interesting to note that the oxidation of Si shows effects of water at below the ppm range!), and even these traces can influence the radiation sensitivity of quartz (25, 26).

3.2.5. <u>Transition metal impurities</u>. For present purposes, the main effect of interest is the range of charge states which these ions can show (why this is possible is another issue, also important). Thus they can act to pin the Fermi level when one tries to change carrier concentrations, or simply as a sink of carriers produced under irradiation. They may act as efficient recombination centres, since both electrons and holes can be trapped with efficiency. Clearly, transition metal impurities can form complexes with other defects too.

4. SURFACES AND SURFACE DEFECT PROCESSES
4.1 Properties of defect-free surfaces
It is easy to give an over-idealised description of surfaces. On the one hand, many experiments are interpreted as if surfaces were chemically pure and crystallographically unique; on the other hand, many theories ignore the substantial surface polarisation and distortion. As a start, therefore, I note some of the important aspects of real crystal surfaces, concentrating mainly on oxides, (refs. 29-31) since these are especially important in practice, but also remarking on other ionic crystals and semiconductors as appropriate.

(1) If one tries to grow microcrystals of oxides so as to produce a high area for study, the morphology depends greatly on how the sample is prepared. MgO grown by decomposition of the carbonate differs in morphology from that grown from the hydroxide; in turn, these differ from MgO smoke from combustion of Mg. Moreover, exposure to moisture can change the morphology.

(2) Which surfaces are obtained is governed mainly by the surface energy. But some faces have an infinite surface energy (examples are the polar 111 surfaces of MgO) and these should not be observed. They can be made to form (e.g. iodine doping of a MnS melt leads to stable 111 facets). However, it is possible to have impurity-stabilised surfaces which would not be stable otherwise; it is also possible to have small areas as stable components, e.g. there could be alternate patches of Mg and O (111) surface (though this has not been demonstrated, and any claims of 111 MgO surfaces should be treated with caution).

(3) The outer layers of ions show both "rumpling", when anions move out

relative to cations, and "relaxation", i.e. a modification of interplanar

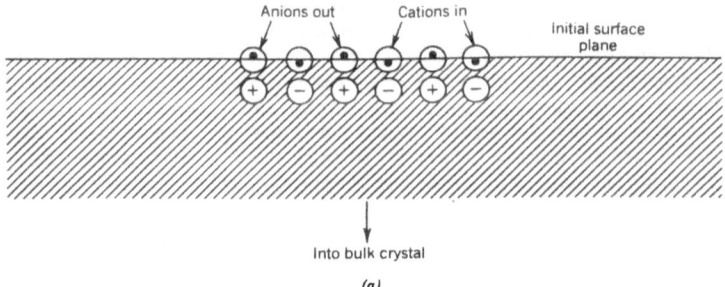

(a)

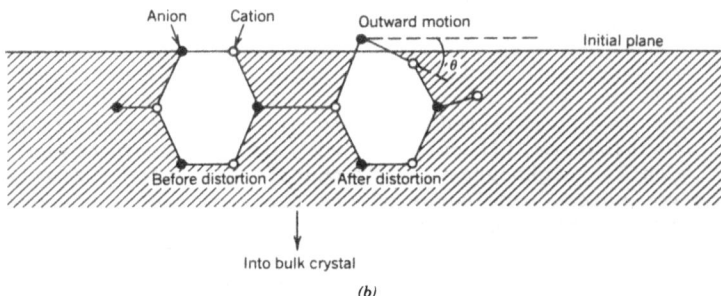

(b)

Figure 7   Surface distortion of polar crystals. (*a*) Rumpling of the (100) NaCl surface. Only the core motion is shown; relaxation altering mean layer spacings is omitted. (*b*) Puckering (buckling) of the (110) zincblende surface. For II–VI and III–V compounds the bond rotation (or tilt angle) θ is usually around 30°.

spacings.   Rumpling is generally called "puckering" for semiconductors, though the behaviour is very similar.   The relaxations are not large for oxides, but they have an important effect on the Madelung potential and on the surface electric field.   The Madelung constant is changed rather little (and that too only for the outer surface layer to a significant extent) but the Madelung energy change is quite high.

(4)   Low-angle surfaces can be regarded as simpler surfaces with steps. Near the steps there are substantial fields and lattice relaxation, and this has consequences for surface processes.   The surface energy depends on the orientation in a way which can be represented by an exponential repulsion between steps with a range of a few Angstroms.

(5)   The concentrations of thermal defects near the surface can be modified, with several consequences.   There can be an excess of species with one charge (e.g. interstitial ions, rather than vacancies) which alters the energy needed to move charges from the interior of the crystal to the outside.   One other aspect which is important is the image interaction: charges near to an interface at which the dielectric constant changes (like a free surface) have altered energies.   In particular, charged defects are repelled from a free surface, but attracted to an interface with a metal, such as in an oxide film on a metal.

4.2 Surface defects and impurities

Here one must start with a caution (ref. 30).   There are many types of signal seen apparently associated with surfaces.   But how do you decide first that it is a surface signal, and secondly whether it corresponds to some defect model?   Normally deciding whether or not it is a surface defect uses one of these approaches: (a) the effect of changing surface area to bulk volume, (b) the possibility of wiping out defect signals rapidly by exposure to gas, (c) the use of a probe which only penetrates a short distance (or which produces electrons which only escape through a few layers of atoms), or (d) the use of tricks to enhance fields at surfaces. Scanning tunnelling microscopy is an especially promising approach, and is dominated by the outer layer of atoms or ions in almost all circumstances. It can be used in several different modes: one can change the sign of the bias to distinguish anionic and cationic species; one can use the force microscope to look at insulating surfaces; one can work in an atmosphere, or indeed in an electrolyte, subject to relatively minor restrictions.

Relatively few defect models are well established (refs. 32,33). Spin resonance has proved the strongest approach so far, but many other methods have been effective. Some of the defects noted are these (ref. 30):

(1) Anion vacancies. The surface F+ (i.e. a surface oxygen vacancy) centre has been identified clearly on MgO. The other charge states occur, one assumes; indeed the Fo state may be associated with the signals attributed to the "low-coordinated oxygens" or with the reflectivity data which correlate with the surface F+ centre. It is important to realise that electrons may be transferred rapidly in small microcrystals. Anion vacancies on strontium titanate or on rutile appear to have electrons associated with Ti neighbours, rather than in the vacancy itself.

(2) Cation vacancies. Analogues of the bulk centres (see above) are seen; again, presumably the hole is on the anion in MgO and on the cation in NiO.

(3) Segregated species. Impurity concentrations at surfaces can be much higher than in the bulk, and this can have dramatic effects on behaviour. Thus Mg segregated to the alumina surface leads to enhanced sintering; Sr and Ba segregated to an MgO surface should inhibit sintering. This control of ceramic processing is of real importance, for whereas one wants very complete sintering for optical ceramics, one wants the high surface area of catalyst substrates to be stable.

(4) Adsorbed species. Clearly, some adsorbed species form strong chemical bonds to the surface. In other cases, there can be binding without chemical interaction, for the polarisation energy and Coulomb interactions may suffice. As examples, consider Ag on silica and the molecular ion $(O_2)^-$ on MgO. For Ag, it seems that the main interaction is of the Si ion with the dipole it induces in the Ag atom, and indeed the Ag atoms are found to lie above Si on the surface. This is essentially the same phenomenon which underlies metal/oxide adhesion when there is no chemical reaction (ref. 34). For the oxygen molecular ion (which appears to be one of the most important species in gas-sensor operation, and perhaps in catalysis too) it is easily seen that the oxygens can lie above two Mg cations, so that a simple Coulomb term dominates. More precise calculations have been made, including those of electron transfer reactions between bulk transition metal impurities and molecular oxygen.

4.3 Removal of Surface Species

Two of the main surface processes are those which add species to the surface (like condensation from the vapour, molecular beam epitaxy, or oxidation) and those which remove species from the surface. Removal includes removal by evaporation, by electrochemical means, by reduction of oxides, or by various stimulated processes. It is these stimulated processes, rather than collisional processes, which are summarised here, since they relate most closely to ion-beam methods.

The stages of stimulated desorption (ref. 35) are basically (a) energy deposition, (b) energy localisation, at least if the process is not just

heating, (c) the several types of atomic processes leading to emission, and
(d) the possible subsequent processes involving the vapour phase, whether
chemical decomposition of fragments, molecule formation, or change of
charge state. These processes may be influenced by behaviour deeper into
the solid (even if only a few layers), such as the accumulation of charged
defects to produce electric fields (an effect absent for metals) or the
segregation of species yielding enhanced surface concentrations.

Two main mechanisms are commonly invoked to described desorption, and
were mentioned briefly earlier because of their analogies with the
radiation damage mechanisms. The Menzel–Gomer–Redhead mechanism invokes
excitation (usually from a valence state) into an antibonding state, for
which the energy surface favours separation of the component ions. This
has been deomonstrated for the desorption of adsorbed species, and it seems
probable that the same mechanism works for the removal of surface ions
following valence electron excitation of alkali halides. The
Feibelman–Knotek mechanism (ref. 36) has been demonstrated for maximal
valence systems, e.g. rutile, in which the 4+ Ti cation has no valence
electrons. If a core electron is removed, the subsequent Auger process in
which an oxygen loses several electrons leaves a positive oxygen at a
surface site, where it is driven off by the strong Coulomb interaction. Of
course, in principle direct core excitation could occur. If the excited
ion is a surface ion (e.g. Cl^- at a KCl surface becoming Cl+), emission
of chlorine is expected. If the Cl^- is in the second layer of ions, it
may be able to drive off the K+ ion immediately outside it (ref. 37).

REFERENCES

1. Hayes W and Stoneham AM: Defects and Defect Processes in Non-Metallic Solids, New York: John Wiley and Sons, 1985.
2. Dreyfus RW (this meeting).
3. Eccles AJ, van den Berg JA, Brown A, and Vickerman JC: Appl Phys Lett 49 188, 1986.
4. Petroff PM and Kimerling LC: Appl Phys Lett 29 461, 1976.
5. Baraff GA and Schluter M: Mat Sci Forum 10-12 293, 1986.
6. Itoh N: Adv Phys 31 491, 1982.
7. Itoh C, Tanimura K and Itoh N: J Phys C19 6887, 1986.
8. Hayes W, Kane MJ, Salminen O, Wood RL, Doherty SP: J Phys C17 2943, 1984.
    Tanimura K, Tanaka T, Itoh N: Phys Rev Lett 51 423, 1983.
    Trukhin AN, Plaudis AE: Sov Phys St 21 644, 1979.
9. Griscom D: SPIE Vol 541 Radiation Effects in Optical Materials, 1985 p53.
10. Stoneham AM: Rep Prog Phys 44 1251, 1981.
11. Itoh N and Saidoh M: Journale de Physique 34(C19) 101, 1973.
12. Itoh N, Stoneham AM and Harker AH: J Phys C10 4197, 1979.
13. Madey TE, Ramaker DE and Stockbauer R: Ann Rev Phys Chem 35 215, 1984.
14. English CA and Jenkins ML: Mat Sci Forum 15-18 1003, 1987.
15. Johnson WL (this meeting).
16. Knotek ML and Feibelman PJ: Phys Rev Lett 40 964, 1978.
17. Steeples K, Dearnaley G and Stoneham AM: Appl Phys Lett 36 981, 1980.
18. Naguib HM and Kelly R: Rad Eff 25 1, 79, 1975; also Kelly R, this meeting.
19. Torres VJB, Masri PM and Stoneham AM: J Phys C20 L143, 1987.
20. Hobbs lW, Hughes AE and Pooley D: Proc Roy Soc A332 167, 1973.
21. Kennedy TA: Mat Sci Forum 10-12 283, 1986.
22. Dean PJ: p147 of "Deep Levels in Semiconductors", Pantelides S(ed), New York: Gordon and Breach 1986.
23. Rochet F, Rigo S, Froment M, D'Anterroches C, Maillot C, Roulet H and Dufour G: Adv in Phys 35 237, 1986.
24. Zimmerman J: J Phys C4 3265, 1971.
25. Martini M, Spinolo G and Vedda A: J Appl Phys 61 2486, 1987.
26. Haliburton LE, Koumvakalis N, Merkes ME and Martin JJ: J Appl Phys 52 3565, 1981.
27. Stoneham AM and Sangster MJL: Phil Mag B43 609, 1980.
28. Fowler WB and Elliott RJ: Phys Rev B34 5525, 1986.
29. Stoneham AM and Tasker PW: to appear in "Oxide Surfaces", Dufour L and Nowotny J(Eds), Amsterdam: Elsevier, 1987.
30. Stoneham AM: to appear in Cryst Latt Def and Amorph Mat, 1987.
31. Stoneham AM: J Am Ceram Soc 64 54, 1981.
32. Stoneham AM and Tasker PW: Mat Res Soc Symp 40 291, 1985.
33. Henrich V: Rep Prog Phys 48 1481, 1985.
34. Stoneham AM and Tasker PW: J Phys C18 L549, 1985.
35. Knotek ML: Rep Prog Phys 47 1499, 1984.
36. Knotek ML and Fiebelmann PJ: Phys Rev Lett 40 964, 1978.
37. Itoh N, Stoneham AM and Harker AH: AERE Report TP1221 (1987) and to be published.

FAST-ION-INDUCED DEFECTS IN SILICON STUDIED BY DEEP LEVEL TRANSIENT SPECTROSCOPY.

A. HALLÉN, P. HÅKANSSON, B. U. R. SUNDQVIST AND E. TILLBERG

DIVISION OF ION PHYSICS, DEPARTMENT OF RADIATION SCIENCES
UPPSALA UNIVERSITY, BOX 535, S-751 21 UPPSALA, SWEDEN

1. INTRODUCTION

Ion beam irradiation of silicon has been used extensively in semiconductor technology, but mainly in implantation of dopant atoms (B, As, P etc.). The dopant atoms substitute the silicon atoms in the lattice and change with their extra electron or hole, the electric conductivity of the material. In the bandgap they appear as shallow levels, lying close to the conduction or valence band edges. This technique uses ions in the keV energy region and the penetration depth is at the most a few μm (1).

Ions in the MeV energy region, i.e. fast ions, have lately also been used for modification of silicon (2). This field does not concern implantation of dopant atoms, but utilizes light ions such as protons and alpha particles, in order to reach as deep as possible in the silicon. Instead of substituting the silicon atoms, these light nuclei disturb the lattice, creating vacancies, divacancies and other types of defects. The more stable defects, or damage, are created towards the end of the track, where nuclear stopping plays a significant role. The defects appear as deep levels in the silicon bandgap, where holes and electrons can recombine, thus shortening their lifetime (3). The major advantage of this technique of reducing the carrier lifetime, is that it enables a localization of the lifetime reducing region by varying only the energy of the bombarding ions. The role of nuclear and electronic stopping in the damage creation mechanism is not yet understood, but the usefulness of this application of ion irradiation is already documented for a number of silicon power devices, such as thyristors (4,5) and power transistors (6).

2. EXPERIMENTAL

A system analogous to a conventional ion implanter, but intended for fast ion beams has been built at The Svedberg Laboratory in Uppsala (7). The system can scan a 5 MeV proton beam over an area of 15×15 cm^2 with a dose variation of less than 5% across the surface. A scale drawing of the set up is presented in figure 1. The ions are supplied by an EN-tandem accelerator which, in combination with a sputter ion source and two duo plasmatrons, enables a large variety of particles and energies to be used. Compared to low energy implanters only low beam currents are available, but since the effects discussed here are seen for doses as low as 10^9 cm^{-2}, this is not a major limitation.

For this investigation clean wafers and diodes made from Float Zone, (111), 200 and 110 Ωcm, NTD (neutron transmutation doped) silicon were irradiated with protons and alphas. The diodes were formed by diffusion of aluminum ($\sim 10^{16}$ cm^{-3}).

231

R. Kelly and M. Fernanda da Silva (eds.), Materials Modification by High-fluence Ion Beams, 231–236.
© *1989 by Kluwer Academic Publishers.*

The irradiations took place at room temperature and in this particular study no annealing was done.

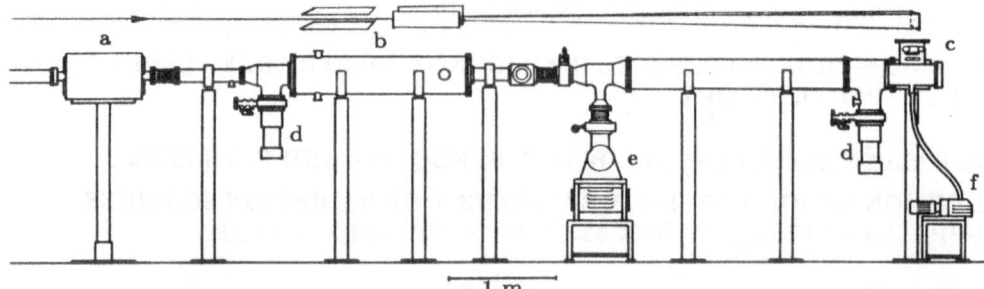

FIGURE 1. Raster scan beam line: MeV ions enter the system from left in the drawing. The ions are then focused by a quadrupole lens (a) and move on into the deflection chamber (b), where an electrostatic field scans the beam in x- and y-directions. The target chamber and vacuum lock (c) are situated to the far right. Also seen are the cryopumps (d), a turbomolecular pump (e) and a roughing pump (f).

3. MEASUREMENTS

Shallow implantations can readily be characterized by sheet resistivity, RBS, SIMS or other surface analysing technique. To analyse the effects of light MeV ions, which penetrate hundreds of μm into the silicon bulk, different measurements are required. In the next section two such methods, which use the change in electric behaviour to monitor the effects of the irradiation, will be described.

3.1 Resistivity profiles

A straight forward way to reach the silicon bulk is to grind the irradiated area at a small angle, projecting the damage to the surface as in figure 2. Then the resistivity can be measured by ordinary two- or four-point probes.

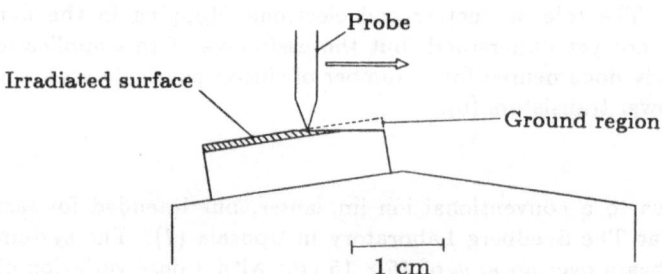

FIGURE 2. Recording of a resistivity profile.

There is no direct way to calculate the number of defects from the resistivity values and the grinding also introduces additional defects besides destroying any processed device. Despite this, resistivity profile measurements give some information of the effects of MeV ions (8).

3.2 Deep level transient spectroscopy

DLTS, introduced by D. V. Lang (9) is a non destructive measurement which uses

Schottky or pn-diodes made on the silicon wafer. The diodes are reversed biased with the voltage V_r and then the depleted region is flooded with carriers by an injection pulse V_i. The pulse duration is sufficiently long to fill all recombination centers with carriers and when the diode again is under reverse bias, these trapped carriers are emitted with a time constant typical of the specific trap at that temperature. The trapped carriers affects the capacitance and as the carriers are emitted, the capacitance returns to its quiescent state (figure 3).

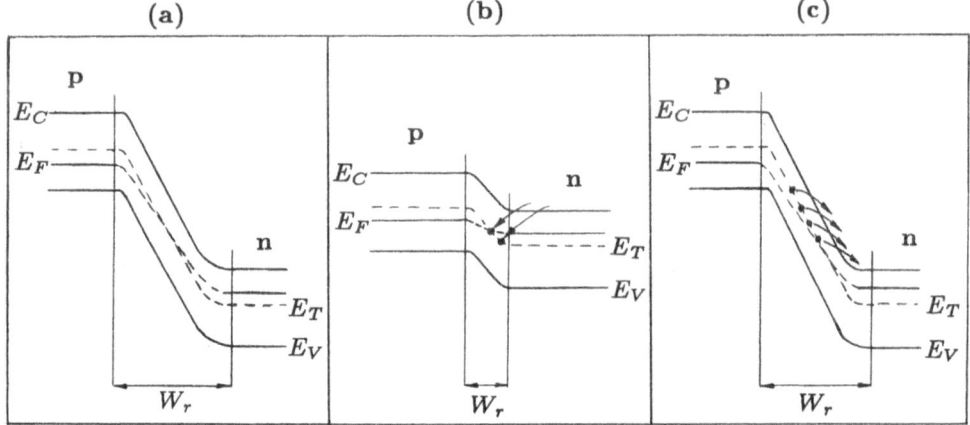

$$(a) \qquad\qquad (b) \qquad\qquad (c)$$

FIGURE 3. Band diagram for a pn-junction under reverse bias (a), during the filling pulse (b) and in the transient state (c). W_r is the depletion region.

By ramping the temperature of the diode the fermi level (E_F) reaches different trap levels (E_T) in the bandgap and every level adds a transient behaviour to the capacitance. Recording of this change in capacitance as a function of temperature gives a spectrum of existing deep levels in the bandgap. From this spectrum the exact bandgap position and emission rates are deduced and the peak magnitude is a direct measure of the trap concentration.

Increasing the reverse bias extends the depleted region further in the n-type silicon and traps can be detected at different depths. With the temperature fixed at the value corresponding to a certain trap, profiling of this trap concentration is accomplished by varying the height of the filling pulse from some steady state reverse bias. There are also ways to separate majority and minority carriers by the settings of V_r and V_i.

4. RESULTS AND DISCUSSION

Resistivity profiles recorded by a two point probe for MeV proton and alpha irradiated silicon showed a nonlinear dose dependence for doses varying between 10^9 cm^{-2} and 10^{13} cm^{-2} (figure 4). At lower doses no increase in resistivity was seen after irradiation and at higher doses the resistivity saturated, along the entire range, around the value of intrinsic silicon ($\approx 10^5 \Omega$cm). Since all defects and deep levels in the bandgap contribute to the change in resistivity, it is impossible to relate any stopping mechanism to the different parts of the resistivity profile. Considering the nonlinear dose dependence, however, it seems unlikely that only nuclear stopping is involved in the defect creation.

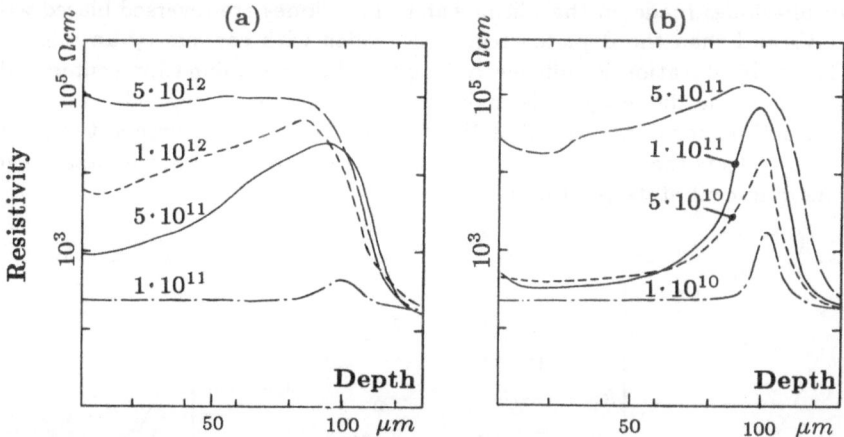

FIGURE 4. Resistivity profiles from (a) 3.2 MeV proton and (b) 13 MeV alpha irradiation of 200 Ωcm silicon for different doses (cm^{-2}).

DLTS plots give more information of the types of defects and figure 5 shows three spectras from proton and alpha irradiated diodes (10). Each spectrum belongs to a diode irradiated according to figure 6. Unirradiated diodes have no pronounced DLTS-peaks. Doses are $5 \cdot 10^{11}$ cm^{-2} for all three cases.

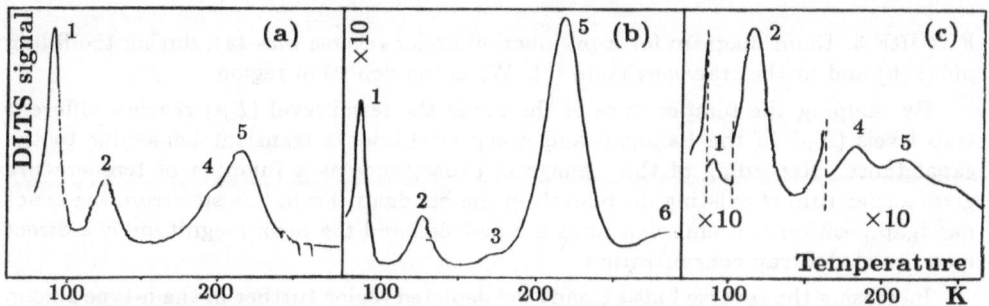

FIGURE 5. DLTS plots. (a) 8 MeV protons, (b) 6.5 MeV protons from the cathode side and (c) 13 MeV alphas. Note that the DLTS-signal in (b) and parts of (c) should be magnified ten times.

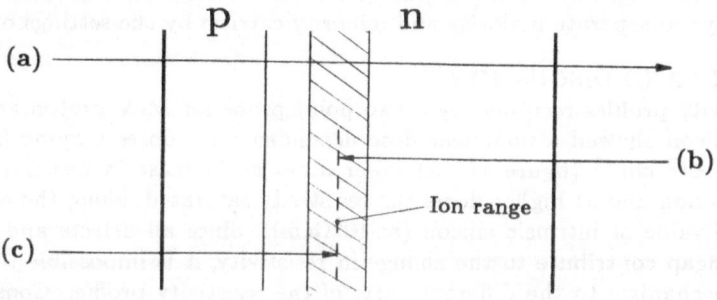

FIGURE 6. The irradiated diodes. The region contributing to the DLTS spectra is shown dashed.

Identification of defects is not straight forward and other types of measurements such as IR-spectroscopy, ESR and EPR together with annealing are needed to give a complete characterization. Here only tentative origins will be assigned to the peaks in figure 5. The peaks 1, 2 and 5 seem to be present after all three irradiations. 1 is possibly due to vacancy/oxygen center (11,12) while 2 and 5 are due to divacancies, double and single negatively charged respectively (13,14). Peaks 3 and 6 are less pronounced, but might be hydrogen-related defects since they only appear in spectra (b) (15). Preliminary calculations of the corresponding bandgap positions for the most significant peaks give:

Peak number	Position
1	$E_C - 0.20 \pm 0.03$ eV
2	$E_C - 0.29 \pm 0.03$ eV
5	$E_C - 0.44 \pm 0.03$ eV

Although the concentration of defects are in the same order of magnitude, peak 1 is dominant in spectrum (a), where electronic stopping plays a more significant role. In spectrum (b) and (c), where effects due to the nuclear stopping region should be enhanced, peak 5 is somewhat more pronounced. A possible interpretation could be that the lower lying level, E_C-0.20 eV, is more associated with electronic stopping, while the deeper level is more related to nuclear stopping.

Depth profiling of these defects has not yet been done but is essential before more definite conclusions can be drawn about the relative importance of the two stopping mechanisms in the defect creation.

5. CONCLUSIONS

The results presented here are still preliminary but shows that DLTS measurements of fast-ion implanted silicon is a promising method to characterize deep defects and such studies may contribute to the understanding of the mechanism for defect creation.

REFERENCES

1. B. G. Streeman: Solid State Electronic Devices (2nd ed.). pp. 131, Prentice-Hall Inc, N. J. 1980

2. J. Bartko and Kuan H. Sun: Reducing the Switching time of Semiconductor Devices by Nuclear Irradiation. U. S. Patent 4 056 408, Nov 1, 1977

3. S. K. Ghandhi: Semiconductor Power Devices. Wiley, New York, 1976

4. D. C. Sawko and J. Bartko: Production of Fast Switching Power Thyristors by Proton Irradiation. IEEE NS-30, No 2, April 1983

5. D. Silbert, W-D. Nowak, W. Wondrak, B. Thomas and H. Berg: Improved Dynamic Properties of GTO-Thyristors and Diodes by Proton Implantation. IEEE, 162-IEDM 6.6, 1985

6. A. Mogro-Campero, R. P. Love, M. F. Chang and R. F. Dyer: Shorter Turn-Off Times in Insulated Gate Tran- sistors by Proton Implantation. IEEE, EDL-6, No 5, May 1985

7. A. Hallén, A. Ingemarsson, P. Håkansson, G. Possnert and B. U. R. Sundqvist: A Fast-ion Beam Raster Scan System: In manuscript. Dept. of Radiation Sciences, Uppsala University, Box 535, S-751 21 Uppsala, Sweden

8. A. Hallén, P. Håkansson and B. Sundqvist: Resistivity Profiles in Proton and Alpha Irradiated Silicon. Internal report, Tandem Acc. Lab. Box 533, 751 21 Uppsala, Sweden, TLU 144/86

9. D. Lang: J. Appl. Phys. 45, 3023, 1974

10. E. Tillberg: DLTS-studier av protonbestrålningsinducerade defekter. Project report for Master of Eingineering Sciences. Kungliga Tekniska Högskolan, Stockholm, 1986. (In swedish)

11. L. C. Kimerling: Radiation Effects in Semiconductors 1976, edited by N. B. Urli and J. W. Corbett. Conference Series No. 31, The Institute of Physics, Bristol, 1977

12. G. D. Watkins and J. W. Corbett: Phys. Rev. 121, p. 1001, 1961

13. A. O. Evwaraye and E. Sun: J. Appl. Phys. 47, p. 3776, 1976

14. K. L. Wang, Y. H. Lee and J. W. Corbett, Appl. Phys. Lett. 33,

15. K. Irmscher, H. Klose and K. Maass: Hydrogen-related deep levels in proton-bombarded silicon. J. Phys. C: Solid State Phys., 17, 1984

LOW ENERGY (300 eV - 10 keV) Ar+ AND Cl+ ION IRRADIATION OF (100) Si

S. KOSTIC, D.G. ARMOUR AND G. CARTER

Department of Electronic and Electrical Engineering, University of
Salford, Salford M5 4WT, UK.

1. INTRODUCTION

The use of reactive ions to etch Si and SiO$_2$ has been shown, under
appropriate conditions, to enable anisotropic and selective etching to be
obtained [1]. In an etching process where reactive ion etching is used,
damage is inevitably produced by high kinetic energy ions impinging on a
surface from a plasma [2]. Ion beam etching offers greater control and can
be employed to shape features that are unobtainable by other etching
techniques; in fact, this process introduces the possibility of controlled
isotropic material shaping [3]. However, the radiation damage produced by
this process is a serious problem.

The aim of the present work is to study and compare the radiation damage
produced in single crystal Si by two distinctly different ion species,
similar in mass yet very different in their chemical nature, the two ion
species in question being Cl+ and Ar+. These ion species were implanted
into Si in the energy range 300 eV - 10 keV, i.e. values applicable to low-
energy ion implantation as well as reactive ion-etching processes. The
resulting disorder generated within the crystal is quantified employing the
well known technique of Rutherford backscattering/channelling. Attempts
have been made to determine etching and disorder mechanisms exploiting a
plasma etching apparatus [4,5]. However, in such systems it is not easy to
control individual parameters such as bombardment species, energy and
target environment independently. The use of a mass analysed ion beam in an
efficiently differentially pumped system is therefore essential if the
fundamental processes associated with dry etching are to be studied
effectively.

2. EXPERIMENTAL

All of the samples used in this study were commercially prepared, n-
type, high resistivity (250 Ωcm), (100) oriented Si single crystals. The
substrates were mounted in a UHV target chamber, base pressure ~7 x 10^{-8}
Pa, such that the ion beam would impinge on the substrate surface at normal
incidence. Some channelling would occur, but as has been demonstrated
previously this effect is negligible with increasing ion fluence [6]. All
the samples were etched in hydrofluoric acid followed by a rinse in
methanol immediately prior to loading in the target chamber through a
differentially pumped target transfer system.

^{40}Ar+ and ^{37}Cl+ ions were generated and accelerated to 10 keV in a
Freeman type ion source. The ion beam was mass analysed and finally
decelerated to energies of 3 keV, 1 keV and 300 eV. The ion flux was kept
constant to 0.1 μA/cm^2 for all implants, and, although the substrates were
maintained at room temperature, no significant temperature increase would
be expected due to the low power input from the irradiation (0.03 - 1
mW/cm^2). The ion beam uniformity was measured with a Faraday cup to be

237

R. Kelly and M. Fernanda da Silva (eds.), Materials Modification by High-fluence Ion Beams, 237–243.
© *1989 by Kluwer Academic Publishers.*

238

better than 5%. The same Faraday cup was employed to measure the ion beam
current before and after implantation. For high dose implants, spot checks
on the current were made during the implant period by moving the Faraday
cup into the beam using the mechanical, scanning arrangement. Continuous
beam sampling was precluded by the geometry of the target and Faraday cup
assembly. The ion fluence was varied in the range 3×10^{12} – 10^{16}
ions/cm². The above UHV ion beam deposition/implantation system overcomes
the complexity of the processes involved in plasma or wide bore, non-mass
analysed systems, where the identity of the bombarding species is virtually
impossible to specify, making it difficult to carry out any definitive
measurements in such systems.

Subsequent to ion implantation, analysis was performed with 2 MeV He⁺
ions and the combined technique of Rutherford backscattering and
channelling. Alignment was in the ⟨100⟩ direction for the channelling
investigations. Both random and aligned incidence spectra were recorded
with normal (or near normal) incidence and two scattering geometries, $\Theta =$
168° and $\Theta = 99°$. In the latter case a much improved depth resolution is
achieved.

In the case of $\Theta = 168°$ scattering geometry and near normally incident
and channelled probing ion beam, the depth resolution is quite poor and
hence, even for the highest ion implant energy and fluence, the aligned
spectrum does not reach the random incidence level. The more reliable and
informative data is obtained with the $\Theta = 99°$ scattering geometry. From
the aligned spectra with this geometry we observe that after a certain
implant fluence the height of the backscattered peak, at the leading edge
of the Si, remains constant with further increase in ion fluence, although
the width of this peak increases with fluence. This invariance of the Si
peak height is used to indicate amorphisation of the Si from the surface
inwards as it is otherwise difficult to obtain "random" spectra for the $\Theta =$
99° scattering geometry. Typical RBS/channelling spectra for the 99°
scattering geometry are shown in Figure 1. The region of the spectrum
corresponding to the position where backscattering from implanted Ar or Cl
occurs is omitted for clarity.

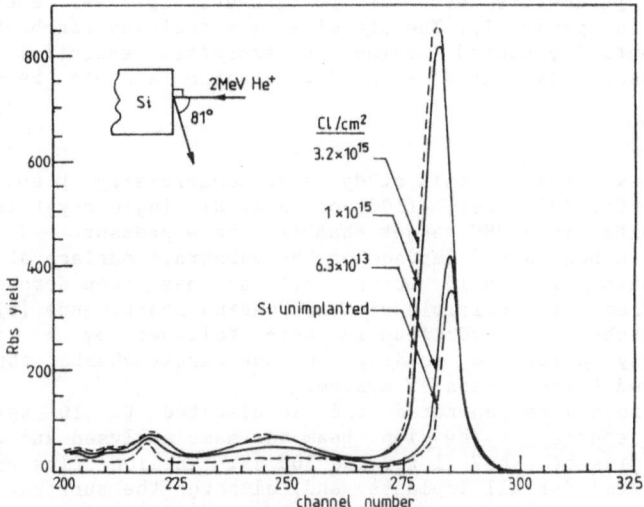

Figure 1. RBS/channelling spectra of 2 MeV He⁺ ions backscattered from
(100) Si implanted with 3 keV Cl⁺.

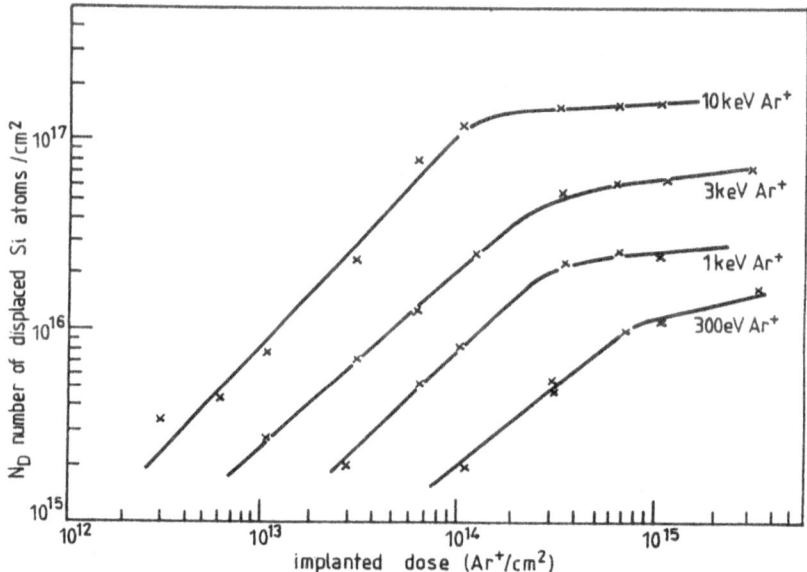

Figure 2. Total disorder, N_D, as a function of incident Ar⁺ ion fluence, Φ, for four different incident ion energies.

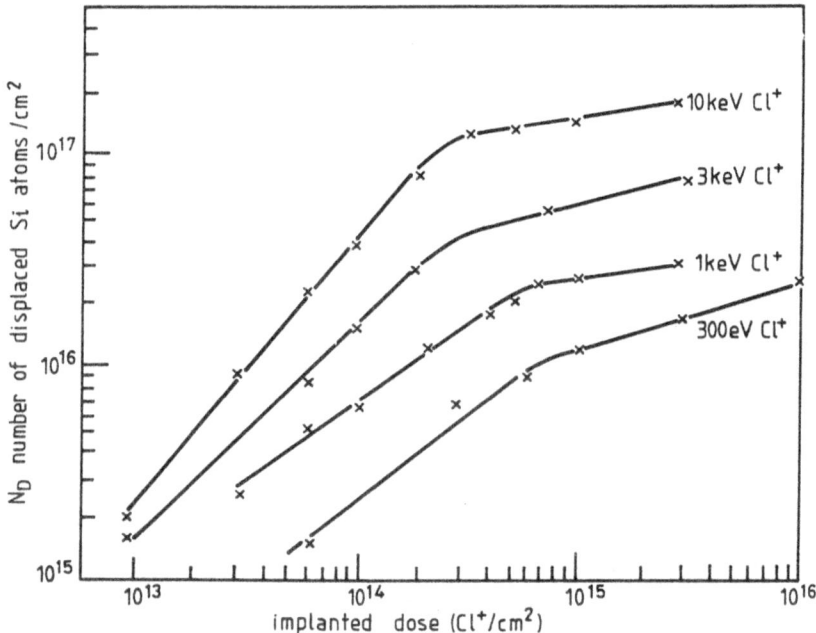

Figure 3. Total disorder, N_D, as a function of incident Cl⁺ ion fluence, Φ, for four different incident ion energies.

240

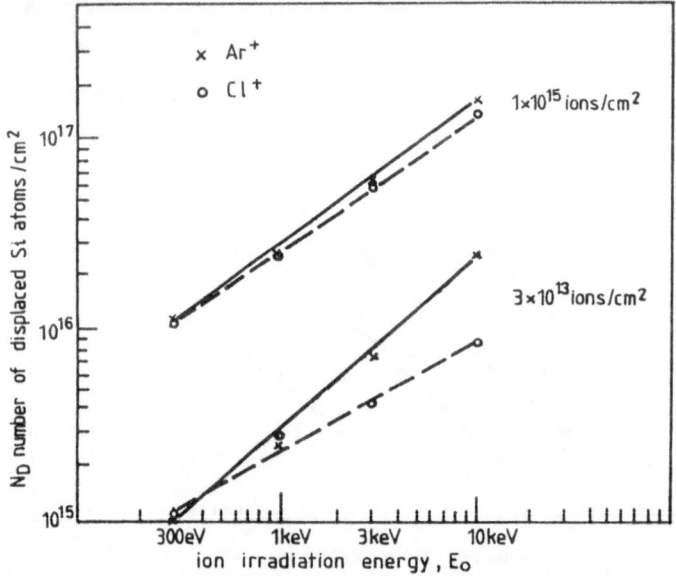

Figure 4. Total disorder, N_D, as a function of incident Ar^+ and Cl^+ irradiation energy for two different ion fluences: 3×10^{13} cm^{-2} and 1×10^{15} cm^{-2}.

The number of displacements per incident ion, N_D*, expected to be produced, on theoretical grounds applicable to linear cascade conditions, can be expressed, using the modified Kinchin-Pease model [10], as

$$N_D^*{}_{K-P} = \frac{0.42 \; \gamma(E_o)}{E_d} \cdot E_o$$

where $\gamma(E_o)$ is the energy fraction deposited into nuclear scattering events, E_o is the incident ion energy and E_d is the effective displacement energy (= 13 eV for Si). The values of $\gamma(E_o)$ were obtained from a TRIM code employing Monte Carlo calculations [11], thus enabling estimations of $N_D^*{}_{K-P}$. The experimental value, $N_D^*{}_{EX}$, can be readily determined from

Figures 2 and 3 at 10% of the saturation level. All the values of N_D*, experimental and theoretical, are tabulated in Table 1 for comparison.

E_o =	10 keV		3 keV		1 keV		300 eV	
	Ar^+	Cl^+	Ar^+	Cl^+	Ar^+	Cl^+	Ar^+	Cl^+
$N_D^*{}_{EX}$	780	210	275	143	77	67	20	32
$N_D^*{}_{K-P}$	278	276	86	86	29	29	9	9

Table 1.

3. RESULTS AND DISCUSSION

From raw RBS data similar to those shown in Figure 1, the number of displaced Si atoms, after each Ar^+ or Cl^+ fluence, was determined by first subtracting the spectrum from an unimplanted substrate and then employing a linear dechannelling correction across the surface peak [7]. Integration of the region corresponding to implanted Ar or Cl enabled the quantities of Ar or Cl retained in the sample to be deduced. This retained dose was measured, although this was only possible for the higher implant fluences.

Figures 2 and 3 present the total disorder N_D as a function of Ar and Cl implant fluence respectively, for all four different implant energies employed. The slopes of all four curves in Figure 2 are quite similar and are all slightly greater than unity, namely ~1.2. For the case of Cl^+ irradiation, from Figure 3, the slopes of the disorder-fluence curves take on values between 1.2 for the highest incident energy ($E_o = 10$ keV) and 0.7 for the lowest incident energy ($E_o = 300$ eV), i.e. decreasing slopes with decreasing energy.

A unity slope behaviour for log N_D = logΦ curves, where N_D is the number of displaced atoms and Φ the incident ion fluence, is normally associated with accumulation of simple defects, which decrease towards a saturation or quasi-saturation level at higher fluences corresponding to the formation of a continuous amorphous layer at some depth within the material [8,9]. In the case of Ar^+ bombardments where a greater than unity (~1.2) slope is observed at all energies, the excessive damage may be attributable either to non-linear cascade conditions or to significant distortion of the lattice due to the incorporation of the gas atoms. For the energy range in question, linear cascade conditions are expected (even down to 300 eV) for both Ar and Cl. The fact that the lower energy Cl curves have slopes below unity, while the collision cascades should remain virtually identical to those for Ar, rules out the cascade itself as the source of the additional damage. Under these circumstances the differences between Ar and Cl as well as the level of damage observed (Table 1) must be associated with the nature of the implanted species and in the way in which it is accommodated in the lattice, unless the Cl bombardment results in additional sputtering due to chemical effects. If this were the case, then the quasi-saturation levels obtained with Cl would be expected to be lower than those for Ar. Since the levels observed (Figures 2 and 3) are very similar, chemical sputtering as an effective annealing process during Cl bombardment can be ignored.

The above assumptions may be emphasized in Figure 4 where N_D is plotted as a function of incident ion energy for a low fluence (3×10^{13} cm^{-2}) and high fluence (1×10^{15} cm^{-2}) case corresponding to the low pre-saturation and quasi-saturation regime respectively. Little or no difference is shown for the Ar^+ and Cl^+ irradiation at a fluence of 1×10^{15} cm^{-2} in the energy range studied here. Indeed, the near unity slope of the log N_D - log E_o curves shows the as-expected behaviour for amorphous region growth with increasing incident ion energy. However, in the low fluence, 3×10^{13} cm^{-2} regime the Cl^+ data exhibits a trend in disorder with increase in irradiation energy which is also linear on the log N_D - log E_o plot but with a lower slope than that for the Ar^+ case. This means that at the low fluences the damage build up with energy increase is slower with Cl^+ than Ar^+. This trend is consistent with the large difference observed in the 10 keV Ar^+ and Cl^+ disorder curves in the low fluence or pre-saturation range.

Although at this stage it is difficult to give a definitive explanation, it is possible, on the basis of these damage build-up curves and the damage behaviour expected from linear cascade conditions, to suggest the most probable mechanisms involved.

For the Ar-Si system the experimental values of N_D in Table 1 are ~ 2.5 times greater than theoretical estimates for all four incident ion bombarding energies employed. The Cl-Si system shows a very different behaviour whereby the ratio of the experimental to theoretical values, $N_D{}^*{}_{EX}/N_D{}^*{}_{K-P}$, increases with decreasing energies and is ~ 1 for the highest E_o case ($E_o = 10$ keV).

At this stage it must be pointed out that the statistical error, from RBS data, is greatest for the low energy low fluence conditions. Also, all the RBS damage data include subtraction of the unimplanted surface peak which amounts to $\sim 10^{16}$ Si atoms/cm^2. This is equivalent to ~ 20 monolayers of the Si substrate and indicates the presence of a native oxide. The low energy (300 eV) irradiation will thus largely be dissipated in this near surface region which is not taken into account (i.e. it is subtracted as the virgin contribution) in the calculation of N_D. This implies that both the total disorder produced by Ar$^+$ and Cl$^+$ in a perfect Si crystal could be much greater than that deduced here and the difference in disorder caused by the 300 eV Cl$^+$ or Ar$^+$ ions could in fact be far greater than already observed. Verification of this is not possible with the RBS technique due to insufficient depth resolution. This uncertainty in N_D is not as pronounced for the higher energy bombardments where the ion penetration and the resulting disorder distribution is greatly in excess of this already disordered surface region, hence yields much more reliable information on disorder data.

In conclusion, it is suggested that in the energy range 300 eV - 10 keV, the disorder-fluence behaviour for chemically active as compared to inert ion species of similar mass in single crystal Si is very much dependent on the solubility of these ions in the matrix as well as the chemical activity of the species and its ability to be accumulated into the lattice structure without causing significant relaxation or distortion. However, once the onset of amorphisation has been achieved these differences become unimportant in the level of disorder produced. In the energy regime studied, sputter yield differences (physical and chemical) do not play a role for fluences up to at least 10^{16} cm^{-2} as far as the crystal disorder level is concerned. However, at still higher fluences, where surface erosion becomes significant, the disorder levels may be greatly reduced with lower incident ion energies, i.e. values applicable to the reactive sputter etching techniques.

Although the present investigations have yielded new and informative data on the fundamental applicability of mainly collisional considerations in determining the extent of low energy induced damage production in Si, it has also been shown that extreme precautions in ensuring initial and continuing surface cleanliness must be taken to unequivocally interpret the results of testing low energy ion implantation since these are influenced by surface conditions.

ACKNOWLEDGEMENTS

The authors are grateful to Dr J. Van den Berg for assistance with low energy ion implantation at the initial stages of this work.

REFERENCES

[1]   Revell P J, Goldspink G F.   Vacuum <u>34</u> 455 (1984).
[2]   Mu X C, Fonash S J, Yang B Y, Vedam K, Rohatgi A, Rieger J.   J Appl Phys <u>58</u> 4282 (1985).

[3] Kay E, "Erosion and Growth of Solids Stimulated by Atom and Ion Beams" (Ed: G Kiriakidis, G Carter, J L Whitton), Martinus Nijhoff Publishers, 247 (1986).

[4] Coburn J W, Winters H F. J Appl Phys 50 3189 (1979).

[5] Horiike Y, Shibagaki M, Kadano K. Jap J Appl Phys 18 2309 (1979).

[6] Kostic S, Begemann W, Abril I, Armour D G, Carter G. Rad Eff 100 1 (1986).

[7] Carter G, Grant W A. "Ion Implantation in Semiconductors", Academic Press, NY (1970).

[8] Thompson D A, Golanski A, Haugen H K, Stevanovic D Y, Carter G, Christodoulides C E. Rad Eff 52 69 (1980).

[9] Webb R P, Carter G. Rad Eff 42 159 (1979).

[10] Sigmund P. Rev Roum Phys 17 823 (1972).

[11] Biersack J P, Eckstein W. Appl Phys A34 73 (1984).

THE CHARGE STATE OF IRON IMPLANTED INTO SAPPHIRE

C. J. MCHARGUE, P. S. SKLAD, C. W. WHITE
Oak Ridge National Laboratory, P.O. Box X, Oak Ridge, TN 37831 USA

G. C. FARLOW
Wright State University, Dayton, OH 45435 USA

A. PEREZ, N. KORNILIOS, G. MAREST
Université Claude Bernard Lyon I, Villeurbanne Cedex, France

1. INTRODUCTION
     The use of ion implantation to modify the near-surface mechanical pro-
perties of ceramics is currently being explored at a number of laboratories.[1-3] The changes in hardness, apparent fracture toughness, and
flexure strength have been qualitatively related to the damage microstructures. However, the variation in mechanical properties with concentration
of implanted ion or between similar concentrations of different chemical
species has not been accomplished because detailed information on the local
defect structures has not been available.
     The nature of defects present in implanted insulating materials
(ceramics) depends upon the residual charge state of the implanted species.
In most studies to date, the implanted species has been a cation; therefore, there must be compensation for an excess positive charge.
     The use of conversion electron Mössbauer spectroscopy (CEMS) has given
insight into the local surroundings of iron implanted into MgO, LiF, and
TiO_2 (ref. 4–6). Iron in as-implanted MgO and LiF was present in three
well-defined charge states, Fe^{3+}, Fe^{2+}, and Fe^0 (metallic). By studying
the variation in the relative amounts of these charge states, these
investigators were able to develop models for the defect structures. The
purpose of this work was to extend such studies to sapphire.

2. EXPERIMENTAL CONDITIONS
     High-purity Al_2O_3 single crystals having <0001> normal to the surface
were given an optical grade polish and then annealed 120 h at 1450°C in
flowing oxygen to remove any residual polishing damage. Following this
treatment the crystals were almost defect free, as determined by Rutherford
backscattering-channeling (RBS) measurements. The minimum yields, X_m
(defined as the ratio of the backscattered yield from a c-axis aligned
crystal to that of a random specimen), were 2% in the aluminum sublattice
and 8% in the oxygen sublattice.
     Crystals were implanted at room temperature using a mass analyzed beam
of ^{57}Fe (160 or 100 keV). The fluences covered the range of 10^{16} to 10^{17}
ions/cm² to give peak iron concentrations of 3 to 30% of cations. The
crystals were implanted with the ion beam ~7° off-normal to minimize
channeling effects. Subsequently, the specimens were analyzed by RBS
using 2 MeV He⁺ ions.
     Specimens for transmission electron microscopy (TEM) were prepared in
both cross-section and plan view by mechanical polishing followed by ion
milling.

R. Kelly and M. Fernanda da Silva (eds.), Materials Modification by High-fluence Ion Beams, 245–254.
© 1989 by Kluwer Academic Publishers.

Mössbauer spectra were obtained using the technique of conversion electron Mössbauer spectroscopy (CEMS).[7] Two important features of CEMS for studies of implanted materials are: (1) CEMS probes only the surface to a depth of 150 to 200 nm, a thickness comparable to the width of the implanted zone; and (2) Al and O produce low photoelectron backgrounds in conversion-electron detection, thus yielding a high signal-to-noise ratio.

In the present study, CEMS spectra were determined at 4 K, 77 K, and room temperature using backscattered geometry. The ^{57}Co source was contained in a rhodium matrix and was mounted on a constant acceleration triangular-motion velocity transducer. The data were folded to produce a constant background. The velocity scale and all data are referred to a metallic α-iron absorber. The spectra were fitted with a computer least-squares procedure with the assumption of Lorentzian shapes of Mössbauer lines.

3. RESULTS

Micrographs obtained by TEM show the as-implanted microstructure to consist of tangled arrays of dislocations extending from the surface to a depth of about 170 nm. The corresponding electron diffraction patterns showed that the implanted zones remained crystalline. No evidence for precipitates was found; however, the large residual stresses and dense dislocation arrays would prevent the detection of very small precipitate particles as well as the determination of the nature of the dislocations.

The RBS spectra show the residual disorder in the aluminum and oxygen sublattices. The value of X_{min} at the position of maximum disorder in the Al-sublattice varied from 0.55 (1×10^{16} Fe/cm²) to 0.82 (1×10^{17} Fe/cm²), Figs. 1 and 2. Note that Fig. 2 shows the effect of the large concentration of iron (~30 cation %) on the RBS spectra of Al. The fraction of substitutional iron (as viewed along the c-axis) was: 0.86 at 1×10^{16} Fe/cm²; 0.72 at 2×10^{16} Fe/cm²; 0.50 at 4×10^{16} Fe/cm²; and ~ 0 at 1×10^{17} Fe/cm². The significance of this measurement in the presence of the disorder at the higher fluences is of course questionable. Nevertheless, the values at the lower fluences may aid in developing a model for the defect structure.

The CEMS spectra measured at room temperature are given in Fig. 3. The spectra consist of the superposition of several overlapping components. The following components were identified on the basis of consistent sets of computer fits for all the spectra: three quadrupole split doublets and one single line. These components can be assigned to a ferric ion (Fe^{3+}), two forms of a ferrous ion (Fe^{2+}_I and Fe^{2+}_{II}) and metallic iron (Fe^0). The Mössbauer parameters (isomer shift, IS; quadrupole splitting, QS; line widths, W; and relative line areas, R) are given in Table 1, and the fluence-dependence of the relative amount of each component is given in Fig. 4.

The ferrous iron, Fe^{2+}, ions are described by two different quadrupole-split doublets characterized by the isomer shifts IS ≈0.7 and 1.15 mm·s^{-1} and the quadrupole splittings QS ≈1.84 and 1.7 mm·s^{-1} for Fe^{2+}_I and Fe^{2+}_{II}, respectively. The parameters for the Fe^{2+}_I component are similar to those for the Fe-O bond in wustite.[8] The Fe^{2+}_{II} component represents a more ionic state and is similar in iron to $FeAl_2O_4$ (ref. 9). Since $FeAl_2O_4$ has a partly inverse spinel structure, the iron likely resides in octahedral sites, consistent with the relatively large value of IS observed. Essentially all the iron resides in the ferrous states at the lower concentrations (fluences) of iron and the relative amount decreases as the implantation fluence increases (Fig. 4), although the total amount in these states continues to increase (Fig. 5).

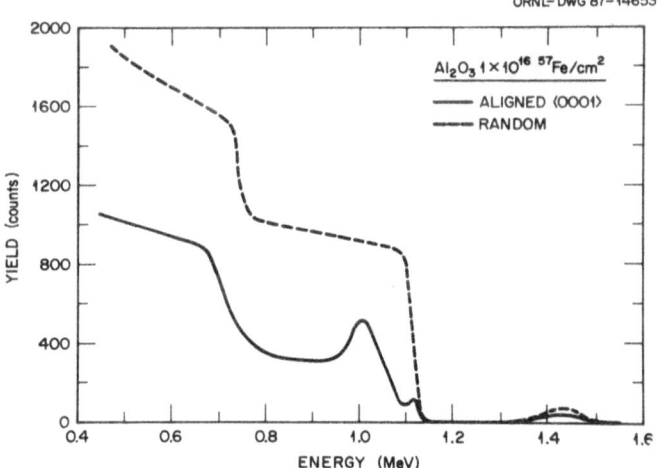

FIGURE 1. Rutherford-backscattering spectra for 2 MeV He$^+$
from α-Al$_2$O$_3$ implanted with 1×10^{16} ^{57}Fe/cm^2
(160 keV) at room temperature.

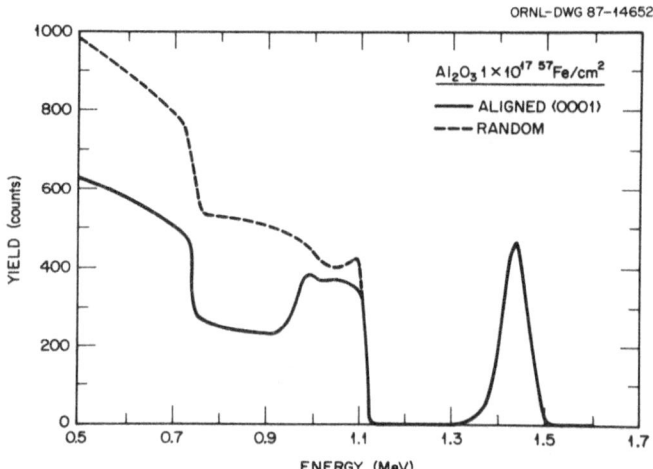

FIGURE 2. Rutherford-backscattering spectra for 2 MeV He$^+$
from α-Al$_2$O$_3$ implanted with 1×10^{17} ^{57}Fe/cm^2
(160 keV) at room temperature.

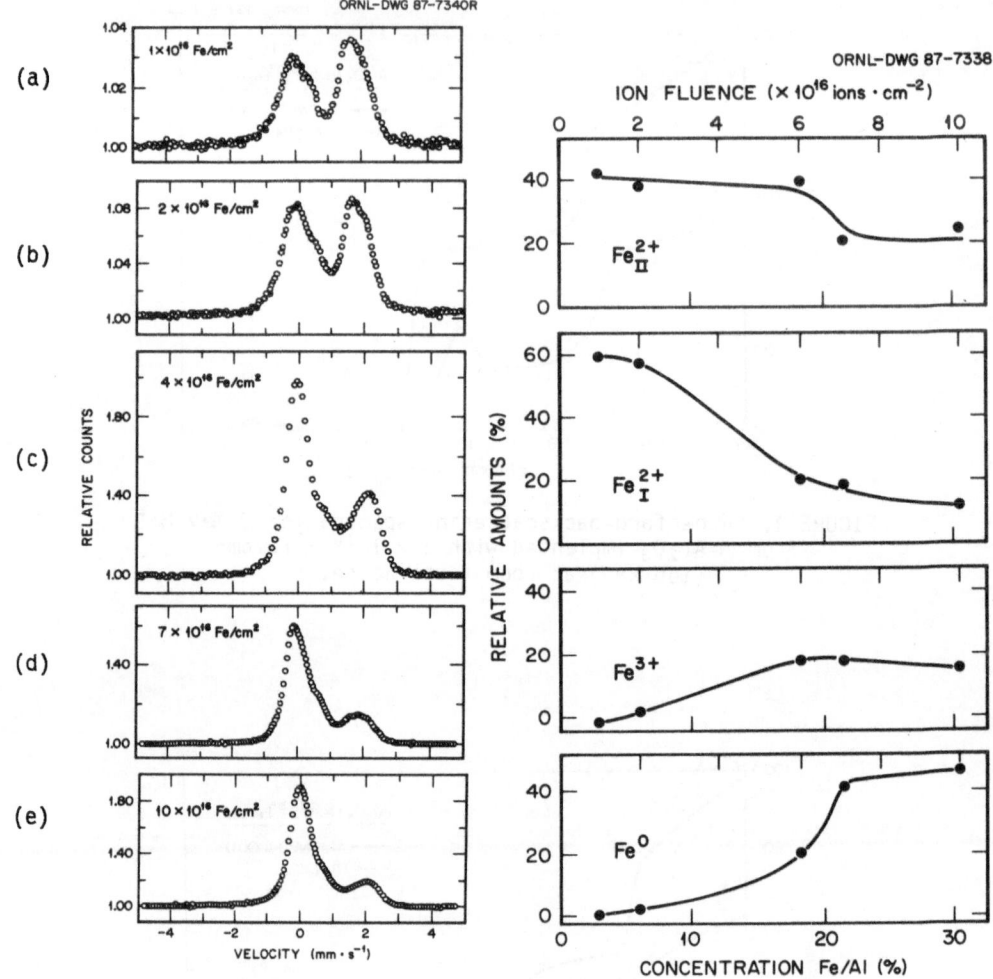

ORNL-DWG 87-7340R

ORNL-DWG 87-7338

FIGURE 3. Conversion electron Mössbauer spectra measured at room temperature on α-Al₂O₃ implanted with ⁵⁷Fe:
(a) 1×10^{16} ions·cm⁻² (160 keV);
(b) 2×10^{16} ions·cm⁻² (160 keV);
(c) 4×10^{16} ions·cm⁻² (100 keV);
(d) 7×10^{16} ions·cm⁻² (160 keV);
(e) 1×10^{17} ions·cm⁻² (160 keV).

FIGURE 4. Fluence (i.e., concentration) dependence of relative amounts of components in Mössbauer spectra of iron-implanted Al₂O₃.

TABLE 1. Mössbauer parameters of components present in as-implanted Al_2O_3

Fluence (ions cm^{-2})			1×10^{16}	2×10^{16}	4×10^{16}	7×10^{16}	1×10^{17}
E (keV)			160	160	100	160	160
Concentration (Fe:Al)			0.03	0.06	0.18	0.21	0.30
Doublet Fe^{2+}_I	IS	(mm·s^{-1})	0.70	0.68	0.69	0.66	0.67
	QS	(mm·s^{-1})	1.80	1.86	1.94	1.96	1.85
	W	(mm·s^{-1})	0.63	0.65	0.54	0.50	0.55
	R	(%)	59	58	21	19	12
Doublet Fe^{2+}_{II}	IS		1.19	1.20	1.14	1.17	1.17
	QS		1.68	1.64	1.94	1.78	1.85
	W		0.66	0.65	0.62	0.55	0.61
	R		41	38	39	20	24
Doublet Fe^{3+}	IS		--	0.24	0.26	0.18	0.22
	QS		--	1.00	1.08	0.97	1.02
	W		--	0.50	0.55	0.38	0.40
	R		--	4	19	18	16
Single Line Fe^0	IS		--	--	-0.09	-0.07	-0.08
	W		--	--	0.48	0.61	0.57
	R		--	--	21	43	48

IS = Isomer Shift
QS = Quadrupole Splitting
W = Line Width
R = Relative Area

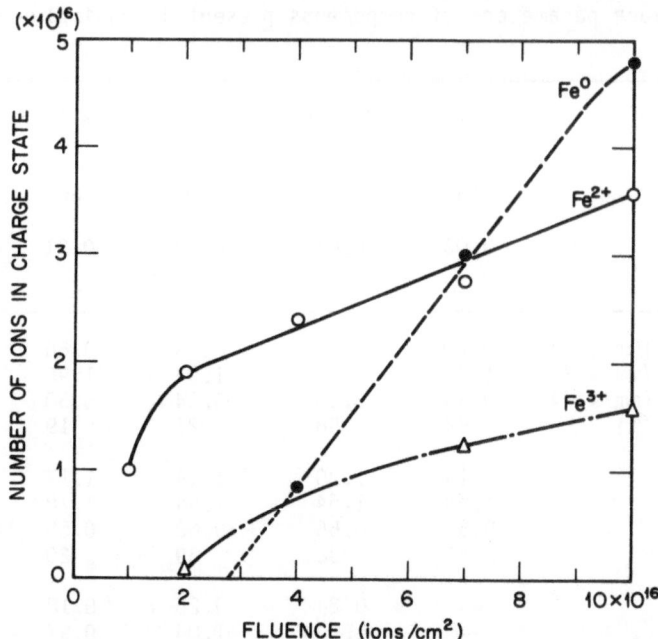

FIGURE 5. Total amounts of each charge state
of iron as functions of fluence.

The ferric iron, Fe^{3+}, is represented by a doublet having the
parameters: IS ≈ 0.22 mm·s^{-1}, QS ≈ 1.0 mm·s^{-1}. These values are comparable
to those of iron substitutionally located in alumina $(Al_{1-x} Fe_x)_2O_3$
(ref. 10) or amorphous Fe_2O_3 (ref. 11) and are indicative of covalent-
distorted octahedral surroundings. The relative amount of this Fe^{3+} component increases from about zero at 1×10^{16} Fe/cm^2 to a maximum of 20% for
the high fluence implants.

The single line with IS ≈ 0 mm·s^{-1} is attributed to small metallic iron
precipitates which behave superparamagnetically. The spectra taken at 77 K
were similar to the room temperature spectra (Fig. 3e) but those taken at 4
K show magnetic splitting. Figure 6 shows the 4 K spectra after Fourier
transform filtering and indicates a superparamagnetic sextet and the Fe^{2+}
doublet which distorts the symmetry of the line shapes. Thus, this component arises from small metallic clusters or precipitates with a size of
~2 nm. The relative fraction of this component increases with fluence and
represents about 48% of the implanted iron at the highest fluence.

In the earlier CEMS studies on iron implanted into nonmetallic crystals,
attempts were made to use a simple statistical model to analyze the fluence
dependence of the charge state.[4],[8] The model indicated the trend of the
data for LiF and showed quite good fit in the case of MgO. A similar
approach was applied to the present data.

The probability of finding various configurations of iron atoms (ions)
was calculated as a function of atom fraction of iron ($\bar{x}$) using the binomial distribution:

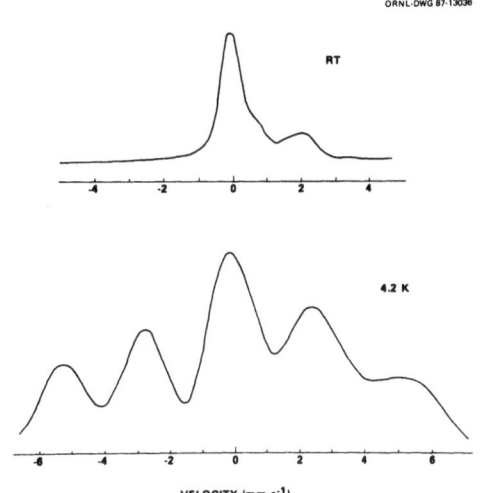

FIGURE 6. Mössbauer spectra obtained at 4 K of specimen
implanted with 1×10^{17} Fe/cm² (160 keV).

$$P_n\ (n,\bar{x}) = \left| \begin{matrix} N \\ n \end{matrix} \right| \bar{x}^n\ (1 - \bar{x})^{N-n}\ .$$

In this expression, N corresponds to the number of neighboring atoms which
can be substituted by iron and which form complexes of n impurity atoms
with the iron probe atom. Thus, the probability for finding an isolated
iron atom (n = 0), two irons (n=1), etc., can be calculated for each iron
concentration ($\bar{x}$) for a volume containing N nearest neighbors of the probe
atom. In this case, the best fit was obtained for N = 13, that is the
number of nearest cation neighbors in 4 coordination shells for an atom
occupying a cation site in the α-Al_2O_3 lattice.

Figure 7 shows the fits for the relative amounts of Fe^{+2}, Fe^{+3}, and Fe^0.
The solid line for Fe^{+2} represents the sum of the probabilities of finding
0, 1, and 2 iron atoms (ions) in addition to the probe iron atom within the
volume defined by four coordination shells of cations. This corresponds to
all configurations that have 1, 2, 3 total iron atoms in this volume. The
curve for Fe^{+3} is the calculated probability of finding exactly four iron
atoms in this volume, and that for Fe^0 represents the probability of
finding five or more such atoms. The points are the observed values of
relative amounts of each state taken from Fig. 4. The agreement between
these calculations and the data is excellent at low to moderated fluences.
At the highest fluence the amount of damage in the Al-sublattice makes the
concept of near-neighbors questionable.

ORNL-DWG 87-12974

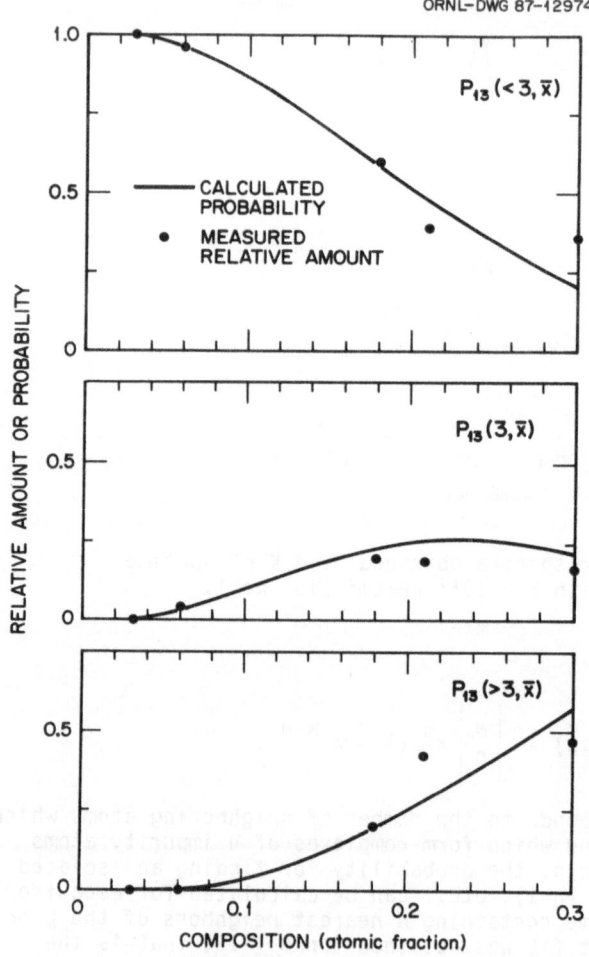

FIGURE 7. Probability for
various iron atom config-
urations in Al_2O_3 lattice
calculated from the bi-
nomial function with N=13.
Points are observed
relative amounts of Fe^{2+},
Fe^{3+}, and Fe^0.

4. DISCUSSION OF RESULTS

The distribution of the charge states of iron implanted into
Al_2O_3 differs significantly from those observed for LiF, MgO, and TiO_2.[4-6]
The high fluence implants in those cases resulted primarily in the Fe^{2+}
state, with up to 20% of the Fe^0 state in the case of MgO.

The probability calculations suggest that the Fe^{2+} state is occupied
when there are 1, 2, or 3 iron ions within four coordination shells. The
single iron corresponds to a single interstitial impurity ion and the
excess positive charge may be compensated by near-by oxygen interstitials
produced by the elastic collisions of the bombarding ions.

The calculated defect energies of Catlow et al.[12] indicate that clustering of defects will dominate upon the introduction of di-valent impurities
into Al_2O_3 in the absence of oxygen exchange with the atmosphere. That
analysis shows that either vacancy compensation (in which two impurity ions
occupy nearest neighbor sites with respect to an anion vacancy) or self-
compensation (in which the charge on a 2+ interstitial is neutralized by

two 2^+ impurity ions on substitutional sites located above and below the interstitial along the c-axis) are energetically preferred.

A consistent scheme for the defect structure involving the ferrous ions is:

One Fe^{2+} in four coordination shells = isolated interstitial

Two Fe^{2+} in four coordination shells = two Fe^{2+} substitutionals plus one oxygen vacancy

Three Fe^{2+} in four coordination shells = two Fe^{2+} substitutionals plus one Fe^{2+} interstitial

The ferric state, Fe^{3+}, appears to be a direct substitution of Fe^{3+} for Al^{3+}, either in the Al-sublattice or within the stoichiometric interstitial loop damage structure proposed by Pells and Stathopoulos.[13] It is not yet clear as to why this configuration is preferred for exactly four iron ions within the four coordination shells.

At high fluences (concentrations) of iron, the probability of forming clusters dominates and the large excess positive charge is avoided by formation of the metallic iron clusters. There is ample opportunity for the iron ions to capture electrons since much of the incident energy is lost by inelastic (electronic) processes.

The Mössbauer line associated with the small metallic iron clusters (or precipitates) has an isomer shift of about -0.08 mm·s^{-1} relative to bulk α-iron (bcc). There are two possible reasons for obtaining such a value rather than IS ≈ 0. The isomer shift for γ-iron (fcc) is also -0.08 mm·s^{-1}, suggesting that the initial configuration of the very small particle could be face-centered cubic. A second possibility is that the clusters are initially under very high isostatic pressure which increases the electron density at the nucleus and thus decreases the isomer shift. Using the expression for pressure $p = 2\Gamma/r$ with $\Gamma = 2065$ mJ m^{-2} for the interfacial free energy for Fe/Al_2O_3 (ref. 14) and $r = 2$ nm, a value of 1.0 GPa is calculated. This value is similar to that deduced from lattice spacing measurements for similarly sized solid xenon precipitates in Xe-implanted aluminum (15). Annealing studies (to be reported elsewhere) show that the value of IS goes to zero as the particles increase in size during annealing.

5. SUMMARY

Several techniques (RBS, TEM, CEMS) have been used to characterize sapphire single crystals implanted with iron at room temperature to fluences of 10^{16} to 10^{17} ions cm^{-2}. At low fluences the as-implanted iron is found mainly in the ferrous state. As the fluence is increased, Fe^{3+} and metallic iron clusters become dominant. There is a strong correlation between the probability of finding specific configurations of iron ions within four cation coordination shells and the relative amount of each charge state observed. The superparamagnetic behavior of the clusters suggest that they are of the order of 2 nm in size but the large amount of irradiation-induced damage and residual stress has prevented their imaging by TEM.

Research sponsored in part by the Division of Materials Sciences, U.S. Department of Energy, under contract DE-AC05-84OR21400 with the Martin Marietta Energy Systems, Inc.

254

REFERENCES

1. McHargue CJ: Nucl. Instrum. Methods Phys. Res. B19/20, 797 (1987).
2. Barnett PJ and Page TF: J. Mater. Sci. 19, 3524 (1984).
3. Hioki T, Itoh A, Noda S, Doi H, Kawamoto J, and Kamigaito O: J. Mater. Sci. Lett. 3, 1099 (1984).
4. Perez A, Marest G, Sawicka BD, Sawicki JA, and Tyliszczak T: Phys. Rev. B 28, 1227 (1983).
5. Kowalski J, Marest G, Perez A, Sawicka BD, Sawicki JA, Stanek J, and Tyliszczak T: Nucl. Instrum. Methods 209/210, 1145 (1983).
6. Guermazi M, Marest G, Perez A, Sawicka BD, Sawicki JA, Thevenard P, and Tyliszczak T: Mat. Res. Bull. 18, 529 (1983).
7. Sawicka BD and Sawicki JA: pp. 139–66 in Topics in Current Physics, Vol. 25, ed. U. Gonser, Springer, Berlin 1981.
8. Greenwood NN and Gibb TC: pp. 249 in Mössbauer Spectroscopy, Chapmann and Hall Ltd., London, 1971.
9. Rossiter MJ: J. Phys. Chem. Solids 26, 775 (1965).
10. Janet C and Gilbert H: Bull. Soc. Fr. Mineral. Cristallogr. 93, 213 (1970).
11. Van Diepen AM and Popma Th. JA: Solid State Commun. 27, 121 (1978).
12. Catlow CRA, James R, Mackrodt WC, and Stewart RF: Phys. Rev. B 25, 1006 (1982).
13. Pells GP and Stathopoulos AY: Rad. Eff. 74, 181 (1983).
14. McLean M and Hondros ED: J. Mater. Sci. 6, 19 (1971).
15. Templier C, Gaboriaud RJ, and Garem H: Mater. Sci. Eng. 69, 63 (1985).

IMPLANTATION AND MIXING

DIRECT AND RECOIL IMPLANTATION, AND COLLISIONAL ION-BEAM MIXING: RECENT
LOW-TEMPERATURE EXPERIMENTS

ALBERTO GRAS-MARTI[1] and UFFE LITTMARK[2]

[1]Health and Safety Research Division, Oak Ridge National Laboratory, Oak
Ridge, TN 37831-6123, USA, and Dept. of Physics, University of Tennessee,
Knoxville, TN 37996.   Permanent address: LCFCA, Dpt. Física Aplicada,
Universitat d'Alacant, Apt. 99, E-03080 Alacant, Spain

[2]IGV-KFA Julich, Association EURATOM/KFA, D-5170 Julich, W. Germany

Table of Contents

1. SCOPE AND OBJECTIVES
    This contribution serves as a bridge between the section, in this
NATO Advanced Study Institute, devoted to fundamentals of ion-solid
interactions, and that of the applications to materials modification.
Theoretical concepts directly relevant to the topics covered in this paper
have been discussed in the preceding lecture. Here we collect and discuss
recent low-temperature (T < 300 K) experimental work on ion and recoil
implantation, and ion-beam mixing, and their relation with collisional
theory.   We shall not include those experiments the interpretation of
which hinges on thermal or temperature-activated processes, like radiation
enhanced diffusion, segregation, etc.   These topics are covered in the
following contributions. Computer simulation experiments are discussed in
previous chapters.   In a recent extensive report, Hubler [HU 87] describes
all aspects of ion-beam implantation processing, from the calculation of
range profiles to the discussion of time and costs for practical
applications of ion implantation.   This reference provides a brief

257

R. Kelly and M. Fernanda da Silva (eds.), Materials Modification by High-fluence Ion Beams, 257–284.
© 1989 by Kluwer Academic Publishers.

introduction to many of the topics discussed in this NATO–ASI. Another general introduction is offered in [TO 86].

We provide here a bibliographic review in tabular format, and discuss briefly recent experimental progress and its relation with existing collisional theory. No specific data or results will be reproduced here. A copy of the (more tutorial and descriptive) notes prepared for the lecture given at the NATO–ASI may be requested from the first author [GR 87a]. Because of space limitations, and also due to the existence of recently published bibliographies, our data bank includes work published after 1983. References in which no attempt is made to compare the measurements with theory have been generally disregarded. We apologize for inadvertent omissions of relevant material. Literature search was mainly based on the "scan–through" approach, with minor computerized data searches. The following journals have been scanned through systematically: Nucl. Inst. and Meth. Section B, Rad. Eff., Rad. Eff. Lett.

References are quoted as follows: [XY MN] is a paper published in 19MN by a first author whose last name starts with XY. [XY MNa,b,...] distinguishes among otherwise ambiguous cases, when necessary.

2. INTRODUCTION

2.1 Experimental Techniques

A recent text [FE 86] presents clearly the variety of techniques that is available and in use to investigate ion–beam bombardment effects. Freeman [FR 86b] has reviewed the historical developments of the machine aspects of the production, transport and use of beams of heavy ions. Ion implanters and ion sources for materials modifications are reviewed in [BY 86], covering the range 1-200 keV in energies and μA to mA beam intensities. A history of commercial implantation is offered in [RO 85]. See also the proceedings of several international conferences on implantation equipment, techniques and applications of accelerators (published in *Nuclear Instruments and Methods*).

2.2 Notation and Acronyms

A list of notations used in this paper, and the usual units of some quantities, follow.

A/B	:	bilayer with film (or layer) A on top of substrate (or film) B
Dt	:	variance of the mixed profile (Å^2)
E	:	ion energy (keV)
E_d	:	minimum energy required to displace a lattice atom (eV)
F_D	:	deposited–energy distribution (eV/Å)
F_R	:	range profile (Å^{-1})
M_1	:	bombarding ion mass (amu)
Y	:	sputtering yield (atoms/ion)
Y_R	:	recoil–implantation yield (atoms/ion)
dY_R/dX	:	recoil–implantation depth profile (atoms/ion/Å)
T_{irr}	:	irradiation temperature (K)
x	:	an axis perpendicular to the target surface
X	:	width of the intermixed region in ion mixing (Å)

X_1	:	film thickness (e.g. in a recoil-implant experiment or in layered-substract mixing) (Å)
$\bar{x}_D$	:	mean damaged depth (Å)
$\bar{x}_R$	:	mean projected range (Å)
$\langle \Delta X \rangle$	:	mean shift of the marker profile in a mixing experiment (ΔX_{peak} is the shift in the peak position) (Å)
Δx_D	:	straggling in deposited energy (Å)
Δx_R	:	range straggling (Å)
N_D	:	number of displaced atoms per ion

<u>Greek symbols</u>

ϵ	:	Lindhard's reduced energy
$\epsilon_\perp$	:	perpendicular strain
ϕ	:	fluence in ions·cm^{-2} (sometimes called dose)
$\dot{\phi}$	:	flux density in ions·cm^{-2}·s^{-1} (sometimes called current density)
Ω_m^2	:	variance of the distribution of marker atoms in a mixing experiment

<u>Acronyms</u> - No distinction has been made among those acronyms corresponding to experimental or simulation techniques, and those acronyms describing physical processes.

AES	:	Auger Electron Spectroscopy
C-e	:	Chemical Etching
CEM	:	Conversion Electron Mossbauer technique
C-V	:	Capacitance-Voltage measurement
CVD	:	Chemical Vapor Deposition
dpa	:	Displacements per Atom
DRM (DRI):		Dynamic Recoil Mixing (Implantation)
EM	:	Electron Microscopy
ERD	:	Elastic Recoil Detection
GS	:	Gibbsian Segregation
GDOS	:	Glow Discharge Optical Spectrometry
IBIEC(IBIIA):		Ion - Beam Induced Epitaxial Crystallization (Interface Amorphization)
IBP	:	Ion-Beam Processing
LEIS (ISS):		Low-Energy Ion Scattering (Spectroscopy)
LHe	:	Liquid-He temperature
LN_2	:	Liquid-N temperature
LSS	:	Lindhard, Scharff and Schiøtt's atomic collision theory
MARLOWE	:	A computer simulation code of collision cascades
MI (MD)	:	Multiple Interactions, or Molecular Dynamics: computer simulation programs of collision cascades
NRA	:	Nuclear Reaction Analysis
RBS(RBS-C):		Rutherford Backscattering Spectrometry (occasionally called BSS); (RBS-Channeling)
RED	:	Radiation-Enhanced Diffusion
RI	:	Recoil Implantation

RIS   :   Radiation-Induced Segregation
SEM   :   Scanning Electron Microscopy
SIMS   :   Secondary Ion Mass Spectroscopy
TEM   :   Transmission Electron Microscopy
TRIM   :   A Monte Carlo simulation program
WSS   :   Winterbon, Sigmund and Sanders' theory of ranges and
              deposited energy
XPS   :   X-ray Photoelectron Spectroscopy
XRD   :   X-Ray Diffraction

3.   DIRECT ION IMPLANTATION

A *range profile* is the probability distribution of rest points for one ion, and an *implant profile* is the resulting concentration profile after a given irradiation fluence.   However, the same notation, $F_R(x)$, for both profiles is used in Tables 1a, 1b, which contain summaries of recent quantitative ion-implantation experiments.

Although this NATO-ASI is tailored towards high-fluence ion–beam effects, low-fluence range and damage experiments will also be covered in this section.   Tracer-level implantation (ϕ $<10^{14}$ ions/cm^2) usually permits a more direct comparison with theoretical predictions of the range distribution, and single-ion profiles constitute necessary input for high fluence calculations.

3.1 Range and Implant Profiles

Because of space limitations, and also because this manuscript is a companion to the preceding paper on collisional aspects of beam-solid interactions, we have restricted the entries in Tables 1a and 1b for range and implants profiles according to the following criteria: range measurements under channeling conditions, e.g. [HA 84], are excluded; we collect only those experimental reports where the profiles are quantified and some comparison with theory is attempted; only experiments carried at low implant temperatures ($<$ 300 K) are included; and, finally, experiments on implantation of exotic targets (e.g., condensed gases) or on beam induced oxidation, have been excluded.   Note that [ZI 85a] contains a review of experimental range data up to ~1984.

Now we discuss briefly the contents of Tables 1a and 1b.   Large-scale efforts to compare range measurements in *homogeneous* targets with theoretical predictions, usually based on the TRIM model, come from the Berlin-Porto Alegre connection [TJ 86, BE 84] and others [KI 84]. Although the agreement between theory and experiments is usually very good (see also [ZI 85a]), even for large off-normal beam incidence [KA 87, DE 84], still many authors voice concerns that range and straggling predictions are too low, especially for range stragglings, and in particular at low energies [MA 84b]. Discrepancies in mean range as large as 20% (or, for the straggling, up to 100%) are reported [CO 86].   A comprehensive study of ranges at low energies is given by [IZ 86].   An area of increasing attention is high energy ($>$ 0.5 MeV) ranges [IN 85, SK 87, JO 87, ...], but existing theory seems to suffice.

An analysis of the sources of range data referenced by the different authors indicates that as many authors use LSS-based predictions as computer codes, most frequently TRIM.   Users of LSS predictions have a wide choice of tabulations that have been referenced in the preceding paper by Littmark and Gras-Marti.   Very few authors [WO 87, HE 84] claim sufficient accuracy in the experimental profiles to enable them to extract spatial moments higher than the second (the straggling).

Implantation in *inhomogeneous* targets is a newly explored area. Simulation models seem to operate adequately in the prediction of range profiles in bilayer [BE 86a, OL 86] or multilayer structures [PI 87]. Other inhomogeneous targets which have been investigated include implantations in a target previously implanted with a different ion [KO 84].

An area of interest within this NATO-ASI is, of course, high-fluence ion implantation. Several authors quoted in Tables 1a, 1b analyze high-fluence data quantitatively. Successful understanding of implant evolution has been achieved [RA 84, FA 85] using models that contain collisional mixing and sputtering as major components, and more comprehensive models exist [BU 85, WA 86b]. In general, however, every group develops its "own" simulation model for implant profile evolution (see tables, label "own" under the heading "Comparison with theory"), with the result that use, or even critical assessment of that approach, by others is difficult. So far, the agreement between experiment and theory of implant evolution is mostly qualitative, and further work is necessary.

For a presentation of some experimental results and comparisons with theory, extracted from the references quoted in Tables 1a and 1b, see the lecture-notes [GR 87a].

3.2 Damage Profiles

The criteria used to compile Table 2 are the same as for Tables 1a and 1b. Furthermore, experiments on the accumulation of damage with increasing ion fluence, and the evolution towards amorphicity, are not included in the compilation; the field will be covered in other contributions in this book.

The experimental analysis of damage profiles is usually carried out using RBS in the channeling mode (RBS-C), which is indicated in Table 2 as RBS for brevity. An area of apparent recent activity is the analysis of the correlation between strain measurements in the irradiated targets [MY 86, PA 87, CH 87, SE 87] and the deposited energy profiles; these profiles are also compared with optical measurements [FA 86a, GO 86, KU 86].

The number of recent experimental publications on energy-deposition profiles is, as seen from Table 2, much smaller than for range and implant profiles. Furthermore, very few authors report comparisons between predicted and measured average energy-deposition depth, not to mention straggling and higher order moments over the damage profiles. In the few instances where this comparison is made, large discrepancies are still found [ED 85, KO 86, ZI 86], and the situation is worse when the measured total damage production rate is compared with the predictions of the Kinchin-Pease model [MA 84a, SL 84 & 85, PA 86a]. However, some positive reports exist [KO 86b], even for varying angle of beam incidence. A need for tabulations of damage parameters for low energy projectiles (< 10 keV) is apparent [KO 86, VI 86].

Recent experimental measurements on metallic targets are almost nonexistent. The main candidates competing with Si as a target in damage measurements are II-VI and III-V compounds. An area that needs further clarification is the origin of the deep damage regions observed frequently in recent experiments [KO 86, ZI 86, GR 87a].

4. RECOIL IMPLANTATION

We collect in Table 3 the references to quantitative experimental studies of recoil implantation (RI), i.e. those making use of the recoil mixing process to implant a given element into a substrate. A film containing the desired species is deposited on a substrate and the layered target is irradiated with a different ion species, usually a noble gas ion. The relation between this topic and what will be termed "low-energy mixing" and "layer mixing" in the following section, is not always clear in the literature (see also the comments in the preceding lecture). Rather than there being a distinction in terms of the underlying mechanisms that produce atomic motion, the difference between those groups of studies resides in their objectives. In the RI technique one is interested in the impurity profile implanted in the substrate, whereas in atomic mixing experiments (see Section 5) one investigates the interface between the film and the underlying substrate, and the atomic redistribution that takes place in both materials as a consequence of the irradiation. In RI experiments, the surface layer (the "film") is usually removed before the resulting profile of recoil atoms is analyzed. The fluences used in RI experiments are generally smaller than those used in typical atomic mixing studies. A review of RI up to ~1984 is given in [KE 84].

Two different modes of RI operation have evolved, the difference being whether the thickness of the film is maintained constant during the irradiation (DRI, or "unlimited supply" mode [AR 84, AH 84]), or not (the usual procedure). Several advantages in the resulting product are reported by the DRI pioneers [AH 84], apart from the theoretical consistency of comparing experimental results with models that assume a constant film thickness.

Few extensive theoretical-experimental studies of RI are reported and usually computer simulations are used for comparison. One exception is [GR 84b], where a semiempirical formula for Y_R is developed and widely tested. Depth profiles of strain in the target appear as an indirect "measure" of the RI process [CE 85]. (See also the relation between damage distributions and strain in Section 3.2.)

Measured recoil-implantation yields are in the range $Y_R \simeq 1$-10 [AR 84, GR 84b], and the depth profiles of recoil implanted species dY_R/dX display an exponential tail [BL 84] at relatively large depths. Agreement between experiments and theory is within a factor 2 [KA 85, ER 85], and there is a clear need for a comprehensive code that simulates the RI process [YA 87].

Clearly, further experimental and theoretical research on the RI technique is needed, not only because it may be a useful technique for many applications [CO 85], but also because many process schedules used in electronics industries (or device manufacture) demand implantation through thin dielectric films [BL 84], and the recoil mixing effect needs to be understood and quantified. At high fluence, primary recoil mixing is a minor component in the atomic relocation process, and the experiment is more properly in the category of ion-beam mixing, which is discussed in the following section.

5. ATOMIC (COLLISIONAL) MIXING

Ion-beam mixing processes are investigated both as inevitable nuisances affecting the depth resolution in profiling and as a useful consequence (important for materials modifications) of the ion slowing-down process in the solid. Roughly, studies can be divided into low and high-energy mixing, in terms of the energy of the ion beam.

5.1 Low ion energies. Sputter depth-profiling

Table 4 refers mainly to sputter depth-profiling analysis of interfaces between two homogeneous media; a few studies of other geometries or substrates include "marker" experiments [MA 86a, TO 86a], multilayer structures [TO 86a, TA 87a], and low-fluence [VA 86] or saturation [VA 84] implants. In all cases, the initial profile is broadened, which is generally referred to in Table 4 as a degradation in the "depth resolution," and occasionally the eventual shift of the profile is also considered. A review of low-energy mixing up to ~1984 can be found in [KI 84b].

Few attempts are made to discuss quantitatively mixing effects and, when these are done, analytical theory is preferred [WI 85c, KI 84b] over computer simulation codes. Very limited use is made, so far, of the models and results developed for ballistic mixing (see the previous chapter), although clear examples of their usefulness in describing the data, with no further ingredients, are found [KI 84b, TS 86].

Distorted profiles with both leading and trailing exponential tails are usually reported [WI 85c, LI 85], with the trailing part being usually correlated with the projected range of the ions [SE 84, AN 86]: the larger the range is, the larger is the observed extent of the mixing in the depth profile. A good depth resolution in sputter profiling requires heavy ions used to sputter at low energies. Usually the angle of incidence of the beam is off-normal at ~40-50°, which reduces the penetration of the ions, and hence increases depth resolution.

The important problem of extracting the original profile by deconvoluting the effects of the analyzing technique, has been successfully attacked in [KI 85a], making use of ballistic mixing concepts and theory in the diffusion approximation.

A few studies of temperature-dependent effects are reported [KI 84a, TO 86a], and of the influence of surface roughness [TO 87a], which is an important factor contributing to the degradation in depth resolution. RED effects have also been demonstrated [VA 86].

5.2 High ion energies.

Tables 5a and 5b are an updated version of similar compilations of experimental studies reported up to ~1984 [PA 85]. The two basic approaches to investigate high-energy ion-beam mixing, namely irradiations of marker samples, or irradiations of layered structures, are tabulated separately. The experimental results reported in a Workshop on Ion Mixing and Surface Layer Alloying [NI 84] have not been included in this review, because most of them appeared in more formal archival sources.

The review by [PA 85] describes systematically the quantities involved, and the essential facts needed, in the analysis of high-energy mixing experiments. There are several other recent and detailed accounts of ion-beam mixing [MA 83, WA 84, AP 84, AV 86, PA 85]. The major points discussed there will not be duplicated here. We only make a few remarks. Tables 5a and 5b do not contain studies performed at high-temperatures because under those conditions the basic concepts needed for the interpretation of the data usually go beyond the ballistic concepts discussed in the previous chapter. Those studies are discussed in other contributions to this NATO-ASI.

In contrast to the situation up to ~1984 [PA 85], and probably due to the larger interest in applications, most efforts have concentrated in the ensuing years on applications of ion-beam mixing to multilayer studies, and not as many studies have been reported on marker experiments. Although the body of knowledge has obviously been enlarged, still many unsolved questions remain, and confident predictions are difficult to make (see the lecture notes for some examples [GR 87a]).

ACKNOWLEDGEMENTS

One of us (AG-M) expresses sincere thanks to the members of the Biological and Radiation Physics Section for the relaxed working atmosphere during his sabbatical leave. The assistance of Drs. I. Abril and R. Garcia-Molina in the literature search has been invaluable. This project is partly funded by the Spanish CAICYT, the US-Spain Committee CCHN, and the Office of Health and Environmental Research, U.S. Department of Energy, under contract DE-AC05-84OR21400 with Martin Marietta Energy Systems, Inc. IBM-Spain kindly contributed to travel expenses. This contribution would not have its present format without the efficient and patient typing of a cumbersome manuscript by Norma Kwaak.

REFERENCES
(See Section 1. for an explanation)

AH 84   Ahmed NAG, Colligon JS, and Hill AE, *Thin Solid Films* **117**, 223.
AH 85   Ahmed NAG, Colligon JS, and Hill AE, *Thin Solid Films* **129**.
AL 85   Al-hashmi SAR and Carter G, *Rad. Eff. Lett.* **85**, 265.
AN 84a   Anttila A, Paltemaa R, Varjoranta T, and Hentela R, *Rad. Eff. Lett.* **86**, 179.
AN 86   Anderle M and Loxton CM, *Nucl. Inst. Meth. B* **15**, 186.
AP 84   Appleton BR, in *Ion Implantation and Beam Processing*, eds. JS Williams and JM Poate, Academic, New York.
AR 84   Argyrokastritis P, Karpuzov DS, Colligon JS, Hill AE, and Kheyrandish H, *Phil. Mag.* **A49**, 547.
AS 86   Ascheron C, Schindler A, Flagmeyer R, and Otto G, *Phys. Stat. Sol.* **a96**, 555.
AV 85   Averback RS and Peak D, *Appl. Phys.* **A38**, 139.
AV 86   Averback RS, *Nucl. Inst. Meth.* **B15**, 675.
BA 85   Balchaitis G, Kameneckas J, Petrauskas G, and Sakalas A, *Phys. Stat. Sol.* **91**, 295.
BA 86   Battaglin B, Boscoletto A, et al., *Rad. Eff.* **98**, 101.
BA 87   Banwell T and Nicolet M-A, *Nucl. Inst. Meth.* **B19**, 704.
BA 87a   Banwell T, Nicolet M-A, Sands T, and Grunthaner PJ, *Appl. Phys. Lett.* **50**, 571.
BA 87b   Battaglin G, Lo Russo S, Paccagnella A, Principi G, and Zhang PQ, *Nucl. Inst. Meth. B* **27**, 402.
BE 84   Behar M, Biersack JP, Fichtner PFP, Fink D, De B Leite Filho CV, Olivieri CA, Patnaik BK, De Souza JP, and Zawislak FC, *Rad. Eff. Lett.* **85**, 117
BE 85   Behar M, Fichtner PFP, Olivieri CA, De Souza JP, Zawislak, FC, Biersack JP, Fink D, and Städele M, *Rad. Eff.* **90**, 103.
BE 85a   Behar M, Fichtner PF, Olivieri CA, De Souza JP, Zawislak FC, and Biersack JP, *Nucl. Inst. Meth.* **B6**, 453.

BE 86   Behar M, Biersack P, Fichtner PFP, Fink D, Olivieri CA, and Zawislak FC, *Nucl. Inst. Meth.* **B14**, 173.

BE 86a   Behar M, Fichtner PFP, Olivieri CA, and Zawislak FC, *Nucl. Inst. Meth.* **B15**, 78.

BH 85   Bhattacharya RS and Rai AK, *J. Appl. Phys.* **58**, 248 and 2798.

BI 80   Biersack JP and Haggmark LG, *Nucl. Inst. Meth.* **174**, 257.

BI 82   Biersack JP and Ziegler JF, in *Ion Implantation Techniques*, edited by H Ryssel and H Glawischnig, Springer, Berlin.

BL 84   Blunt RT, Sweda R, and Sanders IR, *Vacuum* **34**, 281.

BR 87   Brice DK, *Nucl. Inst. Meth.* **B18**, 121.

BU 85   Bubert H, *Mikrochim Acta* **11**, 49.

BU 86   Bussmann U, Hecking N, Heidemann KKF, and Te Kaat E, *Nucl. Inst. Meth.* **B15**, 105.

BY 86   Byers P, Bailey P, Judge PA, and Armour DG, page 15 in [HO 86].

CA 86   Carter WB, page 63 in [HO 86].

CA 87   Carter G, Armour DG, Kostic S, Jimenez-Rodriguez JJ, Karpuzov DS, and Nobes MJ, *Nucl. Inst. Meth.* **B19**, 758.

CE 85   Cembali F, Mazzone AM, and Servidori M, *Rad. Eff. Lett.* **87**, 83.

CH 81   Christel LA, Gibbons JF, and Mylroie S, *Nucl. Inst. Meth.* **182/3**, 187.

CH 84a   Chereckdjian S and Wilson IH, *Nucl. Inst. Meth.* **B1**, 258.

CH 87   Chami AC, Ligeon E, Danielou R, Fontenille J, and Eymery R, *J. Appl. Phys.* **61**, 161.

CH 87a   Cherian S, Reid I, and Gecim HS, *Can. J. Phys.* **65**, 129.

CO 85   Colligon JS, *Mat. Sci. Engn.* **69**, 67.

CO 86   Coghlan WA, Rhee MH, Williams JM, Streit LA, and Williams P, *Nucl. Inst. Meth.* **B16**, 171.

CU 85   Culbertson RJ and Pennycook SJ, *Mat. Res. Soc.* **51**, 357.

DA 86   Davisson CM, *Nucl. Inst. Meth.* **B13**, 421.

DA 87   Daniels LO and Wilbur PJ, *Nucl. Inst. Meth.* **B19**, 221.

DE 84   De Cata L, Williams JS, and Harrison HB, *Nucl. Inst. Meth.* **B4**, 368.

ED 85   Edmond JA, Withrow SP, Kong HS, and Davis RF, *Mat. Res. Soc.* **51**, 395.

ER 85   Erichsen R, Baumvol IJR, and De Souza JP, *Nucl. Inst. Meth.* **B7/8**, 316.

FA 85   Farkas D, Singer IL, Rangaswamy M, *J. Appl. Phys.* **57**, 1114.

FA 86a   Faik AB, Chandler PJ, Townsend PD, and Webb R, *Rad. Eff.* **98**, 233.

FE 86   Feldman LC, Mayer JW, *Fundamentals of Surface and Thin Films Analysis*, North-Holland, New York.

FE 87   Fenn-Tye IA and Marwick AD, *Nucl. Inst. Meth.* **B18**, 236.

FI 84   Fink D, Biersack JP, Städele M, Tjan K, Haring RA, and De Vries AE, *Nucl. Inst. Meth.* **B1**, 275.

FI 86   Fichtner PFP, Behar M, Olivieri CA, Livi RP, De Souza JP, Zawislak FC, Fink D, and Biersack JP, *Rad. Eff. Lett.* **87**, 191.

FI 86a   Fichtner PFP, Behar M, Olivieri CA, Livi RP, De Souza JP, Zawislak FC, Fink D, and Biersack JP, *Nucl. Inst. Meth.* **B15**, 58.

FI 86b   Fink D, Biersack JP, Städele M, Tjan K, Behar M, Fichtner PFP, Olivieri CA, De Souza JP, and Zawislak FC, *Nucl. Inst. Meth.* **B15**, 71.

FR 86b   Freeman JH, *Rad. Eff.* **100**, 161.

FU 86   Furukawa S, Asano T, Fukada T, Ishiwara H, and Tsutsui K, *Mat. Res. Soc.* **54**, 207.

GA 86   Gartner K, Glaser E, Gotz G, and Hehl K, *Nucl. Inst. Meth.* **B15**, 317.

GA 87   Gaboriaud RJ, Grob JJ, and Abel F, *Nucl. Inst. Meth.* **B19**, 648.

GA 87b   Galuska AA, *J. Vac. Sci. Technol.* **B5**, 1.

GI 84   Gill SS, *Rad. Eff. Lett.* **85**, 67.

GN 86   Gnaser H, Bay HI, and Hofer WO, *Nucl. Inst. Meth.* **B15**, 49.

GO 86   Götz G, *Rad. Eff.* **98**, 189.

GR 84b   Grob IJ, Mesli N, Grob A, and Siffert P, *Appl. Phys.* **A35**, 161.

GR 85   Grabowski KS, Correll FD, and Vozzo FR, *Nucl. Inst. Meth.* **B7/8**, 798.

GR 87   Grob A, Grob JJ, and Golanski A, *Nucl. Inst. Meth.* **B19**, 55.

GR 87a   Gras-Marti A, Littmark U, Lecture notes on Direct and Recoil Implantation, and Collisional Ion Beam Mixing: Recent Low Temperature Experiments, presented at this NATO-ASI (ORNL Report No.   ).

HA 84   Hautala M, *Nucl. Inst. Meth.* **B2**, 130.

HA 85   Hahn H, Averback RS, Diaz de la Rubia T, and Oakamoto PR, *Mat. Res. Soc.* **51**, 491.

HA 86   Hagmann D, Steiner D, and Schellinger T, *J. Electrochem. Soc.* No. **133**, 2597.

HA 86b   Hall BO, *Nucl. Inst. Meth.* **B16**, 177.

HE 84   Heidemann KF, Grüner M, and te Kaat E, *Rad. Eff.* **82**, 103.

HE 86   Henriksen O, Laursen T, Johnson E, Johansen A, Sarholt-Kristensen I, and Whitton JL, *Nucl. Inst. Meth.* **B15**, 356.

HI 86   Hirvonen JK, page 49 in [HO 86].

HI 86a   Hirvonen J-P, Nastasi M, and Mayer JW, *Nucl. Inst. Meth.* **B13**, 479.

HO 85   Horino Y, Matsunami N, and Itoh N, *Appl. Phys. Lett.* **47**, 967.

HO 85a   Holland OW, Fathy D, Narayan J, and Oen OS, *Rad. Eff.* **90**, 127 and *Nucl. Inst. Meth.* **B10/11**, 565.

HO 86   Hochman RF, editor, *Ion Plating and Implantation, Application to Materials*, American Society for Metals (1986).

HO 86a   Horino Y, Matsunami N, and Itoh N, *Nucl. Inst. Meth.* **B16**, 50.

HO 87   Horino Y, *Nucl. Inst. Meth.* **B19**, 61.

HU 85   Hung LS and Mayer JW, *Thin Sol. Films* **123**, 135; *J. Appl. Phys.* **58**, 1527.

HU 86   Hughes AE, *Rad. Eff.* **97**, 161.

HU 87   Hubler GK, *Ion Beam Processing*, Naval Research Laboratory report No. 5928.

IN 85   Ingram DC, Baker JA, and Walsh DA, *Nucl. Inst. Meth.* **B7**, 361.

IZ 86   Izsak K, Berthold J, and Kalbitzer S, *Nucl. Inst. Meth.* **B15**, 34.

JA 85   Jaouen C, Riviere JP, Bellara A, and Delafond J, *Nucl. Inst. Meth.* **B7/8**, 591.

JI 84   Jimenez-Rodriguez JJ, Tognetti NP, Marsh T, and Collins R, *Nucl. Inst. Meth.* **B2**, 792.

JO 86   Jorch HH and Werner RD, *Nucl. Inst. Meth.* **B15**, 47.

JO 87   Johnson ST, Cozzolino C, and Williams JS, *Nucl. Inst. Meth.* **B19**, 762.

KA 85   Karpuzov DS, Colligon JS, Kheyrandish H, and Hill AE, *Nucl. Inst. Meth.* **B6**, 474.

KA 87   Kakoschke R, Binder H, Röhl S, Massell K, Rangelow IW, Saler S, and Kassing R, *Nucl. Inst. Meth.* **B21**, 142.

KE 84   Kelly R in *Beam Modification of Materials*, eds. O Auciello and R Kelly.

267

KI 84 Kisielewicz M, *Rad. Eff.* **80**, 81.
KI 84a King BV and Tsong IST, *Ultramicroscopy* **14**, 75.
KI 84b Kirschner J and Etzkorn H-W, *Topics in Current Physics* **37**, 103.
KI 85 Kim S-J, Newlet M-A, Averback RS, and Baldo P, *Appl. Phys. Lett.* **46**, 154.
KI 85a King BV and Tsong IST, *Nucl. Inst. Meth.* **B7/8**, 793.
KI 86 Kim S-J, Nicolet M-A, and Averback RS, *Appl. Phys.* **A41**, 171.
KI 86a Kido Y, Konomi I, Kakeno M, Yamada K, Dohmae AK, and Kawamoto J, *Nucl. Inst. Meth.* **B15**, 42.
KI 86b Kisielewicz M, *Nucl. Inst. Meth.* **B9**, 178.
KI 86c Kiriakidis G, Carter G, and Whitton JL, *Erosion and Growth of Solids Stimulated by Atom and Ion Beams*, NATO ASI, Series E, No. 112, Martinus Nijhoff.
KI 87 Kim S-J, Nicolet M-A, and Averback RS, *Nucl. Inst. Meth.* **B19**, 662.
KI 87b Kido Y, and Kawamoto J, *J. Appl. Phys.* **61**, 956.
KO 84 Kostic S, Jimenez-Rodriguez JJ, Karpuzov DS, Armour DG, and Carter G, *Nucl. Inst. Meth.* **B2**, 182.
KO 86 Kostic S, Begemann W, Abril I, Armour DG, and Carter G, *Rad. Eff.* **100**, 1.
KO 86b Kostic S, Kiriakidis G, Nobes MJ, and Carter G, page 423 in [KI 86].
KU 86 Kulik M and Zuk J, *Nucl. Inst. Meth.* **B15**, 744.
KU 86a Kurup MB, Bhagawat A, and Prasad KG, *Nucl. Inst. Meth.* **B13**, 473.
KW 86 Kwok HL, Lam YW, Wong SP, and Poon MC, *J. Mat. Sci. Lett.* **5**, 633.
LA 85 Land DJ, Simons DG, Brennan JG, and Glass GA, *Nucl. Inst. Meth.* **B10/11**, 234.
LE 85 Lennard WN, Geissel H, Alexander TK, Hill R, Jackson DP, Lone MA, and Phillips D, *Nucl. Inst. Meth.* **B10/11**, 592.
LE 87 Leiberich, *Nucl. Inst. Meth.* **B19**, 457.
LI 85 Littlewood SD and Kilner JA, *SIMS V*,
LI 87 Ligthart HJ, Gerritsen E, Van Den Kerkhoff PH, Hoekstra S, Van de Leest RE, and Keetels E, *Nucl. Inst. Meth.* **B19**, 209.
LI 87b Lilienfeld, *Nucl. Inst. Meth.* **B19**, 1.
LI 87c Lifanova, *Nucl. Inst. Meth.* **B19**, 638.
LI 87d Li W-Z, Kheyrandish H, Al-Tamimi Z, and Grant WA, *Nucl. Inst. Meth.* **B19**, 723.
MA 74 Manning I, and GP Mueller, *Comput. Phys. Commun.* **7**, 85 (Code EDEP-1).
MA 83 Mayer JW, Lau SS, Chapter 9 in
MA 84 Maydell-Ondrusz EA, *Thin Sol. Films* **114**, 357.
MA 84a Macarlay-Newcombe RG and Thompson DA, *Nucl. Inst. Meth.* **B1**, 176.
MA 84b Malherbe JB, *Nucl. Inst. Meth.* **B2**, 774.
MA 85 Maszara WP, Rozgonyi GA, Simpson L, and Wortman JJ, *Mat. Res. Soc.* **51**, 381.
MA 86 Mazzoldi P and Miotello A, *Rad. Eff.* **98**, 39.
MA 86a Macht M-P and Naundorf V, *Nucl. Inst. Meth.* **B15**, 189.
MA 86b Maszara M-P and Rozgonyi GA, *J. Appl. Phys.* **60**, 2310.
MA 87 Mantl, *Nucl. Inst. Meth.* **B19**, 677.
MC 84 McHargue CJ, Sklad PS, Angelini P, and Lewis B, *Nucl. Inst. Meth.* **B1**, 246.
MI 85 Miura T, Hatsukawa Y, Muramatsu H, and Nakahara H, *Nucl. Inst. Meth.* **B9**, 123.

MI 87   Michel AE, Kastl RH, Mader SR, Masters BJ, and Gardner JA, *Appl. Phys. Lett.* **44**, 404.

MO 87   Morgan AE and Maillot P, *Appl. Phys. Lett.* **50**, 959.

MY 86   Myers DR, Picraux ST, Doyle BL, Arnold GW, and Biefeld RM, *J. Appl. Phys.* **60**, 3631.

NA 84   Narayan J, Fathy D, Holland OW, Appleton BR, Davis RF, and Becher PF, *J. Appl. Phys.* **56**, 1577.

NA 85   Namavar F, Budnick JI, and Sanchez FH, *Nucl. Inst. Meth.* **B7/8**, 357.

NA 87   Nastasi M, Hirvonen J-P, Caro M, Rimini E, and Mayer JW, *Appl. Phys. Lett.* **50**, 177.

NI 84   Nicolet M-A and Picraux ST, *Ion Mixing and Surface Layer Alloying. Recent Advances*, Noyes Publications, New Jersey, (Proc. Workshop held May 1983).

OK 85   Okano, *Appl. Surf. Sci.* **22**, 72.

OL 86   Oliviri CA, Behar M, Fichtner PFP, Zawislak FC, Fink D, and Biersack JP, *Rad. Eff.* **98**, 27.

OP 86   Opyd WD, Gibbons JF, Bravman JC and Parker MA, *Appl. Phys. Lett.* **49**, 974.

PA 85   Paine BM and Averback RS, *Nucl. Inst. Meth.* **B7/8**, 666.

PA 85a   Paprocki K and Brylowska I, *Rad. Eff. Lett.* **87**, 63.

PA 86   Parkin DM and Elliot RO, *Nucl. Inst. Meth.* **B16**, 193.

PA 86a   Parikh NR, Thompson DA, and Carpenter GJC, *Rad. Eff.* **98**, 455.

PA 87   Paine BM, Hurvitz NN, and Speriosu VS, *J. Appl. Phys.* **61**, 1335.

PI 87   Picraux ST, Tsao JY, and Brice DK, *Nucl. Inst. Meth.* **B15**, 21.

PO 85   Poker DB, and Appleton BR, *J. Appl. Phys.* **57**, 1414.

PR 86   Prins JF, Derry TE, and Sellschop JPF, *Phys. Rev.* **B34**, 8870.

PR 86a   Prasad KG, Kurup MB, and Bhagawat A, *Nucl. Inst. Meth.* **B15**, 698.

RA 84   Rauschenbach B, Blasek G, and Dietsch R, *Phys. Stat. Sol.* **A85**, 473.

RA 86   Rauschenbach B, *Appl. Phys.* **A40**, 47.

RA 86b   Rai AK, Bhattacharya RS, and Rashid MH, *Thin Sol. Films* **137**, 305.

RA 87a   Rai AK, Bhattacharya RS, Pronko PP, and Tai-il-Mah, *Surf. Interf. Anal.* **10**, 142.

RA 87b   Rauschenbach B and Hohmuth K, *Nucl. Inst. Meth.* **B23**, 323.

RE 87   Reeson, *Nucl. Inst. Meth.* **B19**, 269.

RO 85   Rose PH, *Nucl. Inst. Meth.* **B6**, 1.

RO 86   Roth J, Möller W, Poker DB, and Wittmaack K, *Nucl. Inst. Meth.* **B13**, 409.

RO 86a   Ross GG and Terreault B, *Nucl. Inst. Meth.* **B15**, 61.

SA 84   Saito K and Iwaki M, *J. Appl. Phys.* **55**, 4447.

SA 85   Saito K and Iwaki M, *Nucl. Inst. Meth.* **B7**, 626.

SA 85b   Saris FW, Westendorp JFM, and Vredenberg A, *Mat. Res. Soc.* **51**, 405.

SC 85   Schneider F and Unterricker S, *Phys. Stat. Sol.* **a85**, 455.

SC 86   Scholten D and Burggraaf AJ, *Rad. Eff.* **97**, 191.

SC 86a   Scanlon PJ, Farrell G, Ridgway MC, and Valizadeh R, *Nucl. Inst. Meth.* **B16**, 479.

SE 84   Seah MP and Hunt CP, *J. Appl. Phys.* **56**, 2106.

SE 84a   Seah MP, Mathieu HJ, and Hunt CP, *Surf. Sci.* **139**, 549.

SE 87   Servidori M, *Nucl. Inst. Meth.* **B19**, 443.

SH 86   Shaanan M, Kalish R, and Richter V, *Nucl. Inst. Meth.* **B16**, 56.

SI 84   Singer IL, *Vacuum* **34**, 853.

SK 87   Skorupa W, Wieser E, Groetzschel R, Posselt M, and Buecke H, *Nucl. Inst. Meth.* **B19**, 335.

SL 84   Slater M, Nobes MJ, and Carter G, *Rad. Eff.* **83**, 219.

SL 85   Slater M, Kostic S, Nobes MJ, and Carter G, *Nucl. Inst. Meth.* **B7/8**, 429.

SN 84   Snowdon K, Onsgaard J, and Tougaard S, *Nucl. Instr. Meth.* **B2**, 797.

SO 87   Sood DK, Battaglin G, Kulkarni VN, Lo Russo S, and Mazzoldi P, *Nucl. Inst. Meth.* **B19**, 632.

SR 87   Srinivasan V, and Bhattacharya RS, *Surf. Interf. Anal.* **10**, 131.

SY 86   Syshchenko AF, Anishchik VM, and Komarov FF, *Rad. Eff.* **97**, 111.

TA 84   Takadoum J, Pivin JC, Pons-Corbeau J, Berneron R, and Charbonnier JC, *Surf. Interf. Anal.* **6**, 174.

TA 87   Takadoum J, Pivin JC, Pollock HM, Ross JDJ, and Bernas H, *Nucl. Inst. Meth.* **B18**, 153.

TA 87a   Taylor E, Blanchard B, and Villegier JC, *Surf. Interf. Anal.* **10**, 1.

TA 87b   Tao K, Hewett CA, Lau SS, Buchal Ch, and Poker DB, *Nucl. Inst. Meth.* **B19**, 753.

TJ 86   Tjan K, Fink D, Biersack JP, and Stadele M, *Nucl. Inst. Meth.* **B15**, 54.

TO 85   Tougaard S, Zomorrodian AR, Kornblit L, and Ignatiev A, *Surf. Sci.* **152/153**, 932.

TO 86   Townsend PD, *Contemp. Phys.* **27**, 241.

TO 86a   Tonn DG, Sankey OF, and Tsong IST, *Nucl. Inst. Meth.* **B15**, 193.

TO 87a   Tompkins HG, *Surf. Interf. Anal.* **10**, 105.

TS 86   Tschersich KG, Littmark U, and Fleischhauser HP, *Surf. Interf. Anal.* **9**, 297.

VA 84   Varga P and Taglauer E, *Nucl. Inst. Meth.* **B2**, 800.

VA 85   Van Rossum M, *J. Appl. Phys.* **58**, 1527.

VA 86   Vandervorst W, Shepherd FR et al., *Nucl. Inst. Meth.* **B15**, 201.

VA 86a   Van Ommen AH, Willemsen MFC, Boudewijn PR, and Reader AH, *Mat. Res. Soc.* **54**, 221.

VA 87   Van Rossum M, Cheng Y-T, Nicolet M-A, and Johnson WL, *Appl. Phys. Lett.* **46**, 610.

VA 87   Vasiljev, *Nucl. Inst. Meth.* **B19**, 1987.

VI 86   Vitkavage DJ, Dale CJ, Chu WK, Finstad TG, and Mayer TM, *Nucl. Inst. Meth.* **B13**, 313.

WA 84   Wang Z-L, *Nucl. Inst. Meth.* **B2**, 784.

WA 85a   Wang K-M, Burman C, Lanford WA, and Groleau R, *Rad. Eff. Lett.* **85**, 177.

WA 86   Wang Z-L, Yi L, and Zhang J, *Nucl. Inst. Meth.* **B13**, 453.

WA 86b   Wampler WR and Brice DK, *J. Vac. Sci. Technol.* **A4**, 1186

WA 86c   Wasserman B, *Phys. Rev.* **B34**, 1926.

WE 85   Westendorp JFM, Roe PK, Sanders JB, and Saris FW, *Nucl. Inst. Meth.* **B7/8**, 616.

WE 86   Westendorp JFM, Littmark U, and Saris FW, *Nucl. Inst. Meth.* **B18**, 54.

WE 87   Westendorp JFM, *Nucl. Inst. Meth.* **B26**, 539.

WH 85   Whitlow HJ, Keinonen J, and Hautala M, *J. Appl. Phys.* **58**, 3246.

WI 84   Wilson IH, *Nucl. Inst. Meth.* **B1**, 331.

WI 85   Wittmaack K, *Nucl. Inst. Meth.* **B7**, 779.

WI 85b   Wittmaack, *J. Vac. Sci. Tech.* **A3**, 1350.

WI 85c   Wittmaack K, *Nucl. Inst. Meth.* **B7/8**, 750.

WI 86   Williams JS, *Rep. Prog. Phys.* **49**, 491.

WI 86b   Wilson RG, *J. Appl. Phys.* **60**, 2797.

WI 87   Wilson RG, *J. Appl. Phys.* **61**, 933 and Wilson RG, Jamba DM, Sadana DK, and Hopkins CG *J. Appl. Phys.* **61**, 1355.

WO 87   Wong H, Deng E, Cheung NW, Chu PK, Strathman EM, and Strathman MD, *Nucl. Inst. Meth.* **B21**, 447.

XI 87   Xi X-X, Ran Q-Z, Liu J-R, and Guam W-Y, *Sol. St. Commun.* **61**, 791.

YA 87   Yamamoto Y, Fujima S, Takada H, Segawa Y, Ishibashi K, Shim TE, Itoh T, and Suzuki S, *Nucl. Inst. Meth.* **B19/20**, 392.

YA 87a   Yang GQ, Campisano SU, and Rimini E, *Nucl. Inst. Meth.* **B19**, 623.

ZI 84   Ziegler JF, *Ion Implantation*, Academic Press, New York.

ZI 85   Ziemann P, *Mat. Sci. Engn.* **69**, 95.

ZI 85a   Ziegler JF, Biersack JP, and Littmark U, *The Stopping and Range of Ions in Matter*, Pergamon, New York.

ZI 86   Zinkle SJ, Kulcinski GL, and Knoll RW, *Nucl. Mat.* **138**, 46.

TABLE 1a

SUMMARY OF EXPERIMENTAL MEASUREMENTS OF RANGE AND HIGH-FLUENCE IMPLANTATION PROFILES IN METALS

Ion/Target	E (keV)	ϕ ($\times 10^{16}$ cm^{-2})	T_{irr} (K) (R=room)	$F_R(x)$	$\bar{x}_R$	Δx_R	Comparison with theory	Experimental analysis	Reference	Notes
Ti,N/Fe,steel	40-200	1-100	R	X			Incl. diffusion	RBS,SIMS,AES	SI 84	c)
N/steel	60	20	<423	X	X		LSS	AES	DA 87	d)
Al/Cu	170-257	10-50			X		LSS	RBS,AES,NRA	LI 87	e)
Ti/steel	55,190	5-50		X			own	AES	FA 85	h)
B,P/Ni,Fe	70	0.08-30		X	X	X	LSS	SIMS,GDOS	TA 84	l)
N,B/Fe,Al	30-60	<100		X	X	X	LSS, Coll.mix.	AES	RA 84	m)
B/Cu	40	5,20		X	X	X	LSS	NRA	HE 86	n)
Ta/Fe	150	1-18		X			own	RBS	GR 85	p)
Fe/ZY	15-110	<40	various	X			own	RBS,AES	SC 86	q)
N$_2$/r)	200-400				X		LSS,Monte Carlo	NRA	AN 84a	r)
Ni/TiB$_2$	1000	10	R	X	X		LSS	TEM,STEM,XRD	MC 84	s)
He/a)	35	14	R		X	X	Monte Carlo	NRA	LE 85	a)
Sb/Sn;I/Cu	10-60	10^{-6}-10^{-5}			X	X	LSS	Radiochem.	MI 85	b)
N/f)	800	1.5				X	LSS	NRA	LA 85	f)
N/Cu	0.25-2.5				X	X	LSS,TRIM	AES	MA 84b	g)
review	low				X	X	LSS	RBS	IZ 86	t)
u)/Al,Si.	45-420		100	X	X		LSS	RBS,NRA	KI 86a	u)
Xe/Al,Si	400-1000	0.9			X		LSS	RBS	JO 86	v)
H/Be,C,Si	0.7-2		LN$_2$	X	X	X	Monte Carlo	ERD	RO 86a	i)
Li/j)	50-300	0.01-1	R	X	X	X	TRIM	NRA	TJ 86	j)
He,Li,B,Bi/k)	150-600		R	X	X	X	TRIM	RBS,NRA	BE 86a, FI 86b	k)

Notes to Table 1a (Literature published after 1983)

a) Targets: Al, Ti, V, Ni, Cu, Zn, Zr, Nb, Ag, Sn, Ta, W, Au, Bi; mean range and straggling oscillate with Z_2; experimental errors 5-15% but discrepancies are larger with other reported measurements.

b) Experimental error 20-30%; LSS predictions agree for $0.02 < \epsilon < 0.1$, but are too low for $\epsilon < 0.02$.

c) Claim that predictions of microstructures are needed.

d) Fluence given for N_2^+; rough agreement for mean range; analyze influence of $\dot{\phi}$; diffusion at high $\dot{\phi}$.

e) Channeling tail in the AES profile.

f) Targets: $Z_2 = 6$-90; emphasis on determining electronic stopping power from range data; the puzzle of why ^{14}N (not ^{15}N) experimental range is much smaller than LSS predictions is solved.

g) Large (150%) discrepancy with theory which underestimates range at low ϵ (0.02-0.5).

h) Model includes lattice dilation, sputtering and mixing, carbon gettered from the vacuum system.

i) ERD resolution at the surface ~4 nm. Energy dependence of mean range is extracted from the data, but larger discrepancies in actual values between different theories and experiments.

j) Targets: Si, Ni, Cu, Ge, Be, Al, Ti, V, Fe, Co, Ni, Zn, Se, Zr, Mo, Ag, Cd, In, Sn, Sb, Te, Ta, W, Pt, Au, Tl, Pb, stainless steel, and graphite. Range parameters also compared with theory [BI 82] which predicts 30% larger straggling.

k) Bilayer targets: Al/Ti, Al/V, Au/Si, and others; good agreement with TRIM when profile is convoluted with detector resolution (~280 Å).

l) Theoretical straggling narrower. RBS profiling is used to check SIMS profiles.

m) Good agreement with collisional mixing model of high fluence implantation.

n) Experimental profiles broader and at larger depth than LSS predicts (by a factor of 2).

p) Retained dose higher than simple theoretical model. Collision ion mixing appears to be effective.

q) Good agreement of profile with collection model including experimentally determined sputtering yields.

r) Targets: metals, semiconductors and ionic solids; $\epsilon < 0.01$; no Z_1-range oscillations for metals, contrary to semiconductors.

s) Ions: N, Al, Ar, Mn, Ni, Zn, Te, Xe. Also targets: Al_2O_3, GaAs, GaP. New scaling coefficient gives universal expression for projected ranges.

t) Targets: Al, Ti, Cu, Mo, Hf, W, Steel; ranges also compared with PRAL [BI 82]. Agreement within a few %.

u) ~10% (20-40%) higher range (straggling) than predicted. Good agreement in third moment, where measurable.

v) Microstructure was modified to depths much larger than the mean range.

TABLE 1b

SUMMARY OF EXPERIMENTAL MEASUREMENTS OF RANGE AND HIGH-FLUENCE IMPLANTATION PROFILES IN NON-METALS

Ion/Target	E (keV)	φ (×10¹⁶ cm⁻²)	T_{irr} (K) (R=room)	Quantification $F_R(x)$	Quantification $\bar{x}_R$	Quantification Δx_R	Comparison with theory	Experimental analysis	Reference	Notes
As/Si	50	2	393	X				AES	HA 86	i)
N₂/Si	3	Saturation	R	X				AES	SN 84	j)
P/SiO₂ on Si	~1000	5×10⁻⁵-0.1		X	Peak			SIMS	SK 87	l)
Be/Ge/Si...	70-385	0.05	77, R	X			TRIM [BR 87]	RBS	PI 87	m)
B/Si	5	0.02		X			TRIM	SIMS	MI 84	n)
B,P,As,Sb/Si	400-6000	10⁻³-10⁻²		X	X	X	LSS	Resistance	IN 85	o)
As,Bi/Si	40	0.03-10		X, collected dose			MARLOWE	RBS	CA 87	r)
O₂(+Al)/Si	10,20	2.4-33		X, collected dose			Own	SIMS	WI 85	s)
Al,P/Si	20-600	3×10⁻¹-3		X	X	X	LSS	SIMS,C-V	WI 86b,86,87	t)
N/Si	150-250	25-100	R	X			Monte Carlo	AES	BU 85	u)
As,P,B/Si	1-11 MeV			X	X	X	LSS,TRIM	SIMS	WO 87	y)
As/Si	80	1		X			TRIM	SIMS	KA 87	z)
As,Sb/Si	35,200	0.5-2		X			MARLOWE	SIMS	CU 85	a)
Bi/Al/KCl	300	0.3	R	X			TRIM	RBS	OL 86,BE 86	b)
H/Si	190	0.1-140	250	X	X	X	LSS	c)	HE 84	c)
Ga/Si	15-350				X	X	TRIM	RBS	BE 84	e)
Li/Se,Te	50,300		R		X		TRIM	NRA	FI 84	f)
f)/Si	100-300	0.1-10		X	X		LSS	Various	KI 84,86b	
O/C	1.5,3	0.07-7	R	X			Own	RBS	WA 86b,BR 87	x)
g/Si	10-390	0.1-0.2	R	X	X	X	LSS,TRIM	RBS	BE 85a	g)
Be/GaAs, InSb	40	30		X	X	X	TRIM [MA 74]	RBS	SH 86	h)
As/Si	290	0.5	77	X	X	X	TRIM	NRA	OO 86	k)
B/Si	10	0.3		X	X	X	TRIM	SIMS	SC 86a	p)
As,Sb/Si	60		77-750	X	X	X	LSS,TRIM [BU 80]	NRA	DE 84	q)
Al/Si	100	100		X	X	X	LSS,Marlowe	NRA	NA 85	w)
Pb and Kr/Si	40 and 80	0.5-7.5,0.5		X	X	X	LSS,TRIM	RBS	KO 84	11)
He/Si	0.25-80	0.6-2	300	X	X	X	LSS,TRIM	SIMS	GN 86	ch)
C,N/Si	1-24 MeV	0.03-20	200	X			LSS	Optical	BU 86	ny)

Notes to Table 1b (Literature published after 1983)

a) Annealing studies: profiles of substitutional and total concentration of implants.

b) Bilayer target system. 61 nm Al-film. Comparison with implantations in KCl. Reasonable agreement in range profile.

c) Skewness also evaluated; optically determined range profile (depth profile of complex refractive index).

e) Li mobile after implantation.

f) Ions: O, P, As, Ar, Kr, Xe; profile fitted to Pearson distributions. Oxide growth rate is related to implant parameters.

g) Ions: Bi, Au, Eu, Yb, Sn, Cs. Modern interaction potentials are used; discrepancy for Au at low energies attributed to Z_1-range oscillations. References: BE 84, 85, 85a; FI 86, 86a.

h) NRA resolution ~30 nm; high fluence needed for NRA analysis within reasonable (~3 h) time. Better agreement for GaAs. Deconvolution technique described.

i) $F_R(x)$ determined after 800 C treatment to grow SiO_2.

j) Sharp surface peak; no broad plateau observed; due to RED?

k) Also T_{irr} = 300 C, 600 C. Effect of microstructure on As profile. 20% or 200% larger measured range or straggling at -196 C.

l) Shoulder in profile near surface, as reported by others, not found. Good agreement with TRIM using Brandt-Kitagawa electronic stopping.

m) Multilayered target: Si on Ge and Ge on Si. Interface discontinuities as predicted by theory.

n) Channeling effect on range profile, even with optimum alignment (~12° off axis).

o) Good agreement for mean range, but considerably larger (20-50%) experimental straggling (energies < 100).

p) Glancing-angle NRA increases resolution by ~$(\sin \theta)^{-1}$. 6% or 20% smaller measured range or straggling, than predicted.

q) Implantation at glancing angle down to 86°; good agreement in profile with TRIM. At 0° implants, measured moments 5-14% larger than LSS.

r) Generation of materials standards for surface analytic techniques: amount collected in target, versus fluence.

s) Some broadening due to mixing observed; simple model of ion retention plus sputtering.

t) Four moments extracted from Pearson IV fits to the data: channeling effects studied; straggling 30-60% larger than LSS.

u) Good fit with theory including cascade mixing, sputtering, and change of atomic concentration of the target during implantation.

w) Demonstrate influence of vacuum condition on implant profile and existence of dose rate effects.

x) Sputtering, collection and relaxation is included in the model; good agreement.

y) Also skewness, kurtosis are provided: fit to Pearson-IV distributions. Better agreement with LSS than with TRIM.

z) Different incidence angles: 20-80°. TRIM had to be modified for reflection at grazing incidence: agreement is good.

ll) Good agreement theory - experiment as a function of ϕ.

ch) Good agreement with theory, even below 10 keV.

ny) Ranges determined by optical reflectivity using bevelled samples. Agreement with range theory within 10-20%.

TABLE 2

SUMMARY OF EXPERIMENTAL MEASUREMENTS OF DAMAGE PROFILES

Ion/Target	E (keV)	ϕ ($\times 10^{16}$ cm^{-2})	T_{irr}(K) (R=room)	$F_D(x)$	$\bar{x}_D$	Δx_D	N_D	Comparison with theory	Experimental analysis	Reference	Notes
Cu/Cu	14×10^3	5	373–773	X				WSS	X-TEM	ZI 86	t)
H/Si,SiC	80	2–80	95–800	X				Own	RBS	HA 86b	d)
In/Si	80	10^{-3}–0.05	R	X	X	X	X	WSS,Marlowe	RBS	KO 86b	h)
Ar/Si	5	10^{-2}–1			X	X	X	Marlowe	RBS	KO 86	i)
B,F,P/Si	12–133	10^{-5}–10^{-2}	35–77	X		X	X	WSS	RBS	GR 87	j)
As/SiO$_2$+Si	150	5×10^{-3}		X					RBS	ME 87	
Be/GaAsP/GaP	75	0.1	R	X				TRIM	RBS	MY 86	n)
o)/GaAs	100–500	0.1	R	X				WSS,TRIM	XRD	PA 87	o)
p)/SiC	50–320	2×10^{-2}–0.2	R	X				TRIM	RBS	ED 85	p)
Si/Si	150,300	2×10^{-2}–1	82–296	X				TRIM	TEM	MA 85,86b	q)
Ar/GaP	150	3×10^{-3}		X				TRIM	RBS	KI 87b	
N/SiO$_2$	180–2000	1–15	50,300	X				TRIM	u)	FA 86a	u)
v)/CdS	40–60	0.05–1	50,300		X		X	WSS	RBS,TEM	PA 86a	v)
He,B,Ar/SiO$_2$	35,70,150	10^{-3}–0.2	300	X				WSS	RBS	GO 86	w)
m)/quartz	15–120	6×10^{-4}–24	50,295				X	WSS	RBS	MA 84a	m)
Ar,Ne,H/Si	0.25–1	1–30	R	X		X		Own	RBS	VI 86	c)
Ti,Ar/GaP	200	10^{-3}–0.1	R	X				TRIM	Optical	KU 86	f)
Si/Si	50	10^{-3}–0.2	20	X				WSS	RBS	CH 87	g)
Si/InP	40	10^{-4}–0.1	R	X			X	WSS	RBS	SL 84,85	l)
Si/Si	60–100	5×10^{-4}–0.1	R	X				Monte Carlo	XRD	SE 87	a)
Si/Si	120	0.025	77,R	X				Marlowe,TRIM	RBS,TEM	HO 85a	b)

Notes to Table 2 (Literature published after 1983)

a) Measurements of strain distribution, also compared with calculated recoil-implantation profiles (for Si implantation through oxide layers).

b) For a room temperature irradiation, residual damage increases with $\dot{\phi}$; at cryogenic temperatures, it is independent of $\dot{\phi}$; model as homogeneous nucleation of damage. MARLOWE gives better agreement in profile shape.

c) Model accounts for sputtering; damage is expressed as an equivalent amorphous-Si thickness, and saturates after $2-3 \times 10^{15}$ cm^{-2} Ar or Ne ions.

d) Own model for T-dependent damage accumulation: as T increases saturation takes higher ϕ. Model includes annealing and stabilization of damage by H trapping, detrapping and diffusion.

e) Good fit in profile shape. Threshold damage energy per atom increases with T_{irr} and E. Amorphization studies: theoretical profiles broader and displaced to higher depths.

f) Correlation between optical properties (ellipsometrically determined) and damage profiles. Qualitative agreement.

g) Perpendicular strain profile measurements. The uniaxial elongation profile corresponds to the nuclear energy distribution.

h) Influence of incidence angle on damage profile and moments; reasonable agreement except for straggling, which is difficult to extract.

i) Damage width 50% larger than simulation predicts.

j) Molecular ion combinations were used also, leading to excess damage which increases with the number of atoms in the molecule, and with decreasing E.

l) $\dot{\phi}$ dependence observed; the damaging fluence of the analyzing beam is evaluated; measured damage rate much higher than predicted.

n) Strained layer superlattice target: good agreement in profiles.

o) Ions: He, B, C, Ne, Si, P, Te; strain measurements $\epsilon_\perp \alpha \phi \cdot F_D$. Technique sensitive to low fluences ($< 10^{12}$ cm^{-2}).

p) Ions: Al, P, Si, Si+C; amorphization studies. Significant discrepancies (50%) in peak position of profile.

t) Damage extends to depths 30% greater than predicted; electronic stopping power used in calculations too large?

u) Reflectivity data; F_D versus refractive index profiles, vacancy and impurity distributions.

v) Bi, Kr, Ar, Ne ions: experimental $\bar{x}_D$ much larger than WSS predictions; $dN_D/d\phi$, for $\phi \to 0$, two orders of magnitude smaller than Kinchin-Pease predictions. Measured N_D proportional to energy of RBS beam.

w) Also LiNbO$_3$ was irradiated with 150 keV, $2 \times 10^{14} - 2 \times 10^{16}$ N$^+$/cm^2. Refractive index related to F_D.

m) Ions: Ne, Ar, Kr, Sb, Bi; damage production rates 20 times larger than linear theory predicts; also damage due to RBS beam is analyzed.

TABLE 3

SUMMARY OF RECOIL-IMPLANTATION EXPERIMENTS

Film/Substrate	X_1 (nm)	Ion	E (keV)	ϕ ($\times 10^{15}$ cm^{-2})	Quantification Y_R	dY_R/dX	Comparison with theory	Experimental analysis	Reference	Notes
Au/Si	5-28	Ar	10	0.5-10	X		MARLOWE	RBS,C-e	AR 84,KA 85 CO 85	a)
b)/Si	30	Kr	20-300	0.05-10	X	X	Own	RBS,NRA	GR 84b	b)
Sb/Si	20-140	P	120	0.1-20			Own	C-e,electrical	KW 86	c)
Au/Si,SiO₂		Ar	10	< 100				STEM, RBS	AH 84,85	g)
SiO₂/Si	27-78	Si	100	5		X	Monte Carlo	XRD	CE 85	i)
Al/Si	73-105	Ar	100	0.1-10		X		C-V	PA 85a	j)
Al/GaAs	80	Ar	75	0.1-10				RBS,SIMS,C-e	BA 85	d)
Si/GaAs	95-120	As	150-280	1			Monte Carlo	TEM,SIMS	YA 87	e)
Si₃N₄/GaAs SiO₂	50-150	B,P,Se	60-340	0.1-1	X	X	[CH 81]	SIMS	BL 84	f)
Sb/Si	60	Ar	40-800	0.1-100	X	X	WSS	RBS	ER 85	h)

278

Notes to Table 3 (Literature published after 1983)

a) DRM experiment; 30-50% error in Y_R; Y_R decreases with increasing X_1; $\int d\phi \cdot Y_R$ superlinear in ϕ; theory predicts a peak in Y_R at F_D peak ($X_1 \sim 35\text{Å}$). Good agreement with simulations of transmission sputtering.

b) Films: Sb, Ga, Bi, or B: a semiempirical formula for the recoil yield is developed, including transmission sputtering and sputtering of previously introduced impurities.

c) Sheet resistance of direct P implant versus recoil Sb implant is compared.

d) $x_R + \Delta x_R < X_1$; studies of physical-chemical changes, radiation defects, disorder (with RBS).

e) (Samples are post-annealed.) The carrier concentration generated by recoiled or diffused Si is related to the distribution of implanted As. The simulation is a modified TRIM that deals with all cascade processes of recoils.

f) $dY_R/dX \sim e^{-X/L}$ at intermediate depths. Data supports the model used by [CH 81].

g) DRM experiment: at 10^{17} ions cm^{-2}, film adhesion increases a factor of 30; effects (like film texturing) are specific for the DRM process and do not occur in ordinary RI mode.

h) Sample temperature and implant dose rate do not affect the process. The energy dependence of Y_R follows the predicted amount of energy deposited at the interface. Factor of 2 discrepancy among data from different groups.

i) Strain depth profile agrees with simulated depth profile; oxygen depth profile is calculated.

j) Profiles of electrically active recoil-implanted Al are compared with direct implantation profile of Al.

TABLE 4

SUMMARY OF LOW-ENERGY ION-BEAM MIXING EXPERIMENTS

Top or Marker/ Substrate	X_i (nm)	Ion	E (keV)	T_{irr}(K) (R=room)	Quantification (1)	(2)	(3)	Comparison with theory	Experimental analysis	Reference	Notes
Si/Ti/Si...	30/5/60	Ar	5		X			Coll.+RED	SIMS	KI 84a	a)
Si/SiO$_2$/Si	35(top)	Cs	4-12		X			Own	SIMS	WI 85c	b)
Ta$_2$O$_5$/Ta	29,96	Ar	2		X	X		Own	AES	SE 84,84a	c)
Si/Cr,Ti,Pd/Si	60/2/∞	O$_2$,Cs	10	>120	X				SIMS	LI 85	d)
Al/Nb	~2	Ne...Xe	1.5-5.5		X	X			SIMS,TEM	TA 87a	e)
C/Si,Ni,Cr	~16	Ar	2		X			Collisional	AES	TS 86	f)
Ge/Si,g)	10-20	Ne...Xe	.5-5		X	X	X	Collisional	AES	KI 84b	g)
SiO$_2$/Si	2-100	Ar	5		X	X		Own(mixing)	AES	TO 87a	h)
He/Ta$_2$O$_5$	i)	Ar	1		X	X			AES,ISS	VA 84	i)
Ti,Mo/Si	5-7	Ar	5		X	X	X	Own(mixing)	SIMS	KI 85a	j)
Ag,Mg,W/Si	2-24	Ar	6	80-775	X	X			SIMS	TO 86a	k)
As/Si	0~12	O	1.7,12.5	R, 40	X			Collisional	RBS	VA 86	l)
SiO$_2$/Si	102	Cs,He...Xe	8		X	X			AES,SIMS	AN 86	m)
In,Ni/Cu	50	Ar,O$_2$	2-14	140-450	X	X			SIMS	MA 86a	n)

Notes to Table 4 (Literature published after 1983)

(1) - Profile; (2) - Depth resolution; (3) Shift.
(X_i is the (first) interface depth or the marker width.)

a) Own model including cascade mixing and RED. Use combined diffusion coefficient as fitting parameter. Role of RED shown.

b) Effect of mixing and selective sputtering on the tail. Exponential profile for leading and trailing edges. Characteristic width of tails increases with E.

c) Effect of atomic mixing and electron range. Atomic-mixing-induced profile exponentials are correlated with range of overlayer atoms.

d) Effect of temperature on beam-induced broadening in a marker sample; exponential decay lengths.

e) 20 multilayers; effect on ion mixing of M_1, E. Ion range correlated with the amount of mixing; Xe and Kr ions allow better visualization of the multilayer.

f) Carbonized sample. Profiles modeled entirely on ballistic mixing.

g) Review of depth profiling. Also Ge/Ge and Si/Si. Profiles and profile broadening, modeled entirely on ballistic mixing.

h) Interface width versus sputtered depth is given; ion mixing factor of 10 Å and roughening factor ~1%.

i) Sample previously altered to equilibrium surface concentration by He implantation (0.7-3 keV); altered layer correlates with He range.

j) Deconvolution of atomic mixing effects from SIMS profiles of marker-samples. Good agreement with the original distributions in the as-deposited films.

k) Multilayer structures with markers. Temperature-dependent and independent mixing regimes are identified.

l) As was implanted in Si at 6 keV; broadening of the distribution is larger for larger E; oxygen fluence used: $1-13 \times 10^{16}$ cm^{-2}. At high-energy oxygen irradiation, RED dominates the broadening, even at the lower temperatures; collisional mixing predictions too low.

m) Artifact in SIMS profiling with Cs due to changing Cs concentration at interface (matrix effect); this effect dominates over collisional mixing. For noble-gas profiling, interface broadening increases as ion mass decreases.

n) Marker experiment. Atomic mixing and surface roughness increase with T_{irr}. Exponential tail in profiles. Width of mixing zone comparable to ion projected range.

TABLE 5a

SUMMARY OF HIGH-ENERGY, LOW-TEMPERATURE MEASUREMENTS OF ION-BEAM MIXING OF MARKER SAMPLES

Medium/ Marker	X_o Ion	E (keV)	φ (×10¹⁶ cm⁻²)	T_{irr}(K) (R=room)	$\frac{Dt}{\phi F_D}$ (Å⁵/eV)	Discussion (model)	Experimental analysis	References	Notes
Cu/(a):Zr/(a)	45 Kr	750	0.5-2	7,77	15-50	Spike	RBS at T_{irr}	KI 86,87	a)
Cu/W,Au,Ta	33 Xe	100-400	0.1-0.9	R,10		Collisional	RBS,X-TEM	WE 85,86,87 SA 85b	e)
Pd/Pt	20 Ne...Sm	100-400	0.1-3	12	6-18	Spike+overlap	RBS	FE 87	f)
Si/Al	30 Ne	40	2-10	R	53	Collisional	NRA	WH 85	g)
¹²C/¹³C	F_R ¹²C,D	13-150,	10²-10³	10-1500		RED,TRIM	RBS,SIMS	RO 86	h)
Cu/Au;Ni,Ti/Hf Hf/Ni	Kr	0.5,1,1.7 (x10³)		6,80,R		MI		AV 85	i)
CuEr,NiTi/Ag	30 Kr	10³	1	4,R,400	10-150		RBS	HA 85	j)

Notes to Table 5a (Literature published after 1983)

(X_o is the marker depth in nm)

a) Markers in Cu: Nb, Ru, Ag, In, Sb, Hf, Pt, Au, Bi; no dependence with T_{irr} observed; markers in Zr, Ti, Cr, Fe, Co, Ni, Cu, Hf, W, Au; $Dt/\phi F_D$ higher in Cu than in Zr, explained by thermal-spike mechanism.

e) 10-40 Å markers; measured shifts are independent of F_D but depend on marker thickness; ballistic theory is a factor of 3 below data

f) Ions: Ne, Ar, Ni, Kr, Ag, Sn, Sm. ~5 Å markers; Dt is linear with φ except for Ne and Sm; $Dt/\phi F_D$ data increases with M_1, with the value for Ne close to ballistic predictions.

g) Dt α φ: collisional theory is a factor of 3-6 below experiments. 30 Å markers.

h) Mixing larger than ballistic predictions even in the T-independent regime (10-220K).

i) No temperature dependence at low temperature.

j) Alloy targets, 10-20 Å markers. RED mechanism important in CuEr; Dt α φ. Amount of mixing can be greater in the alloys than in either of the pure components.

TABLE 5b

SUMMARY OF HIGH-ENERGY, LOW-TEMPERATURE MEASUREMENTS OF ION-BEAM MIXING OF LAYERED SAMPLES

Top/Under-layer	X_1 (nm)	Ion	E (keV)	ϕ ($\times10^{16}$ cm^{-2})	T_{irr}(K) (R=room)	Comparison with model	Experimental analysis	Reference	Notes
Cu/Au		Ar,Xe	100-300	0.5-2	R	free energy	RBS,EM	RA 86	c)
Al/Cu_xAu_{1-x}		Xe	320	0.5-1.5	R	free energy	RBS,XRD	YA 87a	
Pb/Ni	~30	Kr	200	0.5-3	R		RBS,EM	SO 87	d)
Fe/Al	30-50	Xe	280	0.25-2	R.		RBS,EM,SIMS	GA 87	f)
Ti/Ni	$\bar{x}_D$	Ar	150	10^{-2}-10	R		EM,AES	SA 84,85	j)
Ni/Ti	50	Au	1000	0.3-2		collisional	RBS,EM	RA 96b	k)
Al/Sb	83	Au	420	0.8	R		RBS	KI 87b	y)
Fe,Ni,Co/Al	8-15	Xe	320	10^{-2}-10^2	77,300		XRD,TEM	JA 85	z)
Ni/Ti,SiC	50	Au;Ni	1-2.5$\times10^3$	0.3-2			X-TEM,RBS	RA 87a	e)
Pt/Ni-alloy	15	Pt	1000	0.2-0.5			RBS,AES	SR 87	g)
Cr/Si	20	Cr	45	0.2-1.5	193-573	collisional	RBS	LI 87d	l)
Cr/Si	6-25	H,He;Ar	1-5;150	1-100	520		RBS	HO 85	m)
Ni/Si(n)	15-70	Si	175	0.05-5				HE 87	n)
Sn/GaAs	20	Ar,Kr	40-50	0.1-1	77-498		RBS,EM	JO 87	p)
Ti,Mg/Si,Ge	Variable	Ar,Kr	240-540	0.5-6	R			TA 87b	q)
Ni/SiO$_2$/Si	Variable	Xe	240-290	10^{-3}-0.1	77-773			BA 87,87a	r)
Ni/SiC	20-50	Xe	350	0.5-5			RBS,EM	NA 84	t)
Ta/Si	50	Sb	180	0.1-1	473-673		a)	GA 87b	a)
Au,V,Cr,Pd/Si	50-150	Si	300-375		77		RBS	PO 85	x)
Ag/glass	25	b)	120-1000	5×10^{-3}-1	R,77		RBS	WA 85a	b)
Ag/Si	22-37	Ar	45	0.5-2		collisional	RBS	JI 84	i)
Sn/Si	30,45	Ar	100-220	0.5-5	R		RBS,CEM	PR 86a	o)
Pd/Si	25/20-10^3	H,He,Ar	2.5-10^3	12		RED	RBS	HO 86a	s)
Ni/SiO$_2$/Si	80	Xe	240-290	10^{-3}-1	298-773	Own	RBS	BA 87	u)
Ni/Si	25-45	Ar	100-400	2-10	R		RBS,AES	CH 87a	v)
W,Pd,Sn/Si		Ar	70-250	0.5-5	R		RBS	KU 86a	w)
Ti,Fe/C/Ti,Fe		Xe	600	0.8-1.7	R		RBS,TEM	HI 86a	a1)
Pt/GaAs	$\bar{x}_R$	Si,Ar,Ge	160	1			RBS	FU 86	a2)
Mo/Si	30	As	50-200	0.2-2	R		RBS,SIMS,TEM	VA 86a	a3)
Fe/SiO$_2$		Ar	100	0.3-13	R		RBS,SEM	BA 87b	a4)
Al/Fe	7-50	Xe	200-330	0.5-3	R		EM,AES	RA 87b	a5)

Notes to Table 5b (Literature published after 1983, and not included in the review by [PA 85]).

a) Experimental analysis: RBS, SIMS, SEM, XRD; amount of mixing $\alpha\ \phi$.

b) Ions: He, Ne, Ar, Kr, Xe; Dt/ϕ versus nuclear energy deposition.

c) Higher mixing rates for Au-rich alloy (x< 0.5). $X\ \alpha\ \phi^{1/2}$.

d) Observed effects of RIS, RED and demixing. No mixing effects observed, in accordance with phase diagrams.

e) $X\ \alpha\ \phi^{1/2}$; amorphization starts at the interface.

f) Crystalline Xe is observed; mixing is blocked after 2×10^{16} ions/cm^2 due to the role of the third phase (Xe).

g) $X\ \alpha\ \phi^{1/2}$; slower mixing than in Pt/Ni indicates influence of chemistry.

h) Underlayers: Ti, V, Cr, Mn, Co, Ni (3d metals); range Xe ~ 50 nm: chemically enhanced diffusion.

i) Diffusional theory of mixing explains major aspects of data.

j) $X\ \alpha\ \phi^{0.75}$; Ar bubbles formed at 10^{17} ions/cm^2.

k) Thermal model predictions above experimental results.

l) $X\ \alpha\ \phi$; collisional theory a factor ~10 too low; ϕ dependence?

m) RBS: depth resolution for Cr layer ~ 200 Å; yield of Cr atoms reacted per ion is proportional either to F_D or to the number of knock-ons.

n) Multilayer structure; mixing efficiency $\alpha\ \phi^{1/2}$, and increases with the number of layers in the samples.

o) Considerable amount of transport of Si atoms into the Sn region. X increases with E, ϕ.

p) Variable dependence of X with ϕ: collisional and RED regimes are identified.

q) Also the reverse top/underlayers, and W markers, were investigated; the inverse Kirkendall effect was identified.

r) Isotropic and anisotropic transport of matter.

s) $X\ \alpha\ \phi$ for low F_D, $X\ \alpha\ \phi^{1/2}$ for high F_D.

t) Thickness of mixed layer increases with ϕ. Comparison with laser mixing.

u) Areal density of Ni incorporated in SiO$_2$ has contributions from R.I. ($\alpha\ \phi$) and from overlapping cascades ($\alpha\ \phi^{1/2}$).

v) Collisional mixing theory in the diffusion approximation. No mixing at 100 keV. $X\ \alpha\ F_D$ at 200-400 keV.

w) Cascade mixing dominates in Pb/Si. $X\ \alpha\ \phi^{1/2}$; more complex process in Sn/Si: no mixing in W/Si.

x) Mixing rate (atoms/ion) ~0.8-2.5; depth of mixing varying as ϕ, not $\phi^{1/2}$, is discussed.

y) Mixing rate (atoms/ion) ~ 13 Al/Xe. ~ 1 Sb/Xe; Au marker defines the interfaces.

z) Resistivity measurements; amorphous phase at 77K in Ni, Co/Al; amorphous phase at 77, 330 K in Fe/Al; influence of F_D.

a1) $X\ \alpha\ \phi$: less efficient mixing of C with Ti.

a2) Amount of GaAs reacted with Pt proportional to the mass of the ion.

a3) $X\ \alpha\ \phi$, E: ballistic mixing dominates at low T_{irr}, and RED at high T_{irr}.

284

Notes to Table 5b (Continued)

a4) $X \alpha A \phi + b \phi^{1/2}$.

a5) Multilayer sample with total thickness $\sim 2\bar{x}_R$.

MIXING BY DEFECT-ASSISTED MIGRATION OF THIN MARKERS IN SOLIDS

R. GARCIA-MOLINA AND I. ABRIL

Departament de Física Aplicada, Facultat de Ciències, Universitat d'Alacant
Apt. 99, E-03080 Alacant, Spain

Ion-beam mixing of layered solid systems has been the subject of increasing interest, both experimentally and theoretically, in the last years, mainly due to the capability of this technique to modify the composition of solids on a very small scale, which cannot be done by other conventional techniques. However, the role played by the different possible processes responsible for the atomic mixing induced by ion-beam bombardment is not yet completely clarified because of the complexity of mixing phenomena. It is known that together with collisional mixing, which occurs in all systems when they are bombarded by ions, also defect-assisted migration of the atoms is present during the mixing process. Much experimental and theoretical work has been done in order to elucidate the relative importance of the different mechanisms that can take place in the mixing process at a microscopic level. Among the different mechanisms proposed to describe mixing phenomena, we have analysed the mixing process that occurs in a damaged region and which may be described by diffusion mathematics. This model was first proposed by Sigmund /1/.

In this paper, we have focused on the mixing of thin markers in solids, because, although bilayer systems are the systems most commonly used for ion-beam mixing studies (due to their interest for potential applications), marker systems are the simplest configurations in order to study the different mechanisms acting during ion-beam mixing. We apply the model proposed by Sigmund /1/, which assumes that a given marker atom may move in the gradient of defects created by the collisional cascade, being trapped in some type of trapping center. From a general evaluation of the relocation profile moments, this model provides gaussian relocation profiles and it also gives some estimates about the broadening of a thin impurity layer beneath a substrate.

Our aim in this work is to consider the dependence of the defect-assisted migration mixing process with the depth where the marker lies and to characterise the spreading of the marker after ion bombardment from the explicit calculation of the relocation profile. In our calculations we describe the initial profile of the thin impurity marker by means of a delta function (typical experimental values of the width are of the order of a few tens of angstroms or less).

The relocation profile $P_\phi(x, \Delta x)$ is defined as the probability density that a given atom originally at x undergoes a relocation within the substrate layer ($x + \Delta x, d(\Delta x)$) after an ion beam, with fluence ϕ and energy E, bombards the system (see Fig.1). It is given within a multiple relocation theory by (see Ref./2/ for definitions)

$$P_\phi(x, \Delta x) = \frac{1}{2\pi} \int_{-\infty}^{+\infty} dk \; \exp\{ik\Delta x - \phi\, \sigma(k)\} \; , \tag{1}$$

R. Kelly and M. Fernanda da Silva (eds.), Materials Modification by High-fluence Ion Beams, 285–290.
© *1989 by Kluwer Academic Publishers.*

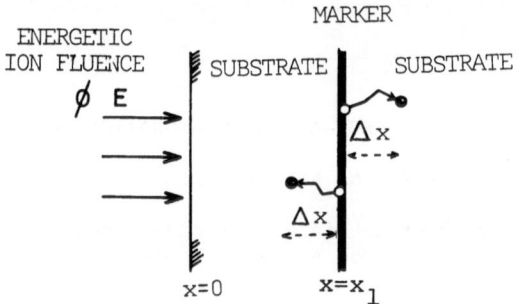

FIGURE   1. Ion-beam mixing parameters. The marker   is located at the depth x_1 .

where $\sigma(k)$ is the transport cross section

$$\sigma(k) \;=\; \int\limits_{(z)} d\sigma(x,z)\,\{\,1-\exp(-ikz)\,\} \;, \qquad (2)$$

calculated from the differential relocation cross section $d\sigma(x,z)$ for a defect-assisted migration mixing process (the defects being created by the collisions induced by the ion-beam bombardment on the solid) /1/. $d\sigma(x,z)$ is given by

$$d\sigma(x,z)= \tfrac{1}{2}\; \sigma(x)\; p\; (\frac{\lambda}{D})^{\frac{1}{2}}\; \exp\{-(\frac{\lambda}{D})^{\frac{1}{2}}|z|\}\; dz \;, \qquad (3)$$

where the parameters are the following /1/: $\sigma(x)$ is the mean area for an individual damage zone facing the beam, p is the probability for an atom within a damaged zone to undergo defect-assisted migration, λ is the trapping probability per unit time for a uniform trap concentration, $D= \nu \ell^2/6$ is an "effective diffusion coefficient", ℓ is the jump distance (a few times the interatomic distance), ν is the jump frequency and z is the relocation distance of an impurity in an individual event. In general all these parameters depend on the depth, as was quoted but not taken into account in Ref./1/. Note that the normalization of the differential relocation cross section, eq.(3), is $\sigma(x)\cdot p$, i.e., the total cross section for migration in the damaged area.

We have calculated explicitly the relocation profile   using the saddle point method /3/   and we get

$$P_\phi(x,\Delta x)\approx \frac{1}{2\sqrt{\pi}}\; \{\sigma(x)\; p\; \frac{\lambda}{D}\; \phi\}^{-\frac{1}{2}}\; \sum_{k_0}\; \frac{(\lambda/D + k_0^2)^{3/2}}{(\lambda/D - 3k_0^2)^{1/2}} \;\cdot$$

$$\exp\{\frac{i\,k_0\,\Delta x}{2}\;(1-\frac{D}{\lambda}k_0^2)\} \;, \qquad (4)$$

where the k_0's are the saddle points, given by the following equation

$$(k_0^2 + \lambda/D)^2 + 2i\sigma(x) p \frac{\lambda}{D} \frac{\phi}{\Delta x} k_0 = 0 \qquad (5)$$

The dependence with the depth where the marker atoms lie initially, x_1, (or where they are relocated after the ion bombardment, $x_1 + \Delta x$) has been incorporated into the mean area $\sigma(x)$, through $\sigma(x) =_\pi \rho^2(x)$, where $\rho(x)$ is the transverse damage straggling at the depth x. We have assumed $\rho(x_1)$ to be proportional to the energy deposited into damage at the depth x_1, $F_D(E,x_1)$, which was taken from Winterbon's book /4/.

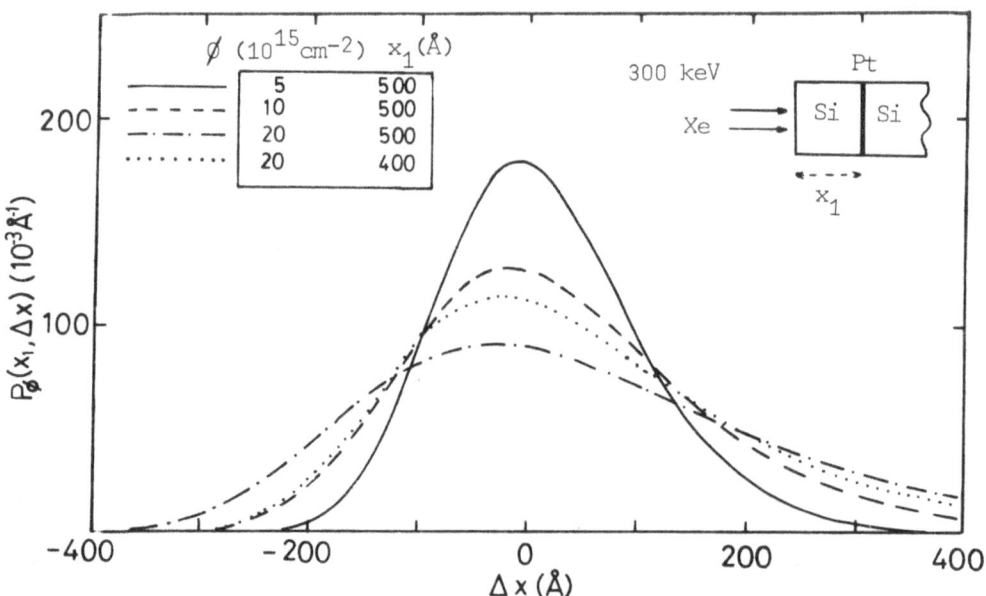

FIGURE 2. *Relocation profile,* $P_\phi(x, \Delta x)$, *for a Pt marker in Si after bombarding by Xe ions of 300 keV, at various marker depths and fluences. The jump distance we have used is $\ell = 4$ Å.*

In figure 2 we have depicted the relocation profile $P_\phi(x, \Delta x)$ corresponding to (300 keV) Xe ⟶ Si/Pt/Si, for various fluences and with the Pt marker located at various depths. The asymmetric relocation profile we obtain is due to the fact that we have taken into account the different value of F_D at the right and at the left of the initial marker position. As the shifts of the marker shown in Fig.2 are small and the relocation profile remains like a gaussian in the vicinity of the original marker position, in all the cases that follows we will assume that $\rho(x_1)$ is proportional to $F_D(E,x_1)$.

In figure 3 we compare the marker broadening, defined as half width at half maximum (HWHM) as calculated from the relocation profile, eq.(4), with experimental data /5/. The data relate to the systems

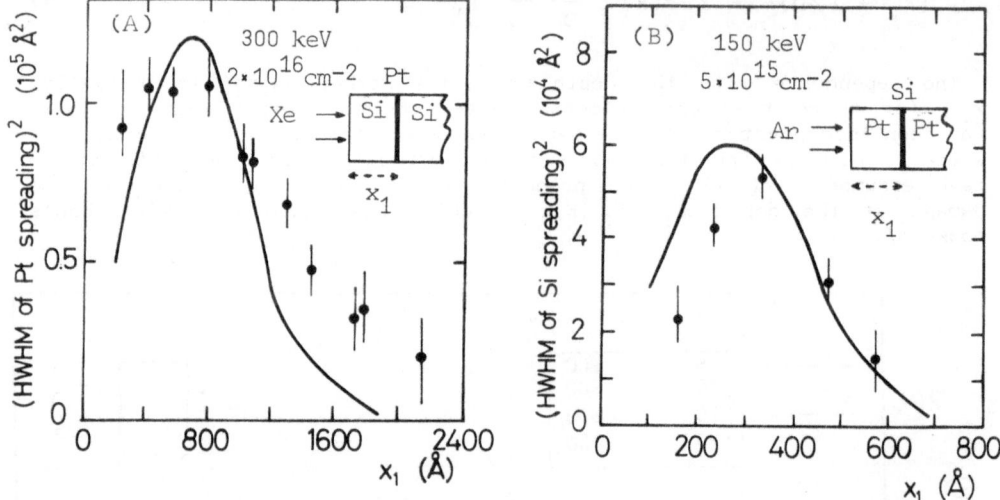

FIGURE 3 . Dependence of the spreading of the marker (induced by the ion bombardment) as a function of the marker depth. The solid curves are the theoretical results calculated from the relocation profile, eq. (4); we have assumed that $\rho(x_1) \propto F_D(E, x_1)$. The dark circles and bars represent experimental data from Ref. /5/. (a) (300 keV, $2 \cdot 10^{16}$ cm^{-2}) Xe → Si/Pt/Si, the jump distance we have used is $\ell = 4$ Å; (b) (150 keV, $5 \cdot 10^{15}$ cm^{-2}) Ar → Pt/Si/Pt, $\ell = 9$ Å.

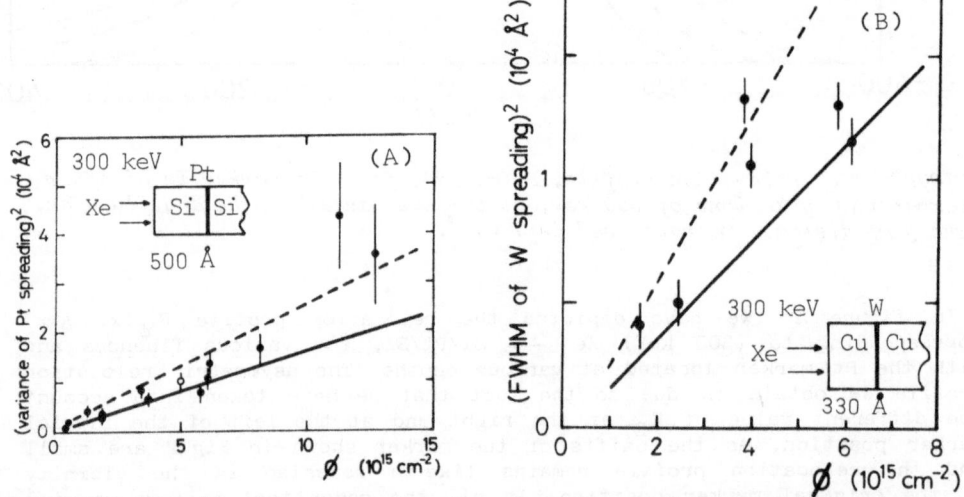

FIGURE 4. Spreading of the marker as a function of ion beam fluence. Lines represent theoretical results and dark circles and bars correspond to experimental data. (a) (300 keV) Xe → Si/Pt/Si ($x_1 = 500$ Å); —— : theoretical result with $\ell = 4$ Å; - - - : theoretical result with $\ell = 5$ Å; experimental data, ♦, were taken from Ref. /6/. (b) (300 keV) Xe → Cu/W/Cu ($x_1 = 330$ Å) ; —— : theoretical result with $\ell = 3$ Å; - - -: theoretical result with $\ell = 4$ Å; experimental data, ♦, were taken from Ref. /7/.

(300 keV, $2 \cdot 10^{16}$ cm^{-2}) Xe $\longrightarrow$ Si/Pt/Si and (150 keV, $5 \cdot 10^{15}$ cm^{-2}) Ar $\longrightarrow$ Pt/Si/Pt as a function of the depth where the marker lies. In figure 4 we have plotted the marker broadening after ion bombardment for two different systems as a function of the ion fluence, and compared it with experimental data /6,7/. It can be noted that the square of the spreading increases linearly with the ion fluence. Figures 3 and 4 show that the agreement between experimental data and theoretical predictions derived from Sigmund's model are fairly good. The jump distance, ℓ , is the only adjustable parameter and corresponds to typical values of the order of a few interatomic distances.

We have also analysed the change in the marker spreading as a function of the ion beam energy, E. For the system depicted in Fig. 4(b) we have obtained that for E=100 keV the FWHM$^2 \simeq 3 \cdot 10^3$ Å^2 and for E=400 keV the FWHM$^2 \simeq 2 \cdot 10^3$ Å^2. These results are in line with the experimental results of Ref./8/, which give a decrease in the mixing when the ion-beam energy increases. In general, the theoretical prediction of the marker broadening as a function of the ion-beam energy depends on the marker position with respect to the mean damage depth.

From our calculations we have, obviously, reproduced the general trends obtained by Sigmund /1/ (such as mixing profile close to a gaussian and broadening increasing linearly with the square root of the ion fluence) and we have made some comparisons between the broadening of initially very narrow layers extracted from our relocation profile calculations and that experimentally observed. These calculations show that: (i) mixing by defect-assisted migration can give account of the observed profile broadenings by itself or combined with the effects of collisional mixing; (ii) the displacement of the peak is very small ($\lesssim 10$ Å) and its motion relative to the surface depends on the position of the marker with respect to the mean damage depth; and (iii) the spreading variation of the marker as a function of the ion energy depends on the position of the marker with respect to the mean damage depth.

Summarising, we conclude that some improvements (such as considering the effects due to the depth at which the marker atoms lie in the substrate) added to Sigmund's model /1/ give theoretical predictions that agree with available experimental data /8-10/. Nevertheless, the complexity of the mixing process include many possible mechanisms whose relative contribution to the total effect remain not completely understood.

We would like to thank. Dr. A. Gras-Marti for useful discussions and a critical reading of the manuscript. Financial support from the Generalitat Valenciana is acknowledged.

REFERENCES
/1/ P. Sigmund, Appl. Phys. A 30 (1983) 43
/2/ P. Sigmund and A. Gras-Marti, Nucl. Instrum. Meth. 182/183 (1981) 25
/3/ P.H. Morse and H. Feshbach, Methods of Theoretical Physics, McGraw-Hill Publ. Co., New York 1953 (p.437)
/4/ K.B. Winterbon, Ion Implantation Range and Energy Deposition Distributions, Plenum, New York 1975, Vol.2
/5/ B.Y. Tsaur, S. Matteson, G. Chapman, Z.L. Liau and M-A. Nicolet, Appl. Phys. Lett. 35 (1979) 825
/6/ B.M. Paine, J. Appl. Phys. 53 (1982) 6828
/7/ J.F.M. Westendorp, U. Littmark and F.W. Saris, Nucl. Instrum. Meth.

B 18 (1986) 54
/8/ R.S. Averback and D. Peak, Appl. Phys. A 38 (1985) 139
/9/ B.M. Paine and R.S. Averback, Nucl. Instrum. Meth. B 7/8 (1985) 666
/10/ D. Peak and R.S. Averback, Nucl. Instrum. Meth. B 7/8 (1985) 561

THE TRIUMF OPTICALLY PUMPED POLARIZED H⁻ ION SOURCE

W.M. LAW*, C.D.P. LEVY, M. McDONALD, and P.W. SCHMOR

TRIUMF, 4004 Wesbrook Mall, Vancouver, B.C., Canada V6T 2A3
*Also at Physics Department, Univ. of Manitoba, Winnipeg Canada R3T 2N2

1. INTRODUCTION

TRIUMF presently uses a Lamb-shift polarized H⁻ source capable of pro-viding ~1 μA of ~75% polarized beam on target. This current is barely ade-quate for a number of approved experiments. Since 1983 an optically pumped source has been under development at TRI- University Meson Facility of Canada (TRIUMF). This is expected to eventually produce intense dc H⁻ beams (~50 μA) at a polarization of ~70%, with an emittance suitable for injec-tion into the cyclotron. Approximately 50% of the dc ion source beam will be accelerated to full energy and extracted. To date, the optically pumped source has been shown to be capable of ~10 μA of H⁻ at a polarization of ~60% within a normalized emittance of 0.4π mm-mrad and at an ionizing field of 1.5 kG. Work has begun on attaching the source to the cyclotron.

Optically pumped H⁻ ion sources have been successfully demonstrated in pulsed operation by Mori et al.[1] at National Laboratory for High Energy Physics of Japan (KEK), using an electron cyclotron resonance (ECR) proton source, and by Zelenskii et al.[2] at Institute for Nuclear Research of Moscow (INR). The TRIUMF source is the only existing dc optically pumped source.

The optical pumping technique as proposed by Anderson[3] is as follows. Circularly polarized dye laser light tuned to the sodium D_1 transition is used to electron polarize ground state sodium atoms in an optically thick vapour. An electron-spin polarized atomic hydrogen beam is produced by passing protons through the polarized sodium vapour, where charge exchange occurs. A diabatic field reversal technique, similar to that used in Lamb-shift sources, transfers the electron polarization to the nucleus. Charge exchange in a second unpolarized alkali vapour cell (negative ion ionizer) yields a nuclear polarized H⁻ beam. Figure 1 shows the layout of the TRIUMF

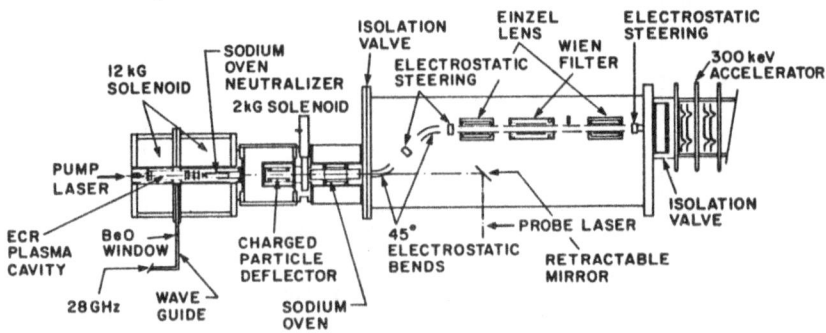

FIGURE 1. Layout of the TRIUMF optically pumped polarized H⁻ ion source.

291

R. Kelly and M. Fernanda da Silva (eds.), Materials Modification by High-fluence Ion Beams, 291–294.
© 1989 by Kluwer Academic Publishers.

source and the optics to match the 5 keV beam to a 300 kV accelerator.

2. ECR PROTON SOURCE

A hydrogen plasma is produced in the multi-mode ECR cavity by up to 400 W absorbed cw microwave power at 28 GHz, from a Varian extended interaction oscillator (model VKQ-2H35F). The extraction electrodes and polarized sodium vapour cell are located in a ~12 kG axial magnetic field. The field has a mirror configuration with a minimum at 8 kG, where the microwave power is fed in radially. Hydrogen gas is fed in through the same wave guide and with a quartz liner in the ECR cavity the proton ratio, $[H^+/(H^+ + H_2^+ + H_3^+)]$, is greater than 0.75. The current density extracted at an energy of 5 keV from the water-cooled acceleration-deceleration type multi-aperture molybdenum electrodes increases linearly with absorbed microwave power to a maximum of 120 mA/cm^2, limited by the available power. Earlier results showed that up to 300 mA/cm^2 could be extracted through a relatively small 2 mm diameter hole.

3. SODIUM POLARIZATION

The sodium atoms in the neutralizer are polarized by circularly polarized light of up to 1 W at 5896 Å from a Coherent CR-599 broadband dye laser, and the polarization is measured using a Faraday rotation technique.[4,5] Narrowing the bandwidth of the pumping light from a nominal 30 GHz to ~6 GHz with an uncoated 0.5 mm thick intra-cavity etalon increases the spectral power density of the laser light within the 3 GHz Doppler width of the sodium D$_1$ transition. The polarization of the atoms depends on the polarization rate due to optical pumping and the depolarization rate due to wall relaxation, diffusion out of the target, and at higher target thickness, radiation trapping.[6,7] Relaxation of electron spin $\langle S_z \rangle$ of optically pumped sodium atoms on copper and dry-film[6] coated walls have recently been measured as a function of magnetic field within the neutralizer at TRIUMF. The polarized atoms relaxed in the dark after the pumping light was suddenly cut off by a mechanical chopper.

It was found that a copper wall became less depolarizing as the magnetic field was increased (Fig. 2). The dry-film became so effective that

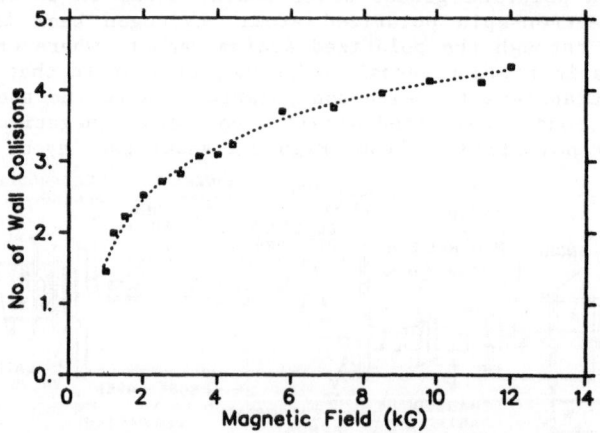

FIGURE 2. Mean number of non-depolarizing wall collisions vs magnetic field, 4.9 mm diameter copper cell.

we were not able to measure the very low depolarization rate, since the observed polarization relaxation was dominated by molecular flow of polarized sodium atoms out of the cell. However, we find that dry-film coatings are rapidly destroyed when running the ion beam.

Further experiments and calculations[7,8] show that a sodium polarization over 90% at a density of 4×10^{12} atoms/cm^3 can be attained with a laser power of a few watts over the sodium absorption bandwidth, even with metallic walls. The TRIUMF source will soon use three broadband dye lasers with total power of ~3 W. The transfer of polarization[9,10] from sodium atoms to hydrogen nuclei has an efficiency of ~60% at a field strength of 12-13 kG. Therefore, a nuclear polarization of ~54% will be achieved in the TRIUMF source. A superconducting solenoid will be used later to increase the magnetic field to 20-25 kG in the neutralizer region, thus increasing the polarization transfer efficiency and the final nuclear polarization.

4. BEAM CURRENT AND EMITTANCE

Using a multi-aperture electrode configuration, 8 mA of proton current was transmitted through the 8 mm neutralizer cell aperture in the absence of sodium vapour. The proton current decreased to near zero as the sodium cell temperature was raised, as an increasing proportion of the beam was neutralized. Other results indicated that a neutral beam current of 220 µA particle equivalent current was present downstream of the ionizer cell, as shown by both secondary electron emission and calorimetry, at a neutralizer cell sodium thickness of 5×10^{13} atoms/cm^2. Given an equilibrium H⁻ yield of 7% at 5 keV in a thick sodium target,[11] such a neutral current corresponds to an H⁻ current of ~15 µA.

The emittance of the H⁻ beam is determined mainly by the emittance growth as the beam leaves the 1.5 kG field of the ionizer cell. The H⁻ beam emittance was measured as a function of ionizer magnetic field using a slit scanner of a type developed at Los Alamos,[12] and the results are shown in Fig. 3. The graph shows that the H⁻ beam has a zero field normalized emittance at the 60% contour level of 0.07π mm-mrad, which is equivalent to that of the H⁰ beam accepted by the ionizer. At 1.5 kG the H⁻ beam has an effective emittance of 0.4π mm-mrad in the transverse planes after leaving the field.

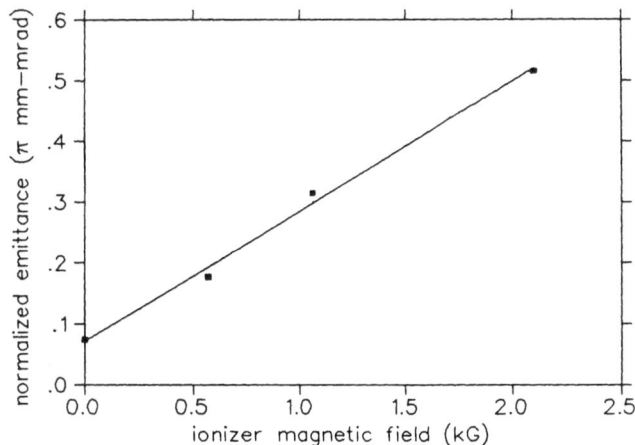

FIGURE 3. H⁻ beam emittance vs ionizer magnetic field.

The focussing elements used to transport the 5 keV H⁻ beam to the 300 kV accelerating column (Fig. 1) have been computer-designed using the results of the H⁻ emittance measurements. A pair of parabolic 45° electrostatic deflectors separates the H⁻ beam from the H⁰ beam not ionized in the ionizer and allows the probe laser light to enter the neutralizer. A Wien filter rotates the nuclear spin of polarized H⁻ by 90° from the horizontal to the vertical plane, parallel to the magnetic field of the cyclotron. Two einzel lenses of focal lengths of 7 cm and 18 cm are used to focus the H⁻ beam into the Wien filter and accelerating column. The geometries of these lenses were designed using particle-tracing computer codes to give an aberration of $\lesssim 5\%$.

CONCLUSION

The TRIUMF source is capable of producing ~10 µA H⁻ ion beam polarized at ~60% within a normalized emittance of 0.4π mm-mrad at the 60% contour level, suitable for injection into the TRIUMF cyclotron. Although the depolarizing effect from wall collisions of dry-film coated wall is small, a problem in cw systems is the rapid destruction of the coating by the ion beam. The depolarizing effect of a copper wall, on the other hand, is reduced at high magnetic fields, improving the performance of the neutralizer. The polarization of the H⁻ beam will be enhanced by using superconducting coils around the neutralizer cell to increase the magnetic field to 20-25 kG. Installation of the ion source on the TRIUMF cyclotron is scheduled to be completed by the end of 1987.

REFERENCES

1. Mori, Y., Ikegami, K., Igarashi, Z., Takagi, A. and Fukumoto, S. American Institute of Physics Conference Proceedings 117, 123 (1984).
2. Zelenskii, A.N., Kokhanovskii, A.S., Lobashev, V.M., and Polushkin, V.G., Nucl. Instrum. Methods, A245, 223 (1986).
3. Anderson, L.W., Nucl. Instrum. Methods 167, 363 (1979).
4. Cornelius, W.D., Taylor, D.J., York R.L. and Hinds, E.A., Phys. Rev. Lett. 49, 870 (1982).
5. Mori, Y., Ikegami, K., Takagi, A., Fukumoto, S. and Cornelius, W.D., Nucl. Instrum. Methods 220, 264 (1984).
6. Swenson, D.R. and Anderson, L.W., Helv. Phys. Acta 59, 652 (1986).
7. Tupa, D., Anderson, L.W., Huber, D.L. and Lawler, J.E., Helv. Phys. Acta 59, 657 (1986).
8. Mori, Y., Takagi, A., Ikegami, K., Fukumoto, S., Ueno, A., Levy, C.D.P. and Schmor, P.W., National Laboratory for High Energy Physics of Japan report 86-2, (1986).
9. Mori, Y., Takagi, A., Ikegami, K., Fukumoto, S. and Ueno, A., J. Phys. Soc. Jpn. 55, 453 (1986).
10. Uegaki, J., Schmor, P.W., and Levy, C.D.P. Phys. Lett. A 115, 216 (1986).
11. Schlachter, A.S., Stalder, K.R., and Stearns, J.W., Phys. Rev. A 22, 2494 (1980).
12. Allision, P.W., Holtkamp, D.B., and Sherman, J.D., IEEE Trans. NS-30(4), 2204 (1983)

SOME HIGH-CURRENT ION SOURCES FOR MATERIALS MODIFICATION

T. TAYLOR

Atomic Energy of Canada Limited
Chalk River Nuclear Laboratories
Chalk River, Ontario, Canada K0J 1J0

1. INTRODUCTION

Ion sources for materials modification have evolved through three distinct generations. The first generation was adopted from research accelerators. These cold-cathode plasma-discharge devices generate beam currents of less than 100 µA. They are suitable for low-dose semiconductor applications such as threshold shifting. The hot-cathode plasma-discharge ion sources, originally developed for isotope separation, comprise the second generation. They produce between 100 µA and 10 mA of beam current and are appropriate for medium-dose semiconductor processes, for example, the formation of sources and drains. The third generation ion sources give beam currents in excess of 10 mA. This technology, transferred from industrial accelerators, has already made SIMOX (Separation by IMplanted OXygen) into a commercially viable semiconductor process and promises to do the same for ion implantation of metals and insulators.

This paper is not a comprehensive treatise on ion sources. There are already a half dozen excellent reviews in the literature. In particular, the recent works of Stephens (1) and Aitken (2) are highly recommended. They complement each other well, the former concentrating on the physics of ion sources while the latter emphasizes the engineering. The focus here is on the third generation technology that will play a key role in the future of ion implantation. But first, to put the material on high-current ion sources into perspective, a brief look at a second generation device is presented.

2. THE FREEMAN SOURCE

Almost every modern commercial ion implanter uses a version of the Freeman source (3) (see Fig. 1) developed at Harwell in the early 1960's for isotope separation. Like all other plasma-discharge ion sources the Freeman source is comprised of four components: a) the feed system, b) the electron generator, c) the plasma generator and d) the extraction system. The operating principles of the Freeman source are relatively simple. A pressure of about 0.1 Pa (10^{-3} torr) is established in the arc chamber either by admitting a gas through a needle valve or by vaporizing a solid in the oven. Electrons are thermionically emitted from the 2 mm diameter tungsten filament by Ohmic heating with a current of 150 A. A voltage of 50 to 100 V applied between the filament and the arc chamber sustains a plasma discharge. An externally generated 10 mT axial magnetic field in combination with the 2 mT azimuthal magnetic field from the filament forces the electrons to follow convoluted trajectories from the filament to the arc chamber efficiently ionizing the gas or vapour. An ion beam is extracted electrostatically from the dense plasma close to the filament through a 2 mm wide slit mounted only 3 mm away from the filament.

R. Kelly and M. Fernanda da Silva (eds.), Materials Modification by High-fluence Ion Beams, 295–301.
© *1989 by Kluwer Academic Publishers.*

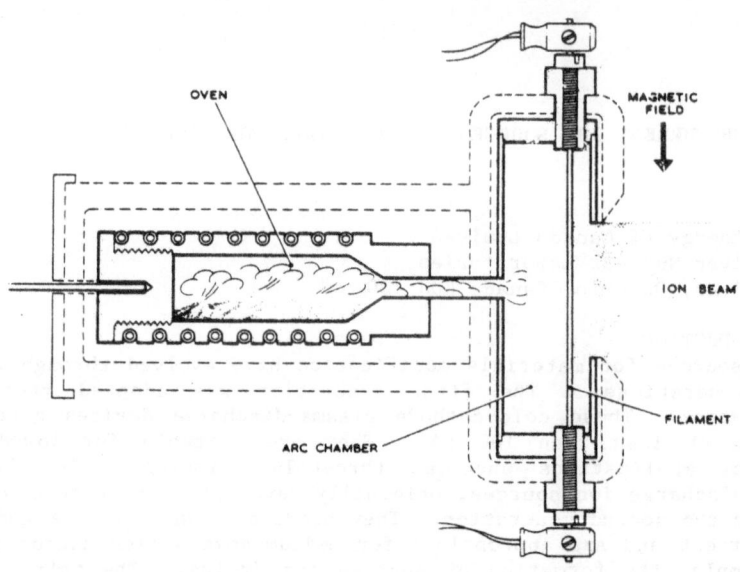

FIGURE 1. The Freeman source sectioned through the length of the slit (from (3)).

The Freeman source is extremely versatile. The oven and the arc chamber can be operated at temperatures up to 1100°C so that virtually any ion can be generated either from the corresponding element or from a compound. Furthermore, compounds can be created in the oven by introducing halogenating agents such as carbon tetrachloride. Even ions of noble metals can be produced by sputtering a target mounted at the back of the arc chamber with inert gas ions. Stable beams of up to 10 mA of P^+, Ar^+, As^+ and Sb^+ and lesser currents of B^+ (from boron trifluoride) are easily obtained. The extraction geometry of the Freeman source is also quite flexible. The length of the slit is virtually unlimited. (Although 40 mm is typical, slits as long as 90 mm are used in commercial implanters.) Moreover, the slit can be curved in the direction of the beam to generate a vertical waist at an analyzing magnet.

In summary, the Freeman source is a simple, reliable source of high quality beams of almost every ion. Nonetheless, the Freeman source cannot generate the beam currents required for the high-dose applications of ion implantation that are now being developed. Some alternative source concepts will now be considered.

3. CHORDIS

CHORDIS (4) (Cold or HOt Reflex Discharge Ion Source), shown in Fig. 2, was developed at GSI Darmstadt for the heavy-ion accelerator UNILAC. Like the Freeman source, CHORDIS is a hot-cathode plasma-discharge device. There the similarity ends. The cathode is comprised of six tantalum helices suspended from a tantalum disk. Reflector electrodes connected to the negative leg of the cathode force the electrons to oscillate, enhancing the ionization efficiency. (The resultant reflex discharge is often

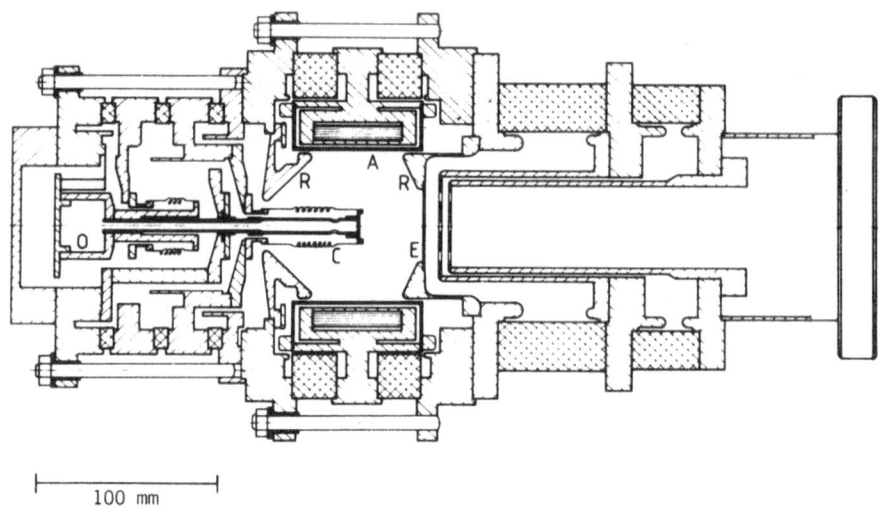

Oven/cathode chamber   Discharge chamber   Extraction system

FIGURE 2.   CHORDIS with anode (A), cathode (C), extraction grid (E),
reflectors (R) and oven (O) (from (4)).

called PIGing after the Penning Ionization Gauge.) The plasma is confined
radially by magnetic cusps generated by eighteen 200 mT samarium-cobalt
permanent magnets.   Probably the most significant departure from the
second generation technology is the extraction geometry.   Assuming that
the plasma temperature and density are fixed, the only way to increase the
beam current from an ion source is to increase the area of the plasma from
which beam can be extracted.   Unfortunately, all else being equal, the
divergence of the beam increases linearly with the width of the extraction
aperture (5).   If low-divergence high-current ion beams are required
multi-aperture extraction is essential. CHORDIS has been operated with up
to seven 0.7 cm diameter apertures and thirteen 0.3 cm diameter apertures
giving a total extraction area of as much as 2.7 cm^2.
    All of the feed systems developed for the second generation ion sources
work equally well with CHORDIS.   High-quality ion beams of about 100 mA
have been produced for a variety of elements. Table 1 gives a sample of
the current densities achieved to date.

4. THE CHALK RIVER DUOPIGATRON
    The Chalk River duoPIGatron, shown in Fig. 3, is a spin-off from the
Atomic   Energy   of   Canada   Limited   (AECL)   spallation   neutron   source
programme.   It is, in essence, CHORDIS with an external electron generator.
    The operating principle of the duoPIGatron was first elucidated by
Demirkhanov et al. (6).   An arc is struck between the filament and the
intermediate electrode.   The   current   flowing   through   the   resistance
associated   with   the   intermediate   electrode   reduces   the   potential
difference between the intermediate electrode and the filament.   The
discharge then propagates through the hole in the intermediate electrode
to the anode. A 150 mT axial field in the region between the intermediate

TABLE 1. CHORDIS beam current densities and feeds for various ions (4).

Ion	Current Density (mA cm^{-2})	Feed
H$^+$	87	Hydrogen Gas
He$^+$	235	Helium Gas
Li$^+$	30	Lithium Vapour
N$^+$	19	Nitrogen Gas
Ne$^+$	60	Neon Gas
Al$^+$	3	Sputtered Aluminum
Ar$^+$	32	Argon Gas
I$^+$	20	Iodine Vapour
Xe$^+$	36	Xenon Gas
Bi$^+$	19	Bismuth Vapour

electrode and the anode pinches the discharge creating a very high electron density. The electrons oscillate (PIG) in the potential well between the intermediate electrode and the plasma aperture plate. The result is exceptionally good arc power and feed efficiency.

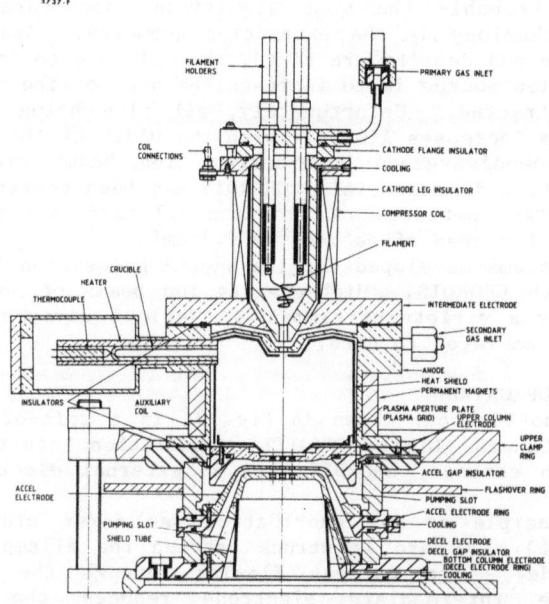

FIGURE 3. A demonstration hot Chalk River duoPIGatron ion source.

Because the duoPIGatron has two chambers, the electron generator can be operated on an inert primary feed when the PIG region is operated on a reactive secondary feed. This significantly enhances filament life. The Eaton NV-200 high-current high-energy oxygen ion implanter (7) uses a source based on this concept under license from AECL.

The various versions of the Chalk River duoPIGatron have been operated with as many as thirteen 0.5 cm diameter extraction apertures with a total area of up to 2.6 cm^2. Representative current densities along with the corresponding feeds are given in Table 2.

TABLE 2. Chalk River duoPIGatron beam current densities and feeds for various ions.

Ion	Current Density (mA cm^{-2})	Feed
H$^+$	213	H$_2$ Gas (8)
He$^+$	122*	He Gas
Li$^+$	17*	Li Vapour with Ar Gas
B$^+$	11	BF$_3$ Gas with Ar Gas (9)
N$^+$	75	N$_2$ Gas (8)
O$^+$	100	O$_2$ Gas with Ar Gas (10)
Ne$^+$	78*	Ne Gas
P$^+$	42	PH$_3$ Gas with Xe Gas (9)
Ar$^+$	113	Ar Gas (8)
Ca$^+$	50*	Ca Vapour with Ne Gas
As$^+$	23	AsH$_3$ Gas with Xe Gas (9)
Xe$^+$	72	Xe Gas (8)

* Preliminary data. Higher current densities are almost certainly obtainable.

Until now we have considered only the quantity of beam that can be extracted from each ion source. In most ion implantation applications, beam quality is at least as important. The quality of an ion beam can be characterized in terms of: a) current and voltage stability, b) charge state and mass distribution, and c) emittance.

Ion sources that rely on high magnetic fields and/or oscillating electrons are susceptible to plasma oscillations which can cause fluctuations in beam parameters. Therefore, the Chalk River duoPIGatron is equipped with an auxiliary coil that shapes the axial magnetic field to suppress plasma oscillations generating stable "hash-free" beams.

Unfortunately, ions of interest often make up only a small fraction of the total beam current. For example, because elemental boron has an extremely low vapour pressure, B$^+$ ions are generated from a boron trifluoride secondary feed with an argon primary feed. A typical mass analysis is shown in Fig. 4. Only 17% of the beam is the desired species. The balance must be magnetically separated.

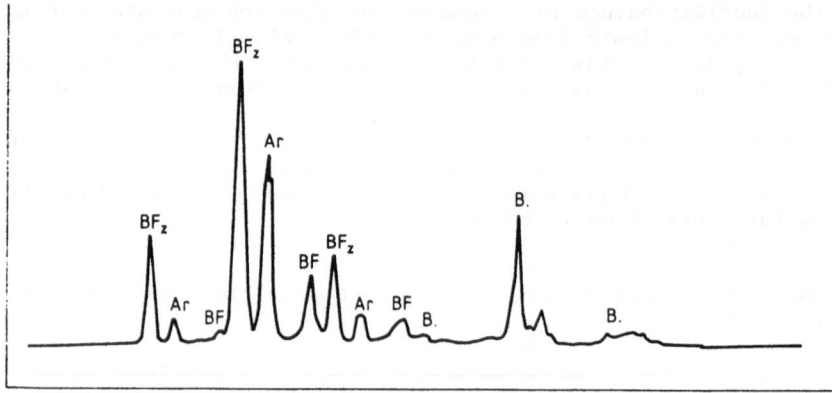

FIGURE 4. Mass analysis for boron trifluoride secondary feed with argon primary feed (from (9)). Each species contributes one peak for each of three apertures.

Usually, the beam from an ion source must be matched to a transport system. A most useful matching tool is the phase-space plot – a graph of transverse position versus angular divergence for a representative selection of ions. Figure 5 shows a typical phase-space plot for a Chalk River duoPIGatron with three in-line apertures. The solid contours at 2% of the maximum beam intensity include 95% of the beam. The dashed contours show where the additional beamlets would have been if a seven aperture extraction system had been used. The emittance is essentially the area of the figure that encloses all of the beamlets, in this case 0.121 π cm-mrad taking into account momentum normalization. (The momentum normalized emittance is especially useful as it is independent of ion energy and mass.) The figure is tilted rather than upright because the measurements were made 56 cm downstream from the ion source.

5. CONCLUSIONS
Third generation ion source technology is capable of delivering a wide variety of high-quality ion beams with currents of hundreds of milli-amperes. Ion sources like CHORDIS and the Chalk River duoPIGatron will be essential to the commercialization of materials modification by high-fluence ion beams.

ACKNOWLEDGEMENTS
The author is indebted to E.C. Douglas, R.G. Maggs, T. Tran Ngoc and J.S.C. Wills for their contributions to the design and testing of the demonstration hot Chalk River duoPIGatron ion source and to M.S. de Jong for many enlightening discussions.

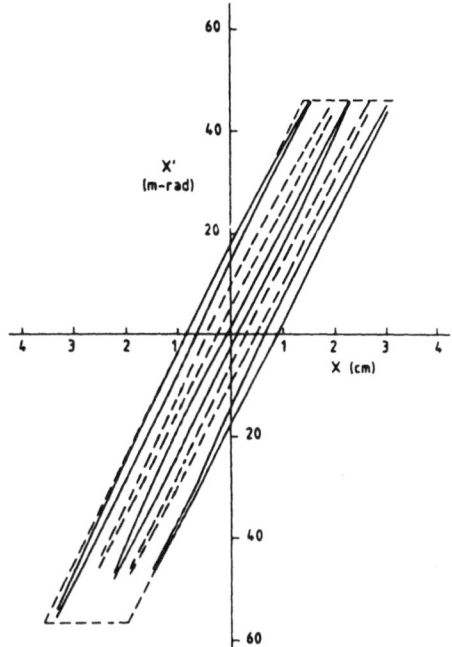

FIGURE 5. Phase space plot for oxygen beam with three in-line apertures (from (10)).

REFERENCES

1. Stephens KG: An Introduction to Ion Sources, in Ion Implantation Science and Technology, JF Ziegler(ed). New York: Academic Press, 1984.
2. Aitken D: Ion Sources, in Ion Implantation Techniques, H Ryssel and H Glawischnig(eds). Berlin: Springer-Verlag, 1982.
3. Freeman JH: A High Intensity Isotope Separator Ion Source, in Proc. 15th Int. Conf. on Ion Sources. INSTN-Saclay, France, 1969.
4. Keller R, P Spädtke and H Emig: Recent Results with a High-Current, Heavy-Ion Source System, Vacuum 36 (1986) 833.
5. Coupland JR, TS Green, DP Hammond and AC Riviere: A Study of the Ion Beam Intensity and Divergence Obtained from a Single Aperture Three Electrode Extraction System, Rev. Sci. Instrum. 44 (1973) 1258.
6. Demirkhanov RA, YuV Kursanov and VM Blagoveshchenskii: High-Intensity Proton Source, Prib. Tekh. Eksp. 1 (1964) 30.
7. Ruffell JP, DH Douglas-Hamilton, RE Kaim and K Izumi: A High Current, High Voltage Oxygen Implanter, Nucl. Instrum. and Meth. B21 (1987) 229.
8. Shubaly MR and MS de Jong: High Current DC Ion Beams, IEEE Trans. Nucl. Sci. NS-30 (1983) 1399.
9. Shubaly MR and RG Maggs: Private Communications, 1985.
10. Shubaly MR, RG Maggs and AE Weeden: A High-Current Oxygen Ion Source, IEEE Trans. Nucl. Sci. NS-32 (1985) 1751.

COMPOSITIONAL AND CHEMICAL CHANGES

BOMBARDMENT-INDUCED COMPOSITIONAL CHANGE WITH ALLOYS, OXIDES, AND OXYSALTS. I THE ROLE OF THE SURFACE BINDING ENERGY

ROGER KELLY

IBM RESEARCH DIVISION, T. J. WATSON RESEARCH CENTER, YORKTOWN HEIGHTS, NY 10598, U.S.A.

1. INTRODUCTION

1.1 The early situation

Bombardment-induced compositional change with alloys, oxides, and oxysalts has been the subject of both experimental and theoretical study for the past two decades. Concerning theoretical work, it is worth pointing out that the emphasis has changed to a remarkable extent as the experimental base has grown. Work written before 1980 tended to emphasize preferential sputtering as triggered by differences of mass, chemical binding, or volatility. Mass and chemical binding were envisaged as governing collisional sputtering, volatility as governing thermal sputtering.

Mass is now regarded to be an important factor in preferential sputtering, thence compositional change, only under near-threshold conditions, as when Ta_2O_5 is bombarded with 1 keV H^+ or He^+ and loses O [1-1], and in isotope sputtering, as when lighter isotopes are lost preferentially with Li, Ti, Ga, and Mo (Fig. 1.1 [1-2]). The role claimed for mass _other than_ near the threshold or with isotopes [1-3] was subsequently shown [1-4] to be largely irrelevant: this was not because the theory [1-3] was wrong but because bombardment-induced Gibbsian segregation [1-4] had been neglected and the latter can cause either a light or heavy species to be lost. (A simplified restatement of the argument of [1-3] about the role of mass is given in [1-5].)

Chemical binding, reflected explicitly in the surface binding energy and ideally proportional to ΔH^a, the heat of atomization [1-6], was at one time the most popular framework for explaining preferential sputtering, thence compositional change. It had the distinct advantage, one lacked by the mass argument, that it gave a nearly perfect correlation: e.g. Au is known to be lost from _near the surface_ of Au-Pd (Figs. 1.2 [1-7] and 1.3) and, at the same time, although it is heavier, it has a slightly lower ΔH^a: $\Delta H^a_{Au} = 3.82$ eV, $\Delta H^a_{Pd} = 3.90$ eV. It is easily shown, however, that chemical binding as manifested in the surface binding energy should normally lead to significantly smaller composition changes than are observed, for example the evolution of $Au_{0.50}Pd_{0.50}$ to $Au_{0.498}Pd_{0.502}$ [1-5] (Sect. 2.5). But as manifested in segregation, chemical binding can lead to very large effects indeed, for example the evolution of $Au_{0.50}Pd_{0.50}$ to $Au_{0.76}Pd_{0.24}$ (Fig. 1.3, curve labeled "annealed"; see also [1-5,1-8]). Nevertheless, there is a hint that with oxides and halides the surface binding energy may play an unsuspected role. As will be shown in Sect. 2.6, this is because

R. Kelly and M. Fernanda da Silva (eds.), Materials Modification by High-fluence Ion Beams, 305–327.
© 1989 by Kluwer Academic Publishers.

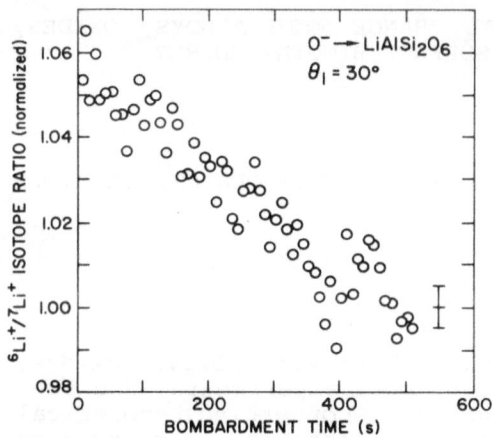

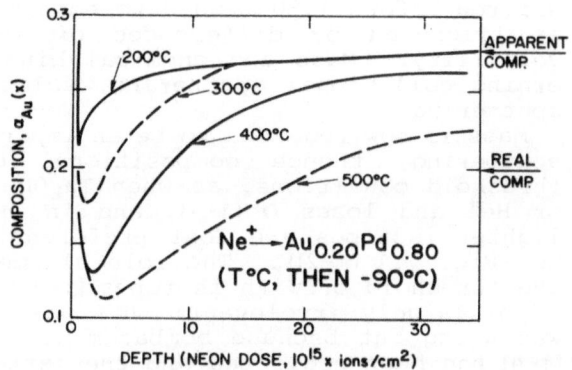

Figure 1.1. Change of the $^6Li^+/^7Li^+$ isotope ratio during the initial sputtering of a fresh $LiAlSi_2O_6$ (spodumene) surface. Normalization is to the steady-state value. The flux density of primary ions, 14.5 keV O^-, was approximately 3.8×10^{14} ions/cm^{2}s, the angle of incidence 30°, and the sputtered depth about 15-25 nm. Initially the $^6Li^+/^7Li^+$ ratio is enriched by a factor of about 1.054 relative to the steady-state value of $^6Li/^7Li \simeq 0.081$. This is an example of preferential sputtering showing a true mass effect. Due to Gnaser and Hutcheon [1-2].

Figure 1.2. Composition-depth profiles for bombarded $Au_{0.20}Pd_{0.80}$ as obtained by first bombarding to steady state with normally incident 2 keV Ne^+ at the indicated temperatures, then quenching to -90°C, and finally profiling with 2 keV Ne^+. Compositions were obtained with ISS, "ion scattering spectroscopy", using the same 2 keV Ne^+ as the probe ion. The reason that the bulk composition implied by the figure, $Au_{0.28}Pd_{0.72}$, disagrees with the stated composition, $Au_{0.20}Pd_{0.80}$, is that the system is subject to on-going Pd loss as a mass effect (cf. Fig. 1). For notation, see Sect. 2.1. Due to Swartzfager, Ziemecki, and Kelley [1-7].

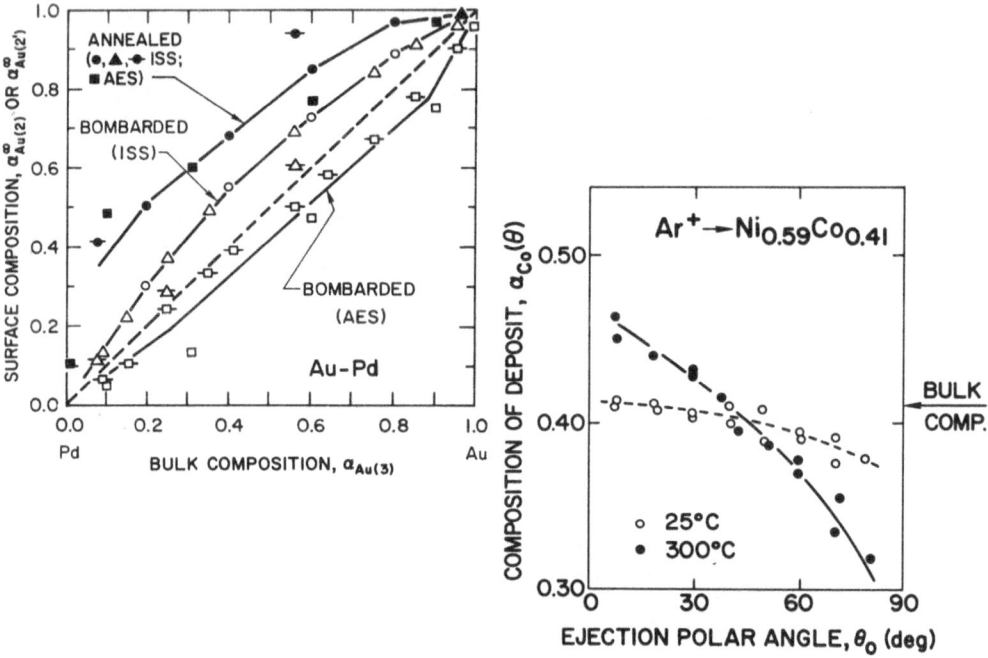

Figure 1.3. Comparison of surface or subsurface compositions with bulk compositions for Au-Pd specimens which have been either annealed or bombarded. The segregating element is seen to be Au and, in addition, the outermost atom layer of a bombarded surface is also Au rich ("ISS"); in agreement with Fig. 2, however, the subsurface region of a bombarded specimen is deficient in Au ("AES", "Auger electron spectroscopy"). This example is generic of many others in which the species which segregates on annealing shows a composition spike followed by severe depletion when the specimen is bombarded. Details are as follows: ● (600°C, Ne ISS [1-7]), ▲ (500°C, Ne ISS [1-38]), ◆ (400°C, He ISS [1-39]), ■ (600°C, AES [1-40]), O (bombarded, Ne ISS [1-7]), △ (bombarded, He ISS [1-41], ◈ (bombarded, He ISS [1-39]), □ (bombarded, AES [1-40]), ⊟ (bombarded, AES [1-42]). A further point "●" which lies in the upper left under the "E" of "bombarded" is not shown. For notation, see Sect. 2.1.

Figure 1.4. Angular variation in the sputtered atom yields for normally incident 3 keV Ar^+ bombardment of $Ni_{0.59}Co_{0.41}$ as determined by X-ray analysis of the collected sputter deposit. Bombardments were carried out at 25 and 300°C, with the angular variation significantly stronger at the higher temperatures. Co is seen to be emitted preferentially normally to the surface. It follows that there is excess Co (i.e. $\alpha_{Co} > 0.41$) in atom layer two and that the species showing bombardment-induced segregation is Ni. Self-consistently, it is also Ni which undergoes thermally activated segregation [1-43]. The segregation ratio, K, can be shown to be 2-2.5 at 25°C and ~60 at 300°C [1-8]. Due to Ichimura et al. [1-19A].

energies calculated for atom removal from perfect surfaces can differ markedly depending on whether one considers an anion or cation, e.g. 12 eV for O in Al_2O_3 and 30 eV for Al in Al_2O_3 [1 – 9]. The binding energy is also believed to play a major role in such cases as the sputtering of Si by a combined flux of gaseous Cl_2 and inert-gas ions [1-10].

The concept of thermal sputtering has been alternately accepted and rejected for 130 years [1-11]. In particular, it was in a phase of acceptance in the decade 1970-1980 for a number of reasons: since the experiments of Nelson [1-12] had not yet been re-interpreted [1-13,1-14], since physically correct theories had just appeared [1-15,1-16], and since thermal effects were being widely advocated to justify the presence of ions and excited states amongst sputtered particles [1-17,1-18]. For example, the present author [1-18] showed that differences in volatility (i.e. thermal sputtering) were able to explain most examples of oxygen loss from oxides, with both the correlation ("yes" or "no") and the required magnitude of the volatility $(10^{2\pm1}$ atm at ~4000 K) being reasonably correct for the systems then known. Since then, many additional oxides have been shown to lose oxygen, including some like Al_2O_3, Nb_2O_5, SiO_2, Ta_2O_5, and ZrO_2 which (unlike the systems known when [1-18] was prepared) have very low volatility. Thermal sputtering, while it may contribute to compositional changes with oxides, is thus not the whole story.

1.2 Bombardment-induced segregation

We have been emphasizing the inadequacies of the older theoretical works, those before 1980, on the subject of what causes bombardment-induced compositional change, with their emphasis on a more or less universal role for preferential sputtering as triggered by differences of mass, chemical binding, or volatility. The first new feature was the realization of a pervading role for bombardment-induced Gibbsian segregation in causing changes with alloys [1-4,1-5]. (For brevity we thenceforth leave out the word "Gibbsian".) Such segregation occurs in nearly all cases in the same sense as equilibrium segregation (Figs. 1.2 and 1.3), though normally has a smaller numerical extent for ambient temperature [1-5,1-8]. It is demonstrated in characteristic composition-depth profiles (Fig. 1.2), in conflicts between ISS and AES analysis of bombarded surfaces (Fig. 1.3), in characteristic results for yield-vs.-angle [1-19] (Fig. 1.4 [1-19A]), in delayed segregation, and in suppressed sputtering [1-19B]. Delayed segregation relates to such systems as a bilayer of Ni on Ag [1-20] or homogeneous Au-Cu (Fig. 1.5 [1-21]), where segregation was found to continue evolving for a few minutes after bombardment ceased. Such experiments show in a beautifully explicit way that chemical effects can occur during the prolonged relaxation that occurs subsequent to particle impact, a subject that was anticipated also in computer simulation [1-22], in work with alkali halides [1-23], and, particularly relevant here, in work on composition changes with oxysalts [1-24,1-25,1-26]. The subject was recently reviewed by Li [1-27], where the point is made that the mass transport seen in Fig. 1.5 is probably bombardment-induced and not the result of thermal activation in the usual sense. Segregation is, we would emphasize, not a type of sputtering, so invoking segregation to explain composition change constitutes a major change.

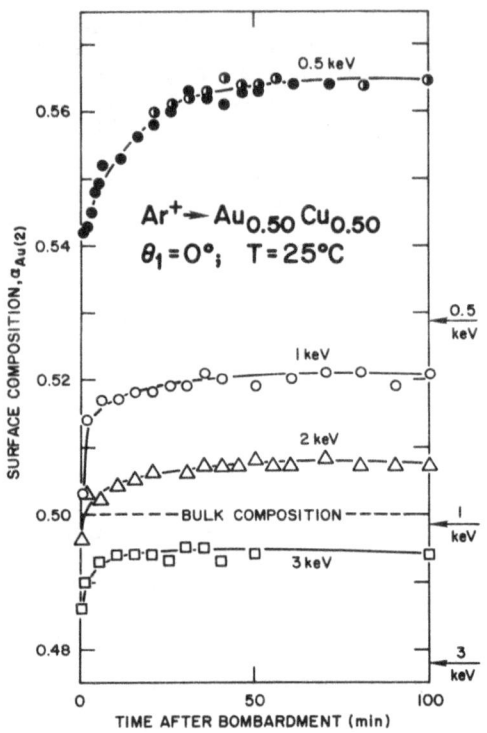

Figure 1.5. Changes in surface composition with time after stopping bombardment of $Au_{0.50}Cu_{0.50}$ at 25°C with normally incident Ar^+ having energies of 0.5-3 keV. Compositions were obtained with low-energy (60-69 eV) AES. The arrows mark the <u>average</u> steady-state compositions achieved during bombardment and differ from what is shown in the figure for t = 0 only because of being averages [1-44]. The composition changes are seen to involve an increase in the amount of Au at the surface and are in accordance with what is found with heated specimens [1-45]. The relevant mass transport is not regarded as thermally activated diffusion but rather the post-bombardment analog of bombardment-enhanced diffusion [1-27]. For notation, see Sect. 2.1. Due to Li, Tu, and Sun [1-21]

Oxides also show segregation but the evidence for it playing a role in composition change was slow to come. In such a case as Na^+ on SiO_2, the segregation of the Na^+ was a response to the charge of the incident particle [1-28], an effect that can occur only under conditions of ultra-high diffusivity. In the case of CaO in MgO, thermally activated segregation requires a very high temperature ($\geq$ 900°C) [1-29], so it is unclear whether the bombardment-induced variety would take place in an unheated target. The first straightforward example, with behavior like that of an alloy, relates to Na in $Na_2O \cdot SiO_2$ [1 - 30] .

1.3 Cascade chemistry

The second new feature in the post-1980 period originated with three separate groups of authors, Christie et al. [e.g. 1-24], Marletta et al. [e.g. 1-25], and Rabalais et al. [e.g. 1-26]. They suggested an important variant of thermal sputtering. Basically, the traditional point of view emphasizes transient vaporization subject to the normal laws of thermodynamics. When several components are present, compositional changes arise due to the loss of the more volatile species. Although this is a strictly surface process, it could influence greater depths if the surface composition change served as a diffusion boundary condition, as when a surface of Ti_2O_3 triggers bombardment-enhanced out-diffusion of O from TiO_2 [1 - 31]. The point of view of Christie-Marletta-Rabalais is that the cascade itself should be regarded as a source of new stoichiometries as the initially disrupted lattice relaxes (cf. Fig. 1.5) and various alternative stoichiometries compete. In effect, a species of what we will term "cascade chemistry" occurs very similar to what is postulated to govern ion-beam mixing [1-32]. The normal laws of thermodynamics, especially the relevance of the free energy, probably do not apply. For example, bombarded $Ca(NO_3)_2$ [1-24] could return to $Ca(NO_3)_2$ itself, but as an alternative could evolve to CaO or Ca_3N_2. The argument goes that $Ca(NO_3)_2$ is unfavored because of requiring a high degree of ordering, whereas between CaO and Ca_3N_2 (i.e. $CaN_{2/3}$) the former is favored energetically (Fig. 1.6 [1-24]). We will show elsewhere a possible way of combining the twin criteria that the phase formed tends to be simple and at the same time to be energetically favored.

1.4 Redistribution

The third new feature in the post-1980 period was the general acceptance that composition change can be caused by point-defect fluxes whether or not sputtering occurs [1-33]. We will term this "redistribution". Actually, this effect has long been known (e.g. [1-34]) and it is surprising that its importance was often neglected. In the most modern simulations, both redistribution and segregation are taken into account separately, as they are indeed unrelated effects [1-33]. To some extent, the terminology has been a problem, as the word "segregation" is frequently used to mean "redistribution", just as "adsorption" is used to mean "segregation".

1.5 Surface binding energies of oxides and halides

It is clear from Section 1.2-1.4 that the role of preferential sputtering (in the formal sense of the term) in causing compositional change is much less important than had once been thought. Exceptions occur in the near-threshold regime [1-1] and

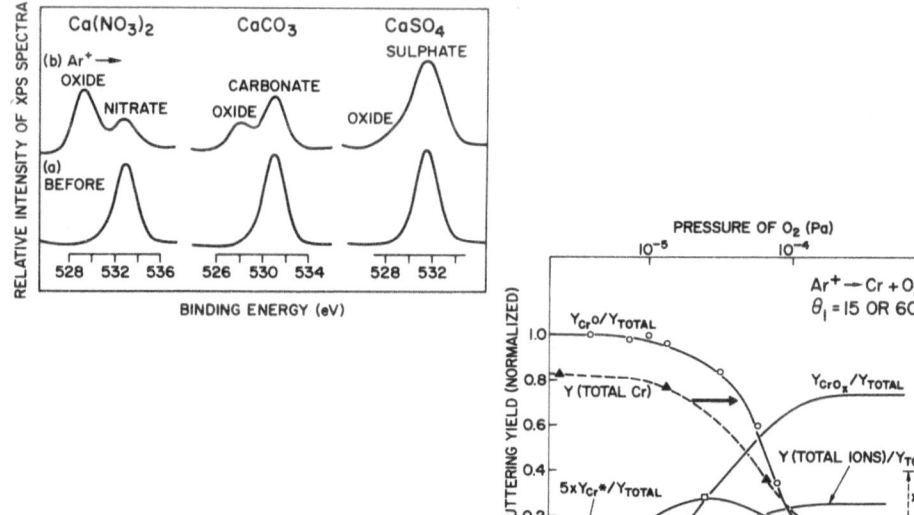

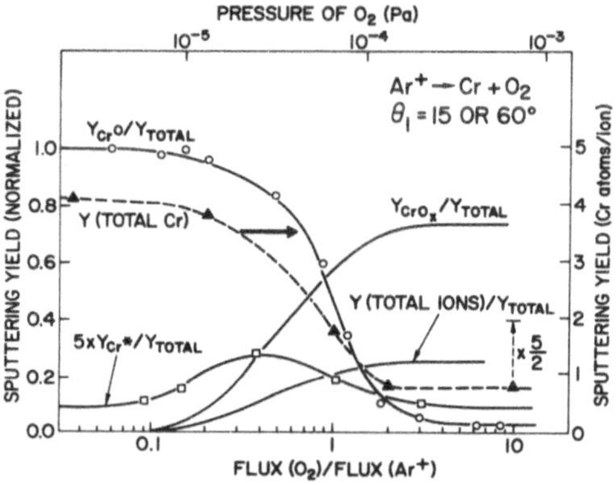

Figure 1.6. High resolution XPS, "X-ray photoelectron spectroscopy", spectra of the oxygen 1s photoelectron peaks from $Ca(NO_3)_2$, $CaCO_3$, and $CaSO_4$: (a) before Ar$^+$ bombardment; (b) after 3 keV Ar$^+$ bombardment to a fluence of 7.5×10^{16} ions/cm^2 at an angle of incidence of 50°. The targets were in the form of anhydrous powders. Similar results were obtained for compounds of Ca, Sr, and Ba, with the fractional conversion to oxide being 0.92-0.94 for nitrates, 0.53-0.57 for carbonates, and 0.40-0.45 for sulphates. We agree with the authors that what is being observed is more nearly a chemical re-arrangement of atoms displaced within each cascade than, for example, a process related to the thermal spike. Whether the re-arrangement occurs on a short time scale (as envisaged by Johnson [1-32]) or on the rather long time scale evident in Fig. 5 is unclear. Due to Christie et al. [1-24].

Figure 1.7. Composition of the sputtered flux as a function of the ratio of the O_2 to Ar$^+$ fluxes or of the O_2 partial pressure for 15 keV Ar$^+$ bombardment of polycrystalline Cr. All curves except that for Y(TOTAL Cr) refer to the left-hand scale and give the fractional composition of the sputtered flux. Cr^0 is the Cr ground state, a$'S_3$, Cr* is the 425.4 nm transition of the state z$'P_4^0$, while CrO$_x$ denotes neutral molecules in general, of which CrO can be expected to be the dominant component. The relative yields other than that for Cr* were obtained by LIF, "laser-induced fluorescence", at $\theta_1 = 60°$, the Cr* yield was obtained by light emission at $\theta_1 = 15°$, while the total Cr yield was obtained by use of a quartz-crystal microbalance and $\theta_1 = 15°$. The unexpected feature is the factor of 130 fall in the Cr^0, a result that has been explained [1-9] in terms of the surface binding energy for Cr^0 being significantly higher than that for O^0. Due to Betz and Husinsky [1-35].

with isotopes (Fig. 1.1), while another exception relates to the surface binding energy of oxides and halides. Thus, there is very recent evidence that the anions and cations in some oxides may have significantly different binding energies. That oxides have somewhat lower <u>overall</u> yields has long been known (Fig. 1.7, "total Cr" [1-35]) but that <u>cation atoms</u> are particularly reluctant to be emitted was realized only when laser-induced fluorescence was applied to the sputter products of systems such as Cr + O$_2$. The yield of Cr° fell by over a factor of 100 as the ratio flux(O$_2$)/flux(ions) increased (Fig. 1.7). Behavior as in Fig. 1.7 was subsequently rationalized by calculations in which energies were deduced using parameters appropriate to point defects in perfect oxide [1-9] or halide [1-36] surfaces. The cation binding energy was found to be unusually high for oxides of group III and beyond [1-9], while a similar effect was shown for halides of group II and beyond [1-36].

Ideally, what follows would be an attempt to give a contemporary review of all aspects of compositional change with alloys, oxides, and oxysalts. For present purposes, however, we will treat only binding energies. For a discussion of mass effects, we refer to [1-1,1-2,1-37], for a discussion of segregation we refer to [1-37A], while for a review of redistribution we refer to [1-33]. Material on "cascade chemistry" is in preparation.

It is worth commenting further on what should already be clear: the concepts of <u>preferential sputtering</u> and <u>compositional change</u> may or may not coincide. The former relates to changes in the outermost atom layers and is correctly attributed to differences of mass, chemical binding, or (in particular cases) volatility. Refs. [1-3] and [1-9], for example, relate to preferential sputtering. Preferential sputtering is, by its nature, an essentially universal effect but it will contribute significantly towards compositional change only when the system, for whatever reason, fails to show segregation, "cascade chemistry", or redistribution. Table 1.1 is an attempt to clarify the difference between preferential sputtering and compositional change.

2. THE ROLE OF THE SURFACE BINDING ENERGY

2.1 <u>Notation</u>

For better or for worse, we will use in what follows the somewhat awkward notation of [1-5,1-37A]. Subscript "o" will designate a property of a cascade particle, "1" a property of the incident particle, "2" a property of a surface atom (atom layer <u>one</u>), "2'" a property of a subsurface atom (atom layer <u>two</u>), and "3" a property of a bulk atom. "α" denotes atom fraction, subscript A, B or (in general) i a component, and superscript "∞" steady state, whence the use of such forms as $\alpha_{A(2)}^{\infty}$. M designates a metal, O oxygen, X a halogen, U the surface binding energy, Y the sputtering yield, Z the co-ordination number, and λ the mean atomic spacing. We note that in [2-1], the sense of "2" and "3" is interchanged.

Atoms and point defects will be designated (as is usual [2-2]) with the entity as the main symbol, the location as subscript, and the charge as superscript. Thus, Mg_{Mg}^{2+} is a normal Mg ion in such a compound as MgO, V_{Mg} is an Mg vacancy without

TABLE 1.1. Mechanisms for compositional change

Effect	Systems where the effect is important	Role in compositional change
Preferential sputtering due to mass differences	(a) The outermost atom layer of nearly all alloys	Normally overwhelmed by segregation or redistribution
	(b) Near-threshold conditions	Loss of lighter species
	(c) Isotope sputtering	Loss of lighter species
Preferential sputtering due to differences in chemical binding	(a) The outermost atom layer of nearly all alloys	Normally overwhelmed by segregation or redistribution
	(b) Oxides of group III and beyond	Loss of O; reduction of the cation (tentative)
	(c) Systems like Si exposed to Cl_2 plus inert-gas ions	Loss of the species with the most altered binding
Preferential sputtering due to differences in volatility	(a) Alloys with a very volatile component	(Untested)
	(b) Alloys in general	Normally overwhelmed by other effects
	(c) Oxides with high decomposition pressures	Loss of O; reduction of the cation (tentative)
Bombardment induced segregation	(a) Almost all alloys	Massive subsurface loss of the segregating species
	(b) The system Na_2O-SiO_2	Subsurface loss of Na_2O
Cascade chemistry	(a) Oxides and oxysalts in general, especially if complex	Loss of C,N,O,S; reduction of the cation (tentative)
Redistribution	(a) Alloys containing Si	Massive subsurface loss of Si
	(b) Alloys in general	Subsurface loss of whichever species is redistributed

associated electrons, Mg_i^+ is an Mg^+ interstitial, e^- is an electron, and h^+ is a hole.

It will sometimes be necessary to distinguish "s", i.e., solid or crystalline, "ℓ", i.e., liquid or amorphous, and "g", i.e., gaseous or sputtered. For example, when $TiO_2(s)$ is bombarded, it evolves first to $TiO_{2-x}(\ell)$ and then to $Ti_2O_3(s)$ by loss of $O(g)$.

2.2 The surface binding energy in cascade sputtering

Let us consider briefly the recoil-density derivation [2-3] of the standard relation for slow collisional (i.e. cascade) sputtering. This approach has the advantage of being somewhat more transparent than that based on transport theory [2-4] but is otherwise equivalent. As seen in Fig. 2.1, an ion is regarded as starting from the reference plane $x = 0$, which is equivalent to the surface. During the penetration of the ion into the target, assumed to be random, a "linear" collision cascade is generated. Let E_1 be the incident energy and $C_n(x)dx$ be the differential depth-distribution function for energy deposited in elastic (nuclear, "n") events. The deposited energy creates recoiling target atoms with energy E_0 and an assumed isotropic motion. If we accept the result that the recoil density has the form appropriate to the value $m = 0$ for the power-law parameter,

$$F(E_1, E_0)dE_0 \sim (\Gamma E_1/E_0^2)dE_0; \quad \Gamma = 6/\pi^2 = 0.608, \tag{2.1}$$

then it follows that the total number of recoils at the surface which would be able to overcome an energy barrier U is just

$$\int_0^{\pi/2} d\theta_0 \sin\theta_0 \cdot \int_{U/\cos^2\theta_0} dE_0 \cdot \Gamma E_1 E_0^{-2} \cdot C_n(0)L_0$$
$$= (1/3)\Gamma E_1 C_n(0)L_0/U,$$

where θ_0 is the polar angle of a sputtered atom, $C_n(0)dx$ has been replaced with $C(0)L_0$, and L_0 is the average depth of origin of sputtered particles. L_0 has been evaluated both theoretically [2-4] and experimentally [1-8,2-5], though unfortunately with somewhat different results [1-8]. It is conventional to introduce a factor of 1/2 to take into account that only outward-moving recoils will be sputtered and to replace L_0 with $3L_0/4$. Taken together, this gives for the cascade sputtering yield, $Y_{cascade}$, the usual [2-3,2-4] form:

$$Y_{cascade} = (1/8)\Gamma E_1 C_n(0)L_0/U. \tag{2.2}$$

The important quantity in Eq. (2.2), in the present context, is U. If it differs for different components of a target, then sputtering will cause a compositional change, though the change will be significant only if the difference in U is sufficient.

2.3 The bulk binding energy, U_b

Eq. (2.1) for the recoil density assumes, amongst other things, that the bulk binding energy, U_b, is zero. The possibility of a non-zero U_b was discussed first by Sigmund [2-6], where the result obtained was stated to be valid only for $E_1 \gg E_d \gg U_b$. We have obtained a very similar result and were able to show that it is more generally valid, namely for $E_1 \gg E_d$ or U_b [2-7]. The form appropriate to $m = 0$ is

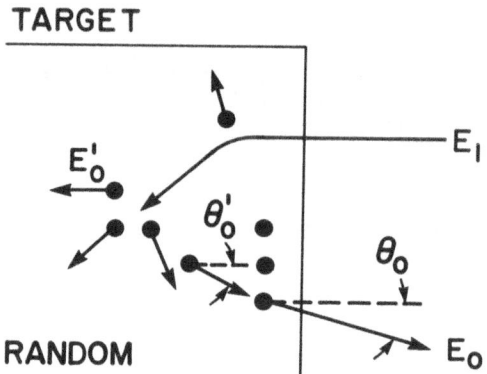

TARGET

E_0'

θ_0'

θ_0

E_1

E_0

**RANDOM
CASCADE**

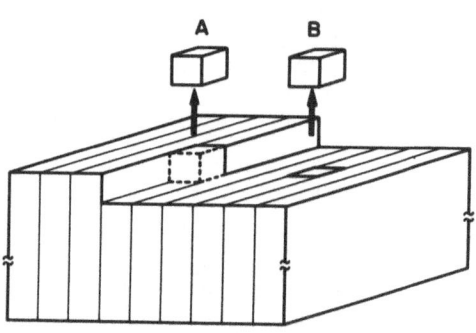

Figure 2.1. Sketch of quantities relevant to slow collisional (i.e. cascade) sputtering. E_1 is the energy of the incident particle, E_0' the recoil energy of a target atom, E_0 the energy of a sputtered atom, θ_0' the polar angle of a moving target atom, and θ_0 the polar angle of a sputtered atom. The incident particle is shown as creating a random cascade and as coupling with the surface atoms only indirectly via this cascade.

Figure 2.2. (A) Sketch of a half-space atom with coordination number $Z_3/2$ which has been emitted by vaporization. Such an atom, which is bound by the cohesive energy or heat of atomization, ΔH^a, is _not_ characteristic of the sputtering process because it is atypical of an undisturbed surface. When a half-space atom is removed, there is no resulting surface vacancy but rather the jog is displaced. (B) Sketch of an in-surface atom with coordination number Z_2 which, being more nearly typical of an undisturbed surface, is here assumed to be relevant to sputtering. If the expulsion is sufficiently _rapid_, such an atom is bound by $(2Z_2/Z_3)\Delta H^a$, thence by a quantity somewhat greater than ΔH^a. A surface vacancy is always formed, i.e. whether the expulsion is rapid or slow.

$$F(E_1, E_o, U_b)\, dE_o \sim \frac{\Gamma(E_1 + U_b)}{(E_o + U_b)\, E_o}\, dE_o, \tag{2.3}$$

where Γ is as before. We are still studying the significance of Eq. (2.3), the main reason for introducing it here being to show that the problem of U_b being non-zero is tractable and U_b should not be perfunctorily overlooked.

2.4 The surface binding energy for alloys

Attempts to define U for metals and alloys showing miscibility have been made repeatedly. With metals, it has been usual to identify U with the cohesive energy, i.e. the heat of atomization ΔH^a. We would point out that, if U_{AA} is the A-A "bond strength" and if Z_3 is the bulk ("3") coordination number, then we have the well-known result [2-8]

$$\Delta H^a = -(1/2)Z_3 U_{AA}, \tag{2.4}$$

which means, effectively, that the use of ΔH^a is equivalent to regarding the sputtered atom as occupying a half-space site (A in Fig. 2.2). Since half-space atoms are atypical of an undisturbed surface as compared with in-surface atoms (B in Fig. 2.2), it follows that ΔH^a underestimates U.

Similar in spirit to the use of ΔH^a with metals, is to describe binding in miscible alloys in terms of the cohesive energies of the pure substances [e.g. 2-9]. However, not only is the wrong type of atom being dealt with, but such an approach neglects the fact that binding in an alloy is governed in all cases by the statistics of site occupancy (e.g. random, ordered, segregated) and in some cases (as when species such as B or Si are involved) by changes in the character of the binding.

An approach which avoids the twin problems of ΔH^a underestimating U and of ΔH^a not being correct in any sense for an alloy, is based on the "quasichemical" variant of thermodynamics [1-5,1-37A]. It is easily applied to both metals and binary alloys as it requires knowledge only of nearest-neighbor "bond strengths", U_{AA}, U_{BB}, and U_{AB}. Self-consistency is achieved if U_{AA} and U_{BB} are defined as in Eq. (2.4). Similarly, U_{AB} is defined in terms of the heat of mixing ΔH_m [2-8]:

$$\Delta H_m \equiv \alpha_{A(3)}\alpha_{B(3)}h_m = \alpha_{A(3)}\alpha_{B(3)}Z_3[U_{AB} - (1/2)(U_{AA} + U_{BB})],$$

where $\alpha_{i(3)}$ is the atom fraction of component i in the bulk. The surface binding energy U now follows as the sum of nearest neighbor energies under the assumption that the typical atom expelled is an in-surface atom (reasonable if L_o in Eq. (2.2) is 0.80 λ [1-8,2-5]) with atom fraction $\alpha_{i(2)}$ and coordination number Z_2:

$$U_A = (Z_2/Z_3)[(1 + \alpha_{A(2)})\Delta H_A^a + \alpha_{B(2)}(\Delta H_B^a - h_m)] \tag{2.5a}$$

for a binary alloy and

$$U = (2Z_2/Z_3)\Delta H^a = (1.42 \pm 0.08)\Delta H^a \tag{2.5b}$$

for a unary system, i.e. $\alpha_{A(2)} = 1$. The numerical factor 1.42 $\pm$ 0.08 applies to the more densely packed surfaces of fcc, hcp, and bcc. In practice, the term h_m is normally unimportant [1-5,1-37A], with Si-containing systems being a marked exception (e.g. Fe-Si [2-10]). Nevertheless, we take it into account (Table 2.2, to follow).

Eq. (2.5) should be acceptable whenever two conditions are met: U describes atom removal from an undisturbed surface and the act of removal is sufficiently rapid that the relaxation of the lattice around the surface vacancy can be neglected. If the act of removal is slow, the relaxation around the vacancy cannot be neglected. As discussed in [6], U is then somewhat smaller than in Eq. (2.5).

Examples of U as given by Eq. (2.5b), as well as experimental values where known, are given in Table 2.1. The agreement between theory and experiment is moderately good, though with experiment typically more similar to ΔH^a than to Eq. (2.5b). The obvious problem is that atom ejection from real systems does not involve ideal surfaces as in Fig. 2.2 and as assumed in Eq. (2.5b). Nevertheless, we conclude that it is realistic to assume a _proportionality_, $U \propto \Delta H^a$, and this will be done in what follows.

2.5 Application to compositional change with alloys

In view of the inverse dependence, $Y \propto 1/U$, shown in Eq. (2.2) it follows that the sputtered flux ratio for a binary alloy is, to lowest approximation,

$$\frac{\text{surface flux}_A}{\text{surface flux}_B} = \frac{\alpha_{A(2)} Y_A}{\alpha_{B(2)} Y_B} = \frac{\alpha_{A(2)} U_B}{\alpha_{B(2)} U_A} \qquad (2.6a)$$

but to higher approximation

$$\frac{\text{surface flux}_A}{\text{surface flux}_B} = \frac{\alpha_{A(2)}(1 - \beta) + \alpha_{A(2')}\beta}{\alpha_{B(2)}(1 - \beta) + \alpha_{B(2')}\beta} \cdot \frac{Y_A}{Y_B}, \qquad (2.6b)$$

where "2'" refers to atom layer _two_ and β is the fraction of sputtering from beyond atom layer _one._ Values of β (the notation $f(\lambda)$ is also used) range from 0.11 to 0.30 with an average of 0.19 [1-8]. If $U_A \neq U_B$ the surface compositions will change until a steady-state ("∞") condition is achieved. That based on Eq. (2.6a) is just

$$\frac{\text{surface flux}_A}{\text{surface flux}_B} = \frac{\alpha_{A(2)}^{\infty} Y_A}{\alpha_{B(2)}^{\infty} Y_B} = \frac{\alpha_{A(2)}^{\infty} U_B}{\alpha_{B(2)}^{\infty} U_A} = \frac{\alpha_{A(3)}}{\alpha_{B(3)}} \qquad (2.7)$$

If U_B/U_A is taken as defined by Eq. (2.5a), then the latter expression and Eq. (2.7) can be solved iteratively for $\alpha_{B(2)}^{\infty}/\alpha_{A(2)}^{\infty}$ (column 6 of Table 2.2).

Table 2.2 (final column) also contains observed values for $\alpha_{B(2)}^{\infty}/\alpha_{A(2)}^{\infty}$, in all cases taken from work using AES or XPS. The outstanding feature is that the predicted compositional changes due to bond-strength effects are without exception very much less than what is observed with AES or XPS. There are two reasons for such a trend:

(i) One relates to the mathematical form of the expression for U_B/U_A. As seen in Table 2.2, the ratio U_B/U_A is significantly closer to unity than $\Delta H_B^a/\Delta H_A^a$, and it is the former which governs $\alpha_{B(2)}^{\infty}/\alpha_{A(2)}^{\infty}$.

(ii) The other reason is a practical problem. Systems such as Ag-Mo or Au-W, which might be expected to show a large com-

TABLE 2.1. Examples of calculated and observed surface binding energies for metals.

Metal	ΔH^a (eV)	$U \simeq 1.42\Delta H^a$ (Eq. (2.5b)) (eV)	U from experiment (eV)	Ref.
Al	3.41	4.8	3.6	2-18
Au	3.82	5.4	...	
Ba	1.89	2.7	2.1[b]	2-19
Ca	1.85	2.6	1.3	2-20
Ce	4.38	6.2	...	
Cr	4.12	5.9	4.2 ± 0.2	2-21
			4.4 ± 0.2	2-22
Cr[a]	>4.12	>5.9	5.1 ± 0.2	2-22
Cu	3.49	5.0	...	
Fe	4.31	6.1	4.3; 5.0[b]	2-23;2-24
Ge	3.88	5.5	...	
In	2.52	3.6	4.0	2-25
Mg	1.52	2.2	...	
Mo	6.82	9.7	...	
Ni	4.46	6.3	...	
Pt	5.85	8.3	...	
Rh	5.73	8.1	8.0; 11 ± 1	2-25; 2-26
Si	4.72	6.7	...	
Sn	3.12	4.4	...	
Th	5.96	8.5	...	
Ti	4.87	6.9	4.6	2-18
U	5.42	7.7	5.4	2-27
W	8.80	12.5	...	
Zn	1.35	1.9	...	
Zr	6.31	9.0	6 ± 2	2-28; 2-29
			6.3	2-30

a) Stainless steel with 17% Cr and the remainder dominantly Fe.
b) Corrected by Garrison et al. [2-31] for geometrical effects [2-12].

TABLE 2.2. Examples of calculated and observed surface compositions for alloys.

System, A-B	ΔH_A^a (eV)	ΔH_B^a (eV)	h_m [a)] (eV)	$\Delta H_B^a/\Delta H_A^a$	$U_B/U_A =$ $\alpha_{B(2)}^\infty/\alpha_{A(2)}^\infty$ from Eqs. (2.5a), (2.7)	$\alpha_{B(2)}^\infty/\alpha_{A(2)}^\infty$ from experiment using AES or XPS
$Ag_{0.50}Au_{0.50}$	2.94	3.82	-0.193	1.30	1.14	2.0
$Ag_{0.50}Pd_{0.50}$	2.94	3.90	-0.216	1.33	1.15	2.2
$Au_{0.50}Cu_{0.50}$	3.82	3.49	-0.197	0.91	0.96	1.0
$Au_{0.50}Ni_{0.50}$	3.82	4.46	+0.295	1.17	1.08	~2.7
$Au_{0.50}Pd_{0.50}$	3.82	3.90	-0.320	1.021	1.010	1.4
$Cu_{0.50}Ni_{0.50}$	3.49	4.46	+0.074	1.28	1.13	1.8
$Cu_{0.50}Pd_{0.50}$	3.49	3.90	-0.461	1.12	1.054	1.6
$Cu_{0.50}Pt_{0.50}$	3.49	5.85	-0.461	1.68	1.27	2.1
$Mo_{0.50}W_{0.50}$	6.82	8.80	+0.084	1.29	1.14	1.4
$Ni_{0.50}Co_{0.50}$	4.46	4.44	~0	0.9955	0.998	1.2
$Ni_{0.50}Pt_{0.50}$	4.46	5.85	-0.362	1.31	1.14	1.6
$Pd_{0.50}Ni_{0.50}$	3.90	4.46	~0	1.14	1.069	1.3

a)Most values of h_m were obtained by averaging values of integral ΔH_m for solid alloys [2-10] using $h_m = \Delta H_m/\alpha_A(3)\alpha_B(3)$. The value for Mo-W is from [2-32].

positional change in view of the highly dissimilar values of ΔH^a, are not miscible.

Nevertheless, the conclusion is inescapable: compositional changes with alloys as measured by AES or XPS are not governed by bond-strength effects.

We note in passing that, even if the numerical value $U \simeq 1.4 \Delta H^a$ was not fully in agreement with experiment (Sect. 2.4), at least the proportionality $U \propto \Delta H^a$ was acceptable. This is a sufficient condition for the arguments made here to be valid.

2.6 The surface binding energy for oxides and halides

Oxides and halides do not permit quite as straighforward a definition of U as do metals and alloys: ionized species and diatomics are emitted to an important extent and, due mainly to the major role of polarization, oxides and halides cannot be described in simplistic terms such as pair-wise interactions [2-11] or quasichemical thermodynamics [1-5,1-37A].

We have proposed [1-9] that a possible description of binding at undisturbed surfaces with oxides and halides is in terms of processes in which individual surface atoms are removed slowly to infinity, using as the basis defect theory of the type pioneered by Norgett and Lidiard [e.g. 2-2]. Slow removal was suggested on the grounds that the characteristic time for electronic relaxation is of order 10^{-15} s and thus distinctly shorter than the sputtering time (e.g. 9×10^{-14} s for a 5 eV Al atom or 22×10^{-14} s for a 5 eV W atom [1-6]). In evaluating the energies, it is convenient to take advantage of the result that, for an oxide or halide, most bulk defects have similar energies to surface defects, the divacancy binding energy apparently being the only exception [1-9].

Cation atom binding

Using MgO as the example, we consider the process in which a neutral Mg atom is removed slowly to infinity:

$$Mg_{Mg}^{2+} = Mg(g) + V_{Mg} + 2h^+ .$$

The energy change is

$$U_{Mg} = E(V_{Mg}) - I_1(Mg) - I_2(Mg) + 2E(h^+), \qquad (2.8)$$

where $E(V_i)$ is the lattice energy for vacancies of type i, i.e. the energy to slowly remove a lattice ion to infinity, $I_n(M)$ is the nth ionization potential of the cation M, and $E(h^+)$ is the formation energy of a hole inclusive of electronic and ionic relaxation.

Anion atom binding

We next consider the process in which a neutral O atom is removed slowly to infinity:

$$O_O^{2-} = O(g) + V_O + 2e^- ,$$

where e^- is a lattice electron and the energy change is

$$U_O = E(V_O) + I_1(O^{2-}) + I_2(O^{2-}) + 2E(e^-) . \qquad (2.9)$$

Here $I_n(O^{2-})$ is the nth ionization potential of O^{2-} and $E(e^-)$ is the formation energy of a lattice electron. In those instances where $E(e^-)$ is the energy of an electron at the bottom of the conduction band, we have $E(e^-) = - |E_c|$, $|E_c|$ being the conduction band width.

Input parameters for calculating U for oxides, as collected from a variety of sources, are given in [1-9], while evaluations of Eqs. (2.8) and (2.9) as well as corresponding equations for MO diatomics and M^+ ions [1-9] are given in Table 2.3. The results show that, if U for an ionic oxide is taken as the lower of that for the metal atom or O atom, then ionic oxides are more tightly bound than the corresponding metals (Table 2.1) by factors of 1-5. This result, which should be manifested in low total yields, is not surprising in view of what has long been known experimentally. What is somewhat unexpected is that, for ionic oxides of group III and beyond, O atoms as well as MO diatomics are predicted to be far more easily removed than metal atoms, so that one can expect abnormally low metal-atom yields in appropriate circumstances as well as essentially universal preferential loss of O for group III and beyond.

2.7 Application to compositional change with oxides

Metal atom yields

Before considering compositional changes with oxides, we will digress in order to discuss metal-atom yields because of the light which is shed on binding energies. While it has been known for some time that total yields were generally somewhat lower for ionic oxides than for metals, whereas partial yields for ions, excited states, and MO diatomics were greatly enhanced, almost nothing was known about neutral, ground state atoms because of the difficulty in detecting them explicitly. This problem was recently overcome by using either laser-induced fluorescence [e.g. 1-35] or secondary neutral mass-spectroscopy [e.g. 2-13]. The example of Cr, given in Fig. 1.7, reveals that metal-atom yields can fall drastically as the oxygenation of the surface increases. This in turn is strong support for the newly proposed oxide binding energies as in Table 2.3.

Preferential effects

Oxygen loss from oxides is well documented, leading either to well-defined changes in composition (Fig. 2.3 [2-14]) or else to an ill-defined state of understoichiometry (Fig. 2.4 [2-15]). We first note that considerations of mass are not useful for understanding such O loss, since it would follow that all oxides (except, e.g., BeO) would behave similarly and reduce to pure metal, at least at the outer surface. This leaves considerations of binding, with which this section is concerned, together with thermal-spike arguments [1-18] and arguments based on "cascade chemistry" [1-24,1-25,1-26].

What has not been generally appreciated is that a properly defined U leads naturally to preferential effects with oxides and halides: as borne out in Table 2.3, for ionic oxides of group III and beyond, the O atom binding is distinctly (up to a factor of 4) smaller than the metal-atom binding. This is the result of the cations, with their greater charge, sitting in a deeper potential well. There should therefore be a universal tendency for O to be lost from appropriate oxides, as when ZrO_2 evolves

TABLE 2.3. Examples of calculated surface binding energies in eV for oxides as in Eqs. (2.8) and (2.9). The input parameters are given in [1-9].

Type of binding energy, U_i	CaO	MgO	NiO	ZnO	Al_2O_3	ThO_2	ZrO_2
Metal atom, U_M	12.3	13.5	7.7	8.0	29.8	~29.7	32.2
Oxygen atom, U_O	~12.2	~13.1	10.9	4.8	~11.8	8.5	<7.1
MO diatomic, U_{MO}	~8.4	~8.2	~6.8	...	~18.2	~18.1	~19.0
Metal ion, $U_{Mg}+$	14.1	15.8	10.8	10.8	28.5	~33.5	35.6

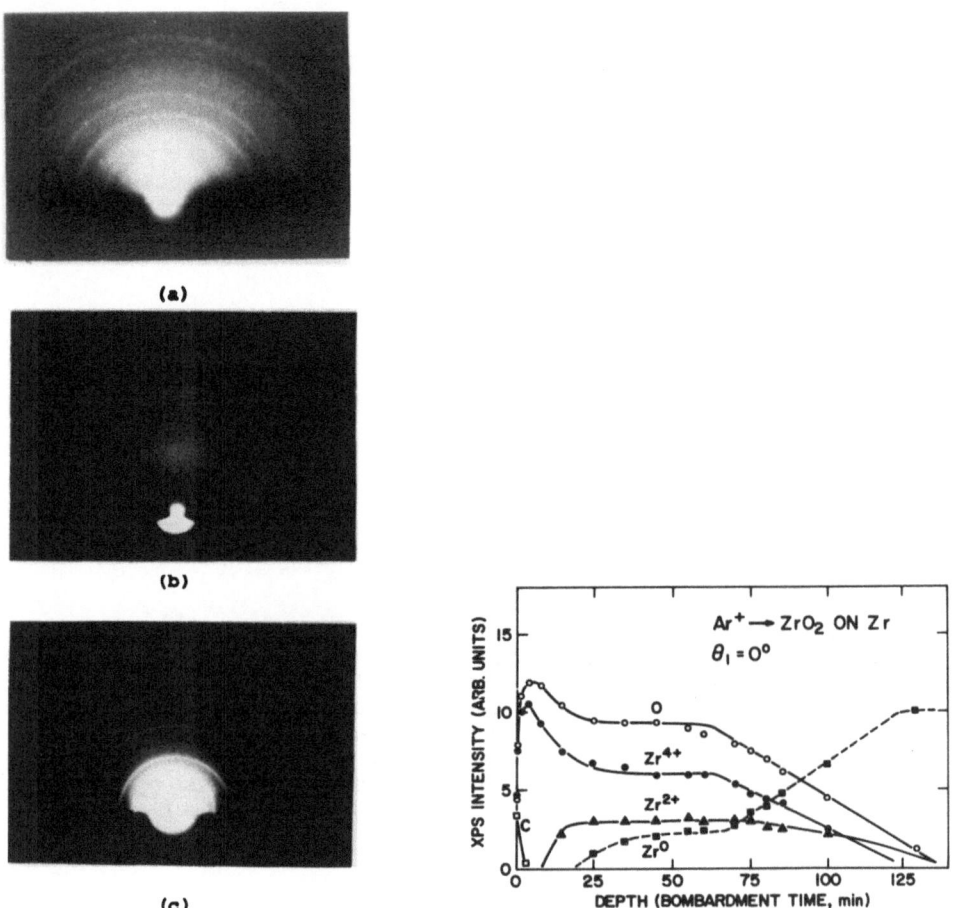

(a)

(b)

(c)

Figure 2.3. Reflection electron diffraction patterns taken at 80 kV for TiO_2: (a) before bombardment; (b) after exposure to 5×10^{15} ions/cm^2 of 30 keV Kr^+ at normal incidence; and (c) after exposure to 6×10^{16} ions/cm^2 of 30 keV Kr^+. The patterns, respectively those of rutile TiO_2, amorphous TiO_{2-x}, and Ti_2O_3, are an explicit indication of preferential O loss which leads to a well-defined change of composition. Due to Parker and Kelly [2-14].

Figure 2.4. Bombardment-induced O loss from a 36 nm anodic film of ZrO_2 on Zr as sensed with XPS while bombarding with 3 keV Ar^+ at normal incidence. The initial rise of the O and Zr^{4+} signals is due to contaminant removal, the changes in all signals at 5-25 min are due to preferential O loss to yield an ill-defined composition $ZrO_{0.6}$, while the more gradual changes beyond 65 min are due to the termination of the anodic film. Due to Hofmann and Sanz [2-15].

324

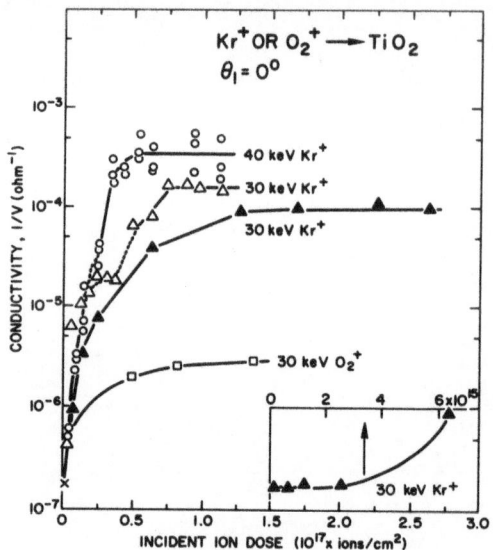

Figure 2.5. Electrical conductivity versus bombardment dose for sintered TiO_2 which is bombarded at normal incidence variously with 40 keV Kr^+, 30 keV Kr^+ in two different experiments, and 30 keV O_2^+. X is a point common to all curves. The inset shows the low-dose region of one of the curves for 30 keV Kr^+. The point is that bombardment of insulating TiO_2 (as well as many other insulators [2-17]) causes an evolution towards a metallic phase, identified in Fig. 2.3 as Ti_2O_3. Due to Parker and Kelly [1-31].

to $ZrO_{0.6}$ (Fig. 2.4). (The latter probably consists mainly of a saturated solid solution of O in α-Zr, namely $ZrO_{0.41}$.) Similar comments apply to TiO_2 but with an important difference: O deficiency is accommodated when sufficiently slight as V_O plus Ti_{Ti}^{3+}, then for greater loss as shear planes, and finally as a lower stoichiometry. Experimentally, bombarded TiO_2 shows a well-defined surface layer of Ti_2O_3, seen both by electron diffraction (Fig. 2.3) and by electron spectroscopy [2-16].

The same argument also explains why TiO_2 evolved to Ti_2O_3 rather than Ti: Ti_2O_3 is metallic (Fig. 2.5 [1-31]) and there is therefore no longer the large difference between U_{Ti} and U_O as for the more nearly ionic TiO_2. This leads to the rule that, if an ionic system has a metallic lower stoichiometry, then this stoichiometry forms under bombardment. Otherwise, there is an ill-defined state of understoichiometry as with ZrO_2.

REFERENCES

[1-1] E. Taglauer, Appl. Surf. Sci. 13 (1982) 80.

[1-2] H. Gnaser and I. D. Hutcheon, Surf. Sci. 195 (1988) 499.

[1-3] N. Andersen and P. Sigmund, Mat. Fys. Medd. Dan. Vid. Selsk. 39, No. 3 (1974).

[1-4] R. Kelly, in: Proc. Symp. on Sputtering, eds. P. Varga et al. (Inst. für Allgem. Physik, T. U. Wien, Austria, 1980) p. 390.

[1-5] R. Kelly and D. E. Harrison, Mat. Sci. Engin. 69 (1985) 449.

[1-6] A. Oliva, R. Kelly, and G. Falcone, Nucl. Instr. Meth. B19/20 (1987) ·101.

[1-7] D. G. Swartzfager, S. B. Ziemecki, and M. J. Kelley, J. Vac. Sci. Technol. 19 (1981) 185.

[1-8] R. Kelly and A. Oliva, Nucl. Instr. Meth. B13 (1986) 283.

[1-9] R. Kelly, Nucl. Instr. Meth. B18 (1987) 388.

[1-10] J. van Zwol, A. W. Kolfschoten, J. van Laar, and J. Dieleman, this volume.

[1-11] W. O. Hofer, in: Erosion and Growth of Solids stimulated by Atom and Ion Beams, eds. G. Kiriakidis et al. (Nijhoff, Dordrecht, Netherlands, 1986) p. 1.

[1-12] R. S. Nelson, Phil. Mag. 11 (1965) 291.

[1-13] R. Kelly, Surf. Sci. 90 (1979) 280.

[1-14] K. Besocke, S. Berger, W. O. Hofer, and U. Littmark, Rad. Effects 66 (1982) 35.

[1-15] R. Kelly, Rad. Effects 32 (1977) 91.

[1-16] P. Sigmund, Appl. Phys. Lett. 25 (1974) 169.

[1-17] A. E. Morgan and H. W. Werner, Anal. Chem. 49 (1977) 927.

[1-18] C. J. Good-Zamin, M. T. Shehata, D. B. Squires, and R. Kelly, Rad. Effects 35 (1978) 139.

[1-19] H. H. Andersen, B. Stenum, T. Sørensen, and H. J. Whitlow, Nucl. Instr. Meth. 209/210 (1983) 487.

[1-19A] S. Ichimura, H. Shimizu, H. Murakami, and Y. Ishida, J. Nucl. Mat. 128/129 (1984) 601.

[1-19B] P. J. Rudeck, J.M.E. Harper, and P. M. Fryer, Appl. Phys. Lett. (in press).

[1-20] J. Fine, T. D. Andreadis, and F. Davarya, Nucl. Instr. Meth. 209/210 (1983) 521.

[1-21] R. S. Li, L. X. Tu, and Y. Z. Sun, Appl. Surf. Sci. 26 (1986) 77.

[1-22] D. E. Harrison and R. P. Webb, Nucl. Instr. Meth. 218 (1983) 727.

[1-23] H. Overeijnder, R. R. Tol, and A. E de Vries, Surf. Sci. 90 (1979) 265.

[1-24] A. B. Christie, J. Lee, I. Sutherland, and J. M. Walls, Appl. Surf. Sci. 15 (1983) 224.

[1-25] G. Marletta, Nucl. Instr. Meth. B (1988).

[1-26] S. Contarini and J. W. Rabalais, J. Electron Spect. and Related Phenom. 35 (1985) 191.

[1-27] R. S. Li, in: Diffusion and Defect Data, ed. F. H. Wohlbier (Trans Tech S. A., Aedermannsdorf, Switzerland) (in press).

[1-28] R. A. Kushner, D. V. McCaughan, V. T. Murphy, and J. A. Heilig, Phys. Rev. B10 (1974) 2632.

[1-29] R. C. McCune and P. Wynblatt, J. Am. Cer. Soc. 66 (1983) 111.

[1-30] A. Torrisi, G. Marletta, A. Licciardello, and O. Puglisi, Nucl. Instr. Meth. B (1988).

[1-31] T. E. Parker and R. Kelly, J. Phys. Chem. Sol. 36 (1975) 377.

[1-32] W. L. Johnson, this volume.

[1-33] N. Q. Lam and H. Wiedersich, Nucl. Inst. Meth. B18 (1987) 471.

[1-34] R. C. Piller and A. D. Marwick, J. Nucl. Mat. 71 (1978) 309.

[1-35] G. Betz and W. Husinsky, Nucl. Instr. Meth. B13 (1986) 343.

[1-36] N. Itoh, A.M. Stoneham, and R. Kelly, work in progress.

[1-37] R. Kelly, in: Ion Beam Modification of Insulators, eds. P. Mazzoldi and G. W. Arnold (Elsevier, Amsterdam, 1987) ch. 2.

[1-37A] R. Kelly and A. Oliva, in: Erosion and Growth of Solids Stimulated by Atom and Ion Beams, eds. G. Kiriakidis et al. (Nijhoff, Dordrecht, Netherlands, 1986) p. 41.

[1-38] P. Biloen, R. Bouwman, R. A. van Santen, and H. H. Brongersma, Appl. Surf. Sci. 2 (1979) 532.

[1-39] G. Hetzendorf and P. Varga, Nucl. Instr. Meth. B18 (1987) 501.

[1-40] A. Jablonski, S. H. Overbury, and G. A. Somorjai, Surf. Sci. 65 (1977) 578.

[1-41] P. Varga and G. Hetzendorf, Surf. Sci. 162 (1985) 544.

[1-42] G. Betz, Surf. Sci. 92 (1980) 283.

[1-43] E. E. Hajcsar, P. T. Dawson, and W. W. Smeltzer, Surf. Interface Analysis 10 (1987) 343.

[1-44] R. S. Li, private communication (1988).

[1-45] G. C. Nelson, J. Vac. Sci. Technol. A1 (1983) 1037.

[2-1] H. F. Winters and P. Sigmund, J. Appl. Phys. 45 (1974) 4760.

[2-2] W. C. Mackrodt, Sol. State Ionics 12 (1984) 175.

[2-3] G. Falcone, R. Kelly, and A. Oliva, Nucl. Instr. Meth. B18 (1987) 399.

[2-4] P. Sigmund, Rev. 184 (1969) 383.

[2-5] B. Jorgensen, M. J. Pellin, C. E. Young, W. F. Calaway, E. L. Schweitzer, D. M. Gruen, J. W. Burnett, and J. T. Yates, this volume.

[2-6] P. Sigmund, Rev. Roum. Phys. 17 (1972) 969. See in particular p. 974.

[2-7] R. Kelly and A. Oliva, in preparation.

[2-8] R. A. Swalin, Thermodynamics of Solids, 2nd. ed. (Wiley, New York, 1972), p. 144.

[2-9] G. Betz, Surf. Sci. 92 (1980) 283.

[2-10] R. Hultgren et al., Selected Values of the Thermodynamic Properties of Binary Alloys (Am. Soc. for Metals, Metals Park, OH, U.S.A., 1973).

[2-11] D. P. Jackson, Rad. Effects 18 (1973) 185.

[2-12] H. L. Bay, W. Berres, and E. Hintz, Nucl. Instr. Meth. 194 (1982) 555.

[2-13] H. Oechsner, H. Schoof, and E. Stumpe, Surf. Sci. 76 (1978) 343.

[2-14] T. E. Parker and R. Kelly, in: Ion Implantation in Semi-conductors and Other Materials, ed. by B. L. Crowder (Plenum, New York, NY, 1973), p. 551.

[2-15] S. Hofmann and J. M. Sanz, J. Trace and Microprobe Tech. 1 (1982-1983) 213.

[2-16] V. E. Henrich, G. Dresselhaus, and H. J. Zeiger, Phys. Rev. Lett. 36 (1976) 1335.

[2-17] R. Kelly, Nucl. Instr. Meth. 182/183 (1981) 351.

[2-18] E. Dullni, Nucl. Instr. Meth. B2 (1984) 610.

[2-19] D. Grischkowsky, M. L. Yu, and A. C. Balant, Surf. Sci. 127 (1983).

[2-20] W. Husinsky, G. Betz, and I. Girgis, J. Vac. Sci. Technol. A2 (1984) 698.

[2-21] W. Husinsky and G. Betz, Nucl. Instr. Meth. B15 (1986) 165.

[2-22] W. Husinsky, P. Wurz, B. Strehl. and G. Betz, Nucl. Instr. Meth. B18 (1987) 452.

[2-23] C. E. Young, W. F. Calaway, M. J. Pellin, and D. M. Gruen, J. Vac. Sci. Technol. A2 (1984) 693.

[2-24] B. Schweer and H. L. Bay, Appl. Phys. A29 (1982) 53.

[2-25] J. P. Baxter, J. Singh, G. A Schick, P. H. Kobrin, and N. Winograd, Nucl. Instr. Meth. B17 (1986) 300.

[2-26] J. P. Baxter, G. A. Schick, J. Singh, P. H. Kobrin, and N. Winograd, J. Vac. Sci. Tech. A4 (1986) 1218.

[2-27] R. B. Wright, M. J. Pellin, and D. M. Gruen, Nucl. Instr. Meth. 182/183 (1981) 167.

[2-28] W. Berres and H. L. Bay, Appl. Phys. A33 (1984) 235.

[2-29] W. Husinsky, J. Vac. Sci. Technol. B3 (1985) 1546.

[2-30] M. J. Pellin, R. B. Wright, and D. M. Gruen, J. Chem. Phys. 74 (1981) 6448.

[2-31] B. J. Garrison, N. Winograd, D. Lo, T. A. Tombrello, M. H. Shapiro, and D. E. Harrison, Surf. Sci. Lett. 180 (1987) L129.

[2-32] S.V.N. Naidu, A.M. Sriramamurthy, and P.R. Rao, Bull. Alloy Phase Diagrams 5 (1984) 177.

INVESTIGATION OF PREFERENTIAL SPUTTERING MECHANISMS BY ANALYSING THE SAMPLE SURFACE AND NEAR—SURFACE REGION WITH AES AND ISS

B. BARETZKY

Max-Planck-Institut für Plasmaphysik, EURATOM Association
D-8046 Garching/München, Fed. Rep. of Germany

Preferential sputtering of compounds and alloys results in a compositional change of the surface and near surface region compared to the bulk composition. This effect can be caused by different mechanisms and the resulting concentration profiles should be different. For segregation-induced preferential sputtering an enrichment in the outermost layer and a loss in the sublayers of the segregating compound is expected /1/, in contrast to collision dominated systems.

Therefore we analysed the surface of ion bombarded compounds and alloys nearly simultaneously with AES (Auger Electron Spectroscopy) and ISS (Ion Scattering Spectroscopy) which have different information depths. We also measured the depth distribution of the near surface region by means of sputter depth profiling.

We compare results from collision dominated systems (such as Ta_2O_5) with those from systems showing segregation (CuLi). The relevance of the various mechanisms is discussed.

1. INTRODUCTION

Ion bombardment of a multicomponent target (compound or alloy) may cause a compositional change of the surface and near surface region due to preferential sputtering /2,3,4,5/. Surface compositional changes under ion bombardment can be ascribed to different reasons: Differences in surface binding energies /6/, large mass differences /5,7,8/ surface segregation /1,4/ and radiation enhanced diffusion /9/. The dominant effect indeed depends on the investigated system.

Bombardment of compounds consisting of components having large mass differences with light ions (H^+ and He^+) at low energies (comparable to the threshold energy of the heavier component) results in a depletion (i.e. . preferential sputtering) of the light component. This can be explained in terms of the different energy transfer in this mass and energy range /5,7,8/. There is also a characteristic depth profile of the altered layer which shows an exponential decrease of the surface-enriched component in the depth until bulk composition is reached /1,10/.

For most multicomponent materials the effect of Gibbsian segregation causes an enrichment of one component in the first atomic layers to minimize the free energy of the system /11/. Bombardment of these compounds or alloys results in sputtering of the segregating component. The Gibbsian segregation tends to compensate this loss in the first layer, which results in a subsurface loss of the segregating component. Therefore the characteristic depth profile of the altered layer consists of an enhanced concentration of the segregating element in the outermost atomic layer compared to the subsurface and its depletion in the layers below compared to the bulk composition, which changes until bulk composition is reached in deeper layers /1,12/.

R. Kelly and M. Fernanda da Silva (eds.), Materials Modification by High-fluence Ion Beams, 329–337.
© 1989 by Kluwer Academic Publishers.

It is the purpose of this paper to discuss the preferential sputtering effect of collision and segregation dominated systems. Therefore we have bombarded the collision dominated system Ta_2O_5 /13,14,15/ and a segregation dominated system, CuLi /16/, with 1.5 keV (1 kev) He^+ and depth profiled with 1 keV Ar^+. The different information depths of AES (Auger Electron Spectroscopy) and ISS (Ion Scattering Spectroscopy) enables us to compare the concentration of the outermost and of the subsurface layers.

2. EXPERIMENT AND COMPUTER SIMULATION

Anodically oxidized Ta_2O_5 samples with an oxide layer of about 350 nm thickness were bombarded with 1.5 keV He^+ at various angles of incidence (e.g. 20^0 and 90^0 relative to the surface). The copper lithium alloy samples with 6 at% and 16.4 at% lithium (6 at% Li alloy and 16 at% Li alloy) had been prepared in the Argonne National Laboratory, USA /17/. They were bombarded with 1 keV He^+ at an angle of incidence $\psi = 60^0$ relative to the surface.

All experiments were performed in an ultrahigh vacuum chamber with a base pressure of about 2×10^{-8} Pa at room temperature. The target surfaces were analyzed nearly simultaneously by means of AES using a retarding field analyser (for details see ref. /13/) and ISS. For the measurement of the low energy lithium ($\sim$ 43 eV) and copper ($\sim$60 eV) signal the electron impact angle was changed from 51^0 (for Ta_2O_5) to 30^0 (for CuLi-samples) and the modulation voltage from $V_{p-p} = 5V$ to $V_{p-p} = 1V$. The information depth is determined by the mean free path of the electrons depending on the electron energy /18/. For tantalum and oxygen and the high energy copper ($\sim$920 eV) Auger electrons the information depth is about 10 Å - 20 Å, whereas for the low energy lithium and copper Auger electrons it is about 4 Å corresponding to 2 atomic layers. ISS was performed by bombarding the target surface with 500 eV (1 keV for Ta_2O_5) He^+ at a fixed impact angle of 70^0 relative to the surface and analysing the He^+ ions scattered at 140^0 with a spherical condensor (for details see ref. /15/). In this energy range the signal arises from the outermost atomic layer. The surface composition is defined by the AES peak-to-peak amplitudes and the maximum value of the scattered intensity of the He^+-ions, and can be normalized to a pure copper standard. The electron beam width of less than 0.5 mm diameter and the analyzing ion beam width of about 0.9 mm diameter are small compared to the sputter ion beam width of about 1.6 mm diameter to ensure that the surface composition is measured in the center of the ion beam spot. The adjustment of the three beams and the measurement of the fluence were performed with a Faraday cup with an aperture of 0.5 mm diameter.

To obtain the compositional depth distribution of the altered layer the helium prebombarded surface was sputtered with 1 keV Ar^+ ions at a fixed angle of incidence $\psi = 30^0$ relative to the surface, because of its smaller preferential sputtering effect.

The experimental results of Ta_2O_5 were compared and normalized with computer simulation calculations, using the binary collision Monte Carlo Code TRIDYN /19/, based on TRIM /20/ (for details see refs. /14,21/).

3. RESULTS

The measurements of tantalum oxide show good agreement between the two different experimental techniques of AES and ISS as well as with the computer simulation: the ratio of the tantalum to the oxygen AES and ISS signals increases with the fluence of the 1.5 keV He^+ bombardment until a steady state is reached. The steady state value and the characteristic fluence to reach steady state increase by a factor of 3 and 90, respectively, by increasing the angle of incidence from 10⁰ to 90⁰, relative to the surface /5,14,15/. Depth profiling of the bombarded surface with 1 keV Ar⁺ ions shows that the depth of the altered layer also increases with increasing impact angle /5,14,15/.

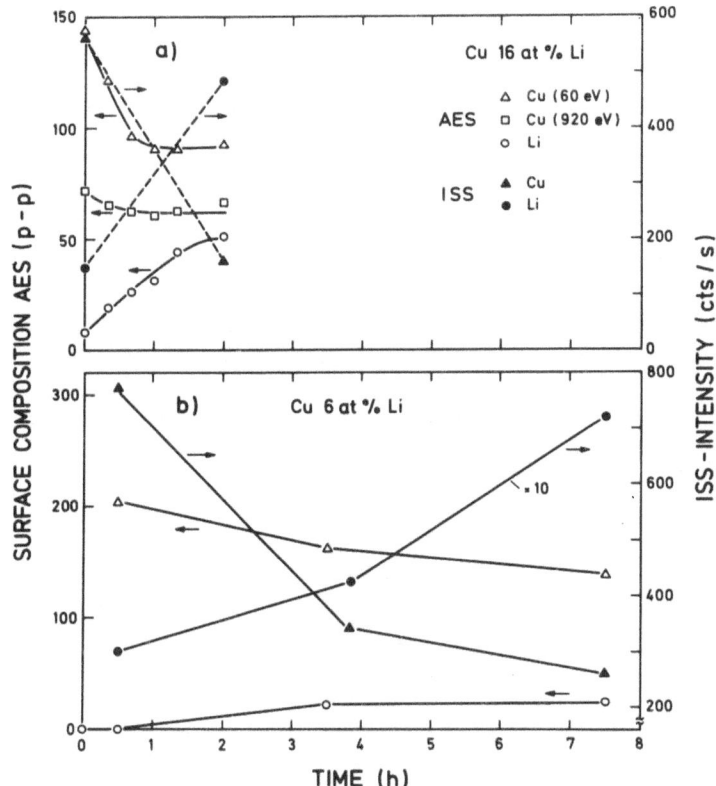

FIGURE 1. Segregation of lithium in CuLi alloys at room temperature: a) Surface composition of the CuLi alloy with 16 at% Li as a function of the time after bombardment, measured with AES and ISS. b) Surface composition of the CuLi alloy with 6 at% Li as a function of time after bombardment, measured with AES and ISS.

Figure 1 shows the segregation of Li in two different copper lithium alloys. The surface was bombarded with 1 keV He⁺ which leads to a removal of lithium. At this stage we assume that the prebombardment has no other effect than cleaning the surface, i.e. the segregation is due to thermally activated mass transport and not due to accumulated point defects. The AES and ISS signals for copper and lithium are plotted in dependence of time after bombardment. The ISS signal which comes from the outermost atomic layer shows an increase of the lithium signal and a decrease of the copper signal. About the same behavior is seen with the AES measurements but the changes are less pronounced because of the larger information depth compared to the segregation layer. The ISS saturation value of the copper signal is 21% of the starting value for the alloy with 16 at% Li, whereas the AES-copper signal changes to 64% of the starting value. The high energy AES copper signal shows an even smaller change to 87%, indicating that the segregation occurs mainly in the first atomic layer. The segregation effect also increases with the lithium concentration, which can be seen by comparing fig. 1a and 1b: the amount of the segregating lithium increases for the alloy with 16 at% Li by a factor of 2.5 for the AES-lithium signal and by a factor of 6.7 for the ISS-lithium signal compared to the 6 at% Li alloy. Supposing an exponential dependence of the segregation flux, i.e. $\exp(-t/\tau)$, we can get a characteristic time constant τ for the segregation process which also decreases with increasing lithium concentration ($\tau = 1 \times 10^2$ sec for the alloy with 16 at% Li and $\tau = 2 \times 10^2$ sec for the alloy with 6 at% Li).

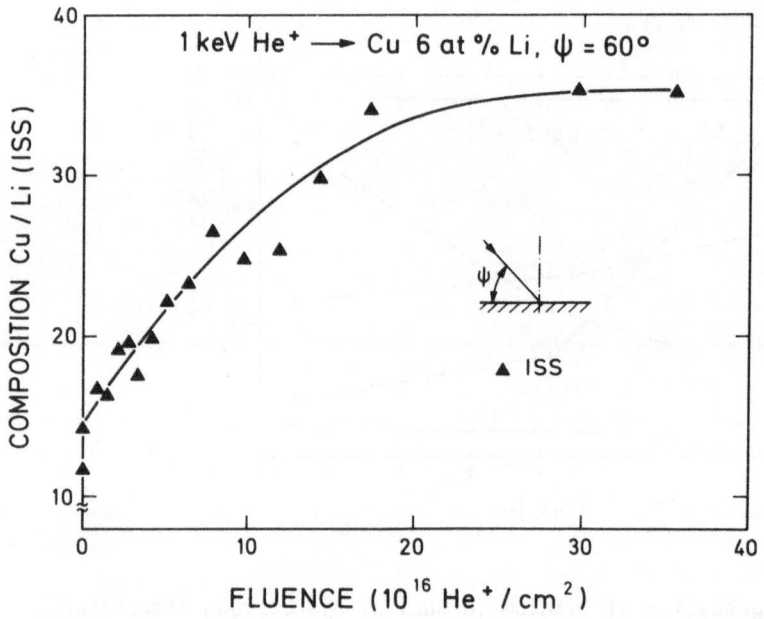

FIGURE 2. Change of the surface composition of the CuLi alloy with 6 at% Li in dependence of the fluence of 1 keV He^+ ions, measured with ISS.

The polished virgin samples were cleaned by means of 1 keV Ar^+ sputtering. Before each bombardment we waited for several hours until the segregation layer was established. Bombardment of the copper lithium alloy with helium and argon ions results in a depletion of the segregating component lithium and therefore an enrichment of copper (see Fig. 2 and 3). Figure 2 shows the increase of the ratio of the copper to lithium ISS signal with the fluence by bombarding the 6 at% Li alloy with 1 keV He$^+$ at an angle of incidence of ψ = 60^0 relative to the surface. Only the ISS results are available in this case. Bombardment of the 16 at% Li alloy with 1 keV He$^+$, ψ = 60^0 also results in an increase of the ratio of both the AES and ISS copper to lithium signal, as seen in Fig. 3.

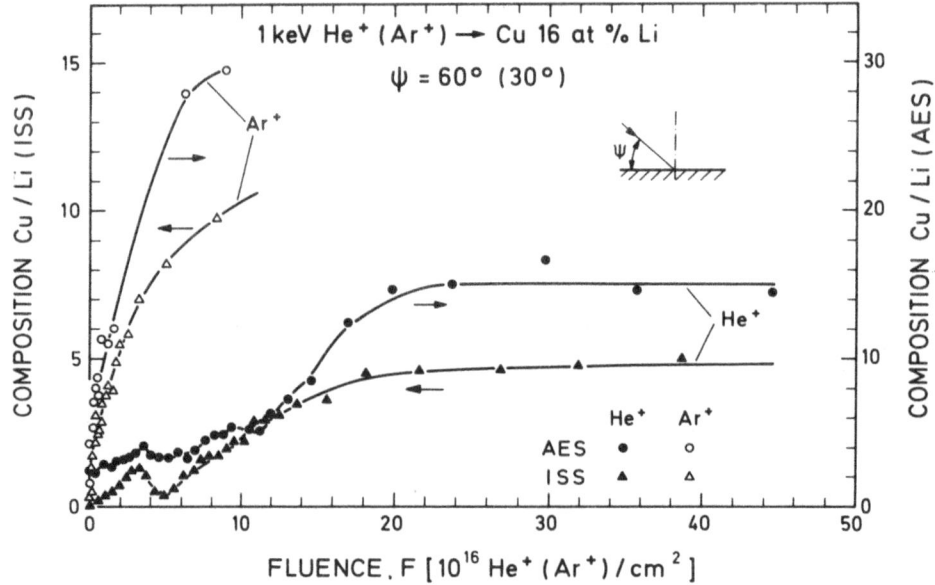

FIGURE 3. Change of the surface composition of the CuLi alloy with 16 at% Li in dependence of the fluence of 1 keV He$^+$ ions and 1 keV Ar$^+$ ions, measured with ISS and AES.

The steady state concentration ratio is lower by a factor of 7 compared to that of the sample with 6 at% Li, which indicates a higher steady state lithium concentration for the alloy with 16 at% Li. Because of the fast segregation process measurement of the depth profile of the altered layer caused by the helium bombardment is impossible. Figure 3 shows the bombardment of the segregated surface of the 16% Li alloy with 1 keV Ar$^+$ ions at an angle of incidence ψ = 30^0 relative to the surface. This "post"-bombardment causes a further increase of the concentration ratio by a factor of 2 which is in contrast to the Ar$^+$-sputtering of tantalum oxide.

4. DISCUSSION
The good agreement between AES and ISS measurements on the one side, and the good agreement between experiment and computer simulation on the other side, demonstrate that in this mass and energy range the preferential sputtering of tantalum oxide is dominated by collisional sputter processes. For the CuLi alloy we have seen that lithium segregates to the surface already at room temperature. The amount of the segregating lithium increases with increasing lithium bulk concentration (Fig. 1).

The striking difference between the sputtering of tantalum oxide and the copper lithium alloys shows up in the sputtering results with Ar^+ ions: We find an increase of the Cu/Li ratio and a decrease of the Ta/O ratio compared to the He^+ bombardment. This different behavior can be explained by the additional segregation process existing for the CuLi alloys. First we make a rough estimate of the depth distribution of the segregating lithium using the different information depths of the ISS and AES signals.

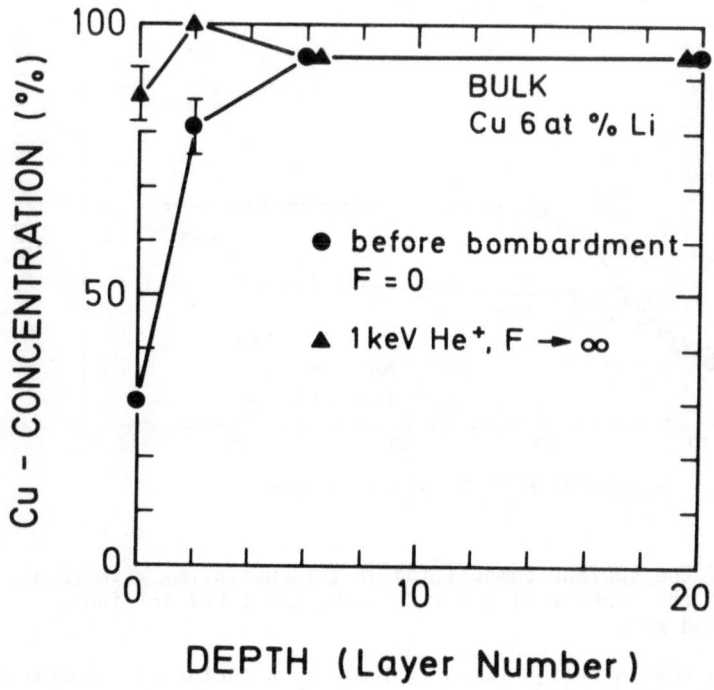

FIGURE 4. Variation of the Cu concentration as obtained from signals arising from various information depths: ISS (top layer), low energy AES (2. layer) and bulk concentration (6. layer and 20. layer). The symbols refer to different pretreatment of the CuLi sample with 6 at% Li: segregation at RT (●), bombardment with 1 keV He^+ to steady state (▲). The ISS signal and the AES signal are nomralized to a pure copper standard.

For this reason we compare the ISS and AES intensities with that of a pure copper standard. Two depth distributions of the normalized copper signals are shown in Fig. 4 for the 6at% Li alloy, one before sputtering and the other after reaching the steady state with 1 keV He^+ bombardment. The composition of the outermost atomic layer is measured with ISS and the AES signal gives us the composition of the second layer. In this case the high energy AES copper signal is not available in contrast to the CuLi alloy with 16 at% Li. There it shows no significant variations and we

assume that the composition in deeper layers (e.g. sixth atomic layer)
corresponds to the bulk composition.

Bombardment with 1 keV He$^+$ leads to a "preferential" sputtering of the
lithium and an increase of the copper signal in the outermost as well as in
the sublayer, from 30% to 87% and from 81% to 100%, respectively. This can
be interpretated by a lithium subsurface loss due to preferential
sputtering dominated by Gibbsian segregation as mentioned by R. Kelly /1/
and N.Q. Lam et al. /12/.

The same behaviour is seen for the 16 at% Li alloy in fig. 5. The
composition in the sixth atomic layer is measured with the high energy AES
copper signal. We assume that the concentration in the twentieth layer
corresponds to the bulk composition. The segregation of lithium is more
pronounced and its layer is thicker than for the alloy with 6 at% Li. The
surface consists nearly of 100% lithium. Sputtering with 1 keV He$^+$ results
in a reduced segregation profile and increases the surface composition to
50% copper. Sputtering with 1 keV Ar$^+$ results in a further enrichment of
the copper concentration to 87%. The steady state surface composition
approximately corresponds to the bulk composition.

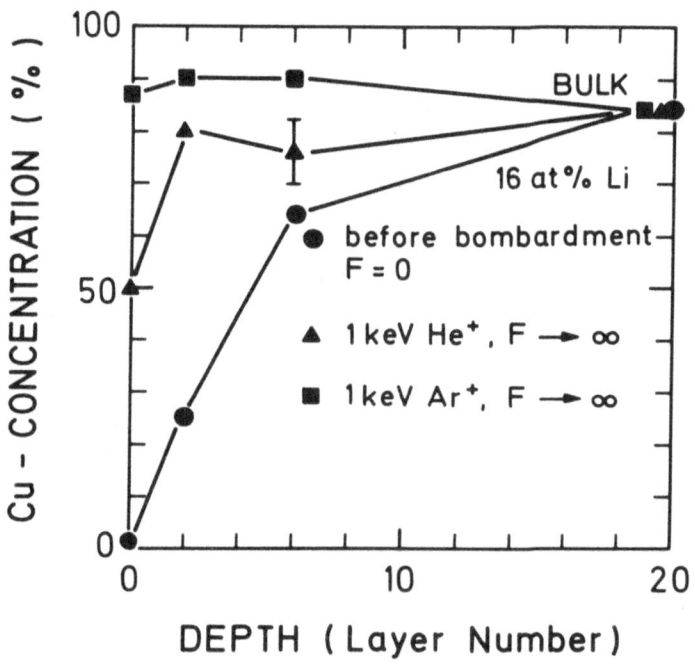

FIGURE 5. Variation of the Cu concentration as obtained from signals
arising from various information depths: ISS (top layer), low energy AES
(2. layer), high energy AES (6. layer) and bulk concentration. The symbols
refer to different pretreatment of the CuLi sample with 16 at% Li:
segregation at RT (●), bombardment with 1 keV He$^+$ to steady state (▲),
bombardment with 1 keV Ar$^+$ to steady state (■). The ISS signal and the AES
signals are normalized to a pure copper standard.

The dependence of the steady state surface concentration on the bulk concentration of the alloy and the sputtering ions can be explained by the main processes involved: the segregation and the sputter process. In steady state the flux of the segregating lithium atoms toward the surface $j_{S,Li} = n_{Li} \cdot (c^{\infty}_{Li} - c_{Li})/\tau$ must be equal to the flux of the sputtered lithium particles $j_{Y,Li} = Y^{c}_{Li} \cdot i_{I} \cdot c_{Li}$. The surface concentration of lithium is determined by this equilibrium and can be calculated by the following equation (which will be discussed in more detail in ref. /22/).

$$c_{Li} = \frac{c^{\infty}_{Li}}{Y^{c}_{Li} \, i_{I} \, \tau \, n^{-1}_{Li} + 1} \quad ; \; c_{Li} + c_{Cu} = 1, \tag{1}$$

with the saturation surface concentration of the segregation c^{∞}_{Li} measured by ISS, the estimated surface density of lithium $n_{Li} = 1 \cdot 10^{15}$ cm^{-2}, the component sputtering yield of lithium Y^{c}_{Li} measured by Laser Fluorescence Spectroscopy /23/, the ion current density i_{I} and the characteristic time constant τ, which is much shorter than the bombardment time.

The surface concentrations calculated with eq. (1) are in good agreement with the experimental results /22/. Therefore the assumption made above that sputtering and (Gibbsian) segregation are the determining processes seems to be justified. There is no evidence from these results for bombardment-induced Gibbsian segregation. The increase of the copper concentration and the decrease of the lithium concentration, respectively, in Fig. 5 is due to the increase of the sputtering yield of Ar^{+} by a factor of 10 compared to He^{+} sputtering, assuming a characteristic time constant for the segregation process independent of the bombarding ions.

For the bombardment of the two different CuLi alloys with He^{+} the component sputter yield is the same. Therefore the surface concentration is dominated by τ : The decrease of τ with increasing lithium bulk concentration results in a higher lithium and lower copper surface concentration for the alloy with 16 at% Li compared to that of the alloy with 6 at% Li.

5. CONCLUSION

We have investigated two different systems, one collision dominated (Ta$_2$O$_5$) and one segregation dominated (CuLi):
- The surface composition and the depth profile of tantalum oxide follows from collisional processes.
- For the copper lithium alloys lithium segregates to the surface at room temperature. Under ion bombardment the surface composition is determined by the balance between the sputtering and segregation process.

ACKNOWLEDGEMENTS
I would like to thank E. Taglauer for many helpful discussions and for critically reading the manuscript. The assistance of D. Mehl with the measurements should also be gratefully acknowledged.

REFERENCES

/1/ R. Kelly, Surf. Interf. Anal. 7 (1985) 1.
/2/ G. Betz and G.K. Wehner, in: Sputtering by Particle Bombardment II,
    ed., R. Behrisch (Springer, Berlin, 1983) p. 11.
/3/ For recent reviews, see e.g. H.H. Andersen, in: Ion Implantation and
    Beam Processing, eds., J.S. Williams and J.M. Poate (Academic Press,
    Sydney, 1984) p. 128; R. Kelly, in: Chemistry and Physics of Solid
    Surface V, eds., R. Vanselow and R. Howe (Springer, Berlin, 1984),
    p. 159.
/4/ R. Kelly, in: Materials Modification by High-Fluence Ion Beams (in
    this volume).
/5/ E. Taglauer, in: Materials Modification by High-Fluence Ion Beams (in
    this volume).
/6/ H.H. Andersen and H.L. Bay, Radiat. Eff. 13 (1972) 67.
/7/ E. Taglauer and W. Heiland, Appl. Phys. Lett. 33 (1978) 950.
/8/ E. Taglauer, Appl. Surf. Sci. 13 (1982) 80.
/9/ R. Bastasz and J. Bohdansky in: Proc. Int. Symp. on Sputtering, eds.:
    P. Varga, G. Betz and F.P. Viehböck (Inst. Allg. Physik, TU Wien,
    1980) p. 430.
/10/ P. Varga and E. Taglauer, Nucl. Instr. & Meth. B2 (1984) 800.
/11/ J.W. Gibbs, Trans. Connecticut Acad. Sci. 3 (1875-76) 108.
/12/ N.Q. Lam and H. Wiedersich, Nucl. Instr. & Meth. B18 (1987) 471-485.
/13/ B. Baretzky and E. Taglauer, Surf. Sci. 162 (1985) 996-1002.
/14/ B. Baretzky, W. Möller, E. Taglauer, Nucl. Instr. & Meth. B18 (1987)
    496-500.
/15/ B. Baretzky, W. Möller, E. Taglauer, in preparation.
/16/ A.R. Krauss, O. Auciello, A. Uritani, M. Valentine, M. Mendelsohn,
    D.M. Gruen, Nucl. Instr. & Meth. B27 (1987) 209-220.
/17/ M.H. Mendelsohn, D.M. Gruen, and A.R. Krauss, J. Nucl. Mater. 141-143
    (1986) 184.
/18/ G. Ertl, H. Küppers, Low Energy Electrons and Surface Chemistry,
    Verlag Chemie, 1974.
/19/ W. Möller and W. Eckstein, Nucl. Instr. & Meth. B2 (1984) 814;
    W. Möller, Nucl. Instr. & Meth. B15 (1986) 688.
/20/ J.P. Biersack and L.G. Haggmark, Nucl. Instr. & Meth. 174 (1980) 257.
/21/ W. Möller, in: Materials Modification by High-Fluence Ion Beams (in
    this volume).
/22/ B. Baretzky, E. Taglauer, in preparation.
/23/ R. Schorn, private communication, in preparation.

HIGH-FLUENCE IMPLANTATION IN INSULATORS,
PART I: COMPOSITIONAL, MECHANICAL, AND OPTICAL CHANGES

Paolo Mazzoldi

Unità CISM-GNSM, Dipartimento di Fisica dell'Università, Via Marzolo 8,
35131 Padova, Italy; and Laboratori Nazionali INFN, Legnaro, Italy

1. INTRODUCTION

Ion implantation introduces impurity or alloying atoms into the surface in a controlled and reproducible manner. The interaction of implanted ions with insulators results in the deposition of energy into electronic processes (ionization and excitation) and into atomic collisional events (Coulomb and hard-sphere interactions). The partitioning of energy into these two categories depends on the velocity of the ion relative to that of the target electrons. Although these interactions are common to all material classes, the utilization of electronic-energy deposition in physical property modification has been found to be significant only in insulators. This sensitivity is well known in the case of organic polymers (1).

The defects which can be formed by ion implantation depend upon the insulator structure and composition. Thus we expect for glasses and ceramics different changes in mechanical and tribological properties, network dilatation, induced optical absorption and luminescence, compositional changes and modifications in the chemical behaviour.

In the first section of this paper we will discuss the modifications induced by ion implantation in the composition of glasses, with particular reference to alkali silicate glasses (2), in the second section the mechanical and tribological properties of ion implanted insulators, in particular glasses and ceramics (3), and in the last one the optical properties.

2. ALKALI DEPLETION IN GLASSES

When alkali silicate glasses are irradiated with electrons (as in Auger Electron Spectroscopy AES) or ion beams their surface composition undergoes marked modifications (see for example refs. 2,4,5). The alkali migration can be correlated to different mechanisms, clearly connected to the different stopping-power regimes of incident particles. Particularly in the electronic stopping-power regime (low-mass particle irradiation), the alkali profile evolution appears to be governed by ordinary and field-assisted diffusion, with an electric field which is a function of the depth (6,7). In the nuclear-stopping-power regime (heavy-ion irradiation) the observed alkali depletion at the surface can be interpreted on the basis of a phenomenological model, which takes into account a preferential ejection of alkali atoms from the surface and an enhanced diffusion over the range of implanted ions (8,9).

R. Kelly and M. Fernanda da Silva (eds.), Materials Modification by High-fluence Ion Beams, 339–356.
© *1989 by Kluwer Academic Publishers.*

Electron-irradiation produces an alkali surface-depletion while an accumulation occurs at a depth roughly corresponding to maximum electron range (9,10). In the case of MeV-proton bombardment, alkali migration occurs from the interior towards the surface (11,12). Examples of radiation effects are shown in Figures 1 and 2 for irradiation of soda-lime glasses with electrons or protons. In both cases, alkali migration starts after a characteristic time interval has elapsed (incubation time). This interval is a function of the target temperature and of the incident beam current (12,13).

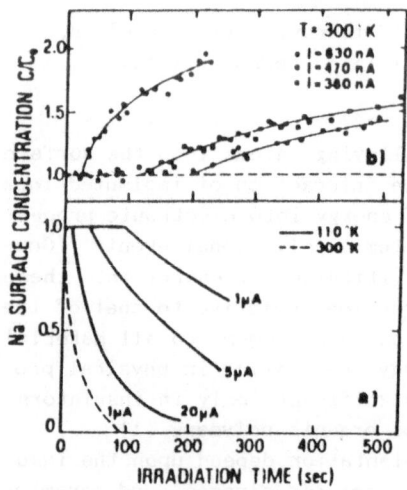

FIGURE 1

(a) Decay of the Na peak-to-peak Auger height at 110 K, normalized to the peak height at zero time, as a function of the time of electron irradiation for beam currents of 1,5,20 µA. A decay curve measured at room temperature and 1 µA is also shown (14). (b) Surface Na concentration at 300 K, normalized to the initial value at zero time, as a function of the time of 600 keV proton irradiation at different currents (15).

Other effects are present in the case of proton or electron irradiation of soda lime glasses: alkali migration towards the surface (proton irradiation) is accompanied by a Ca surface depletion while the Na depletion (electron irradiation) is accompanied by a Ca surface accumulation (see insets in Figure 2).

Heavy-ion implantation first removes alkali ions from the region of the implanted-ion distribution and for higher fluence levels removes alkali from the entire near-surface region to a depth well beyond the ion projected range (4,16-18). This means of altering the near-surface composition was used by Arnold and Peercy (19) to bring about low-temperature crystallization of an $Li_2O \cdot 2 SiO_2$ glass after inert-gas implantation. Measurements (Figure 3) made with elastic recoil detection (ERD) showed that implantation of, e.g., 5×10^{16} 250 keV Xe/cm^2, resulted in a depletion of Li from the region corresponding to the Xe end-of-range depth and a corresponding indiffusion of incumbent surface H, due to contamination. Heating to 500°C resulted in crystalline lithium disilicate formation in the implanted region, whereas no changes were noted in the unimplanted regions. Phase separation can occur in the lithia-silica system as is evident from the phase diagram (20). A loss of ≥ 2 mol.% Li_2O would result in a composition which could phase-separate for temperatures ≤500°C. The depletion of Li noted in Fig. 3 could easily bring about this situation. The phase diagram also makes it clear that crystallization could also be expected when small amounts of Li$_2$O are added to SiO$_2$. This situation occurs when Li ions are implanted into fused silica.

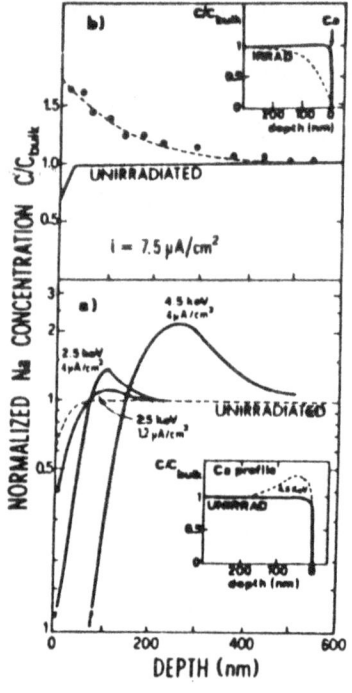

FIGURE 2

(a) Na depth profiles, normalized to the unirradiated bulk value, for 2.5 keV (1.2 µA/cm^2, 4.0 µA/cm^2) and 4.5 keV (4.0 µA/cm^2) electron irradiations. The normalized Ca profiles for unirradiated and 4.5 keV electron irradiated glasses are shown in the inset (9). (b) Normalized Na profiles for unirradiated (—) and 600 keV proton irradiated (●) glass. The normalized Ca profiles for irradiated (--)samples are also reported in the inset (10).

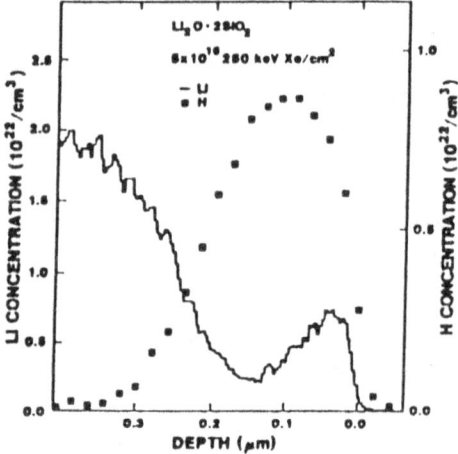

FIGURE 3

Comparison of the Li and H profiles in the near-surface region of Li$_2$O·2SiO$_2$ glass implanted with 5x10^{16} 250 keV Xe ions/cm^2 (19).

Arnold (21) showed that a polycrystalline α–quartz surface on fused silica can be formed by Li-ion implantation at fluences from 5x10^{16} – 1x10^{17}/cm^2 followed by annealing at 750˚C for 30 minutes. Quartz formation was verified by both IR spectra and x-ray diffraction. The phase separation of small Li$_2$O·2 SiO$_2$ particles in an ion-damaged silica host could serve to impede the network SiO$_2$ reformation as the annealing temperatures is increased and thus favor the transformation of SiO_2 tetrahedra into the quartz structure.

From the results reported in the literature for heavy-ion implantation, we can enumerate the following general points:

(1) the thickness of the alkali-depleted layer is a function of the implantation dose, reaching a steady-state value at sufficiently high doses.

(2) The alkali profile, for a fixed irradiation dose, is independent of the incident-ion current density.

(3) The depleted-layer thickness increases with the implantation energy.

(4) The alkali-depletion mechanisms seem to be independent (at least for near-room temperature) of the alkali element present in the glass.

(5) The sputtering cross-section of alkali elements shows an almost linear dependence with the electronic stopping power for a variety of incident ions.

(6) The modified layer is thicker than the calculated incident ion range, in particular for high-energy deposition, indicating a migration of produced defects.

More recent work (22) showed peculiar features in the alkali profiles, after irradiation, which may be explained as a contribution to the diffusion process of an electric field due to incident ion deposited charge. Such effects are even more evident with a higher velocity of incident ion.

Fig. 4 shows the RBS spectrum concerning the scattering from Rb atoms, for an Rb_2O-SiO_2 glass, irradiated at a temperature of 100°C with a 300 keV Ar^+ beam at doses of 4.5×10^{16} and 10.5×10^{16} ions/cm^2. For low-dose irradiation the profile is characterized by three features: 1) a surface peak, 2) a maximum at a depth of about 100 nm, 3) a large depletion at a depth of 300 nm, corresponding to the Ar projected range. At higher implantation doses an Rb segregation occurs at a depth corresponding to the Ar projected range. The surface peak is due to a partial trapping of the alkali element, possibly due to surface-vacancy agglomerates.

The surface alkali aggregate is water-soluble at room temperature, suggesting that it is not bonded to the matrix (23). The high solubility in aqueous environments could drastically modify the surface alkali profile in implanted glass samples.

The experimental data concerning the migration processes have been analyzed in the framework of the thermodynamics of irreversible processes by Miotello and Mazzoldi (5). The calculated diffusion-coefficient values are some orders of magnitude larger than that expected for the ordinary processes in glass of similar composition. In Figure 5 we report an evaluation of the Rb "enhanced" diffusion coefficient, during Ar irradiation of an Rb_2O-SiO_2 glasses at different target temperatures. The Rb self-diffusion coefficient, D_{th}, is also reported for comparison.

Cooperative transport processes are also evident during charged-particle irradiation of glasses explained by using the quoted theoretical treatment (5). Here we report an indicative example of the cooperative effect occurring in the Na-Ca migration in silicate glasses during Auger analysis (Figure 6).

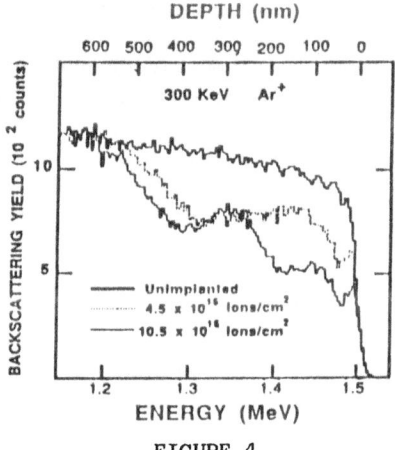

FIGURE 4

RBS spectrum from Rb in glasses irradiated with Ar at 300 keV (24).

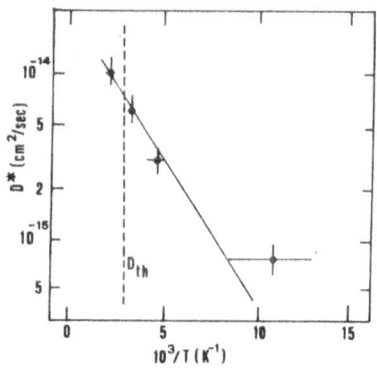

FIGURE 5

Calculated enhanced diffusion
coefficient of Rb in heavy-ion-
irradiated Rb_2O-SiO_2 glasses
(24). D_{th} is the Rb self-diffusion coefficient.

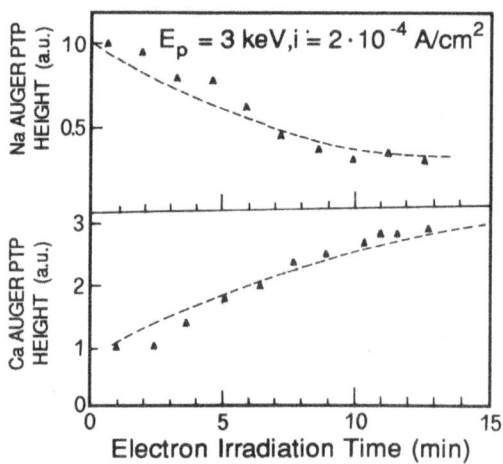

FIGURE 6

Experimental and calculated Na and
Ca Auger peak heights as function
of electron bombardment time (25).

An interesting result emerging from a theoretical analysis of effects
like that reported in Figure 6 is connected to the possibility that high
diffusion coefficients are not accompanied by high ionic mobilities. This
is most probably due to electrodynamical screening effects, possibly connected to local lattice deformations, when charges of different kinds are
involved in the global transport processes (5).

3.MECHANICAL PROPERTY MODIFICATIONS

3.1.Glasses

The defects produced by ion implantation induce volumetric changes and, as a consequence, concomitant modifications in surface stress, hardness and refractive index.

For most simple silicate glasses, implantation results in compaction within the implanted region. This might be expected _a priori_ on the basis of the large specific volume and more open structure of the glassy state relative to that of its crystalline counterpart. On the contrary, crystalline materials generally expand due to lattice displacement.

The implanted layers are shallow, on the order of less than one micron in depth, and the constraint of the undamaged substrate is such that the dilatation is normal to the surface, i.e., $\Delta V/V \simeq \Delta L/L$. The dilatation of the implanted surface also results in a lateral stress which can be tensile, due to compaction, or compressive, due to expansion in the implanted layer. These stresses can be measured directly by photoelastic and interferometric techniques (26-27) or in situ by a cantilever-capacitor measurement (28).

For a comprehension of the volume-change process, it is useful to quote the Shelby (29) experiments. ^{60}Co γ-ray irradiation produced expansion for hydrogen-impregnated silica. In ordinary fused silica which has not been hydrogen-impregnated, there is room for movement of network atoms into the interstices. This movement, resulting in compaction, occurs either when bonds are broken by ionizing radiation or when direct displacement occurs via collisional processes. When H-impregnated silica is irradiated, the hydrogen can react with the broken bonds to form Si-OH and Si-H. The presence of H does not allow densification; instead, the network relaxes and a net expansion results. The addition of network modifiers, such as Na_2O, produces similar results (Figure 7) for high irradiation energy.

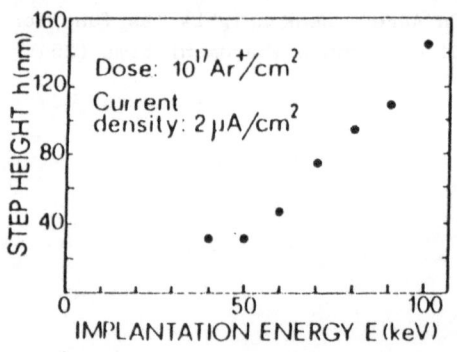

FIGURE 7

The swelling of the implanted region as a function of incident energy at constant dose and current density, in alkali silicate glasses (30).

Compressive or tensile surface stresses depend on whether the dilatation is positive or negative.

Figure 8 shows (28) the integrated tensile stresses, as obtained by cantilever beam technique, for fused silica implanted with various ions. The maximum stress depends on the deposited energy, which, for fused silica and for thermally grown SiO_2 on Si, corresponds to a value of about $2 \cdot 10^{20}$ keV/cm^3. The ion fluences at which saturation occurs are in reasonable agreement with calculations of the fluence at which ion-track overlap takes place (31). The large number of broken bonds in the implanted surface allows a form of plastic flow, or an elastic-plastic transition, to occur, which accounts for the decrease in integrated stress as ion fluence is further increased.

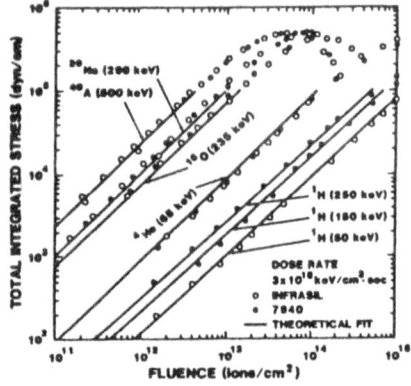

Integrated stress (dynes/cm) versus ion fluence (ions/cm²) incident on fused silica (Infrasil and Corning 7940) for implanted H,He,O,Ne, and Ar at the indicated energies (2).

Figure 9 shows (32) the integrated stress for an alkali-borosilicate glass, irradiated with different ions.

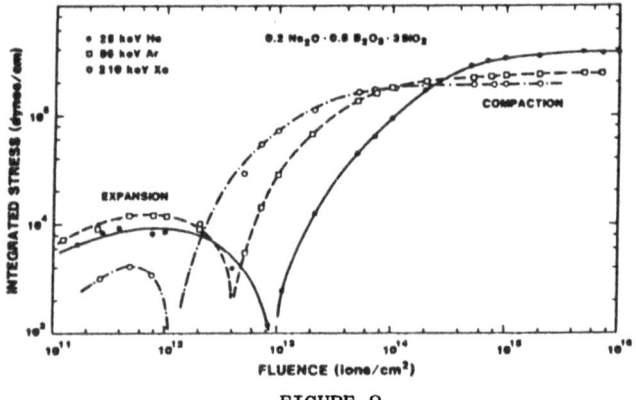

FIGURE 9

Integrated stress (dynes/cm) versus ion fluence (ions/cm²) for He, Ar and Xe incident on 0.2 Na₂O·0.8 B₂O₃ 0·3 SiO₂ (2).

In this case the ion-fluence regime over which expansion takes place has been shown to be related to the alkali content for the reasons given by Selby for H-related expansion. The results for the alkali-borosilicates are different in that the measured stresses scale with energy into electronic processes rather than collision-energy density, i.e., for these glasses, which include nuclear waste-glasses, the efficiency of stress generation is greatest for the ionizing component of the ion energy. Bond-breaking accompanying ion implantation results in changes in surface hardness. This has been evidenced in work by Jensen et al. (33) for a soda-lime-silicate glass implanted with 480 keV H-ions, measuring surface hardness with a Vickers (square) diamond indenter under a 60 g load. The data show a softening (increased length of the indenter trace) with increasing fluence to a maximum value ($\simeq 5 \times 10^{20}$ keV/cm³) followed by an increase in hardness. More significantly, the median-vent surface cracking reflected the stressed condition of the surface, i.e., increasingly longer cracks as the surface tensile stress increased (compaction) to a maximum followed by shorter crack traces at higher fluences. Arnold (31) has done similar experiments on fused silica implanted with 500 keV Ar using a shallow Knoop indenter under a light load (15 g). Under these conditions the hardness measurement reflects the condition of the implanted layer itself. The data show a decrease in hardness on the order of 10% at a fluence level ($\simeq 6 \times 10^{13}$ ions/cm²) which gave a maximum in integrated tensile stress (28). At higher fluence levels the hardness increased to values greater than that of the original glass at an energy deposition value of 3×10^{22} keV/cm³ and then decreased again.

A number of experiments have been reported which show that soda-lime-silica glass or more complex nuclear waste glasses exhibit increased fracture toughness under certain ion fluence conditions.

Figure 10 shows the percentage of cracks (with respect to the number of indentations), which develop after application of a loaded square pyramid indenter on 100 keV-Ar implanted soda-lime glasses, as a function of implantation dose for three applied loads (4). A reduction of crack development is observed for all implantation fluences.

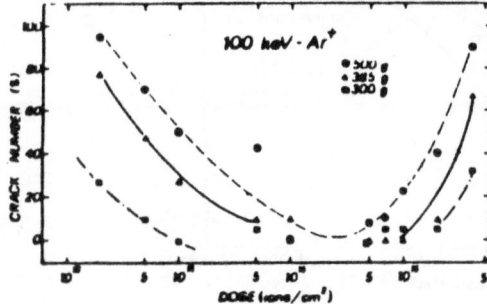

FIGURE 10
Percentage of cracks versus implantation dose for three applied loads (4).

Figure 11 reports the difference, Δc, of the average crack length, c, between the implanted and the unimplanted glasses, at the same indenter load, as a function of the ion dose. A reduction in the crack length, the maximum of which occurred for implantation doses of the order of $5 \cdot 10^{15}$ ions/cm², is evident. A reduction of crack formation is accompanied by a shorter length (surface compression). The thermal stability of the surface mechanical resistance of implanted glass has been studied by performing indentation tests on samples annealed at different temperatures for a constant time (4). The annealing results have been analyzed by studying the temperature dependence of the so-called P^x load, at which the crack percentage reaches the value 50%. Figure 12 shows the parameter P^x using a logarithm scale, as a function of $1/T$, where T is the annealing temperature. Such a behaviour indicates that the process is governed more by implantation-induced defects than the loss of Na, due to Ar-irradiation.

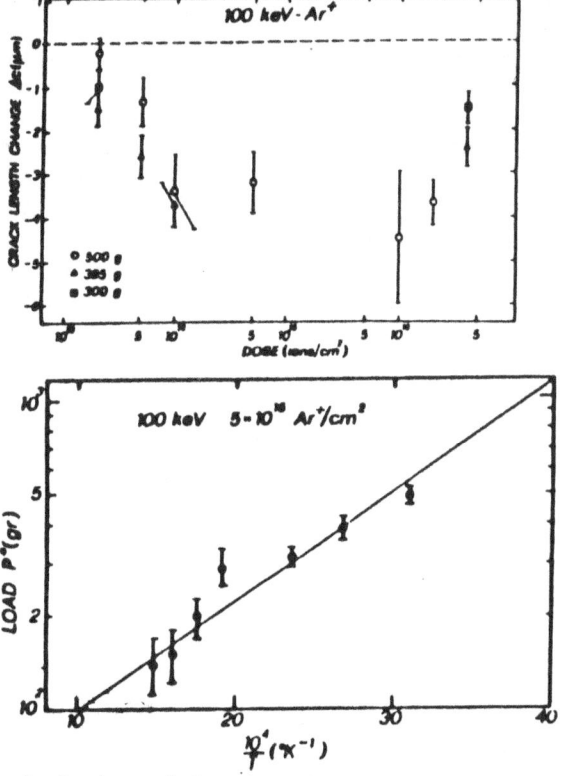

FIGURE 11
Change of radial crack length as function of implantation dose for three applied loads (4).

FIGURE 12
Applied load at which the percentage of cracks reaches a value of 50% versus the annealing temperature (4).

Implants of C and N into soda-lime glasses, in the range of 10^{16}–10^{18} ions/cm^2, show an increase in Knoop indentation hardness which saturates at about 1.5x10^{16} ions/cm^2 or about 10^{22} keV/cm^3 (34). The state of compression in the implanted surface, leading to hardening at high energy-deposition values could arise from several causes; (a) oxygen bubble formation at high fluences, (b) swelling due to the inserting of N or C ions, or (c) a compositional change to a silicon oxynitride or silicon carbide phase. More work is needed in this area to make ion-implantation hardening of glasses a viable technological technique.

3.2.Ceramics

The mechanical properties of ceramics are more sensitive to surface conditions than any other class of structural materials. For high-quality structural ceramics, most failures under applied loads initiate at the surface. Usually, this is due to a tensile component of the applied stress or the propagation of a pre-existing flaw or both. As a consequence, the useful strengths of these materials can often be raised by surface treatments which remove the flaws or reduce their severity or which generate a compressive stress layer at the surface. The severity of flaws can be reduced by increasing the strength and fracture toughness of the host material. Hardness testing on ceramic surfaces present two difficulties: large elastic recovery effects and penetration depth higher than the implanted thickness (3).

In attempts to overcome the problems caused by the penetration depth, equipment employing ultra-low loads (in the μN range) and measuring the displacement as a function of load has been developed (35,36). The application of an ultra-low-load tester to implanted ceramics overcame the problem of depth of impression relative to the depth of implanted layer, but the method of handling the elastic recovery is not yet totally satisfactory (37).

The effect of ion implantation on hardness has been reported for several ceramics, including α-Al$_2$O$_3$, MgO, ZrO$_2$, SiC, Si$_3$N$_4$, and TiB$_2$. Figure 13 shows the Knoop hardness of Al$_2$O$_3$, expressed as the ratio of the implanted to the unimplanted hardness, for different implantation at different temperatures (38).

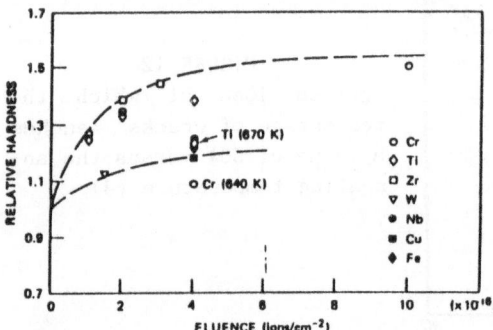

FIGURE 13
Relative hardness of α-Al$_2$O$_3$ implanted with indicated ions (3).

Hardness increase has been found in MgO single crystals, yttrium-stabilized zirconia, TiB$_2$, as well as covalent-bonded ceramics such as α-SiC, Si$_3$N$_4$ and Silicon Aluminum Oxi-Nitride (SIALON) for low-fluence implantation. The formation of an amorphous layer, at higher fluences, decreases the hardness value (3).

Post-implantation annealing modifies the surface hardness of ceramics. Studies on Al$_2$O$_3$ (39-41) and MgO (42) find that the implantation-induced hardness increases begin to recover as the point-defect clusters anneal. Depending on the phase relationships of the implant species and host, the hardness may proceed to that of a solid solution or may show another increase due to the onset of precipitation of one or more phases. Such results indicate that the increase in hardness involves both defect and solid-solution strengthening. Further increases in these properties may occur during post-implantation annealing due to precipitation hardening in appropriate host-implant species combinations.

Fracture toughness is an important mechanical property of ceramic materials. Increases in the indentation fracture toughness have been reported for several ion-implanted ceramics, such as Al$_2$O$_3$, SiC, MgO, yttrium stabilized zirconia, TiB$_2$. These results are attributed to both the surface compressive stress and the change in method of crack nucleation (3).

It should be noted that in all reported experiments the cracks initiated in the unimplanted substrate and propagated toward the surface. Such observations note the influence of the modified surface layer on the process and final stages of crack propagation, but whether or not it is accurate to call this an effect on fracture toughness is not yet resolved.

Increases in the transverse rupture strength for Al_2O_3 implanted with different ions have been reported (3). The increase in strength has been attributed to residual stresses that decreased the net stress concentration at surface flaws under an applied tensile stress.

The presence of a residual compressive stress in the implanted surface layer can explain, at least in part as discussed, the increases in hardness, fracture and flexural strength.

The integrated stress caused by implantation of argon, oxygen, nitrogen, helium, and hydrogen into Al_2O_3 (43,44) has been measured by using the cantilever-capacitor method. The stress initially increased almost linearly with fluence and for each ion except hydrogen exhibited a saturation at high fluences. This saturation has been correlated to the onset of amorphization and the dynamic annealing of defects. A maximum lateral stress of 3.4 GPa was calculated for a fluence of 10^{16} Ar/cm^2 (500 keV).

Using the same technique, the influence of fluence and substrate temperature on residual stress for nickel-implanted Al_2O_3 has been determined (Figure 14). The integrated stress increased monotonically with fluence at 300 K with a broad maximum at $5x10^{15}$ Ni/cm^2. Assuming the thickness of the compressive region to be equal to the range of the implanted ions, a value of 2 GPa was obtained for the maximum compressive stress. Implantation at 100 K caused a rapid increase in integrated lateral compressive stress until the onset of amorphization, at which fluence stress began to drop sharply with further implantation. The decrease in stress at high fluence is attributed to the presence of a weaker, amorphous, phase.

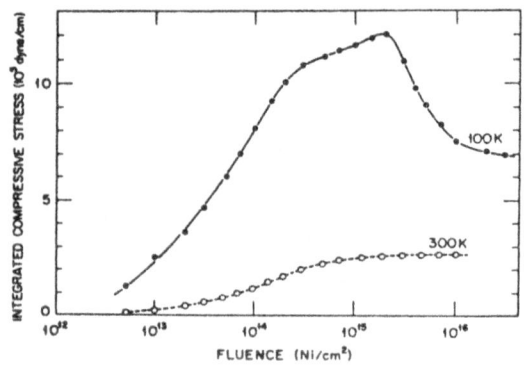

FIGURE 14
Integrated lateral compressive stress as a function of fluence of Al_2O_3 implanted with 300 keV Ni ions at 100 or 300 K (45).

Comparable results have been obtained on TiB_2 implanted with 10^{17} Ni/cm^2 (46). The increase in the hardness and the reduction of subsurface crack propagation in the implanted regions, result in an improvement in the resistance to material removal during wear tests (3).

In summary, the surface mechanical properties of ceramics can be altered

350

by ion implantation, due to the microstructural changes and the residual
stresses produced in the implanted zone. Because of these implanted layers,
the data are generally qualitative in nature and indicate only the direction of the changes.

4.REFRACTIVE INDEX CHANGES IN GLASSES

Ion bombardment of insulators produces refractive-index modifications,
which are determined by many effects (47) such as phase change (e.g.
quartz, LiNbO$_3$), microporosity (e.g. SiC, B$_4$C, Al$_2$O$_3$), colour center introduction (e.g. SiC, MgO), compaction (e.g. vitreous silica) or compositional
changes (e.g. alkali loss in glass).

The optically modified layer thickness corresponds to the incident-ion
range in most of the studied conditions. For low-mass-ion irradiations
(proton, helium or deuteron), the reflection coefficient measurements show
evidence of nonuniformity in the modified surface film because of the large
ion penetration depth and the deposited-energy distribution. However, in
the case of medium-weight ions (neon, argon or krypton), the smaller penetration and the collision-cascade mechanism reduce the nonuniformity in the
optically modified layer (48).

The bombardment of silica glass produces a glass network compaction, due
to radiation damage, and consequently a refractive index enhancement. The
vitreous silica reaches a more ordered and denser state, which should be
present in thermodynamic equilibrium.

In the case of quartz bombardment, the large interstitial and vacancy
concentration, which occurs in the implantation processes, does not allow
the lattice to return to the initial configuration, when most of the primary defects have been annealed. A disordered structure builds up and saturation would be expected for high implantation doses.

Figures 15 and 16 show the changes of refractive index of ion implanted
quartz crystals, fused silica, and LiNbO$_3$ as a function of the nuclear-deposited energy for various kinds of ions.

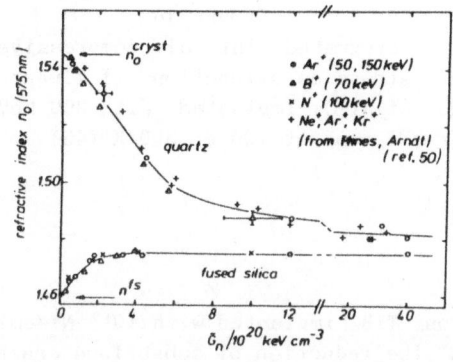

FIGURE 15
Refractive index of ion-implanted
quartz crystals and fused silica as
a function of the nuclear-deposited
energy density G$_n$ for various kinds
of ions (49).

The observed refractive-index modifications in vitreous silica suggested
an analogous behaviour for a glass, in consideration of its disordered
structure. However, Hines (50) showed that 40 keV-Ar$^+$ ion implantation of a

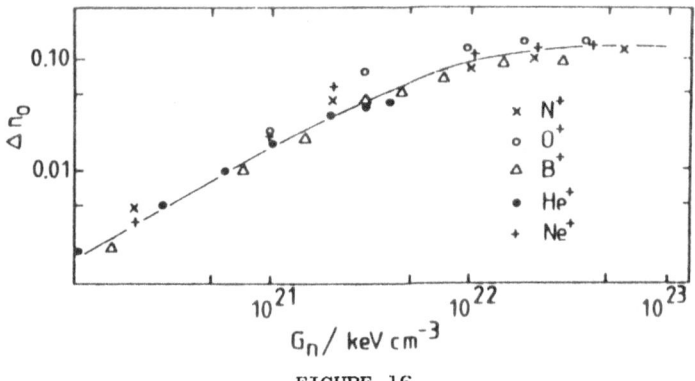

FIGURE 16

Change of the refractive index $\Delta n_o (n_{unimp} - n_{impl})$ of ion-implanted $LiNbO_3$ as a function of the nuclear deposited energy density G_n for various kinds of ions (49).

soda-lime glass caused a decrease of the refractive index in the surface layer (n=1.343 to be compared with n=1.525 for unimplanted glass). Such a result indicates that the changes must be related to the other elements present in glass apart from SiO_2. The refractive index decrease was correlated to preferential removal of sodium from the surface, during the implantation process.

A systematic investigation of the optical modifications induced by Ar-ion implantation in soda-lime glass was performed by Geotti et al. (51). The formation of a sodium-depleted surface layer with antireflective (A.R.) properties was observed for low current densities (up to 2 $\mu A/cm^2$), whereas medium-current densities and high implantation energies produced light-diffusing surfaces.

For 50·keV Ar implantation, at doses lower than 10^{16} ions/cm^2 no significant changes were observed in the reflectance curves. For doses between 10^{16} and 5×10^{16} ions/cm^2 the specular reflectance curves showed an AR effect, which increased with the dose, as shown in Fig. 17. The reflectance curves show an interferential minimum, which becomes deeper and shifts towards higher wavelengths as the dose increases. The thickness and average refractive index of the modified surface layers have been calculated from the position and height of the interferential minima in the reflectance curves.

Figure 18 shows the reflectance modifications, which were observed as the implantation energy was further increased up to 100 keV. The optical thickness increases (the minima shift towards the near IR range) and new extrema points appear. For a discrete, homogeneous, transparent film the interferential maxima are expected to be tangential to the curve of the substrate reflectance. The fact that they lie higher or lower was attributed to the onset of light scattering, particularly in the UV and visible ranges. This hypothesis has been confirmed by the observation of a diffuse-reflectance component in the UV and visible range for implantation energies higher than 60 keV.

Analogous information has been obtained from the optical scattering indicatrix of the implanted surfaces as a function of implantation energy.

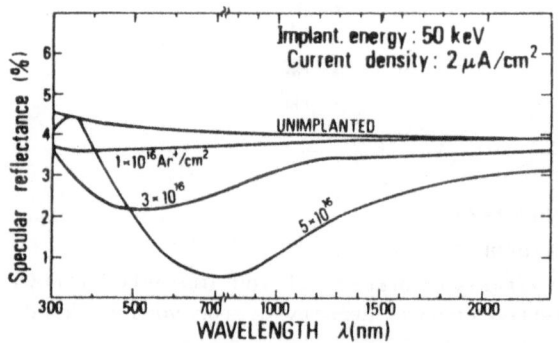

FIGURE 17
Experimental specular-reflectance curves versus wavelengths for 50 keV Ar⁺ irradiation at different doses.

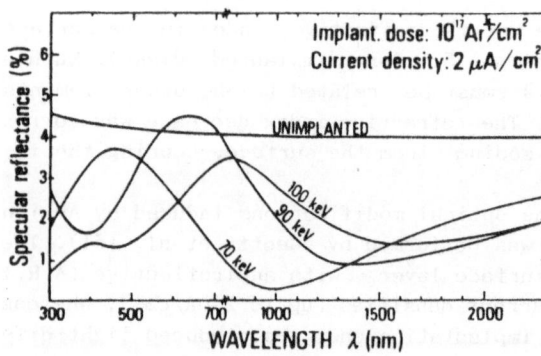

FIGURE 18
Experimental reflectance curves versus wavelength for Ar⁺ irradiations at different energies. The implantation dose was 10^{17} ions/cm².

The thicknesses of the modified layers were evaluated from both nuclear (sodium depletion) and optical measurements. The agreement was satisfactory. Only implantation energies of ≤70 keV are promising for optimizing transmittance since they yield noticeable antireflective effects without significant scattering losses.

Figure 19 compares the solar terrestrial irradiance curve and the spectral reflectance of the 50 keV sample representing the most favorable energy trade-off able to minimize solar-reflection losses. In fact, any change in energy would either further displace the minimum from the solar maximum or produce less-pronounced antireflective effects.

For the best sample, surface-reflectance losses are cut by 66%, with a reduction of almost 22% of total transmission losses with respect to the unimplanted sample (which in these preliminary investigations had an energy absorption of 4.5%). The table shows, for comparison, the performances of samples obtained by vacuum coating the same glass with a single-layer or three-layer antireflective coating.

Thus, leaving aside the cost factor, thanks to the low refractive indices attained, this technique can yield results at least comparable to vacuum-coating. However, the reflectance minima are not as broad as those provided by some leach processes which result in gradient index and are suitable for treating both sides at the same time.

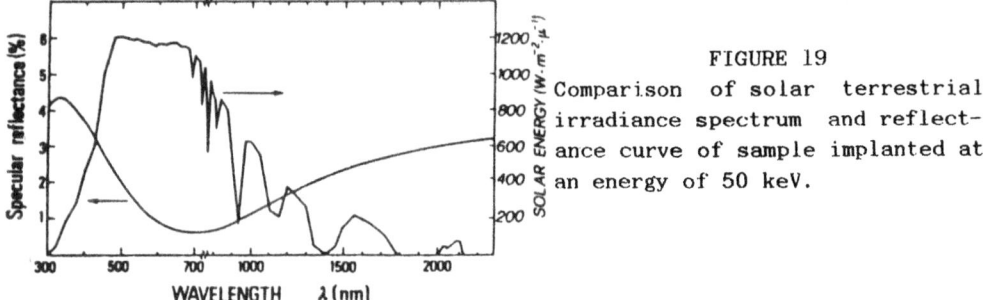

FIGURE 19

Comparison of solar terrestrial irradiance spectrum and reflectance curve of sample implanted at an energy of 50 keV.

Solar Characteristics Obtained for Antireflective Layers Produced on Same Glass by Varied Methods

Sample	First surface solar reflectance (%)	Total solar transmittance losses (%)
Unimplanted sample	4.1	12.0
Implanted at 50 keV, 10^{17} Ar$^+$/cm^2, 2 μA/cm^2	1.4	9.3
Vacuum-coated with MgF$_2$ (100 nm)	2.0	10.0
Coated with antireflective tri-layer	1.5	9.4

Optical slab waveguides can be produced by high-energy implantation of light ions (refs. 52-56). For the fabrication of strip waveguides several methods are possible (Figure 20). The simplest one uses the implantation of high-energy light ions to produce a buried layer of lower refractive index. The lateral boundaries will be produced by means of masks. A strip waveguide with a refractive index value smaller than that of the virgin material will be obtained in the central region.

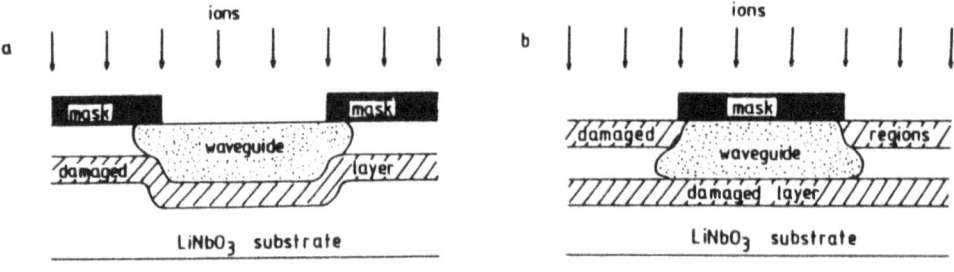

FIGURE 20

Formation of strip waveguides in LiNbO₃ by ion implantation (49).

In the second case two implantation steps are applied: a high energy implantation creates a buried layer with decreased refractive index and a low energy implantation through a mask generates the lateral boundaries of the waveguide. The damaged regions at the surface can be easily etched away.

Other methods for fabrication of strip waveguides make use of combined techniques, i.e. diffusion and ion implantation.

5. CONCLUSIONS

A review, very short due to space limitations, has been presented on the mechanical and optical properties of ion implanted glasses and ceramics. The following paper of J.P. Dran will consider other aspects of ion implantation of insulators.

REFERENCES

1. Venkatesan T, Calcagno L, Elman B S and Foti G: in Ion Beam Modification of Insulators (ed. P. Mazzoldi and G.W. Arnold, Elsevier, Amsterdam 1987) Ch. 8.
2. Arnold G W and Mazzoldi P: in reference 1, Ch. 5.
3. McHargue G J: in reference 1, Ch. 6.
4. Mazzoldi P: Nucl. Instr. Meth. 209/210 (1983) 1089.
5. Mazzoldi P and Miotello A: Rad. Eff. 98 (1986) 205.
6. Miotello A: Phys. Lett. 103A (1984) 279.
7. Miotello A and Mazzoldi P: J. Phys. C15 (1982) 5615.
8. Miotello A and Mazzoldi P: J. Phys. C16 (1983) 221.
9. Battaglin G, Della Mea G., De Marchi G, Mazzoldi P and Puglisi O: J. Non-Cryst. Solids 50 (1982) 119.
10. Battaglin G, Della Mea G, De Marchi G, Mazzoldi P and Puglisi O: Rad. Eff. 64 (1982) 99.
11. Battaglin G, Della Mea G, De Marchi G, Mazzoldi P, Miotello A and Guglielmi M: J. Phys. C: Sol. State Phys. 15 (1982) 5623.
12. Battaglin G, Della Mea G, De Marchi G and Mazzoldi P: Appl. Phys. Lett. 45 (1984) 736.
13. Pantano C G and Madey T E: Appl. Surf. Sci. 7 (1981) 115.
14. Pantano C G, Dove D B and Onode G Y Jr.: J. Vac. Sci. Tech. 13 (1976) 414.
15. Battaglin G, Della Mea G, De Marchi G, Mazzoldi P and Miotello A: in P. Mazzoldi (Ed.), Induced Defects in Insulators, Les Editions de Physique, Les Ulis (France), 1984, p. 165.
16. Arnold G W and Borders J A: J. Appl. Phys. 48 (1977) 1488.
17. Arnold G W and Borders J A: in G. Carter, J.S. Collingon and W.A. Grant (Eds.), Applications of Ion Beams to Materials, Conference Series No.28, The Institute of Physics, Bristol, 1976, p. 121.
18. Borders J A and Arnold G W: in O. Myer, G. Linker and F. Kappeler (Eds.), Ion Beam Surface Layer Analysis, Vol. 1, Plenum, New York, 1977, p. 415.
19. Arnold G W and Peercy P S: J. Non-Cryst. Solids 41 (1980) 359.
20. Tomazawa M: Phys. Chem. Glasses 14 (1973) 112.

21. Arnold G W: Rad. Eff. 47 (1980) 15.
22. Battaglin G, Della Mea G, De Marchi G, Mazzoldi P, Miotello A, Boscoletto A, Tiveron B: Nucl. Instr. Meth. B19/20 (1987) 948.
23. Battaglin G, Della Mea G, De Marchi G, Mazzoldi P, Miotello A and Tiveron B: Rad. Eff. (1986).
24. Battaglin G, Della Mea G, De Marchi G, Mazzoldi P, Miotello A: Nu-. cl. Instr. Meth., in press.
25. Corchia M, De Logu P, Giorgi R, Mazzoldi P, Miotello A: Nucl. Instr. Meth., in press.
26. Primak W: Surface Sci. 16 (1969) 398.
27. Primak W: J. Appl. Phys. 55(4) (1984) 852.
28. EerNisse E P: J. Appl. Phys. 45(1) (1974) 167.
29. Shelby J E: J. Appl. Phys. 50(5) (1979) 3702.
30. Mazzoldi P: Society of Photo-Optical Instrumentation Engineers Vol. 400 (1984) 94.
31. Arnold G W: Rad. Eff. 65 (1982) 17.
32. Arnold G W: Rad. Eff. (1987) to be published.
33. Jensen T, Lawn B R, Dalglish R L and Kelly J C: Rad. Eff. 28 (1976)245.
34. Burnett P J and Page T F, J. Materials Sci. 20 (1985) 4624.
35. Pethica J B, Hutchings R and Oliver W C: Philos. Mag. A48 (1983) 593.
36. Newey D, Pollock H M and Wilkins M A: in V. Ashworth, W.A. Grant and R.P.M. Procter (Eds.), Ion Implantation into Metals, Pergamon Press, New York, 1982, pp. 157-71.
37. Oliver W C, McHargue C J, Farlow G C and White C W, in: Y. Chen, W.D. Kingery and R.J. Stokes (Eds.), Defect Properties and Processing of High-Technology Nonmetallic Materials, in press.
38. Farlow G C, White C W, McHargue C J, Sklad P S and Appleton B R: Nucl. Instrum. Meth. Phys. Res. B7/8 (1985) 541-46.
39. Burnett P J and Page T F: J. Mater. Sci. 19 (1984) 3524-45
40. Yust C S and McHargue C J: in R.F. Davis, H. Palmour III and R.L. Porter (Eds.), Emergent Process Methods for High Technology Ceramics, Plenum Press, New York, 1984, pp. 533-47.
41. Naramoto H, White C W, Williams J M, McHargue C J, Holland O W, Abraham M M and Appleton B R, J. Appl. Phys. 54 (1983) 683-98.
42. Burnett P J and Page T F: in G.K. Hubler, O.W. Holland, C.R. Clayton and C.W. White (Eds.), Ion Implantation and Ion Beam Processing of Materials, North-Holland, Amsterdam, 1984, pp. 401-6.
43. Arnold G W, Krefft G B and Norris C B: Appl. Phys. Lett. 25 (1974) 540-42.
44. Krefft G B and Eernisse E P: J. Appl. Phys. 49 (1978) 2725-30.
45. Hioki T, Itoh A, M. Ohkubo, Noda S, Doi H, Kawamoto J and Kamigaito O: J. Mater. Sci., in press.
46. Lawn B R and Fuller E R: J. Mater. Sci. 19 (1984) 4061-75.
47. Primak W: J. Appl. Phys. 48 (1977), 1556.
48. Hines R L: J. Appl. Phys. 28 (1957) 587.
49. Götz G: in ref. 1, Ch. 10.
50. Hines R L, Arndt R: Phys. Rev. 119 (1969) 623.

51. Geotti-Bianchini F, Lo Russo S, Mazzoldi P, Polato G: Journ. Am. Ceram. Soc. 67 (1984) 39.
52. Destefanis G L, Gailliard J P, Ligeon E L, Valette S, Farmery B W, Townsend P D and Perez A: J. Appl. Phys. 50 (1979) 7898.
53. Götz G and Karge H: Nucl. Instr. Meth. 209/210 (1983) 1079.
54. Townsend P D: Nucl. Instr. Meth. 182/183 (1981) 727.
55. Destefanis G L, Gailliard J and Townsend P D: Rad. Eff. 48 (1980) 63.
56. Naden J and Weiss B: Radio and Electr. Eng. 54 (1984) 227.

HIGH-FLUENCE IMPLANTATION IN INSULATORS.
PART II: CHEMICAL CHANGES

Jean-Claude DRAN

Centre de Spectrométrie Nucléaire et de Spectrométrie de Masse du CNRS,
B.P. n°1, 91406 Orsay, France

1. INTRODUCTION

The aim of this paper is to review some recent works dealing with chemical modifications induced in inorganic insulators by ion bombardment. Obviously chemical transformations always occur during materials processing with high fluence ion beams. However, in spite of their ubiquity, they have been rarely investigated per se and thus have received much less attention than other types of beam induced modifications such as, for instance, electronic, mechanical or optical ones. The only exception is the improvement of the corrosion resistance of metals and alloys induced by ion implantation, which has been the subject of numerous investigations. As the field of ion-induced chemical modifications is exceedingly vast, the present review does not intend to be exhaustive but will be restricted to a few examples which best illustrate this field from the standpoint of both fundamental research and potential applications. The limitation to inorganic insulators may appear arbitrary, but will be justified below.

The chemical transformations involved in ion beam processing of materials can be roughly classified in two different groups: those which are only related to energy deposition and do not depend on the nature of the incident ions (and can thus be produced by chemically active or inert beams as well), and those which strongly depend on the ions, generally because of the formation of chemical bonds between them and constituent atoms of the target. While this last class of transformations can be found during processing of any material, including monoatomic targets such as metals, the first one is far more frequent in polyatomic materials. This particularity explains the special emphasis of the present review to inorganic insulators. A second reason stems from our personal interest in this field: our work in Orsay directly or undirectly deals with chemical changes induced by ion implantation in insulators. Our experiments are not oriented toward materials processing but rather use ion implantation as a practical tool for simulating irradiation processes occurring on natural or technological materials such as extraterrestrial matter exposed to solar wind bombardment or actinide-rich minerals and nuclear glasses subjected to internal irradiation. An additional subdivision among chemical modifications can be made whether the emphasis is on the final chemical composition for itself, or on the chemical reactivity toward particular reagents with special interest to water which plays a dominant role in materials corrosion in natural environment. As compositional changes induced by ion bombardment are treated at length by several other lecturers (R. Kelly, P. Mazzoldi and E. Taglauer), the focus will be given here on the particular ones which involve the dopant in the formation of compounds.

The present review will only deal with some inorganic insulators, mostly simple oxides and mixtures of oxides such as silicate glasses. The paper will be divided into three parts: in the first one will be given the general background of chemical transformations in inorganic insulators. The

357

R. Kelly and M. Fernanda da Silva (eds.), Materials Modification by High-fluence Ion Beams, 357–383.

second one will deal with the changes in chemical composition and will be illustrated by molecular synthesis and astrophysical implications. The third one will be devoted to modifications of chemical reactivity, with special emphasis on the alteration of the aqueous corrosion of glasses and minerals in conjunction to the problem of the long-term durability of nuclear glasses and some geochemical applications.

2. OVERVIEW ON CHEMICAL MODIFICATIONS OF ION-IMPLANTED MATERIALS

Table 1 contains a list of topics selected from the recent literature, which are relevant to compositional modifications induced by ion bombardment in insulating materials. It only intends to show the extreme variety of investigated materials and scientific or technological objectives. Extensive reviews on this issue have been published (1-6).

TABLE 1. List of topics relevant to chemical changes induced by ion bombardment on insulators

Materials	Effect	Application	Laboratory
Silicate glasses	Atomic redistribution	Optics	Padova
Halides Oxides Garnets	Metallic precipitates	Solar energy Magnetism	Lyon
Silicates, Ices Frozen gases	Molecular synthesis	Exobiology	Julich, Orsay Catania, Bell Labs.
Inorganic salts	Radiolysis		Catania, Julich
Organic Organometallic compounds	Decomposition Metal substitution	Labelling	Heidelberg Julich
Oxides Silicates	Noble metal precipitates	Catalysis	Heidelberg
Polymers	Bond breaking Cross linking	Micro-electronics	Many Labs.

Figure 1 illustrates the three different chemical processes associated with ion beam processing of materials. As physicists and chemists frequently label the same process with different names, it is worth giving here the proper correspondence. As stated in the introduction, compounds undergo chemical transformations under bombardment with inert or reactive ions. In the first case, modifications are only due to radiation-induced compositional changes occurring in the collision cascade. This type of transformation is known among the chemistry community under the name of radiation chemistry, mostly investigated with ionizing radiations. In the second case, are superimposed processes involving the hetero-atom coming to rest within a damaged matrix and forming bonds with constituent atoms. Such processes are equivalent to the so-called "hot atom" chemistry induced by the recoil atom during nuclear reaction or radioactive decay. It should be pinpointed that

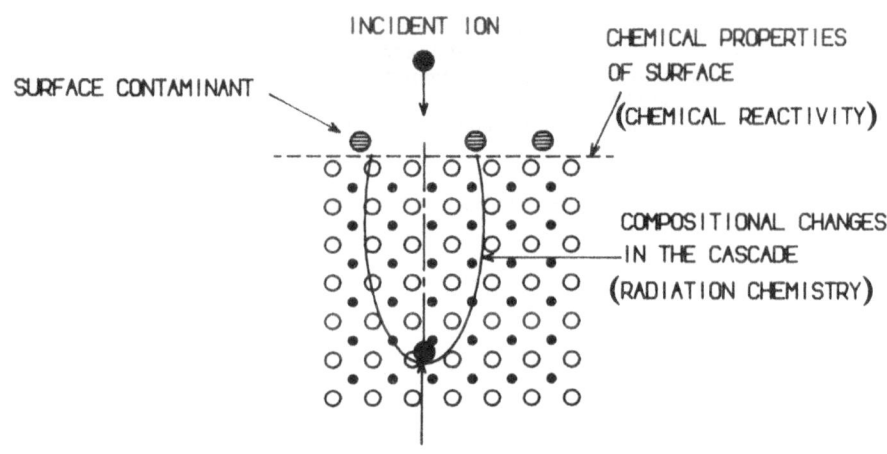

INCIDENT ION

SURFACE CONTAMINANT

CHEMICAL PROPERTIES
OF SURFACE
(CHEMICAL REACTIVITY)

COMPOSITIONAL CHANGES
IN THE CASCADE
(RADIATION CHEMISTRY)

BOND FORMATION IF REACTIVE SPECIES
(HOT ATOM CHEMISTRY)

FIGURE 1. The three types of chemical transformations induced by ion bombardment.

surface contaminants (most often O or C) knocked on by incident inert ions can also form bonds with matrix constituents. As compositional changes are treated by other lecturers (R. Kelly and E. Taglauer), only bond formation will be described. The third chemical aspect of ion bombardment of materials stems from the modified chemical properties of the surface, that is its chemical reactivity with respect to particular external reagents.

As shown by Roessler (3), bond formation by ion implantation is always accompanied by radiation chemistry which can interfere with it and thus modify primary bonds. This situation mostly occurs because experiments dealing with compound formation under implantation of stable ions, generally rely on in-situ analysis by optical spectroscopy which necessitates relatively high doses ($>10^{14}$ ions/cm^2). In contrast, those based on the nuclear recoil technique to implant radioactive ions suitable for high sensitivity radiochemical methods, can avoid such a difficulty, as the corresponding dose can be as low as 10^8 ions/cm^2.

In summary, we will review two broad categories of chemical changes induced by high fluence ion beams on insulators, namely changes in chemical composition related to bond formation between the implant and matrix constituents, and changes in chemical reactivity; for the first category, the interest is on the final chemical state of the system target-impurity at a microscopic level, whereas the second one corresponds to a macroscopic change of the chemical behavior with respect to a specific reagent and does not necessarily involve significant compositional modifications. In the following chapters some examples will be reported which are relevant to these categories of chemical modifications.

3. CHANGES IN CHEMICAL COMPOSITION : COMPOUND FORMATION WITH INCIDENT ION

As stated in the introduction, compound formation between a chemically reactive impurity and target constituent elements is not specific of insulators but occurs as well in metallic targets. An example is the formation of surficial nitride layers on nitrogen implanted metals which

are at least partly responsible for their improved mechanical properties
(7). The specificity of insulators (and semiconductors) in this area is the
possibility to use in-situ optical spectroscopic techniques to evidence
compound formation within the implanted layer. In table 2 are listed
typical insulator-incident ion systems where the creation of chemical bonds
has been evidenced (8-20).

TABLE 2. Evidence for bond formation in inorganic insulators under light
          ion implantation

Target	Ion	Bond	Techniques	References
SiO_2	H	O-H	IRS	8
SiC	H	Si-H C-H	IRS	9-10 11-18
Si_3N_4	H	Si-H N-H	IRS	14 14
Al_2O_3	H	O-H	IRS	9
TiO_2	H,D	O-H,O-D	IRS,OA	15-18
MgO	H C	O-H O-C	SIMS	19
Garnet $Y_3Fe_5O_{12}$	H,D	O-H,O-D	IRS	20

IRS: INFRA-RED SPECTROMETRY
OA: OPTICAL ABSORPTION
SIMS: SECONDARY ION MASS SPECTROMETRY

3.1. Hydrogen implantation

Numerous studies on hydrogen implantation have been performed, mainly
in connection with the question of hydrogen interaction with the first wall
of fusion reactor (21). Hydride formation has been shown to be an efficient
trapping mechanism (labelled chemical trapping) while evaporation of
volatile hydrides constitutes an erosion process (chemical sputtering).
Table 2 summarizes the main results relevant to formation of chemical bonds
in hydrogen implanted insulators. Optical spectroscopic studies on H (or D)
implanted oxides such as SiO_2, Al_2O_3 or TiO_2 have revealed the occurrence
of OH (or OD) bonds. The general trend is that the bonding probability
which is low at low ion doses, increases with the dose up to a maximum and
then decreases at high doses (10^{17}-10^{19} H cm^{-2}) due to a competition with
H_2 release. It seems that a condition for compound formation is a certain
amount of lattice disorder, whereas large damage concentration reduces
compound formation and favors clustering of the implanted ions.

Figure 2 shows the evolution of the infra-red spectrum of TiO_2 implanted with increasing doses of 50 keV deuterons (18). The absorption band
near 2510 cm^{-1} is characteristic of O-D bonds. Its amplitude increases
linearly with the ion dose up to a saturation value at $3x10^{17}$ ions cm^{-2},
which corresponds to a local D concentration of $3x10^{22}$ cm^{-3}. A correlated

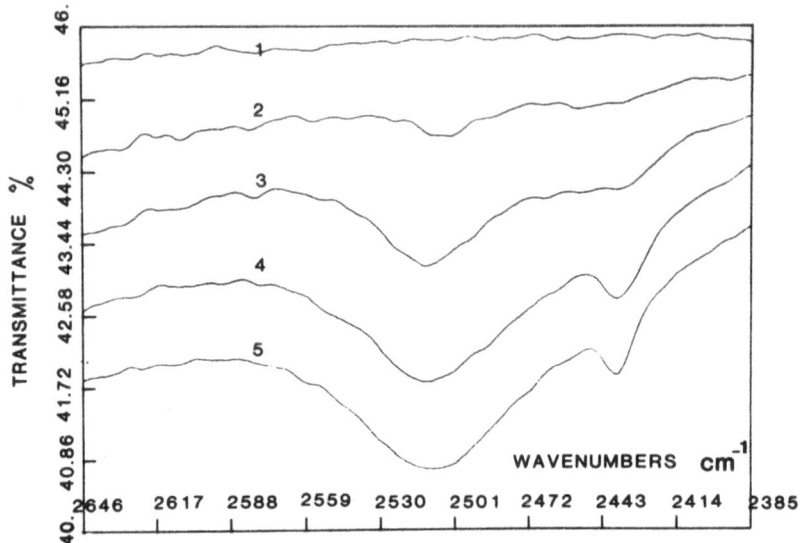

FIGURE 2. Infrared transmission spectra of rutile implanted with 50 keV D at increasing doses (18): (1) before implantation; (2) 4×10^{16} cm^{-2}; (3) 1.6×10^{17} cm^{-2}; (4) 5×10^{17} cm^{-2}; (5) 10^{18} cm^{-2}.

valence change of titanium is observed in order to preserve the electric neutrality. The Ti^{3+} ions produced are detected by their optical absorption near 1 µm. Similar studies have been performed with several other ceramics, such as SiC and Si$_3$N$_4$. For the first material, infra-red spectroscopy reveals the formation of Si-H and C-H bonds and in some instance distinct hydrocarbons (9-13). For the second one, both Si-H and N-H bonds have been evidenced (14).

In a recent study dealing with H implantation on MgO (19), indirect evidence of an O-H bond has been obtained from the stability of the depth profile of implanted H against thermal annealing, determined by SIMS.

3.2. Carbon implantation

Compound synthesis induced by C implantation in SiO$_2$ has been evidenced by means of in-situ infra-red spectrometry (22,23). Use of ^{13}C as a projectile enables to distinguish synthesized CO$_2$ from residual atmospheric one. Characteristic spectra are shown on figure 3. The main results are the following: (i) both CO and CO$_2$ are synthesized; their absorption bands differ from those of gaseous species both in shape (single band instead of double one) and position (slight shift toward lower wavenumbers). (ii) The amount of synthesized compounds depends on the defect concentration. Above the amorphization dose (5×10^{15} C cm^{-2}), about 50% of the implanted carbon has combined with oxygen to form either carbon monoxide or dioxide. For lower doses, the fraction of combined carbon is less than 10%, but can reach the same 50% value if, prior to C implantation, the target has been amorphized with, for example, a high dose of He ions (figure 4). (iii) Below a fluence of 10^{16} cm^{-2}, the major product is CO$_2$, with no detectable CO. When increasing the fluence, the amount of CO$_2$ remains constant but its production yield (number of molecules per incident C ion) decreases down to 5%, whereas that of CO increases up to 50%.

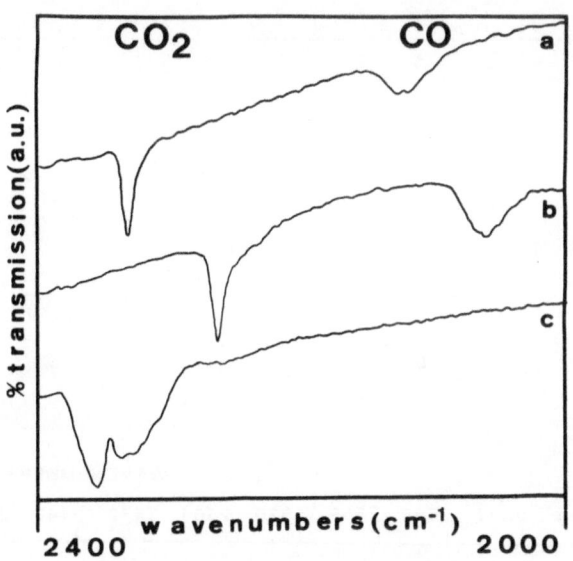

FIGURE 3. Infra   spectra of a silica film (23); (a) implanted with
5×10^{16} cm^{-2} ^{13}C;   implanted with 5×10^{16} cm^{-2} ^{12}C; (c) unimplanted
showing the doub: nd of atmospheric CO_2.

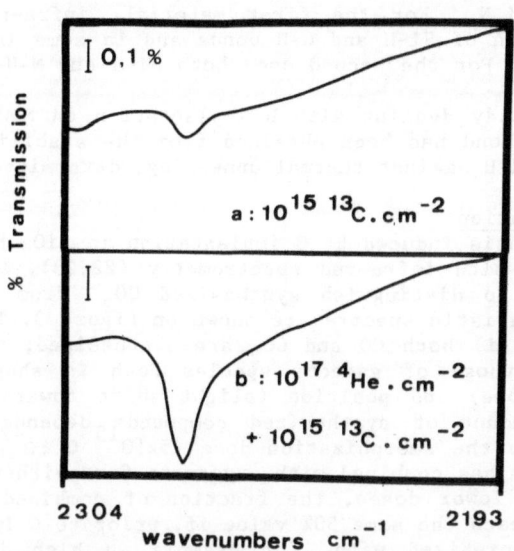

FIGURE 4. Infrared spectra of a ^{13}C implanted silica film showing the
effect of predamaging with He ions.

3.3. Molecular synthesis by successive C and H implantations

Successive implantations of C and H (or H and C) on SiO_2 lead to quite interesting features (23): (i) when H is implanted after C, the resulting I-R spectrum exhibits the same CO_2 band but a much reduced CO one (figure 5d).(ii) Such a CO disappearance does not occur when He is implanted instead of H or when H is implanted prior to C. Consequently, the CO destruction is not the result of radiation damage induced by the second implantation but rather of chemical reaction with H. This is supported by the observation of an absorption band at 3050 cm^{-2} characteristic of the C-H bond.

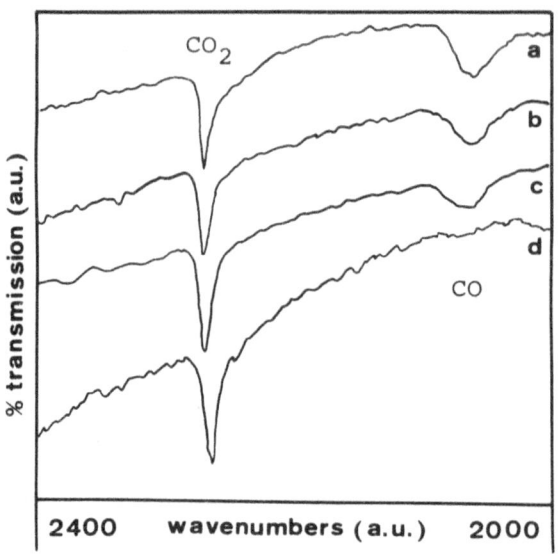

FIGURE 5. Infrared spectra of silica films subjected to various implantations (22): (a) $5x10^{16}$ cm^{-2} ^{13}C; (b) same as (a) + 10^{17} cm^{-2} He; (c) $5x10^{17}$ cm^{-2} H + $5x10^{16}$ cm^{-2} ^{13}C; (d) $5x10^{16}$ cm^{-2} ^{13}C + $5x10^{17}$ cm^{-2} H.

3.4. Astrophysical implications of molecular synthesis

Laboratory experiments on the synthesis of small molecules induced by ion implantation on insulating materials have been performed on a great variety of targets, assumed to be the main constituents of planetary surfaces, cometary materials or interstellar grains. One major aim of such experiments was to evaluate the efficiency of particle-solid interactions as compared to that of particle-gas interaction to synthesize the interstellar molecules, the number(over 60) and complexity (up to 13 atoms) of which constantly grow, as a result of the improvement of observational techniques. In fact, ion implantation processes likely occur in various astrophysical sites: (i) Within the solar system. Solar wind particles of about 1 keV/amu, mostly constituted of protons and helium, bombard unshielded planetary surfaces and interplanetary grains. Particles accelerated in planetary magnetospheres may also irradiate circumplanetary grains either in rings or on satellite surfaces; for instance the jovian satellite Io is subjected to a flux of 1 MeV protons of the same intensity as that of the solar wind on the moon surface. (ii) Outside the solar system. All types of stars appear to emit a stellar wind which can be for the massive ones much more intense and energetic than the solar wind. In addition some of them

364

likely expel solid grains within their expanding envelope (T. Tauri, novae, supernovae).

Most of the experiments have been performed on ices at liquid nitrogen temperature, in pure forms or in mixtures; such ices include H_2O, NH_3, CO_2, CH_4, etc. It is out of the scope of this article to review these works which are summarized in a contributed paper of these proceedings (25). I will only give an example of such studies, which deals with the identification of CO_2 synthesized by solar wind implantation in lunar dust grains (figure 6). A small amount (~ 1 mg) of the finest fraction (grain size ~ 1 μm) of a lunar dust sample has been imbedded into a KBr pellet and compared with similar pellets incorporating crushed quartz either unimplanted or subjected to implantation with ^{12}C ions. Spectrum (a) refers to unimplanted quartz and contains the absorption band of atmospheric CO_2, (b) is implanted quartz and exhibits the sharp peak of synthesized CO_2, and (c) is the spectrum of lunar dust showing a similar CO_2 band ascribed to solar wind induced synthesis.

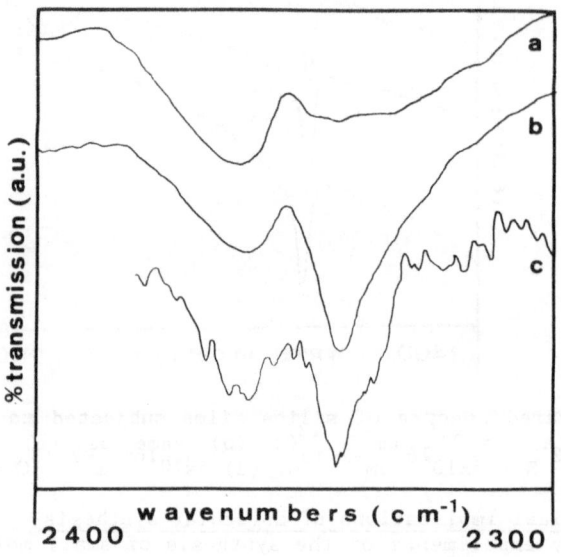

FIGURE 6. IR spectra of silicate grains (22). (a) unimplanted quartz grains observed without purging the spectrometer and showing the double band of atmospheric CO_2; (b) quartz grains implanted with ^{12}C showing the single band of synthesized CO_2 superimposed on the previous one; (c) lunar dust grains exhibiting a similar feature due to synthesis by the solar wind.

4. CHANGES IN CHEMICAL REACTIVITY : VARIATIONS OF CORROSION PROPERTIES

Ion implantation has been extensively used to modify the corrosion properties of metals (26). In the case of insulators, only a limited number of examples can be found scattered in the literature, mostly connected to ion-beam microlithography (see section 4.2.). Moreover, there is a marked difference between the two classes of materials from the standpoint of the expected effect of ion bombardment: indeed, whereas ion-beam processing of metals is oriented toward an increase in their corrosion resistance, on the contrary, ion bombardment of insulators is aimed at an enhancement of their corrosion rate. As most of the investigations in this field deal with silica-based materials in connection with aqueous corrosion problems, it is necessary to distinguish two different modes of corrosion, which can be labelled for sake of simplicity, etching (i.e. matrix destruction) and leaching (i.e. hydration and elemental release).

4.1. Variation of etch rate

Most studies of etch rate variation induced by ion implantation rely on experiments illustrated by figure 7. Polished sections partially coated with a resist are irradiated through a mask, etched with a proper reagent and then probed with a stylus device in order to measure the two step-heights h and h*, corresponding to the unirradiated and irradiated zones. Typical kinetic curves are shown and allow to measure the etch rates V and V*, the enhancement factor $K = V^*/V$ and the etchable range R_E i.e. the thickness of the layer which has an enhanced etchability.

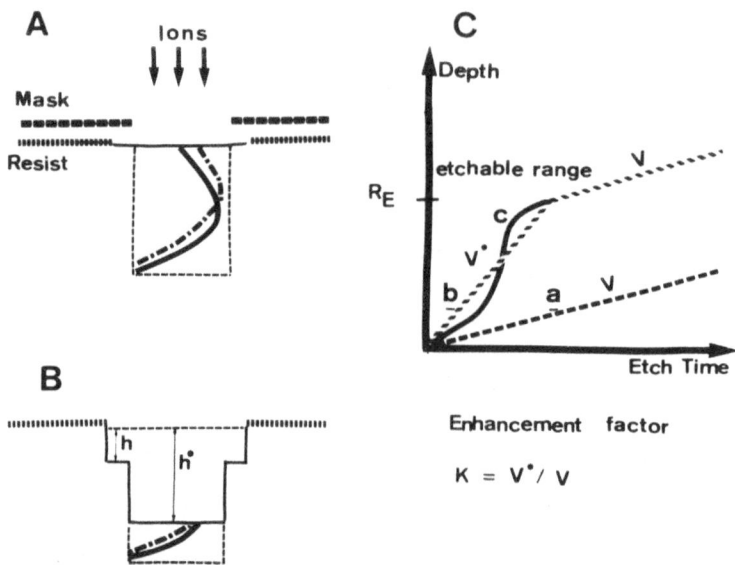

FIGURE 7. Principle of measurement of beam-enhanced etch rate.
In (A) is shown the damaged layer resulting from the implantation through the mask, with the implantation (solid) and damage (dotted) profiles. In (B) is shown the hole obtained after partial etching of the damaged layer. (C) represents the variations with etch time of the two step-heights h and h* (see text); straight line (a) shows the variations of h (etch rate V), curves (b) and (c) those of h*, with respectively a constant and a variable V*.

In table 3 are reported the main data found in recent literature and relevant to etch rate enhancement induced by ion bombardment in insulators. Most of the quoted works deal with light ions in the energy range 10-100 keV/amu where electronic interaction governs the slowing down of the incident ions in the solid. Vitreous silica is the most studied material; its etch rate with dilute HF is only moderately increased, for example by a factor of ~5 under B implantation (30), and C, N or O implantation (33). The etch rate of the implanted layer has been found to be constant along the range of the ions and correlated to the electronic stopping power (33).

TABLE 3. Examples of studies of ion-enhanced etch rate in inorganic insulators (classified in chronological order).

Target	Ion energy keV	Fluence ion cm^{-2}	Etchant	K	References
Mica Al_2O_3	Kr 1.7-8.4	$>10^{12}$	HF NaOH		27
Zircon WO_3	Kr 10	$<5\times10^{15}$	KOH		28 29
SiO_2	B 33-165	$10^{12}-10^{16}$	HF	5	30
Garnet	H, He 200	$<4\times10^{16}$	H_3PO_4	2000	31
Garnet	H,Ne,Ar,Xe 200	$>10^{12}$	H_3PO_4	1000	32
SiO_2	H 20-60 C,N,O 125-300	$>10^{12}$	HF	2 5	33
Si_3N_4	H,He,B,P,Ar 1-25	$10^{14}-5\times10^{17}$	HF	15	34
$LiNbO_3$	N 150-300	$10^{15}-3\times10^{16}$	HF	15	35
SiC	Cr 95-280	$6\times10^{15}-3\times10^{16}$	$K_3Fe(CN)$ $+KOH$	4	36
Silicate glasses	Pb,H,He 50-570	$>10^{12}$	NaCl	50	37,38

Another important material is garnet; its etch rate in H_3PO_4 is, in contrast, drastically enhanced, by a factor up to 2000 (31,32). Among other interesting features, it is worth noting a linear dependence with the irradiation dose and a correlation with the nuclear stopping power (31), as well as the occurrence of a critical fluence which depends on the atomic number of the ion (32). A previous study (27), based on Kr ions in the energy range 0.02 to 0.1 keV/amu, where atomic collisions are the dominant process of energy loss, also reported a critical fluence $\sim 5\times10^{12}$ ions cm^{-2}.

4.1.1. <u>Dose dependence</u>. We have investigated in detail the dose dependence of the etch rate enhancement factor K, by implanting a great variety of ions in different amorphous and crystalline silicates, with special emphasis to muscovite mica. This layered silicate has highly anisotropic etching properties, the etch rate in a direction perpendicular to the sheets being very small, about 1000 times smaller than that on a parallel direction. Such etching properties allow to visualize individual tracks of low energy heavy ions (39), as shown in figure 8. Consequently the experimental procedure can be simplified, as it is no longer necessary to protect part of the unirradiated zone with a resist for etch rate measurement. A typical experiment, based on the irradiation of a grid-

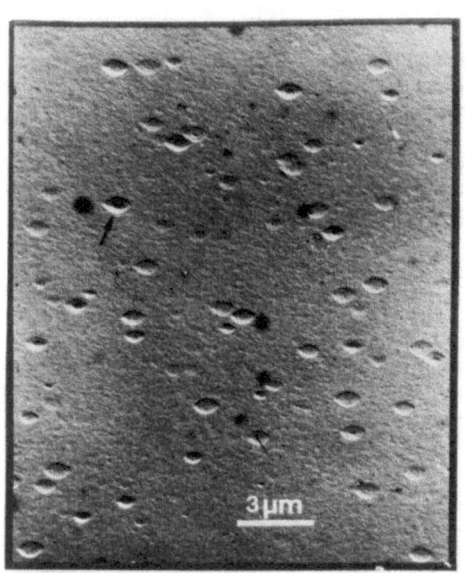

FIGURE 8. Transmission electron micrograph of etched α-recoil tracks in muscovite mica (36).

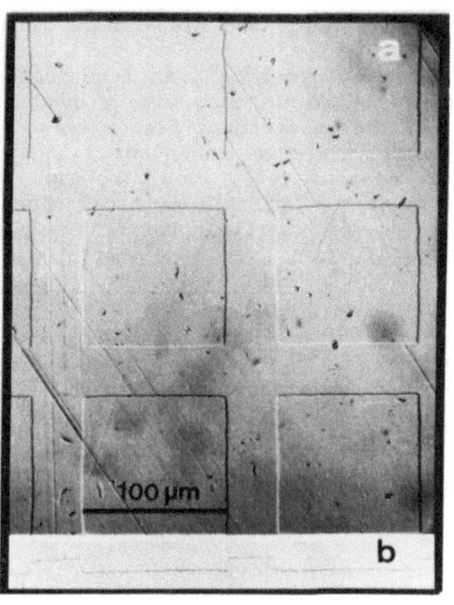

FIGURE 9. Interference contrast optical micrograph and associated step-height profile of an etched mica sample after implantation through a grid.

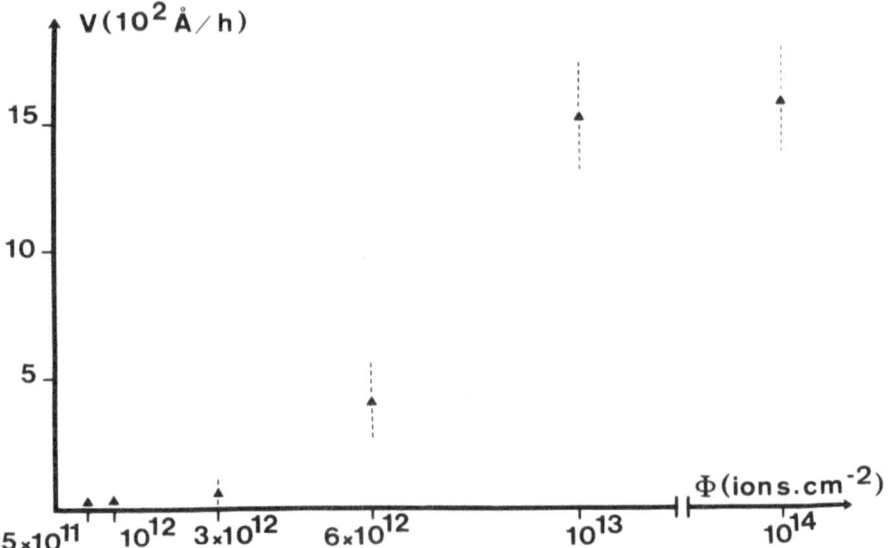

FIGURE 10. Dose dependence of the etch rate (in 4% HF at 30°C) of mica implanted with 207 keV Pb.

covered sample and step-height measurement after etching, is illustrated by
figure 9.

When plotting K in function of the ion fluence $\emptyset$, we have observed a
sigmoidal variation, with a sudden increase at a critical value $\emptyset_c$ (figure
10). The enhancement factor depends on the solid-solution system and is the
greater the more resistant is the solid towards dissolution and the milder
the etchant (40). For 1 keV/amu lead ions, $\emptyset_c$ is ~ 5×10^{12} ions cm^{-2}. We
have shown that this critical fluence strongly depends on the atomic number
of incident ions but not on their energy in the range 0.3-3 keV/amu investigated.

4.1.2. Interpretation of the critical fluence. This particular dose
dependence has been interpreted as a percolation phenomenon, on the basis
of a Monte Carlo model of etching (40), the main features of which can be
summarized as follows:
a) the volume of the solid is divided into elementary cubic cells; b) one
assumes that the collision cascade induced by each incident ion generates a
damaged island of a given shape. Such an island has an enhanced intrinsic
etch rate (v_{i^*}) when compared to that of the undamaged material ($v_i = k_1 V$).
k_1 is therefore the asymptotic value of the etch rate amplification factor,

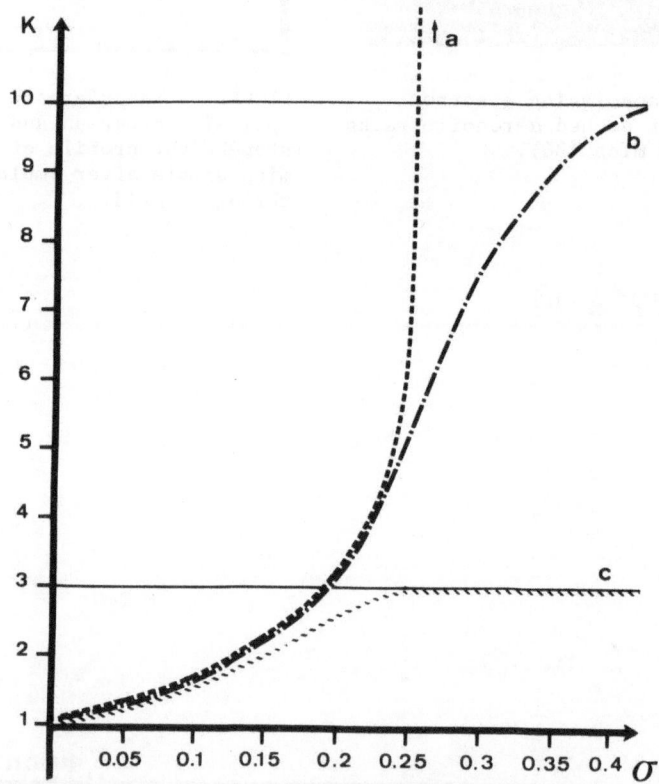

FIGURE 11. Theoretical variations of the enhancement factor K of the etch
rate versus the degree of occupancy σ of damaged islands obtained with the
Monte Carlo code for 3 values of the k_1 parameter (see text); (a): 100;
(b): 10; (c): 3.

reached when the degree of occupancy σ of the damaged islands is equal to 1. These islands are then randomly distributed in the solid volume by means of the Monte Carlo code, their number being proportional to the fluence of the incident ions. c) When going from the external surface of the solid, which contains the layer of elementary cells directly in contact with the etchant, the code acts by removing successive layers inwards to the bulk of the solid in the following way: each elementary etching step consists in removing from the layer in contact with the reagent all the undamaged cells and k_i successive damaged ones; the etching process goes on until the consumption of all available adjacent damaged cells. d) The average etch rate of the irradiated solid V and thus the K-factor can then be evaluated as a function of σ. One computes the average number of cells stripped during an elementary etch step of duration Δt, when an equilibrium has been reached after the removal of the first uppermost layers.

The main results of these calculations are reported on figure 11 which shows the theoretical variations of the mean amplification factor K of the implanted layer with the degree of occupancy σ of the damaged islands, for several values of the intrinsic amplification factor k_i. For any value of k_i sufficiently high (>10), one notes the sharp increase of K when σ is about 0.25. For smaller value of k_i, the theoretical growth curve does not show evidence of a marked discontinuity.

4.1.3. Relationship to defects. One interesting aspect of ion beam enhanced etchability is the possibility to infer information on the defect structure of the implanted layer. This is particularly useful in the case of amorphous materials for which direct defect characterization is not feasible by current techniques such as RBS associated with channelling. A pioneering work by Reid and Kelly (41) was aimed at measuring the thickness of amorphized layers by assessing the depth where the etch rate of the implanted material retrieves its original value. A good correlation between the etch rate variations with depth and the damage distribution has been found in $LiNbO_3$ implanted with N ions (35).

In an attempt to get some insight on the defect structure of implanted silicates, we have investigated the influence of thermal annealing at increasing temperatures on two parameters characterizing their etching behavior, namely the etch rate and the etchable range. The resulting variations for muscovite mica are reported on figure 12 and are in agreement with the percolation model described above. Indeed the low dose target (10^{13} cm^{-2} Pb) appears very sensitive to annealing whereas the high dose one (10^{15} cm^{-2}) is nearly unsensitive. This is due to the fact that in the former, the defect concentration remains above the percolation threshold at all temperatures, while in the latter, it rapidly decreases below the threshold. Incidentally, it should be pinpointed that for all materials investigated, the R_E value for heavy ions largely exceeds the theoretical damage range, at least by a factor of 2. This range excess issue will be discussed later.

4.2. Applications to microlithography: wet chemical etching

The interest in the ion beam modifications of the etch rate of materials was initiated by their potential applications as microlithographic techniques for the microelectronic industry. In table 4 are listed those techniques involving the use of ion beams (42,43). The first three are based on the use of a mask constituted of a fine layer of material called resist (aimed at resisting to a subsequent processing step applied to the underlying material). The resist is generally a polymer, but can be in certain circumstances inorganic. Lithographic procedures involve two steps, the drawing of a primary pattern on the resist and its transfer to the sub-

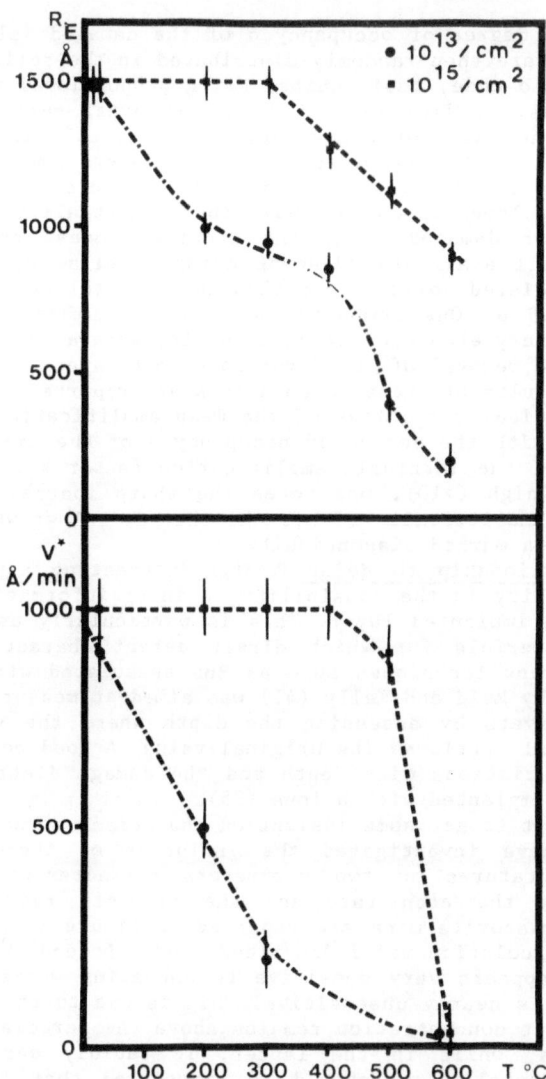

FIGURE 12. Effect of thermal annealing on the etchable range R_E and the etch rate V^* in Pb-implanted mica etched with 4% HF at 30°C.

strate. The primary pattern can be realized by various techniques, but most often by exposure of the resist to visible or UV light. Transfer of the pattern relies on several techniques, including those involving ion beams listed above, and based on chemical etching (in wet or dry environment) or sputtering.

The resist is defined with respect to the effect of exposure to photons or charged particles as positive or negative, depending on whether the exposure enhances or decreases its etch rate. The main property of resists which conditions their use is their sensitivity to radiations, defined as the dose necessary to induce a given change in the etch rate. This property

TABLE 4. Different techniques of microlithography using ion beams

---

1. Wet chemical etching
   - Ion beam enhanced etching
   - Ion beam inhibited etching

2. Plasma etching
   - Ion beam assisted

3. Reactive ion beam etching

4. Maskless ion beam etching

---

explains the preference to polymers which are highly sensitive to radiations. The pattern transfer requires maximum anisotropy and selectivity of the etching. Because of these requirements the present trend is to rely on reactive ion beam etching, instead of wet chemical etching. However, the use of wet etching on inorganic resists can be necessary in particular cases, and thus it is noteworthy to briefly report some relevant examples taken from recent literature.

Some examples of procedures involving ion-beam enhanced etching of insulating materials such as silica or garnet, have been already included in table 3.

Examples of ion-beam inhibited etching can also be quoted. Implantation of silica with a dose of 100 keV N_2^+ ions greater than 2×10^{15} cm^{-2} considerably reduces its etch rate in buffered HF (44). Annealing at 1400 K in a nitrogen atmosphere during 1/2 h suppresses the implantation effect. The origin of this protective role of implanted nitrogen is not clear yet.

4.3. Variations of hydration rate

In the case of silicate glasses, aqueous corrosion generally proceeds via a marked surficial hydration, particularly for acidic or neutral pH solutions characteristic of natural environments. Therefore, the amount and penetration depth of incorporated hydrogen constitute good corrosion monitors. Both quantities are obtained by profiling the near-surface region of the sample by means of resonant nuclear reaction analysis (RNRA) or elastic recoil detection analysis (ERDA). The interest in ion bombardment effects on hydration of glasses derives first from technological applications such as, for example, the environmental corrosion of beam processed optical fibers, or the durability of nuclear glasses, discussed in section 4.4. An additional reason stems from the potential use of incorporated hydrogen as a means to decorate ion-induced defects (45). The principle of this approach is shown on figure 13, where it is compared to that based on etch rate variations with depth. The first one (dissolution curve) relies on the stepwise removal of the investigated material and the characterization of the new surface and the second (defect decoration) on the diffusion of an active species and its fixation on defects. Most of the studies related to this issue deal with vitreous silica which only exhibits a very small and surficial hydration before ion bombardment.

4.3.1. Hydration kinetics and dose dependence. Ion-induced enhanced hydration has been investigated on silica samples implanted with a wide variety of ions, with particular emphasis on lead ions.

Hydration kinetics has revealed a dependence on the dose of 1 keV/amu lead ions (46), as shown on figure 14. For the highest dose (10^{15} cm^{-2}) and long leach times (> 1 day), one notes a flat H-profile extending to

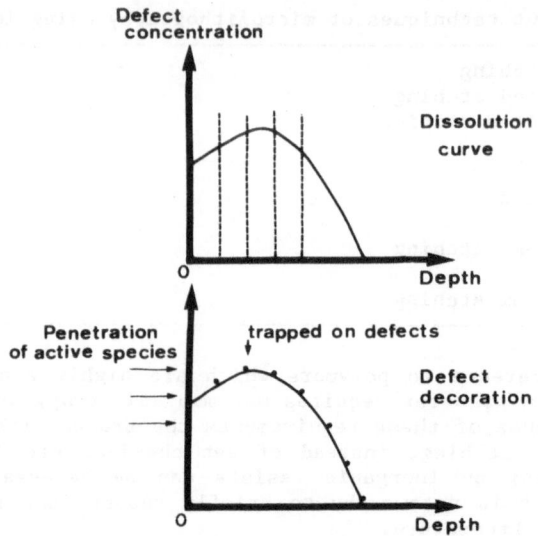

FIGURE 13. Corrosion properties of implanted materials as indirect means to study their defect structure: dissolution curve and defect decoration.

~ 1500 Å, which has to be compared to the R_p ~ 600 Å; this profile then steeply decreases to the bulk H-concentration in less than ~ 500 Å. The H-profile clearly reaches a saturation value for a leach time of 1 day. Pursuing the leach test up to 13 days, reduces the thickness of the H distribution, D_H (depth at half maximum), probably as a result of dissolution of the hydrated layer. Incidentally, this method allows the evaluation of the rate of matrix destruction. On the contrary, the 10^{13} cm^{-2}-target shows a gaussian- like profile even for the longest investigated leach time (1 day) with a maximum at about 500 Å. Hydration apparently reaches a saturation value of approximately 6×10^{21} H cm^{-3}, corresponding to a stoichiometry approaching 8 SiO$_2$ - 1 H$_2$O.

The variations with dose of the amount of incorporated hydrogen (figure 15) show the occurrence of a critical fluence (47) which is one order of magnitude higher than that measured for matrix destruction (for instance ~ 5×10^{13} ions cm^{-2} for lead ions). This peculiarity has been also reported for more complex silicate glasses (48).

The tentative interpretation of the above features was as follows. Hydrogen (in an unspecified form H$^+$, H$_3$O$^+$, H$_2$O), while penetrating in the material at an enhanced diffusion rate because of the heavily damaged structure, would react with a chemically reactive defect (possibly a non bridging oxygen) thus decorating the defect distribution. In fact the maximum of the quasi-gaussian hydration curve, observed at low dose, is very close to the calculated range of the incident ions (~ 600 Å).

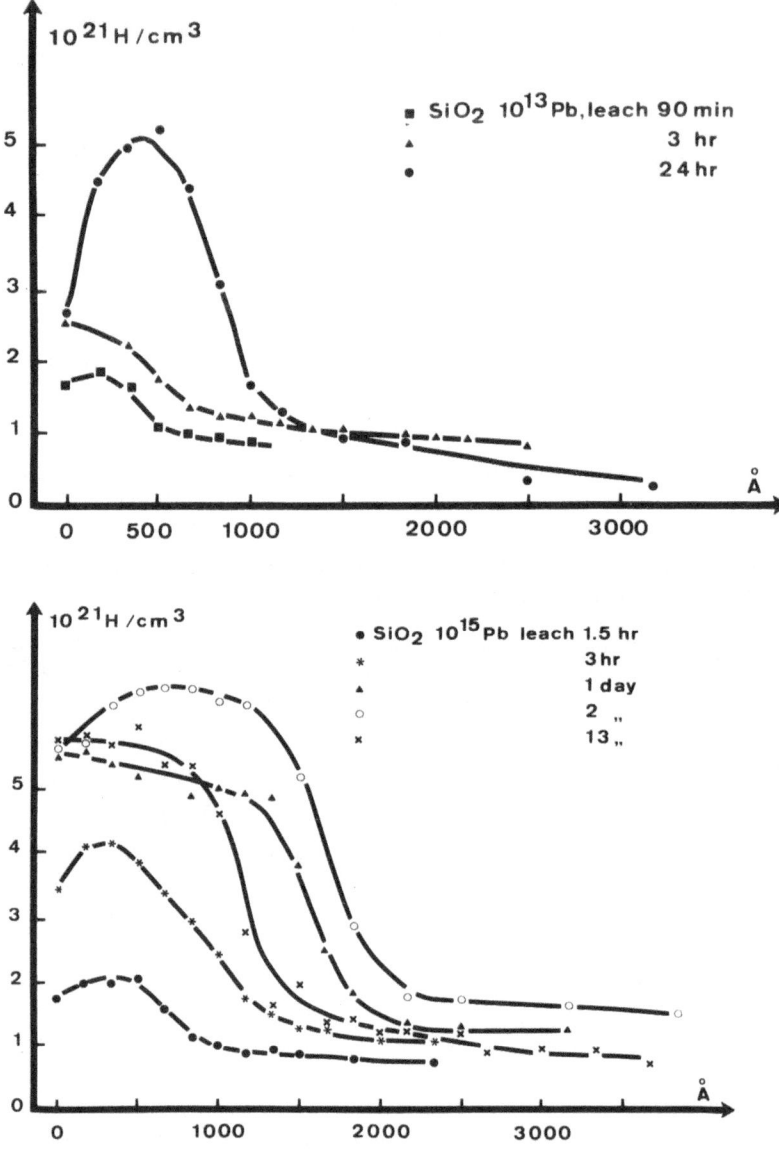

FIGURE 14. Hydrogen depth profiles of vitreous silica implanted with two doses of 207 keV Pb ions and leached in 100°C water for increasing times.

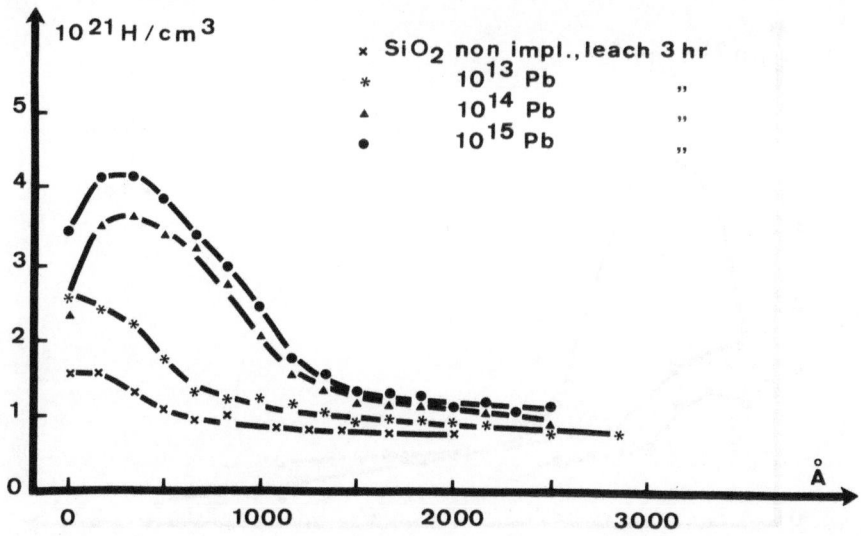

FIGURE 15. Hydrogen depth profiles of silica implanted with increasing doses of 207 keV Pb and leached in 100°C water.

4.3.2. <u>Range excess</u>. Burman et al. (49,50) were the first authors to report a range excess on the hydration of ion implanted silica. They observed that the maximum hydration depth was over twice the ion range (figure 16).

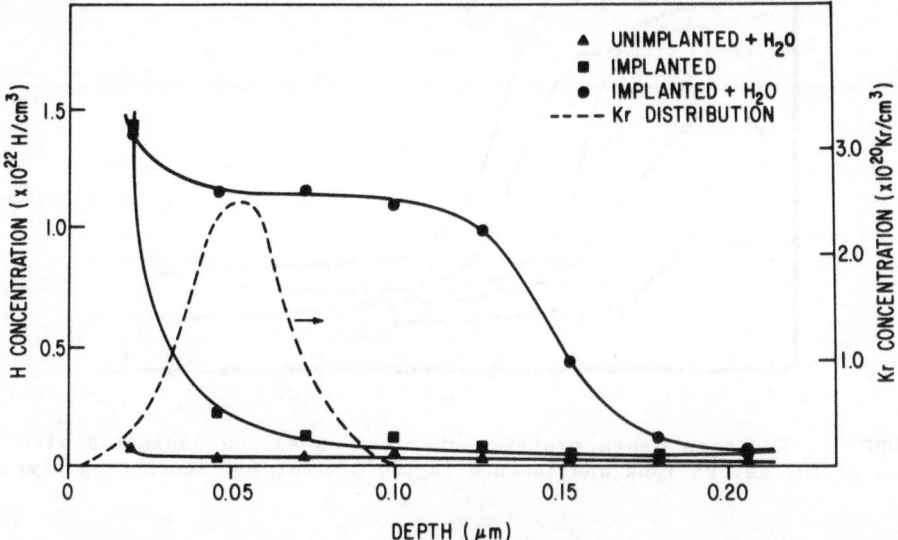

FIGURE 16. Hydrogen depth profile of vitreous silica implanted with 120 keV Kr ions and leached in 90°C water for 48 h (50).

Della Mea et al. (51) have related the enhanced H penetration to the tail of the damage distribution, in agreement with an earlier interpretation of the range excess observed upon etching Kr-implanted WO_3 (29). In fact, for implantation doses 1 or 2 orders of magnitude larger than the critical dose, the defect concentration exceeds the critical value at depths much larger than $R_D + \Delta R_D$ (respectively damage range and range straggling).

4.3.3. <u>Thermal stability of the hydrated layer</u>. The fact that hydrogen atoms are strongly bonded to the silica lattice is supported by results of post-leaching thermal annealing experiments (51). Heating the hydrated samples for 2 h at temperatures up to 300°C does not significantly alter the H-profiles (figure 17). The H-profile shows moderate modifications at 400°C (~ 20% of hydrogen being released), whereas it is markedly altered at 500°C. Similar results have been obtained with a soda-lime glass where hydrogen is much less strongly bonded to the network, its profile being already significantly modified after heating at 200–300°C.

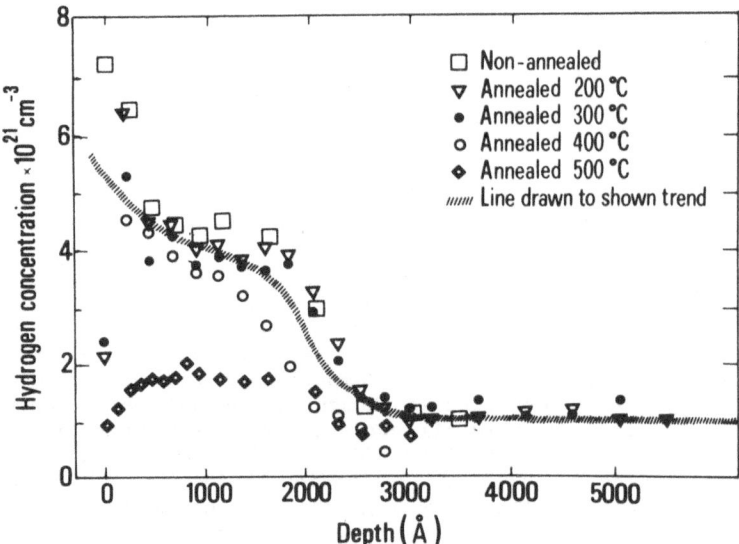

FIGURE 17. Evolution of the hydrogen depth profile of silica implanted with 10^{15} cm^{-2} 207 keV Pb ions, leached 24 h in 100°C water and subsequently heated at increasing temperatures.

4.3.4. <u>Relationship to defects</u>. The effect of thermal annealing on the hydrogen uptake has been investigated in the range 200–900°C on Pb implanted silica samples to identify the role of the different types of defects susceptible to be present (51). At low ion dose (10^{13} cm^{-2}) the annealing produces a regular decrease of both the maximum hydrogen concentration, C_{max}, and the thickness D_H of the H-profile. In contrast, the 10^{15} cm^{-2} targets are not sensitive to annealing up to 400°C, at which temperature only a slight decrease in C_{max} and D_H can be noted (figure 18). Nevertheless, there is a sharp reduction of these quantities between 400 and 500°C and a further decrease of C_{max} (but not D_H) at 600°C. These results have been interpreted in terms of the different densities of point defects and clusters of defects which are postulated to be present at both fluences.

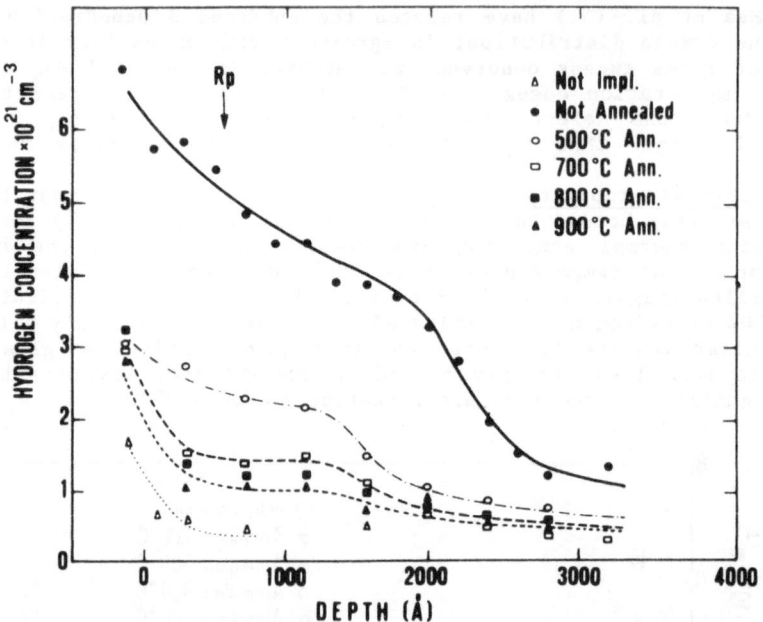

FIGURE 18. Effect of thermal annealing on the hydration of silica implanted with 10^{15} cm^{-2} 207 keV Pb ions and leached 24 h in 100°C water.

4.4. Implications for the long-term durability of nuclear glasses

The relationship between the durability of nuclear glasses and ion bombardment effects on insulators may appear not straightforward. It stems from the following facts: (i) nuclear glasses are subjected to self-irradiation with different particles among which the recoil nucleus associated with α-decay of actinides is thought to be the most damaging for the glass network (52); such a recoil nucleus can be simulated by bombardment with heavy ions of energy of about 100 keV. (ii) The property of nuclear glasses which rules the release of radionuclides is its resistance to aqueous corrosion. It is therefore legitimate to check whether or not this property is affected by the accumulation of radiation damage induced by bombardment with α-recoil nuclei. Before describing the relevant simulation experiments it is necessary to briefly describe the properties of nuclear glasses. For more detailed informations, the reader should refer to comprehensive reviews (52,53).

The sources of nuclear wastes include all the facilities of the nuclear fuel cycle, and in particular the fuel reprocessing plants in countries which have adopted this option for retrieving uranium and plutonium from the spent fuel. For an optimum management, nuclear wastes are separated into several categories on the basis of their radioactivity level and incorporated into solid matrices. The so-called high-level waste (HLW), are constituted of nitric solutions produced in the reprocessing plants and containing high quantities of fission products and actinides. After drying and calcination they are imbedded within a matrix of borosilicate glass. The choice of such a matrix is conditioned by several constraints: (i) it must accommodate a wide variety of oxides. (ii) Its fabrication does not need a too high temperature which can produce radioactive contamination by volatile fission products. (iii) The production must be feasible at an industrial level. This is actually the case in several countries and for example in France where a vitrification plant has operated since 1978 at Marcoule. An alternative HLW storage matrix has been proposed, namely a synthetic ceramic labelled "synroc" (for synthetic rock), constituted of three main mineral phases incorporating specific categories of radio-nuclides (54). These minerals are titanates and have been chosen for nuclear waste encapsulation because they occur in nature with high content of actinides and exhibit a great corrosion resistance.

The most widely accepted strategy of nuclear waste disposal relies on their burial in a deep geological repository which constitutes the ultimate barrier against dissemination of radioelements toward the biosphere, the other barriers being the waste solid form itself, the metallic container and the so-called engineered barrier which is the cavity filling material, mostly constituted of clay minerals. Because of the presence in such wastes of radiotoxic elements of very long periods, and particularly the α-decaying transuranic ones, confinement of radioactivity should be guaren-teed for time spans exceeding several thousand years. The most likely scenario for radioactivity dissemination implies water intrusion into the repository and aqueous corrosion of nuclear glasses. The above considera-tions thus explain why one is so much concerned by the resistance of nuclear glasses against corrosion and by the potential influence of self-irradiation. However, in addition to damaging the glass structure by atomic displacements, self-irradiation can induce several other types of detrimen-tal effects, listed on table 5 and linked to ionizing radiations.

TABLE 5. Potential radiation effects in a nuclear waste repository
----

1. Damage on glass network
   - Displacement damage by alpha-recoils → changes in density,
     etchability, mechanical properties, recrystallization.
   - Ionization damage by alpha, beta, gamma radiations → radiolysis.

2. Radiolysis of ground water
   (mostly by gamma radiation)
   - decreases pH → increases etch rate;
   - oxidizes heavy elements → increases their mobility.

Radiation effects due to α-decay in nuclear glasses have been assessed by means of two types of simulation experiments, one based on bombardment with external beams of heavy ions and the other relying on the incorpora-tion of α-emitters of relatively short periods, as compared with the actual transuranic radionuclides of concern. The use of external ion beams has been criticized on grounds of both the anisotropic irradiation and the extremely high dose rate (10^8 to 10^{10} times the actual value in real nuclear glasses). For both types of experiments, two major difficulties hamper a reliable prediction of the long-term durability of nuclear glasses with respect to aqueous corrosion: (i) a significant parameter which weights the progress of the corrosion reaction should be defined. For instance, it turns out that the weight loss which potentially measures the amount of destroyed matrix is not a good monitor. In fact the release rate of radionuclides of concern appears as the parameter of most interest. So is also the hydration rate which is thought to be the first step of corro-sion. (ii) The kinetic law for such a corrosion is quite difficult to establish and thus it is nearly impossible to extrapolate laboratory data of short-term leach tests (at most a few years) to actual corrosion processes occurring during the disposal period (up to a million years). With these constraints in mind, let us go through the main results of the two types of simulation reported above.

In an early work (37), we investigated the effect of lead ion bombardment on the dissolution rate of various glasses with a concentrated sodium brine; this rate was measured by a step-height technique. In all cases we found an enhanced etch rate for an ion dose above the same critical value as reported in section 4.1. However, the maximum enhancement factor markedly depended on the glass composition, being of the order of 20 for pure silica and only 2-3 for simulated nuclear glasses. Arnold and coworkers (48,55) evaluated the effect of ion bombardment on the corrosion of a simulated nuclear glass by comparing the H depth profiles on samples unimplanted and implanted with increasing Pb ion doses. The results (figure 19) clearly show a threshold dose dependence, with a critical dose of about 5×10^{13} ions cm^{-2}, i.e. the same value as that we have reported for hydration of Pb-implanted silica. Both the penetration depth and the total amount of hydrogen is markedly increased.

Leach tests conducted with borosilicate glasses doped with short-lived α-emitters also indicated a moderately enhanced leach rate for the incorporated radionuclide. For example a glass doped with ^{238}Pu only exhibited a two fold increase after a dose of α-decay of 5.6×10^{18} g^{-1} i.e. largely exceeding that corresponding to the critical fluence of Pb ions (10^{18} g^{-1}).

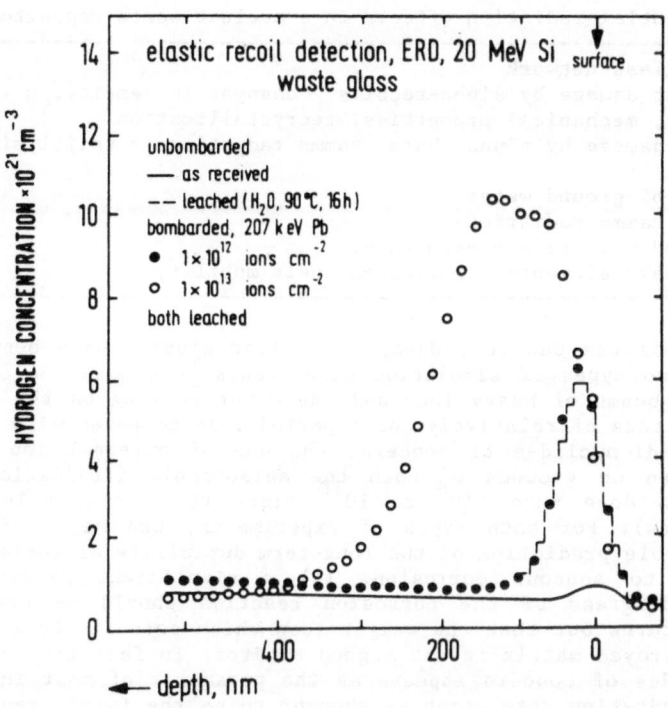

FIGURE 19. Hydrogen depth profiles obtained by ERDA on nuclear glass samples implanted with 207 keV Pb and leached in 90°C water (55).

The mechanical properties of nuclear glasses, such as hardness and fracture toughness could also influence their durability, for example by increasing the specific surface area as a result of intense fracturing, thus enhancing the surface exposed to water attack. Similarly density changes could induce detrimental stresses. Radiation effects on these properties have been investigated mostly with doping techniques. The results generally indicate an increase of the fracture toughness with damage level and very small density changes (<0.6%), with the occurrence of either swelling or shrinkage, depending on the glass composition (53).

In contrast, radiation effects associated with α-decay seem much more severe on crystalline waste forms. Swelling exceeding 5% has been observed on the zirconolite component of synroc after self-irradiation with incorporated α-emitters. Evidence of a marked increase in leach rate has also been reported.

Electron microscope studies have evidenced the formation of oxygen bubbles under electron bombardment as a result of radiolytic decomposition of the glass associated with ionization processes (56). However, the dose necessary (> 10^{12} rads) excluded the occurrence of such a phenomenon at a significant level, under the sole action of β-decay of fission products incorporated in the nuclear glass. In contrast, recent results (57) indicate a much greater yield of gamma radiation for producing such oxygen bubbles (threshold dose of about 2×10^8 rads). Current investigations are intended to assess the possible consequences on the glass durability.

4.5. Implications for uranium geochemistry and prospecting

It is worth mentioning in this review an original application of ion bombardment enhanced reactivity to the field of Earth Sciences. It stems from the consideration that in actinide- rich minerals, the accumulation of radiation damage associated with α-decay should produce a much larger increase in their corrosion rate than that predicted for nuclear glasses and therefore could be responsible for an enhanced release of uranium from rocks. Such an argument sheds new light on the question of the origin of uranium in ore bodies associated with crystalline rocks, which is still debated.

Exploitable concentrations of uranium which occur during the geochemical cycle of this element, require several favorable conditions among which the leachability of uranium is a key parameter. These conditions can be fulfilled in crystalline formations, which are the direct source of numerous uranium ore bodies. The postulated mineralization mechanism is shown on figure 20: it is based on the solubilization of uranium at the hexavalent state, the transport and the precipitation of this element in a reducing trap. However, not all crystalline formations are mineralized in uranium and it is therefore necessary to use various guidelines for uranium prospecting. When assessing the respective contribution of major and minor minerals on the basis of their dissolution rates, the possible radiation-enhancement of these rates due to α-decay should not be overlooked. For studying this effect, Petit et al. (58) have implanted a variety of minor minerals (allanite, monazite, zircon, sphene, apatite, xenotime, uraninite, thorianite) with low energy lead ions (~ 1 keV/amu) with doses ranging from ~ 10^{10} to ~ 10^{15} ions cm^{-2}. They have evidenced a marked radiation-enhanced dissolution for most minerals and etching conditions at the same critical fluence $\emptyset_c$. Indeed for monazite which is particularly corrosion resistant, amplification factors of at least ~ 10^4 have been measured in acidic solutions.

Simple calculations show that the critical dose of α-recoils is quite easily reached in minor minerals of crystalline rocks, providing that they

380

have both sufficient age and concentration of uranium and thorium. For
example, for monazite with a U + Th concentration ~ 2% an age of ~ 300
million years allows the system to reach $\emptyset_c$. Due to this phenomenon,
although originally highly insoluble in usual environmental conditions,
these minerals could be a source of uranium for ore bodies associated with
crystalline formations. Therefore, the possible occurrence of such a
radiation-enhanced dissolution could be used as a guideline for the pro-
spection of favorable crystalline rocks with respect to the formation of
uranium ore bodies.

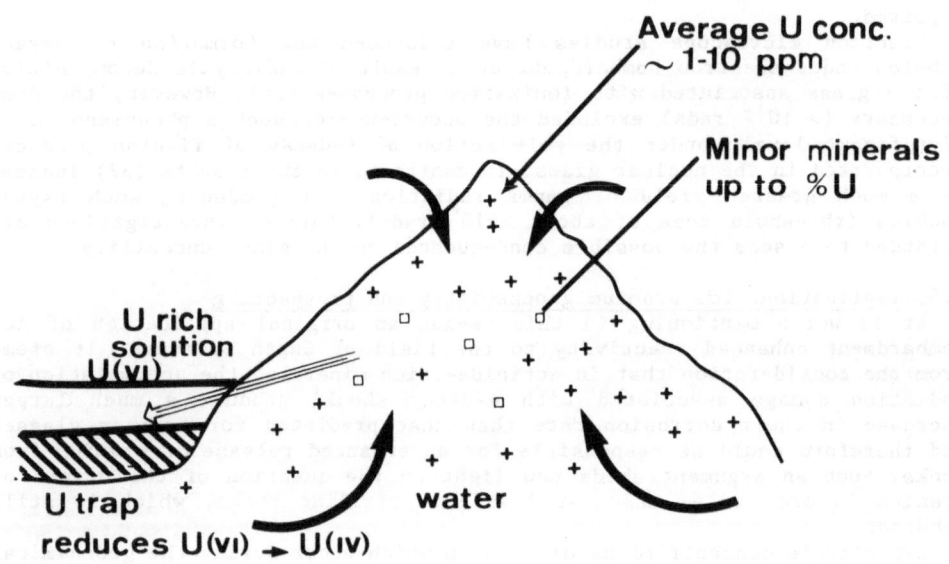

FIGURE 20. Principle of uranium ore formation based on radiation enhanced
release from minor minerals of granite.

5. SUMMARY AND CONCLUSIONS

Ion-beam processing of materials exhibits several chemical aspects which can be roughly classified in changes of their chemical composition or of their chemical reactivity. Such beam induced chemical effects have been particularly investigated on inorganic insulators, due in great part to the wide variety of analytical techniques which can be implemented on such materials.

In the first part of this review, I have described some recent works dealing with compound formation with light reactive incident ions such as hydrogen and carbon and based on in-situ infrared spectrometric measurements. The dose dependence of the yield of synthesized compounds indicates a marked influence of defect creation. These experiments have some astrophysical relevance, in particular on the origin of extra-terrestrial organic molecules.

The second part is devoted to studies of ion-beam modifications of the corrosion properties of various materials, with emphasis to silicate glasses. In the first type of corrosion regime of these materials (etching), characterized by matrix destruction, the rate enhancement is strongly dependent on the solid-etchant system. The influence of various implantation parameters has been investigated. In particular, a significant range excess has been evidenced for heavy incident ions. In addition, the sigmoidal dose dependence of the etch rate has been interpreted by means of a Monte Carlo code, as a percolation phenomenon induced by damage accumulation. Possible implications on microlithography have been briefly discussed. The second corrosion regime (leaching), where hydration is the dominant process, has been illustrated by the behavior of ion-implanted vitreous silica. A link between hydrogen depth profile and damage distribution has been demonstrated and could provide a useful means for damage study in amorphous materials (defect decoration). The same range excess has been evidenced for heavy ions but the critical dose for the enhancement of the hydration rate is apparently an order of magnitude higher than for the etch rate. The consequences of these data for the important issue of nuclear waste disposal have been summarized and an original application to an aspect of uranium geochemistry has been inferred, namely the formation of uranium ores by radiation enhanced release of this element from specific minerals.

This review did not intend to be exhaustive; its sole ambition was to discuss on a new basis the chemical aspects of ion implantation on inorganic insulators and to show the links existing among studies which have quite different objectives.

REFERENCES

1. G.K. Wolf: Treatise Mat. Sci. and Technol. 18 (1980) 373, J.K. Hirvonen ed., Academic Press.
2. G.K. Wolf: Rad. Effects, 48 (1980) 237.
3. K. Roessler: Rad. Effects, 505, 1986.
4. G.K. Wolf and K. Roessler: in "Ion beam modification of insulators", P. Mazzoldi and G.W. Arnold ed., Elsevier, (1987) pp. 558.
5. A. Pérez: Nucl. Instr. Meth., B1 (1984) 621.
6   A. Pérez and P. Thévenard: in "Ion beam modification of insulators", P. Mazzoldi and G.W. Arnold ed., Elsevier, (1987) pp. 156.
7. A. Antilla, J. Keinonen, M. Uhrmacher and S. Vahvaselka: J. Appl. Phys., 57 (1984) 1423.
8. F. Rocard and J.P. Bibring: Phys. Rev. Lett., 48 (1982) 1763.
9. D.M. Gruen, R. Varma and R.B. Wright: J. Chem. Phys., 64 (1976) 5000.
10. D.M. Gruen, B. Siskind and R.B. Wright: J. Chem. Phys., 65 (1976) 363.
11. R.B. Wright, R. Varma and D.M. Gruen: J. Nucl. Mater., 63 (1976) 415.
12. W.J. Choyke, L. Patrick and P.J. Dean: Phys. Rev. B, 10 (1974) 2554.
13. R.B. Wright and D.M. Gruen: Rad. Eff., 33 (1977) 133.
14. H.J. Stein: Appl. Phys. Lett., 32 (1978) 379.
15. R. Roy and R.T. Greer: Solid State Commun., 5 (1967) 103.
16. B. Siskind, D.M. Gruen and R. Varma: J. Vac. Sci. Technol., 14 (1977) 537.
17. M. Guermazi, P. Thévenard, P. Fais, M.G. Blanchin and C.H.S. Dupuy: Rad. Effects, 37 (1978) 99.
18. M. Guermazi, P. Thévenard, R. Brenier, J.P. Thomas and J.M. Mackowski: Nucl. Instr. Meth., B19/20 (1987) 912.
19. U. Knipping, I.S.T. Tsong and G.W. Arnold: Rad. Effects, 97 (1986) 209.
20. G. Marest, A. Perez, P. Gerard and J.M. Mackowski: Phys. Rev. B., 34 (1986) 4831.
21. S.T. Picraux: Treatise Mat. Sci. and Technol., 18 (1980) 135, J.K. Hirvonen ed., Academic Press.
22. Rocard: Thèse de doctorat d'Etat, Université de Paris XI (1986).
23. J.P. Bibring, J. Chaumont and F. Rocard: Nucl. Instr. Meth., B1 (1984) 628.
24. J.P. Bibring and F. Rocard: Rad. Effects, 65 (1982) 399.
25. J. Benit, J.-P. Bibring and F. Rocard: these proceedings.
26. V. Ashworth, R.P.M. Procter and W.A. Grant: Treatise Mat. Sci. and Technol., 18 (1980) 175, J.K. Hirvonen ed., Academic Press.
27. C. Jech: Phys. Stat. Sol., 21 (1967) 481.
28. C. Jech and R. Kelly: J. Phys. Chem. Solids, 31 (1970) 41.
29. N.Q. Lam and R. Kelly: Can. J. Phys., 50 (1972) 1887.
30. F.N. Schwettmann, D.J. Dexter and D.F. Cole: J. Electrochem. Soc., 120 (1973) 1566.
31. W.A. Johnson, J.C. North and R. Wolfe: J. Appl. Phys., 44 (1973) 4753.
32. T. Tsurushima and H. Tanoue, in O. Meyer, G. Linker and F. Kapler (Ed.), Ion beam surface layer analysis Vol. 2, Plenum, New York, 1976, pp.685.
33. A.P. Webb, A.J. Houghton and P.D. Townsend: Rad. Effects, 30 (1976) 177-182.
34. P.D. Parry and S.P. Bristol: J. Vac. Sci. Technol., 15 (1978) 664.
35. G. Gotz and H. Karge: Nucl. Instr. Meth., 209/210 (1983) 1079.
36. C.J. McHargue, M.B. Lewis, J.M. Williams and B.R. Appleton: Mat. Sci. Engin., 69 (1985) 391.
37. J.-C. Dran, M. Maurette and J.-C. Petit: Science, 209 (1980) 1518.

38. J.-C. Dran, M. Maurette, J.-C. Petit and B. Vassent, in: J.G. Moore (Ed.), Scientific Basis for Nuclear Waste Management, Vol. 3, Plenum, New York, 1981, pp. 449–456.
39. J. Borg, J.-C. Dran, Y. Langevin, M. Maurette, J.-C. Petit and B. Vassent: Rad. Effects, 65 (1982) 413.
40. J.-C. Dran, Y. Langevin and J.-C. Petit: Nucl. Instr. Meth., B1 (1984) 557.
41. I. Reid and R. Kelly: Rad. Effects 17, (1973) 253.
42. O. Auciello in: Ion bombardment modifications of surfaces, O. Auciello and R. Kelly ed., Elsevier (1984), p. 435.
43. I. Adesida: Nucl. Instr. Meth., B7/8 (1983) 923.
44. Z. Znamirowski and J. Martan: Nucl. Instr. Meth., B7/8 (1985) 920.
45. G. Della Mea, J.-C. Dran, J.-C. Petit, J. Chaumont, G. Bezzon and C. Rossi-Alvarez, in: P. Mazzoldi (Ed.), Proc. Mat. Res. Soc. Symp. Induced defects in insulators, Les [ditions de physique, Paris, 1984, pp.135–140.
46. G. Della Mea, J.-C. Dran, J.-C. Petit, G. Bezzon and C. Rossi-Alvarez: Nucl. Instr. Meth., 218 (1983) 493.
47. J.-C. Petit, Th]se de doctorat d'Etat, Universit[de Paris XI (1982).
48. G.W. Arnold, C.J.M. Northrup and N.E. Bibler, in: W. Lutze (Ed.), Mat. Res. Soc. Symp. Proc., Vol. 11, Elsevier, New York, 1982, pp. 357.
49. C. Burman, Wang Ke-Ming and W.A. Lanford, in: S.V. Topp (Ed.), Mat. Res. Soc. Symp. Proc., Vol. 6, North Holland, New York, 1982, pp. 641.
50. C. Burman and W.A. Lanford: J. Appl. Phys., 54 (1983) 2312.
51. G. Della Mea, C. Rossi-Alvarez, G. Mazzi, G. Bezzon, J. Chaumont, J.-C. Dran, M. Mendenhall and J.-C. Petit: Nucl. Instr. Meth., B19/20 (1987) 943.
52. W.G. Burns, A.E. Hughes, J.A.C. Marples, R.S. Nelson and A.M. Stoneham: J. Nucl. Mater., 107 (1982) 245.
53. Hj. Matzke in "Ion beam modification of insulators", P. Mazzoldi and G.W. Arnold ed., Elsevier, (1987) pp. 501.
54. A.E. Ringwood: Mineral. Mag., 49 (1985) 159.
55. G.W. Arnold and J.-C. Petit: Nucl. Instr. Meth. 209/210 (1983) 1071.
56. J.F. De Natale and D.G. Howitt: Rad. Effects, 74 (1983) 151.
57. W.J. Weber: Nucl. Instr. Meth. (1988) in press.
58 J.-C. Petit, J.-C. Dran and Y. Langevin, Geochim. Cosmochim. Acta, 49 (1985) 871.

ELECTROCHEMICAL AND CORROSION BEHAVIOUR OF ION AND LASER➤BEAM MODIFIED
METAL SURFACES

P.L. BONORA[*], L. FEDRIZZI[**]

[*] University of Trento, Faculty of Engineering, Material Branch, and IRST
[**] Institute of Scientific and Technological Research (IRST)

1. INTRODUCTION

Engineering components must usually possess appropriate bulk
characteristics together with surface properties necessary for their
efficient behaviour over the required lifetime. In particular, corrosion
and wear resistance, together with the frictional properties, are surface
characteristics of great importance in the engineering applications. But it
is often impossible to obtain the necessary bulk and surface properties if
the component is constructed from a single material, and where such a
solution is possible, it is usually uneconomic. Thus an interesting
solution is to coat the bulk material with another one able to provide the
required surface characteristics. There are many different methods for
surface modification or, alternatively, coating of materials. Ion
implantation has proved to be very effective in the production of compounds
with predetermined composition, whose characteristic are interesting for
the improvement of surface properties (1,2,3).

Ion implantation may improve the resistance of metals to wear and
corrosion both :
1) by the formation of stable or metastable chemical compounds localized in
   a thin external layer or
2) by the so-called "radiation damage" which is mainly active in lowering
   the potential gradients between different grains and between grain and
   grain boundaries as an effect of subdivision into substructures, up to a
   quasi-amorphous inert surface.

Laser irradiations are also able to modify metal surface behaviour. In
particular the electrochemical parameters can be drastically changed by the
formation of external amorphous layers or by the formation of a rough and
thermally stressed surface as a consequence of melting and resolidification
(4,5).

In order to get thicker modified layers and higher implanted atom
concentrations, new processes are being developed, which involve both film
deposition and ion implantation. One of such processes is ordinary ion
mixing, where a coating is first applied and subsequently bombarded with an
ion beam (6,7,8). This process can be further extended by multiple
sequential deposition/implantation steps thus making the depth of the
treated region nearly independent of the projected range of the implanted
ions. Moreover, the ion beam plays a role in increasing the adhesion and

R. Kelly and M. Fernanda da Silva (eds.), Materials Modification by High-fluence Ion Beams, 385–402.

homogeneity of the deposited film, and in converting it into a chemically different material (9).

The electrochemical and corrosion behaviour of metal surfaces modified by direct energy beams is a research subject widely studied in our laboratory during the last years using surface microanalysis techniques, scanning, electron microscopy (SEM) observations and electrochemical techniques (10,11,12,13,14,15,16). Such last methods have contributed immensely to the understanding of all types of aqueous corrosion. In fact the possibility of measuring and controlling the electrochemical potential and the current density of a metal electrode, offers a fast and reliable approach for characterizing a corroding system (17,18,19).

Now we are going to explain our results obtained on iron and steel by using ion and laser beams. The changes in the corrosion and in the corrosion fatigue behaviour of iron as well as in its catalytic properties as a consequence of ion implantation are considered. The electrochemical behaviour of laser irradiated and ion mixed metals in terms of changed corrosion rate, corrosion morphology, passivability or breakdown of passivity is also explained.

2. RESULTS

2.1. Corrosion behaviour of implanted iron

Many experimental results (10,11,13) have shown that ion implantation improves the corrosion resistance of metals : lowering the corrosion rate both in acidic and in near neutral solutions, modifying the corrosion morphology from a localized form of attack towards a more uniform dissolution, and hindering pitting corrosion under passivating conditions, by improving the protective properties of the surface oxide layers.

Nitrogen-implanted iron at different implantation doses was studied in sulphuric acid at two pH values (0 and 3.5), under controlled fluid-dynamic conditions (Fig. 1 and 2). Anodic current displays in both environments lower values for the implanted samples, with a minimum at the 10^{15} N $\cdot$ cm^{-2} dose. A possible explanation for such a behaviour is : the ion implantation "per se" causes a rearrangement of the metal surface structure, leading to a lowering of the electrochemical potential difference between grains and grain boundaries and hence to a more uniform and not intergranular attack. In neutral electrolytic solutions, and in air, many experimental results support the assumption that nitrogen implantation causes the growth of thin and protective iron oxide surface layers; in neutral media the comparison of breakdown times of passive films in the presence of chloride ions shows that the induction times before passivity breakdown are 10 up to 100 times longer on implanted samples than on pure iron; pits are shallow, randomly nucleated and distributed.

But our experimental results show the possibility of a radiation damage effect superimposed on a chemical one.

In order to separate completely the chemical modifications from the physical ones, Ar$^+$ implantation into ARMCO iron at different doses was also performed. The samples implanted with 10^{15} Ar$^+$ ions/cm^2 show a very low initial reactivity in a neutral Na$_2$SO$_4$ solution, but for all the treated

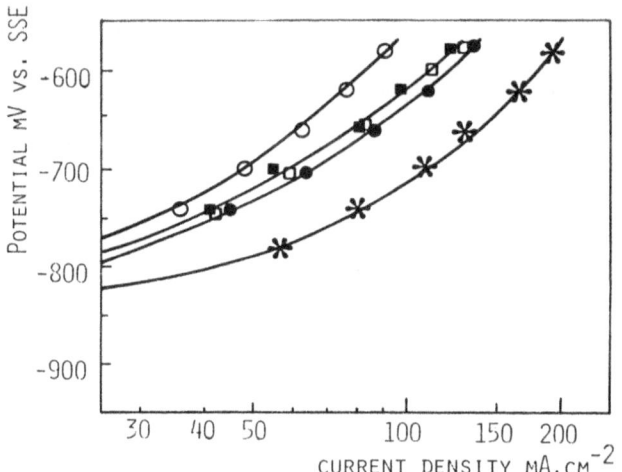

Fig. 1 : Potentiokinetic polarization curves in 1M H_2SO_4 (pH=0.03) on
rotating disk electrode (SSE = saturated sulphate electrode)
(*) unimplanted ARMCO iron, (■) $5.10^{14} N^+/cm^2$
(o) $10^{15} N^+/cm^2$, (▫) $10^{16} N^+/cm^2$, (●) $10^{17} N^+/cm^2$

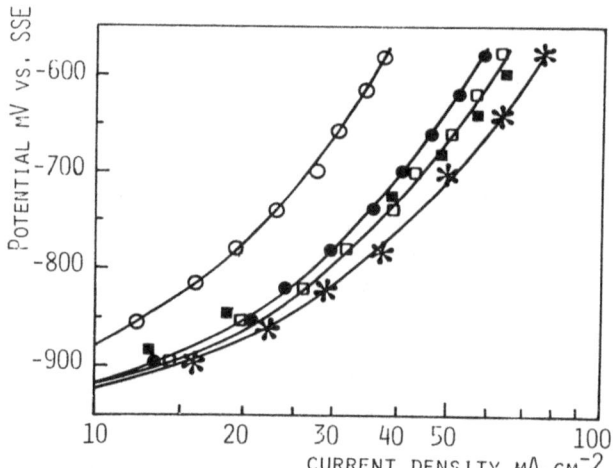

Fig. 2 : As in Fig. 1 but in 1M Na_2SO_4 (pH = 3.5)

samples, after about 10 h, the corrosion rates were the same (Fig. 3).

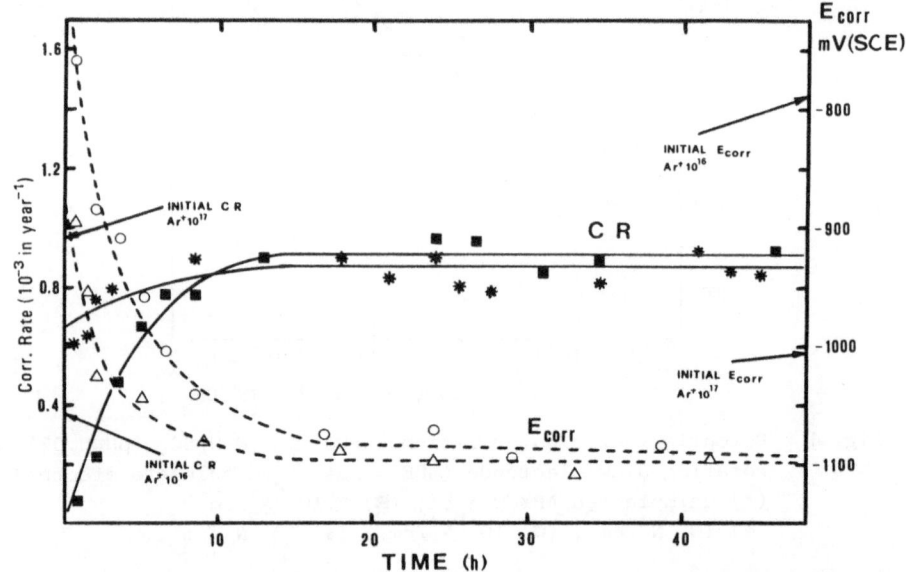

Fig. 3 : Plot of calculated corrosion rate (C.R.) vs. time (——) and E_{corr} vs. time (---) for untreated ARMCO iron (*,Δ) and for iron implanted with 10^{15} Ar^+/cm^2 (■,o) in a 0.3% Na_2SO_4 solution.

The improvement in the initial electrochemical behaviour shown by the samples implanted with lower doses seems to be due to more stable oxide layers which have been "ion beam mixed" by Ar^+ ions rather than to structural modifications which are very small because of the weak action exerted by this ion.

These conclusions are confirmed by pitting induction time measurements (Fig. 4). Under the same experimental conditions the pitting induction time is markedly longer for nitrogen implantation. Such results confirm the critical importance of the choice of the implanted element in order to obtain an improved corrosion behaviour for what concerns both the structural modifications and particularly the formation of compounds which can alter the chemical surface reactivity.

2.2. Corrosion fatigue behaviour of implanted steel

The corrosion fatigue behaviour of nitrogen implanted steel in a neutral solution of lower conductivity (0.3% Na_2SO_4 in pure water) has been investigated utilizing an oscillating cantilever beam method, by means of

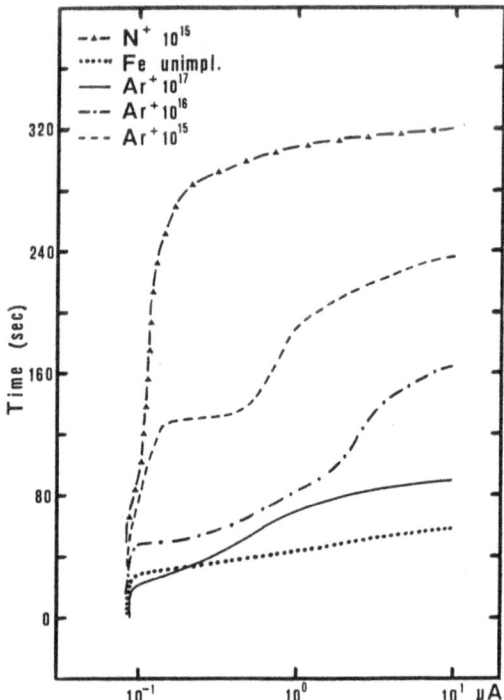

Fig. 4 : Induction time to pitting in a borate buffer solution (pH=8.7) in the presence of 5000 ppm Cl⁻.

electrochemical measurements and SEM techniques (20). In this testing device the specimen acts as a cantilever beam shaped as a rectangular blade. Its free end is submitted to a sinusoidal displacement and consequently the section close to the mounting, which is the implanted area, is subject to the highest sinusoidal stress. Contrary to the previous example, in this case the surface treatment worsens the behaviour of iron. In fact, ion bombardment lowers the fatigue corrosion life, as is shown in Table I, where the number of cycles before cracking is listed, at two frequency values and three implantation doses. The effects of increasing the implanted dose are quite evident, as well as the pure mechanical action of the higher frequency imposed, which reduces, by almost one order of magnitude, the service life extent.

Detailed information on the electrochemical behaviour of implanted and untreated samples can be obtained from Fig.5 where the values of the corrosion current (I_{corr}) calculated through polarization resistance (R_p) measurements carried out during cyclic loading are plotted. It is possible to discern three zones :

- first zone, corresponding to stagnant conditions, where the corrosion rate is higher for pure than for implanted iron;

- second zone, starting from the instant at which the cyclic load was applied to the first 10^4 cycles; the same behaviour as in the first zone is found;
- third zone, where I_{corr} is always higher for implanted than for pure iron; the time to breaking is consequently shorter.

Materials	16.6 Hz	R	3.3 Hz	R
Fe	100,000		800,000	
10^{15} N^+ cm^{-2}	43,000	43 %	400,000	50 %
10^{16} N^+ cm^{-2}	35,500	35 %	300,000	38 %
10^{17} N^+ cm^{-2}	26,500	26 %	270,000	33 %

Table I : Cycles before cracking of implanted and untreated iron canti-lever beams under cyclic loading at two frequencies ; R represents the percentage diminution of the corrosion fatigue life of implanted samples with respect to untreated ones.

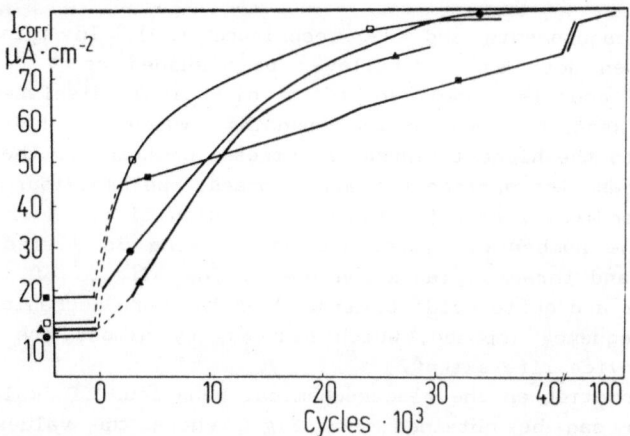

Fig. 5 : Calculated I_{corr} values vs. number of cycles in a 0.3% Na_2SO_4 solution for : (■) untreated iron, (●)10^{15}, (□)10^{16}, (▲)10^{17} N^+/cm^2 implanted iron.

The electrochemical behaviour shown in the first zone is in agreement with the previous example; a highly protective and homogeneous layer following the implantation treatment reduces the corrosion current.

A possible working hypothesis to explain the successive complex behaviour is as follows : the hard and brittle surface layers, caused by implantation, are considerably nobler, from an electrochemical point of view, than the bulk material as shown by the lowest initial corrosion rate; but they fracture in a random manner under cyclic loading causing a large number of microcracks. The SEM observation confirms such an idea. A sharp difference in crack topography emerges : cracks on pure iron are lined up parallel to the stressed section, while the implanted iron specimens show a random distribution of fracture paths. So the steeper increase of implanted steel corrosion rate may be related to the rapid increase of reactive area, as a consequence of the widespread opening of microcracks on the hard and brittle implanted layers. Furthermore, these layers are cathodic with respect to untreated inner ones; the crack propagation is hence accelerated electrochemically and the service life is consequently shortened.

2.3. Electrocatalytic properties of implanted iron

Recently the interest of some researchers has been devoted to the evaluation of the electrocatalytic properties of materials, the surfaces of which have been modified by ion implantation techniques. Our work studied the behaviour of ARMCO iron implanted with different doses of boron ions (2.10^{15} - 4.10^{16} B^+/cm^2) in a 1M H_2SO_4 solution, as a support for hydrogen evolution reaction (HER) (21).

We used a chemical species which shows no special catalytic property towards hydrogen evolution reaction, with the purpose of checking the possible effects due to the radiation damage by itself. On the basis of previous research work, carried out by Caprani et alii (22), we utilized a rotating disk electrode for our experimental measurements on HER. We tried to show the influence of ion implantation on the amount of the reactional area and the distribution of the active sites on it. For what concerns the theoretical model and the experimental details on which we based our experimental work, see the specific bibliography(22,23).

Figure 6a and b, shows, as an example, the electrochemical impedance data (presented as Bode and Nyquist plots) obtained for untreated and 10^{16} ions/cm^2 implanted ARMCO iron at the imposed potential of -1100 mV vs. saturated sulphate electrode (SSE). The calculated charge transfer resistance (R_t) values are, in the two cases, 160 $\Omega \cdot cm^2$ and 23 $\Omega \cdot cm^2$ respectively. Figure 7 represents a plot of the calculated R_t values. The lowering of R_t with a sharp minimum at the 10^{16} dose, is a clear sign of the trend of the reduction reaction rate constant. This is the evidence of a peculiar catalytic activity for redox reaction due to ion implantation.

Fig. 8a,b,c, are related to the experimental results obtained by rotating disk electrode measurements and are concerned with the non diffusive contribution to the total cathodic current amount (A). These figures show how a better tendency for proton reduction is connected to the modifications induced on the surfaces by ion implantation; furthermore it

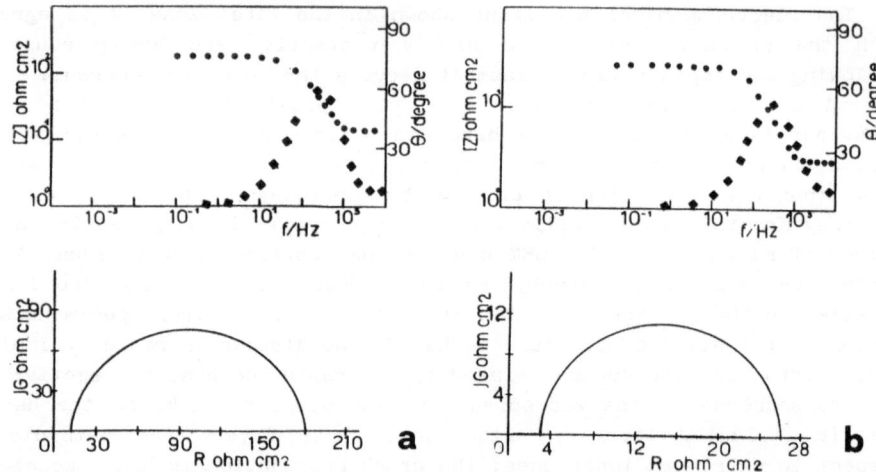

Fig. 6 a and b : Bode and Nyquist plots for unimplanted and $10^{16} N^+/cm^2$ implanted iron samples in 0.5 M H_2SO_4 solution. Z is the electrochemical impedance; jG and R are respectively the imaginary and real part of the electrochemical impedance.

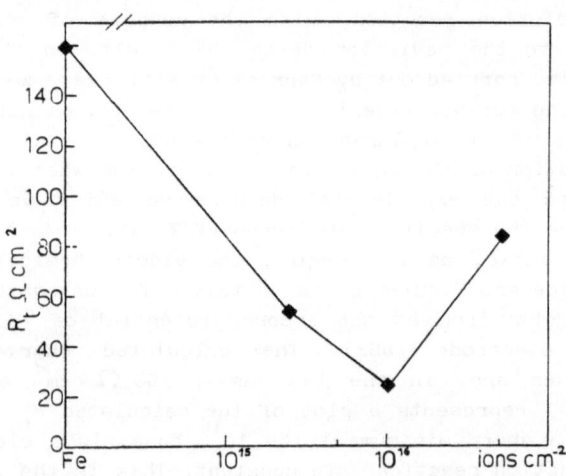

Fig. 7 : Charge transfer resistance (R_t) values as a function of implantation dose.

is possible to point out an increasing reactivity which is closely related both to the temperature and the overvoltage. As a matter of fact the value of A, which only depends on H_2 evolution, is progressively growing as the implantation dose increases at the lower reaction temperature. This value is found to be three times higher for the samples implanted with a dose of $4.10^{16} B^+/cm^2$ than for the untreated ones.

If we consider that the implanted species (boron) seems to have no particular catalytic properties for the HER, it is possible to assume that the cause of such a particular behaviour, pointed out by electrochemical tests, is only due to the radiation damage. The bombarded surface is subjected, during such a treatment, to modifications due both to its pseudo-amorphization and/or homogeneization and to the structural damage of the substrate, both of which are able to promote the adsorption of the H_{ads} (weakly bonded) and consequently to support the whole reduction reaction. Furthermore the mathematical elaboration of electrochemical data gave the main and most interesting topographic parameters listed in Table II, where $(1-\psi)$ represents the surface area covered by adsorbed hydrogen species weakly bonded (H_{ads}) and ρ is the number of active sites per surface unit. The values of $1-\psi$ for unimplanted ARMCO iron samples are quite similar to the ones obtained by Caprani. For the implanted samples the main result is the significant increase of the value of $1-\psi$, closely related to the increasing implantation dose : with respect to the unimplanted ARMCO iron we noted a three—fold increase. The effect of ion implantation is particularly noticeable at the lower temperatures.

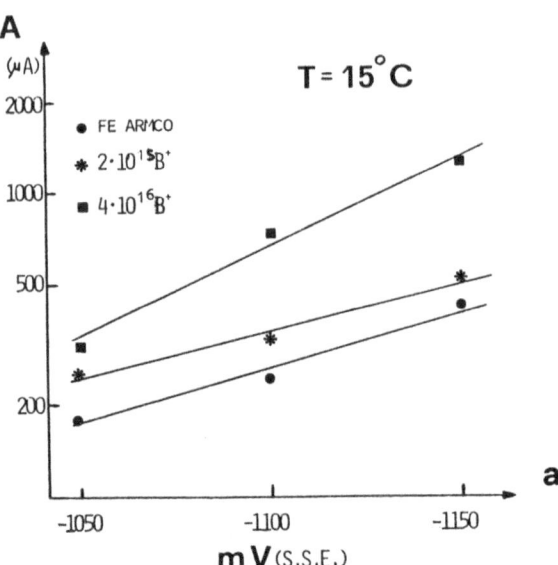

Fig. 8 a,b,c : Values of A (non-diffusive component of cathodic current) at different temperature conditions.

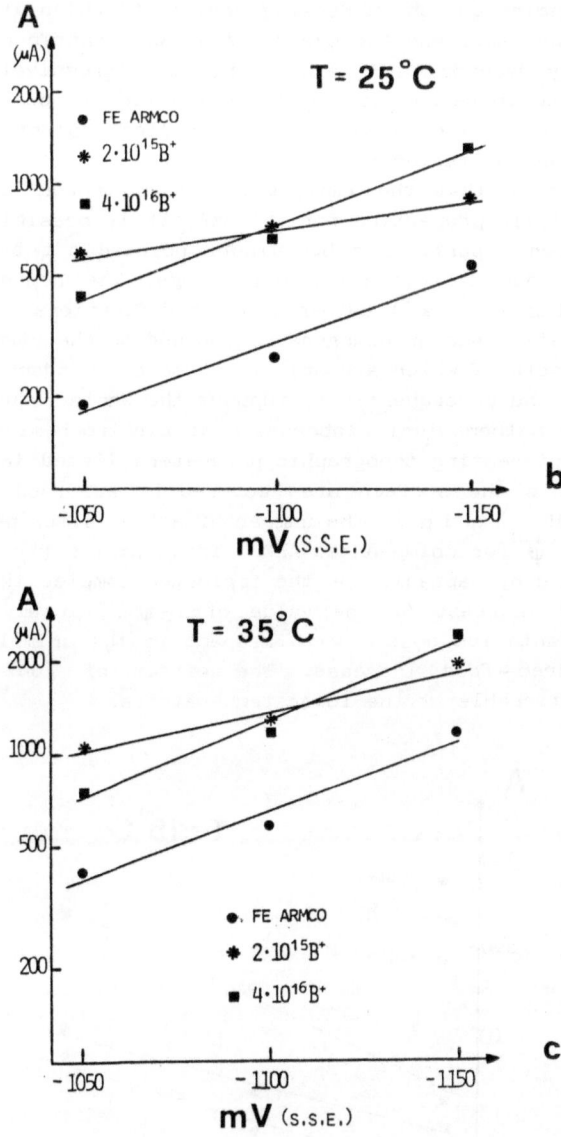

A further result of our experimental work is the set up of ρ values concerning the arrangement of the active sites. It is noticeable that, especially at high implantation doses, a marked compactness of the active sites takes place, pointed out by the ρ values, which result one or two orders of magnitude lower with respect to the ones calculated for pure ARMCO iron (22). It is well known that Ion Implantation introduces interstitial atoms in the near surface metal layers. This fact is likely to hinder hydrogen diffusion into the metal and together with the calculated

Temperature potential (SSE)		$(1 - \psi) \cdot 10^2$			ρ		
		A	B	C	A	B	C
15°C	- 1050	1.84	2.57	3.27	53.3	45.7	1.8
	- 1100	2.51	3.44	7.48	22	37.25	0.037
	- 1150	4.31	5.33	12.75	1.7	36.1	0.016
25°C	- 1050	1.88	5.93	4.34	263	1.36	0.68
	- 1100	2.68	7.27	6.62	61.8	329	0.24
	- 1150	5.59	9.09	12.9	2.46	0.66	0.011
35°C	- 1050	4.26	10.8	7.44	4.62	0.21	0.14
	- 1100	6.1	13.04	11.6	1.05	0.02	0.12
	- 1150	11.9	20.5	24.2	0.06	0.003	0.018

Table II : A=unimplanted ARMCO iron; B=$2.10^{15} B^+/cm^2$; C=$4.10^{16} B^+/cm^2$
Topographic parameters $(1-\psi)$ and ρ obtained by elaboration of
the experimental results in terms of chemical desorption mechanisms ($2H_{ads} \rightarrow H_2$). $(1-\psi)$ is the surface area covered by
adsorbed hydrogen species weakly bonded; ρ = number of active
sites per surface unit.

compact topography of the active sites, supports the experimental evidence
of an enhanced bubble evolution. As a matter of fact it is interesting to
report that the lowest ρ value is found with the sample implanted with a
$2.10^{15} B^+/cm^2$ dose, on which the current was hardly recorded at low rotation
speed, because of gas evolution.

2.4. Localized corrosion of passivating alloys

The behaviour of stainless steel after ion implantation or laser
irradiation has been widely studied (24,25). Under our experimental
conditions which have been described in detail elsewhere (26) the surface
finish after laser irradiation was poor because of craters, thermal fatigue
striations and an increased surface area. Pit nucleation is therefore
markedly enhanced. The AISI 304 stainless steel behaviour in an environment
inducing passivity breakdown (a deaerated solution of 0.05 N Na_2SO_4 plus
0.1N NaCl) was studied. In this case the unfavourable surface finish and
the increase in the surface area play a critical role in promoting pit
nucleation.

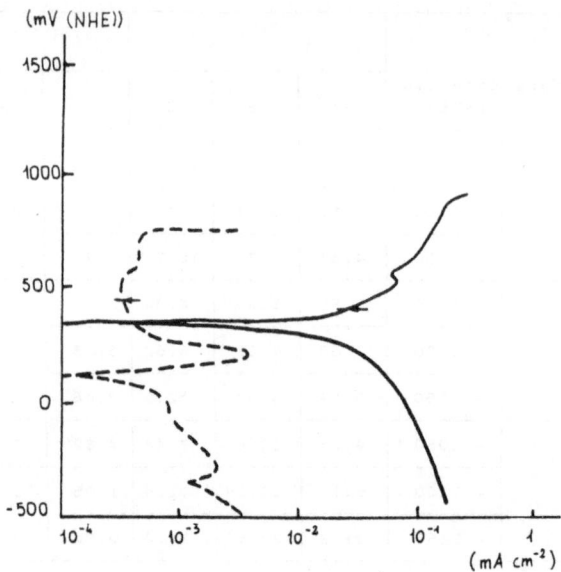

Fig. 9 : Potentiokinetic polarization curves of AISI 304 stainless steel
rotating electrodes (2000 rpm) in a solution of 0.05 Na_2SO_4 plus
0.1 N NaCl --- untreated sample, —— sample irradiated in air
with a pulsed ruby laser (5.6 Jcm^{-2}; 16 ns). NHE = normal hydrogen electrode.

The potentiokinetic polarization curves in Fig. 9 show in fact that the
laser irradiated sample has very high cathodic and anodic current densities
and does not undergo passivation. The zero current potential value is quite
different from that of the untreated alloy, because of larger variation in
the polarization of the electrode reactions. However, the potential range
in which pit nucleation and breakdown are possible is almost the same in
the two cases.

The attack morphology is able to explain the electrochemical behaviour.
Figures 10a,b show the localized attack on the untreated and laser
irradiated samples respectively. The sharply localized pit in Fig. 10b with
complete protection of the surrounding areas and the diffuse passivity
breakdown on the untreated alloy are quite evident. At a higher
magnification, Fig. 11, a thick quasi-amorphous near-surface layer, which
overlays a cavity, can be seen. The surface layers in many cases remain
uncorroded while the attack proceeds inside the cavity. A possible
explanation for this type of behaviour is the following : the pitting
attack nucleates on the surface where there are craters and holes caused by
laser melting. As soon as one nucleation site undergoes autocatalytic pit
growth, the crystalline unmodified underlying layers which are more anodic
with respect to the near surface metal provide intense cathodic protection
on the surrounding surface area, which remains uncorroded as can easily be
seen. Thus in this case the surface treatment promotes localized

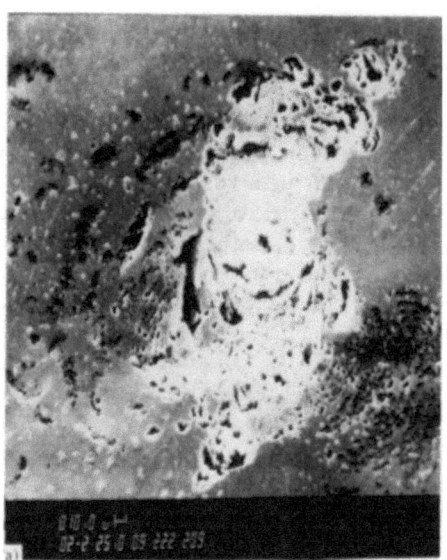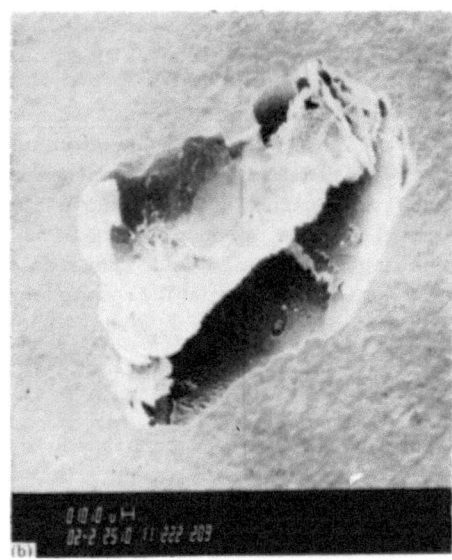

Fig. 10a,b : Pits on (a) untreated and (b) laser irradiated AISI 304
stainless steel samples.

Fig. 11 : Pit on laser irradiated AISI 304 stainless steel sample, showing uncorroded surface layers.

instead of generalized attack. The same behaviour has previously been observed with less marked features on laser irradiated aluminum (26).

2.5. Corrosion behaviour of nitride coated iron surfaces

In the last years some new techniques have been developed in order to improve, in terms of reduced thermal oxidation or wear or corrosion rate, the results obtained by ion-implantation. Ion beam mixing has proved to be effective in the production of highly adherent surface coatings with interesting chemical and physical properties. On the other hand electrochemical tests performed to control the corrosion behaviour of such films have pointed out typical problems due to the film thinness or defectiveness.

In order to overcome such difficulties, in our laboratories have been studied some new coating methods such as Reactive Ion Beam Enhanced Deposition (RIBED), consisting of simultaneous or sequential deposition/implantation steps, or such as high thickness multilayer deposition (about 100 nm) followed by a high energy implantation. In these ways we have obtained, on pure iron samples, well adherent and hard surface layers consisting of boron or chromium nitrides. These materials, particularly suitable for anti-wear aims, have been characterized from an electrochemical point of view in order to illucidate their corrosion behaviour which is of great importance in the case of mechano-chemical attack.

The films were produced by :

i) a sequence of 4 to 25 steps each one consisting of a 10 nm Cr deposition, followed by a N_2 implantation to a dose of 8.4×10^{16} ions/cm^2

ii) 4 sequential steps of a 27 nm B deposition, followed by a N_2 implantation to a dose of 1.4×10^{17} ions/cm^2.

The apparatus used to perform RIBED experiments was described elsewhere (15). Fig.12 shows the potentiodynamic polarization curves, performed in 1M NaCl solution, of RIBED obtained coatings and of pure Fe, and in Table III some electrochemical parameters are summarized, deduced from Fig. 12 and polarization resistance (Rp) measurements.

Both the treated samples compared to pure iron and to multilayer coatings reveal a nobler free corrosion potential (E_{corr}) and a reduced corrosion current (i_{corr}), about 2 orders of magnitude lower. The i_{corr} values were obtained by Tafel slopes (βa, βc) and Rp measurements performed as described by D.A. Jones (27). It is interesting to note the trend of the curves which show a significant inhibition of the cathodic reactions. This fact can justify the very small i_{corr} data observed. Polarization resistance measurements were also performed in a 0.3% Na_2SO_4 solution as a function of time in order to study the materials behaviour in a neutral environment (Table IV).

All the samples considered reveal an improved initial behaviour with respect to iron, particularly the RIBED obtained coatings, which remain effective over a long time range. These measurements, as well as

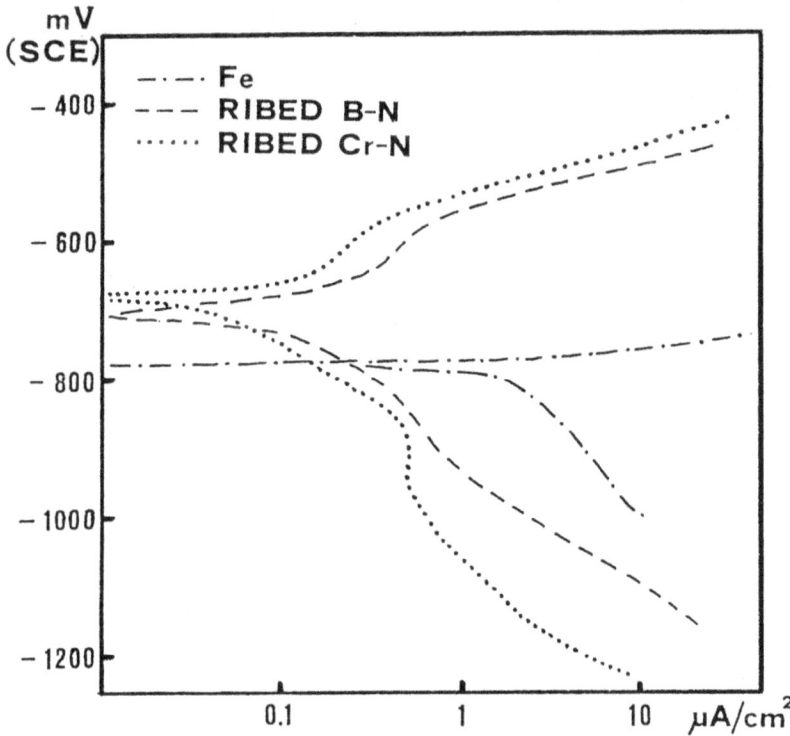

Fig. 12 : Potentiodynamic polarization curves in 1M NaCl (pH=4) deaerated solution. SCE = saturated calomel electrode.

electrochemical impedance data, point out a better behaviour of the Cr-N
coatings with respect to the B-N coatings. This fact can be justified by
SIMS and AES analysis which demonstrate the presence of an outer chromium
layer able to improve the corrosion resistance of the whole coating.

The protective effectiveness of the RIBED obtained Cr-N coatings was
also tested by evaluating the susceptibility to Cl⁻ attack. The samples
were passivated in a borate buffer solution (pH=8.7); 5000 ppm of Cl⁻ ions
were then added to the solution and at the same time the samples were
anodically polarized in order to induce pitting. The chromium deposit shows
no pitting attack at the imposed potential of -350 mV(SSE) contrarily to
all the other tested samples (Table V). To induce a pitting attack, the
polarization potential must be shifted towards more anodic values. At -100
mV (SSE) we can observe a pitting induction time of about 58 minutes (i.e.
about twice the maximum value obtained with the previously studied
samples).

S A M P L E	E_{corr} (mV)	Rp (kohm)	βa (V)	βc (V)	i_{corr} ($\mu A/cm^2$)
ARMCO Fe	− 780	24.5	0.05	0.165	1.36
Fe/B/Fe .... on Fe impl. with N_2^+	− 820	20	0.04	0.16	1.39
Fe/B/Fe .... on Fe impl. with Kr^+	− 790	14	0.065	0.10	2.44
RIBED obtained BN ON Fe	− 720	650	0.065	0.16	0.062
RIBED obtained Cr_2N on Fe	− 670	720	0.07	0.17	0.06

Table III : Free corrosion potential and instantaneous corrosion rates in 1M NaCl solution, deaerated (samples area 0.5 cm^2).

The reactive ion beam enhanced deposition technique is able to produce surface coatings with an improved corrosion behaviour with respect to pure iron. This behaviour is related to the presence of low reactivity nitride layers (BN or Cr_2N) and, in the case of RIBED obtained Cr-N coatings, to the presence of an external easily passivable pure Cr layer. On the other hand

S A M P L E	TIME (h)							
	0	1	2	3	5	8	24	50
ARMCO Fe	5.06	5	4	3.84	–	–	3.1	
Fe/B/Fe ... on Fe impl. with N_2^+	–	25	15.7	10.1	–	4.1	2.9	
RIBED obtained BN on Fe	90	67.5	–	–	10	7	6	
RIBED obtained Cr_2N on Fe	191	73	40	29	21.3	17.4	16.5	13

Table IV : R (kohm) measurements vs. immersion time in a 0.3% Na_2SO_4 solution (area = 0.5 cm^2).

S A M P L E	ARMCO Fe	N^+impl.Fe	Fe/B/Fe.on Fe impl.with N_2^+	RIBED obtained Cr_2N on Fe
t_{ind}(min) at-350 mV(SSE)	0.5	5	26.6	no pitting
t_{ind}(min) at-100 mV(SSE)	-	-	---	58

Table V : Pitting induction time after passivation in a borate buffer solution (pH=8.7) with an addition of 5000 p.p.m. Cl^- ions.

the presence on the surface of faults or defects can have a deleterious
effect on the protective action inducing a galvanic couple between the
substrate (anode) and the more noble coating (cathode) so accelerating a
localized corrosion.

Therefore it is very important to optimize all the physical parameters
which influence the structure of our samples in order to obtain well
adherent and more compact and reproducible coatings.

REFERENCES

1. V. Ashworth, W.A. Grant, R.P.M. Procter and E.J. Wright, Corros.Sci. 18
    (1978) 681
2. H. Ferber and G.K. Wolf, in Ion Implantation into Metals, V. Ashworth
    et al (eds.), Pergamon Press, Oxford, 1982, p.1
3. W.K. Chan, C.R. Clayton, R.G. Allas, C.R. Gossett and J.K. Hirvonen,
    Nucl.Instr.Methods 209/210 (1983) 857
4. P. Mazzoldi, G. Della Mea, G. Battaglin, A. Miotello, M. Servidori and
    E. Jannitti, Phys. Rev. Lett. 44 (1979) 88
5. P.L. Bonora, M. Bassoli, P.L. De Anna, G. Battaglin, G. Della Mea and
    P. Mazzoldi, Mat. Chem. 5(1980) 73
6. G. Dearnaley, Thin Solid Films 107 (1983) 315
7. C. Weissmantel, Proc. of 3rd Int. Conf. on Solid Surfaces (Vienna,
    1977) p. 1533
8. G. Dearnaley, P.D. Goode, F.J. Minter, A.T. Peacoock and C.N. Waddell,
    J. Vac. Sci. Technol. A3 (1985) 2684
9. M. Elena, L. Fedrizzi, V. Zanini, M. Sarkar, L. Guzmàn and P.L. Bonora,
    Nucl.Instr. and Meth. B 19/20 (1987) 247
10. P.L. Bonora, M. Bassoli, G. Cerisola, P.L. De Anna, P. Mazzoldi, S. Lo
    Russo, I. Scotoni, C. Tosello and M. Maja, Mater. Chem. 4 (1979) 17
11. P.L. Bonora, G. Cerisola, C. Tosello and S. Tosto, in Ion Implantation
    into Metals, V. Ashworth et al (eds), Pergamon Press, Oxford, 1982,p. 1
12. P.L. Bonora, G. Cerisola, L. Fedrizzi and C. Tosello, Mat. Sci. Eng. 69
    (1985) 283
13. P.L. Bonora, M. Bassoli, G. Cerisola, P.L. De Anna, S. Lo Russo, P.
    Mazzoldi, B. Tiveron, I. Scotoni, C. Tosello and A. Bernard, Nucl.
    Instr. Methods 182/183 (1981) 1001
14. F. Marchetti, L. Fedrizzi, F. Giacomozzi, L. Guzman and A. Borgese,

Mat. Sci. Eng. 69 (1985) 289

15. L. Guzmàn, F. Giacomozzi, B. Margesin, L. Calliari, L. Fedrizzi, P.M. Ossi and M. Scotoni, Mat. Sci. Eng. 90 (1987) 349

16. L. Fedrizzi, L. Guzmàn, A. Molinari, S. Girardi and P.L. Bonora, Nucl. Instr. Meth. B7/8 (1985) 711

17. M.G. Fontana and N.D. Greene, "Corrosion Engineering", (Mc Graw-Hill, N.Y. 1978)

18. H.H. Uhlig, "Corrosion and Corrosion Control", John Wiley, New York, London, Sydney, Toronto

19. L.L. Shreir, "Corrosion" Newnes Butterworths, London 1979

20. P.L. De Anna, G. Cerisola, P.L. Bonora, S. Lo Russo, P. Mazzoldi, I. Scotoni, C. Tosello, Werkstoffe und Korrosion 31 (1980) 783

21. P.L. Bonora, G. Cerisola, L. Fedrizzi, S. Lo Russo "Electrochemical Methods in Corrosion Research" Ed. M. Duprat, Material Science Forum Vol. 8 (1986) 23

22. A. Caprani, M. Keddam, P. Morel, 3 Congrès Int."Hydrogène et Matériaux", Paris C2 (1982) 251

23. V.G. Levich, "Physicochemical hydrodynamics" Prentice Hall Englewood Cliffs, N.Y. 1962

24. C.R. Clayton, W.K. Chan, J.K. Hirvonen, G.K. Hubler and J. R. Reed, "Fundamental Aspects of Corrosion Protection by Surface Modification" Washington, DC, 84-3 (1984) 17

25. V. Ashworth, W.A. Grant, A.R. Mohammed and R.P.M. Procter, Proc. 7th Int. Congr. on Metallic Corrosion, Association Brasileira de Corrosion, Rio de Janeiro 1978

26. G. Cerisola, P.L. De Anna, P.L. Bonora, M. Bassoli, G. Battaglin, and A. Carnera, Proc. 2nd Int. Congr. on Heat Treatment on Materials 1982, Associazione Italiana di Metallurgia, Florence 1982

27. D.A. Jones, Corrosion 39 (1983) 444.

STRUCTURAL CHANGES

ION-IRRADIATION INDUCED PHASE CHANGES IN METALLIC SYSTEMS

William L. Johnson

W. Keck Laboratory of Engingeering, California Inst. of Tech., Pasadena, CA
91125

1. INTRODUCTION

It is well known that irradiation of thin metallic film structures by
high-energy ion beams produces a variety of metallurgically interesting
modifications of the atomic structure and morphology of the irradiated
material. The modifications include changes in the composition profile of
the film structure due to ion mixing, creation of nonequilibrium defect
populations within the phases present, and even changes in the phases
themselves. The latter type of modification is the primary concern of this
paper. In order to address the question as to why phase changes occur, one
must in fact also develop an understanding of the evolution of composition
profiles and defect distributions within the phases present since these are
the factors which control the thermodynamic stabilities of the competing
phases. This subject can be naturally divided into two cases according to
whether the initial configuration of the film structure is chemically
homogeneous or inhomogeneous. In the former case, ion irradiation does not
alter the overall composition profile. Neglecting effects such as
preferential sputtering of one component from a free surface, the
composition of the film remains uniform throughout the sample during
irradiation, and ion mixing produces only local concentration fluctuations.
In the latter case of an initially inhomogeneous material (e.g. a bilayer
or multilayer film with two or more components), the overall composition
profile evolves under irradiation and the phases present are subjected
changes in average composition.

The effect of composition changes on phase stability can be best
discussed within the framework of free-energy diagrams and phase diagrams
in the composition/temperature (c,T) plane. The basic principles used are
outlined in section 2 below. The extension of the equilibrium diagrams to
the case of nonequilibrium phases will be covered, and it will be shown
that nonequilibrium phases may be either metastable or unstable. Section 3
deals with the case of ion mixing in a bilayer diffusion couple. A
phenomenological model developed by the author and colleagues (1) is used
to describe the evolution of the composition profile as a result of ion
mixing. This model includes material-dependent parameters such as chemical
heats of mixing, and the cohesive energies of the phases present. It treats
the case of ion mixing at low temperatures where the experimental mixing
efficency is observed to be temperature independent. The stability of the
initial phases under the influence of the evolving composition profile is
discussed. It is shown that under these conditions, one expects to form
either extended solutions of the original phases or an amorphous phase. The
mechanism by which an amorphous phase forms is discussed in some detail.
Section 4 deals with the case of an initially homogeneous intermetallic
phase subjected to irradiation. In this case, the homogeneous phase is
altered by irradiation-induced disorder in the form of vacancies,

R. Kelly and M. Fernanda da Silva (eds.), Materials Modification by High-fluence Ion Beams, 405–420.
© *1989 by Kluwer Academic Publishers.*

interstitial atoms, and anti-site defects (chemical disorder). This again alters phase stability and can result in phase transformations. Formation of amorphous phases under such conditions will be discussed in detail.

2. EQUILIBRIUM AND NONEQUILIBRIUM THERMODYNAMICS

The equilibrium thermodynamics of a two component system at constant pressure and fixed temperature can be best represented in terms of the composition dependence of the Gibbs free-energy function of each phase. From a knowledge of these functions at different temperatures and pressures, one can construct equilibrium phase diagrams. Fig.1a and 1b show schematic free-energy curves at two different temperatures for a binary system composed of two metals, A and B, having structures α and β. One intermetallic compound having structure γ is assumed. Fig.1c shows the corresponding equilibrium phase diagram with equilibrium two phase fields determined by the common tangents bounding the curves from below as shown in Fig. 1a and 1b.

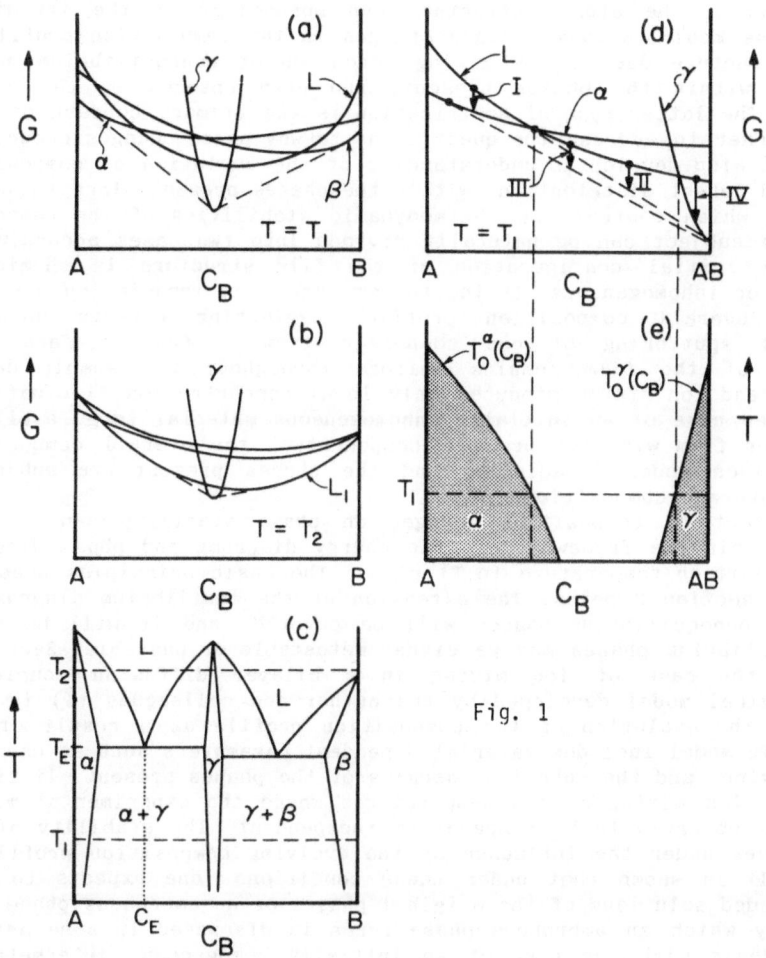

Fig. 1

In the case of ion beam mixing, we will see that the evolution of a collisional cascade into an energy spike produces atomic displacement and replacement events which mix two metals in very short time scales ($\lesssim 10^{-10}$s) under very energetic conditions (2,3). Following the mixing process, energy is transported to the surrounding material and the excited region of the cascade is quenched to ambient temperature in a very short time scale of $\lesssim 10^{-9}$s. The composition profile created by the cascade mixing is "frozen" into the material at ambient temperature where further evolution of the profile is suppressed for low enough ambient temperature. Under these conditions, it will be useful to consider a type of constrained equilibrium whereby further chemical segregation within a mixed region is suppressed. This leads one to consider a nonequilibrium situation where the local composition of a phase is dictated by the energetic mixing process. To decide which phase is more stable at a specific composition, one examines the composition dependence of the free energy of each phase at the ambient temperature. Fig 1d shows that at temperature T_1, the α-phase has lower free energy up to a critical concentration c_B^* where its free energy crosses that of the liquid phase. This leads one to define a so-called $T_o^\alpha(c_B)$ curve which delineates the boundary separating the regions of relative stability for a single α-phase material and a single liquid phase material. Such curves were originally introduced by Cahn (4) and Massalski (5) to describe regions of a binary phase diagram in which partitionless crystallization could occur in an undercooled melt during rapid solidification. The analogy with the kinetics of mixing and quenching in a collision cascade is noteworthy. Fig.1e illustrates the T_o-curves of both the α-phase and the γ-phase for the present case. This figure can be called a polymorphous phase diagram in the sense that it depicts a constrained equilibrium under conditions where the local composition is fixed. In a polymorphic phase diagram, it is often the case that for certain composition ranges, the liquid phase is the most stable single material at low temperatures as seen in Fig. 1e. In fact, liquids are known to configurationally freeze below a glass transition temperature T_g (6). As such, the glassy or amorphous phase may be the most stable phase in certain ranges of temperature and composition lying outside the T_o-curves of the competing crystalline phases. This fact will later be seen to provide one criteria for predicting ion-mixing induced amorphization.

Having introduced the T_o-curve concept, one might now ask what happens when a phase is driven compositionally outside of its T_o-curve. For example, one can ask what happens to an α-phase (Fig.1e) at temperature T_1 and compostion outside the $T_o^\alpha(c_B)$ curve. Such a phase can thermodynamically transform without long range chemical diffusion to an amorphous phase of the same composition. Such a transformation can be termed a polymorphic phase change. Whether or not this amorphization transformation occurs depends on kinetic factors. In the following section, attention will be focussed on phase changes which take place during and following the evolution and subsequent quench of a collisional cascade occurring in a local region within the sample. The detailed nature of how a collisional cascade evolves both spatially and temporally, ultimately becoming a thermal spike, has been recently discussed and reviewed in ref. (2). Space precludes a detailed discussion here. For present purposes, it is important to recognize that a phase transformation (e.g. to the amorphous state) could occur either during cascade evolution (in the prompt regime) or after the reestabishment of thermal equilibrium following the cascade (a delayed transformation). If the ambient temperature is very low, a delayed transformation may be kinetically severely hindered. The author has discussed the nucleation of an amorphous phase from a nonequilibrium crystalline phase under low—temperature isothermal conditions (6). Like ordinary melting at the equilibrium melting temperature,T_m, polymorphous melting along the T_0—curve generally involves a nucleation barrier. This barrier becomes smaller as a crystalline phase is carried progressively beyond its T_0-curve. There in fact exists a second type of curve in the (c,T) plane for a given crystalline phase which is defined by the condition that the nucleation barrier for melting or amorphization vanish. The author has termed this the $T_\mu(c)$ curve. A crystalline phase at its T_μ-curve can undergo a massive and spontaneous transformation to the liquid or amorphous phase with no nucleation barrier. That such a curve exists can be seen at least in the case of an elemental metal by examining the temperature dependence of the elastic shear modulii extrapolated above the thermodynamic melting point. Tallon and Wolfenden have examined this problem for the case of many elemental metals and several congruently melting compounds (7,8). For cubic metals (e.g. Al,Cu, etc.) they find that the shear modulus $\mu_2 = (C_{11}-C_{12})/2$ vanishes at roughly 1.7 times the thermodynamic melting point of the metals. This establishes where the T_μ curve should intersect the temperature axis above a pure component in a binary system such as that in Fig.1e. Fig. 2 illustrates the T_μ curve of the α-phase of Fig.1.

Fig. 2 Illustration of the T_μ and T_0—curves of the α-phase previously considered in Fig.1 together with the features of the equilibrium phase diagram. The script S and L refer to the equilibrium solidus and liquidus curves of the α-phase. The cross hatched region is the equilibrium α-field, the shaded areas represent the regions of metastable α-phase inside and outside of the T_0-- curve. A homogeneous single phase of α is unstable outside the T_μ-curve.

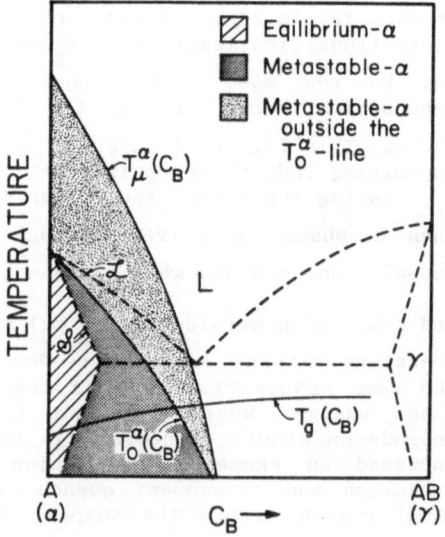

It also shows the T_0-curve, as well as the original equilibrium phase diagram (dashed lines). Finally, we see in Fig.2 that the α-phase can be characterized by it equilibrium phase field, a metastable phase field (outside of equilibrium but inside the T_0-curve), and limit of metastability (the T_μ- curve) outside of which the α-phase will spontaneously transform to an amorphous or liquid phase with no nucleation barrier. The T_μ-curve may thus be important in determining whether or not a delayed crystal—to—amorphous transformation can occur at low ambient temperatures. We will return to this point in the following section.

3. COMPOSITION PROFILES AND PHASE CHANGES DURING ION MIXING OF BILAYERS

The previous section discussed the composition dependence of phase stability in terms of phase diagrams appropriate to ambient temperatures which lie below or near the melting points of crystalline solids. During the evolution of a collisional cascade, individual particles move with kinetic energies far higher that those associated with low temperature phase diagrams. In a purely ballistic model of a cascade (2,9), atomic displacements and replacements are induced by binary collisions involving an energetic particle in motion and a stationary target atom of the lattice which is subsequently set into motion. In the early stages of cascade evolution, one has a relatively small number of high energy particles moving in a matrix of relatively stationary atoms residing on a lattice. As further generations of collisions occur, more particles are set into motion, the energy of a typical moving particle decreases, and the density of moving particles (compared to stationary particles) in a locally excited region increases. At a later stage, termed the "cool-down phase" by Averback and Siedman (2), local regions develop in which all of the particles are excited and have kinetic energies distributed according to a Maxwell-Boltzmann distribution. At this point, these regions can be said to characterized by a temperature. The local temperature profile then evolves according to an ordinary heat—flow equation to ambient temperature. In this paper, we will assume that mobile interstitials created during the cascade migrate and annihilate with vacancies or sinks during the late stages of and following the cool-down stage. The remaining vacancies are then configurationally frozen at the final ambient temperature. The thermal spike model first introduced by Seitz and Koehler (10) and later developed by Vineyard (11), gives a useful description of this cool—down phase within local regions. Recently, Cheng et al. (12,13) have shown that a collisional cascade can be viewed as statistical fractal tree having a fractal dimensionality which depends on the nature of the interaction potential of the colliding particles. Using power law potentials of the form

$$v(r) = constant/r^{1/m} \qquad (1)$$

and an idealized collisional cascade in which each collision has a fixed scattering angle equal to the average scattering angle, they showed that the fractal dimensionality of the tree is simply $d_f = 1/(2m)$. From this, they argued that a collisional cascade can only become space-filling if m<1/6, that is when d_f becomes greater than the dimensionality,d=3, of the Euclidean space in which the cascade is embedded. In practice, a potential with m<1/6 is characteristic of ion/atom potentials when kinetic energies lie below the ˜100 eV range. At very high energies (˜10-100 keV), potentials with m=1/2 have been found appropriate. An actual collisional cascade is thus an inhomogenous fractal with d_f increasing as the cascade evolves. Space filling (required for a thermal spike) can only occur in the late stages of the cascade where particle kinetic energies lie near or below the ~100 eV range.

For the present purposes, we are interested in the process of atomic mixing. In particular, it is important to know whether atomic mixing occurs predominantly in the early ballistic cascade or in the later cool-down phase where the thermal-spike model is applicable. This question can be answered by appealing to experimental observation. In a series of experiments, Cheng, Van Rossum, Nicolet, and the author clearly demonstrated that, in most cases of interest, the majority of atomic mixing occurs during the cool-down phase of the cascade (1,14) and involves particles with kinetic energies of the order of 1 eV. To deduce this, they carried out mixing experiments at low ambient temperature (77K) on a series of bilayer films in which the top layer was chosen to be either Au or Pt and the bottom layer was chosen to be a 3d-transition metal (Ti, Cr, Fe, Ni, Cu). The bilayers were all irradiated with 600 keV Xe^{++} ions to induce ion mixing at the interface. Since the masses and atomic numbers of Au and Pt are nearly identical as are those of the 3d-transition metals, ballistic mixing at high energies should be nearly identical for all of the bilayers. On the other hand, the various pairs of metals (e.g. Au/Ti, Pt/Ni) have very different chemical heats of mixing which range form near zero (Pt/Ni) to very large negative values (~1eV per atom at the equiatomic compostion) for Au/Ti and Pt/Ti. The experiment showed that metal pairs with large chemical heats of mixing were mixed much more efficiently by the ion beam. Fig. 3 (taken from ref. 14) shows this result in the form of a plot of mixing efficiency vs. heat of mixing.

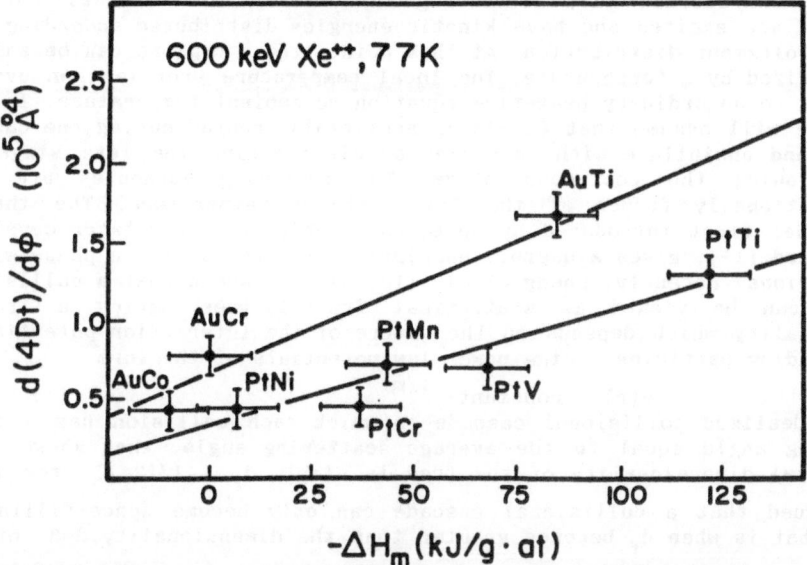

Fig. 3 Correlation between the mixing rate d(4Dt)/dφ and Miedema's heat of mixing for bilayers irradiated with 600 keV Xe^{++} ions. (see ref. 14)

In a second series of experiments, the same authors (15) demonstrated that the cohesive energy of the metals being mixed also plays an important role. Metals having a higher cohesive energy undergo fewer atomic displacements during the cool down phase. Averback and Siedman (2) argue that most mixing occurs by point defect migration and that the activation energy for such defect migration roughly scales with the cohesive energy of the lattice. The chemical effect of Fig.3 can be explained if it is assumed that the mixing process is a chemically biased diffusion process as first described by Darken (16). In this interpretation, the slope of the lines in Fig. 3 yields the effective temperature at which the atomic mixing process took place. This effective temperature turns out to be about 10^4 K and corresponds to particle kinetic energies of 1 eV. In other words, using the known energy scale for chemical heats of mixing, and the observed chemical biasing of the mixing process, one can directly determine the kinetic energy of the particles at the time when most of the mixing takes place. This shows conclusively that mixing occurs in the cool-down phase and further combined with the above discussion of space filling implies that most atomic mixing occurs under near thermal-spike conditions. Using the Darken analysis together with a thermal-spike description of atomic displacement jumps during the cool-down phase, the author and colleagues derived a phemonenological equation for the effective diffusion constant in ion mixing which can be written as follows (1)

$$\frac{d(4Dt)}{d\phi} = \frac{K_1 \, \epsilon^2}{\rho^{5/3} \, (\Delta H_{coh})^2} \left[1 - K_2 \, \frac{\Delta H_m}{\Delta H_{coh}} \right] \qquad (2)$$

where ΔH_m is the heat of mixing of the equiatomic alloy, ΔH_{coh} is the average cohesive energy of the two metals being mixed, ϵ is the energy deposition per unit path length of the projectile ion, ϕ is the dose, and ρ is the atomic density of the target material. The parameters K_1 and K_2 are phenonemological constants which were determined from fitting to data taken from mixing about 20 binary systems to have values of $K_1 \cong 0.0037$ nm and $K_2 \cong 27$. Fig. 4 demonstrates the degree to which eqn. (2) can account for the observed ion mixing efficiency in the 20 different binary systems.

This shows that by suitable choice of only two parameters, K_1 and K_2, one can account quite well for the variation of mixing efficiency (actual mixing efficiencies vary by more than an order of magnitude) in the 20 binary systems. The diffusion constant obtained from eqn. (2) for a given binary system at a given dose rate $d\phi/dt$, can be put into the ordinary Fick equation to give the time evolution of the concentration profile in a binary system. It is noteworthy that when ΔH_m is positive and sufficiently large so that the term in the bracket of eqn. (2) is negative, then D must be negative. In other words, one predicts uphill diffusion (decomposition) during ion irradiation of binary systems characterized by a large positive heat of mixing. It has in fact been observed that such binary systems as Cu/W which exhibit liquid miscibility gaps (and large positive heat of mixing) cannot be mixed under ion irradiation (18,19). This again shows that most atomic mixing takes place at relatively low particle kinetic energies during the cool-down phase of the cascade.

Having developed an understanding of the mixing process which leads to

412

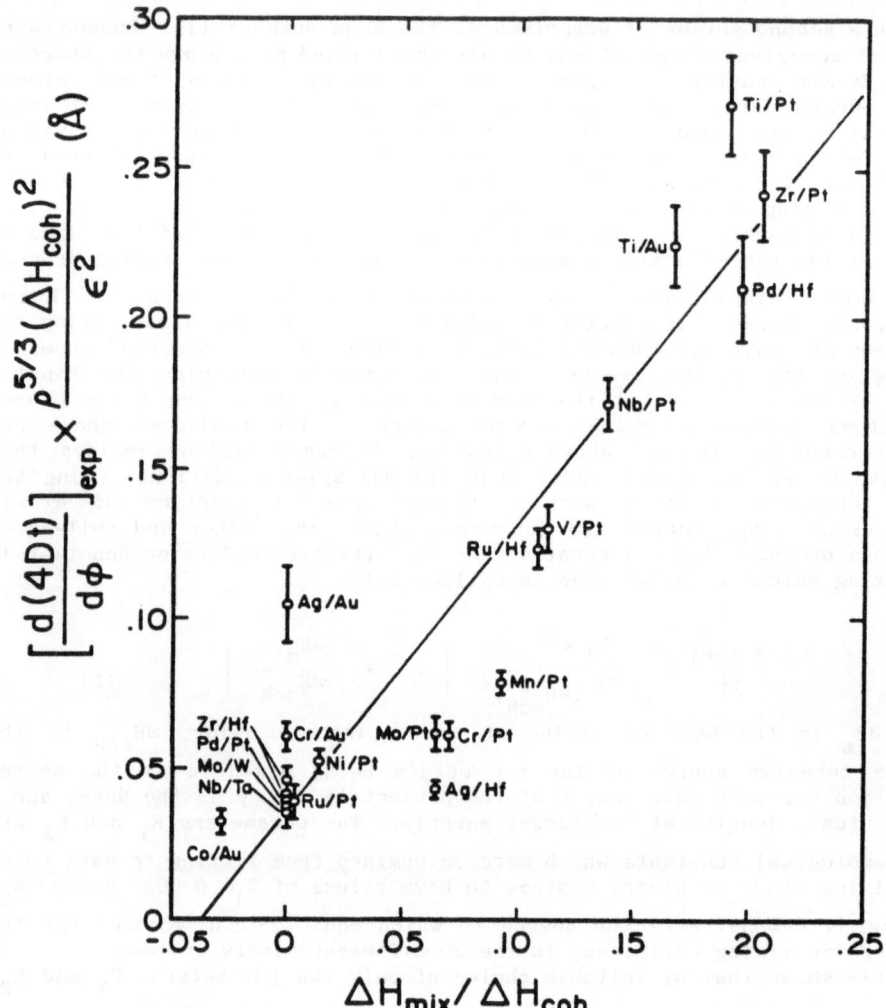

Fig. 4   Comparison of eqn. (2) with experimental results of mixing efficiency on 20 binary metal systems. All of the bilayers were mixed using 600 keV Xe^{++} ions. The straight line represents a least squares fit based on eqn. (2) and determines the value of the constants K_1 and K_2. See ref. 17 for details.

evolution of the composition profile, one is now in a position to ask whether compositionally driven phase changes take place during ion mixing. We have already seen that systems characterized by a large positive heat of mixing should not mix. Systems with less positive heats of mixing should mix slowly, while system with zero or negative heats of mixing should mix rapidly. In a more exact treatment of the diffusion problem, ΔH_m should be treated as composition dependent. This means that the effective D for mixing should depend on composition. This will cause the composition profile to deviate somewhat from a complementary error function, which is the appropriate solution for D independent of composition. This detail will be ignored in what follows. It will be assumed for simplicity that each cascade produces mixing in a local region. The combined effect of many overlapping cascades occuring in a spatially and temporally uncorrelated fashion will be to make the bilayer composition profile evolve along the direction normal to the interface separating the layer. Composition fluctuations along the direction of the interface (as well as normal to it) arising from the Poisson statistics of a cascade occurrence in space will be ignored as a first approximation.

As pointed out earlier, a phase transformation could initiate either during the cascade (prompt) or during the time intervals between cascades (delayed) at ambient temperature. Both cases are discussed with respect to transformation to the amorphous state. It is assumed for the case of delayed transformations that the ambient temperature is sufficiently low so as to suppress vacancy migration and that defect recombination in the late stages of the cool-down phase is complete. Under these conditions, when two metals are mixed, the composition profile will evolve according to eqn. (2). This will lead to mixing of A atoms into the B phase and vice versa. The result will be to produce solid solutions of the two metals. During a single cascade event, local regions having a spatial extent of a few tens of angstoms (2) will be excited, mixed, and undergo a thermal-spike type of cool-down phase. The duration of this thermal spike is of order 10^{-9} s. During this period the effective local temperature drops from 10^4 K to ambient temperature. The local cooling rate will thus be exceedingly high ~ 10^{13} K/s. For a phase transformation to occur during the prompt regime requires that it be characterized by very fast kinetics. For a new phase which requires nucleation (i.e. a phase which has an interfacial surface tension with the parent crystalline phase), not only must the local region be of dimension larger than the dimension of a critical nucleus (typically ~ 2 nm), but the nucleation event must take place in tens of picoseconds. This precludes the nucleation of nearly all new crystalline phases. Amorphization or melting, on the other hand, may not require nucleation if the system is driven to instability (locally having a temperature and composition lying outside of the respective T_μ-curve). Melting might in fact occur. A melting transition would require a sufficiently spatially extended thermal spike at a sufficiently high temperature to carry the material locally outside of the T_μ-curve. This is most likely to happen in metals which have very high atomic number together with a relatively low melting point. The high atomic number will result in well-developed thermal spikes (2,12,13), the low melting point will mean that the system remains outside of the T_μ-curve far into the cool-down phase of the thermal spike. Metals such as Pb, Bi, Au, and Hg would be good candidates. It is noteworthy that Au/Ag bilayers are the only example of bilayers with mixing efficiency substantially above that predicted by eqn. (2), (14-17), and that cascades in Au exhibit anomalously large numbers of atomic displacements (2). Both of these observations are consistent with the enhanced mixing which might be associated with local melting. In the extreme case of mixing of low melting point high-Z metals, amorphization may actually occur by prompt melting followed by a sufficiently rapid quench to configurationally freeze that material locally into the glassy (amorphous) phase.

At this point, one may ask if the amorphous transformation may occur in the delayed regime. According to the above discussion of the T_o and T_μ-curves, amorphization could also occurs without nucleation as a massive transformation at low temperatures. When the solid solutions produced by mixing are driven outside of the T_μ-curve at low ambient temperature, such catastrophic amorphization should in fact occur. Since we have already argued that nucleation of new crystalline phases should be very difficult in the prompt regime, and should be suppressed altogether at low ambient temperatures due to the nucleation barrier, it follows that ion mixing will likely produce either extended solutions or an amorphous phase in the absence of RED. As such, we can predict whether or not a given ion mixing experiment will lead to amorphization by examining the T_o and T_μ-curve of the parent phases being mixed. If there exists a range of compositions lying outside of the T_μ-curve of both parent solution phases, then amorphization should occur whenever the mixing governed by eqn. (2) forces the material into this composition range. Otherwise, the likely result of mixing will be to produce only extended solid solutions. The only exception to this rule would be the case where the metastable extended solutions locally melt during the prompt regime and then fail to recrystallize during the cool down phase. This, as seen above, should only happen for metals with high atomic number and relatively low melting points.

Table I gives a list of results of mixing experiments performed on a variety of binary systems in the bilayer configuration. The results show several interesting features which generally confirm the points made in the above discussion. First, one notes that amorphization occurs generally below the characteristic temperature at which RED becomes appreciable. Thus for the case of Zr/Ru bilayers, amorphization is possible below 410K, but not above (20). For Ti/Ru, amorphization is not possible at room temperature, but becomes possible at 77K. Second, one notes that for metals which have restricted mutual solubility (e.g. Zr/Ru, Co/Mo, Cu/Ta, Au/Ti, Au/V) in the equilibrium phase diagram (24), and therefore also plunging T_o and T_μ-curves separated by a gap in the polymorphous phase diagram (see Fig.1), one always tends to form an amorphous phase by mixing to compositions lying near the center of the phase diagram. Third, for systems which form extended solid solutions with no gap between the T_o-curves of the parent solutions (e.g. Ag/Cu, Mo/Nb), one does not observe amorphization. Finally, metals which do not mix at all (due to a large positive enthalpy of mixing, ΔH_m) do not form amorphous phases by ion beam mixing.

Table I   Amorphous Alloys formed by ion mixing of thin film diffusion couples.   Effects of substrate temperature, positive heat of mixing, etc.

Metals Mixed	Irradiation	Substrate Temperature	Amorphous Alloys Formed [a]	Not Amorphous (Comments)	References
Ru-Ti	600 keV, Xe^{++}	77 K	$Ru_{50}Ti_{50}$ [a]	$Ru_{25}Ti_{75}$, $Ru_{75}Ti_{25}$	20
Ru-Ti	600 keV, Xe^{++}	300 K	None	$Ru_{75}Ti_{25}$, $Ru_{50}Ti_{50}$, $Ru_{25}Ti_{75}$	20
Ru-Zr	600 keV, Xe	77 K	$Ru_{25}Ti_{75}$, $Ru_{50}Ti_{50}$	$Ru_{75}Ti_{25}$ [b]	20
Ru-Zr	600 keV, Xe^{++}	370 K	$Ru_{25}Ti_{75}$, $Ru_{50}Ti_{50}$		20
Ru-Zr	600 keV, Xe^{++}	410 K	None	$Ru_{25}Ti_{75}$, $Ru_{50}Ti_{50}$, $Ru_{75}Ti_{25}$	20
Ru-Mo	300 keV, Xe^{+}	RT	$Mo_{55}Ru_{45}$		21
Ru-Si	300 keV, Xe^{+}	RT	$Ru_{45}Si_{55}$		21
Ni-Ti	300 keV, Xe^{+}	RT	$Ti_{50}Ni_{50}$		21
Ni-Mo	300 keV, Xe^{+}	RT	$Ni_{65}Mo_{35}$, $Ni_{50}Mo_{50}$, $Ni_{35}Mo_{65}$		21
Cu-Ag	100-300 keV, Xe^{+}	RT, 77K	None	Forms continuous solid solutions[c].	22
Co-Mo		RT	$Co_{65}Mo_{35}$, $Co_{35}Mo_{65}$		21
Cu-W	400 keV, Ar^{+}	RT	None	Metals do not mix.[d]	23
Cu-Ta	100-300 keV, Xe^{+}	RT	$Cu_{50}Ta_{50}$		22
Au-V	100-300 keV, Xe^{+}	RT	$Au_{50}V_{50}$		22
Au-Ti	300 keV, Xe^{+}	RT	$Au_{35}Ti_{65}$, $Au_{60}Ti_{40}$		21
Co-Au	100-300 keV, Xe^{+}	RT	$Co_{50}Au_{50}$		22
Al-Ni		RT		$Ni_{50}Al_{50}$	24
Al-Nb	300 keV, Xe^{+}	RT	$Al_{55}Nb_{45}$		21
Mo-Nb	300 keV, Xe^{+}	77 K	None	Forms continuous solid solutions	21

RT - room temperature

(a) $Ru_{50}Ti_{50}$ forms the "CsCl"-type structure (with T_{00} T_{m}) in equilibrium. See Section IV.B. in text.

(b) $Ru_{75}Ti_{25}$ contains an hcp solution together with an amorphous phase after mixing to high dose (5 x 16 ions/cm^2) at 77 K.

(c) The Cu-Ag system has a slightly positive heat of mixing - extended solutions remain metastable (see refs. 2 and 40).

(d) Cu-W This system exhibits liquid immiscibility and is characterized by a large positive heat of mixing (see refs. 39, 102, and 105). Ion irradiation to high dose does not produce mixing.

4. HOMOGENEOUS INTERMETALLIC PHASES UNDER IRRADIATION - AMORPHIZATION AND OTHER PHASE CHANGES

For phases which are initially chemically homogeneous, ion irradiation will not modify the overall composition profile, at least as a first approximation. Exceptions to this include the case of migration of point defects to sinks during RED, preferential sputtering of one species from a surface, etc. (26,27). These topics are adequately treated elsewhere and will not be dealt with here. Instead, attention will be focussed on the role of a random distribution of point defects on the stability of crystalline phases.

With the exception of Ga (28), elemental metals are not observed to become amorphous under ion irradiation, even when the irradiation is carried out at cryogenic temperatures. Damage to the lattice of an elemental metal is in the form of self-interstitials and vacancies. The interstitials are mobile, even at relatively low temperatures and are found to migrate to surfaces and sinks (2). Vacancies are relatively immobile at low temperatures and therefore tend to accumulate with progressive ion dose. The author (6) has estimated that a vacancy concentration of 1 % would be required to render an elemental metal unstable against amorphization. Such vacancy concentrations are apparently not in general reached during cumulative damage induced by ions. The reason seems to be that vacancies tend to collapse into dislocation loops and voids before reaching this concentration. The metal Ga occurs in several allotropic modifications of low crystallographic symmetry. Futher, Ga has an unusually low melting point (29 C) and a very small heat of fusion. As discussed by the author (6), this implies that the amount of stored energy nec̃essary to induce amorphization is comparatively low for Ga. This may explain the observed amorphization of Ga at low temperatures.

More interesting is the case of intermetallic compounds which are frequently observed to undergo radiation induced amorphization. For intermetallic compounds, there are additional mechanisms for storing enthalpy in the lattice during ion irradiation. The most important is chemical disorder. Just as in the case of ion mixing, ion induced cascades cause atomic displacement and replacement events in intermetallic compounds. Again, the displacement process in the prompt cascade is followed by a very rapid quench to ambient temperature. The chemical disorder produced by the cascade is then configurationally frozen into the sample. In the simple Bragg-Williams theory of ordered compounds, the chemical enthalpy associated with ordering can be written

$$H = H_o S \tag{3}$$

where S is the Bragg-Williams order parameter and H_o is the chemical ordering energy. Ion irradiation tends to disorder the atoms on the sublattices of the intermetallic compound crystal structure. Like the case of ion mixing, this can be characterized as a high temperature random walk occuring among sublattices of the ordered compound. The author and others (6,29) have described this process as a competition between disordering in the prompt cascade (including the cool-down phase) and reordering due to atomic migration at ambient temperature. This can be written as

$$\left.\frac{dS}{dt}\right]_{irr} + \left.\frac{dS}{dt}\right]_{ord} = \left.\frac{dS}{dt}\right]_{tot} \tag{4}$$

where the first term is represents the rate of cascade-induced disorder and the second term represents reordering at ambient temperature (RED) or possibly reordering during the cool-down phase of the cascade. In a simple treatment of this problem, the author (6) showed that whenever an intermetallic compound has an ordering energy H_o which is comparable to the heat of fusion of the compound at the melting point, then ion irradiation at temperatures sufficiently low to suppress reordering kinetics, should result in amorphization. The criteria for amorphization can be alternatively stated as follows. Whenever an ordered intermetallic compound retains chemical long range order up to its melting point, it should be possible to induced amorphization by ion mixing at low temperatures. This follows from the fact that when a compound remains ordered to the melting point, then the ordering energy should be at least as large as the heat of fusion (6). This criterion in fact underestimates the potential of the crystal to store excess enthalpy since it is assumed that all of the enthalpy is stored in the form of chemical disorder. There will in fact be contributions arising from other point defects (e.g. vacancies etc.). These will contribute to the stored enthalpy. When the total stored enthalpy in a crystal at low temperatures equals the enthalpy difference between the crystal and the amorphous phase (i.e. the heat of crystallization of the amorphous phase), then the crystal can transform polymorphically to the amorphous phase. This can be conveniently thought of in terms of the T_o-curve concept already introduced. For the case of a disordered crystal, one might plot a T_o-curve as a function of disorder (for example as a function of S in the case of chemical disorder. For T=0, a critical value of S= S_{crit} exists below which the crystal has higher internal energy than the amorphous state. This defines the point at which the $T_o(S)$-curve goes to zero temperature. It can in fact be argued (30) that the T_μ-curve should not extent beyond this limit. Thus at low temperatures (where the question of configurational entropy can be neglected), a crystal with enthalpy higher than that of the amorphous phase should be unstable with respect to amorphization.

Table 2 contains a list of intermetallic compounds which have been observed to tranform to the amorphous state under irradiation by both ions and electrons. Electron irradiation is of interest since it involves single isolated displacement event with no cascade evolution. Nevertheless, electron irradiation still produces atomic site disorder in the intermetallic compounds. Luzzi et al. (31-33), have in fact shown that amorphization occurs when the Bragg-Williams order parameter of Cu_4Ti_3, $CuTi_2$, and NiTi, is decreased below a critical value. They in fact observed a critical temperature range (~250 K), above which thermally activated reordering kinetics prevent the critically disordered state from being achieved. Above this temperature, no amorphization is achieved. This is a clear example in which chemical disordering is responsible for the crystal-to-amorphous transformation. For both the cases of ion and electron irradiation which are shown in Table 2, one observes that whenever an intermetallic compound remains ordered up to its melting point, amorphization is observed at low temperatures. Exceptions in the table are the cases of Al alloys irradiated at room temperature. It is known that room temperature is inadequate to suppress thermally activated defect migration in Al base alloys (34). A more detailed comparison of ion induced cascade amorphization in intermetallic compounds must await futher detailed studies of the type done by Luzzi et al. for electron irradiation where measurements were made over a range of temperatures, and measurements of S

Table II. Intermetallic compounds which undergo irradiation induced amorphization. Effects of ambient temperature, T_{00}, irradiation type, etc.

Compound	Ta- Ambient Temperature (K)	Irradiation	$T_{00}(C)$	$T_m(C)$	Result	References
Cu_4Ti_3	>260	2 MeV, electrons	>Tm	920**	C	31
	<175	2 MeV, electrons	>Tm		A	
$CuTi_2$	>265	2 MeV, electrons	>Tm	990**	C	32
	<265	2 MeV, electrons	>Tm		A	
NiTi*	275	2 MeV, electrons	>Tm	1310	A	33
	290	2 MeV, electrons	>Tm		C	
AlCo*	RT	500 keV, Xe++	>1300	1645	C	35
AlNi*	RT	500 keV, Xe++	>Tm	1638	C	25
AlIr*	RT	500 keV, Xe++	?	?	C	35
AuZn*	RT	500 keV, Xe++	>Tm	725	C	35
RuZr*	370	600 keV, Xe++	>Tm	2100	A	20
	410	600 keV, Xe++			C	
RuTi*	77K	600 keV, Xe++	>Tm	2120	A	20
	300K	600 keV, Xe++			C	
SiRu*	473K	500 keV, Xe++	>Tm	1800	C	35
	RT	500 keV, Xe++			A	
FeTi*	RT	2.5 meV, Ni+	>Tm	1317	A	36
FeAl*	RT	2.5 MeV, Ni+	700	1260**	C	36
CoTi*	RT	500 keV, Xe++	>Tm		A	37
	160K	2 MeV, electrons		1350	C	38

RT - Room Temperature

* - "CsCl" - type intermetallic compounds

** - peritectic melting

were simultaneously conducted. Such studies should serve to confirm the role played by chemical disordering in the amorphization of crystalline intermetallics.

5. Conclusion

In conclusion, we have briefly surveyed the results of studies of transformations from the crystalline to the amorphous state induced during ion mixing and during irradiation of intermetallic compounds. It has been seen that the kinetic constraints imposed by cascade evolution impose severe constraints on the nucleation of new phases. By introducing the concept of T_0 and T_μ-curves, we can establish the conditions under which a crystal—to—amorphous transformation can occur spontaneously with little or no kinetic hinderance. These concepts have allowed us to develop criteria for predicting when ion mixing or irradiation of compounds can lead to amorphization. On the other hand, we have also seen that, in the absence of radiation enhanced diffusion due to defect migration, it is generally very difficult to nucleate new intermetallic crystalline phases. In summary, one might claim that ion irradiation at low temperature produces extended solid solutions, disorders homogeneous crystalline phases, and produces amorphous materials when the competing crystalline alternatives become unstable.

REFERENCES

1. W.L. Johnson, Y.T. Cheng, M. Van Rossum, and M-A. Nicolet, Nucl. Inst. & Meth. B7/8, 657 (1985)
2. R.S. Averback and D.N. Siedman, Mat. Sci. Forum, 15–18, 963 (1987)
3. see for example, J.R. Beeler Jr., Phys. Rev., 150, 470 (1966)
4. J.W. Cahn, in Rapid Solidification Processing, ed. by R. Mehrabian, B.H. Kear, and M. Cohen, (Claitors Press, Baton Rouge, LA), p.24 (1980)
5. T.B. Massalski, Proc. IV Int. Conf. on Rapidly Quenched Metals, ed. by T. Masumoto and K. Suzuki, (Japan Inst. of Met., Sendai), p.203 (1982)
6. W.L. Johnson, Prog. in Mat. Sci., Vol.30, p.81 (1986)
7. J.L. Tallon and A. Wolfenden, J. Phys. Chem. Sol., 40, 831 (1979)
8. J.L. Tallon, Phil. Mag., 39, 151 (1979)
9. P. Sigmund, Appl. Phys. Lett., 25, 169 (1974): also Appl. Phys. Lett., 27, 52 (1975)
10. F. Seitz, and J.J. Koehler, Sol. St. Phys., ed. by F. Seitz and D. Turnbull, Vol.2 (Academic Press, New York, 1956)
11. G.H. Vineyard, Radiat. Effects, 29, 245 (1976)
12. Y. T. Cheng, W.L. Johnson, and M-A. Nicolet, Phys. Rev. Lett., 58, 2083 (1987)
13. Y.T. Cheng, in this proceedings
14. Y.T. Cheng, M. Van Rossum, M-A. Nicolet, and W.L. Johnson, Appl. Phys. Lett., 45, 185 (1984)
15. M. Van Rossum, Y.T. Cheng, M-A. Nicolet, and W.L. Johnson, Appl. Phys. Lett., 46, 610 (1985)
16. see for example, P.G. Shewmon, Diffusion in Solids, (McGraw Hill, New York, 1963) p. 126
17. T.W. Workman, Y.T. Cheng, W.L. Johnson, and M-A. Nicolet, Appl. Phys. Lett., 50 , 1485, (1987)
18. Z.L. Wang, J.F.M. Westendorp, and F. Saris, Nucl. Inst. Meth. 209/210, 115 (1983)

420

19. Y.T. Cheng et al. have observed demixing of amorphous Ag/Fe film prepared by cosputtering and subsequently ion irradiated, unpublished results

20. Y. T. Cheng, W.L. Johnson, and M-A. Nicolet, Mat. Res. Soc. Symp. Proc. , Vol. 37, 365 (1985)

21. B.X. Liu, W.L. Johnson, and M-A. Nicolet, Appl. Phys. Lett., 42, 45 (1983); also B.X. Liu, E. Ma, J. Li, and L.J. Huang, Nucl. Inst. Meth. B19/20, 682 (1987)

22. B.Y. Tsaur, S.S. Lau, and J.W. Mayer, Appl. Phys. Lett., 36, 823 (1980)

23. see references 18 and 19

24. L.S. Hung, M. Nastasi, J. Gyulai, and J.W. Mayer, Appl. Phys. Lett., 42, 672 (1983); see also Ph. D. thesis by M. Nastasi, Cornell Univ. (1985)

25. T.B. Massalski, Binary Alloy Phase Diagrams, (American Metals Society, Metals Park, Ohio, 1986)

26. see for example, G. Martin and P. Bellon, Mat. Sci. Forum, Vol. 15-18, 1337 (1987)

27. see other paper presented in this proceedings under the topic of sputtering

28. R.S. Averback and D. Peak, MRS Sym. Proc., in press (1986)

29. R.H. Zee and P. Wilkes, Phil Mag., A42, 463 (1980)

30. W.L. Johnson, Proc. VI Int. Conf. on Rapidly Quenched Metals, Montreal, Canada, (1987), in press

31. C.E. Luzzi, H. Mori, H. Fujita, and M. Meshi, Acta. Met. , in press (1986)

32. D. E. Luzzi, H. Mori, H. Fujita, and M. Meshi, Proc. MRS Sym. A. , Beam Interactions with Solids, 1986, in press

33. H. Mori, H. Fujita, and M. Fujita, Japan J. Appl. Phys., 22, L94 (1983)

34. Y.T. Cheng, X.A. Zhao, T. Banwell, T.W. Workman, M-A. Nicolet, and W.L. Johnson, J. Appl. Phys. , 60, 2615 (1986)

35. M. Nastasi, D. Lilienfeld, H.H. Johnson, and J.W. Mayer, J. Appl. Phys. , in press (1986)

36. J.L. Brimhall, H.E. Kissinger, and L. A. Charlot, in Metastable Materials formed by Ion Implantation, ed. by S.T. Picraux and W.H. Choyke, (Elsevier, New York, 1982) p.243

37. L.S. Hung and J. W. Mayer, Nucl. Inst. Meth. B7/8, 676 (1985)

38. H. Mori, H. Fujita, M. Tendo, and M. Fujita, Scripta Met., 18, 783 (1984)

THE TOPOGRAPHY OF ION-BOMBARDED SURFACES

IAN H. WILSON

DEPARTMENT OF ELECTRONIC AND ELECTRICAL ENGINEERING,
UNIVERSITY OF SURREY, GUILDFORD, SURREY GU2 5XH, ENGLAND

1. PREAMBLE

The nature of the response of a surface to ion beams is strongly materials dependent. As will become clear here, and can also be determined from other recent reviews of this topic (1 to 4), very few general laws can be applied to the subject as a whole and it therefore does not lend itself to a grand analytical theory. One needs to select well defined areas in order to begin to accurately quantify the effects. The simplest situation one can visualize is that of a monoisotopic, amorphous, conducting solid bombarded with a collimated, spatially and temporarily uniform ion beam. One then chooses to neglect all atom fluxes other than those of the beam and primary (i.e. as a direct result of the incident ion collision cascade) sputtering. Even then iterative numerical methods still have to be used, except in the most trivial cases, to predict the evolution of surface contours with increasing fluence.

In order to describe the zoology of observed phenomena we need to classify them. One could devise classification schemes based on materials, for example: metal/insulator or element/mixture/alloy/compound/polymer, or crystalline/amorphous or nuclear stopping/electronic stopping. These will inevitably emerge later but here we will discuss processes.

I will now try to classify the atom fluxes into and out of an element of surface.

1.1 Atom fluxes

The various atom fluxes are shown schematically in Fig. 1, we will now discuss each in turn.

i) The incident ion beam. This is usually assumed to be uniform, but in general it will have imperfect collimation ($\Delta\Theta$) and spatial non-uniformity (Δx). For example a scanned gaussian beam involves both Δx and $\Delta\Theta$. On the microscopic scale, the statistical nature of the collision process must not be neglected. Craters and asperities on the scale of the collision cascade may be formed in metals, and most certainly in covalent materials. An example is the ion explosion spike that can occur in insulators due to intense ionisation during electronic stopping (1). These microscopic features can serve as nuclei for other processes.

ii) The sputtered flux. This is usually assumed to be dominated by prompt collisional events in the cascade. However the surface stopping power and surface binding energy are both affected by composition. This will be altered by the implant species, bulk and surface atom transport. Implantation of an active species can create an altered surface layer of very low binding energy

R. Kelly and M. Fernanda da Silva (eds.), Materials Modification by High-fluence Ion Beams, 421–466.
© *1989 by Kluwer Academic Publishers.*

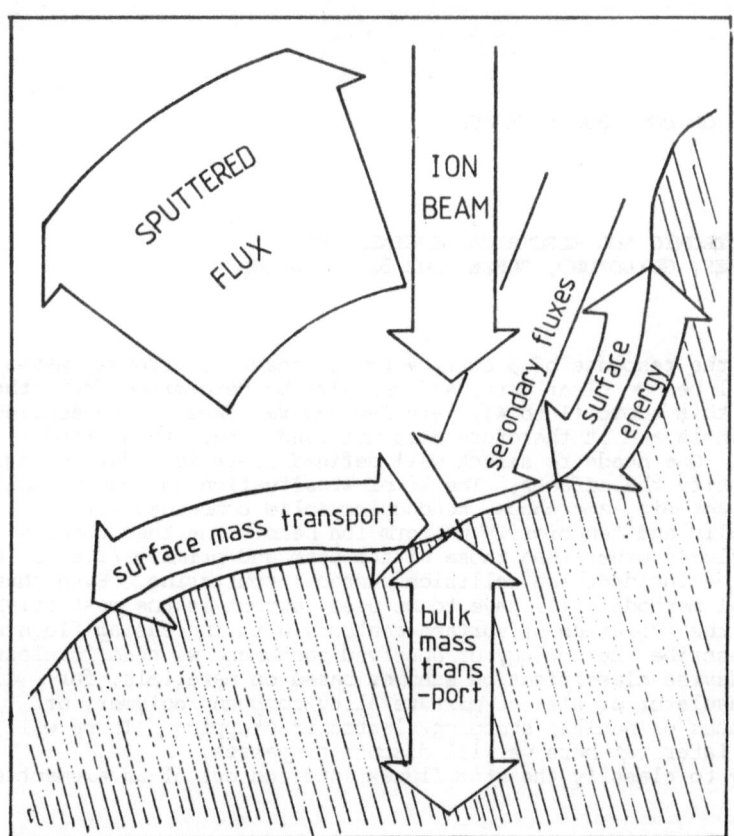

Fig. 1 Atom fluxes to and from an element of surface during ion bombardment.

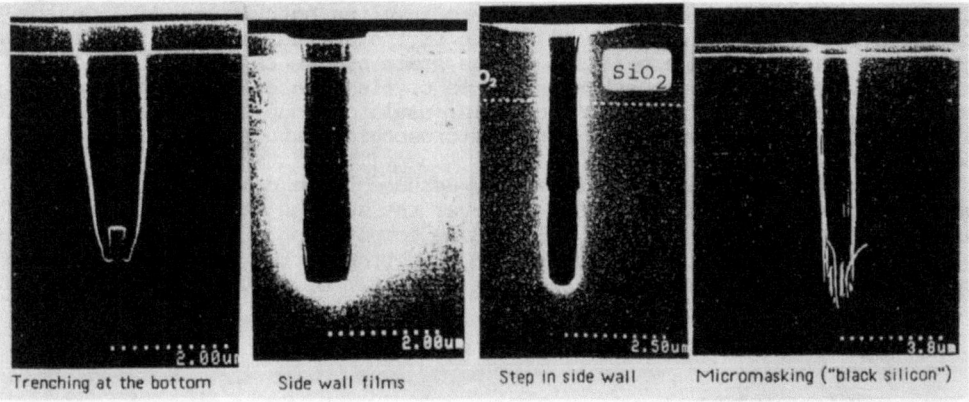

| Trenching at the bottom | Side wall films | Step in side wall | Micromasking ("black silicon") |

Fig. 2. Problems encountered in deep-trench ion milling of Si.

which can result in a dramatic increase in sputtering yield. This is the area of so-called 'chemical sputtering'. Often delayed effects (thermal and diffusional) and electronic stopping (ionisation) are largely ignored, although all can lead to erosion or growth. Electrons will sputter alkali-halides (62). The most important consequence of collisional sputtering is that as the angle of ion incidence Θ is tilted away from the normal (Θ=0), energy is deposited closer to the surface, and therefore the sputtering yield, S(Θ), increases. This increase is not monotonic for single crystals because channelling of the incident ion will depress the yield in low index directions.

iii) <u>Secondary fluxes</u>. These can be sub-divided into two main types, both geometry dependent: a) Primary ion reflection, which occurs if the beam is incident on another part of the target, for example the side of a masking layer, at close to grazing incidence. Reflected ions loose little energy and so will add to the erosion caused by the primary ion beam. This is termed 'flux enhancement'. b) Sputtered atoms. These will arrive from all parts of the surface that can be seen from our element of surface. In general the average energy of sputtered atoms is ~10eV so they will not cause further sputtering but their energy imparts high sticking probability. This is termed 'redeposition'. However close to grazing incidence the yield of high energy sputtered atoms may be sufficient to give net erosion.

iv) <u>Surface mass transport</u>. If we have more than one species on the surface (for example when the surface is exposed to flux of evaporated atoms at the same time as the ion beam) then diffusion may take place. The driving forces that operate mainly are concentration gradients and chemical potentials. The former gives a more homogeneous surface whilst the latter lead to segregation, e.g. nucleation and growth of a new phase at steps on the surface, with consequent spatial variation in sputter yield. Asperities, cones, pyramids and whiskers can be formed this way.

v) <u>Surface energy</u>. This is a third driving force for diffusion, which occurs to minimise surface energy by minimizing surface area. This leads to rounding of sharp points, filling-in of small craters, flattening of small asperities and the formation of bulging tops to large asperities. Anisotropy in the surface energy of single crystals manifests itself in the formation of facets. Twisting and plastic deformation of asperities may also occur in order to reduce surface stress.

vi) <u>Bulk mass transport</u>. Large point defect fluxes are created by ion bombardment. These can lead to segregation to/from the surface, altering the surface binding energy. Segregation can lead to the formation of precipitates, colloids, bubbles and blisters. Voids may be formed in situations where the collision cascade density is very high. The damage density may be sufficiently high to trigger an amorphous ⇄ crystalline phase change with consequent modification of the sputtering yield.

1.2 <u>This chapter</u>

In this chapter I have separated the effects into primary, secondary and tertiary. Primary sputtering deals with the simplest situation where only i) and ii) operate. Here we are concerned with describing evolution with increasing dose from an original profile considering only the variation of sputtering yield with angle of ion incidence (S(Θ)).

Secondary effects add to primary sputtering. We will be considering in turn secondary fluxes (iii), surface energy (v) and spatial variation of sputter yield (S(x)) due to dislocations, grain boundaries and single ion impact. Tertiary effects include some of the consequences of forming multi-element or multi-phase surfaces as described in (iv) and (vi).

Figure 2 shows an example of some of the effects that are of interest to the microelectronics industry (5). This illustrates some of the problems encountered in deep trench etching of silicon for the fabrication of 3D capacitor DRAM cells. The third dimension is being exploited due to the pressure to increase integration density, and ion etching is the technique that makes it possible. The stepped side wall may arise from the formation of stable edges during the primary erosion process, although redeposition must be an important phenomenon in such deep trenches as evidenced by side wall films. The 'trenching' i.e. flux enhancement effect is clear. So is the effect of spatial variation of sputter yield in the so-called micromasking (black silicon) effect - impurities (residues of carbon from resists?) form low yield regions, effectively very small masks, on the exposed surface creating a forest of small asperities.

1.3 Experimental methods

The scanning electron microscope is almost universally used. The resolution of modern instruments is constantly improving, but despite makers claims it is difficult to get resolution better than 0.5mm in the image at 20×10^3 magnification. This is equivalent to a resolution of 250 Å. One can improve on this by careful alignment, etc., and by placing the specimen as close to the objective lens as possible but the practical resolution limit is ~100 Å.

One can vary the viewing angle in SEM and so take stereo pairs, determine cone angles, and get contrast for low aspect ratio features by extreme oblique incidence. This facility is little used. Also the computer methods of image analysis now available would give a statistical survey of the surface but these are neglected at present. We have found it essential to make a map on the surface using a labelled TEM grid as a mask during the first bombardment. Then each feature can be given a map reference which can be returned to after further treatment. An example of this on a single crystal Cu surface (surface orientation (530)) is shown in Fig. 3.

Resolution of atomic scale topography is now possible using the scanning tunnelling microscope (STM). Hillocks 50 Å diameter and 10 Å in height have been observed on the surface of Si subjected to high doses (up to 1×10^{19} ions cm^{-2}) of low energy (700 eV) Ar^+ (16). No explanation of this phenomenon is given (oxide formation is not thought to be the cause). Could it be surface energy minimisation by wrinkling?

More remarkably STM has very recently been used to observe the effects of single ion impacts (84). Shown in Fig. 4 is an STM micrograph of a crater left after the impact of a 100 keV Ge ion on a Si (100) surface. It is thought that amorphized or high defect density regions oxidise and/or etch slightly more rapidly during oxide removal in a buffered HF etch. This leaves a trace of the cascade in the silicon surface.

Transmission electron microscopy (TEM) is also approaching atom resolution, but is yet to be used extensively to study topography except by the use of surface replicas. However in certain special circumstances the dynamics of atomic movements at and above surfaces can be observed (7).

The third atomic resolution technique is field ion microscopy, where atomic arrangements on an emitter tip can be observed. Individual cascade events have been seen, also surface energy effects. The restricted range of suitable emitter materials limits the use of this technique.

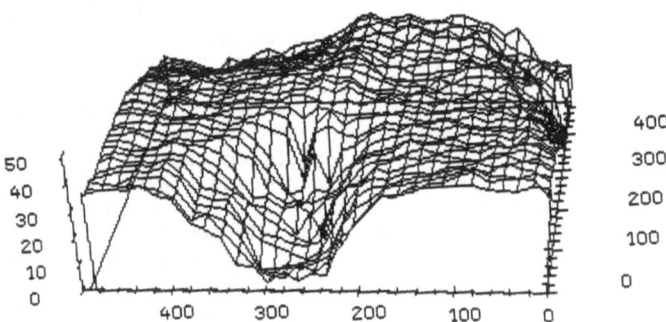

Fig. 3 Mapping the surface. A coded TEM grid is used as a shadow mask.
Cones on the Cu surface (orientation (530)) can be given grid references
within each unique 100 μm square.

Fig. 4 Scanning tunnelling microscope picture of the crater left by a single
ion impact (100 keV Ge$^+$ → Si (100)) and chemical etching to remove surface
oxide (ref. 84). All units in Å.

2. PRIMARY EROSION

In this section we consider the evolution of surface contours under the action of bombardment by a collimated ion beam. The only phenomenon considered is the variation of sputtering yield with angle of ion incidence $S(\Theta)$.

As this phenomenon is central to all approaches mentioned in this section we will first spend some time in discussion of $S(\Theta)$.

2.1 The variation of sputtering yield with angle of ion incidence

Not only is $S(\Theta)$ important in determining how surface topography changes during sputtering by ion bombardment (1,8 to 11), but it is a sensitive test of theories, particularly the anisotropy of momentum, of the collision cascade.

There has been little effort addressed to predicting $S(\Theta)$ theoretically (13) except by Monte-Carlo simulations (14 to 16). It is usual to assume (17) that energy deposited in a layer close to the surface determines the yield and that $S(\Theta)$ turns over with a peak $\hat{\Theta}$ = 60° to 80° solely because of the increasing influence of ion reflection from the inclined surface.

Many aspects of sputtering are satisfactorily described by linear transport theory (18). The Boltzmann equation is solved for a linear cascade assuming a chaotic model and, particularly, an isotropic flux of low energy recoils. The deposited energy is determined, and from this back-sputtering yields. Convenient assumptions have to be made regarding the low energy cut-off, and the surface binding energy. Conventionally the theory is tested by prediction of the variation of sputtering yield with ion energy (13). The maximum in the yield curve is attributed to a maximum in the surface deposited energy. However molecular dynamics simulations (19) have shown that for large ions the yield peaks at an energy where the surface deposited energy is monotonically increasing. This discrepancy is attributed to anisotropy in the momentum distribution within the collision cascade. At energies above the peak in yield the momentum of the primary knock-ons is increasingly directed away from the surface, i.e. towards the direction of ion incidence, with a consequent reduction in the transverse momentum which is responsible for generation of the low energy (higher generation) part of the cascade close to the surface.

Transport theory has been modified in order to determine the anisotropy within the collision cascade (20,21). Calculation of the momentum distribution is substantially more complex than calculation of the energy deposition profile so analytic interactability makes comparision with observables difficult. General predictions can be made with regard to the range of validity of the isotropic model, the energy spectrum of the sputtered particles and particulary the angular distribution of sputtered atoms.

It is surprising that the theory of $S(\Theta)$ has been little studied as one can see intuitively that, as the angle of ion incidence is rotated away from the surface normal, different parts of the collision cascade will come close to the surface and therefore influence the sputtering yield. Indeed, in the presence of anisotropy, high energy components will increasingly be lost from the cascade, with a consequential effect on cascade development. It would seem reasonable that by 'sectioning' the cascade this way one could determine experimentally the momentum distribution by some form of deconvolution. Isotropic energy deposition (18) predicts a $\cos^{-x}\Theta$ dependence ($1 \leq x \leq 1.7$, depending on ion/target mass ratio). Experimental results (13) do not show a consistent trend in agreement with this theory.

Recently a simple approach to the determination of the momentum distribution in the collision cascade has been tried (22 to 25). The geometrical distribution of successive generations of knock-ons is considered. Features such as the peak in the $S(\Theta)$ curve at $\hat{\Theta}$ can be predicted from consideration of anisotropy alone.

Anisotropy effects are particularly important for sputtering at ion energies in the region of and beyond the peak in the sputtering yield curve, so the results of this work are particulary relevant to ion energies of the order of 100 keV.

The geometric model is illustrated in Fig. 5. An ion incident at Θ to the normal of a plane surface penetrates to the 'sputtering depth' X_s where an energetic primary recoil R_0 is generated. The primary generates secondary and higher generation recoils R_1 etc. at the end of its path. The primaries are anisotropic in ϕ, the majority being forward directed. In the simplest model the secondaries etc. are assumed to be distributed isotropically. Thus any one primary generates a spherical volume of secondaries etc. and the cone of primaries at angle ϕ to the direction of ion incidence will generate a toroidal volume of secondaries. The area of intersection of the torous with the surface gives a measure of the contribution to the sputtering yield of primaries inclined at ϕ to the direction of ion incidence. A simple polynomial is used to describe the distribution of primaries with ϕ. Single term distributions of the form ϕ^n for n = 1, 2 and 3 are shown in Fig. 6. Clearly a 'theory' with so many adjustable parameters (X_s, R_1, R_0, ϕ^n) can be used to 'fit an elephant and make the tail wag'. However R_1 and R_0 can be determined by other means leaving just two adjustable parameters. Also the purpose of this work is to demonstrate the importance of anisotropy rather than predict accurately experimental results. Three experimentally determined curves have been fitted: 50keV Ar^+ bombardment of Au and 100keV Ar^+ bombardment of Ge and Si. The results are shown in Figs. 7a to c. The results for Au and Ge are fitted with a primary distribution $\alpha\phi^3$, not too well in the case for Ge; higher orders are probably needed and indeed the Si curve is fitted with ϕ^4. Ion reflection has been included in these results, taken from TRIM (Monte Carlo) simulations. The results for Au and Ge are shown in Fig. 8, and the result of sec Θ combined with ion reflection is shown as the dashed line in Fig. 7a. Sec Θ tends to ∞ much more strongly than reflection tends to 0 and so no turn-over is seen. Anisotropy resulting in incomplete development of the cascade, i.e. loss of primaries from the surface, must account for the relatively low angle of maximum sputtering yields ($\hat{\Theta}$).

The $S(\Theta)$ curve for single crystals is not monotonic as can be seen in Fig. 7d for 20keV Ar^+ on Cu(111) (63). Sharp dips in yield occur in the low index directions due to channelling of the incident ion which reduces the energy deposited close to the surface.

2.2. Profile evolution – analytical theory

Stewart and Thompson (26) used simple geometrical arguments to consider the movement of inclined planes and their point of intersection assuming a smooth $S(\Theta)$ curve with one peak ($\hat{\Theta}$), e.g. an amorphous solid. Their arguments have recently been well summarised by Wehner (4) as follows:

"Assuming normal ion incidence on a macroscopically flat part of a target surface and neglecting redeposition effects one can readily conclude that every contour would remain unchanged if the sputtering yield $S(\Theta)$ (atoms/ion) would be independent of Θ, because every surface element recesses in ion beam direction by the same amount, no matter what its local impingement angle is. If however $S(\Theta) > S(0°)$ with a maximum at $\hat{\Theta}$ between 0° and 90°, every slope with $\Theta \leq \hat{\Theta}$ as shown schematically in Fig. 9(a) would not only recess in beam direction faster than the flat areas but move, as shown, laterally towards the elevated side because d'>d. The thickness d" removed perpendicular to the slope is d" = d' $\cos\Theta$, which can be smaller or larger than d depending on whether $S(\Theta)$ is smaller or larger than $S(0°)/\cos\Theta$. Every slope with $\Theta = \hat{\Theta}$

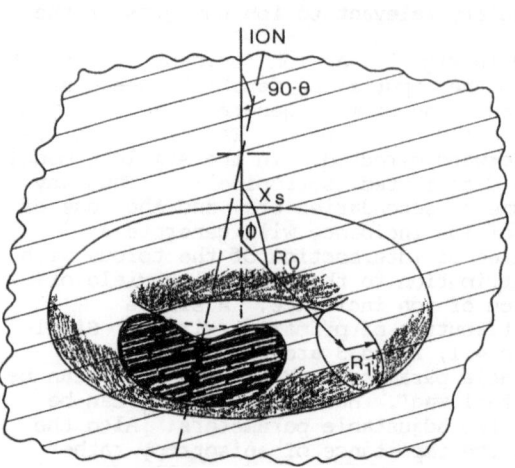

Fig. 5. Parameters for the geometric model for S(Θ).

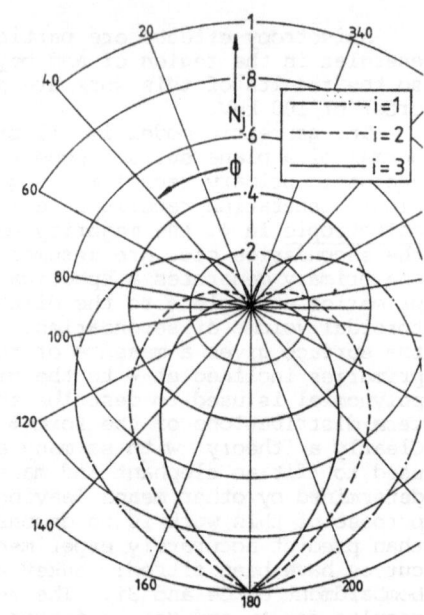

Fig. 6. Angular distribution of primary knock-ons, $N_j = c\phi^i$.

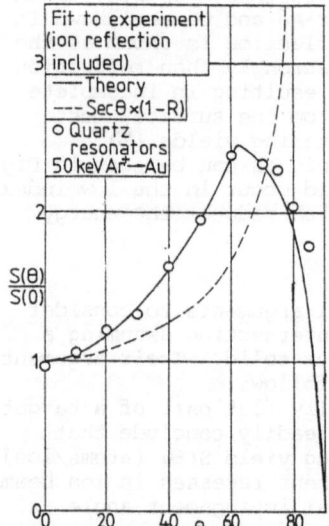

Fig. 7. Variation of sputter yield with angle of ion incidence Θ.
a) 50keV Ar$^+$ → Au (polycrystalline).

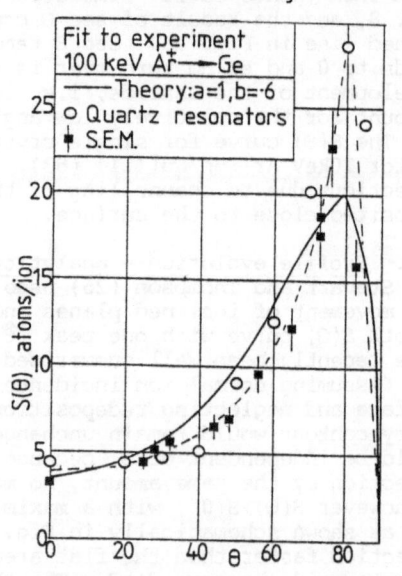

b) 100 keV Ar$^+$ → Ge: $a = \langle R_0 \rangle / \langle X_s \rangle$
$b = \langle R_1 \rangle / \langle X_s \rangle$

would move towards the right in the sketch with maximum speed but the slope itself would remain unchanged. A quite different situation arises when $\Theta > \hat{\Theta}$ such as schematically shown in Fig. 9(b). Because at the upper corner a microscopically small area must have the most favorable attack angle $\hat{\Theta}$, this gradually widens until the whole slope has changed to $\hat{\Theta}$. A very steep slope with $S(\Theta) < S(0°)$, if no deposition effects took place, could move towards the lower side (or to the left in Sketch 9(b)) although starting at its upper corner it would again convert to the $\hat{\Theta}$ slope."

This simple two dimensional theory was used to suggest a reason for cone formation by slopes at $\hat{\Theta}$ meeting, although such a cone would rapidly disappear with a rate of erosion faster than the surrounding $\Theta = 0$ surface. Depressions in the surface should get wider with increasing dose. Both phenomena have been observed, but Stewart and Thompson were the first of many to point out that other experimental factors may be at play. For example surface diffusion, and secondary fluxes. The major problem of this approach when applied to metal targets is the crystalline nature of the target leading to more than one local maximum in $S(\Theta)$. This problem has not yet been fully addressed.

A quite different approach that in principle is more generally applicable to three dimensional surfaces under erosion and growth has been developed by Carter and co-workers over many years (1,2). The concept used is that of a surface as advancing wave. The motion of a surface point of orientation Θ can be described by a propagating wave equation in the variable Θ such as:

$$\left.\frac{\partial \Theta}{\partial t}\right|_x = -\frac{J}{N}\frac{dS(\Theta)}{d\Theta}\cos^2\Theta\left.\frac{d\Theta}{dx}\right|_t , \qquad\qquad 2.1$$

where J is the flux density and N the atomic density.

This presents difficulties in solution as the effective wave velocity varies with Θ, but simplifications allow prediction of $\hat{\Theta}$ cones for example. However, formalisms for the study of space time developing wave fronts with nonconstant wave velocity have been developed (27) to facilitate solution for a number of cases where effects in addition to $S(\Theta)$ occur. The normal method of solution is by determining the trajectories in space and time of elements of the surface. This is the method of characteristics, which are manifest in geometric optics as the optical rays. By this method the wave front (i.e. the surface in our case) can be reconstructed at successive intervals of time. The method of characteristics has been applied by Frank to the disolution of crystals (28,29) and by Luke to geomorphological development by erosion (30,31). For sputter erosion it is usual to compute the trajectories of elements of the surface of constant orientation. For an element of surface at a certain orientation this orientation is reproduced at successive stages of erosion and follows a well defined path at a fixed speed, v_Θ.

If two trajectories (i.e. characteristics) intersect then all elements at values of Θ between the two are lost and an edge (i.e. discontinuous change in Θ) is formed. The total change in orientation from any point on an initial profile to any point on the sputtered profile is given in one dimension by:

$$\delta\Theta = \left.\frac{\partial\Theta}{\partial t}\right|_x \delta t + \left.\frac{\partial\Theta}{\partial x}\right|_t \delta x$$

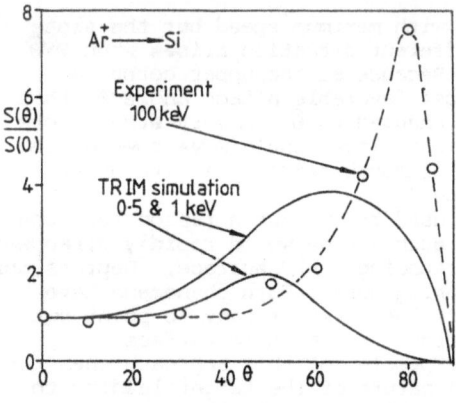

Fig. 7. continued   c) Ar⁺→Si.

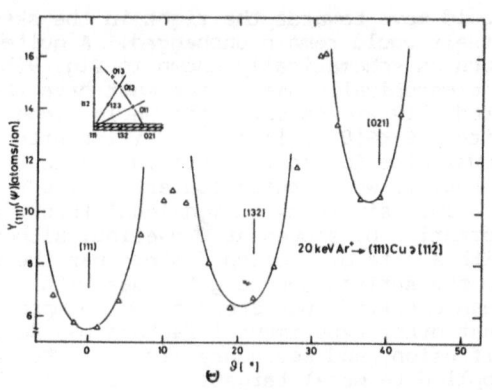

d)   20keV Ar⁺→Cu(111).

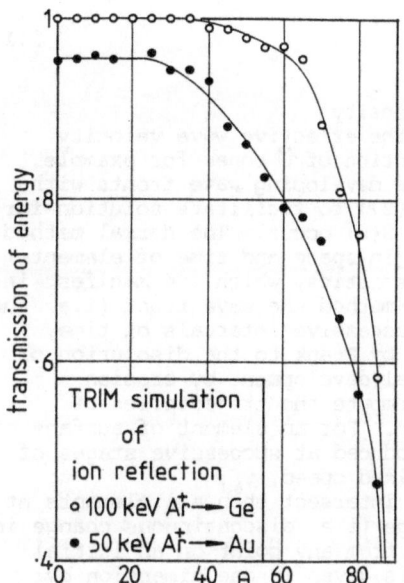

Fig. 8. Transmission of energy
into a surface versus angle of
ion incidence Θ. Monte Carlo
simulation.

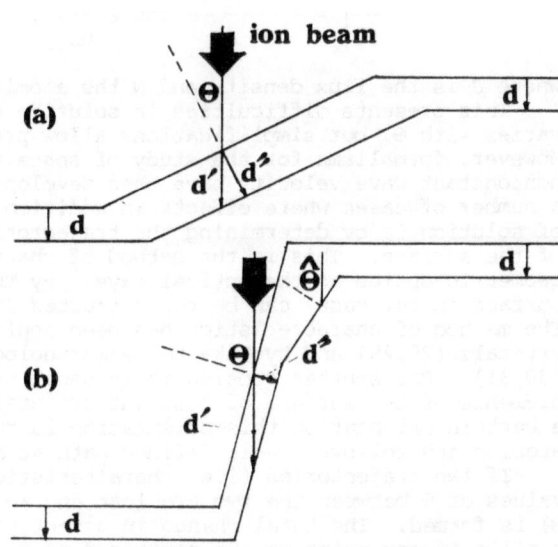

Fig. 9. Analytical theory of
profile evolution. a)   For
$S(\Theta) > S(0)$ and $\Theta < \hat{\Theta}$; b) $\Theta > \hat{\Theta}$

if $\delta\Theta = 0$ then:

$$\frac{dx}{dt}\bigg|_\Theta = - \frac{\partial\Theta}{\partial t}\bigg|_x \bigg/ \frac{\partial\Theta}{\partial x}\bigg|_t$$

and from equation 2.1, for constant Θ:

$$\frac{dx}{dt}\bigg|_\Theta = \frac{J}{N} \cos^2\Theta \, S(\Theta) = v_x,$$

where v_x is the velocity in the x direction of the element at Θ.

By analogy $\dfrac{dy}{dt}\bigg|_\Theta = v_y, \quad \dfrac{dz}{dt}\bigg|_\Theta = v_z$

and $\qquad\qquad v_x{}^2 + v_y{}^2 = v_z{}^2 = v_\Theta{}^2$

We will now briefly look at methods of solution by use of characteristics.

2.3   Charactertistics – graphical methods

This method was developed from the ideas of Frank (29,30) by Barber et al. (32) and an example of one method is shown in Fig. 10 from the work of Witcomb (33). Frank's two theorems state;
1. The locus of an elemental area of crystal surface with a particular orientation is a straight line during etching (assuming that the etch rate is only a function of orientation).
2. The trajectory of this elemental area is parallel to the normal of the polar diagram of the reciprocal of the etch rate at the point of similar orientation (defining the etch rate as measured normal to the actual crystal surface).

In the case of ion etching a polar diagram of the $S(\Theta)$ curve is used, or rather the inverse of this, which is called the 'erosion-slowness' curve. Examples are shown in Fig. 10. The formation of surfaces at $\hat\Theta$ can be seen to occur more rapidly for the element with the more sharply peaked $S(\Theta)$ curve.

A second method, applied chiefly by Nobes and colleagues (1,2) is the use of a characteristic velocity component plot as a cursor. This is illustrated in Fig. 11. The vector from the origin to any point on the plot represents the constant orientation velocity (v_Θ) and the associated recession angle (β). The cursor is derived from an $S(\Theta)$ curve (in this case a simple cosine series). The radius of curvature at any point on the curve is equal to the rate of change of the radius of curvature of the bombarded profile at the orientation Θ. The discontinuities in the curve Θ_{s1} and Θ_{s2} represent the inflection points on the $S(\Theta)$ curve. The peak sputtering yield angle $\hat\Theta$ is on the y axis as $dS/d\Theta = 0$. The development of edges may be directly inferred from the diagram, as illustrated in Fig. 12. Once sections such as XY in Fig. 12 form they would increase in size in proportion to erosion time.

An advantage of this technique is that it can be applied to situations where factors that effect net sputter yield in addition to orientation must be taken into account. We will see examples of this later. To date only two dimensional solutions have been attempted by graphical methods.

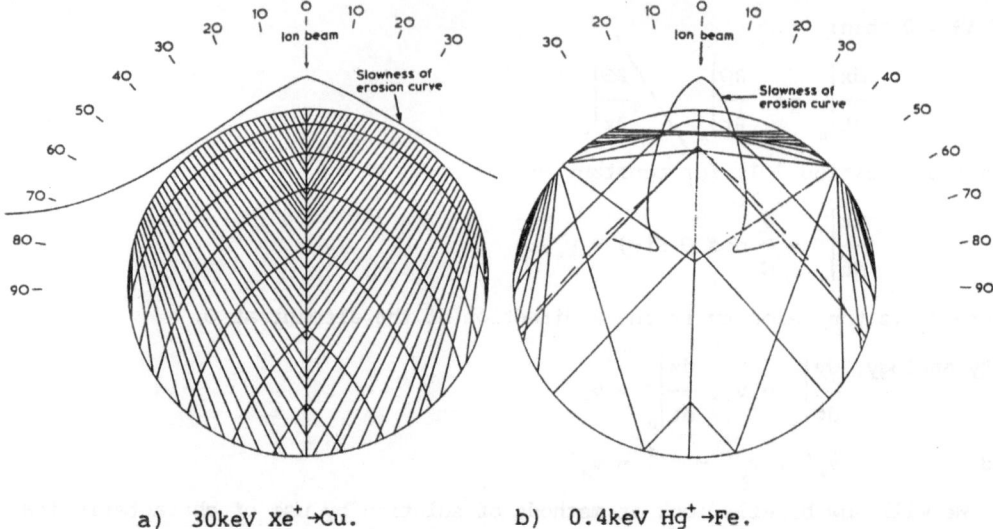

a)   30keV Xe⁺→Cu.   b)   0.4keV Hg⁺→Fe.

Fig. 10   Prediction of the evolution of spheres during ion bombardment by geometric construction from Franks theory of crystal disolution.

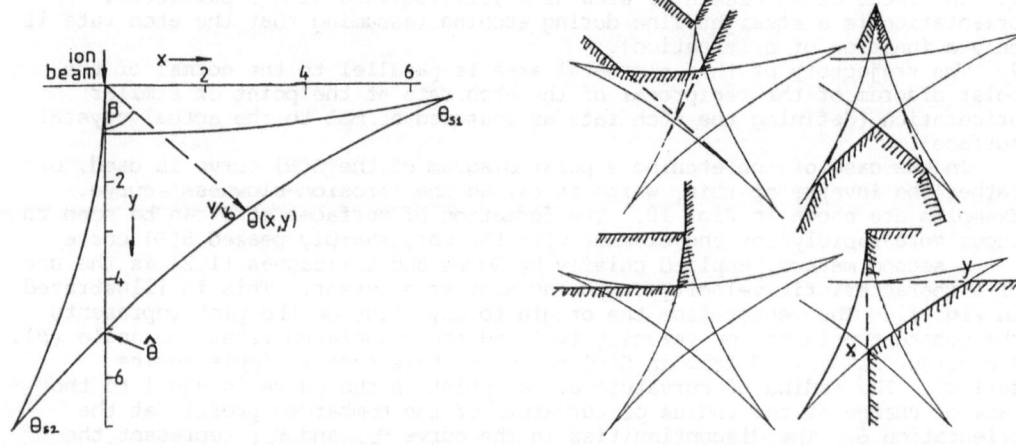

Fig. 11 The continuous line is a plot from $\Theta = 0°$ to $\pi/2$ of the normalized recession velocity (v_θ/v_o) of an element at Θ versus recession angle β. v_o is positive for $\Theta < \hat{\Theta}$, negative for $\Theta > \hat{\Theta}$ and zero at $\Theta = \hat{\Theta}$ (dS/dΘ=0). Cusps at Θ_{s1} and Θ_{s2} represent inflection points on the $S(\Theta)$ curve.

Fig. 12 The etching of plane-bound edges can be predicted by aligning the full cursor ($0°<\Theta<\Pi$) with the beam, apex placed at the edge in the initial (upper) profile. This region will erode to a shape similar to part of the cursor.

2.4   Characteristics - computational methods

In principle, if all the fluxes to each element of surface illustrated in Fig. 1 can be quantified, then an iterative computer code can be used to reconstruct the surface profile in three dimensions. Many codes of this type are used today in the semiconductor industry for modelling ion milling (for example ref. 34). Wilson et al. (35) have used such a code to determine the shift of each segment of surface along its normal in the presence of redeposition of material from the bombarded surface.

Ducommun et al (36) used a non iterative computer code based on constant orientation trajectories to predict the formation and motion of edges on surfaces due to the presence of inflection points (Θ_{s1} abd Θ_{s2} in Fig. 11) on the $S(\Theta)$ curve. They also successfully simulated the development of edges on silicon as a result of low energy (1 keV) argon bombardment.

Smith and co-workers (e.g. ref. 2) have fully developed the computation of characteristics using the basic equations developed by Carter (1). They have given careful consideration to the effects of intersection of characteristics, here I quote from ref. 2:

"Computationally, the characteristic method is fairly straightforward to programme. Its drawback, however, is that it can produce a large set of unwanted data after the characteristic intersections have indicated the formation of edges. Once this has occurred, all subsequent information obtained by integration along the characteristics must be discarded, since the uniqueness of the solution breaks down. If such points were included, the surface would appear to consist of cusps and folds but physically such points have no meaning and must be deleted from the final result. Although it is computationally possible to include pieces of computer code in the programmes to deal with these spurious points, it has been found to be more convenient to delete them manually from the data sets after the eroded surfaces have been calculated."

"Just as edges can form from an initially smooth surface, so can smooth surfaces form from initial gradient discontinuties in a surface. This is computationally more difficult to deal with. It can arise, for example, in pattern delineation where an initial angular profile is subject to further erosion by ion beams. The motion of these angular points requires a special consideration in the computation."

This method has been developed to deal with erosion of crystalline materials (where the complex form of $S(\Theta)$ leads to facet formation), time dependent beam orientation, non-uniform ion fluxes, multiple ion beams, redeposition and ion reflection. Erosion in three dimensions can be determined, as illustrated in Fig. 13, for an amorphous material with uniform flux and $S(\Theta)$ appropriate for silicon. An elliptical hummock erodes to form a sharp ridge which then reduces in size to leave a flat surface.

An alternative method to the kinetic wave formulation has been proposed by Marsh and Collins (37). This is based on a construction analogous to that of Huygens for an optical wave front. It is formally equivalent to the kinetic wave method in the limit of vanishingly small dose increment, but confers some advantages in stability on the resulting computer program. The formation of a facetted valley is shown in Fig. 14 (the angle for peak sputtering yield is here called Θ_p). In its original form spherical Huygens wavelets, radius proportional to v_θ, were used. This leads however to problems at corners and a more proper approach is now being used where the wavelets take the form of the cursor shown in Fig. 11.

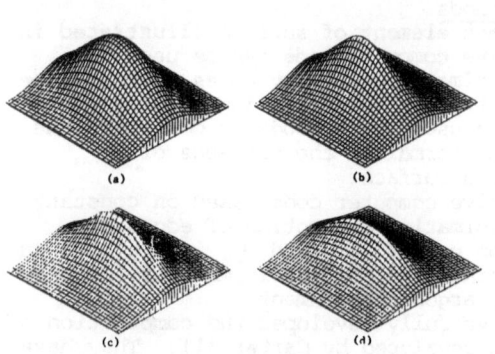

Fig. 13 Three dimensional computation of the erosion of an ellipitical hump (equal dose intervals).

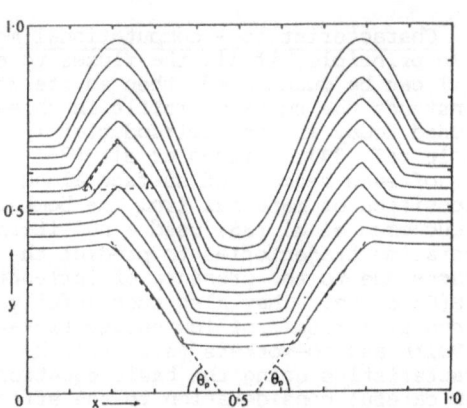

Fig. 14 Use of Huygens construction to compute erosion of valley edged by mountains.

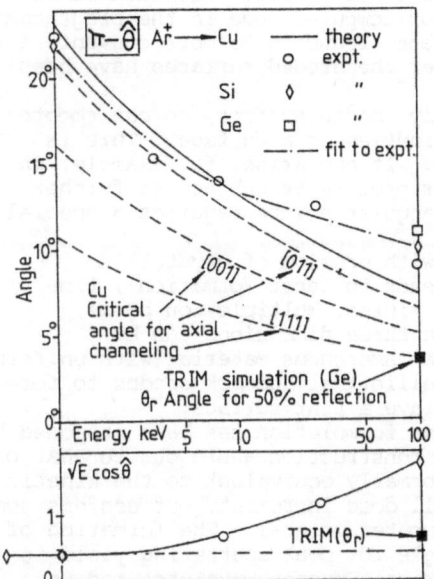

Fig. 15. Experimental and theoretical plots of π-θ̂ versus ion energy compared with axial channelling angles and critical angle for ion reflection (TRIM). Bottom curve: ion velocity normal to the surface at θ̂ versus ion energy.

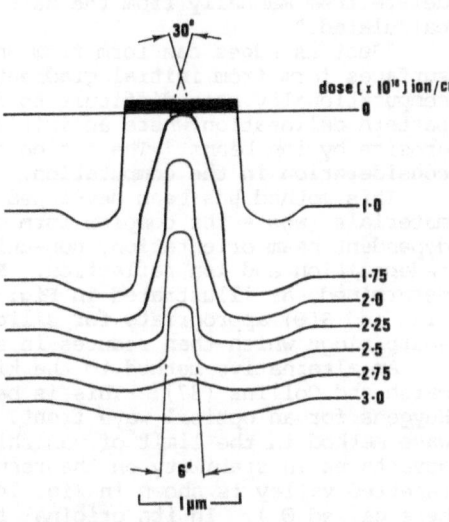

Fig. 16. Trenching or dig-out a) 40keV Ar⁺→GaAs. Profile evolution below a mask from SEM observation of one feature.

3. SECONDARY EFFECTS

With the understanding that $S(\Theta)$ always operates we will now consider four phenomena that can assume importance in determining the shape of ion bombarded surfaces: The secondary fluxes due to reflected ions and sputtered atoms, surface energy, and the variation of sputtering yield with position on the surface, $S(x)$.

3.1 Ion reflection

Ion reflection, which implies non–penetration of the lattice, is analogous to channelling. The ion is reflected by the potential barrier presented by the sheet of atoms comprising the target surface. The theory of the penetration of charged particles through crystals can therefore be applied. This was first attempted by Thompson (26) and later by Witcomb (38) using the Lindhard theory (39). Problems encountered included the choice of potential and the interatomic separation in the surface. Assuming that ion reflection alone was responsible for the turnover in the $S(\Theta)$ curve Witcomb found that he could successfully predict values of $\hat{\Theta}$ for low energies ($\leq$30keV). He used a Thomas–Fermi potential and had to assume the closest packing lattice spacing. His theoretical prediction for Ar^+ on Cu is shown in Fig. 15.

The critical axial channelling angle appropriate for low energy (1 to 50 keV) Ar^+ ions incident on Cu monocrystals is also plotted in Fig. 15 (40). Also shown are values of $\pi-\hat{\Theta}$ determined experimentally by us (41) and other workers (38). The theoretical values, as one would expect, follow the general trend of axial channelling. However channelling would have to start at angles considerably less than $\hat{\Theta}$ in order to create the turn over in yield. Also it is known experimentally that critical angles for planar channelling tend to be at least a factor of 2 less than those for axial channelling. These two facts seem to indicate that ion reflection alone cannot account for the relatively high values of $\pi-\hat{\Theta}$. The experimental results for $\pi-\hat{\Theta}$ show an energy dependence in the range 1 to 5 keV similar to the theoretical prediction but the drop in $\pi-\hat{\Theta}$ flattens out at higher energies. This would support the idea (18,25) that the sputtering yield at high energies and oblique incidence is considerably reduced by incomplete development of the cascade, particularly by loss through the surface of primary knock–ons. A TRIM simulation of the variation of ion reflection with Θ is shown in Fig. 8. In both examples there is a large disparity between the critical angle where 50% of the beam is reflected (84.3° for 50 keV $Ar^+ \rightarrow Au$ and 87° for 100 keV $Ar^+ \rightarrow Ge$) and the experimental value of $\hat{\Theta}$ shown in Fig. 8 (65° for $Ar^+ \rightarrow Au$ and 80° for $Ar^+ \rightarrow Ge$).

Also shown in Fig. 15 is a normalized plot of the velocity perpendicular to the surface at $\hat{\Theta}$ (proportional to $\sqrt{E} \cos\hat{\Theta}$) versus energy E for Cu and Si. Naively one would expect this figure to remain constant, but the value steadily increases at energies $\geq$3keV. This seems to support the incomplete cascade model. The critical angle determined by TRIM for Ge (Fig. 8) agrees well with the low energy values for Si and Cu.

What is incontrovertible is that enhanced erosion occurs at the foot of steep slopes. This can be seen in Fig. 2 for Si, Fig. 16a for GaAs (42), Fig. 16b for Cu and Fig. 16c for CdTe (43). Two facts have emerged from experimental observations: The 'trenching' or 'dig–out' phenomenon occurs only at the foot of very steep slopes ($\Theta > \hat{\Theta}$). And the magnitude of the effect is proportional to sputtering yield. An energetic sputtered atom can sputter several further atoms, whilst it will only deposit one. Note (See Fig. 16a) that the trench forms below the masked column. It does not deepen on further

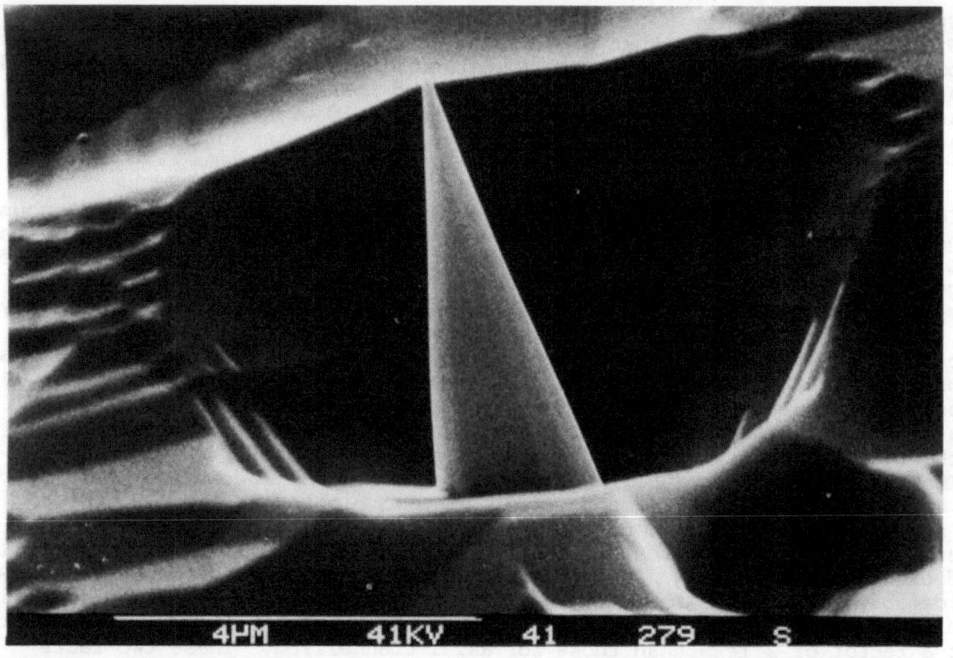

Fig. 16. (continued)b) 100keV Ar$^+$→Cu(530). Pyramid formed from a masked column (cultured cone).

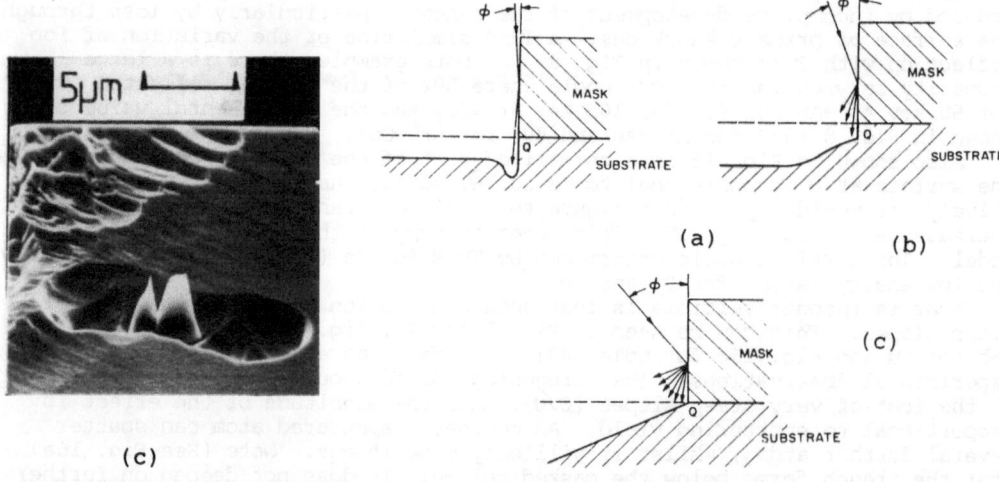

(c)

Deep trenching around an asperity in Ar$^+$ bombarded (40keV)CdTe. This material has a high sputtering yield and the large trenching effect dominates topography as the asperities disappear fast.

Fig. 17. Change in erosion behaviour with angle of ion incidence to mask edge. a) grazing incidence, b) ϕ~20° c) ϕ>30°

bombardment when the walls of the asperity are less steep. In fact the trench widens as would be predicted by primary erosion theory. Also it is known from ion etching of semiconductors, and illustrated in Fig. 17 (44), that trenching only occurs with grazing incidence on mask walls, for $\Theta \leqslant 80°$ sideways etching of the mask and redeposition on the substrate can lead to a build-up of material at the foot of the slope. The important question is therefore: how much is the 'trenching' due to reflected incident ions and how much is due to secondary sputtering caused by high energy sputtered atoms (e.g. escaped primaries)? To answer this question one must know how the energy distribution of sputtered atoms varies with Θ, particularly for $\Theta > \hat{\Theta}$. To my knowledge this has not yet been done experimentally. What could quite easily be done is a Monte Carlo simulation.

Ion reflection will be very sensitive to the nature of the surface. Planar channelling assumes a perfect crystal, i.e. atomically flat planes, but in reality with crystalline targets one may have a stepped surface, unless the whole slope happens to be a low index facet, and with amorphous materials the slopes may be wrinkled or humped. I feel that in many situations secondary sputtering may be at least as important as ion reflection for angles of incidence between the sputter yield peak and the critical planar channelling angle Θ_c ($\hat{\Theta} < \Theta < \Theta_c$) (43). Ion reflection will only predominate when the transverse velocity is insufficient for penetration of the lattice. However with a real surface with atomic steps a head-on collision is still possible at grazing incidence.

3.2 Redeposition

If an ion-bombarded surface is convoluted then certain areas of the surface will intercept the flux of atoms sputtered from other areas. If the sticking probability is non-zero then the movement of any elements of surface will be a balance between the removal of atoms by sputtering and redeposition from other parts of the surface.

In the case of chemical sputtering the sputtered species are mainly molecules with low surface binding energy, the sticking probability will be very low and one does not expect redeposition to play a significant role. In most cases of physical sputtering the sticking probability is expected to be close to unity, therefore consideration of redeposition effects is necessary. This is clearly demonstrated by ion etching at normal incidence where, as illustrated in Fig. 18, material is sputtered from the substrate onto the sidewalls of masks.

Both analytical and numerical methods have been applied to this problem and examples of these models will be described below. The analytical approach has the advantage that one can be certain that no artefacts are introduced, but at present it has only been applied to analysis of the flux arriving at one surface element from the rest of the surface in a few geometrical configurations. The numerical approaches all have the advantage that the evolution in surface topography can be predicted under the combined effects of normal sputtering, sputtering due to ion reflection from other parts of the surface and redeposition. The analytical approach may be useful in determining the flux of particles arriving at the new surface after each iteration.

In most cases it is assumed that the flux of sputtered atoms follows a cosine distribution with the maximum flux normal to the surface and zero flux in the surface plane. It is well known that distributions may be over- or under-cosine depending on ion energy, mass, angle of incidence and target mass. Indeed, the distribution may be asymmetrical about the local surface normal, with a distribution similar to that of specularly reflected light.

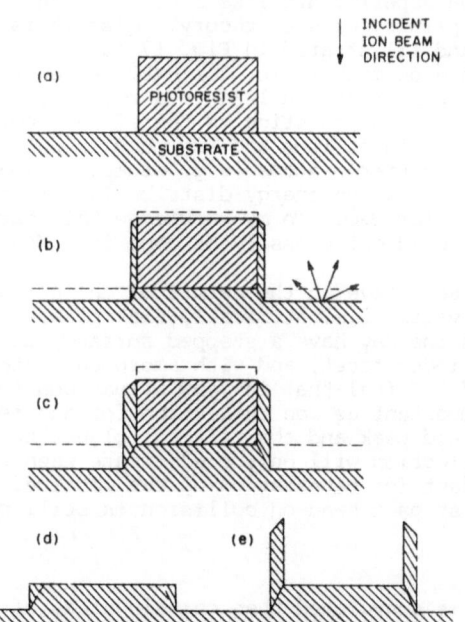

Fig. 18. Redeposition of substrate material onto mask walls. After removal of photoresist, deposit may tear off (d).

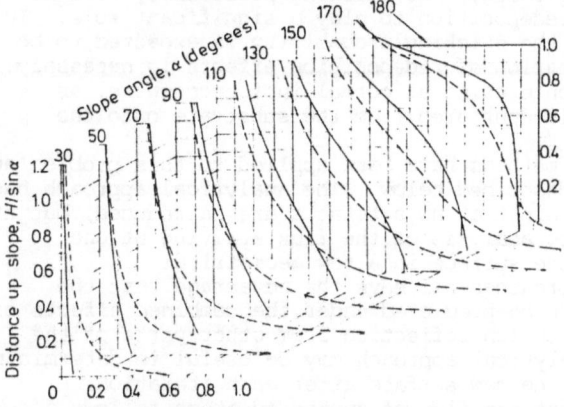

Fig. 20. Analytical determination of the redeposition flux density for cosine (continuous line) and isotropic (dashed line) distributions onto a linear slope, angle α (defined in Fig. 21).

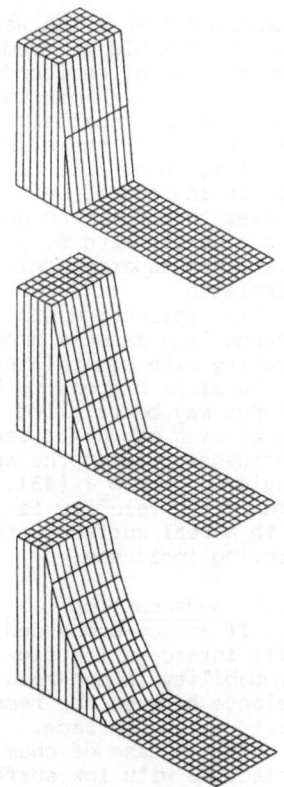

Fig. 19. Computed (method of characteristics) redeposition onto mask walls for overcosine (top), cosine (middle), and undercosine sputtered flux distributions.

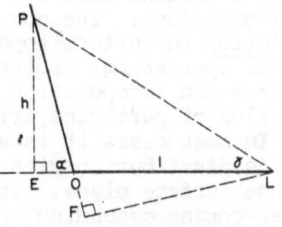

Fig. 21. Geometry for redeposition onto a linear slope.

Theoretical models have yet to take this into account. In particular, the variation of flux distribution with angle of ion incidence may have important consequences in the evolution of surface topography.

3.2.1. Numerical methods

An early iterative technique was that of Lehmann et al.(45). This allows for the inclusion of different mechanisms contributing to the development of surface topography under ion bombardment. In the initial treatment sputter etching arising from the direct ion beam, redeposition from the bombarded surface and redeposition from the volume above the surface were considered. (The last mechanism is thought only to be important in the case of plasma etching.) In more recent treatments a term describing sputter etching arising from ion reflection is added.

The physical quantities that enter this model are; $S(0)$ the sputtering yield per unit time for $\Theta_i = 0$, $S(\Theta_i)$ and $R(\Theta_i)$, the variation of sputtering yield and reflection ratio with ion incidence angle for component i. A cosine distribution of sputtered material is assumed. It is normal to use empirical expressions, such as cosine series, that best fit the experimental $S(\Theta)$ and $R(\Theta)$ relationships (if they are known). It is important to use realistic values for these functions.

The method of constructing new surface profiles from constant orientation trajectories or characteristic lines (The application of this method to sputter erosion has been described above) has also been applied to redeposition (46). A general integral for the flux arriving from an eroding surface at another surface has been derived, and evaluated exactly for redeposition from a horizontal plane (groove bottom) onto a vertical plane (groove wall) as this is of interest for microelectronics applications. Numerical calculations were performed to illustrate the effects of varying the ratio between groove height and groove width on the build-up of material on the groove wall, and the effects of under and over cosine flux distributions. The effects of flux distribution are illustrated in Fig. 19 (the horizontal scale is 10 x the vertical scale).

3.2.2. Analytical methods

Analytical expressions have been obtained in two dimensions for the flux density redeposited onto an arbitrary profile $y = f(\xi)$ from a horizontal emitter (47).

Solutions have been found for linear and sinusoidal receiving surfaces, where the linear surface can be at any angle to the emitter and the aspect ratio of the sinusoidal asperity can be varied to simulate a sharp spike or a low mound. The effect of the emitter distribution of sputtered atoms has been investigated by determining the flux density for both cosine and isotropic distributions, the latter being an extreme case of 'under-cosine' emission, relevant to low energies.

The results for redeposition from a flat surface onto a linear slope for cosine and isotropic distributions is shown in Fig. 20 for various values of α, the slope angle (defined in Fig. 21 and where H is height (h) over emitter length (l)). The results are qualitatively in agreement with those of Smith et al. (46). Deposition rate is very high close to the foot of the slope for an isotropic flux distribution. The curves for $\alpha > 90°$ do not take into account masking effects but they serve to show that the general trends continue.

The case of cosine emission onto linear slope was found to have a particularly simple solution for the normalized flux density arriving at the slope, F/n, namely:

$$\frac{F}{n} = \frac{1}{2} \{1 - \cos(\alpha-\gamma)\}$$

where the angles are defined in Fig. 21. For any value of α the deposition rate at any point P versus γ, the angle subtended by the end of the emitter at L, takes the form of a partial cardioid.

Fig. 20 shows that isotropic emission near the foot of a linear slope gives rise to a deposition rate (F/n) iso which is much greater than for cosine emission (F/n)cos. In the cosine case there is a lack of flux along directions close to the emitter axis and this suppresses the build-up evident in the isotropic case. Both types of emitter produce a reduction in (F/n) with increasing distance up a linear slope. However, the rate of reduction is less for cosine than for isotropic emission. Increasing remoteness from the emitter is counteracted, to some extent, by increasing luminance to give a deposition rate which is less height dependent. The curves for slope angles $\alpha<70°$ show that (F/n)cos is less than (F/n)iso if the distance up the slope is smaller than about one emitter length. However, there is a point, on each curve, above which (F/n)cos > (F/n)iso.

In Fig. 22 is shown a construction of the 'first burst' redeposition onto a sinusoidal asperity for a small emitter (dashed line, maximum height $\underline{a} = 10$ l) and a large emitter (continuous line $\underline{a} = 10^{-3}$ l). Unity sticking coefficient is assumed as is azimuthal symmetry in the distribution of sputtered atoms. Remember also that the solutions are only two dimensional and therefore can only indicate trends for three dimensional features. The effects of redeposition must in reality be greater for convex features as each element of receiver can see more emitter. Two aspect ratios are shown simulating a low bump and a cone-like asperity. The fall off in redepositon for the small emitter is very rapid due to the shadowing effect of the curved profile. In the case of a large emitter the bump tends to flatten out whilst the redeposition onto the cone-like asperity is almost independent of height except at the apex, which would tend to flatten as the diameter of the asperity increases. Similar effects have been observed experimentally (48, 49).

3.3. Surface energy

The effects of the surface diffusion of impurities on ion bombardment topography, particularly their role in generation of asperities, has been discussed by Rossnagel et al. in Refs. 1 and 2 and by Wehner in Ref. 4. As a foreign species is present in these cases I classify this phenomenon as a tertiary effect (which will be discussed in Section 4). However even on a pure surface, diffusion, enhanced chiefly by ion impact heating, can modify the shape of surfaces in order to minimise surface energy. Two examples are rounding or flattening of small asperities and pits in order to minimise surface area and faceting which arises from the anisotropy of surface energy in single crystals. Two other means of reducing surface energy are more mechanical in nature, namely plastic deformation and twisting to a minimal configuration.

3.3.1. Surface stress

The surface of a solid constitutes a discontinuity in the atomic structure and so there are unique forces associated with the boundary. We begin by distinguishing between the Helmohltz free energy γ per unit area of crystal surface and the surface tension s per unit length within the boundary. The

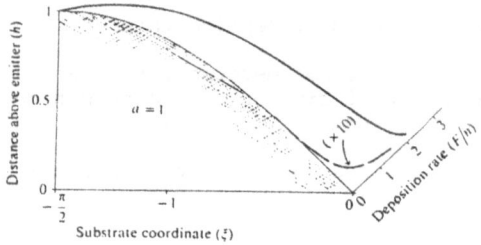

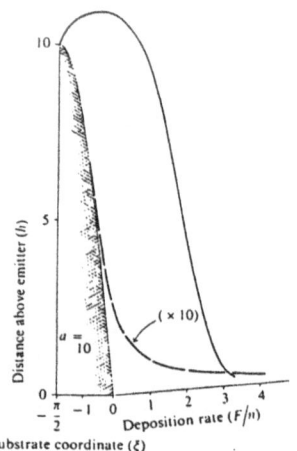

Fig. 22. Redeposition onto
sinusoidal asperities.
Full line:large emitter
($\underline{a}$ = 10^{-3} 1). Dashed
line:small emitter
($\underline{a}$ = 10 1), where a is the
maximum height of the asperity.

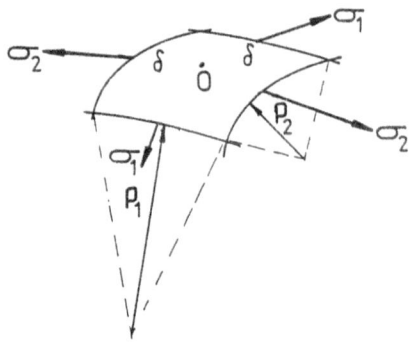

Fig. 23. Surface stress.
σ_1 and σ_2 are the orthogonal
stresses on element of surface
of area A, and side δ. ρ_1
and ρ_2 are the radii of
curvature at origin 0.

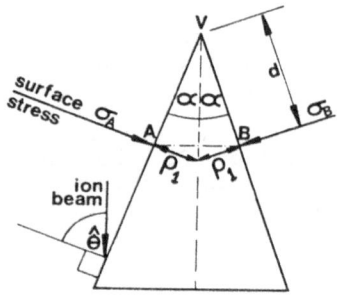

Fig. 24a. Principal section
through a right circular cone.
σ_A = σ_B local normal
surface stress.

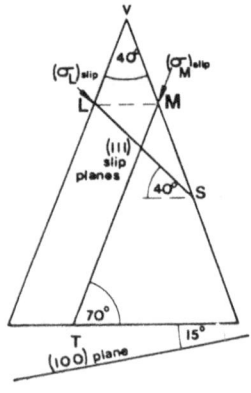

Fig. 24b Principal section
through a right circular
on a (530) surface. (111) slip
planes shown. (σ_L) slip and
(σ_M) slip resolved shear stresses.

work of Shuttleworth (50) has proved invaluable in understanding this distinction.

Consider the thermodynamically reversible isothermal deformation of a crystal face (Fig. 23) having initial area A. Let infinitesimal increases dA_1 and dA_2 be associated with orthogonal stresses σ_1 and σ_2 in the plane. If we do not admit any change in the crystal volume then the work done by these stresses is equal to the change in the Helmholtz free energy of the surface:

$$\sigma_1 \, dA_1 + \sigma_2 \, dA_2 = d(A\gamma_A), \tag{3.1}$$

where γ_A is the surface free energy per unit area.

i.e.
$$\gamma_A = E_A - TS_A$$

E, T and S have their usual thermodynamic meanings. Shuttleworth's analysis showed that for an isotropic substance or for a crystal face with n-fold symmetry ($n \geq 3$) all normal components of the surface stress equal the surface tensions and equation (3.1) takes the form

$$s = \frac{d}{dA}(A\gamma) = \gamma + A\frac{d\gamma}{dA} \tag{3.2}$$

where $dA = dA_1 + dA_2$.

It should be noted that a change of shape at constant area does not necessitate the expenditure of work since in such a case the shear components of the surface stress are zero. We thus arrive at the following physical interpretation of the surface tension of a solid. It is the work required for unit increase in area by means of equal stretching in all directions.

In terms of the strain, ε, equation (3.2) appears as

$$s = \gamma + \frac{d\gamma}{d\varepsilon} \tag{3.3}$$

In a solid under conditions of insignificant atomic mobility the number of surface atoms does not change under deformation. Then γ is a function of area and $d\gamma/dA$ (or $d\gamma/d\varepsilon$) is not zero. Thus in general γ, the specific surface free energy, and s, the surface tension, are not the same.

We shall be concerned below with applications of these considerations to the surfaces of small cones (~1 μm). When a surface element is planar the surface tension can be balanced by external forces applied at the periphery parallel to the surface. However, when the surface is curved a pressure must act normally to the surface in order to maintain equilibrium. Consider the small curvilinear square of side δ, centre 0, shown in Fig. 23. The square lies within the surface of a crystal. Let the principal radii of curvature at 0 be ρ_1 and ρ_2. The surface tension forces acting on the two pairs of sides have components $s \, \delta^2/\rho_1$ and $s \, \delta^2/\rho_2$ parallel to the normal at 0. If the pressure difference across the surface is Δp then

$$s \, \delta^2 (1/\rho_1 + 1/\rho_2) = \Delta p \delta^2$$

i.e.
$$\Delta p = s(1/\rho_1 + 1/\rho_2)$$

This is the known Laplace fundamental equation.
Using equation (3.3) we can relate Δp to $\gamma(\varepsilon)$:

$$\Delta p = (\gamma + d\gamma/d\varepsilon)(1/\rho_1 + 1/\rho_2)$$

From here on we shall treat Δp as a local normal stress (σ) acting in the vicinity of each point of a surface, and write:

$$\sigma/s = (1/\rho_1 + 1/\rho_2) \qquad (3.4)$$

Let us now consider the case of a right circular cone. The curve of constant stress is a circle lying parallel to the x-y plane with its centre on the cone axis. Consider a principal section (Fig. 24a) through the apex V of a right circular cone, half-angle α. We wish to estimate the local surface stress (σ_A or σ_B) acting normally at points such as A or B located at a distance d down the generators from V. The principal radius in the plane of the normal section through A is given by

$$\rho_1 = d \tan \alpha$$

From equation (3.4) (Putting $1/\rho_2 = 0$) we can write

$$\sigma_A = \sigma_B = s/(d \tan \alpha) \qquad (3.5)$$

which is the normal stress at every point on the circle.

We now examine the specific case of ion bombardment of copper for which there are some data available. Firstly we require a feasible value for the surface tension s, as defined in equation (3.3). Various studies of gas ion bombardment effects on copper indicate that a reasonable value for s would be about 2 Jm^{-2}.
Numerical data are also required for the cone half angle α. We shall assume that a value of $\alpha = 20°$ is appropriate for ~2 keV (Fig. 15) although for energies ~ 100 keV a value of 10° appears more appropriate. Equation (3.5) now reduces to

$$\sigma_A = \sigma_B = 5.5/d \qquad (3.6)$$

which allows us to estimate the stress on a copper cone. If d is expressed in meters then σ_A is expressed in Nm^{-2}. For example, if $d = 1~\mu m$ then σ_A is 5.5 x $10^6~Nm^{-2}$.
i) Orientation dependence. Consider a right circular conical copper crystal with its base situated on a {100} plane. The close packed (i.e. 'slip') planes are of the {111} type and are inclined at 54.7° to the {100} plane. The shear stresses within the four {100} planes are identical and so a stress anisotropy which could result in slip is unlikely in this case.
Fig. 24b shows a cone with its base situated on a plane which makes an angle of 15° with {100}. Two of the four {100} slip planes are shown intersecting the principal plane of the cone through L and M. The angle between the normal section and the (111) plane passing through L is 20°. At M the corresponding angle is 50°. If we choose L and M to be such that the normal stresses σ_L and σ_M act along the <111> directions within the {111} planes then we can write the resolved shear stresses within these slip planes as:

$$(\sigma_L)\text{slip} = \sigma_L \cos 20° = 0.94\sigma_L \text{ along AS}$$

$$(\sigma_M)\text{slip} = \sigma_M \cos 50° = 0.64\sigma_M$$
$$= 0.64\sigma_L \Big| \text{ along MT.}$$

Substituting for σ_L and σ_M from equation (3.6) we have,

$$(\sigma_L)\text{slip} = 5.2/d$$
$$(\sigma_M)\text{slip} = 3.5/d$$

On the basis of this simple model there is an asymmetry in the resolved shear stresses amongst the {111} planes touching any circle of constant stress. Thus if the local surface stress at some point on the circle were greater than the critical shear stress (σ_{crit}), slip in a preferred direction might be expected. If, for example, slip were to occur parallel to LS the cones would develop a correspondingly bent appearance. This condition could be expressed by the inequality:

$$(\sigma_L)\text{slip} > \sigma_{crit}, \tag{3.7}$$

where (σ_{crit}) is the critical shear stress for copper. There should be some value of d below which equation (3.7) would be operative.

ii) The critical shear stress. Data on unbombarded crystals of different purities and orientations have been obtained by Rosi (51). For 99.999% pure crystals he quotes a mean critical shear stress of 6.5×10^5 Nm^{-2}. For OFHC (99.98%) copper the mean value was found to be 9.4×10^5 Nm^{-2}. It is well known that the displacement damage which results from exposure to particle beams (neutrons, electrons or ions) may give rise to a very considerable hardening effect in copper, leading to an increase in σ_{crit} of an order of magnitude. We shall suppose here that σ_{crit} is increased from 10^6 to 10^7 Nm^{-2} and we can see that this occurs for $d < 0.5 \mu$m for $\alpha = 20°$, or $d < 1\mu$m for $\alpha = 10°$.

3.3.2. Experimental verification of the theory of plastic deformation
In order to test the theory of bending by surface-induced plastic deformation cones were formed on single crystal copper (42). Surface orientation was (530). To provide a coordinate reference mesh against which individual cones could be identified and located repeatedly in the SEM, standard TEM copper grids with labelled 100μm square cells were bonded at their periphery to the sample surface. The grids acted as shadow masks which transferred a complete image of the grid and cell identification symbols in deep relief (10μm) on the sample surface.

Prior to bombardment the grid-covered surfaces were seeded with alumina particles. Their masking action yielded a culture of cones having their axes parallel to the incident (normal) beam direction. The argon ion beam current was 10μA cm^{-2}, energy 100 keV and dose 8×10^{18} ions cm^{-2}.

It was found that spontaneous deformation occurred only rarely. An increase of surface stress by hydrostatic pressure (~10^4 atmospheres) led to deformation of a few cones. However the most dramatic effect occurred as a result of exposure to a 25 kHz, 50W (acoustic) source for 5 minutes. More than 80% of the cones exhibited some degree of deformation. An example of one deformed (originally upright) cone is shown in Fig. 25 where one can see an abrupt change in curvature and flattening typical of plastic deformation.

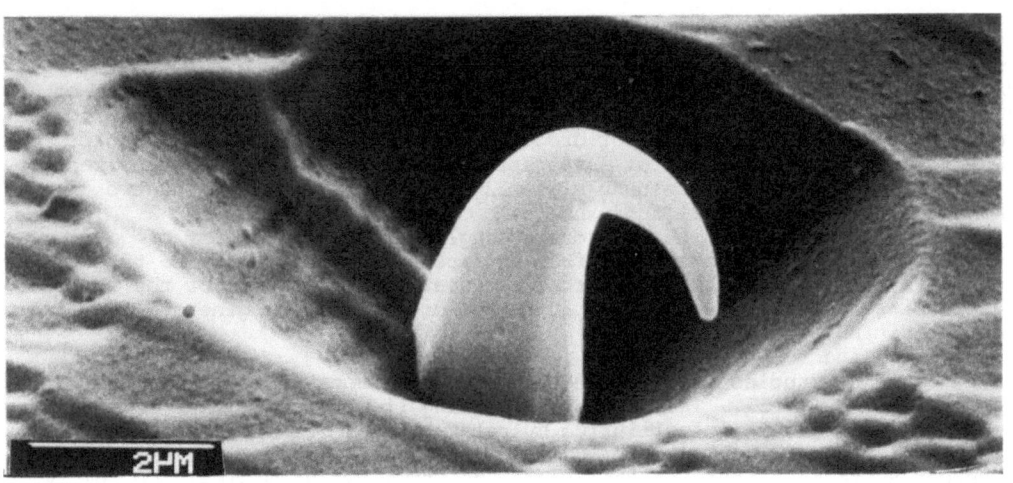

Fig. 25.   Acoustically deformed pyramid on Cu (530).

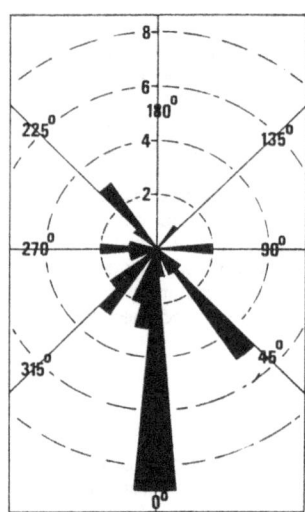

Fig. 26.   Polar histogram of pyramid droop directions. Number of pyramids corresponds to radial distance.   Interval 10°.

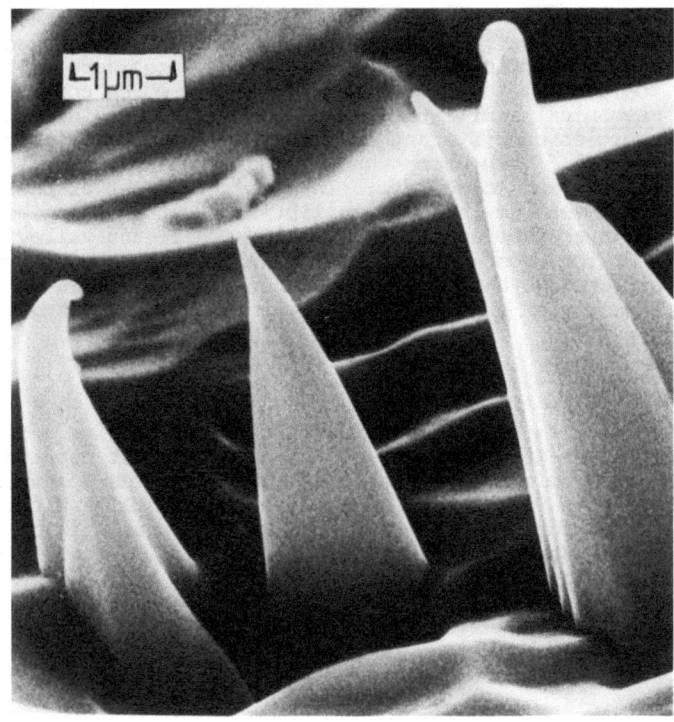

Fig. 27.   Twisted pyramids on Cu (530) (same scale as Fig. 25).

Analysis was made of the relative directions within the surface plane into which 38 cones drooped. All the cones were twisted to some extend so the direction of bending was taken as that of the thickest section. Results of this analysis are summarised in Fig. 26, which is a polar histogram giving the number of cones with direction of droop lying in a given ten degree sector. There is a strong indication of an eightfold symmetry amongst the observed directions of droop with a strong bias (22 out of 38 cones) towards two adjacent directions (labelled 0° and 45°).

The results on direction of droop strongly support the theory of spontaneous deformation resulting from surface stress but cannot be taken as conclusive proof. The observed forced deformation is a result of the combined effects of surface stress and cavitation and/or dislocation generation during vibration in an acoustic field.

3.3.3. Twisting

Careful SEM studies (41) have detected the presence of twisted cones illustrated in Fig. 27. This is not surprising if we suppose that a deformed cone tends to mechanical equilibrium by relaxation of its surface stress (52). From equation (3.4), the equilibrium condition $\sigma = 0$ implies that $(1/\rho_1 + 1/\rho_2)$ is zero. Surfaces satisfying this condition are known as minimal and the two elementary surfaces in this class are the catenoid and the helicoid (Fig. 28). It is unlikely that asperities could achieve a helicoidal geometry near their bases. However for points remote from the base we propose that the observed twisted forms represent an attempt to assume a helicoidal shape. Whether this is by plastic deformation or by surface diffusion is an unresolved question.

3.3.4. Minimum energy morphologies

Up to this point we have been concerned with cases where surface stress may drive changes in cone morphology and where observed distortions are compatible with relaxation of surface stress towards zero. The relaxation process leads the cone to adopt some modified shape which corresponds to a trend towards mechanical equilibrium. We now turn to the influence of the relative free energies of different crystal faces on the equilibrium shape of a small crystal. Much of our understanding in this area is due to the work of Herring (53,54) and co-workers. In particular, out of their considerations of the Wulff theorem (55) on equilibrium crystal shapes, criteria for the stability of a planar face against convoluted ("hill-and-valley") topography emerged. In addition, conditions were clarified under which crystalline forms might be expected to have smoothly rounded surface zones as well as sharp cornered zones. The surface free energy of a crystal is given by:

$$E = \int_S \gamma(\underline{n}) \cdot \underline{dA},$$

where γ $(=\gamma(n))$ is the specific free energy (i.e. the Helmholtz free energy) of the element of surface $\underline{dA}$ and n is the unit outward normal at dA. The term 'energy minimum' is taken to imply that the condition $\delta E = 0$ holds true. The question "what shape does a small crystalline particle adopt under the constraint $\delta E = 0$?" was answered by Wulff without substantial justification. His construction was later put on a firmer theoretical basis for the two dimensional case by Burton et al. (56) and by Herring (54). An example of a typical polar plot of γ and the Wulff construction derived from this is shown in Fig. 29.

Faces having low surface free energy lie closer to the crystal centre and occupy a relatively larger area than those having high γ values.

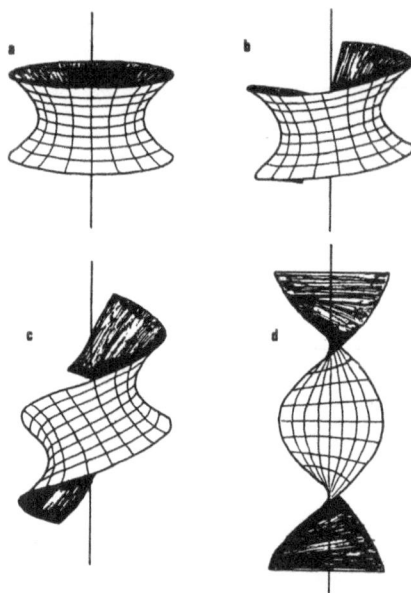

Fig. 28. Minimal surfaces, ($\rho_1 = -\rho_2$). A catenoid (a) transforms into a helicoid (d) via stages b and c.

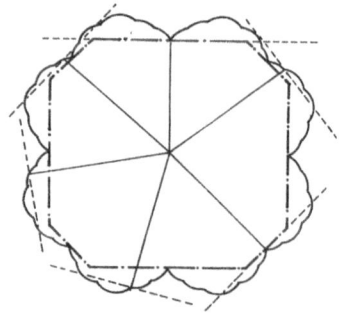

—— Polar plot of surface free energy
- - - Samples of planes normal to radius vectors of this plot
━▾━ Equilibrium polyhedron (Wulff solid)

Fig. 29. Anisotropy of surface free energy in crystals. The equilibrium polyhedron (Wulff solid) is that which is accessible from the origin without crossing planes drawn normal to, and at the end of, the radius vectors of the free-energy plot.

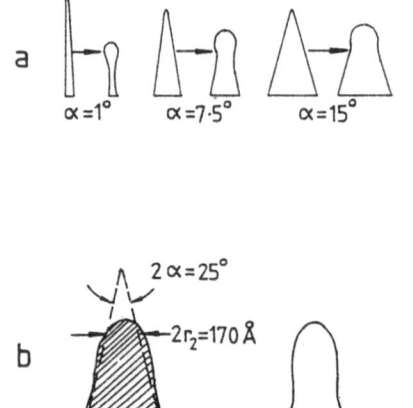

Fig. 30. Rounding of tips of Au dendrites at elevated temperatures.

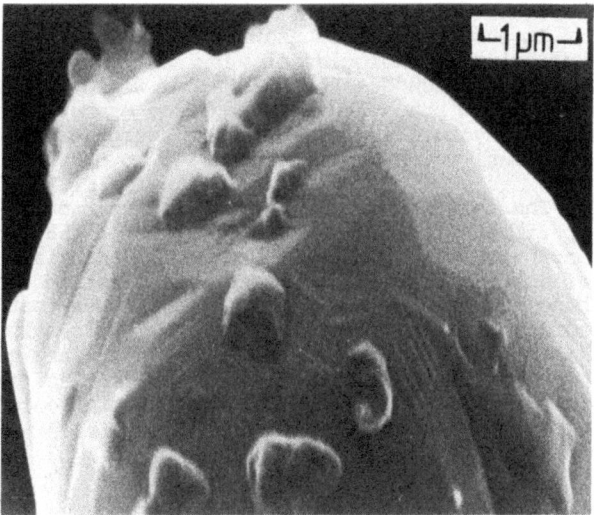

Fig. 31. Facets on the rounded tip of a Cu cone after 1 hr at 900°C.

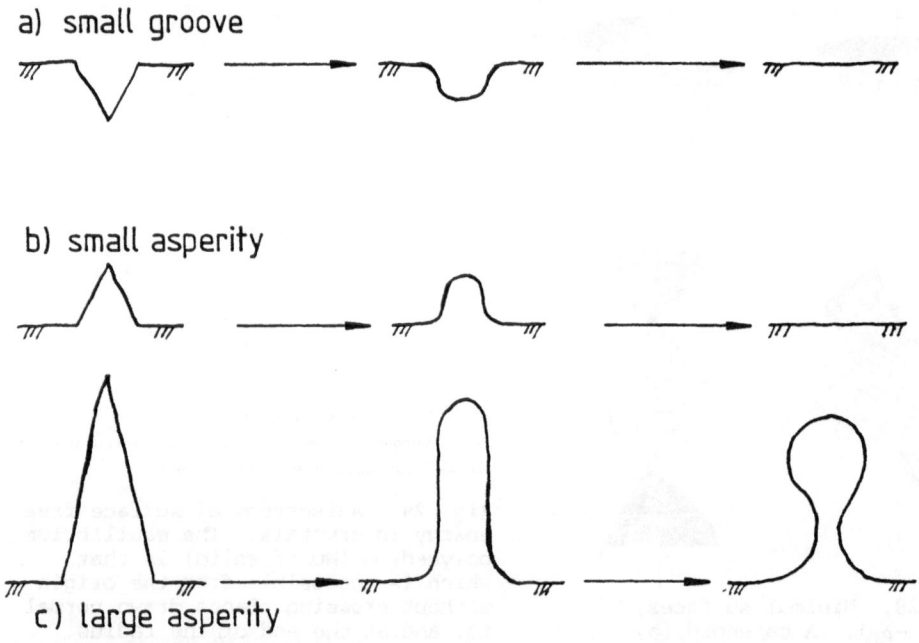

a) small groove

b) small asperity

c) large asperity

Fig. 32. Smoothing and rounding of surface features to minimise surface energy by diffusion.

It should be noted that Wulff construction is a basis for studying equlibrium morphologies of very small crystals only. Larger crystals (having dimensions much greater than ~1μm) would only show <u>trends</u> towards an equilibrium shape because the large number of elementary atomic transport events involved would require an energy expenditure (at ordinary temperatures and times) which is large compared to the reduction in surface energy.

The role of surface energy minimization in relation to ion bombarded asperities has been discussed by Carter (58) and by Chadderton (59). Their conclusions are that sputtering with associated secondary effects (such as ion reflection) and surface energy minimization can contribute to the ultimate shape of the asperity. Chadderton argues that the ion beam defines a primary geometry (the cone) with a characteristic half-angle α determined by the angle θ at which $S(\theta)$ is a maximum. The secondary geometry (faceted cone or "pyramid") is then determined by the surface energy of the bombarded solid. Sputtering, via the associated atomistic redistributions, acts so as to enchance surface diffusion and is thus the impetus for morphological change.

It is probable that whilst this is a sound basis for understanding
morphological evolution it is still somewhat simplistic. We would suggest that
both surface energy and surface stress play a role. It has long been known
that shape changes, particularly at elevated temperatures involve
redistribution of surface atoms and hence depend on the kinetics of surface
diffusion. This was clearly recognized in the work of Mullins (57) in his
treatments of thermal etching and groove evolution. A flux J of surface
material is associated wih the gradient of the chemical potential due to the
curvature gradient

$$J = -B\nabla(1/\rho)$$

where B is a constant defined by

$$B = \frac{D\gamma\Omega^2 d}{kT}, \qquad D = D_0 e^{-Q/kT}$$

D is the surface self diffusion coefficient, γ the specific surface free
energy, Ω the atomic volume, d the areal density of surface atoms and k and T
have their usual meanings. Q is the activation energy for surface self
diffusion.

The effect of the surface flux is to smooth out surface irregularities so
that a V-shaped groove is converted to a U-shaped profile. By analogy a small
asperity would become rounded, and eventually disappear. Large pointed
asperities (i.e. large compared with the characteristic diffusion length) may
develop a large radius tip or 'knob'. These effects are illustrated
schematically in Fig. 32. The flux J of surface atoms is such as to lead to a
large local radius of curvature.

Drechsler et al. (60) have applied these ideas to a study of dendritic gold
crystallites heated at temperatures up to 670K (i.e. half the melting point).
This work is of interest because the dendrites had shapes and sizes similar to
the cones which appear after high dose ion bombardment of seeded metal
surfaces. It was proposed that electron microscopy measurements of tip radii
on the dendrites provide a sensitive determination of D down to 10^{-13} cm^{-2} s^{-1}.
The expression relating D to tip radius, r_2, is

$$D = \frac{r_2^4 \, kT}{\Delta t \, A_\alpha \, \gamma\Omega^{4/3}},$$

where Δt is the time for which the surface is annealed at temperature T(K). A_α
is a constant which depends on dendrite angle α and is given by

$$r_2^4 - r_1^4 = A_\alpha \, B(t_2 - t_1)$$

where r_1 is the radius prior to annealing and $t_2 - t_1 = \Delta t$. Shapes of a tip
before and after annealing are shown in Fig. 30 together with the geometrical
significance of the quantities α and r_2.

It is interesting to compare this work with some recent results (61)
obtained on annealed copper cones at Surrey. Fig. 33a shows the surface of a
polycrystalline copper sample where cones have been cultured by sputter

450

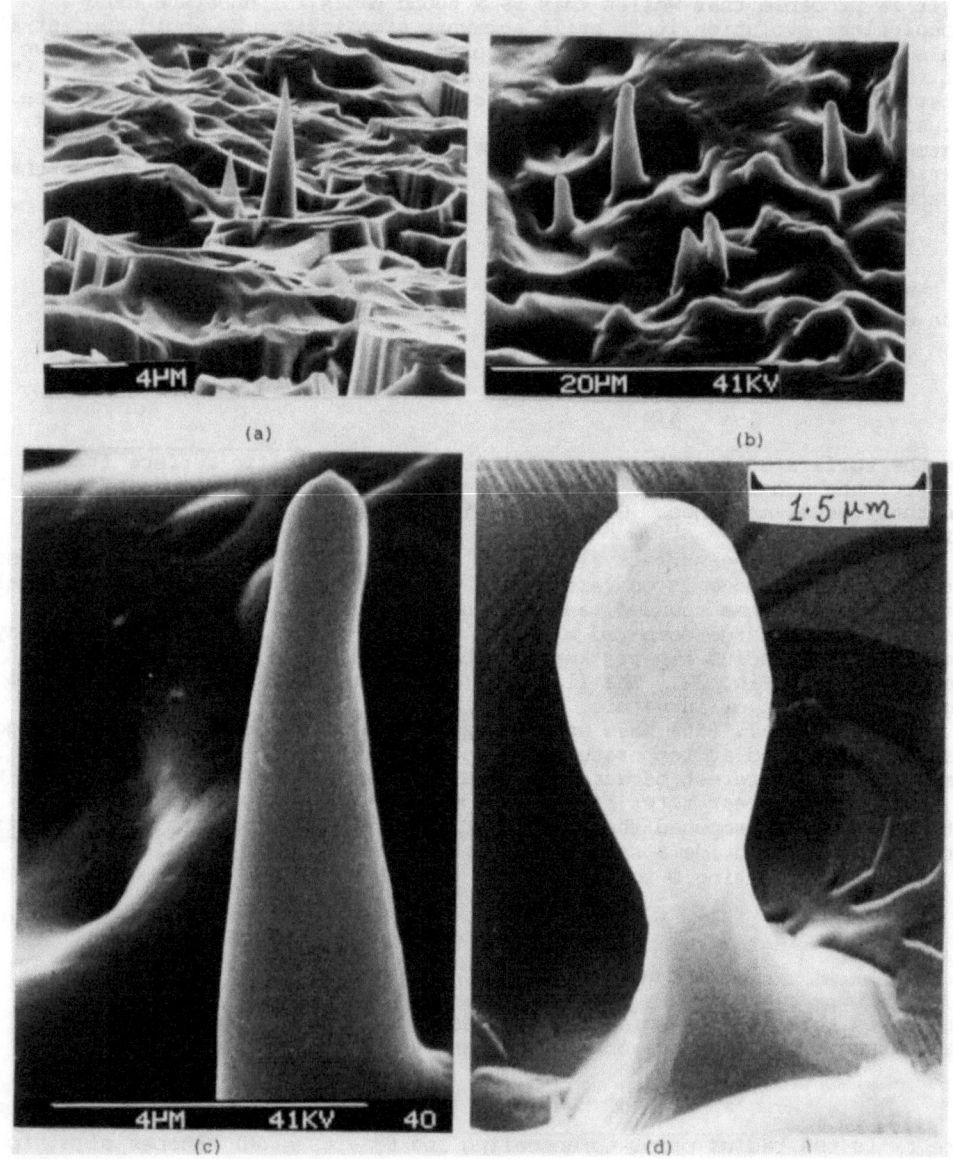

Fig. 33. Cultured cones on polycrystalline Cu. a) As bombarded (100keV Ar$^+$); b) and c) after 20 min at 500°C; d) after 1 hour at 900°C.

etching of the seeded surface as described previously. The sample was heated to 500°C for 20 minutes and the resulting forms were found to be as shown in Fig. 33b. A more detailed picture is given in Fig. 33c. Considerable rounding of all features is evident with the cones tending to become columns with rounded tops. Heating to 900°C for 1 hour (Figs. 33d and 31) leads to formation of a large radius knob at the top of the column with a reduction in radius (i.e. a local waisting) near the base. The knob is bounded by crystalline facets as may be seen in the higher magnification micrograph of Fig. 31.

It can be seen from the micrographs shown here that under the experimental conditions employed, smoothing of surface irregularities does occur. However, for high aspect ratio features such as cones, the tip may tend to a spherical shape (knobbing) in order to minimize surface area, and develop minimum energy low index crystal facets (i.e. tending towards a Wulff solid). The small protrusions that develop at or near the tip of the asperities remain unexplained, but may be a manifestation of crystal growth.

3.4. Variation of sputtering yield with position on the surface

I would like to leave the question of impurities (implanted and contaminating) to Section 4 for reasons outlined at the beginning of Section 3.3.

We are therefore left to consider matrix inhomogeneities such as dislocations and grain boundaries, and effects arising from the finite extent of a single ion impact.

3.4.1. Grain boundaries

As we know from measurements of $S(\Theta)$ for crystalline metals such as those illustrated in Fig. 7d, the sputtering yield varies with orientation with sharp minima in the low index directions due to channelling of the incident ions. The effect of this on polycrystalline specimens is that grains of different orientation will etch at different rates. The early results of Stewart (62) show this rather well for Ar^+ on polycrystalline Al at 5.1 keV (Fig. 34).

With increasing dose the difference in yield results in a step forming between grains, often with sub-grain structure being revealed. Note also that the exposed grain boundaries rotate away from the vertical and become broader. This behaviour is as would be predicted for primary erosion by the method of characteristics as we can see in Fig. 35 (2). The exposed grain boundaries sometimes become very rough, having a faceted appearance. Some grains develop a high density of etch pits whilst other do not. The roughening of the grain boundaries to form a faceted surface appears to be due to elongation of etch pits in the direction of ion incidence.

3.4.2. Dislocations and etch pits

Etch pits are formed in a similar way to chemical etching. They occur where extended defects such as dislocations and stacking faults meet the surface (2,64). In the region of lattice strain around a dislocation the atomic surface binding energy is less and therefore the sputter yield is enhanced. The method of characteristics has been used to predict the surface evolution due to the presence of area defects. The results are shown in Fig. 36 for a monotonic $S(\Theta)$ curve. Although in a real crystal there will be structure on the $S(\Theta)$ curve which may change the detailed development of the surface, the qualitative results, that an etch pit is formed, must be valid.

When surfaces are irradiated at oblique incidence corrugations, terraces or ripple structures are often formed. These were observed on metals in what is probably the earliest paper on ion bombarded topography, by Cunningham et

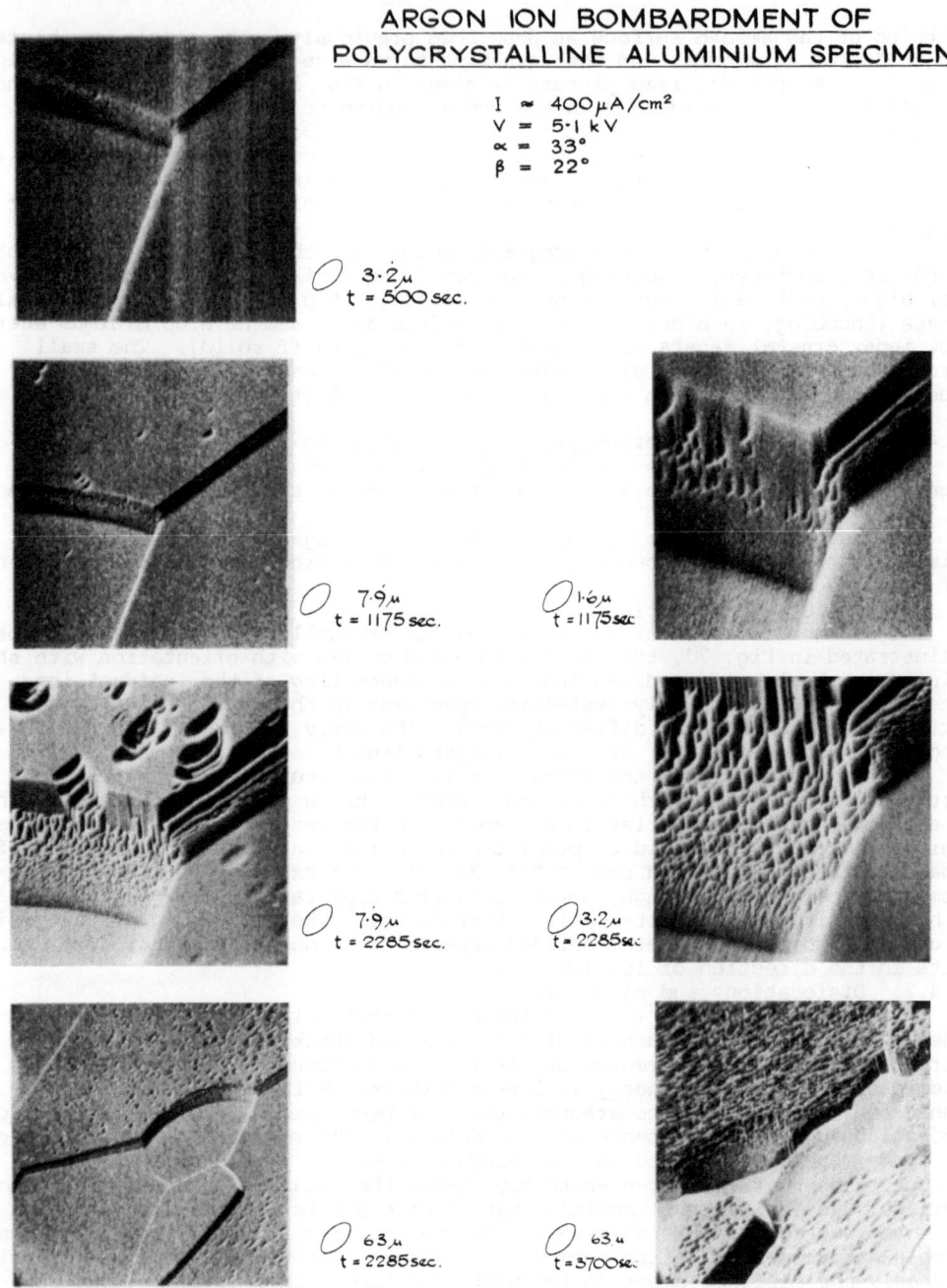

Fig. 34.   Grain–boundary relief.

al. in 1960 (70). More recently similar features have been seen on silicon surfaces (71). Aligned terraced structures are thought to form as a result of the overlap, extension and susequent modification of etch pits. A fine structure of asperities and pits on silicon is seen, possibly arising from gas bubble formation or cascade density effects (72).

The sensitivity of etch pit formation to orientation in metals could simply be due to the preferred planes for condensation of point defects created by radiation damage. The dislocation structures are crystallographically oriented and therefore the nature of their intersection with the surface must vary radically with orientation.

Certain surfaces become very heavily faceted and it has been found that this is accompanied with an anomalously large sputter yield. An example is a Cu (100) surface bombarded with 20 keV Ne incident about 6° from the [100] direction (64). The geometry of the etch pits that form in such a surface is shown in Fig. 37. It is thought that the nucleation of dislocations (initially loops) is stimulated by a perpendicular focusing mechanism producing deep radiation damage. Sufficient momentum is transferred in the [$\bar{1}$10] direction to start a focusing replacement sequence. This leads to an enhancement in the [1$\bar{1}$0] Wehner spot in the angular distribution of sputtered material. Whitton has also found an enhancement of yield in one [110] direction (2). He compared a copper (11 3 1) surface that is flat with one of the same orientation covered with pyramids. The latter showed the enhanced yield. The pyramids sit in etch pits with a (100) base and have facets of the (110) type, so the geometry is similar to that of the etch pit shown in Fig. 37. Whitton proposes that the enhancement in yield is due to the presence of focussed collision sequences using a simple geometrical argument.

The pyramids are observed to be cut from convoluted edges of etch pits. The floor of the etch pit has a (100) orientation and the intersection of the pyramid facets with the base plane is <100> and <110>. The formation of pyramids seems to be part of the process of etch pit formation and therefore related to defect structures formed by radiation damage. The (11 3 1) surface on which these etch pits form is close to <001> and provides high near-surface energy deposition and easy formation of low index surfaces in the pits and on the pyramids.

Kelly and Auciello (73) make the general claim that, in the absence of impurities, bombardment-induced pyramids on pure Cu owe their origin to surface roughness. This is one plausible explanation of the nucleation of pyramids on (11 3 1) surfaces. What does remain unclear is the reason why these pyramids do not appear to erode faster than the normal (11 3 1) surface. The evolution of one feature with ion dose has not yet been studied. However if one takes this as fact, the answer must lie in the peculiarities of the local sputtering yield peak, crystal and ion beam orientation to give a dose-invariant profile. Primary erosion theory cannot help in predicting this until realistic $S(\Theta, \phi)$ (i.e. in 3 dimensions) relationships are available.

3.4.3 Single ion impact

The sputtering process is discontinuous in character. Each ion initiates an atomic collision cascade. The statistical nature of the sputtering process is such that the sputtering yield can vary greatly from one event to another (65). This can have important consequences with regard to surface topography for dimensions similar to that of the atomic collision cascade.

A particularly violent event may leave behind a crater either by sputtering a large number of atoms or by sublimation from a spike or hot spot.

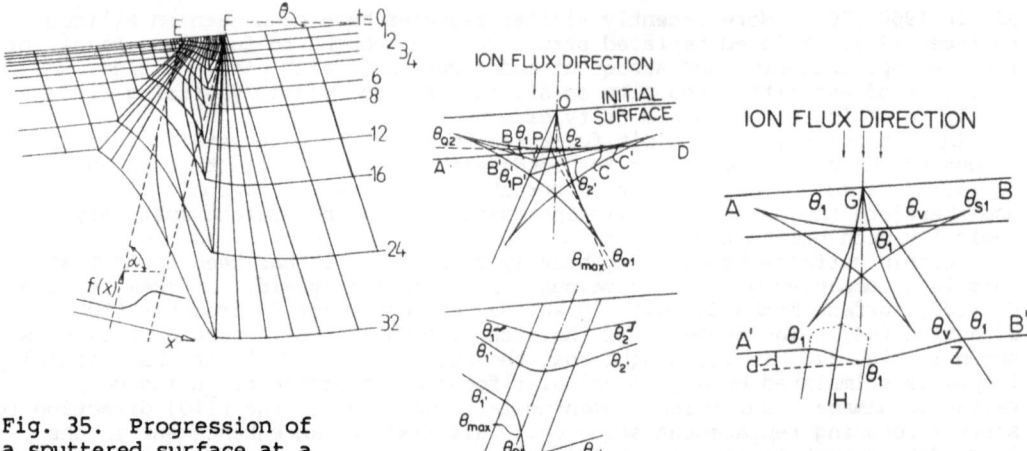

Fig. 35. Progression of a sputtered surface at a grain boundary by the method of characteristics.

Fig. 36. Use of velocity cursor to predict: (a) equilibrium profiles within an area defect; (b) complete surface evolution in the vicinity of a weak area defect.

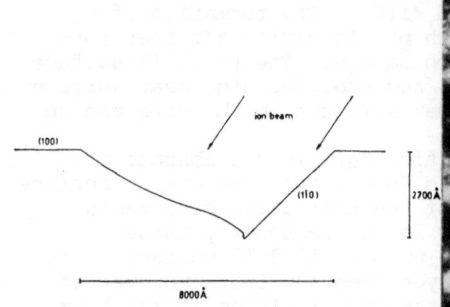

Fig. 37. Facets/etch pits formed on Cu (100) when bombarded 6° from [110] with 20keV Ne$^+$.

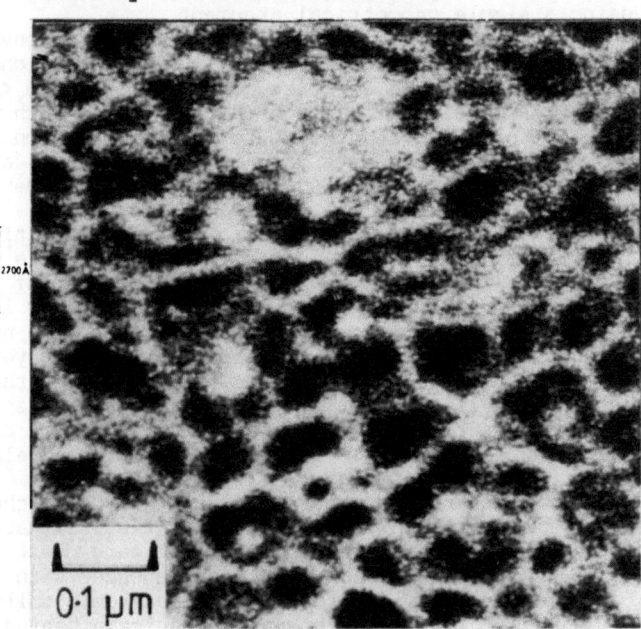

Fig. 38. Shallow indentations, possibly resulting from single ion impact. Ge$^+$ (100keV) bombardment of Ge above the amorphous/crystalline transition temperature.

The variation of sputtering yield with position leads to the prediction (65) that in absence of an additional surface smoothing mechanism such as diffusion, flat surfaces are unstable against roughening over the dimensions of the collision cascade. This instability can lead to enhancement of small irregularities of atomic dimensions such as the single impact crater mentioned above or some other irregularity such as a region of high sputtering yield due to lattice damage or small island of impurities.

Shallow indentions have been observed as a result of germanium bombardment (100 keV) of germanium at temperatures in the region of the amorphous/crystalline transition (350 to 400°C) (68). The craters, shown in Fig. 38 are 500 Å in diameter and 50 Å deep. If they are formed by a single ion, then 9000 atoms would have to be removed by sputtering in a single event. This seems improbable as this is greater than the upper limit of the average number of displaced atoms in the cascade (6000 atoms). However, a small crater could be broadened and deepended by subsequent sputtering as predicted by Sigmund in his theory of surface roughening (66).

Smaller craters that almost certainly arise from single ion impacts have been seen using high resolution transmission microscopy of gold films bombarded with bismuth ions (67). In some cases a cap is seen on one side of the crater. It is suggested that the cratering in this case is due to sublimation of atoms from a local hot spot created by a spike mechanism. When the energy is deposited deeper than a critical distance below the surface a microexplosion occurs with formation of the cap by elastic deformation.

Sigmund predicts that at temperatures below that where appreciable migrational smoothing takes place, the top of ridges and cones will sputter at a lower rate simply because the energy is deposited below the surface and therefore 'down-slope'. Erosion at the foot of slopes will be enhanced for the same reason.

We have therefore three scales of topography development (illustrated in Fig. 39):

Atomic (1 to 10 Å) impact craters → Stable cascade asperities (~100 Å) → SEM observable features (>1000 Å). This provides a plausible evolutionary path but has yet to be verified experimentally.

Kubby and Siegel (69) claim to have seen evidence for cascade effects. They have conducted a TEM study of tungsten emitter wire tips eroded to a conical shape by ion milling (3–15 keV, Ar^+). At the tip of the cone they observe a rounded protuberance (~250 Å radius) which they claim to be evidence to support the predictions of Sigmund coupled with rounding due to radiation enhanced migration. However if we recall Figures 30 and 31 we can see that the tip profile may be entirely determined by surface energy driven 'knobbing' of the tip.

Kelly and Auciello (73) suggest that if fluctuations in the sputtering process create surface roughness this would stabilize tips of asperities if $\Theta > \theta$, and so act as a pyramid generating mechanism. Similar roughening would de-stabilize pits leading to their growth. The pits seen in Ge (68) and Si (72) could arise from this mechanism. In the case of Si bombardment was below the amorphous/crystalline transition temperature. I would like to suggest that local amorphous zones are created by those cascades which deposit most of their energy close to the surface. This would provide a localized variation in sputter yield and therefore a roughening mechanism.

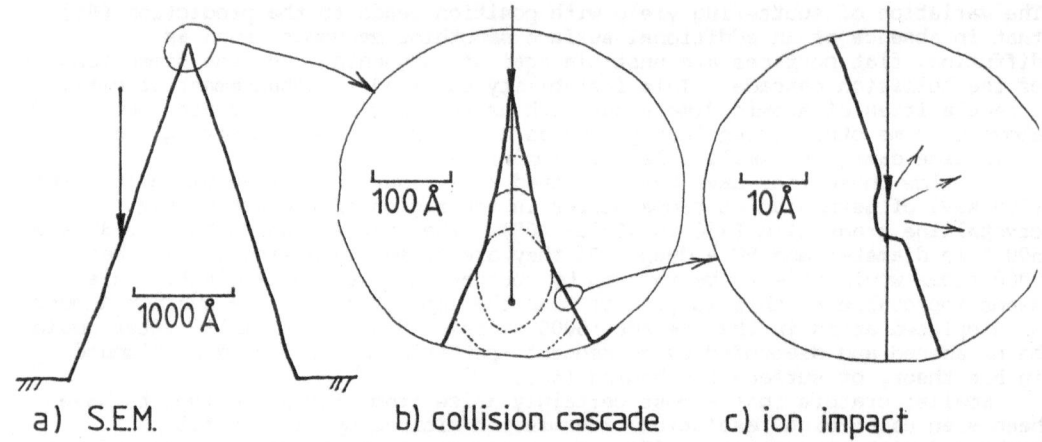

a) S.E.M.   b) collision cascade   c) ion impact

Fig. 39. The scales of topography development.
c) may be stabilized by b) so generating a).

4.   TERTIARY EFFECTS

The multitude of effects that arise from impurities or the aggregation of implanted ions and point defects are here divided into three categories: multicomponent surfaces, surface diffusion of impurities and bulk diffusion of implant or point defects.

4.1. Multicomponent surfaces

Regions of foreign material on the surface will lead generally to variation of sputtering yield and consequent generation of asperities. These will evolve into cones or pyramids (see Fig. 16). This is a masking effect, the asperities do not grow, indeed generally they shrink as they erode faster than the surrounding surface.

The masking effect may arise from dust (crud cones) (9), from deliberate seeding of the surface with submicron ceramic particles (cultured cones) (42), or from material arriving at the surface by evaporation or by sputtering from hardware, particularly the ion beam defining aperture (deposit cones). An example of the latter is shown in Fig. 40. Ge^+ (100 keV) bombardment of Ge results in a swollen cellular structure (68) (more of this in Section 4.3), however the small (500 Å diameter) spheres standing on columns are Fe/Cr deposited from the beam defining aperture. A Ge aperture was subsequently used. It is clear that unless an experimenter takes great care to ensure that the ion beam cannot sputter contaminants onto the target surface, the topography he observes may arise from extrinsic effects.

Multicomponent targets often roughen under ion impact. For example, the structure of a binary eutectic alloy may be revealed by etch rate differences (41). A finely divided sputter resistant phase will generate asperities.

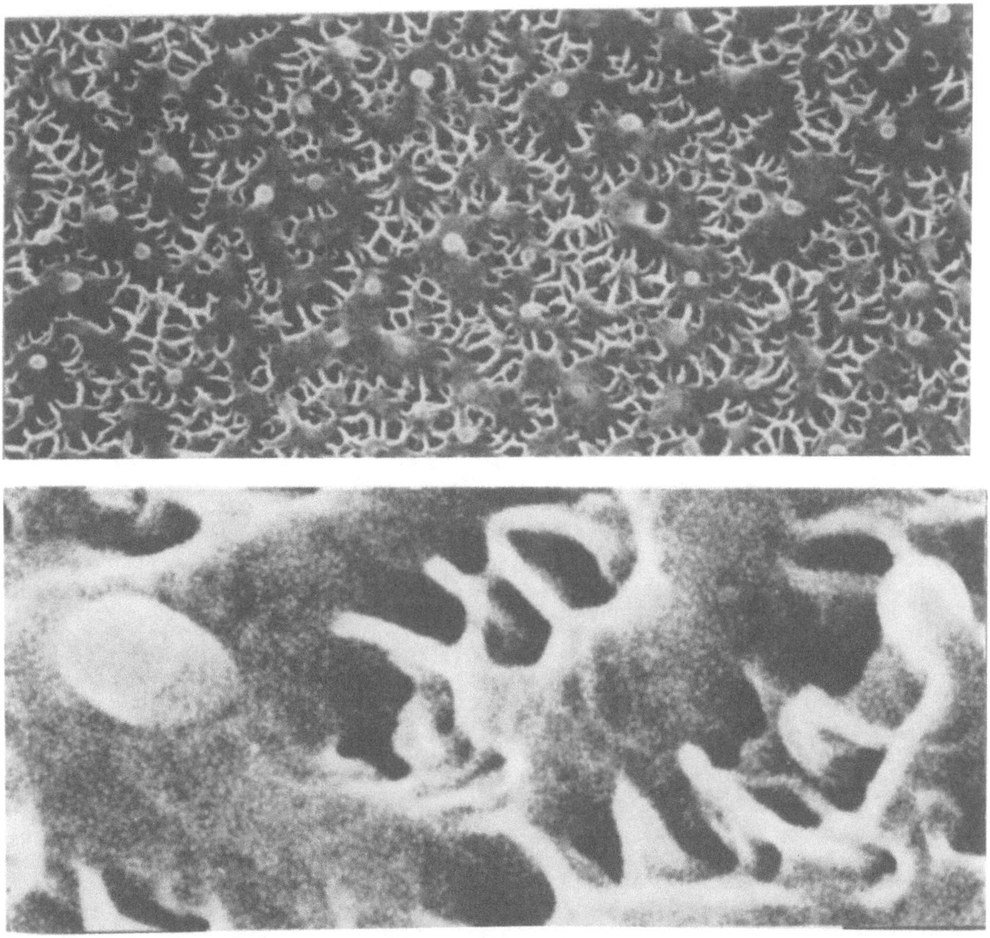

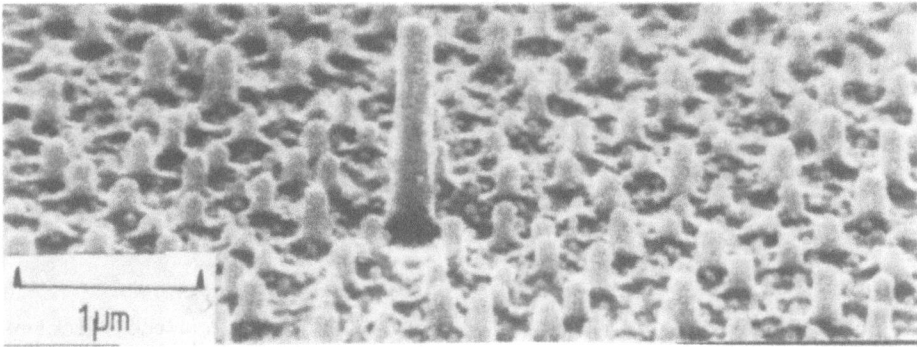

Fig. 40.   500 Å diameter Fe/Cr spheres deposited onto the surface of Ge by sputtering from a stainless-steel beam-defining aperture.

1μm

Fig. 41.   Cones formed by segregation of In to the surface of InP during 40keV Ar[+] bombardment.

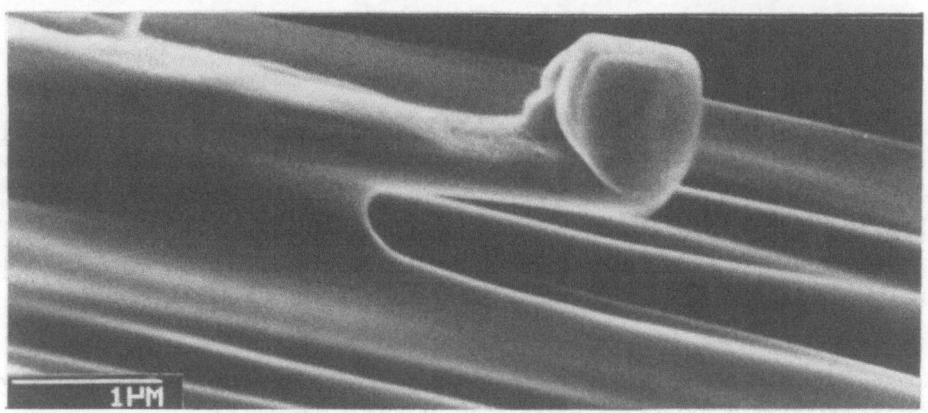

Fig. 42. Asperities formed by small Al_2O_3 particles acting as masks. Top: beam incident (left to right) at 76°. Bottom: beam incident (right to left) at 83° from surface normal. Ar^+ (100keV) on Ge.

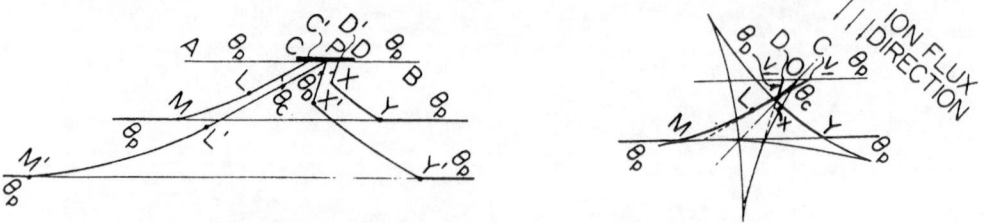

Fig. 43. Predictions of surface profile near a small mask and the velocity cursor used to construct the profile.

Surface segregation of one component of a compound, combined with aggregation, leads to etch rate differences. For example, In clusters form on the surface of InP (43,74), these generate asperities of the type shown in Fig. 41.

Asperities can take various forms. On cooper alone we have perfect pyramids (2), mushrooms (possibly as a result of flux enhancement and redeposition from secondary suttering) (1), and rather indeterminate shapes on faceted bases (75). The interplay of the secondary effects discussed in Section 2 determines the way the surface evolves after the initial masking. The mask itself erodes to present a moving mask edge to the surface (as illustrated in Fig. 17b), thus further complicating the picture.

One interesting feature that occurs on cultured asperities generated by oblique ion bombardment is a hump, or filling-in at the foot of the asperity on the side making an acute angle with the general surface. These are called 'latent planes' by Lewis et al. (76). In Fig. 42 Ar^+ (100 keV) bombarded Ge is shown, $\Theta = 76°$ in the upper micrograph and 83° in the lower. Alumina particles are used as masks. The acute angle notch is filled with a rounded contour and a bulge runs 'up-beam' from the foot. The latter effect is particularly pronounced for 76° incidence. The method of characteristics predicts the development of such a feature on the 'up-beam' side of a mask edge (Fig. 43)(1). The feature follows the shape of the polar diagram cursor placed at the mask edge. Redeposition must contribute to the rounded filling in of the notch.

4.2. Surface Diffusion

If a deposited or segregated species is free to migrate over the surface, atoms may be trapped and nuclei form at steps or irregularities in the surface and therefore localized growth can take place. Under ion bombardment there must be a dynamic balance between erosion and growth. Ever since the first proposal that cone formation is primarily explained by crystal growth (77) this subject has raised unnecessary controversy. Authors tended to imply that their models were generally applicable rather than more properly confining themselves to the narrow special circumstances where their predictions are valid. Growth phenomena are often observed, a recent example is Cs^+ bombardment of GaSb (78). Low doses (5×10^{15} ions cm^{-2}) create a swollen filamentary cellular structure on the surface. This does not occur for Ar^+ or 0^+ bombardment and is interpreted as Cs nucleated microcrystalline growth of GaSb. However Cs^+ is a very heavy ion, with a dense collision cascade, so one cannot rule out point defect aggregation effects. This will be discussed in Section 4.3. The surfaces look very like those seen on Ge^+ bombarded Ge (68).

Growth in In rich regions of InSb after Ar^+ bombardment has also been observed (43). In this case the features have the appearance of whiskers (probably of indium). This is shown in Fig. 44.

Whiskers have been observed on ion bombarded metal surfaces. Some, like that on top of a cultured asperity on a Cu (530) surface shown in Fig. 45, seem remarkably resistant to ion-beam erosion. The whisker (confirmed as Cu by EMA) has grown well above the unbombarded surface seen in the background. Whisker-like features have been seen in low mass materials with hexagonal structure (Be, Mg and graphite) (4). Initiation and survival is believed to be associated with second phase formation with residual contaminants. Rapid initial growth is followed by slower diffusion fed growth, with only whiskers aligned within 10° of the beam surviving (79).

Wehner has undertaken a detailed study of whisker growth as a result of seeding with evaporated or sputtered atoms (4). He finds that the critical flux for creation of seed cones can be very low (e.g. one Mo seed atom per 500 sputtered target Cu atoms). In every case where seed cones were formed the

460

Fig. 44. Whiskers growing in In-rich areas of 40keV Ar^+ bombarded InSb. Shadow-masked beam edge, top right.

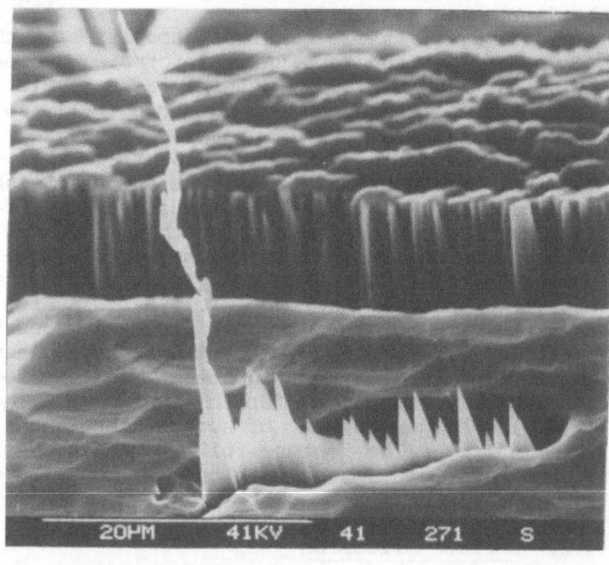

Fig. 45. Large Cu whisker grown during cultured cone formation by 100keV Ar^+ bombardment of Cu (530). The shadow-masked beam edge can be seen at the top of the micrograph.

(a) 4×10^{15} ions cm^{-2}.

(b) 5×10^{16} ions cm^{-2}.

(c) 2×10^{17} ions cm^{-2}.

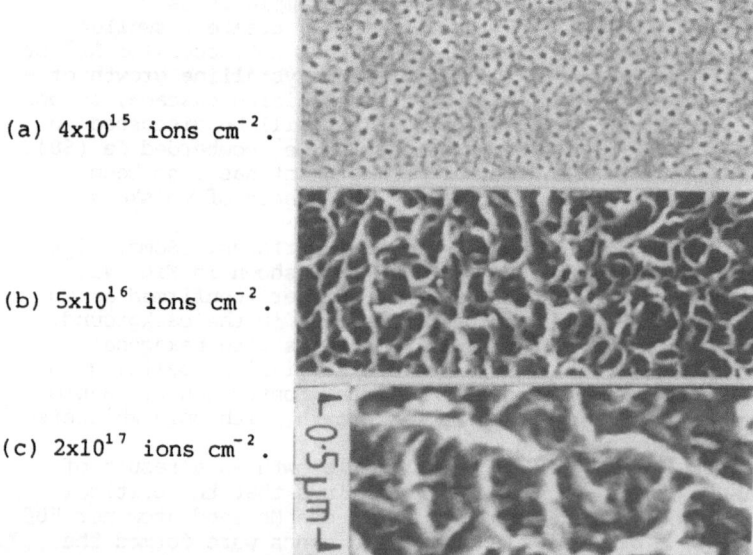

Fig. 46. 50keV Ge^+ bombardment of Ge. An array of holes coarsens into a cellular structure with increasing ion dose.

seed metal had a higher melting point than the substrate. This distinguishes
seed cones from deposit cones.

The relative sputter yield was unimportant. Seed cone formation depends
on surface mobility so necessary conditions include elevated temperatures or
energetic (>1 keV) bombarding ions, and clean surfaces (no oxide or nitride).
At close to sputtering threshold energies whisker growth is seen. It is
assumed that the whiskers are composed of the substrate material. The seed
metal acts to initiate growth. Under higher energy bombardment the whiskers
convert to seed cones by a combination of erosion and redeposition. Wehner
concludes that this behaviour is general for metals, this is where the
controversy exists.

4.3 Bulk diffusion

Sputtering results from the fraction of depostied energy that arrives back
at the surface. The bulk of the bombarding ion energy is deposited below the
surface as are the implanted ions.

At temperatures below those necessary for appreciable diffusion, point
defects and small point-defect clusters will be formed and the implanted
species will be frozen into a metastable state. This will result in a density
change and an overall swelling or compaction of the surface determined by the
nature of the interatomic bonding. In the absence of other effects the
topography will remain flat except for the boundary between the bombarded and
unbombarded surface.

However, if diffusion (perhaps radiation enhanced) can occur the defects
and/or the implanted species can aggregate and this may produce effects that
completely dominate the surface topography. For example, bubbles, blisters
and voids can be formed with the appropriate combination of ion, target,
temperature, ion dose and ion energy.

4.3.1 Voids

Voids produced by neutron bombardment of metals have been a topic of
intensive study by the fission reactor community for some time. One of the
more relevant matters in ion implantation has been the discovery that the
voids can be formed in ion implanted semiconductors when subjected to very
high rates of energy deposition (68). It appears that in this case the voids
do not need to be stabilized by the presence of a gaseous species. The effect
of 100 keV germanium ion bombardment of single crystal germanium is shown in
Figs. 40 and 46.

Holes first appear in the surface at relatively low doses, for example 2 x
10^{15} ions/cm^2 for 50 keV Ge ions. As the ion dose increases the holes grow to
form a cellular structure which coarsens until a dose of ~2 x 10^{17} ions/cm^2 is
reached, where a dynamic equilibrium is established. Cavities 200-400 Å in
diameter were observed below the surface at doses just below the threshold for
the development of any surface topography. The altered layer thickness was
found to increase monotonically from twice the projected range at threshold to
six times the projected range for a dose of 4 x 10^{17} ions/cm^2.

It is proposed that cavities are formed by the coalescence of small voids
(up to 60 Å in diameter) created by the dense collision cascades of the
incident germanium ions.

Voids have also been observed by TEM in 200 keV N^+ implanted at ZnSe (80).
The voids form after annealing at 700°C at the peak of the damage profile.
They are dodecahedra with {110} crystal faces. Size ~300 Å along one edge.

4.3.2 Bubbles and blisters

Gas bubbles occur when ions of insoluble species such as hydrogen, helium and the other inert gases are implanted to depths where diffusion to the free surface is improbable. This topic is of vital interest to those developing materials for the first wall of fusion reactors.

One possible effect on surface topography is that once sputtering exposes a bubble, a cavity is formed on the surface. However, there is evidence (81) that in the case of metals, at temperatures where bubbles (and/or voids) form, surface diffusion is so rapid that the cavity rapidly fills in. The main effect of bubbles will be seen when conditions are suitable for blister formation.

The models for blister formation in metals fall into two categories. One is the gas-pressure model, the other is the integrated lateral stress model (appropriate references are given in Reg. (82)).

The gas-pressure model assumes that bubbles form due to the build-up of excess gas pressure. Blisters form when the inter-bubble separation is sufficiently small in the region of maximum concentration for fracture to be initiated.

The integrated lateral stress model assumes that large lateral stresses in the implanted layer lead to elastic instability and buckling of the implanted surface. Proponents of this model point to the relationship between the most probable blister diameter D_{mp} and the blister skin thickness. They find the D_{mp} is proportional to $t^{1.5}$ in agreement with their model.

However, careful experiments (82) have established that the exponent of t can vary between 0.85 for vanadium and 1.25 for beryllium with nickel and niobium inbetween. It has also been established that the blister skin thickness correlates well with the region of maximum bubble concentration. This region may be deeper than the projected range for low ion energies; for example with 20 keV He^+ irradiation of nickel at 500°C the projected range is 500 Å, but the maximum bubble density is at a depth of 1300 Å. This evidence strongly supports the excess gas pressure model for the systems studied.

Blistering does not appear to be a transient phenomenon. In fact a continuous exfoliation process occurs, fifteen equal thickness exfoliated layers having been seen in He^+ (100 keV) bombarded 316 stainless steel (450°C) (82).

It has been proposed (72) that voids or Ar gas bubbles are responsible for nucleation of etch pits in Ar^+ bombarded Si. A detailed TEM study of low energy Ar^+ (5 keV) bombarded Si reveals a dense shallow layer of bubbles of ~10Å, radius (81). These trap ~50% of the Ar and are remarkably resistant to annealing to 1100°C, only the matrix Ar being released. It is thought that the bubbles are stabilized by the presence of recoil implanted oxygen. It seems very reasonable that these bubbles will generate surface roughness once they are intercepted by the receding, sputter etched, surface.

5. CONCLUSIONS

5.1. Analytical models

These have been developed to describe variation of sputtering yield with angle of ion incidence ($S(\Theta)$), primary erosion, redeposition and surface energy effects. They serve to illustrate the physics and predict trends. In general, evolution of surface contours with increasing etch time (dose) cannot be determined except in simple cases.

5.2. Geometrical models

The model developed for $S(\Theta)$ illustrates the importance of cascade anisotropy but is in need of further development in order to be predictive. Cursor methods applied to primary erosion allow easy prediction of edge development in two dimensions. Addition of extra processes and single crystal $S(\Theta)$ is possible but this makes cursor complex and interpretation difficult.

5.3. Numerical computation

Monte-Carlo ion-impact simulation codes are useful in prediciting $R(\Theta)$ (ion reflection) and $S(\Theta)$ for amorphous targets. Codes based on the characteristics method are usefully predictive of surface contour evolution in three dimensions under the action of primary erosion, redeposition and spacial and temporal variation in flux and yield. Huygen's contruction using the cursor to construct the eroded surface is probably the most convenient method. Single crystals are difficult because of the scarcity of information of $S(\Theta$ and $\phi)$, the sputtering yield variation with both polar and azimuthal angle.

5.4. Experimental

In addition to the requirement for more data on $S(\Theta, \phi)$ for single crystals more data is needed on ion reflection, particularly the energy dependence, in order to indicate the importance of incomplete development of the collision cascade. More data on the energy and angular distribution of sputtered atoms at close to grazing incidence is also needed to quantify the relative importance of ion reflection and energetic sputtered atoms in the flux enhancement effect.

The new high resolution techniques need to be applied to clarify surface roughening and coarsening mechanisms and their role in pit and pyramid generation. There is almost no experimental data on this topic.

Computer image analysis needs to be applied to the evolution of surface features in order to obtain data which is statistically significant, as a reliable test of theory.

6. ACKNOWLEDGMENTS

I would like to thank Brenda Dore and Kathy Woodford for a painstaking and professional job in preparing this manuscript in the face of almost insuperable difficulties created chiefly by myself. I must also thank George Carter for encouragement and Roger Kelly for the invitation to participate. Much of the work reported here was done in collaboration with my friend and colleague Jeff Belson, without whom very little of the theory would have been developed.

464

REFERENCES

1. 'Ion Bombardment Modification of Surfaces' editors O. Auciello and R. Kelly, Elsevier (Amsterdam, 1984), particularly chapters 3 (R. Kelly), 5 (G. Carter and M. J. Nobes), 6 (I. H. Wilson, J. Belson and O. Auciello) and 7 (R. S. Robinson and S. M. Rossnagel).
2. 'Erosion and Growth of Solids Stimulated by Atom and Ion Beams' editors G. Kiriakidis, G. Carter and J. L. Whitton, particularly pages 70 (G. Carter), 103 (M. J. Nobes), 121 (R. Smith and M. A. Tagg) and 181 (S. M. Rossnagel).
3. Carter G. Katardjiev IV, M. J. Nobes and J. L. Whitton: in press.
4. G. K. Wehner: J. Vac. Sci. Technol A3 (1985) 1821.
5. W. Beinvogl and U. Wagner: Proc. Int. Symp. on Trends and New Applications in Thin Films, Strasburg (March 1987), published by Societe Francaise du Vide (Paris).
6. R. M. Feenstra and G. S. Oehrlein: J. Vac. Sci. Technol. B3 (1985) 1136.
7. J. O. Bovin, R. Wallenberg and D. J. Smith: Nature 317 (5 Sept. 1985). 47.
8. A. D. G. Stewart and M. W. Thompson: J. Mat. Sci., 4, (1969) 56.
9. I. H. Wilson and M. W. Kidd: J. Mat. Sci., 6, (1971) 1362.
10. G. Carter, M. J. Nobes and K. I. Arshak: Wear, 53, (1979) 245.
11. R. Smith and M. Walls: Phil. Mag. A, 42(2), (1980) 245.
12. D. E. Harrison Jr.,: Rad. Effects, 70, (1983) 1.
13. H. H. Andersen and H. L. Bay: 'Sputtering by particle bombardment 1', Ed. R. Behrisch, Springer-Verlag, Berlin, (1981), 200.
14. M. Hou and M. T. Robinson: Appl. Phys., 371, (1978), 17.
15. L. G. Haggmark and J. P. Biersack: J. Nucl. Materials, 103 & 104, (1981), 345.
16. M. T. Robinson: J. Appl. Phys., 54(5), (1983), 2650.
17. P. D. Townsend, J. C. Kelly and N. E. W. Hartley: 'Ion Implantation, Sputtering and Their Applications', Academic Press, London, (1976).
18. P. Sigmund: Phys. Rev. 184, 383 (1969), See also P. Sigmund in 'Sputtering by Particle Bombardment I': Topics in Applied Physics, Vol. 47, R. Berisch (Ed), (Springer-Verlag, 1981).
19. D. E. Harrison Jr.: Radiation Effects 70, 1, (1983).
20. J. B.Sanders: Thesis, Univ. Leiden (1968).
21. N. E. Roosendaal, U. Littmark and J. B. Sanders: Phys. Rev. B26, 5261 (1982).
22. I. H. Wilson, S. Chereckdjian and R. Webb: Nucl. Instrum. and Meth. B7/8, 735-741 (1985).
23. R. P. Webb and I. H. Wilson: paper A 4.2, Spring Meeting of MRS (1985).
24. R. P. Webb, C. Jeynes and I. H. Wilson: Nucl. Instrum. and Meth. B13, 449-452 (1986).
25. I. H. Wilson and R. P. Webb: Radiation Effects 99 (1986) 281-291.
26. A. D. G. Stewart and M. W. Thompson: J. Mat. Sci. 4 (1969) 56.
27. G. B. Whitham: 'Linear and Non Linear Waves', Wiley J (New York 1974).
28. F. C. Frank: 'Growth and Perfection of Crystals', (Wiley J, New York) 1958, p.411.
29. F. C. Frank: Zeits. Phys. Chem. Neue. Folge, 77 (1972) 84.
30. J. C. Luke, J. Geophys Res., 77 (1972) 2460.
31. J. C. Luke: Zeits. Geomorph. Supp., 25 (1976) 114.
32. D. J. Barber, F. C. Frank, M. Moss, J. W. Steeds and I. S. T. Tsong: J. Mater. Sci. 8 (1973) 1030.
33. M. J. Witcomb: J. Mater. Sci. 10 (1975) 669.

34. D. W. Younger and C. M. Haynes: J. Vac. Sci. Technol. 21 (1982) 677.
35. I. H. Wilson, S. S. Todorov and D. S. Karpuzov: Nucl. Instrum. and Meth. 209/210 (1983) 549.
36. J. P. Ducommun, M. Cantagrel and M. Moulin: J. Mater. Sci. 10 (1975) 52.
37. T. Marsh R. Collins: Radiat. Eff. 99 (1986) 171.
38. M. J. Witcomb: Radiat. Eff. 27 (1976) 223.
39. J. Lindhard, M. Scharff and H. E. Schiott: Matt. Fys. Medd. Dan. Vid. Selsk. 33 (1963) No. 14.
40. M. T. Robinson in 'Sputtering by Particle Bombardment 1', ed. R. Behrisch, Springer-Verlag, Berlin (1981), p.91.
41. N. Bibic, I. H. Wilson and T. Nenodovic: J. Appl. Phys. 53 (1982) 5250.
42. I. H. Wilson and J. Belson: Phil. Mag. A47 (1983) 351.
43. I. H. Wilson: Radiat. Eff. 18 (1973) 95.
44. L. F. Johnson in ref. 1, p.366.
45. H. W. Lehmann, L. Krausbauer and R. Widmer: J. Vac. Sci. Technol., 14 (1977) 281.
46. R. Smith, S. S. Makh and J. M. Walls: Phil. Mag. A47 (1983) 453.
47. J. Belson and I. H. Wilson: Nucl. Instr. Meth. 182/183 (1981) 275.
48. W. Hauffe: Proc. 1st Int. Conf. Ion Beam Modification of Materials (Budapest, 1978) p.1079.
49. O. Auciello and R. Kelly: Rad. Effects, 66 (1982) 195.
50. R. Shuttleworth: Proc. Phys. Soc. A 63 (1950) 44.
51. F. D. Rosi: Trans. A. I. M. E., 200 (1954) 1009.
52. J. Belson and I. H. Wilson: Physics Letters, 90A (1982) 141.
53. C. Herring: 'Structure and properties of solid surfaces' eds. R. Gomer and C. Smith.
54. C. Herring: Phys. Rev., 82 (1951) 87.
55. M. Von Laue; Z. Kristallogr, 105 (1943) 124.
56. W. K. Burton, N. Cabrera and F. C. Frank: Phil. Trans. A., 243 (1951) 40.
57. W. W. Mullins: J. Appl. Phys., 28 (1957) 333.
58. G. Carter, J. S. Colligon and M. J. Nobes: Rad. Effects, 31. (1977) 65.
59. L. T. Chadderton: Rad. Effects Lett., 50 (1979) 23.
60. M. Drechsler, J. J. Metois and J. C. Heyraud: Surf. Sci., 108 (1981) 549.
61. I. H. Wilson (unpublished).
62. D. J. Elliott and P. D. Townsend: Phil. Mag. 23 (1971) 249.
63. Ph. J. J. Elich, H. E. Roosendaal, and D. Onderdelinden: Radiat. Eff. 14 (1972) 93.
64. H. E. Roosendaal: 'Sputtering by Particle Bombardment 1', ed. R. Behrisch, Springer-Verlag (Berlin, 1981) p.249.
65. J. E. Westmoreland and P. Sigmund: Rad. Eff., 6 (1971) 187.
66. P. Sigmund: J. Mater. Sci., 8 (1973) 1545.
67. W. Jager and K. L. Merkle: 9th Int. Congress on Electron Microscopy, Toronto, 1 (1978).
68. I. H. Wilson: J. Appl. Phys., 53 (1982) 1698.
69. J. A. Kubby and B. M. Siegel: Nucl. Instrum. and Meth. B13 (1986) 319.
70. R. L. Cunningham, P. Haymann, C. Lecomte, W. J. Moore and J. J. Trillat: Appl. Phys. 31 (1960) 839.
71. M. J. Nobes, G. W. Lewis, G. Carter and J. L. Whitton: Nucl. Instrum. and Meth., 170 (1980) 363.
72. W. Begemann, S. Kostic, J. J. Nobes, G. W. Lewis and G. Carter: Rad. Effects Lett. 87 (1986) 229.
73. R. Kelly and O. Auciello: Surface Science, 100 (1980) 135.
74. O. Wada: J. Phys. D., 17 (1984) 2429.
75. J. A. Floro, S. M. Rossnagel and R. S. Robinson: Radiat. Effects lett. 68 (1982) 57.

76. G. W. Lewis, J. S. Colligon and M. J. Nobes: J. Mater. Sci., 15 (1980) 681.
77. R. S. Gvosdover, V. M. Efremenkova, L. B. Shelyakin and V. E. Yurasova: Radiat. Effects, 27 (1976) 237.
78. Y. Homma: J. Vac. Sci. Technol: A5 (1987) 321.
79. R. S. Robinson and S. M. Rossnagal: J. Vac. Sci. Technol., A1 (1983) 1398.
80. J. S. Vermaak and J. Petruzello: J. Appl. Phys. 55 (1984) 1215.
81. R. S. Nelson, D. J. Mazey and J. A. Hudson: Proc. Brit. Nuclear Energy Soc. European Conf. on voids, Reading UK, eds. S. F. Pugh, M. H. Loretto and D. I. R. Norris (1971).
82. S. K. Das, M. Kaminsky and G. Fenske: J. Nucl. Mater. 76/77 (1978) 215, 247 and 256.
83. U. Bangert, P. J. Goodhew, C. Jeynes and I. H. Wilson: J. Phys. D. 19 (1986) 589.
84. I. H. Wilson and I. S. T. Tsong in preparation.

CULTURED BLISTERS

S. CHERECKDJIAN and I.H. WILSON

APPLIED MATERIALS, IMPLANT DIVISION, HORSHAM, W. SUSSEX, U.K.

1. INTRODUCTION

The Surface Acoustic Wave (S.A.W.) resonator (1) is a high Q device and is susceptible to manufactural variations. It is constructed by using thin film aluminium deposition, to realise the transducers and distributed reflector banks, on a piezoelectric substrate (typically ST quartz).

Substrate inhomogeneities and uncertainties in the thin film structures may result in a low yield of devices within specification. High energy incident ions have been used to advantage in the trimming of S.A.W. device performance (2).

This work has resulted from the SEM examination of the aluminium quartz interface after oxygen ion bombardment. Molecular oxygen was implanted at energies between 100 and 150 keV and doses from 2×10^{17} to 2.8×10^{18} oxygen atoms cm^{-2} into thin aluminium films evaporated onto fused quartz samples. Bombardment by oxygen ions gave essentially a clean system (3) in the quartz, while being reactive in the aluminium film (4, 5). Fused quartz substrates were used to eliminate radiation damage swelling.

2. EXPERIMENTAL

Aluminium steps 300nm thick were evaporated onto polished fused quartz substrates. After oxygen implantation the specimens were coated with a layer of sputtered gold to prevent the build up of charge on the exposed quartz surface and analysed by electron microscopy. Surface texture and blister dimensions were measured before and after molecular oxygen implantation.

Larger specimens were analysed with X-ray Photon Spectroscopy (XPS) for chemical information from the blistered surfaces. XPS was performed in the Department of Metallurgy at the University of Surrey. The chamber pressure was better than 1×10^{-9} torr during analysis.

To investigate oxygen ion trapping in the aluminium by Rutherford backscattering (RBS) the specimens were prepared from vitreous carbon blanks. These were mirror finished on a diamond lap, ultrasonically cleaned and finally vapour cleaned in alcohol to ensure that the surface was free of contaminants. Aluminium films were evaporated to the required thickness with the aid of an AT-cut crystal monitor and frequency meter. The background pressure during evaporation was 4×10^{-6} torr. Rutherford backscattering measurements enabled the trapping efficiency and stoichiometry of the implant oxygen atoms to be determined. The analysis was undertaken using an incident beam of 1.5 MeV He^+ ions. The beam divergence was less than $0.2°$, the spot diameter was 1 mm, the scattering angle was $150°$ (laboratory coordinates) and the silicon surface barrier detector resolution was 13 keV (fwhm). Typically, the beam current was 6 nA and the charge collected was 10 μC with a chamber pressure of 1×10^{-6} torr. The energy calibration (3.09 keV channel^{-1}) of the 800-channel analyser was achieved with a thin sputtered gold film on an unimplanted silicon specimen.

The ion bombardment was carried out using a 500 keV heavy ion accelerator with a sector magnet for mass analysis. Beam uniformity was better than 1%, beam convergence was less than 0.2%. The ion dose

467

R. Kelly and M. Fernanda da Silva (eds.), Materials Modification by High-fluence Ion Beams, 467–476.
© 1989 by Kluwer Academic Publishers.

was monitored by measuring the charge collected by the target. An electrostatic field was used to suppress secondary electrons. The vacuum during ion bombardment was better than 4×10^{-6} torr.

3. RESULTS

3.1 S.E.M. observations

Oxygen bombarded S.E.M. specimens displayed blisters in the transition region between the aluminium and the fused quartz. This is shown as areas 'a' and 'b' in plates 1 and 2. Area 'a' is the edge of the aluminium film and 'b' is the quartz originally below the aluminium film but now exposed by lateral sputter etching.

Unimplanted aluminium showed no visible topography but,after oxygen implantation, it was noted that the surface developed aluminium oxide nodules with a mean diameter of 180 ± 80nm. This area is shown as region 'c', in the plates. Any sub-micron surface scratches in the fused quartz, region 'd', also took various surface forms after implantation. One of these scratches provided a quartz marker for an edge so that the area of interest could be stripped and reinvestigated. One particular area is shown as a montage in plate 3. The surface texture can be seen to increase with bubble proximity in the blistered area.

To encourage blister formation the aluminium layer thickness was reduced to 141nm and 100nm in two samples. These thicknesses correspond to the projected ranges for 150keV and 100keV, O_2^+ ions respectively. The different evaporated aluminium layers were obtained by careful monitoring of a crystal resonator during deposition and the thicknesses confirmed by talystep measurements. Plates 4 and 5 demonstrate visually the drastic effect that varying the deposited aluminium thickness has on the blister density. When the oxygen range coincides with the thickness of the aluminium layer, blisters are seen over the entire aluminium surface. The blister density reduces when the projected range is greater than the interface depth, indicating very few trapping sites for implanted oxygen atoms residing in the quartz. Lateral diffusion must provide a release mechanism for reducing the implanted oxygen concentration under the interface level. Table 1 shows the distribution of blisters for four different deposited aluminium thicknesses and two oxygen ion energies. The greatest blister activity occurs when the ion range equals the evaporated aluminium layer thickness and this results in the highest calculated (PRAL (6) range tables) implanted oxygen to aluminium ratio at the interface.

A partially exfoliated 3 μm radius blister is shown in plate 6. This larger blister can be seen to be a coalescence of smaller bubbles. The bubble rupture seems likely to have originated from a fracture propagating along the z axis (normal to the substrate surface) from the upper point of the internal bubble blister towards the cavity surface.

3.2 RBS and XPS measurements

These measurements are comprehensively presented in a previous paper (7) and some of the results are summarised below. Figure 1 shows an unimplanted RBS profile (solid line) and the profile obtained after 100keV, O_2^+ implantation (dotted line).

It is observed that each spectrum will require the subtraction of the unimplanted background to obtain the true oxygen profile. Angular RBS located the first small background peak at the carbon-aluminium interface while the second was found to be at the surface. These

Oxygen dose (atoms cm^{-2})	O$_2^+$ energy (keV)	PRAL range (nm)	Al thickness deposited (nm)	Calculated O/Al ratio at interface	Average no. of blisters (per 10μm^2)	Mean diameter (nm)	Standard deviation (nm)
4 x 10^{17}	100	107.6	58	0.31	26	620	200
			100	0.58	115	580	220
4 x 10^{17}	150	168.4	61.3	0.10	20	909	180
			141	0.42	138	682	200

Table 1: Blister distribution versus aluminium thickness

small surface peaks correspond to oxygen from the surface native oxide, and an oxygen signal from the C-Al interface. This interface peak can be attributed to oxygen gettering by aluminium during the initial stages of evaporation. From the area under the oxygen profile, the oxygen concentration was calculated. Figure 2 shows the variation of the calculated oxygen concentration with the implanted dose. The curve indicates a trapping efficiency of 100% until a stoichiometric aluminium oxide layer is formed.

The amount of oxygen in the implanted layer increases with the implanted dose and reaches a saturation. Saturation of the oxygen trapping occurs at a dose of 9×10^{17} atoms cm^{-2}. The uncertainty bars on the points were calculated from the standard deviation of the mean value of the data, collected for that dose.

Blistered specimens were analysed by XPS after stripping the aluminium surface by immersing in hot concentrated sodium hydroxide. The aluminium content was found to be zero. The Auger parameter (8) for silicon was calculated in all the specimens and found to be 225.2 eV, consistant with the SiO_2 state. Therefore, no stable aluminium mixing with the quartz substrate was detected. Prior to stripping, the aluminium signal comprised of metallic and oxidized aluminium. It was observed that for large implanted oxygen doses the metallic aluminium signal was depleted until only the trivalent aluminium signal remained.

4. SUMMARY

The development of these blister structures is closely related to the range and the trapping mechanism, at the aluminium-quartz interface, of the implanted oxygen ions and to the properties of the gas-solid system. The very low diffusion of the oxygen atoms in the solid-solid interface and the low sputtering yield result in a very high concentration of implanted gas ions.

In the early stages of the interaction of single oxygen atoms the interface trapping sites predominate. At an intermediate dose clustering of the trapped gas atoms occurs leading to gas bubble formation and interface swelling. With higher oxygen doses the amount of trapped oxygen saturates. As the gas concentration at the interface reaches a critical value of 0.1 to 1 gas atom (9) per target atom, the surface deforms into blisters due to the internal gas pressure and lateral compressive stress.

Compressive stresses (7, 10) from the mismatch of aluminium oxide (synthesis) and aluminium crystal structures have been measured. This and reactive ion mixing at the interface must be considered as important factors in the growth of these blisters. These additional factors will compete with the internal gas bubble pressure and material constants as the surface tends to a minimum energy configuration.

The rupture of larger blisters is predicted by theory (11). In this case R_p is less than $\sqrt{3}$ b (b = bubble radius), therefore the tensile stress along the circular blister cavity, which is directly proportional to the internal bubble pressure, is less than the tensile stress in the z axis. Thus a rupture would be expected to propogate along the z axis towards the cavity surface. This assumes that the two dimensional solution holds true in the three dimensional case. Obviously no provision for additional compressive stresses or reactive ion mixing have been included in this model.

5. CONCLUSION

The blistering mechanism seen for thin aluminium films on quartz depends upon the presence of an interface. The blisters seen after oxygen bombardment form at the aluminium-quartz interface which acts as a sink for the implanted oxygen atoms. Blisters are always seen near the edge of the aluminium layer where the layer thickness is of the same magnitude as the range of the implanted oxygen. The quantity of oxygen trapped at the interface will be the greatest where the interface depth equals the oxygen projected range. Thus the maximum blistering occurs when the depth of the interface equals the projected range of the incident ions.

Larger blisters were found to be a coalescence of smaller bubbles. For larger blisters a fracture is predicted by theory to propogate along the z axis towards the cavity surface.

REFERENCES

1. D.W. Parker, Electronics and Power, May (1977), 389.
2. S. Jones and I.H. Wilson, Electronic Letts, October (1979), Vol. 15, No. 21, 683
3. S.S. Gill and I.H. Wilson, Thin Solid Films, 55 (1978), 435.
4. J.G. Perkins, Phd., Univ. of Surrey, Dept. of Electrical Eng. (1971).
5. R.G. Musket and D.W. Brown, IBMM, Cornell, USA (1983).
6. J.P. Biersack, Z. Phys. A-Atoms and Nuclei 305, 95-101 (1982)
7. S. Cherekjian and I.H. Wilson, Nucl. Ins Meth. (1984), B1, 258-264
8. R.H. Wert and J.E. Castle, Surf. Anal., 4, 2, (1972), 68-75).
9. J. Roth, Inst. Phys., Conf. Ser. No. 28, 1976, P280-292.
10. D.J. Arrowsmith, E.A. Culpan and R.J. Smith, Proc. Symp. Anod., Alum., Univ. Aston, Birmingham, April 1976.
11. O. Auciello, Rad. Effects (1976), Vol. 30, 11-16.

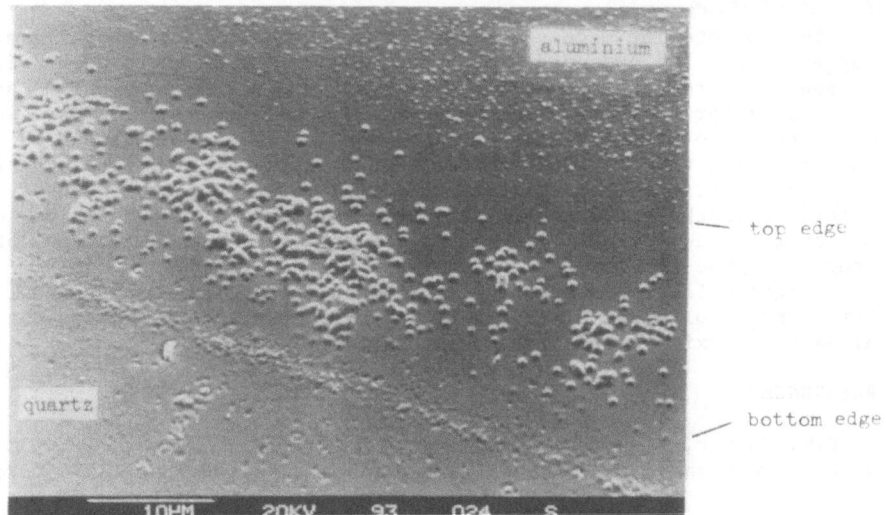

Plate 1   Aluminium(300nm) step on fused quartz after bombardment with 100keV O_2^+ ions $(4 \times 10^{17}$ oxygen atoms $cm^{-2})$.

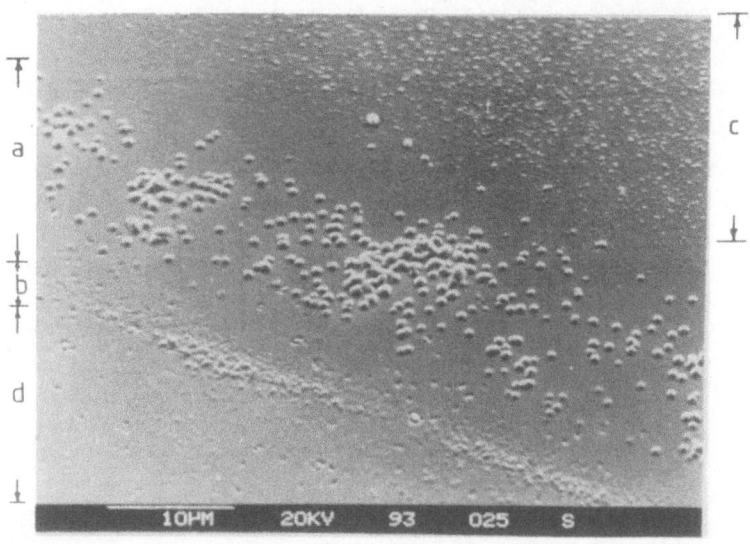

Plate 2   Continuation of the above step.

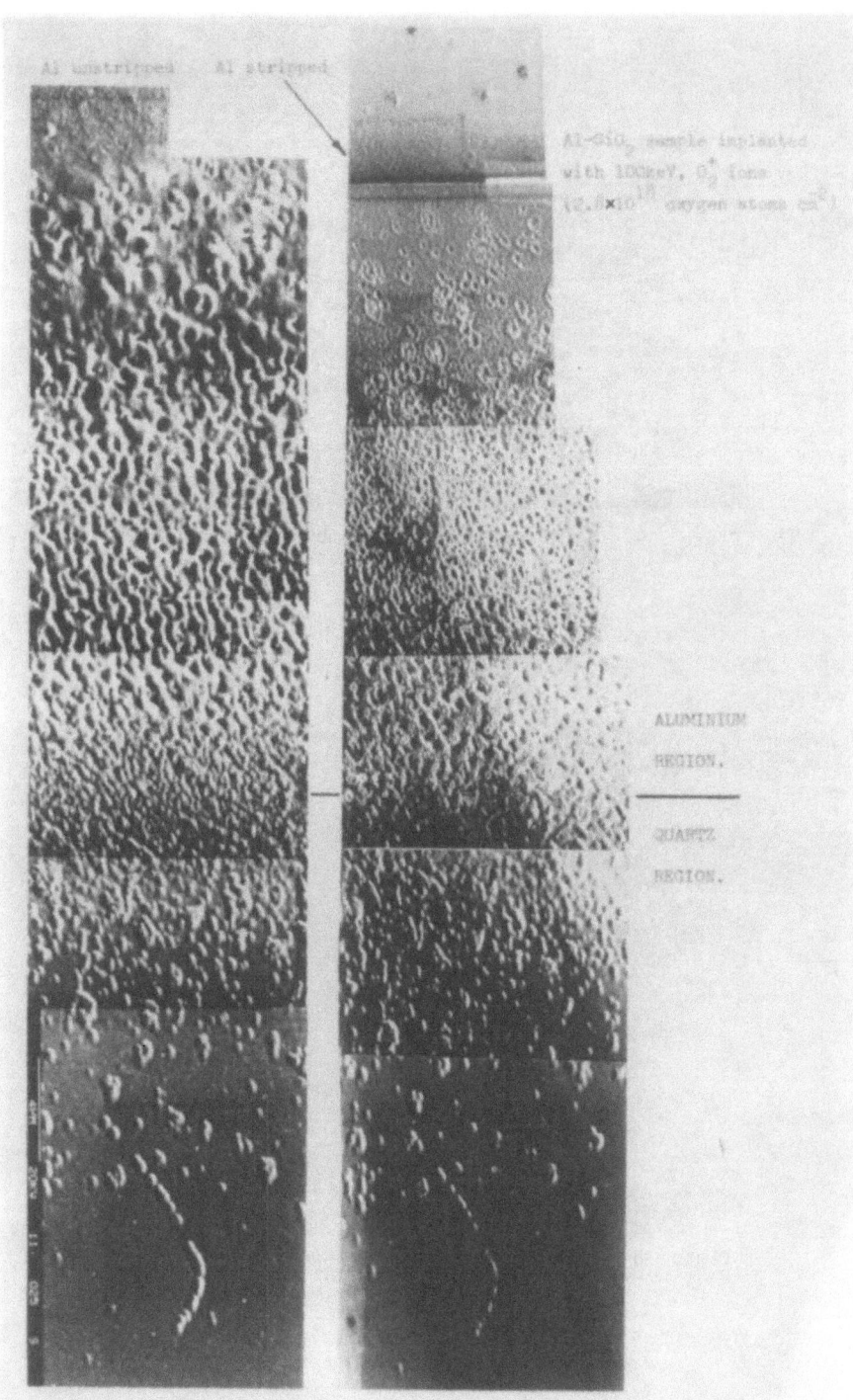

Al unstripped   Al stripped

Al-SiO$_2$ sample implanted
with 100keV, O$_2^+$ ions
(2.8×10^{18} oxygen atoms cm^2)

ALUMINIUM

REGION.

QUARTZ

REGION.

Plate 3   Aluminium-quartz edge, before and after stripping.

474

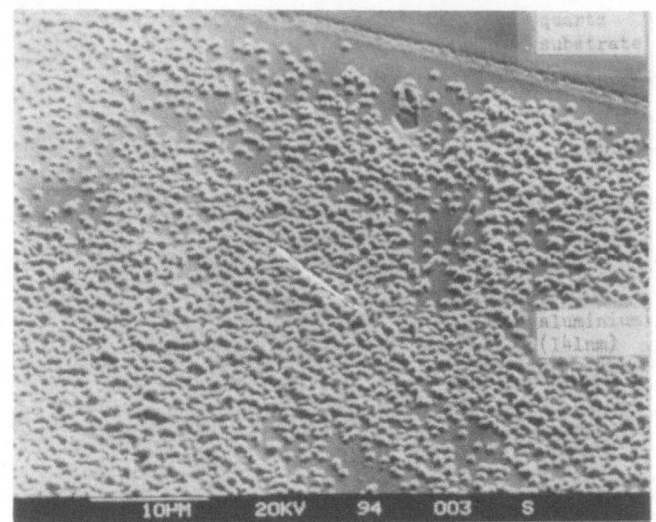

Plate   4   Al(141nm) on quartz bombarded with 150keV
O_2^+ ions (4×10^{17} oxygen atoms cm^{-2}).

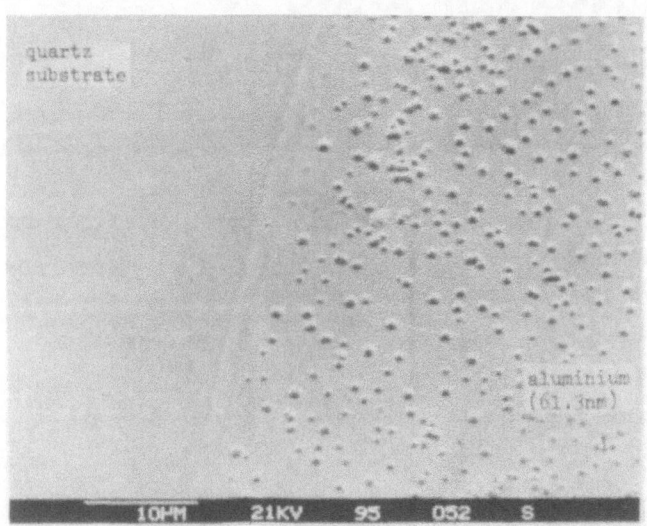

Plate   5   Al(61.3nm) on quartz bombarded with 150keV
O_2^+ ions (4×10^{17} oxygen atoms cm^{-2}).

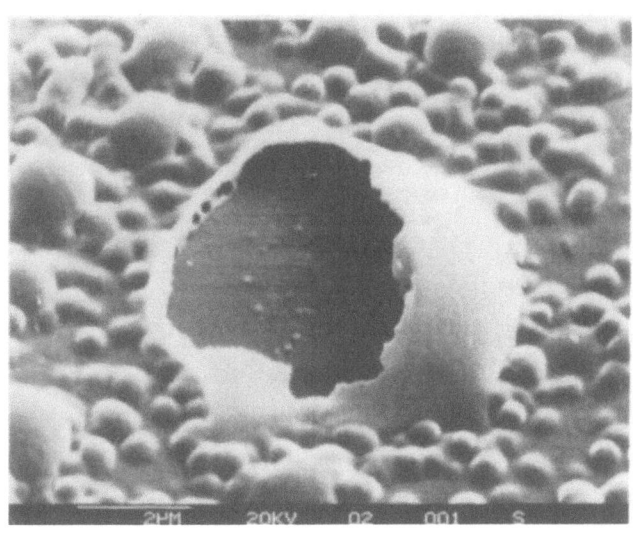

Plate   6   Al(141nm) on quartz bombarded with 150keV O_2^+

ions (4×10^{17} oxygen atoms cm^{-2}).

The S.E.M micrograph shows a ruptured blister

made up of a coalescence of smaller bubbles.

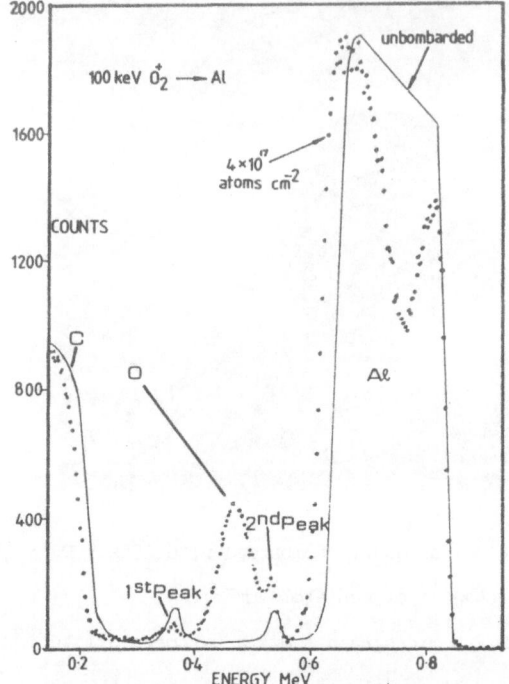

Figure 1: Rutherford backscattered spectra

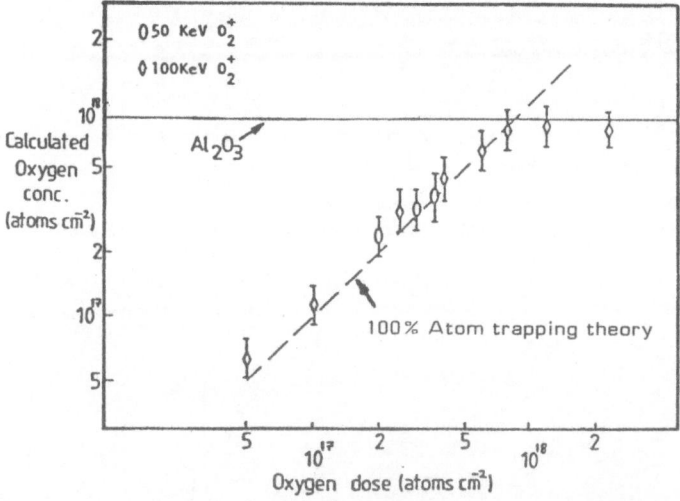

Figure 2: Calculated oxygen concentration versus
oxygen atom dose

ELECTRONIC CHANGES

ELECTRONIC PROPERTIES OF ION-IMPLANTED METALS

Harry Bernas
Centre de Spectrométrie Nucléaire et de Spectrométrie de Masse
91405 ORSAY FRANCE

The starting-point of these lectures is basically the same as that of the other lectures at this Institute: all the properties we shall be interested in are directly related to those two major effects of ion implantation: atomic displacements due to collision cascades, and compositional changes due to the addition of the implanted species into the host. Both of these effects lead to major modifications in the structural properties of metals. Typical examples are collision-induced disordering (or amorphization) of alloys, or phase transformation induced by the combined effects of heterogeneous species, or implantation and radiation-induced (or enhanced) diffusion. Such structural effects form the subject-matter of most publications in the field as well as the main content of many courses in this book. Our point of view here will be complementary to those.

Firstly, we shall go over some rather basic relations between electronic and structural properties. Our purpose here is to emphasize the fact that the nature and properties of chemical bonding determine many features of the implanted alloy's evolution, i.e. that electronic properties underly most of the thermodynamic properties which are so exhaustively studied today.

This is important for a very practical reason: ab initio electronic structure calculations have now reached a level of sophistication and precision where they can predict a number of important ground—state macroscopic properties of compounds, as well as - in many useful instances - their relative phase stability at a given composition. The very fact that structural studies of implanted or ion-mixed alloys demonstrate the importance of such equilibrium thermodynamical quantities as the mixing enthalpy or the Gibbs free energy points to the significance of the underlying electronic properties. It should encourage us to exercise the predictive powers of recent electronic structure calculations in designing ion-beam experiments.

In the second part of these lectures we shall take the reverse approach and discuss the use of ion-beam techniques as a specific tool to study important problems in electronic properties of metals. Examples are: the density of states for amorphous metals; the amorphization mechanism and its effect on magnetism; Anderson localization and the metal-insulator transition; the electronic structure of quasicrystals and very recent work on high-Tc superconductors.

There is one very important area which will not be directly discussed here. That is the crucial field for applications: how to tailor materials (notably metals) by ion—beam processing so that they provide specific transport, optical, magnetic, etc... properties. Partial reviews of several such applications are given in the literature, and in the context of this Institute I preferred to emphasize the more basic features. But I cannot

R. Kelly and M. Fernanda da Silva (eds.), Materials Modification by High-fluence Ion Beams, 479–506.

resist the pleasure of referring the interested reader to the handsome work of   White and co-workers [1], which provides an example of the way in which refined structural studies can lead to the preparation of a system (a buried layer of perfect epitaxial single crystal $CoSi_2$ in this instance) which has remarkable electronic properties opening new avenues for both basic studies and applications.

1. HOW ELECTRONIC PROPERTIES SHOW UP
1.1 What is meant by electronic properties?
We first consider the so-called "ground-state" properties, i.e. those which relate to the system at $T = OK$. Without going into any detail, let us just list a few important features.

The basic one is bonding. The nature of the interatomic forces, the pair and multiple atom interactions - central, covalent, directional, ionic...- are crucial in determining the structure of the basic clusters in all materials. Ab initio analytical and molecular dynamics calculations can now predict, in many cases with reasonable accuracy, the topological arrangement and stability of the clusters which may be formed by bringing together atoms of a given element. This clearly affects the nucleation process in a multi-element medium, and consequently also affects the composition and structure of the precipitates formed.

The introduction of "defects" by ion–induced collisions is a well-known effect in crystalline metals. The displaced atoms perturb the periodic potential set up by the ordered array of metal atoms: this enhances conduction electron scattering (Fig. 1a). As long as the total number of defects is weak, the resistivity increment is linear (i.e., proportional to the number of stable defects). At high defect

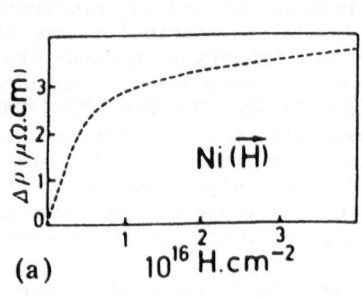

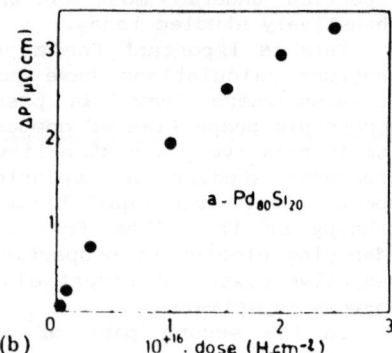

FIGURE 1. Resistivity—fluence dependence of a) crystalline Ni and b) amorphous $Pd_{80}Si_{20}$ films irradiated with protons. Note that main features of both curves are very similar. This example [see J. Less. Common Metals, 129, 1 (1986)] is typical.

concentrations it saturates, because new collisions annihilate as many defects as they create (the analysis of the resistivity saturation provides a measurement of the defect recombination volume). Let us also recall [2] that measurements of the low-temperature resistivity after isochronal anneals at increasing temperature provide information on the nature and evolution of radiation damage. These effects are interpreted in terms of defect (vacancies, interstitials, and clusters of each) evolution. However, irradiation experiments performed in glassy metals (Fig. 1b) have been interpreted in essentially the same way. Now, in a rather covalent system such as, say, a metal-metalloid amorphous alloy, the effect of irradiation (especially when "electronic stopping"- which is specifically due to inelastic interactions of the incoming ion with host atom electrons- is concerned) may be to break bonds. The system then minimizes its energy locally (long range interactions are reduced in the absence of long range order) by reorienting bonding angles. Such changes in configurational energy could also account for the conductivity changes observed when some amorphous alloys are irradiated with GeV ions [3]. This might be more consistent with available knowledge on the structure of the amorphous state than the "point defect creation" approach used so far.

More generally, as long as the interatomic forces are non-central, bonding is naturally the source of anisotropic effects, which influence phase formation via nucleation and via changes in the configurational energy. An example of such effects is the oft-discussed evidence for varying "degrees of amorphousness" in glassy metals. Another example, in a different context, is the driving force for phase formation in artificial multilayers or superlattices.

When metal atoms are brought together to form a solid, the discrete energy levels of the atomic orbitals broaden to form bands. These bands determine the main ground−state properties of the metal, i.e., its cohesive energy, heat of formation, and lattice expansion as well as phase stability. Which crystallographic phases may form at a given composition; what is the relative stability of the various phases that may exist at that composition; and even, in some important cases such as the A15 superconductors, how lattice distortions (and hence phase transitions) are related to the band structure: such are some of the important questions to which electronic structure theorists are now providing answers.

Before giving some examples, let us recall very briefly that bonding and band-structural properties are experimentally determined by a broad variety of techniques. Some of these (notably the various photoemission techniques or soft X-ray spectroscopy) are sensitive to all - or nearly all - the band structure, including in some instances the empty states above the Fermi level. They may also discriminate the partial densities of states (DOS) corresponding to the s, p, d... orbitals; in this way, the amplitude of hybridisation may also be determined. Other techniques are only sensitive to the DOS at the Fermi level: this is the well-known case of static susceptibility and specific heat measurements. An especially sensitive technique is electron tunneling [4] or the proximity effects in superconductors, which provide direct information on both the electronic structure and the phonon structure. Finally, other electric transport measurements provide less direct (but much more easily obtained) information on phenomena occurring in the immediate vicinity of the Fermi level. Photoemission and transport techniques are relatively easy to apply to thin films, and hence have been used to determine properties of implanted alloys. The examples given in Section 2 will show how such measurements provide information on the basic electronic properties (see also the reviews referenced in that Section. The other techniques are far more difficult to implement in the case of usual ion beam-treated samples because of insufficient thickness or total number of atoms. There are several cases, however, where they have been used and in which they have displayed their power (see Ref. 7 for example).

1.2 Macroscopic manifestations of ground-state electronic properties

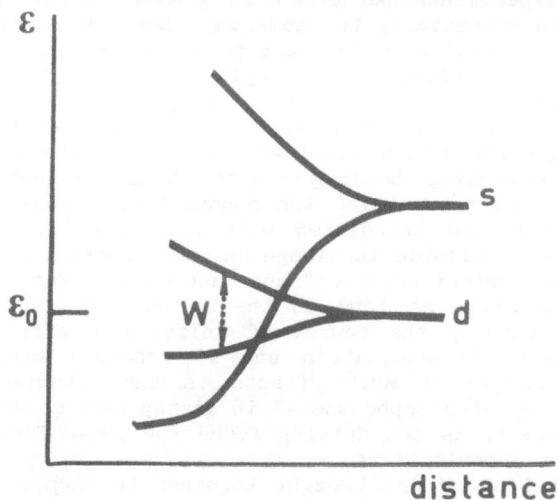

FIGURE 2. Schematics of band formation.

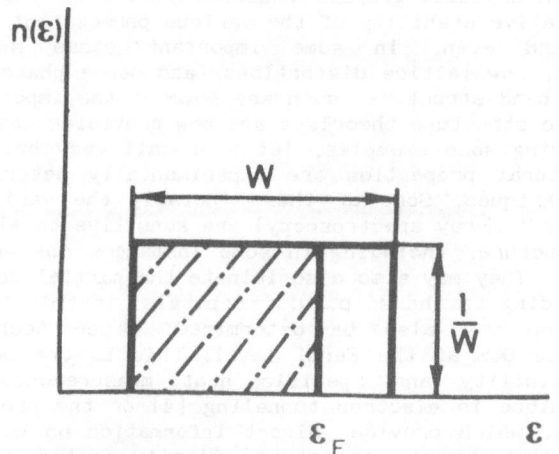

FIGURE 3. Schematics of band filling.

Example 1: The cohesive energy of d-band metals [8,9]

The cohesive energy E is just the difference between the total energy of the sum of individual atoms and the total energy of the metal. As shown in Figure 2, the s and d states of the atom broaden into bands, the d band (consisting of 10 states) being quite narrow. In the simplest model, the d band filling is (Fig. 3):

$$N_e^d = N_{at}^d \times 10 \int_{-\frac{W}{2} + \varepsilon_o}^{\varepsilon_F} \frac{-dE}{W}$$

and the cohesive energy is just:

$$E_c = \int^{\varepsilon_F} n(\varepsilon) \, (\varepsilon - \varepsilon_o) \, d\varepsilon \qquad\qquad (Eq.\ 1)$$

so that:

$$E_c \approx \frac{W}{20} n \, (10-n) \qquad\qquad (Eq.\ 2)$$

where n is the number of electrons per atom and W the band width. Note that similar calculations apply to the formation energy of the metals.

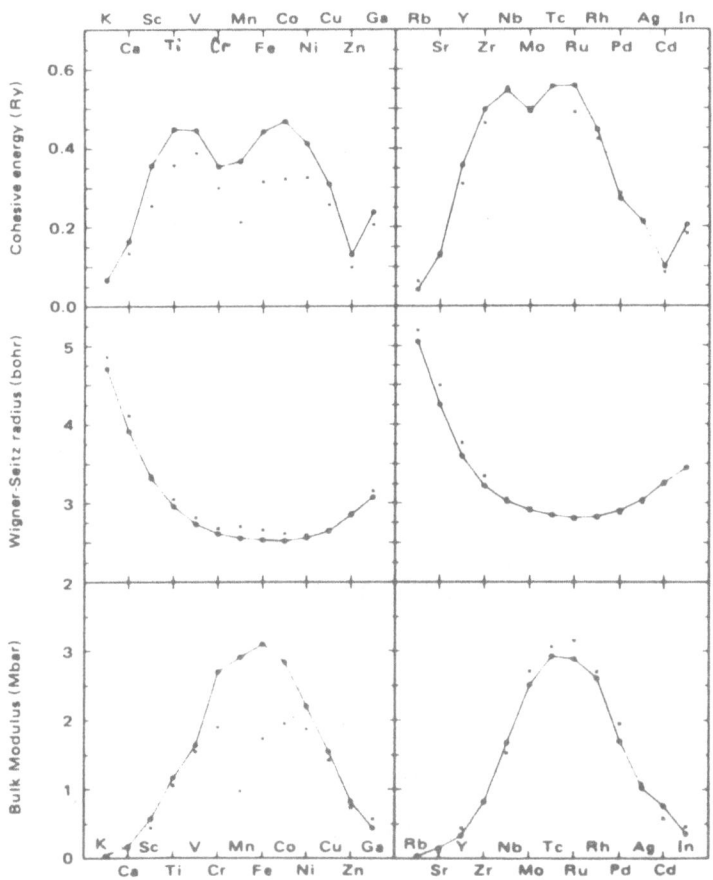

FIGURE 4. *Ab initio self-consistent calculations [9] of 3d and 4d metal ground-state properties (cohesive energy, atomic density, bulk modulus).*

The result of evaluations based on <u>ab initio</u> energy-band calculations is shown in Figure 4 [9] for the <u>3d</u> and <u>4d</u> metals. These calculations also provide the lattice constants (related to the bonding energy which is the difference between the cohesive energy and the repulsive energy) and the bulk modulus (the derivative of the bonding energy with respect to the volume). The difference between the calculated and measured values for the <u>3d</u> series are essentially due to the effect of magnetism in the middle of the series, an effect that may also be included in the calculation.

<u>Example 2</u>: Possible structures of <u>d</u> band intermetallic compounds.

Since the early days of metallurgy, empirical rules have been used to describe general trends for segregation or order in alloys (e.g. the Hume-Rothery rules or the Miedema rules). Figure 5 shows an example of the "structural maps" that may be obtained by correlating such variables as the average number $\bar{N}$ of electrons per atom and the electronegativity difference $\Delta\phi$ (or the valence difference ΔN). Such structural maps may also, as above, be calculated ab initio from interatomic potentials using appropriate techniques [10-15]. The point I wish to make here, however, just concerns the significance of the variable being used. $\bar{N}$ is a measure of the <u>d</u> band filling, and hence directly related to the <u>average potential</u> of the compound; $\Delta\phi$ or ΔN, on the other hand, are a measure of the potential <u>fluctuations</u> from one component atom to another. Thus, a structural map such as this correlates an average with a <u>local</u> parameter. In this context, it is interesting to note that the more or less "modified" Darken-Gurry plots used in some analyses of the stability of ion-implanted solid solutions are a correlation between the electronegativity and the atomic radius (but <u>not</u> the Wigner – Seitz cell size), i.e. between two local parameters. As a result, such correlations may be empirically useful, but provide very little physical insight.

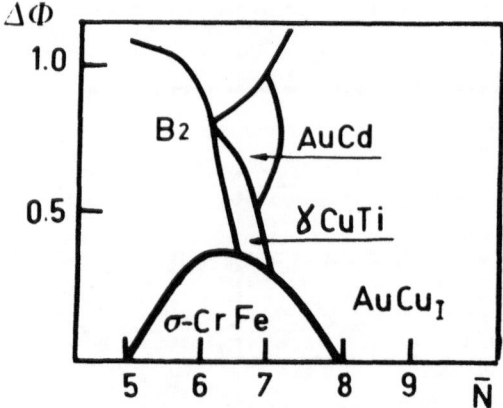

FIGURE 5. Structural map, indicating which is the stable ordered phase, for <u>d</u> band intermetallics. The contours are deduced from energy band calculations [10]. Experimental results (not shown here) are in very good agreement.

Example 3: Long-range order in d band alloys

Consider the dispersion curve $\varepsilon(k)$ of a disordered alloy. It is well-known that ordering of this alloy may introduce a new (higher) periodicity in $\varepsilon(k)$ as shown in Figure 6. If the Fermi level falls in the region where the levels cross and interact, then the total energy of the ordered state may be smaller than that of the disordered state, with the DOS at the Fermi level of the ordered state being lower than that of its disordered counterpart. Hence, the alloy gains energy by ordering.

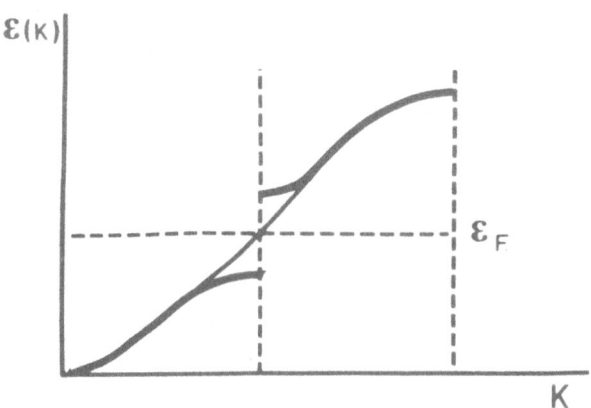

FIGURE 6. *Change in periodicity of dispersion curve due to ordering.*

Similar results are obtained (particularly in low-dimensional systems) by relatively simple distortions of the lattice (Peierls distortion; band and Jahn-Teller effect in the A15 superconductors; see also the tetragonal to orthorhombic transition in the $YBa_2Cu_3O_7$ high T_c superconductors). One consequence of this effect may be important in ion–implantation studies: if a system is only quasi-stable, the energy (as well as the compositional change) provided by ion implantation may be sufficient to drive it into an ordered state, rather than into the disordered or even amorphous state one would naïvely expect if only ballistic effects were considered.

On the experimental side, we note that such structural changes may be detected via modifications in the electronic properties. For example, the ordering effect discussed above leads to a reduction in the specific heat; it also leads to a sudden reduction in the normal resistivity of the alloy as the system orders, since there is a significant reduction in conduction electron scattering when the scattering potential is periodic.

Example 4 : "Short-range" order in amorphous metals

According to the Boltzmann equation [13] and to its extensions to weakly disordered systems [14], the resistivity of a metallic system is directly related to the topological "short-range" order (see Section 1.3). "Short range" means here that we are concerned with the first few

486

neighbour shells around a given atom. As a result, the residual resistivity (and, as we shall see later, the resistivity temperature-dependence) is sensitive to changes in the system's configurational energy.

An example of this sensitivity is the variation in residual resistivity of amorphous $Pd_{80}Si_{20}$ as hydrogen ions are implanted into the alloy. A quasi-linear increase is found until (after having practically doubled) the resistivity saturates at a hydrogen-to-metal ratio of one [18]. This result suggests that the implanted H atoms occupy the quasi-interstitial octahedral sites around the Pd atoms in the amorphous lattice. The correlation between the resistivity change and the H content is the same whether the hydrogen is implanted into, or desorbed from the sample: this confirms the interpretation.

Another example is the change (Fig. 7) in resistivity of the amorphous TeAu alloy after He$^+$ irradiation at 10K and annealing at increasing temperatures: the resistivity clearly detects changes in the configurational energy in the amorphous alloy (whose composition and long-range order remain unchanged).

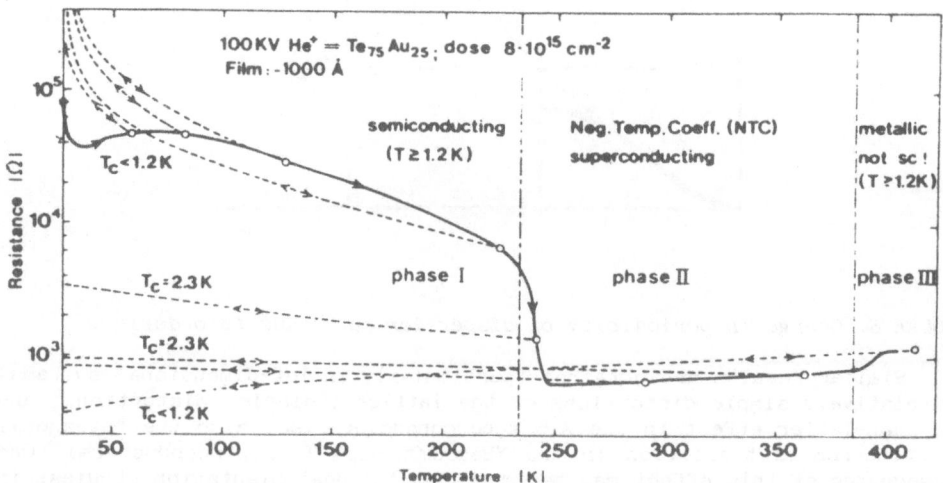

FIGURE 7. Resistivity annealing behavior of $Te_{75}Au_{25}$ after low temperature irradiation with He$^+$ ions. Double (single) arrows indicate reversible (irreversible) changes. From B. Stritzker and J. Meyer, Nucl. Instr. Meth. 183, 965 (1981).

1.3. Dynamic properties

These involve electron motion (transport properties) and the interaction of electrons with the excited states of the lattice or of the electron gas. We have discussed elsewhere [5] the information that may be obtained on implanted alloys and compounds from the temperature-dependent part of the resistivity (electron-phonon interaction), as well as from tunneling measurements on implanted films. The latter provides a powerful means of studying both the electron-phonon interaction and DOS-related electron scattering matrix. It is important at this stage to emphasize that these analyses of the transport properties are restricted to what is known as the "weak scattering" limit, i.e. those conditions under which the Boltzmann transport equation holds for the conductivity [13]. Bloch wave functions (i.e. a periodic potential) are assumed for the electron, whose motion is purely ballistic. Then, according to the Einstein equation:

$$\rho = \frac{m}{ne^2 \tau} \qquad \text{(Eq. 3)}$$

where τ is the elastic scattering time, n being the number of electrons (whose mass and charge are respectively m and e). In a near-ideal crystal, this leads to:

$$\rho \approx \int_0^{2\pi} T(\theta) \, [1-\cos\theta] \, \sin\theta \, d\theta \qquad \text{(Eq. 4)}$$

where $T(\theta)$, the scattering matrix, depends on local screening properties as well as on the electron-phonon interaction. Ziman showed [14]that under certain conditions, it is possible to define an average scattering potential even in an amorphous or liquid system, and to maintain Bloch wave functions which scatter in it. This leads to a resistivity:

$$\rho \approx \int^{2k_F} a(k) |U(k)|^2 k^3 dk \qquad \text{(Eq. 5)}$$

where $a(k)$ is the structure factor of the alloy (as determined by X-ray scattering for instance) and $|U(k)|^2$ the pseudopotential scattering matrix. In this way, the Boltzmann treatment may be extended; it does not, however, allow us to account for those very many instances in which the pseudopotential approximation fails, when strong scattering from random potentials occurs. We shall return to this situation -which is particularly important for implantation-induced strong disorder- in a subsequent Section.

Studies of the normal and extraordinary Hall effect (see below), are also well-adapted to implanted films, and provide access to both magnetic ordering and to lattice disorder effects. Techniques which are sensitive to the generalized succeptibility (relaxation via hyperfine interactions; dielectric constants measurements...) have been used, but are outside the scope of these lectures.

As an aside to these "dynamic" electronic properties, one is tempted to speculate on the possibility of analyzing the details of ion stopping effects on the target. Basically, there is the question of how the energy deposited by the incoming ion is transferred to the various reservoirs, and specifically how it is transferred from the electron bath (plasmons, excitons) to the lattice (phonons). Analyses based on thermal diffusivity provide reasonable numbers, but the underlying physics is somewhat obscure since the electron bath and the phonon bath are just not "on speaking terms", as the relaxation experts say. Modern femtosecond laser techniques using time-resolved analysis of recombinations (in Si) or reflectivity (in metals) may provide interesting information on the energy transfer mechanisms involved. Anyway, it seems likely that the problem is one of relaxation rather than of heat dissipation. In fact, perhaps the analysis in terms of thermal diffusion works simply because the electron-phonon relaxation occurs via the only available electron states, whose population is ~dn/dμ (where μ is the chemical potential), which is directly related to the specific heat...

2. USE OF ION BEAMS IN METAL ELECTRONIC PROPERTY STUDIES

For reasons that are partly historical there is a slight tendency, among physicists working with ion—beam techniques, to neglect some of the problems in basic (as well as applied) solid state physics that can be approached in a novel way by using ion implantation or irradiation. In other words, what follows is an invitation to cross the uncertain borders of metallurgy and to poach on new grounds, just as happened 20 years in semiconductor physics. Our examples are mainly from work performed by the Orsay group and collaborators for obvious reasons of familiarity: there is no pretense at completeness.

2.1. Densities of states in implanted amorphous Ni-P alloys

The electronic structure of the crystalline transition metal-metalloid compounds is quite well understood [9] thanks to elaborate band structure calculations. Cluster calculations [15-17], based on the existence of identical chemical short—range order (CSRO) in the amorphous compound and in the corresponding crystalline phases, have confirmed the effects of CSRO and reproduce both the DOS (as determined via photoemission) and -in simple cases- the structure factors.

The introduction of metalloid atoms into transition metals (e.g. P into Ni) leads to (i) lattice expansion, producing a narrowing of the d band and a corresponding reduction in cohesive energy and (ii) hybridisation of the metal d orbital by the s and p orbitals of the metalloïd, leading to very large modifications in the nature and shape of the DOS and a raising of the Fermi energy. The calculations referred to here -which are in good general agreement with photoemission experiments [for a more detailed discussion, see e.g. Ref. 18]- are performed for amorphous systems in which the CSRO is identical to that of the isocompositional crystal. Can ion implantation produce an amorphous system at the same composition with a higher configurational energy? Will this correspond to a different CSRO? Or -since the DOS is also sensitive to differences in the structure at distances beyond the immediate neighbours via the partial DOS of the p orbitals- are there changes at medium range, relative to the most stable amorphous structure?

In order to study this problem, Belin et al. [19] performed soft X-ray emission and absorption spectroscopy (SXS) experiments. The specific advantage of this technique is to provide not only total, but also partial, DOS of both occupied and empty states: the existence of hybridisation is thus directly determined and compared to theoretical calculations. Their study, performed on P-implanted Ni (producing amorphous $Ni_{80}P_{20}$) and on electrodeposited amorphous Ni-P at similar compositions, reveals several interesting features. Firstly the expected hybridisation of s and p states with the d states, which lower the hybridised pd bonding states relative to the metal d band, are clearly found. So are the hybridised antibonding states around the Fermi energy. In fact, the detailed structure of the DOS is in excellent agreement with theoretical calculations [17]. The specific advantage of ion-implanted samples here is that they are far more homogeneous than their counterparts producted by electrodeposition or splat- cooling. Secondly, there is indeed a difference between implanted and electrodeposited or crystalline samples (Fig. 8). The difference is rather small, and only affects a narrow part of the phosphorus-boron sp states. But this is probably a very significant difference. These states, which are hybridised with the d states at the top of the band, are partially delocalised: they do not reflect the CSRO, which is presumably unchanged as expected, but are sensitive to longer-range effects. Thus, some form of "medium-range order" is different in the implanted (relative to the other) amorphous alloy, affecting the dsp hybridisation in a restricted part of the total DOS. But this modified DOS is just at the Fermi level. As a result of changes in the d-hybridisation at the Fermi level, we must expect the magnetic properties of the implanted amorphous Ni-P systems to be somewhat different from those of the amorphous systems prepared by nearer-to-equilibrium techniques: the hybridisation is larger (so presumably the magnetic component is lower) in the implanted compound. We shall see the consequences of this in the next Section.

Another interesting effect pertaining to the DOS of implanted compounds is the result of Pivin et al. [20] on Ni-B. A comparison of the surface DOS (measured in metastable desorption spectroscopy) and that of the bulk for a-Ni75B25 and for c-Ni3B showed significant surface reconstruction even in the former case.

Both of these results should encourage studies of implanted alloy structures via DOS measurement techniques.

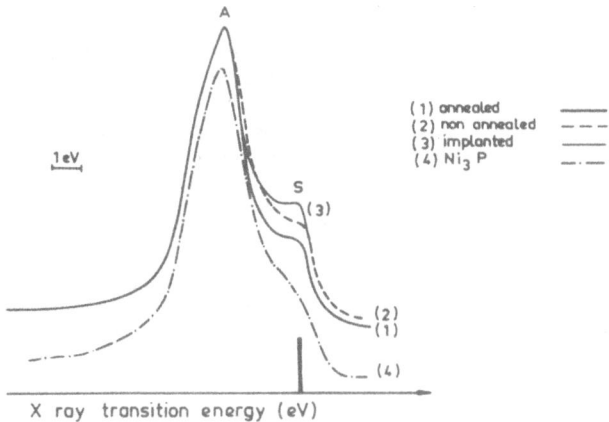

FIGURE 8. Experimental phosphorus p-like DOS in crystalline Ni3P and in amorphous Ni80P20 prepared by P ion implantation (curve 3) or by electrodeposition. The effect of annealing the latter (without crystallization) is also shown. The vertical bar locates the Fermi level.

2.2. Electronic properties of the crystalline - amorphous transition

A structural study, via transmission electron microscopy and channeling [21], of implantation-induced amorphisation in Ni-P revealed that the amorphisation process occurs in the following way. The implantation has two effects: random displacements via atomic collisions, and P concentration variation. Dividing the sample into elementary volumes v_c at the implantation profile depth, the "amorphous fraction α" of the sample can be accounted for by assuming that each volume v_c goes amorphous when the number N of P atoms it contains is larger than a critical number N_c:

$$\alpha = \sum_{N=N_c}^{\infty} (N)^N \exp(-N)/N! \tag{Eq. 6}$$

Averaging over the sample, this leads to a critical concentration (P/Ni atomic ratio) whose value may be directly deduced from the experimental curve (Fig. 9a). The value of v_c -which is the only parameter in the fit- is found to be about $2-4 \times 10^{-21} cm^3$ for Ni-P as well as for other alloys which have been studied so far by several groups (and which may be analysed in the same way). It is very interesting to note that the value of v_c corresponds very nicely to the size of the minimum cluster over which stable chemical short—range order can be defined (as seen, e.g., in EXAFS experiments or cluster calculations).

Thus, in the P concentration range between ~8% and ~16%, the implanted Ni-P system is an inhomogeneous superposition of amorphous cluster and crystalline (low concentration) Ni-P alloy zones. It is interesting to study the electronic properties of such a system as shown in Ref. 22. The conduction properties of this inhomogeneous medium were analysed using the effective medium theories developed by Bruggeman [23] and Landauer [24]. In these theories the conductivity (and the normal Hall constant) are known analytical functions of the conductivity (and the normal Hall constant) of each component as well as of the volume fraction of each component. Since the conductivities and Hall coefficients of the separate components are determined by independent measurements at P concentrations below 8% and above 16%, and since the variation of the amorphous fraction is known also (Fig. 9a and Eq. 6), there are no free parameters in the analysis. Figure 9b shows the result, which is indeed consistent with the conclusions of Ref. 21 and confirms the amorphisation mechanism.

A second conclusion was obtained from a study of the extraordinary Hall effect in the same system (Fig. 9c). This concerns the magnetism of the amorphous cluster, or of the implanted amorphous system. Magnetic studies of bulk (classically prepared, via e.g. splat-cooling) amorphous Ni-P showed that the ferromagnetic-to-paramagnetic transition occurs at a phosphorus concentration of ~18%. A simple analysis of the extraordinary Hall effect results obtained on the implanted system shows that the amorphous clusters are no longer magnetic above an average P concentration of 13%. This difference obviously warrants further investigation. One explanation could be related to the small cluster size (the moment of Ni atoms in such a cluster being smaller than that of the Ni atoms in the crystalline sheath), but a more plausible one is directly related to the preceding Section. The DOS of the implanted Ni-P phase is different from that of the corresponding "nearer to equilibrium" system as revealed by the SXS experiment: the effect of larger pd hybridisation at the Fermi level is to reduce the magnetic moment of the system. This conclusion is impossible to check directly, since there is no way (even at very low temperature) to prepare a homogeneous amorphous Ni-P system at a concentration of 13%. The consistency of all these results does, however, suggest that this second explanation is the correct one. If so, ion implantation has provided us with a means to relate magnetism to be degree of "amorphousness".

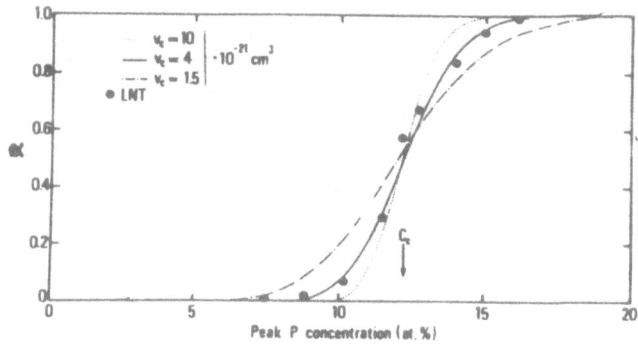

FIGURE 9. a) Metalloid concentration dependence of amorphous fraction α for P-implantated Ni (at 80 K, in the absence of radiation-enhanced diffusion). Curves are calculated, using Equation 6, with different critical volumes vc of the amorphous clusters.

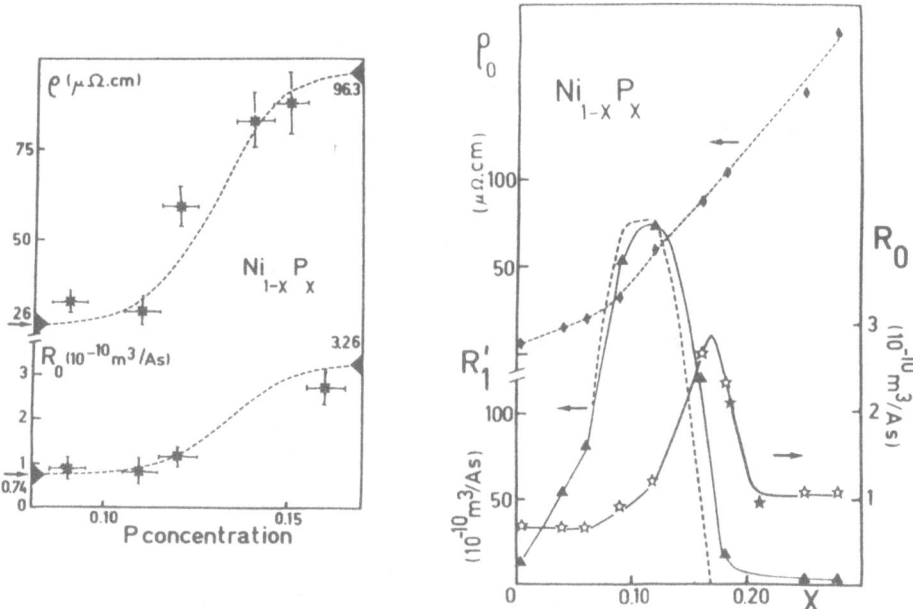

FIGURE 9. b) Comparison of experimental values of the resistivity ρ and the normal Hall coefficient Ro with those calculated from an effective medium theory; c) Residual resistivity ρo, normal (Ro) and extraordinary (R1) Hall coefficients plotted vs x. Solid lines are to guide the eye. The dotted line is a calculated variation (see Ref. 22) for R1 in the range 0.08<x<0.17.

2.3. Electronic properties of disordered metals

The title of this section, in fact, also applies to the two preceding ones; our purpose here is, however, to introduce more general features. As mentioned in Section 1.3, the conduction properties of highly disordered metals cannot always be accounted for in terms of the weak-scattering assumptions underlying the Boltzmann equation. The importance of this problem became increasingly clear in the 1970's: its study has had a major impact in making solid-state physics a branch of science that now studies real (i.e. "dirty", disordered) rather than ideal (i.e., "perfect") materials.

The discovery of the "Mooij correlation" [23]was a significant step. When analyzing the resistivity of disordered or amorphous alloys, of thin film alloys, and even of a number of high-resistivity ordered alloys, an overall correlation was found between the resistivity temperature-dependence and the absolute value of the resistivity at low temperature (i.e., the residual resistivity). There is no explanation of this result within the framework of weak-scattering transport theory. Moeover, the result has been confirmed in hundreds of cases, for materials with widely differing electronic structures and properties which cannot be drawn into the framework of a single model. As seen in Figure 10, the effect is significant for materials whose resistivity is in the 10-300 $\mu\Omega$.cm range. An ideal way of producing such resistivity values in a controlled way is ion implantation. Figure 11 shows an example of this for the Ni-P system, in which the study was performed systematically over the entire composition range where the amorphous structure is produced. [Note, incidentally, that the preparation technique, implantation or other, does not influence the values of ρ or β, which only depend on the P composition. This is not in contradiction with the results of the preceding Sections, even according to weak-scattering theory: the resistivity (Equation 5) depends essentially on the short-range order around the Ni atoms. You will remember that the properties discussed in 2.1. and 2.2. were more sensitive to the medium-range order]. The problem posed by the Mooij correlation is to relate the effects of elastic electron scattering (which affects the residual resistivity) to those of inelastic electron scattering via, e.g., phonons (which affects the resistivity temperature dependence). This relation is now discussed within the framework of Anderson localisation. We shall only indicate the terms of the approach, and refer to excellent reviews for more details [24]-[27].

The emphasis will be on an experiment in which ion implantation is used to demonstrate in a new way the most crucial property of a so-called "weakly localized" system, and to obtain detailed information on the system's evolution.

2.3.1. "New physics" in strongly disordered metals:

As mentioned in Section 1.3, the Boltzmann equation for transport describes the ballistic propagation of electrons in a periodic potential (Bloch waves), and accomodates weak scattering (due to low-level disorder) as a reduction in the electron's mean free path. The phase of the Bloch wave changes upon scattering, but the wave function remains infinitely extended. As shown in recent years (see reviews), in "very random" potentials (strong scattering) the low-energy part of the electron wave function becomes localized: the high-energy part propagates by diffusive motion. This is not accounted for by the Boltzmann equation. Strong disorder also leads to

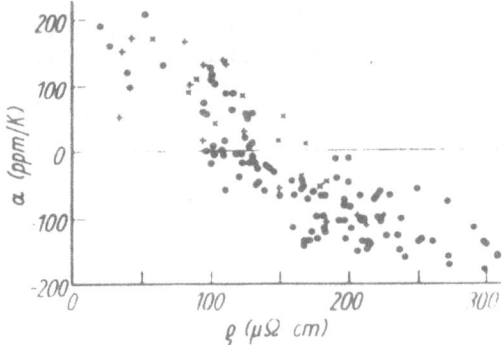

FIGURE 10. Temperature coefficient of resistivity $\alpha=(\rho^{-1})(d\rho/dT)$ versus resistivity ρ, measured around 300 K. Includes data from amorphous alloys, bulk alloys, and crystalline or amorphous thin films [23].

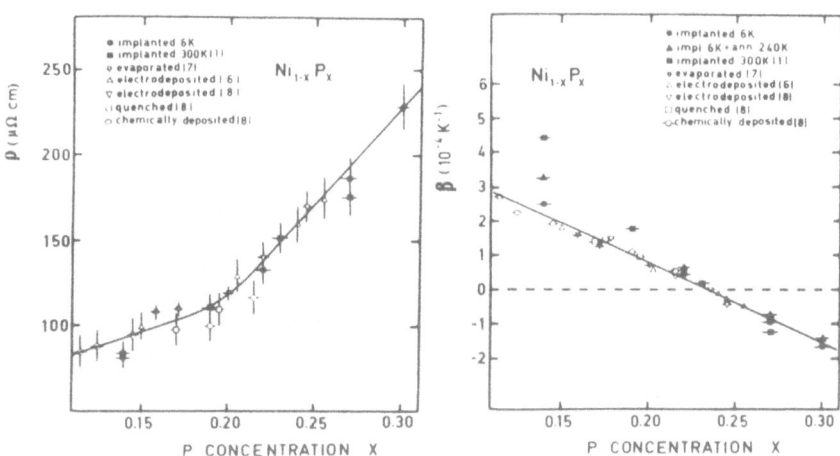

FIGURE 11. Resistivity (L.h.s.) and resistivity temperature coefficient β (r.h.s.) versus phosphorus concentration for Ni1-xPx alloys prepared by different techniques, including ion implantation. From L. Thomé, A. Traverse and H. Bernas, Phys. Rev. B28, 6523 (1983).

strong electron-electron interactions, and hence to the breakdown of standard Fermi-liquid theory in a highly disordered metal.

An illuminating analysis of localization was given by G. Bergmann [25] who showed that it is essentially a quantum interference effect. Consider an electron propagating from A to B in a highly disordered metal, with distance AB >>1, the elastic mean free path defined previously. The probability for an electron to go from A to B is, classically, equal to the sum of probabilities over all possible paths. In quantum mechanics, it is the square of the sum of all amplitudes. As long as there is no phase relation between the amplitudes involved, the classical and quantum probabilities are equal. However, consider the special case where path AB includes a closed loop. Time reversal conservation tells us that any two wave functions must be in phase when they meet at the crossing-point of the loop. Hence the quantum-mechanical probability of describing the loop in two opposite directions is exactly twice the classical probability, and phase coherence is conserved over all loops that may be described from any such crossing-point. The higher the disorder level, the larger the number of loops: disorder maximizes the quantum interference process. Note that this is a backscattering interference process: localization (which is just what confinement in loops leads to) minimizes forward scattering interference and maximizes backscattering interference. Also, localization is stronger when the loops are larger (i.e., maximum elastic scattering). This is the case at zero temperature. When the temperature is raised, inelastic scattering processes (electron-phonon or electron-electron collisions for example) come into play and reduce the size of our loops. Since they correspondingly reduce localization, they enhance the conductivity. Thus the resistivity decreases as the temperature increases in highly disordered metals. This is a qualitative explanation of the "Mooij correlation" shown in Figure 10.

2.3.2. How to approach the metal-insulator (Anderson) transition

A major breakthrough in the theory of solids came about as the relation between the effects of disorder and the existence of the metal-insulator transition became clear. One of the main ingredients was the emergence of a scaling theory, which was combined by Anderson and collaborators [27] with perturbation theory to analyze the scaling behaviour of the conductivity over the entire range between the perfect conductor and the insulator. Their result is shown in Figure 12. A crucial point in this Figure is the fact that, as shown by Thouless, the transition from metal to insulator may occur in one and the same material, just by changing the size L of the measured sample and that the scaling renormalization only depends on the sample's conductance, which in turn is the only relevant parameter characterizing the disorder. Thus the essential properties of localized systems depend neither on the origin nor on the nature of disorder, nor on the details of the material's electronic structure: this accounts for such universal features as the Mooij correlation. It also suggests that where ion implantation or irradiation changes the level of disorder in a metal (producing a change in the conductivity), some of the basic properties of the material will change in a way that is not specific to the mechanism by which the conductivity change has been induced. Conversely we have here a technique with which systematic studies of the most general localization properties may be performed, since there is no simpler (and more specific) way to control the disorder.

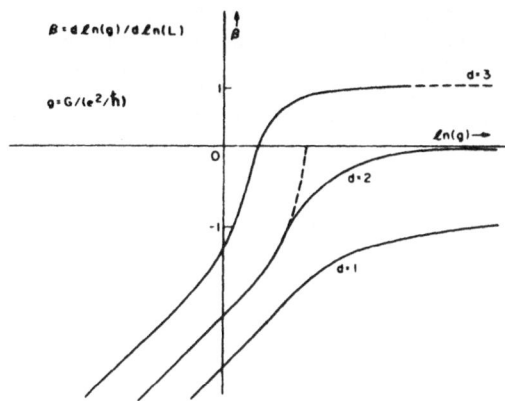

FIGURE 12. The scaling function β(g) vs the dimensionless conductance g for different dimensions. The scaling function β(g)=d(lng)/d(lnL) allows us to deduce the conductance of a sample of size $(2L)^d$ from that of a sample whose size is L^d, where d is the dimensionality (e.g. for a cube, d=3).

The scaling function β(g) in Figure 12 depends on the dimensionless conductance $g=G(e^2/h)^{-1}$, where G is the inverse of the sample resistance (note that we need a dimensionless variable, but one that does depend on sample size). The curves correspond to one-, two- and three-dimensional (3D) solids respectively. The upper r.h.s. of the figure shows that for large sizes and large conductances the system almost follows Ohm's law g(L)=σL, where σ is the conductivity. As the conductivity decreases, it is seen that the 1D and 2D systems are always insulating, while β(g) crosses zero for 3D systems: this is the point at which the 3D system experiences a metal-to-insulator transition. [Note that the meaning of the slope in the lower l.h.s. of the figure is that the conductance varies like g(L)≈exp(-L/ξ), i.e., the electron wave function is strongly localized with a "localization length" ξ, so that if one measured the conductance for a uniform sample over a size L<ξ one would find it to be metallic, while a measurement of the same sample over a size L>ξ would show it to be insulating]. Now consider a 3D metal with a microscopic conductance g_0 over a size L<ξ. Around the crossing-point where β(g)=0, the macroscopic conductivity σ varies like:

$$\frac{d\,\sigma_\infty}{\sigma_\infty} \propto \varepsilon^{-\nu}\,\frac{dg_0}{g_0}$$

where $\varepsilon = \dfrac{g_0 - g_c}{g_c}$ 　　　　　　　　　　　　　　　　　(Eq. 7)

g_c being the microscopic conductance at the crossing-point. The critical exponent ν characterizes the conductance behaviour near the transition, as in other critical phenomena; its value is expected to be around unity here [24].

In order to demonstrate the validity of this analysis, a number of experiments were performed (see Ref. 24 for reviews). Such experiments had two prerequisites: (i) produce a system in which the disorder level is high enough to bring its conductivity in the vicinity of g_c; (ii) then vary in a controlled way the microscopic conductivity and measure the system's properties, notably the macroscopic conductivity as it goes through the metal-insulator transition. The first condition was satisfied by using systems, such as highly doped semiconductors or metastable amorphous metallic compounds, in which the high disorder level is related to the chemistry of the system; the second condition usually implied changing either the chemistry (e.g. the alloy composition) or the geometry of the system which, as a result, was no longer self-similar. Both situations make comparisons with theory somewhat indirect.

These difficulties were circumvented by combining two tricks [28]. The first one consisted in producing a randomly percolating metal-insulator mixture. In the present case, Al and Ge were evaporated simultaneously on a hot substrate: the components coalesce separately (pure Al and pure Ge) in granular form, with the grains connected among themselves on a microscopic scale (a few hundred Angströms), so that a self-similar geometry is produced on a macroscopic scale. Adjusting the Al-to-Ge ratio so that the volume concentration of Al is ≥0.15 guarantees the existence of a 3D infinite cluster in the percolating structure. It has been shown that the (purely classical) conductance of the infinite cluster in a percolating structure depends on the volume ratio of the components (which does not affect the geometry since the system is self-similar). This purely geometrical trick provides a way of approaching the metal-insulator transition for the infinite cluster conductance, while the microscopic conductance (on a scale of a few hundred Angströms) remains that of the pure Al metal. The second trick was to reduce – in a controlled way – the microscopic conductance without changing the sample composition or geometry. This was obtained by low-fluence hydrogen implantation. In order to assess the conductivity changes on the microscopic scale, H^+ implantations were performed in pure Al films whose thickness was comparable to that of the percolating grain diameter. The comparison of the changes wrought by the same implantation on the microscopic and macroscopic scales is shown in Figure 13a. It is clearly seen that a small change ($\approx \rho \times 2.5$) in the microscopic resistivity leads to the Anderson transition ($\rho \times 5000$) on the macroscopic scale. Moreover as seen in Figure 13b, the temperature dependence of the infinite cluster conductivity confirms the insulating behaviour. Note that Ge indeed behaves as an insulator in these low-temperature measurements (all residual resistivities are measured at 4.2K). Also, these measurements allow the relation between the macroscopic and microscopic conductances to be deduced quantitatively: the result provides the critical exponent $\nu = 0.98$ in excellent agreement with localization theory. Finally, it should be stressed that this has provided a unique demonstration of the Anderson transition in a pure metal for which other properties (e.g., the detailed band structure) may be calculated rather exactly.

I would like to emphasize how attractive this area of physics should be to "ion beam" specialists. It includes some very basic problems to which they may provide a new line of attack. Among these, are the features of the various scattering processes which affect localization (see Section 2.3.1.). We have, using ion beams, an excellent way to control and adjust the relative amplitudes of elastic or inelastic scattering. In the latter case, it is easy to introduce and vary the amount of spin-flip (magnetic)

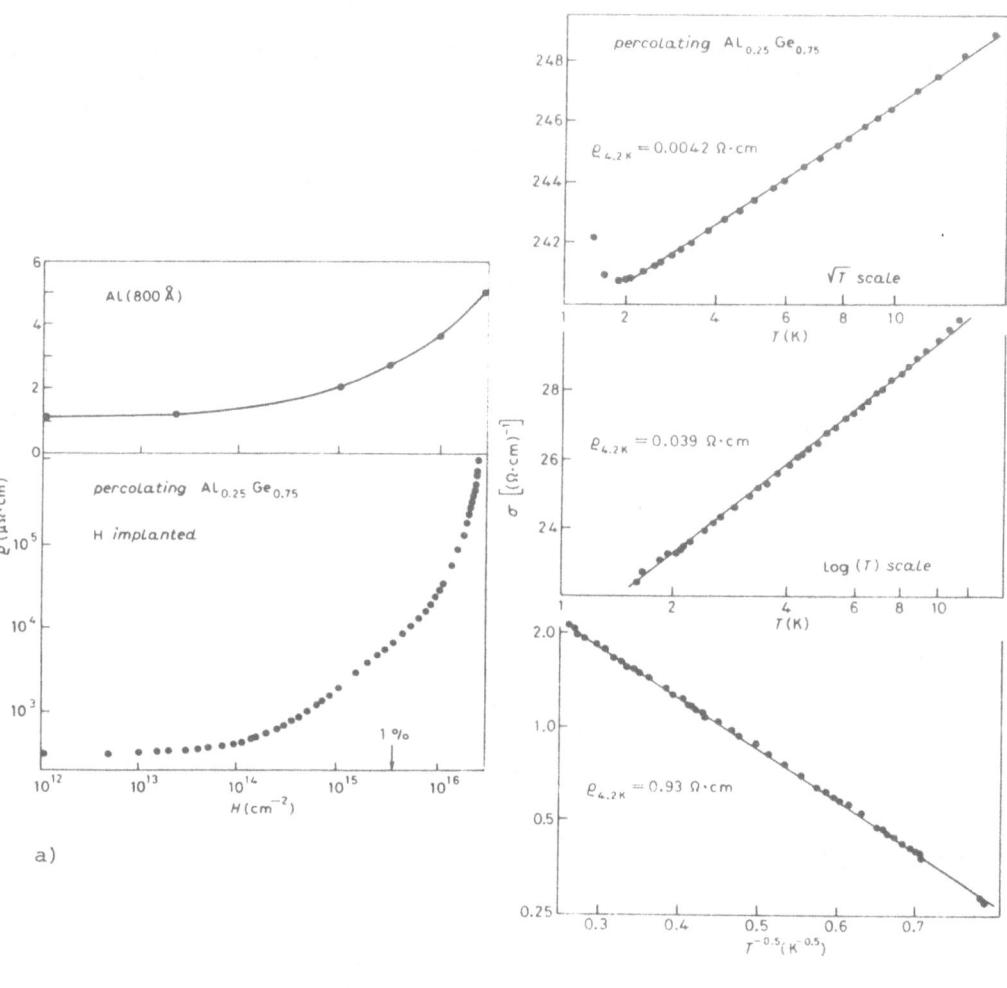

FIGURE 13. a) Residual resistivity dependence measured at 4.2 K vs, the hydrogen fluence. Notice that the resistivity scale is linear for the pure homogeneous Al sample, while it is logarithmic for the percolating Al0.25Ge0.75 sample. An atomic H concentration of 1% corresponds to about $3.4 \times 10^{16} H \cdot cm^{-2}$. Note the dramatic difference in vertical scales.
b) Temperature dependence of the conductivity for the percolating Al0.25Ge0.75 films at different implantation steps. The upper curve shows that the sample is still metallic with a T-dependent resistivity. The two lower curves (corresponding to increasingly large residual resistivities) display the lnT and exponential behaviours that characterize the transition to the insulating state.

impurity or spin-orbit (heavy) impurity scattering, etc... These are only obvious examples. More generally, the characteristic features of a disordered material, as opposed to those of an "ideal" one, relate to both the lattice structure and electron scattering properties. Here as in Part 1 the nature of structural order (e.g.: chemical short-range order in amorphous alloys; periodic or random order in crystalline alloys; medium or long-range order in all cases) is essential to phase stability as well as to the existence and nature of phase transitions. The very basic new feature is that overall the electron scattering properties depend much more on the degree of disorder than on the details of the system's electronic structure: systems with vastly different electron densities (dirty metals, doped semiconductors, granular metals or metallic ceramics such as the high-T superconductors) have similar or even identical scaling properties for some variables (notably transport). Also, the high disorder level leads to strong electron-electron interactions which modify the density of states considerably, particularly near the Fermi level. As a result, when constructing the thermodynamic functions (Gibbs free energy, heat of formation...) for a disordered system, the usual "one-electron" density of states (i.e., for non-interacting electrons) may no longer be appropriate, affecting predictions regarding stability or phase changes.

2.4 Transport properties of quasicrystals

Since their discovery in 1984, quasicrystals have posed a series of exciting problems [29,30]. Most of the work so far has concentrated on structural properties, for obvious reasons related to the nature of the quasicrystal structure and its complexities. A secondary reason may be more prosaic: quasicrystal alloys produced by such usual techniques as rapid quenching are as a rule fairly inhomogeneous, containing varying amounts of pure Al and other crystalline Al-Mn alloys or compounds as well as, say, the icosahedral $Al_{1-x}Mn_x$ with $x\sim 0.1-0.2$. These inclusions may sometimes be accounted for in structural studies, but they preclude high-quality electronic structure studies in general. It was interesting to see that ion beam mixing of multilayers could produce films that were both quasicrystalline and uniform when adequate conditions were satisfied (composition ratio of alternating layers, substrate temperatures, mixing rates, etc) [31]. This is the best way known so far to produce truly uniform samples with controlled composition. But very little work has been reported on their electronic properties, while there are a number of results on presumably less homogeneous samples prepared by rapid quenching.

Why study the electronic properties of quasicrystals? Basically in order to understand how the quasicrystal structure affects the electronic density of states, the transport and/or magnetic properties which all depend on the lattice symmetry properties of the material. Conversely, it would be important to understand (for the same reasons as discussed in Part 1) what special features of the electronic DOS lead to the relative stability of icosahedral order. Remember the main structural properties of quasicrystals: (i) orientational order exists, but there is no translational order. There is, however, a form of "long-range order" since the Fourier transform of the structure function has a discrete spectrum; (ii) there is self-similarity in the lattice symmetry (golden number); (iii) there is a quasi-lattice of defects (twins) which is also self-similar, with the same properties. It is important to note that, after all the structural work performed to date, the symmetries of the quasicrystal structures are clear, but that the atomic arrangements are not.

at all unequivocal (see Section 3 of Ref. 30 for a discussion).

The existence of quasi-periodic potentials in quasicrystals should lead to complex band structures, in which gaps such as that of Figure 6 should open all over the energy spectrum. The inherent disorder necessary to relate the icosahedra among themselves is expected to wash out most of such structures. On the other hand, two features have been demonstrated experimentally on rapidly quenched (hence possibly non-uniform) quasicrystal systems such as $Al_{1-x}Mn_x$. The first is that the resistivities of these alloys (even when their coherence extends over hundreds or thousands of Angströms) are very large- typically 150 to $1000\mu\Omega.cm$, corresponding to electron mean free paths that are of the order (or smaller than) the interatomic distance, if an interpretation in terms of purely elastic Bloch wave scattering is used. Such a result suggests that electron localization be considered; this was studied theoretically by Sokoloff [32], who showed that electron scattering from the quasiperiodic lattice alone does not contribute at all to the resistivity. The second feature is that, while Mn is non-magnetic at low temperatures in crystalline Al-Mn alloys, a fraction at least of the Mn atoms bear moments in the icosahedral phases with compositions between 14 and 23 at %. This result has several consequences. First, it suggested that the special "disorder" induced by the quasiperiodicity could lead to incoherent electron scattering from the strong Mn atom scatterers, thus producing a virtual bound state due to s-d scattering at the Fermi level [33], and consequently large resistivities as well as positive resistivity temperature coefficients for high-resistivity alloys [34] [35]. Secondly, it may open a new line of research in magnetism: that of a special type of spin-glass, with the magnetic moments sitting on a quasiperiodic lattice.

Where does ion implantation or ion beam mixing come in? As mentioned above, they lead to the production of homogeneous films [as checked by transmission electron microscopy and more recently by high resolution microscopy [36]]. As shown in Figure 14, systematic studies of the concentration-dependence of various physical properties are then possible. Resistivity studies are among the easiest, and they yield interesting information here. First, the concentration-dependence of the low temperature resistivity is in line with the magnetic effects expected from Mn^{34}. Also, the absolute values of the resistivity for, e.g., the 20 and 22% Mn samples, preclude explanations in terms of the virtual bound state alone: the latter would only contribute, according to [33], less than $200\mu\Omega.cm$ to the resistivity. The existence of defects in the three-dimensional tiling of the quasicrystals seems to be a crucial factor in the resistivity [32]; it may also be important for the magnetic properties if the existence of a magnetic moment on the Mn atoms is related to its being at such a defect site rather than in the basic icosahedra of the quasi-lattice [30]. In a word, defects seem to play an essential rôle in the electronic properties of quasicrystals, and it is rather obvious that implantation and irradiation experiments can provide useful information again by introducing controlled disorder into the quasicrystal (as well, perhaps, as decorating the icosahedra with light, low energy-implanted, atoms). Finally, let us mention that an SXS measurement of the electronic DOS of $Al_{1-x}Mn_x$ quasicrystals is presently underway [37] using ion beam-mixed multilayers. This experiment will provide a definitive test of the existence of a virtual bound state in the system.

500

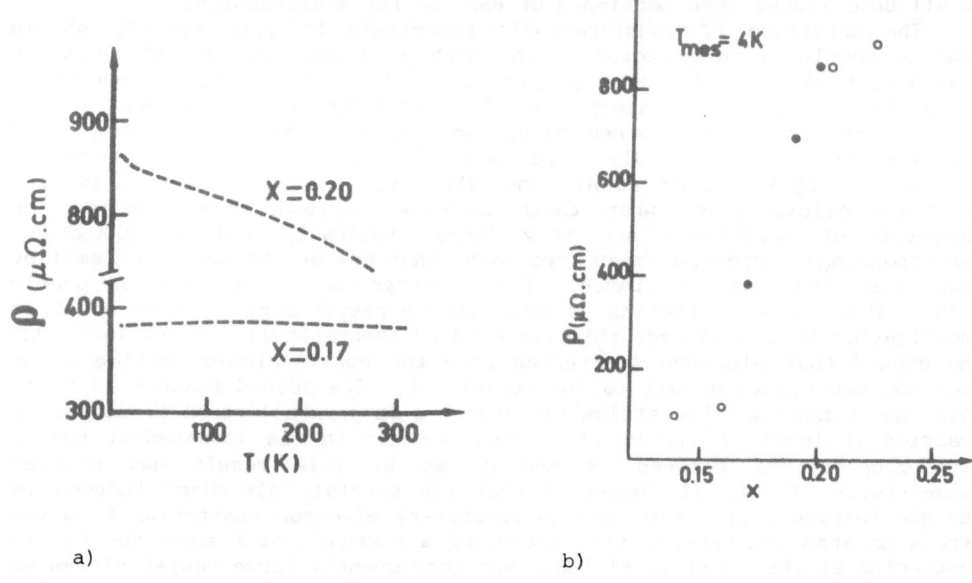

FIGURE 14. a) Typical resistivity temperature-dependence for quasi-crystalline $Al_{1-x}Mn_x$ films produced by ion beam mixing [from Ref. 36]. b) Compositional dependence of resistivity values at 4 K for quasi-crystalline $Al_{1-x}Mn_x$ samples prepared by ion beam mixing (filled circles) [see Fig. 14a] and rapid quenching (open circles) [from Refs. 34-35]. Note very large values at higher concentrations (evidence for magnetic contribution).

2.5 Brief notes on high-T$_c$ superconductors

In view of the rapid development of this field, any attempt at a review would be futile. My propose is only to indicate references for the newcomer to "standard" superconductivity theory and experiment, and to summarize some results comparing the "new" superconductors with the "old" ones. References 38-40 and 26, among many others (and more or less in that order), may prove useful as summaries of the pre-1986 state of superconductivity. The last reference also contains a discussion of the relation between superconductivity and localization, a question that may prove to be important in the high-T$_c$ ceramic oxides because of their inclination to become semiconducting and/or insulating as the oxygen content is reduced or as their structure changes.

The basic problems today are (i) whether or not Bardeen-Cooper-Schrieffer (BCS) theory accounts overall for the existence of superconductivity in the ceramic oxides and (ii) if so, what is the mechanism inducing superconductivity? We recall that BCS theory postulates a retarded attractive interaction between electrons just at the Fermi surface, so that a small fraction of these electrons form pairs ($-\underline{k}^\uparrow$, $\underline{k}_\downarrow$), producing a new ground state for the system, at a lower energy. The corresponding system has a single (macroscopic) wave function – all the electron pairs being correlated, there is an energy gap 2Δ between the new, superconducting ground state and the normal states of the metal. Pairing is thus the essential requirement according to BCS theory. The spatial extension of a pair ξ, called the coherence length, may be quite large and always includes many other pairs which all interact (with the same momentum). The density of states for one-electron excitations out of the ground state is $N(E)=E(E^2-\Delta^2)^{-1/2}$ for energies above the gap energy. Thermally excited electrons lead to occupied k and spin-states, so that the potential energy is increased as well as the kinetic energy. This, as well as the entropy term, leads to the gap temperature-dependence and to a critical temperature T_C (limit for superconductivity) which is shown to be

$$k_B\,T_C = h\,\omega_C\,\exp[-1/g] \qquad\qquad (Eq\ 8)$$

The parameters ω_c (excitation frequency) and g (coupling constant) depend on the mechanism which couples the electrons, i.e., on the excitation inducing the superconductivity. BCS postulated the electron-phonon interaction which does account for superconductivity in essentially all pre-1986 systems and according to which the gap Δ and T_c depend on the Debye temperature (ω_c is then the phonon frequency, leading to an $M^{-1/2}$-dependence of T_c), the electronic density of states and an electron-phonon interaction matrix. When the material has a T_c that is not smaller than, say, about 15% of the Debye temperature, the BCS formulation is modified for strong electron-phonon coupling, which includes renormalization effects due to Coulomb repulsion and retardation of the interaction [see Chapter 10 in Ref. 38]. The ratio $2\Delta/k_BT_c$ varies, for electron-phonon interactions, from 3.52 (BCS, weak coupling) to about 4.3 (strong coupling). A basic feature of all superconductors (besides their perfect conductivity) is their perfect diamagnetism: they exclude an applied flux as long as the applied field is above some critical value H_c. Correspondingly, they will accomodate currents (without applied voltage because all condensate wave functions are in phase) up to intensities which produce the critical magnetic field.

   As of now, the following comparison can be drawn between "classical" and ceramic oxide superconductors.

   a. pairing: this basic point is well-established by dc and Josephson effect experiments, which demonstrate flux quantization in terms of h/2e [41].

   b. coherence length: varies from several hundred Å to about a micron in usual superconductors, while it is only 3-50 Å in $(La_{2-x}Sr_x)_2\,CuO_4$ and $YBa_2Cu_3O_7$! Note that the total number of carriers is only about $10^{21}/cm^3$ in these materials so that, considering their high T_c and correspondingly large number of pairs, the proportion of electrons which contribute to superconductivity in the ceramic oxides is about 100 times larger than in the A15 compounds.

   c. energy gaps have been found in the oxides. The gap energies are large, as expected, but accurate measurements of $2\Delta/k_BT_c$ are still hampered by tunneling barrier control problems or by difficulties in interpreting infrared spectra measurements (see below).

   d. the isotope effect seems to be very small in the oxides: whereas $T_c\sim M^{-\alpha}$ with $\alpha=0.5$ according to BCS, $\alpha=0.16$ in $(La,Si)_2\,CuO_4$ and $\alpha=0.02$ in $YBa_2Cu_3O_7$ as measured on the $^{16}O-^{18}O$ difference.

e. both critical fields and critical currents reveal strong anisotropy in oxide single crystal measurements [42], indicating strong one-and/or two-dimensional effects (note that the structure of these materials is strongly anisotropic).

f. recent nuclear quadrupole resonance and NMR measurements indicate that the energy gap is also anisotropic, with $2\Delta/k_BT_c\sim8$ along the c-axis and ~2.4 in the Cu phases for $YBa_2Cu_3O_7$. There are also indications for antiferromagnetism [43].

g. an important point in the oxides studied so far is the sensitivity of superconducting properties to the crystal structure and - within, say, the orthorhombic structure for $YBa_2Cu_3O_7$ - to the oxygen content [44]. On the other hand, for $YBa_2Cu_3O_7$ there is little difference when Y is replaced by most rare earths (whereas magnetic impurities such as the rare earths are very deleterious to superconductivity in BCS materials).

Ion implantation and irradiation effects in other than the ceramic high-T_c oxides were reviewed previously [see Refs. 5,45,46 for example]. With the exception of low-temperature-implanted metal hydrides or of high temperature-implanted Nb carbides or nitrides, ion beam techniques do not usually lead to ordered compounds. Their main effect is again, as in the previous Section, to produce disorder (combined with a compositional change in special cases) in a controlled way. Now the disordering of superconductors, and its relation to the superconducting properties, is definitely an important problem which is directly related to the problems discussed in Section 2.3. As shown by Anderson [38], as long as the disorder is weak it should not affect T_c because the pair wave function is invariant under time-reversal. Thus, unless the electron-phonon interaction is affected, disordering should have little or no influence on T_c. However, even close to the "weak disorder" regime (such that $k_Fl>>1$, where k_F is the Fermi momentum and l the elastic mean free path), large reductions are found in the T_c of such important superconductors as the A15 compounds. A fundamental point is that the T_c reductions are correlated to the residual resistivities of the material, and are independent of the way in which a given resistivity was obtained (introduction of impurities, cold-working, irradiation...). This has been a much-studied question [see Refs. in [5,46]], to which two main (very different) answers have been proposed. One is related to disorder-induced smearing of sharp peaks at the Fermi level in the electronic DOS, an effect that modifies the electron-phonon interaction. The other is in the framework of strong disorder-induced localization [47] and suggests that the strong electron-electron interactions renormalize the repulsive Coulomb interaction, thereby sharply reducing the electron-phonon coupling constant [the electron-electron interactions can also modify pairing, as well as the Coulomb interaction]. Systematic studies of compounds and alloys with sharp (weak) structure in the DOS, and low (high) disorder levels have still not provided an adequate understanding. If the correlation is indeed due to many-body effects, the corresponding characteristic energies may be of the order of the Fermi energy or the Debye temperature (if they are related to the electron-phonon interaction) as noted in [47]. Then a close scrutiny of the T_c-disorder dependence in the oxide superconductors should be very revealing since in the latter (i) the effects of disorder on the conductivity are much larger, even leading to a metal-insulator transition; (ii) the existence of a metal-insulator transition allows a determination of critical parameters that depend on the electron-electron interaction features; (iii) the nature of the "disorder" may be basically different according to whether it is due to disordering of the perovskite structure (by irradiation for example) or whether it is due to a change in the oxygen content (or occupation ratio of the available sites) in the structure. And obviously, the nature of the superconductivity pairing mechanisms should also play a rôle.

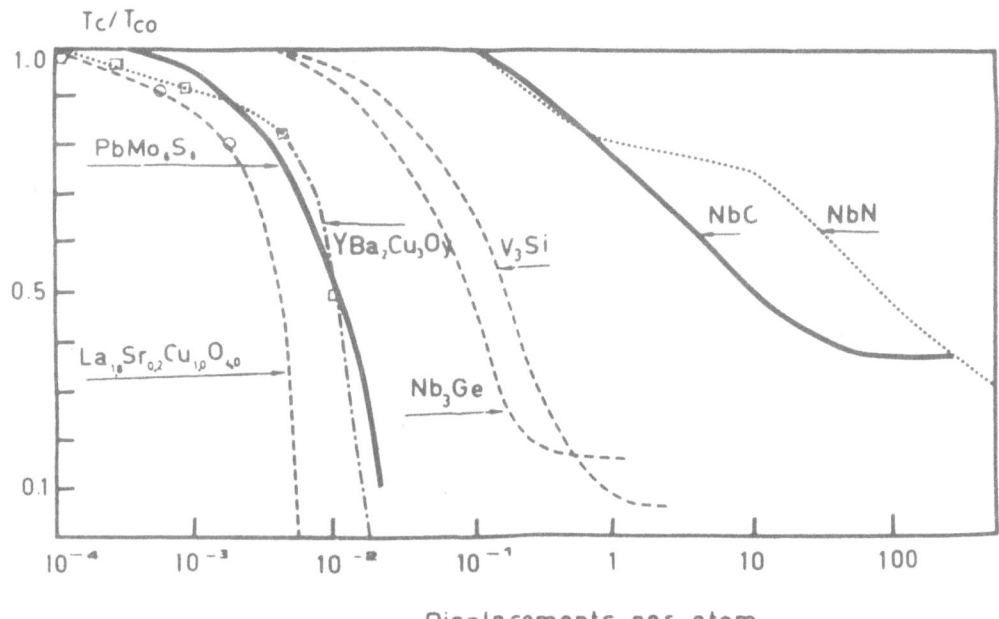

FIGURE 15. Effect of ion irradiation on superconducting critical temperature for various types of superconductors including the high Tc oxides La1.8 Si0.2 Cu O4 and YBa2 Cu3 O7. The critical temperature Tc after irradiation is normalized to the pre-irradiation value Tco. Abcissae are in displacements per atoms in order to allow comparisons among experiments with different ion beams and energies. Results from Refs. 48 and 50.

The first studies of ion-irradiated oxide superconductors [48-50] have shown that their sensitivity (in terms of a Tc-drop per number of displacements per atom) is about 100 times larger than that of the A15 compounds (Fig. 15). Electron irradiations show similar effects [51]. Transmission electron microscopy studies of the damage produced in such irradiation experiments are in progress [52]. The latter are important to identify defects and possible phase transitions, notably those induced by irradiation-induced changes in the oxygen concentration.

Let us finally mention an obvious field of investigation, which is outside of our scope : it concerns the preparation of thin film ceramic oxide films, an all-important asset to high-resolution studies of superconductivity (notably via tunneling and proximity studies, but also -as shown recently- via critical field and current studies), and to the most immediate applications. Whether ion beam mixing, combined with

504

adequate temperature control and annealing techniques, can contribute to the making of better films still remains to be seen at this writing.

ACKNOWLEDGMENTS: It is a pleasure to acknowledge the contribution of colleagues at Orsay and elsewhere with whom I have collaborated, discussed, argued, and generally been friends with. Among them Bill Grant, to whose memory this paper is warmly dedicated.

REFERENCES

1. White A., Short K.T., Dynes R.C., Garno J.P. and Gibson J.M.: Appl. Phys. Lett. 50, 95 (1987).
2. e.g. Vacancies and Interstitals in Metals, ed. Seeger A., Schilling W., Schumacher D. and Diehl J.: North Holland, Amsterdam (1970).
3. Audouard A., Balanzat E., Fuchs G., Jousset J.C., Lesueur D. and Thomé L.: Europhys. Lett. 3, 327 (1987).
4. e.g. Wolf E.L.: Rep. Progr. Phys. 41, 1439 (1978).
5. Bernas H. and Nedellec P.: Nucl. Inst. Meth. 182/183, 845 (1981).
6. Bauriedl W., Ziemann P. and Buckel W.: Phys. Rev. Lett. 47, 1163 (1981).
7. Hitzfeld M., Ziemann P., Buckel W. and Claus H.: Phys. Rev. B29, 5023 (1984).
8. Friedel J. in The Physics of Metals, ed. Ziman J.M., Cambridge University Press, London (1968).
9. Williams A.R., Gelatt C.D. and Janak J.F.: Proc. of Symp. on Theory of Alloy Phase Formation, ed. Bennet L., AIME, New York (1979); Gelatt C.D., Williams A.R. and Moruzzi V.L.: Phys. Rev. B27, 2005 (1983) and refs. therein.
10. Bennet L.H. and Watson R.E.: Phys. Rev. B18, 6439 (1978).
11. Pettifor D.G.: Chap. 3 in Physical Metallurgy, ed. Cahn R.W., Haasen P. (North Holland, Amsterdam, 1984).
12. Gautier F.: High Temperature Alloys: Theory and Design, ed. Stiegler J., AIME (1985) p. 264.
13. Ziman J.M.: Electrons and Phonons (Oxford University Press, 1960).
14. Ziman J.M.: Adv. Phys. 16, 551 (1967). See also Glassy Metals, ed. Guntherodt H.J. and Beck H. (Springer, Berlin, 1981).
15. Fairlie R.H., Temmermann W.N. and Gyorffy B.L.: J. Phys. F, 12, 1614 (1982).
16. Fujiwara T.: J. Non Cryst. Solids, 61-62, 1039 (1984).
17. Khanna S.N., Ibrahim A.K., Mc Knight S.W. and Bansil A.: Sol. State Comm. 55, 223 (1985).
18. Bernas H., Traverse A. and Janot C.: Amorphous Metals and Semiconductors, ed. Haasen P. and Jaffee R.I. (Pergamon, New York, 1986).
19. Belin E., Traverse A., Szasz A. and Machizaud F: J. Phys. F, 17, 1913 (1987).
20. Pivin J.C., Perreau J., Reynaud C., Takadoum J. and Boiziau C.: J. Non-Cryst. Solids, 86, 161 (1986).
21. Cohen C., Benyagoub A., Bernas H., Chaumont J., Thomé L., Berti M. and Drigo A.V.: Phys. Rev. B31, 5 (1985).
22. Traverse A., Paumier E., Nedellec P., Bernas H., Dumoulin L. and Chaumont J.: Phys. Rev. B37, 2495 (1988).
23. Mooij J.H.: Phys. Status Solidi A17, 521 (1973).
24. Lee P.A. and Ramakrishnan T.V.: Rev. Mod. Phys. 57, 287 (1985).
25. Bergmann G.: Phys. Reports 101, 1 (1984) and Physica 126, 229 (1984).
26. see Percolation, Localization and Superconductivity, ed. Goldman A.M. and Wolf S.A. (NATO ASI, Vol 109, Plenum, New York, 1983).
27. Abrahams E., Anderson P.W., Licciardello D.C. and Ramakrishnan T.V.: Phys. Rev. Lett. 42, 673 (1979).
28. Nedellec P., Traverse A., Dumoulin L., Bernas H., Amaral L. and Deutscher G.: Europhys. Lett. 2, 465 (1986).
29. e.g. Proc. Workshop on Aperiodic Crystals, J. Physique 47 C-3 (1986).
30. Henley C.: Comments Cond. Matt. Phys. 13, 59 (1987).

31. Knapp J.A. and Follstaedt D.M.: Nucl. Inst. Meth. B19, 611 (1987); see also Lilienfeld D.A., Hung L.S. and Mayer J.W., MRS Bulletin, Feb. 16, 1987, p. 31.
32. Sokoloff J.B.: Phys. Rev. Lett. 57, 2223 (1986).
33. Berger C., Pavuna D. and Cyrot-Lackmann F.: J. Phys. C-3, 489 (1986).
34. Fukamichi K., Masumoto T., Oguchi N., Inoue T., Sakabua T. and Todo S.:J. Phys. F 16, 1659 (1986).
35. Kimura K., Hashimoto T. and Takeuchi S.: J. Phys. Soc. Japan 55, 1810 (1986).
36. Traverse A., Portier R., Dumoulin L., Ceriani-Sebregondi P., Lachaud Y., Cohen-Bastié P.: Rev. Met. Fr. 9, 450 (1987).
37. Traverse A., Belin E., Dumoulin L. and Senemaud C.: comm. Workshop on Quasicrystals, Grenoble, March 21-25, 1988 (to be published).
38. Superconductivity, ed. Parks R.D. (Dekker, New York, 1969).
39. de Gennes P.G.:Superconductivity of Metals and Alloys (Benjamin, New York, 1966).
40. Butler W.H. in Treatise on Mat. Sci. Tech., 21, ed. Fradin F. (Academic, New York, 1981), p. 165.
41. Estève D., Martinis J.R., Urbina C., Devoret M.H., Collin G., Monod P., Ribault M. and Revcolevschi A.: Europhys. Lett. 3, 1237 (1987).
42. Dinger T.R. et al.: Phys. Rev. Lett. 58, 2687 (1987) and Worthington T.K. et al.: Phys. Rev. Lett. 59, 1160 (1987).
43. Warren W.W. et al.: Phys. Rev. Lett. 59, 1860 (1987).
44. Burger J.P. et al.: J. Phys. 48, 1419 (1987).
45. Meyer O.: Rad. Eff. 48, 51 (1980).
46. Rullier-Albenque F.: Ann. Chim. Fr. 12, 573 (1987).
47. Anderson P.W., Muttalib K.A. and Ramakrishnan T.V.: Phys. Rev. B28, 117 (1983).
48. Geerk J., Li H.C., Linker G., Meyer O., Politis C., Ratzel F., Smithey R. Strehlau B., Xi X.X. and Xiong G.C.: 8th Int. Symp. Plasma Chemistry, Tokyo, Sept. 1987.
49. White A.E., Short K.T., Jacobson D.C., Poate J.M., Dynes R.C., Mankiewicz P.M., Skocpol W.J., Howard R.E., Anslowar M., Baldwin K.W., Levi A.F., Kwo J.R., Hsieh T. and Hong M.: Phys. Rev. B37, 3755 (1968).
50. Clark G.J., Le Gouës F.K., Marwick A.D., Laibowitz R.B. and Koch R.H.: Appl. Phys. Lett. 51, 1462 (1988).
51. Stritzker B., Zander W., Dworschak F., Poppe U. and Fischer K., comm. Mat. Res. Soc. Meeting, Boston, Nov. 5-10 (1987).
52. CSNSM, Orsay - IBM Yorktown collaboration, unpublished 1987-1988.

MECHANICAL CHANGES

TRIBOLOGY OF IMPLANTATION BILAYERS

J.C. PIVIN

C.S.N.S.M., Bâtiment 108, BP 1, 91406 ORSAY CEDEX, FRANCE

1. INTRODUCTION

The mechanical behaviour of implantation films must be analysed in terms of bilayer rheology (laws of mechanical behaviour). It is in itself a complex domain of mechanics, and the models are subject to argument. Indeed, even in the case of a film and a substrate with constant compositions and a well-defined interface, their respective contributions to elastic and plastic deformation are different and vary with the mode of deformation. Scale effects must also be taken into account when considering submicroscopic films.

On the other hand, tribology also takes into account thermodynamical, chemical and metallurgical parameters to interpret the friction properties of a system as a whole. Thus, it is not surprising that the numerous studies on implanted surfaces do not lead to straightforward conclusions, more especially as few of them were comprehensive enough. In many of them, the effect of a given implantation on the wear and friction resistance (or on the lifetime in fatigue) was stated only in very specific conditions. One can distinguish between alloying effects of ion implantation and structural modifications. Alloying affects the basic properties of the crystal: elasticity, cohesion, mobility of planar defects, and finally its surface electronic structure, which determines the reactivity with the atmosphere or the friction counterpart (adhesion). Radiation damage and phase changes act more particularly on the modes of gliding and climbing of dislocations, and fracture mechanisms. For instance ion implantation is a well suited process for obtaining layers of metastable phases: martensites, quasicrystals and metallic glasses, which can play a role of barrier to the propagation of cracks. Precipitates (and segregations) will rather hamper their initiation and have a pinning effect on dislocations, more especially if they present a particular epitaxial relation with the matrix.

Instead of reviewing all experiments on friction or fatigue of implanted surfaces and their possible applications, the purpose of this paper will rather be to analyze a few cases in the light of basic data on the rheology and tribology of bilayers. The most recent studies on ion implantation effects on friction, wear, and fatigue of metals and ceramics are reviewed in (1).

2. TRIBOLOGY OF BILAYERS
2.1. Rheology
    2.1.1. Static contact. For purposes of simplification we will only consider the contact between a planar surface and surfaces exhibiting rotational symmetry: a flat punch, a sphere or a sharp cone. These

R. Kelly and M. Fernanda da Silva (eds.), Materials Modification by High-fluence Ion Beams, 509–533.

configurations are typical models of contacts between single asperities of solids, and other geometries such as a two dimensionnal edge or a pyramid (commonly used in indentation tests) can be derived simply. Let's first consider the static contact between bulk materials and suppose that the flat surface is much softer than its counterpart (indentor). As the force applied normally to the interface increases, the plane will undergo only elastic deformation (i.e. fully reversible) until a critical stress, Y, of plastic flow is attained locally. This flow stress is associated with a critical stress, k, of shear or slip along crystallographic planes at the submicroscopic level: k = Y/2 (Tresca criterion of plasticity) or k = Y/√3 (von Mises criterion). Now, the pressure, P, at the interface and the stress tensor in the indented solid are generally not uniform. They exhibit the symmetry of revolution shown on figure 1 for P. The mean pressure of indentation P_m is also given on the figure for the two extreme cases of a fully elastic or of a fully plastic contact (2): it takes values which are direct measures of the Young's modulus E (elastic coefficient of proportionality between applied stress and deformation), Poisson's ratio ν (elastic coefficient of compressibility) and hardness H. Only in the case of the cone are the Hertzian pressures of elastic contacts independant of the load. In the case of plastic contacts, the mean pressure is by definition the hardness, related to k by a factor c, which varies with the shape of the indentor and interfacial friction but remains within the range P_m/k = 5 to 6 for blunt indentors.

Indeed, the field of plastic deformation depends critically on the angle of indentation θ and on the frictional shear along the interface which modifies the inclination of slip lines with respect to this interface (Figure 2). The slip-line field (S.L.F.) can be constructed by drawing lines parallely to the directions of principal shearing stress in each point, i.e at 45° to the direction of principal direct stress. Their angle α with the interface changes from 45° to a lower value as the exercized stress is no longer normal to this interface. In the case of a sharp cone, indenting a fully plastic solid, the stress is infinite at the tip and the material flows preferentially upward at the periphery of the indentor (Figure 1), giving the typical shape of a Prandl's fan to the plastified zone shown on figures 1 and 2 (3-5). But a cap of undeforming material appears under the indentor when θ and the friction coefficient increase (as α becomes lower than zero and the material behaves as part of the indentor in the triangle ABD), or when the indented solid exhibits an intermediate elastoplastic behaviour (2,6). For θ values over 60° the plastified zone tends to take an hemispherical shape as for a sphere or a punch.

This led Marsh (7) and Johnson (8) to propose a simpler model of the field of deformation than those which may be simulated with the S.L.F. theory (3,4), finite-element models (5), or simplified kinematic models (9). They likened the deformation to that induced by the expansion of an hemispherical cavity under hydrostatic pressure in an elastoplastic solid, for which the problem had already been solved.

The pressure at which a plastified sphere forms is (for ν = 0.5) (8,10):

$$P_m/Y = (2/3) (1 + [\ln E] \times [\cot g\ \theta]/3Y)$$

For a spherical indentor cotg θ is replaced by a/R, where a is the radius of the area of contact, and plastic deformation begins at a depth d = a (a being equal to the Hertzian radius of contact for this limit, i.e $[(3/4)F_N \times R \times (1-ν^2)/E]^{1/3}$ for P_m = (2/3) Y.

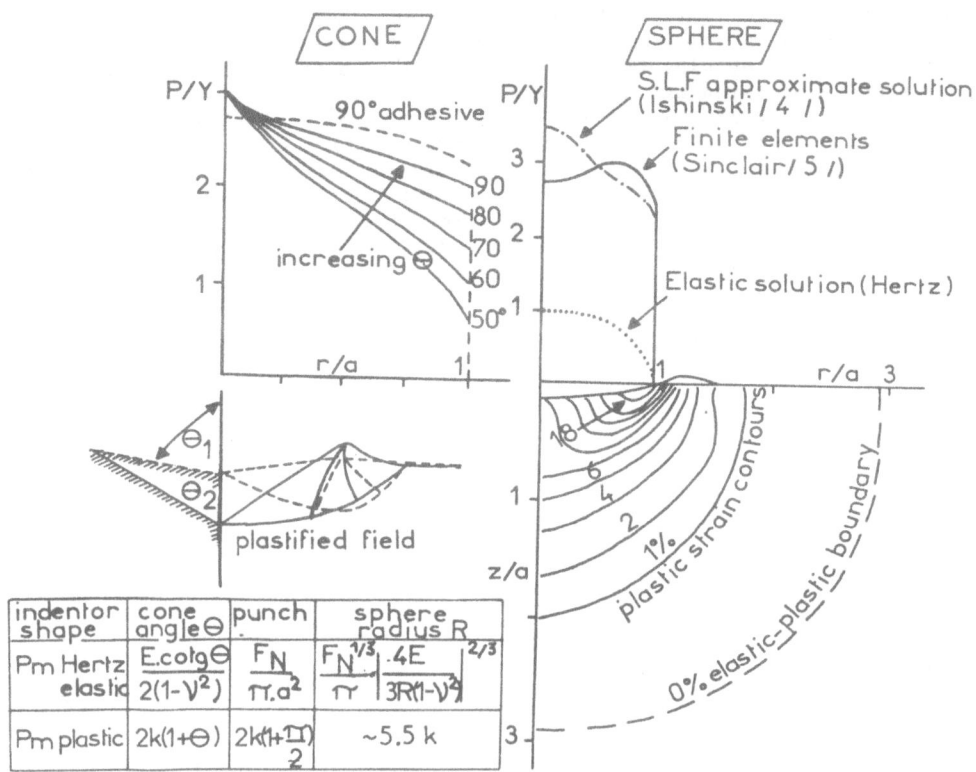

FIGURE 1. Variations of the contact pressure P, normalized to the flow
stress Y of the indented solid, as a function of the distance r from the
center of the contact and depth z, normalized to the radius of contact a.
The shape of the plastified volume and values of P/Y given for the cone
are the exact solutions of a slip-line field (S.L.F.) calculation
(Lockett/4), in the case of a fully plastic contact without friction at
the interface. The comparison of plastic fields of deformation for two θ
values gives evidence of a greater upward flow and of a higher
peripherical pile-up as θ decreases. The strain contours shown for the
sphere were calculated with a finite elements model (Sinclair/5) for a
contact without friction and a/R = 0.3.

512

There is no upper limit to its expansion. In reality it begins for $P_m = Y$, and the contact pressure becomes equal to the hardness 3Y when the plastified zone attains the free surface (d = 3a in the model) (3,4,11). A corrected expression was proposed (12):

$$P_m/Y = 1.1 + 0.58 \text{ x } \ell n \text{ E x} [\cot g \ \Theta]/2.3Y$$

for $P_m < 3Y$. At intermediate pressures there is a good agreement between fields of plastic deformation observed experimentally (8) and the model's predictions, which thus allow us to account for the accomodation of stress by elastic deformation around the plastified core (8,11,13). But the influences of the indentation angle, friction coefficient and work-hardening (foot-note 1) of real materials can only be predicted by using the slip-line field theory. These last two factors induce an increase of P_m and diminish the upward flow at the periphery of the indent as shown on figures 1 and 2.

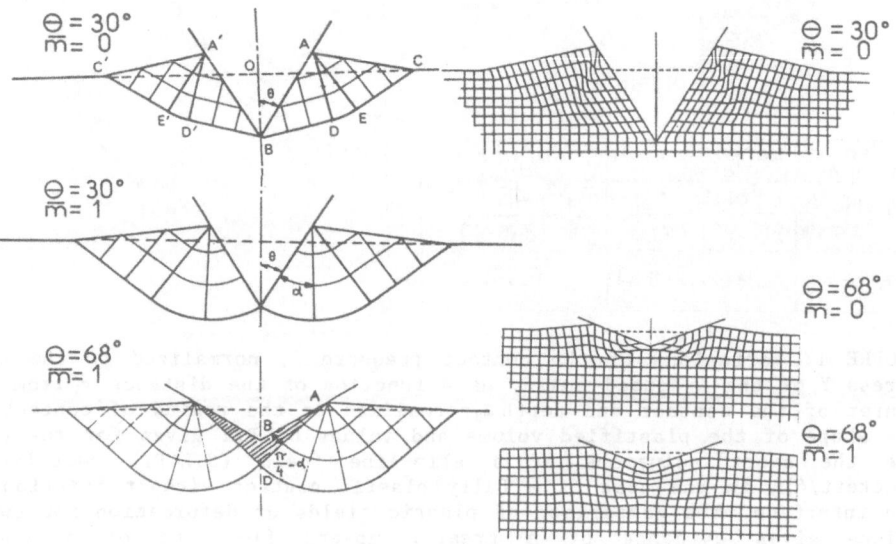

FIGURE 2. Slip-line fields and deduced deformations of square grids for two different semi apical angles of indentation, without friction ($\overline{m}$ =0) or without slip ($\overline{m}$ = 1).

FOOT-NOTE 1. Work hardening can be defined as the increase of Y with the amount of strain, up to its ultimate value equal to H/3 in the fully work hardened, i.e. fully plastic solid.

FOOT-NOTE 2. Brittleness can be defined as the ratio of the work of plastic deformation to the work of elastic deformation, which is equal to the ratio of the plastic indentation depth d_p (real penetration depth) to the elastic recovery in depth during unloading d_e, if there is no recovery in the width of the indent (14). Indentation tests with continuous recordings of depth during loading and unloading allow to measure these works, together with d_p and d_e.

Now if we consider the case of a bilayer, consisting for simplificity of two fully plastic materials A and B, the shape of the indent and the measured hardness depend on the respective hardnesses of A and B and on the thickness h of upper layer A (9). The changes of the plastic field with the ratio of flow stresses $\lambda = Y_A/Y_B$ and $\overline{m}$ are shown on Figure 3. The chosen angle of indentation (35°) corresponds to that of submicroscopic tests we performed in collaboration with H.M. Pollock on implantation films (14). When the ratio of hardness increases, the critical depth at which the substrate begins to contribute to plastic deformation decreases (table 1). As simultaneously, the hard film hampers the upward flow of material, it sinks into the substrate. This sinking is enhanced when the indentation angle or the friction coefficient increases, and can become larger than the penetrated depth. This fact accounts for the suppression of abrasion by ploughing and cutting in sliding contacts. But if the film is too brittle (foot-note 2) to sustain the same plastic deformation as the substrate, it will crack. On the other hand, soft films can absorb the plastic deformation until they are penetrated, and even later on in cases of sharp indentors such as those used in our indentation experiments.

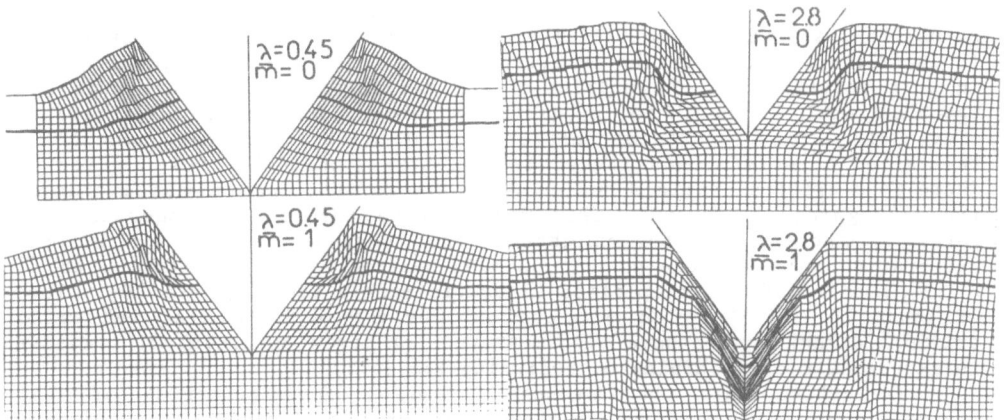

FIGURE 3. Deformations of square grids calculated with a simplified kinematic model of plane strain indentation of bilayers (9,15), shown for d/h = 2.8 in all cases. Whatever d/h and $\overline{m}$, the plastified thickness remains about equal to d in the case of the soft film (λ = 0.45). On the contrary, the hard film (λ =2.8) sinks into the substrate without being penetrated and a dead zone forms under the indentor for $\overline{m}$ = 1.

TABLE 1: calculated critical ratio d/h (experiments)

λ		5	4	3	2	1	0.5	0.1
$\theta=35°$	$\overline{m}=0$	0.70 (0.66)	0.78 (0.74)	0.94 (0.82)	1.06	1.14	1.18 (1.05)	1.30 (1.25)
	$\overline{m}=1$	0.37	0.45	0.53	0.65	0.86	0.98	1.10
$\theta=68°$	$\overline{m}=0$			0.08	0.20	0.60	0.90	1.15
	$\overline{m}=1$				0.12	0.25	0.33	0.66

Note also that in the case of hard films, as the indentation angle increases, the critical depth/thickness ratio at which the plastic deformation becomes characteristic of the bilayer instead of the upper film, decreases also, and tends to zero for spheres with radius R larger than h (otherwise it is between 0.1 and 0.2 for a sphere).

2.1.2. <u>Dynamic contact</u>. As previously we will first consider the contact between an undeformable rider and a fully plastic surface. The resistance to motion (force F_T or shear stress T) depends on the force F_N applied normally to the interface (or applied stress N), the gliding speed V_g, the angle of contact and the adherence coefficient $\overline{m}$. As long as T remains lower than Y, models of abrasion and experiments (16,17) show that T follows the simple laws of proportionality to N and $\overline{m}$ proposed by Coulomb and Tresca (18) respectively: its upper limit is $\overline{m} \times Y$, i.e. $\overline{m} \times$ k/2 or m k/$\sqrt{3}$. Variations with V_g are observed only in cases of materials sensitive to the strain rate: viscoplastic solids or in the case of creep at high temperature.

As already stated for a static contact, when N exceeds Y, an increase of $\overline{m}$ favours plastic deformation under the indentor with respect to the upward flow of material. But in a sliding contact the plastified volume deforms toward the surface because of the tangential component of the stress tensor, T. It is a basic law of mechanics that the concentration of stress in a smaller volume facilitates mechanical instabilities: striction in ductile materials, cracks in brittle ones. Thus, the machining of chips in soft materials or the pulling-off of particles in hard materials are enhanced with respect to ploughing without wear, when $\overline{m}$ increases (inversely, the "Rehbinder effect" of lubricants corresponds to a relocation of strain on a larger depth). Correspondingly, for a given $\overline{m}$, there is a critical angle θ_c under which the deformation mechanism changes from ploughing to cutting. This fact can be intuitively understood in terms of increase of the height of the peripherical rim of plastic deformation with decreasing θ, as well as the increase of θ_c with $\overline{m}$. S.L.F. models (19) or simplified kinematic models (20-21) give θ_c values of about 70 to 75° for $\overline{m}$ = 0 and 85° for $\overline{m}$ = 0.9. Similar values of critical attack angles have been measured for tools or grains of abrasive in experiments of machining or polishing (22,23).

Note here that the hardness is not a good criterion of wear resistance since a soft and fully plastic material would theoretically suffer no wear at all but just ploughing, for θ values lower than θ_c. In reality new asperities, due for instance to wear particles scratching the surface, always form. Moreover the accumulation of strain in the plastified layer induces local instabilities, leading to another wear process by fatigue and delamination at the interface with the elastic hinterland. It proceeds from the initiation of cracks at structural defects (described in next paragraph) and of their extension in an oblique direction, as the mean component of stress exercised in their vicinity by the rider becomes tensile (at the end of contact in this region) (24-26). The changes in abrasion mechanism of metals with their hardness are summarized in figure 4 (27). A transition from ploughing to cutting is observed for steels of increasing hardnesses (but given composition), which induces a relative loss of wear resistance. In some cases the wear rate can even become proportional to the hardness.

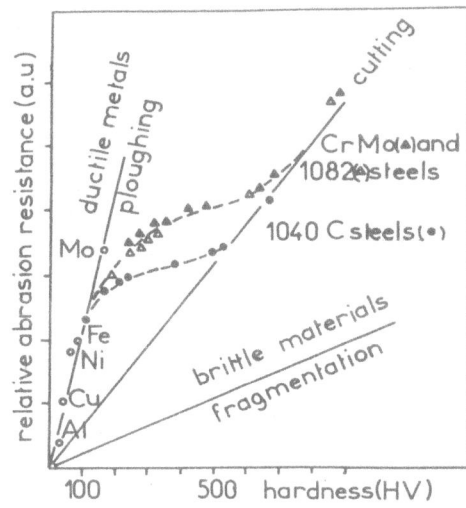

FIGURE 4. Relative wear
resistance of pure metals
and 3 steels as a function
of bulk hardness (27).

Now the plastic deformation of a soft film on a hard substrate can be compared to that of a wrinkled carpet. There is a critical range of thickness h (decreasing with θ and $\bar{m}$) for which the field of plastic deformation extends as far as the interface but remains confined in the film, such that the wrinkle is pushed by the indentor over an undeformed floor without too much frictional resistance. The latter decreases with h and with the ratio of flow stresses λ of film and substrate (see figure 5) (21) until a minimal value of h for which the height t of the wrinkle becomes too high. For this critical ratio $(h/t)_c$, about equal to 0.25 whatever $\bar{m}$ and λ (below 0.8), the mechanism changes from ploughing to cutting of the soft film. Note also that the attack angle for cutting decreases with λ, and thus for a given λ the thickness of a soft coating can be optimized to lubricate a contact without wearing it. The concept is the same as that of the boundary thickness of fluid lubricants allowing us to decrease the attack angle in forming processes (28). Such an optimum was also noticed experimentally by Bowden and Tabor for indium films deposited on tool steel (figure 5c (29)).

Similarly an optimum range of thickness can be defined for hard coatings on soft substrates, in order to suppress wear without increasing friction. As for soft films, the upper limit is that for which the mechanical behaviour is intrinsic to the film ($\theta = \theta_c$ intrinsic), and the critical lower limit $(h/t)_c$ is of order 0.25. Here it is correlated to the increase of T/N.

A last remark concerning the abrasion mechanism of fully plastic bilayers is that shear can occur either in the film or in the substrate, but also at the interface, according to their respective k values. Scratch tests performed in well defined conditions on hard films (with measurements of $\bar{m}$ and control of θ) allow us to characterize the interface toughness by measuring the minimal load F_N for which decohesion occurs, without penetration of the indentor into the soft substrate (30).

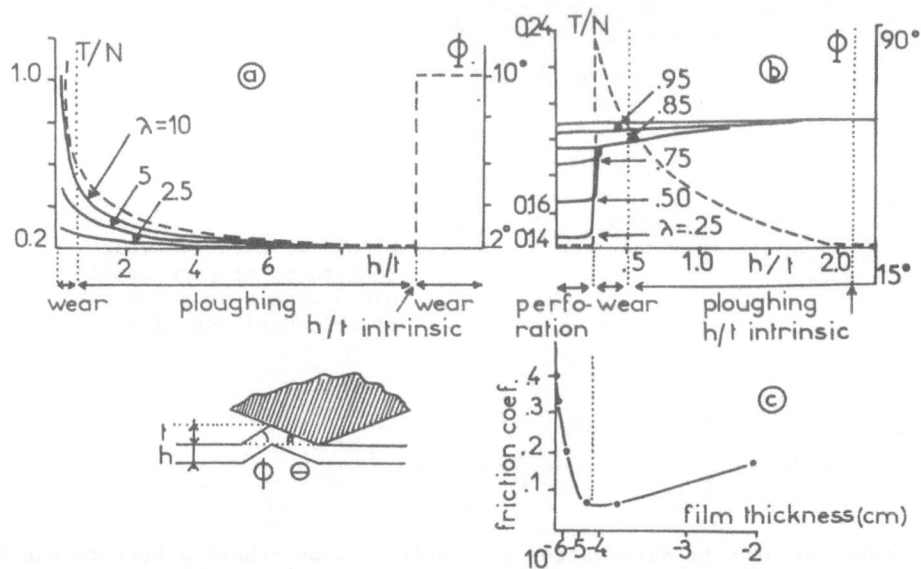

FIGURE 5. a) and b) variations with h/t and λ of the friction coefficient T/N (———) and of the angle Φ (--) made by the rim with the initial surface in front of the rider, calculated by means of a simplified kinematic model of the tangential indentation of bilayers. The results presented are for a bidimensionnal wedge (plane strain) with an attack angle of 8° and $\overline{m}$ = 0.25; c) experimental variations for thin In films on steels.

I will conclude this overview of the rheological properties of bilayers with basic statements on the influence of surface roughness on friction and wear. Its first effect is to diminish the area of contact, such that the applied stress will be locally much higher than the mean pressure. The Newtonian law of variation of F_T with F_N will not be characteristic of the intrinsic elastoplastic properties of the material, but of the collapse and cutting yields of asperities (17). There have been few attempts to correlate them with the statistical distribution of heights and slopes of asperities. On the other hand, these factors are taken into account in optimizing the retention of lubricants in contacts during forming processes (28,31). For instance the surface of rollers is treated by laser in order to give them a roughness profile consisting of plateaux and pockets (statistically distributed, in equal proportions and sizes), such that the lubricant can flow between plateaux but also that its thickness remains slightly over the boundary value upon them in order to avoid seizure events (31). Similarly the efficiency of abrasives is governed by their statistical distribution of grain size and angles and by a well controlled lubrication of the contact (32). Such considerations might at first appear inadequate in a lecture on implantation films, but the lubricant or abrasive nature of the wear products (the so-called "third body") are determined by similar considerations. They form a mixture of abrasive grains of the hardest material in a fluidized bed constituted by particles of metal or oxide abraded from the softer one.

The closer the thickness of the third body will approach to the mean surface roughness, the more efficient will be the abrasion of both surfaces by the abrasive grains , i.e as these grains will not be embedded in one of the surfaces or rolled by the lubricant. On the other hand, if particles of the soft film are transfered onto the hard surface, their shearing will depend on the extent of the plateaux.

2.2. Tribochemistry and metallurgy

2.2.1. Temperature and structural transformations. Part of the energy introduced in the tribosystem is transformed into heat. In most cases of contacts between metals, the mean temperature of both surfaces does not increase significantly, but the flash temperature on asperities can exceed a thousand degrees. This is attested by measurements of infrared emission and contact resistance (29), as well as by structural transformations implying thermal diffusion or quenching from high temperatures (33). For instance the formation or reversion of martensite commonly observed in steels means that the temperature has attained values of the order of 300°C or 800°C. Simple models based on the equation of heat diffusion (33-35) were proposed to estimate flash temperatures as a function of F_N, V_g, and thermal diffusivities. But they assume that all the energy Q is dissipated as heat, on a area of real contact A which is generally unknown ($Q = \overline{m} \times F_N \times V_g \times A$). In fact, the part of Q which is dissipated in both bodies depends not only on their diffusivities but also on the nature of superficial films resulting from tribochemistry (oxides for instance), and part of Q is used for deformations and structural transformations. However, the interest of calibration curves of temperature as a function of the model's parameters which were performed on martensitic steels, is to show that the critical parameter is a normalised value of V : $V = V_g.a/K \sim 10^2$, where a is the apparent Hertzian diameter of contact and K the thermal diffusivity.

The changes of structure which are generally observed in metals depend on their stacking fault energy and on their melting temperature. The more the cross slip and climbing of dislocations occur preferentially to planar slip and piling up, the easier is their rearrangement into a cellular structure (recovery) (36-38). But metals which have a low melting temperature recrystallize instead: for instance Al when compared to Cu in the same conditions of friction (39) in spite of the higher stacking fault energy of Al. When recovery cells are formed, their size and shape varies with depth as strain: near the surface they are elongated in the sliding direction and much smaller. Their size (varying with depth from 5 to 50nm in Cu, Al or steels) determines that of wear particles when the wear mechanism is delamination. But when the latter remains the pure cutting of asperities, the third body consists of the finest cells. In such a case, the formation of cells instead of martensite laths or twins in steels, improves their wear resistance (40). Another structural transformation, reducing the abrasion rate and friction coefficient, is the development of a superficial texture parallel to planes of easy glide, because it diminishes the energy of deformation of asperities. This is the reason for the lubricating character of MoS_2 (41). We could observe the development of a similar (0002) texture, characteristic of hexagonal systems, in the case of Ti during running-in of the surface (figure 6). The topic of textures developed during friction tests was reviewed in reference(42).

518

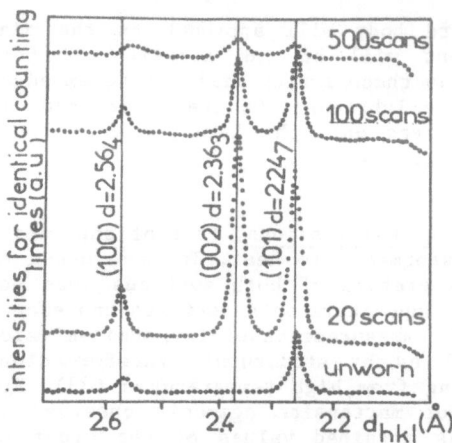

FIGURE 6. Partial diffractograms recorded at grazing incidence of X-ray on a Ti surface (wavelength 1,57Å, incidence angle 0.3° giving a diffracting thickness of 90 nm in the case of Ti (43)), after an increasing number of cycles of a carbon steel ball scanning over the surface under a contact pressure of 10 MPa. The observed transformation accounts for much higher values of local stresses at asperities since the Ti elastic limit is 100 MPa.

2.2.2 <u>Adhesion and reactions with environment</u>. The Dupre energy of adhesion w between two solids is the difference between surface energies γ_1 and γ_2 in vacuum (half the cohesion energy, which can be measured in cleavage experiments or deduced from binding energies) and the interfacial energy γ_{12}. The latter calculation is more difficult, and its experimental determination requires: i) to measure simultaneously the tension exercized to separate the solids and the real area of contact, and ii) to ensure that the rupture mode was effectively an adhesive rupture at the interface instead of a cohesive rupture within the softer material. These tasks are only achieved in contact experiments between a submicroscopic tip and a flat surface in ultra high vacuum, with the measurement of the contact resistance and of the tip radius before contact, by means of its field ion emission, combined to a control of surfaces cleanliness by field ion microscopy and A.E.S (44). The adhesion tensions were also calculated in this particular case of a spherical indentor and criteria of transition were defined as a function of E,Y,w,R and F_N (12). According to these calculations, the tension depends on F_N only in the case of an adhesive rupture after plastic loading. One could thus distinguish between the mechanisms of adhesive wear (with adhesive or cohesive rupture, or spalling of the film in bilayers) by varying the load and shape of the rider. But in macroscopic experiments the area of contact varies also with F_N.

However a cohesive rupture implies a transfer of material, either at the rear of the rider or on both sides of the track if the rider is softer than its counterpart, which can be observed by S.E.M. An adhesive rupture will rather lead to a wear process by cutting of protuding material during following runs or by fatigue and delamination. Whatever the rupture mode, the friction coefficient has a mean value near one and exhibits fluctuations characteristic of the stick and slip process, while it is more constant in the case of abrasion. On the other hand, rims of plastic deformation can form at the periphery of the contact even in the case of Hertzian contacts, if the tension of adhesion exceeds the yield stress of the softer solid (the compression becomes zero at the periphery). Thus, a soft or poorly adherent film may have been entirely stripped from its substrate after a few runs, hampering the study of its intrinsic abrasion resistance even under gentle conditions of loading.

Adhesive events will only occur when T exceeds the flow stress of oxide films (or other products of tribochemistry) which generally form on at least one of the surfaces. Thus, the most simple way to suppress adhesive wear is to attempt to favour the development of an adherent film with a boundary thickness and a low friction coefficient (not necessarly hard). Thick $FeO+Fe_3O_4$ films formed on steel surfaces, or $TiO+TiO_2$ formed on titanium alloys for instance, play this role (41,45,46). Other oxides, such as Fe_2O_3, $Fe(OH)_x$, $Ti(OH)_x$, Cr_2O_3 or Al_2O_3 are too thin to protect efficiently the surface and in some cases are abrasive. As will be shown in the next section, improvements in seizure resistance by ion implantation are generally due to an improvement of the adherence of lubricating oxides: mechanical anchoring or suppression of the substrate delamination by fatigue.

As a last remark it can be stated that there is no particular mechanism of wear resulting from tribo-oxidation. What is called oxidative wear is in fact either the self-abrasion of hard oxide films or the fatigue and delamination within the substrate or at the interface.

3. THE MOST CHARACTERISTIC EFFECTS OF ION IMPLANTATION ON TRIBOLOGICAL PROPERTIES OF METALS

3.1. Metalloid solid solutions of hard compounds (C,N,O,B)

3.1.1. <u>Structure of implanted layers</u>. Compared to other implant species, the influence of nitrogen implantation on tribological properties of iron, steels or titanium alloys has been the most explored for two reasons: i) nitridation or cementation are processes known since antiquity to harden iron ; ii) the easy availability and non-dangerous nature of this gas allow the development of unsophisticated implantation machines for the treatment of large areas under high ionic currents and at low cost. In this field of application ion implantation appeared as an alternative to CVD or PVD processes for obtaining more adherent coatings and vary their composition and structure.

Phase changes induced by N implantation in iron and steels have been extensively studied by T.E.M. (47-49), R.B.S. and N.R.A. in channeling incidence (50), Mössbauer spectroscopy (51), and more recently by X-ray diffraction at grazing incidence (40,52,53). They depend strongly on the initial crystallographic structure and on the nature of metallic additions. In the case of pure iron or of carbon steels, fine precipitates of the quadratic α'-martensite and of the hexagonal $\mathcal{E}$-Fe_2N nitride form simultaneously for concentrations exceeding 5 atomic percent at R_p (40,54-56). The solubility of nitrogen in α-iron is indeed much lower than in martensite or in γ-iron (0.4 atoms % instead of 10%). But the martensitic transformation must also be associated with local stress. It was also observed in a Fe-17Cr-7Ni austenitic steel implanted with rare gas or iron (57), and in Fe-10Ni with a more metastable $\alpha 2$-martensite structure or in Fe-18Cr-10Ni austenitic steel implanted with N (as before for more than 5%N at R_p) (40). In other cases, the reversion of the α' martensite into austenite was observed, maybe because of an overheating of the surface during implantation (58). The nitrides formed in presence of Cr and Ni are generally $\mathcal{E}$-$(Fe, Cr, Ni)_2N$, but Cr_2N or CrN were also observed in some cases (when the alloys contained also C, stabilizing this phase) (53,59).

Their small size (2-10nm for implantations at room temperature) and the existence of epitaxial relations with austenitic matrices give them an optimum pinning effect on dislocations, hampering their piling up in cells or in slip-bands at the surface, which constitute sites for the initiation of cracks. Thus an improvement in resistance to fatigue and wear by delamination is expected compared with that of CVD or PVD coatings with coarser structures. A reduction of abrasive wear was effectively observed for nitrided and cemented surfaces of steels submitted to a further implantation of N inducing a refinement of precipitates (60). In the case of a simple implantation, as the N concentration is increased up to 30%, the precipitates grow slightly (about 30nm) and coalesce into an homogeneous nitride film.

Structural changes in titanium and its alloys are more simple because of the higher solubility of N in the α hexagonal lattice of this metal: 15 atoms %. In most studies on implanted surfaces, solid solutions were effectively observed up to about this concentration, and thereafter precipitates of the δ-TiN cubic nitride. The formation of an intermediate compound ϵ-Ti$_2$N would have been expected for concentrations in the range 15 to 30%, if overheating during implantation (studied samples were generally thin foils for TEM observations) (61), contamination by carbon and oxygen which stabilize the δ-phase (62), and radiation-enhanced diffusion had not promoted the formation of the most stable nitride in the phase diagram. We have effectively been able to detect the ϵ-nitride by X-ray diffraction at grazing incidence on bulk samples implanted under a good vacuum (63). The surfaces were implanted with ions of several energies in order to form homogeneous films of thickness about 400nm. In such conditions the superficial contamination in C and O, due to recoil implantation and to the high affinity of titanium for these elements, did not contribute significantly to structural analysis. Phases formed in the homogeneous part of the films corresponded to the phase diagram; the Ti$_2$N nitride was also present in less implanted regions (between 400 and 500nm) of films containing 30% to 60%N in their homogenous part.

Now the role played by these various structures in fatigue or wear processes must be analyzed as a function of external conditions, expecially in the case of wear.

3.1.2. <u>Abrasion resistance</u>. Review papers on the tribological properties of N implanted steels (48,64-66) state that the friction coefficient does not change significantly, whatever the steel structure and the N fluence, but that the abrasion resistance increases by a factor of 20 to 100 in specific cases.This beneficial effect was attributed to :
i) An interstitial and precipitation hardening, which, as expected, could be detected by indentation experiments at the submicroscopic level (67). As already stated, the precipitates have the ideal range of size to pin dislocations. But precipitates (or segregations) can also have an opposite effect on the work-hardening of superficial layers, by promoting planar slip as against cross slip. This argument can be refuted by noting that implantation films are always fully work hardened by radiation damage. However the effect of N implantation on wear is more durable that the lifetime of the film. This led several authors to propose a second hypothesis.
ii) The diffusion of N along extended defects formed during friction, and also in the bulk in the case of austenitic matrices. Several authors tried to evidence this diffusion by recording depth profiles of concentration in wear tracks (either via sputtering associated with SIMS, XPS or AES, or

via nuclear resonances) (51,54,60,68). N was detected in the track after
complete wear of the implantation film. However the lateral resolution of
the technique used did not permit them to distinguish whether the
remaining amount of N was in fact due to diffusion or encrusted
precipitates. A combination of techniques, including use of a
scanning-tribometer specially designed in order to wear homogeneously a
large area, and of ion microscopy to analyse the tracks (resolution
0.5μm), allowed us to conclude the opposite. The widening of the N profile
was less than 5nm after wear of half the implanted thickness, in the case
of both pure iron and of a 18-10 austenitic steel implanted with 10%N on
300nm. In our experiments the surfaces were worn by SiO_2 and by steel
balls under a pressure of 1GPa, without lubrication, but one may argue
that these conditions of friction were not severe enough. However Fayeulle
(40) came to the same conclusion from TEM observations on an austenitic
steel, chosen for its low resistance to abrasion. Implanted foils (∅= 3mm,
thickness ≃ 0.2mm) were attached to a pin continuously in contact with a
rotating disk (both made of a carbon steel), in order to thin them
homogeneously in lubricated conditions of friction. Only a few nitride
particles remained encrusted in the foil after the abrasion of a thickness
corresponding to the initial film of $(Fe,Cr,Ni)_2N$.

iii) The initiation of a different mechanism of substrate work-hardening
(40,64),when the surface is protected by a nitride film during running-in;
the mechanism then perpetuates after wear of this film. Fayeulle's
observations by X-ray diffraction at grazing incidence and by TEM are
effectively accounted by such an effect (figure7). In the case of
unimplanted surfaces or low implantation fluences (5.10^{16} $N.cm^{-2}$, 40keV),
martensite precipitates were observed in the most highly strained zones of
the tracks (scars) and twins around them, right from the beginning of
friction. After prolonged tests, the work-hardened thickness of material
was over 10μm, but strain remained heterogeneous within it. The formation
of this hard but brittle structure was hampered when the surface was
implanted with a fluence sufficiently high to build an homogeneous film of
nitride. It was replaced by fine cells, constituting a more homogeneously
strengthened underlayer. After wear of the nitride film, which proceeded
from local recrystallization along scars, cracking and crushing of
particles, the strained layer remained limited to a few micrometers and
constituted of the same cells.

The role played by the hard layer of nitride seems thus essentially
to accomodate stress during running-in and to avoid the plastification of
too large a thickness of the soft substrate. The martensitic
transformation might also be beneficial if it occurs homogeneously over a
small depth, as does the recovery under nitride films. Other authors
(69,70) noticed that martensite films with a limited thickness, forming
during friction on sufficiently hard austenitic steels, give them a better
wear resistance than that of ferritic or fully martensitic alloys with
comparable hardness. But when the transformed thickness is too large (or
in fully martensitic alloys), cracks appear rapidly within the film and
its rheological behaviour is unstable. No wonder also that no effect of
nitrogen implantation was observed either on the hardness or on the wear
mechanism and wear rate of various martensitic steels (64-66,71,72).

Similarly, there is an optimal range of thickness of the nitride
films formed by implantation into titanium, and for a given thickness
there is an optimal ratio of nitride over substrate hardness. We were able
to observe that, when the mean contact pressure exercised in friction

522

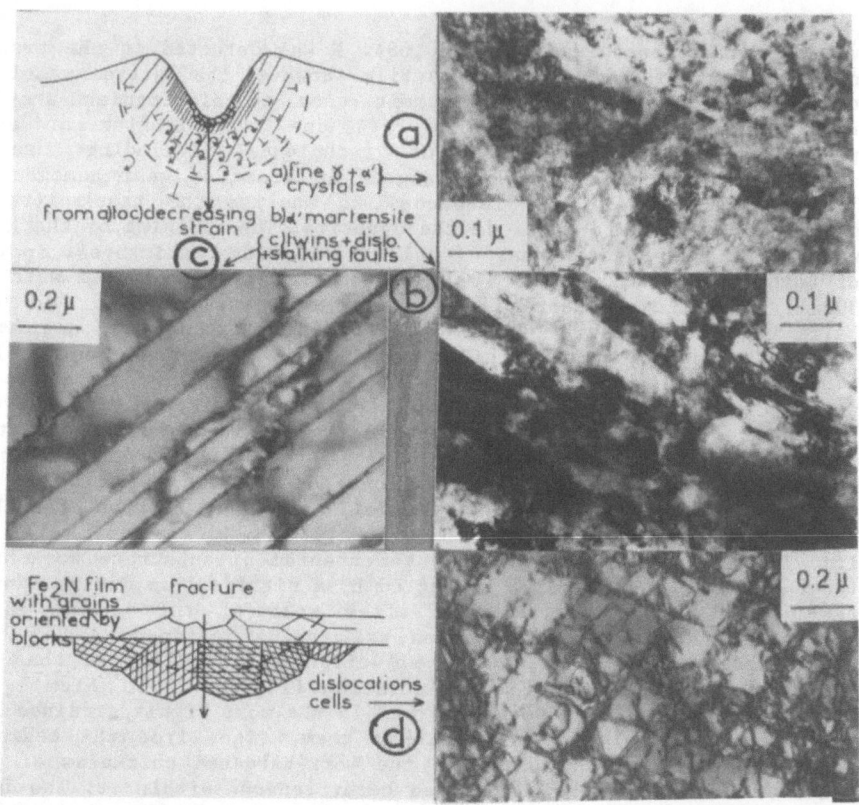

FIGURE 7. T.E.M. micrographs around a scar in (a to c) a Fe–18Ni–10Cr stainless steel after sliding 45min under a load of 0.5N, (d) a surface implanted with 2.10^{17} N.cm^{-2} after the same test.

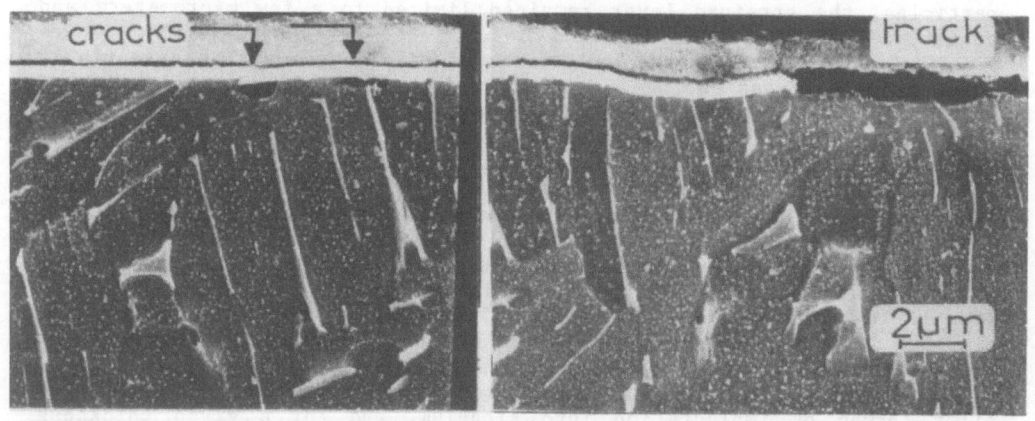

FIGURE 8. S.E.M. micrographs of a cross section in $TiAl_6V_4$ implanted with 44% N over a depth of 400nm, after 4000 cycles of a carbon steel pin, rotating on the surface under a contact pressure of 10MPa.

tests exceeded a threshold slightly above the flow stress of the substrate, ζ-TiN films cracked because they could not sustain the deformation imposed by the substrate. They were finally crushed over large areas and the surface then suffered an adhesive wear as severe as an unimplanted surface (details on friction conditions and on the wear process are given in next paragraph). On the contrary, softer films of ε-Ti$_2$N had a protective effect in the same conditions of friction. We do not yet observe if the work-hardening process of underlying layers was different in this last case, but a change of structure of the implantation film was detected by X-ray diffraction at grazing incidence (see short contribution). The Ti$_2$N film partially oxidized into ζ-Ti-N-O, which was probably in the form of precipitates within ε-Ti$_2$N since depth profiles showed an oxidation through the whole thickness of the film. This transformation seems to have promoted the film adherence. Note here that the generally received idea of an intrinsic adherence of implantation films, associated with a diffuse interface with the substrate, has no meaning when they exhibit a continuous and different structure. Local cracks along the interface are for instance observed on figure 8 in the case of a ζ-TiN implantation film on a TiAl$_6$V$_4$ substrate, in the vicinity of a wear track (within the track itself, the whole film was crushed).

Oliver et al. (73) already observed that the relative increase of wear resistance of Ti or TiAl$_6$V$_4$ implanted with C or N did not depend only on the hardening factor of the surface. Indeed the latter was the same for Ti and for the alloy implanted with 8.10^{17} N/cm^2 of 100 keV (45% N at R$_p$). But above a critical contact pressure which was in the ratio of the substrate flow stress the implantation film quickly experienced breakthrough (and did not improve wear at all). They found also that C and N were equally effective in hardening and improving the abrasion resistance of pure Ti, but that curiously C did not change either the hardness or the wear kinetics of TiAl$_6$V$_4$. On the contrary the experiments of Vardiman (74) and Singer (75) accounted for relative increases of wear resistance 10 to 30 times higher in the case of TiAl$_6$V$_4$ implanted with C than in the cases of N or B implanted at similar fluences and energies. They noted also that i) the increase attained its maximum as the C,N or B fluence was sufficiently high to form a continuous TiC, TiN or TiB$_2$ film, and that ii) the efficiency of implantation at lower fluences can be enhanced by annealing treatments at controlled temperatures, promoting a further precipitation of the implanted element without coarsening of the precipitates. Such treatments,or implantations at high temperatures (76), can even provide the surface of TiAl$_6$V$_4$ with an higher wear resistance than gas nitriding or TiN coatings, in mild conditions of friction : not too rough surfaces, with a lubrication of the contact, and under pressures lower than the beackthrough threshold.

The greatest improvements in the resistance to bulk fatigue of steels and titanium alloys were also obtained when the density of fine carbides or nitrides precipitates was at most. Increase in the lifetime by a factor of 2 were already observed for surfaces hardened by radiation damage (Ne implantation) or N in solid solution. It can be considered as an important effect of a surface treatment on this bulk mechanical property, since ion implantation is only expected to influence the crack nucleation at the surface. However a further increase by a factor of 100 was obtained upon aging, in the case of saturated solid solutions of N into Fe (77), which was associated with the precipitation of a high density of coherent $Fe_{16}N_2$ particles (78). Similarly, it was found that carbon was more effective than nitrogen at comparable fluence (increase by a factor of 6 instead of 2) because of the higher density of precipitates (79).

The durable improvement of wear and friction performances (80) of tools made of WC+Co composite materials when implanted with C or N is also due to a different tribological transformation, perpetuating after wear of the implantation film. The development of a lubricant film of Co embedding the WC grains and avoiding their pulling off, is most probably helped by the ion mixing of constituents of both phases at interfaces and by the transformation of the fcc-Co binder into an harder martensite (and an amorphous Co-W-C alloy near interfaces) as attested by T.E.M. observations (81).

3.1.3. <u>Adhesion and oxidation resistance</u>. Fayeulle (40) selected mild carbon steels (0.05%C + 0.20% Si + 0.38%Mn or 0.42%C + 0.24%Si + 1.1%Mn + 1.0%Cr for both the implanted plane and cylindrical rider), exhibiting intrinsically a poor resistance to oxidation, to investigate the effect of N implantation on the tribo-oxidation of iron based alloys in air, as a function of implantation conditions. He did not observe any change of friction coefficient or of wear kinetics for implantations at room temperature of fluences up to 2.10^{17} $N.cm^{-2}$. A thick Fe_2O_3 film was detected by X-ray diffraction at grazing incidence after friction. Its wear by crushing was heterogeneous and the friction coefficient exhibited large fluctuations around 0.8.

On the other hand, the formation of Fe_2O_3 was delayed when N was implanted at 100° or 150°C, and even after complete wear of the nitride, the surface remained smooth and the friction coefficient more constant (a little lower also: 0.7). The wear rate was diminished by a factor of 2. This difference of behaviour was related to a change of texture and a higher homogeneity of the Fe_2N film. At 100-150°C the basal planes (002) of the hexagonal nitride were parallel to the surface, giving it a minimal reactivity and shear resistance since they are the planes of dense packing. On the contrary, for films formed at lower or higher temperatures (200°C), they were oriented exactly in the perpendicular direction.Another important factor tending to improve the wear resistance after removal of the nitride was the smoothness of the worn surface, which corresponded to an increase of contact area. It is sometimes supposed that the oxide growth rate is proportional to this area and thus decreases with N implantation in steels because the surface hardening reduces the deformation of asperities (60,82). But a better distribution of the transfered energy and thus a lower temperature at asperities appears more plausible. It is also evident that the flatter the surface, the more cohesive the oxide film, and the more constant its thickness. As in the case of pure abrasion, the durable effect of ion implantation here also is to promote a change in the tribo-metallurgy and tribo-chemistry of the substrate.

This fact was also verified for titanium and its alloy $TiAl_6V_4$ implanted with nitrogen (83). Figure 9 shows the variations of friction coefficient registered in a pin on rotating disk test (with a rider made of 0.4%C + 0.25%Si + 1.1%Mn + 1.0%Cr steel, exercizing an Hertzian pressure of 10MPa) in air.

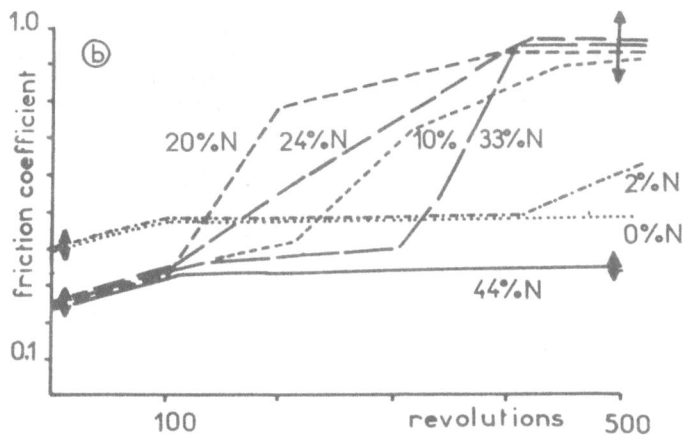

FIGURE 9. Variations of the friction coefficient with sliding distance, for a rotation speed of the pin of 10 revolutions/min.

Surfaces covered by a Ti_2N or TiN film exhibited initially a lower friction coefficient and a higher wear resistance than pure Ti or $TiAl_6V_4$, or than surfaces implanted with only 10%N in solid solution. The greater their stoechiometry and hardness, the longer this effect lasted. Hardnesses, measured by means of a submicroscopic indentation test, were about 9GPa for 10%N, 12GPa for 15 to 40%N, and 18GPa for 45 to 55%N (in comparison with 2.5 GPa for pure Ti). The hardest and least reactive TiN films did not react with atmospheric oxygen as shown by SIMS profiles performed in the same tracks (with the help of the high resolution optics of the used CAMECA IMS 3F ion microscope (84)). This was confirmed by NRA analysis on large and homogeneous tracks obtained with our scanning-tribometer under the same pressure. However nitride grains pulled off at asperities acted as an abrasive powder, the friction coefficient increased, and the film finally destroyed itself. Therafter, the surface suffered the same tribo-oxidation and abrasive wear by a cutting process as the unimplanted surface (NRA and RBS analysis allowed to estimate the oxidized thickness as 3–5µm and the mean oxygen content as 10%). Ti_2N films were only slightly oxidized during the first stage of wear (less than 1% 0 dosed by RBS and NRA) under this pressure, but being less hard than TiN ones they were abraded more quickly. The most durable effect of ion implantation in these mild conditions of friction was observed for a saturated solid solution of 10%N into α-Ti. Despite the fact that its oxidation kinetics was very comparable to that of pure Ti, abrasion was suppressed by an hardening effect of nitrogen incorporated into the oxide. Depth profiles of N within the tracks and in the as-implanted surface remained similar after a number of revolutions (or scans) for which Ti_2N and TiN films were already partially or entirely worn.

The good friction resistance of a Ti-N-0 (amorphous) film, formed by recoil mixing of 0 during implantation of N, has already been observed by Fleche et al. (62). Hutchings et al. (47) also reported that tribo-oxidation films formed on TiAl$_6$V$_4$ implanted with increasing N fluences at a single energy were in all cases f.c.c Ti-N-O solid solutions of same composition, providing the surface with the same reduction of friction coefficient and wear rate (by two orders of magnitude) in comparison with an unimplanted surface. The only effect of increases of N content at R$_p$ from 10% to 60% was to permit the re-formation of the oxi-nitride for a longer time whenever it was locally crushed. In our experiments, the formation of such a compound was not detected by X-ray diffraction at grazing incidence (on large tracks worn with the scanning tribometer), because the tests were performed in air instead of in a lubricant containing only a partial pressure of oxygen.

The tribo-oxidation and abrasion mechanisms of such films depend critically not only on the kinetics of oxygen diffusion but also on the contact pressure.

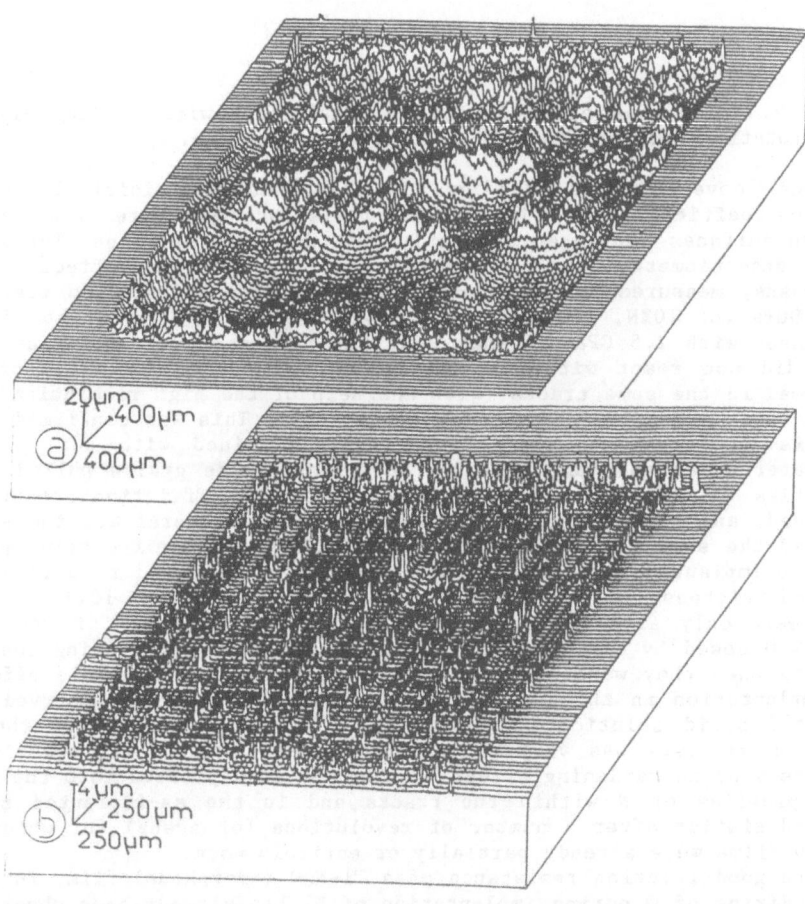

FIGURE 10. Three-dimensional recordings of the surface topography by means of a stylus technique after 500 scans of the rider under a pressure of 800 MPa (with a space between scanned lines of 500μm and an Hertzian diameter of contact of 450μm) on: a)Ti-10%N, b)Ti-25%N implantation films. For most tracks in Ti-35% to 55%N the topography was the same than in case a.

When the contact pressure of our friction tests was increased up to 300MPa and then to 800MPa, the hardening effect of N in the oxide film formed on Ti-10%N was no longer sufficient to avoid a significant plastic deformation of the substrate. The wear process changed from a shearing limited to asperities to a delamination of the interfacial iron-titanium oxide layer over large areas, followed by severe adhesive events (figure 10a). The Ti-10%N film was worn for a number of cycles much smaller than TiN or Ti_2N ones. We already mentioned that Ti_2N films partially oxidized into Ti-N-O under such pressures and exhibited a slower abrasion kinetics than TiN films (figure 10b). NRA analysis of ^{18}O in tracks and profiles using the resonances at 898keV of ^{15}N (p, $\alpha\gamma$)^{12}C and at 629 keV of ^{18}O (p,α)^{15}N, on Ti-50%N samples enriched superficially in ^{18}O have shown that (figure 11):
i) ^{18}O did not diffuse in the bulk of the TiN film, but ^{16}O of the atmosphere could diffuse throught it;
ii) there was a partial loss of nitrogen content per unit area (in the case of the figure, but in most cases the film was entirely worn after 300 to 500 scans under a pressure of 800MPa, while a Ti_2N film was not worn at all), without widening of the N profile.
Both points i) and ii) account for the existence of cracks and local removal of the nitride, permitting the diffusion of oxygen through short paths. RBS profiles of the titanium depth-distribution indicated that the amount of oxygen dissolved in the underlying substrate was of about 10% over more than a micrometer, as for an unimplantated surface.

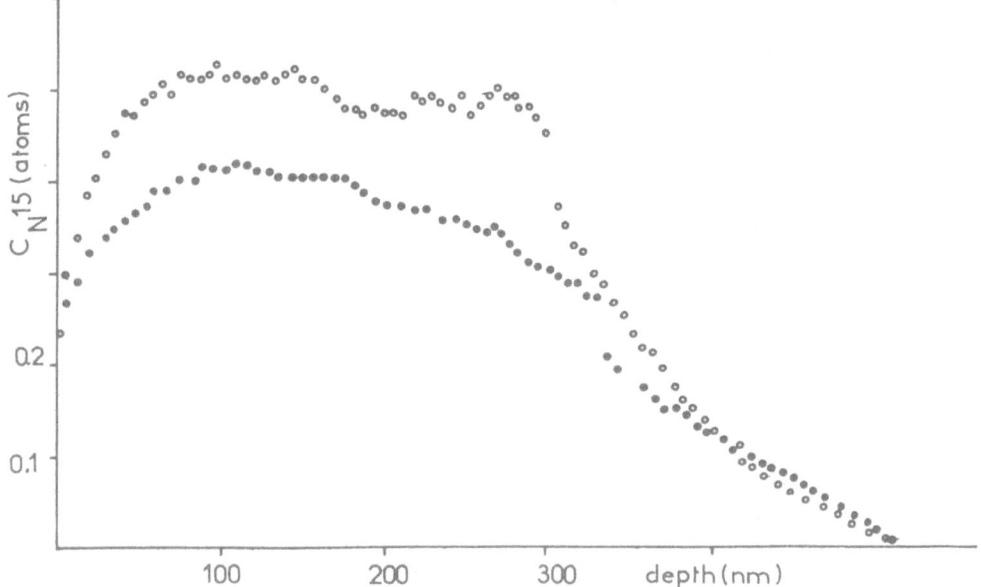

FIGURE 11. Depth profiles of N concentration in Ti implanted with a nominal dose of 2×10^{18} $^{15}N.cm^{-2}$ (E = 20,50,80,145,190 keV), (o) after implantation (1.47×10^{18} N, 3×10^{16} O dosed by NRA), and ($\bullet$) after 2000 scans of the rider under a pressure of 300 MPa (9.4×10^{17} N, 5.7×10^{17} O). The target density was assumed to be 9.5×10^{22} atoms.cm^{-2} to establish the scale of concentrations. The depth scale was defined on basis of a TiN stopping power calibrated on a film with a well known thickness and stoechiometry.

Such a dissolution contributes only a little to surface hardening (as shown by submicroscopic indentation tests in wear tracks) but it hampers the building-up of a fine substructure of work-hardening. Indeed the insertion of oxygen atoms in (004) planes between basal planes of α-Ti diminishes the possibilities of cross-slip and climbing of dislocations.

Additional experiments were performed on Ti-45%N-10%O to Ti-55%N-10%O implantation films with an homogeneous composition over 400nm, in order to confirm the optimum wear resistance of oxi-nitrides. They were less resistant to abrasion than the Ti-N-O film formed during friction on ξ-Ti-25%N (of mean composition Ti-25%N-10%O measured by RBS and NRA), and no more resistant than the unoxidized Ti-45%N to Ti-55%N films. They had an intermediate hardness. These results illustrate once more the strong dependance of the rheological properties of implantation bilayers upon the ratio of film/substrate hardness, and on their intimate substructure. This is not analysed here but a partial recrystallization of the nitrides during friction is expected as for ξ-Fe$_2$N films; in the case of ξ-Ti$_2$N films it is attested by their partial transformation into δ-Ti-N-O.

3.2. Amorphous phases

Metallic glasses are generally harder than the metallic components but less brittle than the crystalline alloys of the same composition (85,86). They are also less reactive. Both properties are due to the strong binding energy between metal and metalloids (Si,P,B,C) constituting most of them, and to the delocalization of extended defects under the form of Somigliana dislocations (87). These characteristics account for the durable improvement of seizure resistance observed for Ni-B and Ni-P amorphous films formed by implantation of B,P into Ni (88), or Fe-Ni-B, Fe-Ni-Si by implantation of B, Si into a 18-10 stainless steel (40,89), or Fe-Ti-C (64,90-92), Fe-Mo-S (93) by Ti+C, Mo+S into austenitic and martensitic steels. The effect of crystallization on hardness, reactivity and adhesive wear was studied in the case of Ni-B films (88,94-97). When annealed they were quickly crushed and the surface suffered the same adhesive wear as unimplanted nickel. On the other hand the durability of the improvement of seizure resistance depends on the matrix. For instance boron implantation into pure iron leads to the formation of an amorphous iron boride, which is harder but more brittle than nickel or mixed borides: no significant improvement of wear was noticed in this case (98). This durability is also expected to be canceled by a crystallization of the film during friction, because of the poor thermal stability of most metallic glasses (T$_c$ in the range 200-600°C). Such a crystallization was effectively observed by Fayeulle in the case of Fe-Si (40). Note that in this case the beneficial effect of the film was perpetuated, as for Fe-N, by a work hardened layer of austenite + martensite + Fe$_3$Si encrusted grains. We observed only a transient and very superficial crystallization of Ni-B films in similar conditions of friction (97, and short contribution in this book), may be because of a higher ductility of the nickel boride allowing a better flattening of asperities.

Other authors observed an increase of a factor of 2 in the lifetime in fatigue or erosion-cavitation of the same Ni-B/Ni bilayers with respect to the unimplanted subtrate (48,99,100). This factor is much more important than for Cu-B/Cu bilayers of same composition, strongly disordered by ion implantation but not amorphous (increase of only 10%). Both improvements were attributed to the blocking effect, either of the amorphous or of the disordered film, on dislocations moving towards the surface. Indeed, observations by SEM of the surface relief of fatigue consisting of slip bands have shown that the greater the implanted dose, the less was the amount of bands, and correlatively the higher the fatigue lifetime (since slip bands are known as preferential sites for the initiation of cracks).

In recent experiments on Al_6Mn films in the glassy, quasicrystalline or fully recrystallised state,[6] obtained by ion irradiation of Al-Mn multilayers deposited on a SiO_2 glassy substrate, we were also able to observe that both glass and quasicrystal exhibited a better resistance to seizure than the crystal (coefficient of 0.2 instead of 0.8 after running-in) in friction tests with a steel rider. On the other hand, they were less resistant to pure abrasion, as shown by friction tests with a glass rider. Their hardnesses have not yet been compared.

3.3. Metallic solid solutions

Few studies in this field have been carried out. They account for inprovements of seizure resistance:

i) either by formation of a solid lubricant in the film during implantation, such as in the case of Mo and S simultaneously implanted in stainless steels (93);

ii) or its formation from implanted elements during friction, for instance by oxidation of Sn into SnO_2 (60,101,102);

iii) or the mechanical anchoring of the matrix oxide by precipitates, for instance of SiO_2, Cr_2O_3, TiO_2 in steels implanted with Cr, Si or Ti (103).

The case of Sn is interesting because its implantation or recoil implantation by mixing (60) promotes the formation of intermetallic compounds: $FeSn_2$, FeSn, Ni_3Sn_2 and TiSn in steels, and SnCu in copper alloys which are being developed to improve the strength and ductility of bulk materials (60,102). Part of the compound is oxidized during friction, forming a lubricant over a hard multilayer, which remains adherent when the surface is submitted to a mild wear test with additional lubrication of the contact (102). In tests of adhesive wear resistance of implanted iron against an unimplanted iron pin in air, a transfer of Sn onto the pin was also observed followed by its transformation into SnO_2. It was noted that if this oxide remained a sufficient time on both surfaces, it allowed their work-hardening and then played the role of a fluid lubricant in the contact (60,102).

Decreases of a factor of 10 in the wear rate were measured for implanted doses 10 times smaller than for nitrogen ions ($10^{16} cm^{-2}$ instead of $10^{17} cm^{-2}$) in lubricated conditions (104), and decreases of a factor of 3 in unlubricated wear tests for iron implanted with 10^{17} Sn cm^{-2} while N implantation had no effect (105).

4. CONCLUSION

The industrial applications of ion implantation have, so far, not matched the amount of research work on structural transformations and tribological effects. Except for the use of N implanted moulds, tools, dies and drills in production lines for plastic or steel components, only isolated applications have been performed on military equipments of strategic importance. Ion implantation remains an expensive treatment when compared with PVD or laser processes. However the recent development of ion beam mixing and of combined evaporation with a low fluence implantation may lead to new openings. For instance such techniques could be used to refine the grain of coatings, amorphize them superficially or improve their adherence by mixing interfacial layers with the substrate. Nearly nothing has been done in the field of ceramico-metallic or plastico-metallic bilayers.

530

Beside this technological research, ion implantation remains an interesting way to vary at will the structure, stoichiometry, hardness and thickness ratios in pure bilayers. Fundamental research on their rheology and diffusion processes during friction (with implantation of selected isotopes and nuclear techniques) may help in the development of new coatings by other techniques. These can be alloys of rare elements, or metastable alloys, for which ion implantation provides a simpler means of obtaining a given composition.

REFERENCES

1. C.J. Mc Hargue, International Metals Reviews 31 (1986) N°2; 49
2. D. Tabor, in "Microindentation Techniques in Materials Science and Engineering", Blau and Lawn ed., ASTM STP 889 (1986) 129
3. R. Hill, The Mathematical Theory of Plasticity, Clarendon Press, Oxford (1950)
4. K.L. Johnson, in "Contact Mechanics", Cambridge University Press (1985)
5. G.B. Sinclair, P.S. Follansbee, K.L. Johnson, Int. J. Solids Structures 20, N°1 (1984) 81 and 21, N°8 (1985) 865
6. L.E. Samuels, T.O. Mulhearn, J. Mech. Phys. Solids 5 (1957) 125.
7. D.M. Marsh, "Plastic Flow in Glass", Proceedings of the Royal Society (London), A279 (1964) 420
8. K.L. Johnson, "The Correlation of Indentation Experiments", J. Mech. Phys. Solids 18 (1970) 115
9. D. Lebouvier, P. Gilormini, E. Felder, 4th EUROTRIB Proceed., S.F.T. ed., Elsevier, II (1985) 5.3.7.
10. C.J. Studman, M.A. Moore, S.E. Jones, J. Phys. D, Appl. Phys. 10 (1977) 949
11. D. Tabor, 'The hardness of Solids", Proceedings of the Institute of Physics F, Physics in Technology 1 (1970) 145
12. D. Maugis, H.M. Pollock, Acta. Met. 32 n°9 (1984) 1323
13. L.E. Samuels, T.O. Mulhear, "The Deformed Zone Associated with Indentation Hardness Impressions", J. Mech. Phys. Solids 5 (1957) 125
14. J.D.J. Ross, H.M. Pollock, J.C. Pivin, J. Takadoum, Thin Solid Films 148 (1987) 171.
15. J.C. Pivin, D. Le Bouvier, H.M. Pollock, submitted for publication in Thin Solid Films
16. P. Gilormini, E. Felder, Wear 88 (1983) 195
17. H. Steffensen, T. Wanheim, Wear 43 (1977) 89
18. P. Baque, E. Felder, J. Hyafil, Y. D'escatha, in "Mise en Forme des Metaux " Dunod ed. 2 (1973)
19. J.M. Challen, P.L.B. Oxley, Wear 53 (1979) 229
20. B. Avitzur, C.K. Huang, Y.D. Zhu, Wear 95 (1984) 77
21. D. Lebouvier, thesis, Nice (1987) - and C.E.M.E.F. internal report, Sophia Antipolis, 06500 Valbonne, France.
22. R.L. Aghan, L.E. Samuels "Mechanisms of abrasive polishing, "Wear 16 (1970) 293
23. T.O. Mulhearn, L.E. Samuels "The abrasion of metals: a model of the process", Wear 5 (1962) 478
24. M. Kaneta, Y. Murakami, H. Yatsuzuka, 4th EUROTRIB Proceed. S.F.T. ed., Elsevier, II (1985) 5.4.10
25. S. Jahanmir, N.P. Suh, Wear 44 (1977) 17

26. J.R. Fleming, N.P. Suh, Wear 44 (1977) 39
27. M.J. Murray, P.J. Mutton, J.D. Watson, J. of Lubrification Technol. 104 (1982) 9
28. F. Delamare, M. de Vathaire, J. Kubie, J. Lubrification Technol. (Trans ASME) 104 (1982) 538 and 545
29. F.P.Bowden, D.Tabor, "The friction and Lubrication of Solids" Oxford University Press, London and New-York (1954-1964)
30. P.A. Steinmann, P. Laeng, H.E. Hintermann, Le Vide, Les Couches Minces, Special n° "Winter school on Sputtering and its Applications" supplement to n° 224 (Nov. Dec. 1984) 260 - and Oberflache - surface 23 n° 4 (1982) 108
31. E. Felder, J. Mech. Theo. et Appl., soumis pour publication.
32. A. Ura, Y. Yamamoto, A. Nakashima, 4th EUROTRIB Proceed, S.F.T. ed., Elsevier II (1985) 5.3.1
33. S.C. Lim and M.F. Ashby, Acta, Met. 35, n°1 (1987) 1
34. J.F. Archard, Wear 2 (1958) 438
35. T.F.J. Quinn, Tribology International 16 (1983) 257 and 305
36. J.P. Hirth, D.A. Rigney, Wear 39 (1976) 133
37. D.A. Rigney, W.A. Glaeser, Wear 46 (1978) 241
38. S.G. Caldwell, J.J. Wert, R.W. Carpentier, in "Wear of Materials", S.K. Rhee, A.W. Ruff and K.C. Ludema ed., ASME, New-York (1981) 63
39. C.M. Rao and T.H. Kosd, in "Wear of Materials", K.C. Ludema ed. ASME, New-York (1985) 364
40. S. Fayeulle, these de doctorat d'Etat, Université Claude Bernard Lyon I, Lyon, France (1987)
41. Y. Mizutani, in "Fundamentals of Tribology", N.P. Suh and N. Saka ed., MIT Press, Cambridge, Mass (1980) 223
42. J.P. Hvith, D.A. Rigney, in "Dislocations in Solids", F.R.N. Nabarro ed., North Holland Pub. Co. (1983) chapter 25
43. F. Pons, S. Megtert, M. Pequignot, D. Mairey, C. Roques-Carmes, J.C. Pivin, Acta Crystal. accepted for publication.
44. H.M. Pollock, P. Shufflebottom, J. Skinner, J. Phys. D: Appl. Phys. 10 (1977) 127 - S.K. Chowdhury, N.E.W. Hartley, H.M. Pollock, M.A. Wilkins, J. Phys. D: Appl. Phys. 13 (1980) 1761
45. J.L. Sullivan, in "Tribology- Fifty Years On", Institution of Mechanical Ingeneers, London (1987) 283 - and J.L. Sullivan, T.F.J. Quinn, D.M. Rowson, Tribology International 13 (1980) 153 - Wear 65 (1980) 1 - Wear 94 (1984) 175
46. J.W. Jones, J.J. Wert, Wear 32 (1975) 363
47. R. Hutchings, W.C. Oliver, Wear 92 (1983) 143, Mat. Sci. Engineer 69 (1985) 129
48. H. Herman, Nucl. Instr. Meth. 182-183 (1981) 887
49. D.C. Kothari, M.R. Nair, A.A. Rangwala, K.B. Lal, P.P. Prabhawaaklar, P.H. Raole, Nucl. Instr. Meth. in Phys. Res. B7-8 (1985) 235
50. J.L. Whitton, G.T. Ewan, H.M. Ferguson, T. Laussen, I.V. Mitchell, H.H. Plattner, M.L. Swanson, A.V. Drigo, W.A. Grant, J. Mat. Sci. Engineer (1985)
51. G. Marest, S. Skovtarides, T. Barnavon, J. Tousset, S. Fayeulle, M. Robelet, Nucl. Instr. Meth. 209-210 (1983) 1063 - Mat. Sci. Engineer 69 (1985) 531
52. N. Moncoffre, M. Brunel, P. Deydier, J. Tousset, Surf. Interf. Anal. 9 (1986) 139
53 Y. Arnaud, M. Brunel, A.M. Becdelievre, M. Romand, P. Thevenard, M. Robelet, Appl. Surf. Sci. 26 (1986) 12

532

54   M. Amara, Doctorat d'Etat thesis, IPN Université Claude Bernard, Lyon I, Lyon, France (1986) - N. Moncoffre, ibid (1986) - T. Barnavon, ibid (1985)

55.   B. Rauschenbach, A. Kolitsch, Phys. Stat. Sol. (a) 80 (1983) 211

56.   E. Johnson, U. Littmark, A. Johansen, C. Christoloudiles, Phil. Mag. A 45 (1982) 803.

57.   E. Johnson, A. Johansen, L. Sarholt-Kristensen, L. Graaboek, N. Hayashi, I. Sakamoto, Nucl. Intr. Meth B (1987) to be published

58.   R.G. Vardiman, R.N. Bolster, I.L. Singer, in "Metastable Materials Formation By Ion Implantation", S.T. Picraux, W.T. Choyke ed., Elsevier, Amsterdam (1982) 269

59.   F.G. Yost, S.T. Picraux, D.H. Follstaedt, L.E. Bre, J.A. Knapp, Thin Solid Films 107 (1983) 287

60.   I.J.R. Baumvol, "Ion Implantation Metallurgy", chapter of a book not yet published, Instituto de Fisika, Universidode Federal do Rio Grande do Sul, Brasil

61.   K. Hohmuth, B. Rauschenbach, Mat. Sci. Engineer 69 (1985) 489

62.   D. Fleche, J.P. Gauthier, P. Kapsa, J. Microsc. Spectrosc. Electron 10 (1985) 219 - Proc. Eurotrib 85, Societe Française de Tribologie ed., Elsevier Pub. I (1985) 1.4

63.   J.C. Pivin, F. Pons, J. Takadoum, H.M. Pollock, G. Farges, J. Mat. Sci. 22 (1987) 1087

64.   I.L. Singer, in "Ion Implantation and Ion Beam Precessing of Materials", G.K. Hubler, O.W. Holland, C.R. Clayton, C.W. White ed., North Holland, New York (1984) 585. Appl. Surf. Sci. 18 (1984) 28

65.   G.K. Hubler, F.A. Schmidt, Nucl. Instr. Meth. B 7-8 (1985) 151

66.   G. Dearneley, Nucl. Instr. Meth. B 7-8 (1985) 158

67.   H.M. Pollock, I.L. Singer, in "Ion Implantation Metallurgy", C.M. Preece, J.K. Hirvonen eds., The Metallurgical Society of AIME, Warrendale, Penn. (1980) 103

68.   C. Fu-Zhai, L. Heng-De, Z. Xiao-Zhong, Nucl. Instr. Meth. 209-210 (1983) 881

69.   N. Jost, I. Schmidt, in "Wear of Materials 1985", K.C. Ludema, ASME, New York (1985) 205

70.   C. Allen, A. Ball, B.E. Protheroe, Wear 74 (1981-1982) 287

71   R. Hutchings, W.C. Oliver, J.B. Pethica, Nucl. Instrum. Meth. 209-210 (1983) 995 - Met. Trans. 15A (1984) 2221

72   I.L. Singer, R.N. Bolster, C.A. Carosella, Thin Solid Films 73 (1980) 283 - Appl. Phys. Lett. 36 (1980) 208

73   W.C. Oliver, R. Hutchings, J.P. Pethica, E.L. Paradis, A.J. Shuskus, Mat. Res. Soc. Symp. Proc. 27 (1984) 705

74.   R.G. Vardiman, Mat. Res. Soc. Symp. Proc. 27 (1984) 699, Elsevier Science Publishing .

75.   R.N. Bolster, I.L. Singer, R.G. Vardiman, Proc. Int. Conf. on Metallurgical Coatings, San Diego (March 1987), to be published in Surface and Coatings Technology

76.   R. Martinella, S. Giovanardi, G. Chevallard, M. Villani, Proc. SMI2B, Mat. Sci. Engineer 69 (1985) 247.

77.   W.W. Hu, H. Herman, C.R. Clayton, J. Kazubowski, R.A. Kant, J. Hirvonen, R.K. Mac Crone, in "Ion Implantation Metallurgy", C.M. Preece and J.K. Hirvonen ed., Metallurgical Society of AIME (1980)92

78   H. Herman, in Proc. Conf. "Ion Plating and Allied Techniques - IPAT 79", CEP Consultants (1979) 255

79.   R.G. Vardiman, R.A. Kant, J. Appl. Phys. 53 (1982) 690

80.   G. Dearneley, Rad. Eff. 63 (1982) 1

81. J. Greggi, R. Kossowsky, in "Science of hard materials" R.K. Viswanadhan et al ed., Plenum Press, New-York (1983) 485 - J. Greggi, Scr. Metall 17 (1983) 765
82. P.D. Goode, A.T. Peacock, J. Asher, Nucl. Instr. Meth. 209-210 (1983) 925
83. F. Pons, J.C. Pivin, G. Farges, J. Mat. Research 2, n°5 (1987) 580-companion paper submitted to Thin. Sol. Films (1987)
84. G. Slodzian, Ann. Phys. 9 (1974) 591 - NBS Spec. Publ. 427, Proceeding of the Workshop on SIMS and IMMA, Gaithersburg 33 (1974)
85. H. Kimura, T. Masumoto, F.S. Spaepen, A.I. Taub, in "Amorphous Metallic Alloys", F.E. Luborsky ed., Butterworths, Monographs in Materials, London (1984) chapter 12 and 13, pp 187 and 231
86. J.C. Pivin in "Les Surfaces en Metallurgie", Ann. Chim. Fr. 11 (1986) 45
87. J.C. M. Li, W. Benoit, G.P. Johari, V. Vitek, J. Perez, in "Plastic Deformation of Amorphous and Semi-crystalline Materials", B. Escaig and C. G'Sell ed., Les Editions de Physique, Les Ulis, France (1982) pp 29, 65, 109, 143, 265
88. J. Takadoum, J.C. Pivin, J. Chaumont, C. Roques-Carmes, J. Mat. Sci. 20 (1985) 1480 - and Proc. Eurotrib. 85, Societe Française de Tribologie, Elsevier
89. I. Singer et al., unpublished work on Fe-Ni-B
90. M. Hirano, S. Miyake, J. Tribology 107 (1985) 467
91. F.G. Yost, L.E. Pope, D.M. Follstaedt, J.A. Knapp, S.T. Picraux, in "Metastable Material Formation by Ion Implantation", S.T. Picraux, W.J. Choyke ed., North Holland, New York (1982) 261
92. D.M. Follstaedt, F.G. Yost, L.E. Pope, in "Ion Implantation and Ion Beam Processing of Materials", G.K. Hubler, O.W. Holland, C.R. Clayton, C.W. White ed., North Holland, New York (1984) 655
93. N.E.W. Hartley, Tribology Int. 68 (1975)
94. J. Takadoum, J.C. Pivin, H.M. Pollock, J.D.J. Ross, H. Bernas, Nucl. Instrum. Meth. in Phys. Research B18 (1987) 153
95. J.C. Pivin, H.M. Pollock, in "Tribology- Fifty Years On", Institution of Mechanical Engineers, London (1987) 179
96. J.C. Pivin, J. Perreau, C. Reynaud, J. Takadoum, J. Non-Crystalline Sol. 86 (1986) 161
97. J. Takadoum, Doctorat d'Etat Thesis, Orsay (1986)
98. E.J. Knystautas, A. Singh, M. Fiset, Proc. IBMM 86, Nucl. Instrum. Meth. Solid State Commun. 52, n°5 (1984) 491
99. C. M. Preece, E.N. Kaufman, A. Staudinger, L. Buene, Proc. Mat. res. Soc. Ann. Meeting, Cambridge, Mass (1979)
100. K. Hohmuth, E. Richter, B. Rauschenbach, C. Blockwitz, Proc. SM2IB, Mat. Sci. Engineer 69 (1985) 191
101. E.F. Finkin, Mat. Engin. Applic. 1 (1979) 154
102. I.J.R. Baumvol, J. Appl. Phys. 52 (1981) 4583 - and J. Appl. Phys. (1984) - Phys. Stat. Sol.(a) Appl. Research 67 (1981) 287 - Nucl.

ADHESIVE AND ABRASIVE WEAR STUDY OF NITROGEN IMPLANTED STEELS

K. Kobs*, H. Dimigen*, R. Leutenecker**, H. Ryssel***

* Philips GmbH Forschungslaboratorium Hamburg,
  Vogt-Koelln-Str. 30, 2000 Hamburg 54, FRG
** Fraunhofer Institut für Festkörpertechnologie,
   Paul-Gerhardt-Allee 42, 8000 München 60, FRG
*** FhG Arbeitsgruppe für integrierte Schaltungen,
    Artilleriestr. 12, 8520 Erlangen, FRG

1. INTRODUCTION

The use of ion implantation for improving the tribological properties of metals has been studied for more than ten years. Various ion species have been employed for reducing friction and wear and especially nitrogen implantation in steels seems to be a suitable method to improve the wear behaviour (1-8) and is approaching industrial application (9). However, more wear investigations are needed before N implantation becomes a real industrial application. The present work is an attempt to contribute to a better understanding of the influence of N implantation on adhesive and abrasive wear.

2. EXPERIMENTAL PROCEDURE

The wear tests were performed on a ball-on-disc tester under oscillating conditions. To simplify matters the wear between the wear partners under unlubrating conditions is defined as "adhesive wear". It should be mentioned that besides adhesion other wear mechanisms like tribooxidation, abrasion, and fatigue can be of importance.

Abrasive wear conditions were adjusted by bringing a suspension consisting of grinding material and glycerol between the ball and the disc. Under these conditions abrasion (scratching and cutting by the particles of the grinding material) plays the dominant role and the definition "abrasive wear" is accurate.

The implantation energy was 100 keV and the dose was $(4-5) \times 10^{17} cm^{-2}$ for all implants. The temperature was 620 K for the stainless steels and only 470 K for the low alloyed steels to avoid effusion. A detailed description of the experimental conditions is given in (4).

3. RESULTS AND DISCUSSION

3.1 Adhesive wear

In the adhesive wear mode eight different steels were investigated and it was found that a wear reduction by nitrogen implantation is strongly dependent on the steel composition. Steels with a low chromium content, the 100Cr6 (AISI 52100) bearing steel, 42CrMo4 (AISI 4140) heat treated steel, 9S20 free cutting steel and even the X34CrAlS5 nitriding steel

R. Kelly and M. Fernanda da Silva (eds.), Materials Modification by High-fluence Ion Beams, 535–540.
© 1989 by Kluwer Academic Publishers.

536

yielded less or no improvement, whereas the stainless steels
- X12CrS13 (AISI 416), X12CrNiS188 (AISI 304 S), X10CrNiTi189
(AISI 321), X90CrMoV18 (AISI 440B) hardened and unhardened
and a chrome plated steel - showed a reduced wear rate by
more than two orders of magnitude. These results are a con-
firmation of our previous statement (4) that the chromium
content plays an important role for the marked wear reducing
effect due to a precipitation hardening by finely dispersed
Cr-N or Fe-Cr-N precipitates (10), which could be proved in
the meantime by X-ray photoelectron spectroscopy and transmission electron microscopy (11).

Due to the fact that the implantation depth is only some
tenth of a micrometer the question arises, whether the wear
reduction is limited by the applied load. Because steels with
a lower bulk hardness should be more critical, this effect
was investigated on the unhardened ferritic X90CrMoV18 steel
and the austenitic X10CrNiTi189 steel; the ball material was
100Cr6 (AISI 52100) steel. The applied load was varied from
0.14 to 13.2 N. We found that the considerable wear reduction
occured only up to a certain load. With increasing dose, the
critical load increased, probably because the protecting layer was thicker. This assumption agrees with the observation
that for the hardened implanted X90CrMoV18 steel the critical
load reached a value of more than 13.2 N, which means that
the hardened matrix provides a sufficient load capacity.

The previously pre-
sented results were ob-
tained by performing the
wear experiments under
laboratory atmosphere.
Dearnaley indicates in
(12) an interrelation
between wear and oxida-
tion in ion implanted
metals. A direct means
for influencing the wear
mechanism is to vary the
ambient atmosphere. For
these investigations the
X10CrNiTi189 steel,
which showed the highest
improvement in wear by
N implantation (dose
$7 \times 10^{17} cm^{-2}$), was used.
In fig. 1 a comparison
of the wear behaviour
under dry nitrogen at-
mosphere and under labo-
ratory atmosphere is
shown. The considerable
wear reducing effect
completely disappeared
under dry nitrogen at-
mosphere. In contrast to
this at a relative humidity of 80% nitrogen, the wear reduction was observed again. Referring to this behaviour, it

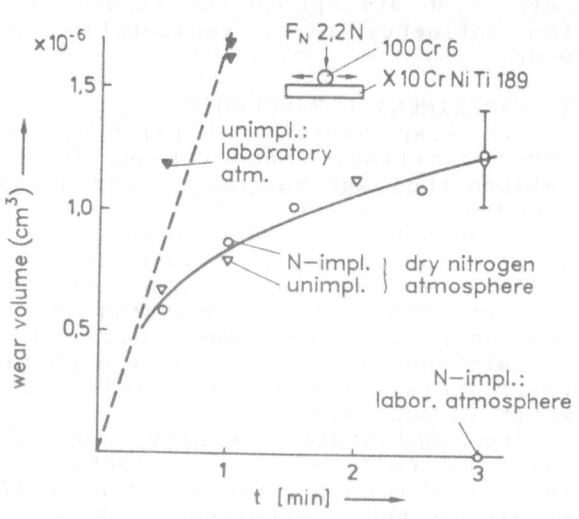

FIGURE 1.
Wear behaviour of the X10CrNiTi189
steel unimplanted and N implanted
($7 \times 10^{17} cm^{-2}$) under laboratory
and dry nitrogen atmosphere

seems that the humidity plays the dominant role. But further
wear tests (13) revealed that oxygen determines substantially
the wear reduction, since under dry oxygen also nearly 'zero
wear' was obtained. This is a clear proof that in addition to
the surface hardening the nitrogen implantation changes the
tribochemical state. A transformation of the severe form of
wear to a milder form of oxidative wear occures. This may be
the reason, why Iwaki et al. (14) found that the wear particles for the unimplanted steels consist of metal mixed with
oxides, in contrast to the implanted steels, where the majority of the particles were made of iron oxides. A further indication is given by Hale et al. (15), who obtained for both
nitrogen implantation and low temperature oxidation of a
SAE 3135 steel the same reduction in wear.

3.2 Abrasive wear

For abrasive wear tests various grinding materials were
used: silicon carbide, corundum, garnet, and quartz with a
mean grain size of 17 μm. The unhardened and hardened
X90CrMoV18 and the chrome plated steel were investigated
(load 0.5 N, ball material X40Cr13 stainless steel). In
fig. 2 the volumetric wear of the unhardened steel after

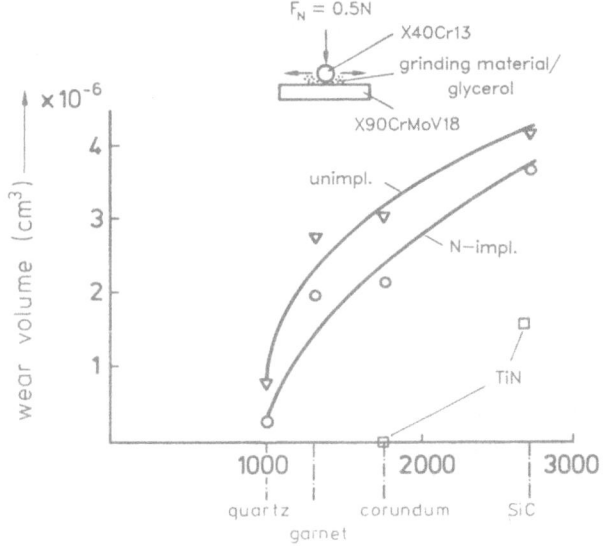

FIGURE 2.
Wear volume of
the unhardened
X90CrMoV18 steel
before and after
N implantation
as a function of
the grain hardness.
For comparison the
wear volume of a
TiN (CVD) layer
is shown.

1260 cycles (180 seconds) as a function of the applied grain
hardness is shown. The implantation yielded only a slight reduction in wear. With increasing the grain hardness, the relative wear reduction decreased. This behaviour was observed
also for the hardened steel and the chrome plated disc, but
the improvement was even lower. Chemical vapour deposited

(CVD) titanium nitride layers are much more effective. The reason for the slight wear reduction by the nitrogen implantation could be clarified by measuring the wear kinetics, which is shown in fig. 3. For the first cycles the wear rate was reduced by a factor 3, independent of the grinding material.

The inflexion point, which characterizes the transition to the wear rate of unimplanted material, is determined by the

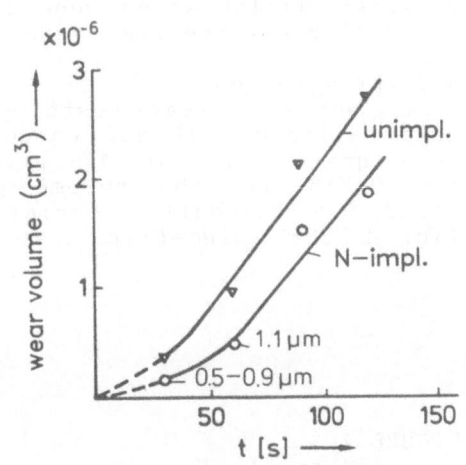

FIGURE 3.
Wear behaviour of the unhardened X90CrMoV18 steel for various grinding materials. At the inflexion point the maximum wear depth is presented.

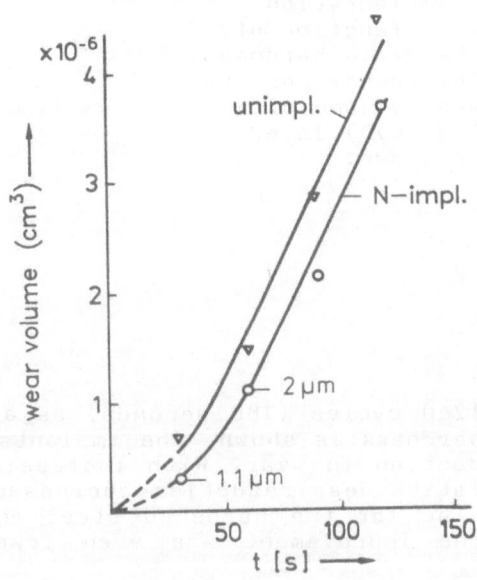

grain hardness. It should be pointed out that the maximum wear depth at the inflexion point was the same for all grinding materials and amounted to 1 μm for the X90CrMoV18 steel and 0.4 μm for the chrome plated disc. This indicates a connection to the implantation depth, because it was found that the range of the nitrogen ions in chrome is lower than in the X90CrMoV18 steel (11).

It should be mentioned that, in abrasive conditions, no nitrogen could be detected by SIMS in wear tracks of 2 μm depth, in contrast to the case of adhesive wear (4). This means that a migration of nitrogen did not occur during the abrasive wear tests, probably caused by the scratching and cutting mechanism of abrasive wear.

4. CONCLUSIONS
It has been shown that nitrogen implantation decreases both the adhesive and abrasive wear. Besides the implantation conditions the wear reduction, which is gained, depends strongly on the steel composition. Only steels with a high chromium content showed the beneficial effect.
1) In conditions of adhesive wear this effect can be more than two orders of magnitude if
   a) the uncritical load, which depends on the implantation dose and bulk hardness, is not exceeded and if
   b) the ambient atmosphere contains oxygen, so that an oxide layer can be formed, which leads in combination with the surface hardening by the N implantation process to a mild oxidative wear. Under dry nitrogen atmosphere no wear reduction was observable.
2) In conditions of abrasive wear a 3-fold wear reduction by the N implantation occured, but only to a wear depth of approximately 1 μm for the X90CrMo18 stainless steel and 0.4 μm for the chrome plated disc. Up to this depth, the reduction in the wear rate is nearly independent on the grinding material. In contrast to the case of adhesive wear no nitrogen could be detected in a wear track of 2 μm depth.

ACKNOWLEDGEMENTS
The authors are indebted to Mrs. C. Witt for assisting the wear measurements and to Mr. H.J. Tolle for the SIMS analysis.

REFERENCES
1. G Dearnaley, in H. Ryssel and H. Glawischnig (eds.), Proc. 4th Int. Conf. on Ion Implantation: Equipment and
2. Techniques, in Springer Ser. Electrophys. 11 (1982) p 332 S J Charter, L R Thomson and G Dearnaley, Thin Solid
3. Films 84 (1981) p 355 C A dos Santos and I J R Baumvol, in H. Ryssel and H. Glawischnig (eds.), Proc. 4th Int. Conf. on Ion Implantation: Equipment and Techniques, in Springer Ser. Electrophys. 11 (1982) p 447

4. H Dimigen, K Kobs, R Leutenecker, H Ryssel and P Eichinger, in G.K. Wolf, W.A. Grant and R.P.M. Procter (eds.), Proc. Int. Conf. on Surface Modifications of Metals by Ion Beam, Heidelberg, 1984, in Mater. Sci. Eng. <u>69</u> (1985) p 181

5. F Z Cui, H D Li and X Z Zhang, in B. Biasse, G. Destefanis and J.P. Gaillard (eds.), Proc. 3rd Int. Conf. on Ion Beam Modification of Materials, Grenoble, 1982, in Nucl. Instrum. Methods 209-210 (1983) p 881

6. K Yu, H D Li, X Z Zhang and J H Tian, Nucl. Instr. Methods 209-210 (1983) p 1063.

7. M Iwaki, in G.K. Wolf, W.A. Grant and R.P.M. Procter (eds.), Proc. Int. Conf. on Surface Modification of Metals by Ion Beams, Heidelberg, 1984, Mater. Sci. Eng. <u>69</u> (1985) p 211

8. A Cavalleri, L Guzman, P.M. Ossi and I. Rossi, Scr. Metallurgica <u>20</u> (1986) p 37

9. P Shioshansi, Thin Solid Films <u>118</u> (1984) p 61

10. F G Yost, S T Picraux, D M Follstaedt, L E Pope and J A Knapp, Thin Solid Films <u>107</u> (1983) p 287

11. R Leutenecker et al., to be published

12. G Dearnaley, Nucl. Instr. Methods <u>B 7/8</u> (1985) p 158

13. K Kobs et al., to be published

14. M Iwaki, T Fujihana and K Okitaka, in G.K. Wolf, W.A. Grant and R.P.M. Procter (eds.), Proc. Int. Conf. on Surface Modifications of Metals by Ion Beams, Heidelberg, 1984, in Mater. Sci. Eng. <u>69</u> (1985) p 211

15. E B Hale, R A Kohser and R A Reinhold, Appl. Phys. Lett. <u>49</u> (1986) p 447.

EFFECT OF α-RECOIL DAMAGE ON THE ELASTIC MODULI OF ZIRCON AND TOURMALINE

H. ÖZKAN

Department of Physics, Middle East Technical University, Ankara, Turkey

ABSTRACT

   The elastic moduli of natural radiation-damaged zircons and thermal neutron irradiated tourmalines have been studied. A systematic   and very marked decrease of the elastic moduli with radiation damage has   been observed. The bulk moduli obtained for the damaged zircons and tourmalines were compared with the theoretical estimates of bulk moduli for irradiated materials containing spherical and nonspherical distributions of defects, displacement cascades, voids and cracks.

1. INTRODUCTION

   In an earlier paper[1] we have reported that the elastic constants and the lattice parameters of several natural zircon crystals have systematical and very marked differences due to the transformation of zircon to the amorphous (metamict) phase caused by the radioactive uranium and thorium impurities. The energetic α-particles and recoil atoms, released by the disintegration of the radioactive impurities, cause radiation damage and transform zircon into a low-density amorphous form.
   Similar α-recoil damage may be induced in boron-containing crystals by thermal neutron irradiation due to ${}_5B^{10}(n,\alpha)\ {}_3L^7$ nuclear reaction[2] . The two energetic fragments released by this reaction also produce lattice defects and may transform the boron compounds into low-density amorphous form[3] . In a recent publication we have also reported the effect of thermal neutron irradiation on the structure and elastic moduli of boron-containing tourmaline crystals[4] . A brief summary of the two previous reports about the effect of α-recoil damage on the elastic moduli of zircon and tourmaline is presented here.

2. ELASTIC DATA OF THE DAMAGED ZIRCONS

   Elastic wave velocities of seven different zircon samples of densities from 4.70 to 3.90 g/cm³ , covering the whole density spectrum for the transformation of zircons into amorphous form, have been measured by the ultrasonic pulse-echo method. The details of sample preparation and the experimental techniques are described in reference 1.
   The measured elastic wave velocities and the calculated elastic moduli are plotted versus density of the samples in Fig.1 and 2. A glance at Fig.1 and 2 reveals a number of interesting features in the behaviour of the elastic moduli of the zircon samples. All the longitudinal and shear moduli, with the exception of C_{66} , decrease systematically and markedly with the radiation damage and approach the saturation values of 154 GPA * and 49 GPA, respectively.

* GPA is gigaPascal.

R. Kelly and M. Fernanda da Silva (eds.), Materials Modification by High-fluence Ion Beams, 541–545.

The largest decrease occurs in C_{33} (69 %), which is the largest modulus for undamaged zircons[5]. The smallest shear modulus C_{66}, which also has anomalous pressure derivatives[6], does not decrease with decrease of density but remains essentially constant approaching the same saturation value for the shear moduli. The systematic decrease in the elastic moduli as well as in the elastic anisotropy indicate that as the radiation damage increases zircon crystals gradually become isotropic.

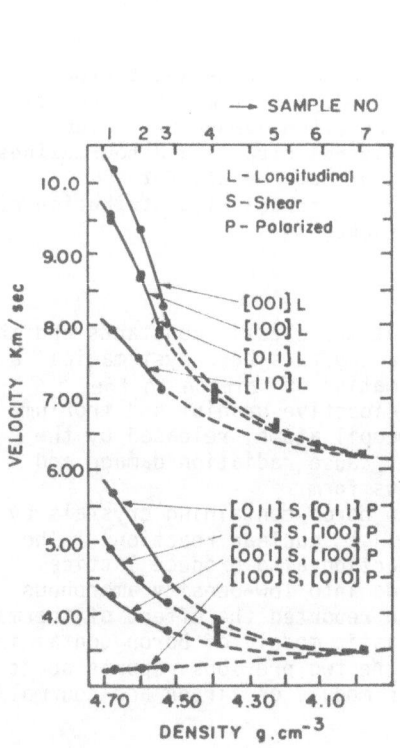

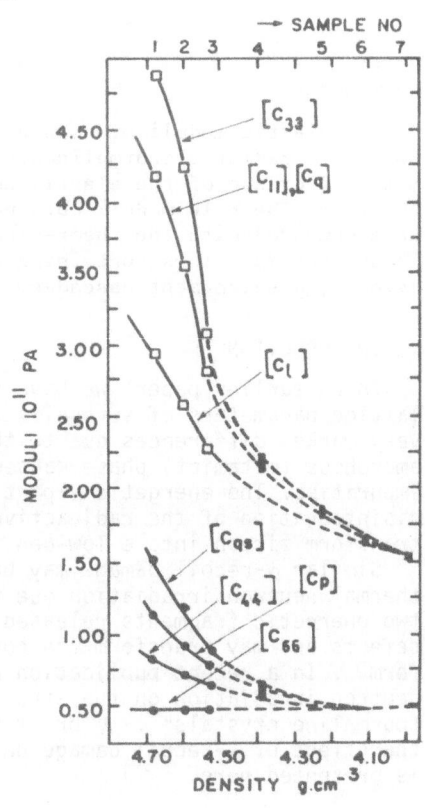

Fig.1. Elastic wave velocities versus density of zircon samples. The solid bars indicate points obtained from several different directions of the unoriented zircon samples 4-7.

Fig.2. Elastic moduli versus density for zircon samples. C_ℓ and C_q denote [110] and [011] longitudinal moduli, C_p and C_{qs} denote [011] shear moduli polarized along [100] and [0$\bar{1}$1], respectively.

As seen in Fig.2, the curves for the longitudinal and shear moduli versus density of zircon show a sigmoid pattern: the rates of decrease of the elastic moduli are initially less rapid, then decrease more rapidly from densities 4.60 to about 4.50 g/cm³. After densities of 4.50 g/cm³ the rates of decrease again reduce gradually. A similar behaviour was also observed in the x-ray diffraction patterns of the zircon samples[1]. The peak positions (2θ values) decrease relatively slowly at first then decrease more rapidly for densities of 4.60-4.50 g/cm³. After densities of 4.50 g/cm³ the decrease in peak positions were relatively small, though the peaks are rapidly broadened and reduced in amplitude.

The observations of the elastic moduli and x-ray diffraction of the damaged zircons at different density regions qualitatively confirm the suggestions[7,8,9] that the transformation of zircons into amorphous form occurs in there different stages as discussed in ref.1.

The average longitudinal and shear wave velocities as well as the bulk and shear moduli of the zircon samples were also calculated from the measured elastic wave velocities. We have found that the bulk and shear modulus of zircons decrease up to, respectively, 61 % and 56 % with the transformation of zircons into amorphous form. The corresponding decrease in the average longitudinal and shear wave velocities are 31 % and 28 % , respectively. All the damaged zircons, except the completely amorphous ones (ρ < 4.0 g/cm^2), recrystallize at temperatures above 900°C and their elastic wave velocities and elastic moduli recover into those of undamaged zircons on heating. The completely amorphous zircons also recrystallize into ZrSiO$_4$ under prolonged annealing at relatively higher temperatures (above 1100°C).

3. ELASTIC DATA OF THE DAMAGED TOURMALINES

Oriented single crystals and powders of tourmaline have been irradiated with thermal neutrons up to total dose of 7.6 x 10^{18} n/cm^2. Elastic wave velocity and x-ray diffraction measurements have indicated that tourmaline crystals are rather stable up to total dose of 8.0 x 10^{17} n/cm^2. Detectable damage in tourmalines started to appear at 1.0 x 10^{18}n/cm^2, with rather rapid decrease of elastic wave velocities at fluences of 2.9 x 10^{18} n/cm^2 and 4.0 x 10^{18} n/cm^2. Wave velocities could be measured even for the most severely damaged zircons, but similar measurements could not be performed for the irradiated tourmalines probably due to non-uniform distribution of microcracks and flaws generated at higher neutron fluences. Nevertheless, we obtained evidence that considerable decrease takes place in the bulk moduli of tourmalines similar to that of the metamict zircons (see Fig.3). We suggest that the large decrease in the elastic moduli of the damaged zircons and tourmalines may be a general characteristic of the crystalline-amorphous transition of such materials

Although tourmalines becomes rather brittle at higher neutron fluences the x-ray diffraction patterns indicate that they essentially preserve the short range order and remain crystalline up to total dose of 7.6 x 10^{18} n/cm^2.

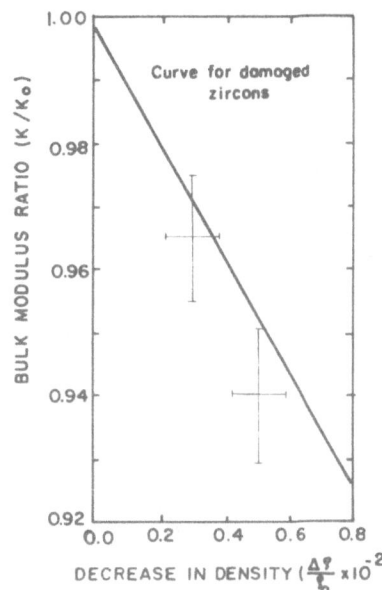

Fig.3. Normalized bulk moduli (K/K_0) versus fractional decrease in density for tourmaline and the corresponding curve for the damaged zircons.

4. DISCUSSION

Mackenzie[12] derived a theoretical expression for the bulk moduli of polycrystalline materials containing spherical pores in terms of Poisson's ratio (α) and porosity (w), defined as the fractional decrease in density:

$$\frac{K}{K_0} = (1 + \frac{3}{2} \cdot \frac{1 - \sigma}{1 - 2\sigma} \cdot \frac{w}{1 - w})^{-1}. \tag{1}$$

Afzali and Nasser[13] have reported a theoretical estimate of the bulk moduli of irradiated materials containing spherical voids in terms of void volume fraction f(= w).

$$\frac{K}{K_0} = (1 + 0.753 \frac{1 + \sigma}{1 - \sigma} f)^{-1} \tag{2}$$

Equations (1) and (2) agree with each other for low f. On the other hand, it has been suggested that voids, pores or cracks in the form of disks and ellipsoids will effect the elastic properties much more than spherical pores[14]. Warren[15] has calculated the effect of pore shape on compressibility and reported curves of normalized bulk moduli versus porosity for pores of oblate spheroids of different axial ratio (a/b, see Fig.4). The experimental bulk-modulus data of the damaged zircons and tourmaline differ from the theoretical estimates involving spherical pores (Eqs.(1) and (2), represented by the upper curve in Fig.4), but follow closely the curve for a/b = 10 involving non-spherical pores and voids. This is an expected result since the displacement cascades and disordered regions in the damaged zircons and tourmalines are not spherical in shape[16]. The dimensions of such elongated cascades, voids and cracks in damaged zircons have to be very small indeed since they are frequently perfectly transparent showing no scattering phenomena, neither of light nor of low-angle x-rays.
However, the dimensions of some of the extended defects generated in the irradiated tourmalines may be larger and they may not be uniformly distributed as evidenced by the brittle nature of the irradiated specimens[4].

Considering the similar effect of α-recoil damage in natural zircons and thermal neutron irradiated tourmalines it was estimated that tourmaline crystals may also be transformed into amorphous form at higher fluence of $\sim 10^{20}$ n/cm^2, just as heavy-ion bombardment causes many silicates to become non-crystalline[17,18]

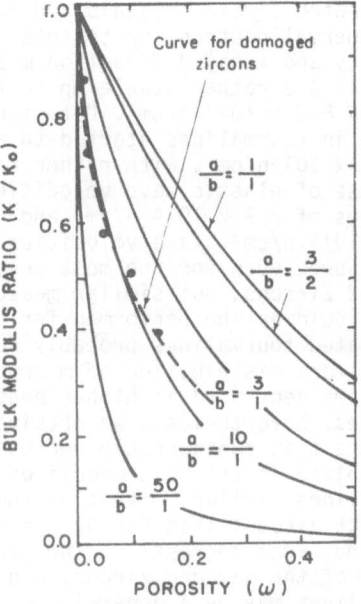

Fig.4. Curves of K/K$_0$ versus porosity for various "oblate-spheroids" of different axial ratio (a/b). The curve for a/b=1 agrees well with Eqs.(1) and (2) for low porosities. The data for zircons and tourmaline differ from the theoretical estimates involving spherical pores, but closely follow the above curve for a/b = 10.

REFERENCES

1. Özkan, H., J.Appl. Phys. 47, 4772 (1976).

2. Billington, D.S., and Crawford, J.H., Radiation Damage in Solids. Princeton Univ. Press, Princeton, N.J. (1961).

3. Birtcher, R.C., and Blewitt,T.H., Rad. Effects 57, 33 (1980).

4. Özkan, H., Rad. Effects 102, 31 (1987).

5. Özkan, H., Cartz, L., and Jamieson, J.C., J.Appl. Phys. 45, 556 (1974).

6. Özkan, H., Jamieson, J.C., Phys. Chem. Minerals 2, 215 (1978).

7. Holland, H.D., Gottfried, D., Acta Cryst. 8, 291 (1955).

8. Bursill, L.A., and Mclaren, A.C., Phys. Status Solidi 13, 33 (1971).

9. Jech, C., and Kelly, R., J.Phys. Chem. Solids 31, 41 (1970).

10. Tatlı, A., and Özkan, H., Phys. Chem. Minerals 14, 172 (1987).

11. Lewis, M.F., and Patterson, E., J.Appl. Phys. 44, 10 (1973).

12. Mackenzie, J.K., Proc. Soc. London B63, 2 (1950).

13. Afzali, M., and Nasser, S.N., J.Nucl. Mater. 87, 175 (1979).

14. Walsh, J.B.,   J. Geophys. Res. 70, 381 (1965).

15. Warren, N., J.Geophys. Res. 78, 352 (1973).

16. Yada, K., Tanji, T., Sunagawa, I., Phys. Chem. Mineral 7, 47 (1981).

17. Cartz, L., and Fornelle, R.A., Rad. Effects 41, 211 (1979).

18. Cartz, L., Karioris, F.G., and Fornelle, R.A., Rad. Effects 54, 57 (1981).

DEPTH-RESOLVED INVESTIGATION OF STRUCTURAL TRANSFORMATIONS AND HARDNESS VARIATIONS IN IMPLANTED FILMS ON BULK SAMPLES

J.C. PIVIN, F. PONS and H. BERNAS

C.S.N.S.M., BP1, 91406 Orsay, France

1. INTRODUCTION

X-ray diffraction experiments at grazing incidence were performed on implanted or irradiated surfaces with a depth resolution of 50 nm, using the intense X-ray source provided by the Orsay synchrotron in conjunction with a new position sensitive detector and a specially conceived instrumental assembly (1). The incidence angle of the monochromatic X-ray beam is first set to a value very near the angle of specular reflection, then increased step by step so that increasing thicknesses of the sample contribute to diffraction.

The mechanisms of various types of transformations could thus be studied on bulk polycrystalline samples, avoiding some technical problems encountered in T.E.M. or R.B.S. investigations (overheating and contamination of thin foils during implantation, preparation of single crystals) and giving complementary information on the nature and depth distribution of phases. Examples presented here are the amorphization of implanted metals (Ni implanted with B, P) or of irradiated alloys with suitable composition (Ni_3B, $CuZr_2$), the successive phase changes in titanium implanted with increasing concentrations of nitrogen, and structural transformations induced by surface heating and stresses during friction in these systems.

These data were correlated to hardness depth-profiles (with comparable resolution) obtained by means of a submicroscopic indentation test conceived by Pollock et al.(2). They allowed us to analyze the rheological and tribological properties of the films in terms of kinematic models of plastic flow in bilayers (3) and of specific deformation mechanisms of the studied phases (4,5).

2. STRUCTURE OF AS-IMPLANTED AND OF WORN SURFACES

2.1 <u>Titanium surfaces implanted with N ions</u>: Ions were implanted at 3 different energies , then in a second series of implantations at 5 energies, in order to obtain a nearly homogeneous depth profile of N concentration over a range of depth of 450nm. The latter was checked by R.B.S and N.R.A (figure 1). The combined results of N.R.A and S.I.M.S. analysis (4) have shown that the contamination of films by recoil implantation of C and O was lower in our case than in other studies of the TiN implantation system, and thus could not affect the structure and hardness changes. The amounts of C and O were less than 2 and 10×10^{16} atoms.cm^{-2} for the highest concentration of 55% N (instead of 1 and 5×10^{16} atoms.cm^{-2} in the natural contamination film on unimplanted surfaces) and the mean concentrations beyond the superficial oxide were of the order of 0.1% and 0.5%.

R. Kelly and M. Fernanda da Silva (eds.), Materials Modification by High-fluence Ion Beams, 547–556.
© 1989 by Kluwer Academic Publishers.

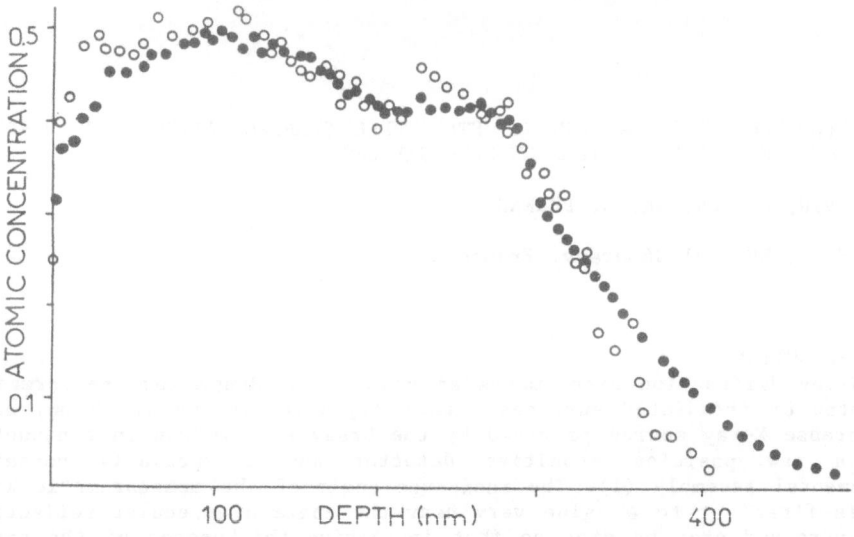

FIGURE 1. (●) N.R.A. profile of C_N, using the resonant reaction $^{15}N(p,\alpha\gamma)^{12}C$ at 898 keV, and (o) R.B.S. profile of $(1-C_{Ti})$, in a titanium sample implanted at 5 energies of 190, 145, 80, 50 and 20 keV with a nominal dose of 1.50×10^{18} $^{15}N.cm^{-2}$. The purity of the film is demonstrated by the near identity of the two profiles thus obtained (1.35×10^{18} N, 5×10^{16} O, 1×10^{16} C cm^{-2} were dosed by N.R.A.).

For implanted concentrations up to 15% N, the only phase identified by X-ray diffraction at grazing incidence was a solid solution α-$Ti_{1-x}N_x$ (with a swelling of the titanium lattice). Then, for N concentrations in the range 15 to 30% N, a single phased film of Ti_2N was formed over a depth of 300 nm, as shown by diffractograms of figure 2a. Crude phase profiles were also established by deriving the contributions of slices of equal thicknesses ΔZ to a given diffraction peak. But they must be considered as very qualitative, because the absorption and reflection coefficients of X-rays are intricate functions of the incidence angle, of the N content and of surface contamination. For instance the real thickness of the Ti_2N film of figure 3a, measured directly on a cross section observed by S.E.M., is 400 nm instead of 250 nm. However such profiles illustrate simply how phases are distributed in multilayers such as those formed on surfaces implanted with 30% to 55% N (figures 2b and 3b). For these concentrations, a single phased film of δ-TiN is formed in the homogeneous part of the film (of thickness 400nm measured on a cross section), and ε-Ti_2N in less implanted regions (figure 3b). Note that other peaks of this phase were indexed in addition to the shallow (111) structure observed in figure 2b.

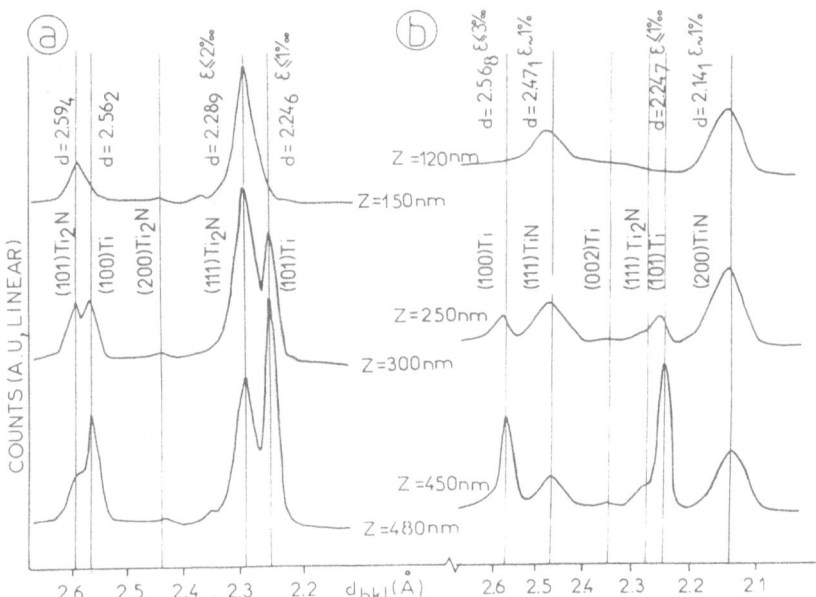

FIGURE 2. Partial diffractograms of Ti surfaces implanted with a) 25% N (7.2x10^{17} N.cm^{-2}), b) 44% N (1.4x10^{18} N.cm^{-2}), recorded with 0.157 nm X-rays at incidence angles of 0.5, 1.5, 2.5°.

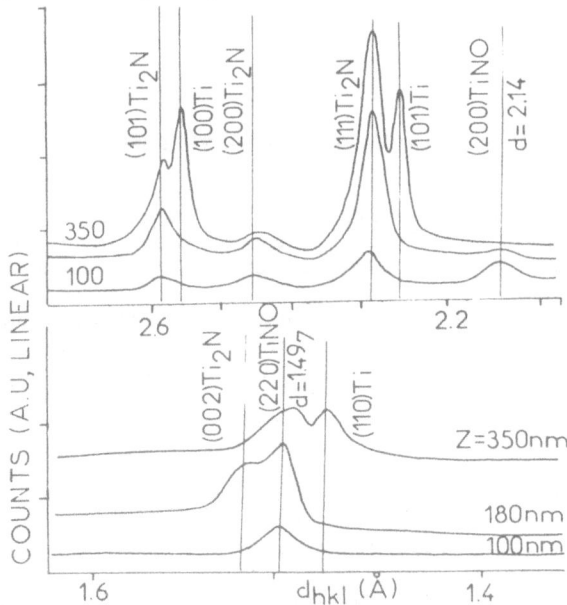

FIGURE 4. Partial diffractograms, recorded with 0.157nm X-rays at increasing incidence angles, on a Ti-25% N film after friction with a carbon-steel ball under a pressure of 800 MPa.

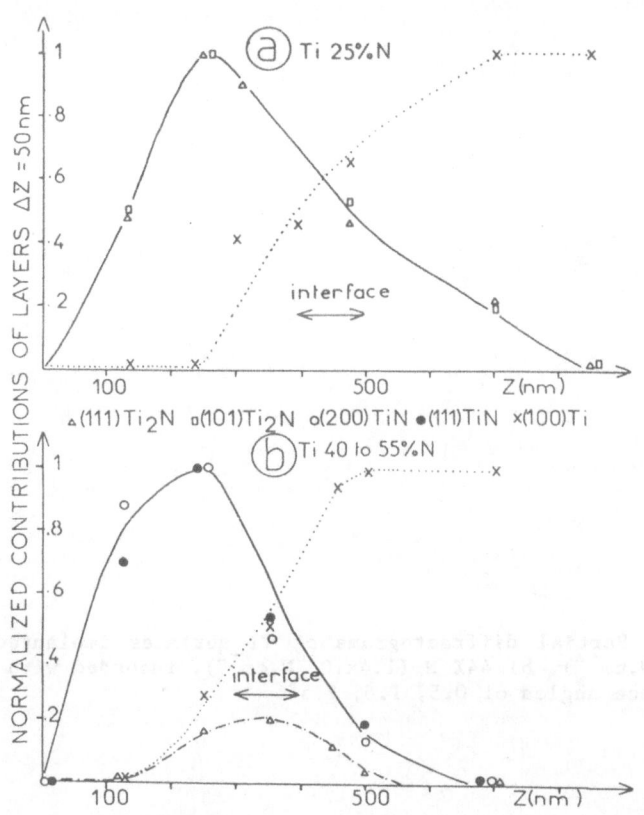

FIGURE 3. Qualitative depth-distributions of phases in the same films as for figure 2.

X-ray diffractograms could also be registered on surfaces worn homogeneously on sufficiently wide areas by means of a specially conceived scanning-tribometer. The combination of both techniques allowed us for instance to demonstrate:
-the development of a superficial texture in titanium during running in (see review paper on the rheology of bilayers in this volume).
-the partial oxidation of Ti_2N into δ -Ti(N,O) in severe conditions of friction (figure 4). Such an oxidation was not observed for TiN films, which explains the difference in the wear resistance of both types of films (lower for TiN- see review paper and ref 4).

2.2. Nickel surfaces implanted with B or P ions: As in the case of titanium, B ions were implanted at several energies in order to form homogeneous films, and their composition was checked by R.B.S, N.R.A and S.I.M.S. (5).

In this case the X-ray diffraction technique has given two useful data. First, it permitted to corroborate the amorphous nature of the film and the disordering kinetics established previously by R.B.S in channeling incidence on single crystals (6).

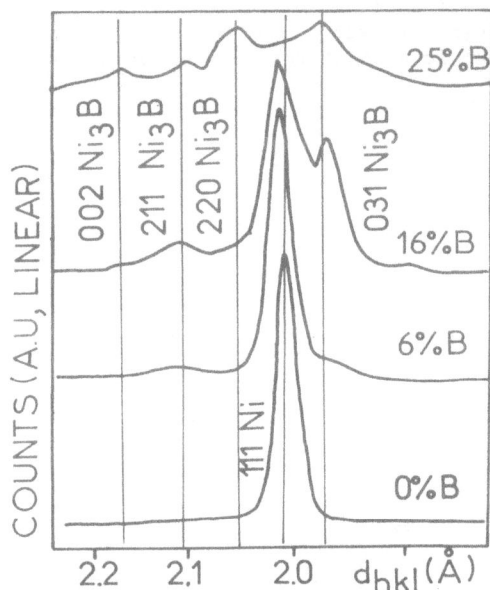

FIGURE 5. Partial diffractograms recorded with 0.190nm X-rays at grazing incidence (0.4° giving a diffracting thickness of 60nm in Ni), on surfaces of pure Ni polycrystals. The latter had been polished electrolytically and then implanted at room temperature with increasing concentrations of boron over a depth of 300 nm (0, 2.4, 5.2, 8.2×10^{17} ions $.cm^{-2}$)

A second interesting result is the identification of Ni_3B crystallites in the outer part (50 nm) of films formed on surfaces implanted with 6% to 25% B at room temperature (figure 5). This compound has not been detected in the main part of the films, which were completely amorphous for B concentrations over 19% (lower diffractogram of figure 7), nor in the outer part of films formed by implantation at low temperature. The presence of Ni_3B is explained by the combined effects of a superficial over-heating during ion implantation and of the diffusion of radiation defects towards the surface. This experimental result does not contribute significantly to the understanding of amorphisation mechanisms by ion implantation, but shows how carefully the structural analysis performed on very thin films must be interpreted. Another example is the discrepancy between our results on Ti surfaces implanted with N and other observations by T.E.M, which all concluded that TiN precipitates preferentially to Ti_2N at intermediate concentrations (4).

For all the systems studied (TiN, NiB, NiP), a swelling of the host lattice of the order of 1% has been measured when the concentration of metalloid in oversaturated solution was at its maximum. Then it decreased when compounds were formed. In the particular case of Ni implanted with B or P at low temperature, this release was observed just at the threshold of concentration for which the disordering kinetics (established by means of R.B.S, T.E.M or X-ray diffraction) exhibits a marked inflexion point. The latter curve could be fitted by a Poisson law, interpreted as being the probability of formation of amorphous clusters in a matrix exhibiting local fluctuations of concentration (the adjustment parameter of this model is a critical volume whose size corresponds to the mean size of ordered clusters in amorphous alloys) (6). The X-ray data corroborate this hypothesis, since a continuous decrease of the lattice swelling would have been observed if the transition from a Ni disordered lattice to an amorphous phase had occured more progressively in the whole implanted volume. Moreover, a decrease of intensity of all diffraction peaks of the Ni matrix was observed until the threshold, then an increase which was particularly marked for (111) and (222) reflections (table 1). Dislocations and stacking faults are just in these planes of dense packing, and a nucleation of the amorphous clusters on these defects would account for this observation.

TABLE 1

P concentration	0	10	12	14	25
amount of disorder X measured by R.B.S.	0	0.1	0.5	0.8	1
Strain % measured on reflections (111), (200), (311).	0	$+3.7\pm1.0$	$+4.9\pm1.0$	0	0
Reflection peaks integrals (111) (222) (200) (220) (311)	10^4 7×10^3 4×10^3 5×10^4 10^3	8×10^2 10^2 8×10^2 2×10^3 2×10^3	10^3 10^3 8×10^2 4×10^3 5×10^2	1.5×10^3 2×10^3 10^2 10^2 10^2	10^2 10^2 10^2 10^2 10^2

Similarly it has been proposed (7) that the amorphization by ion irradiation of crystalline compounds having an amorphous counterpart proceeds from the coalescence of amorphous clusters. These clusters would form after a critical number of d.p.a in overlapping cascades. The disordering kinetics of Ni_3B (001) single crystals by ions of different masses (H, B, Ni, Kr) has been established by R.B.S. and the amorphous nature of fully disordered films was confirmed by X-ray diffraction. As predicted by the model, no threshold of the density of deposited energy per incident ion was observed. Whatever the mass of irradiating ions, the kinetics was the same Poisson function of the number of d.p.a. It appeared interesting to check the validity of the model on bulk specimens of an intermetallic compound for which other kinetic laws have been found (using TEM). Boules of pure polycrystalline $CuZr_2$ could be prepared by melting under a good vacuum, but they exhibited a strong (103) texture hampering a complete study of the amorphization kinetics by X-ray diffraction. However comparable critical values of 0.02 and 0.03 d.p.a were measured for the amorphization of $CuZr_2$ and Ni_3B (figure 6). The diffuse line characteristic of a glassy structure was not detected after irradiating $CuZr_2$ at LN2 temperature with 1.12×10^{13} Ar.cm^{-2} (at 3 energies of 180, 60 and 20 keV with doses in the ratio 2.5, 1 and 1 respectively), while it was the only feature in the outer 70nm of the surface for a dose of 2.25×10^{13} Ar.cm^{-2} and over a depth of 150nm for a dose of 4.5×10^{13} Ar.cm^{-2}. The mean number of d.p.a calculated with the T.R.I.M. program on the same depths are respectively of 1.8 and 0.9, 4.8 and 2.4 for each dose(figure 6b).

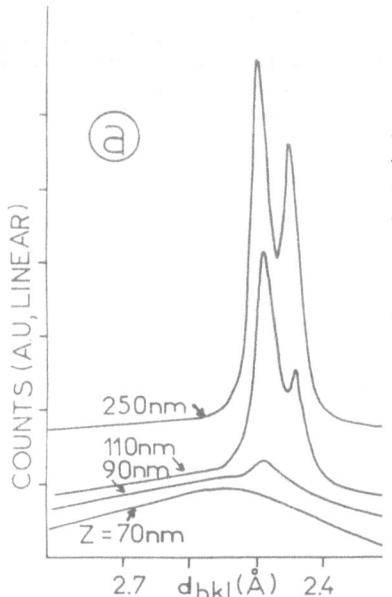

FIGURE 6. a) partial diffractograms around the (103) line of the CuZr$_2$ tetragonal compound, recorded after irradiation with 2.25×10^{13} Ar.cm^{-2}. The splitting of the line (2.50 and 2.46Å) observed at large probed depths is due to the texture of the unirradiated alloy.b)T.R.I.M. calculation of the number of displacements per Cu+Zr atoms (Ed=19 eV for Cu and 40 eV for Zr) for the same irradiation dose (●). Also shown are the normalised contributions of slices ΔZ of equal thickness to the diffractogram for irradiation doses of 2.25×10^{13} (◆) and 4.5×10^{13} (◇) (... crude profiles)

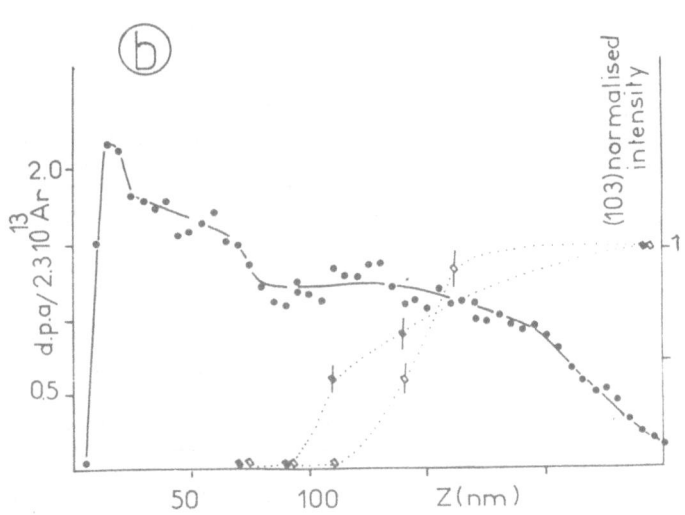

The crystallization of these amorphous alloys during friction is expected since they exhibit a poor thermal stability (T$_c$ = 300°C). NiB implantation films and chemical deposits were submitted to various friction tests: polishing with or without lubrication, dry friction with steel balls under increasing pressures. In most cases, encrusted grains of the abrasive or steel particles transferred onto the surface were the only diffracting phases. A superficial crystallization was detected under the highest pressure with a steel rider, during the running-in stage (upper diffractogram of figure 7). However, the Ni_3B crystallites formed at highly stressed asperities were removed from the track after only 5 more scans of the rider. Thus, the mean temperature of the surface had not increased up to 300°C even during this severe test, because these Ni-B amorphous alloys are very ductile and adhesion resistant (contrarily to their crystalline counterparts) (5).

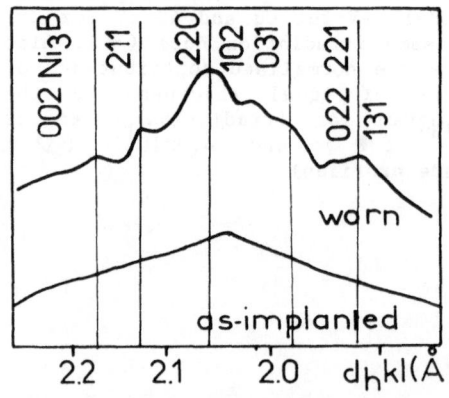

FIGURE 7. Partial diffractograms recorded with 0.154 nm X-rays at an incidence angle of 2° (probed Z = 300nm) on a Ni-25%B film before and after friction with a carbon-steel ball under a pressure of 1GPa (5 scans with a spacing of 500µm between scanned lines).

3. CORRELATION BETWEEN THE HARDNESS AND THE STRUCTURE OF FILMS

Indentation tests were performed with a trigonal diamond. Diamonds can be cut with a very sharp tip when having this shape (their sharpness was checked by S.E.M observations of indents), and the semi-apical angle of indentation is 35°. Thus, the plastified thickness remains about equal to the indented depth as shown by calculations as well as by the measured variations of hardness with depth in submicroscopic bilayers (3). The test procedure consists in varying progressively the applied load with a constant loading rate, and to record the indentation depth Z during loading and unloading (in order to correct the data from the elastic recovery during unloading). From these data are derived the variations with depth of the hardness and of the elastic modulus, as discussed in (3).

The minimum thickness for which the intrinsic hardness of a film can be measured was estimated to be 50 nm. For instance figure 8 shows that a change of hardness can be detected after amorphization by ion irradiation (90 keV B^+ ions to a fluence of 3.5×10^{14} cm^{-2}) of the top 60 nm of a Ni_3B single crystal. The hardness, I_p, of the amorphous layer is equal to that measured with the same technique on thicker implanted films formed by ion implantation at several energies in Ni (described previously) and on thick chemical (electroless) coatings, which gives confidence in the result.

Values of hardness measured on Ni-B and Ti-N films are summarized in table 2. They can be compared to the results of Vickers tests which were performed on the thicker films. They show that the hardness of films increases with the insertion of metalloids in the host lattice, then with the formation of new phases. In the case of Ni-B films the hardness increase with the B concentration was perfectly correlated to the amount of amorphous clusters for C_B over 2.5%. Thermal relaxation of the amorphous films (leading to an increase of density and suppression of part of the chemical disorder) induces a further increase of hardness. Crystallized films are also harder but more brittle, as evidenced by instabilities recorded during loading in indentation tests, and also by their poorer abrasion resistance (4). Single crystals appear also harder at the submicroscopic level than at the microscopic one. This is due to the effect of the number of mobile defects in the plastified field when the size of the indent becomes of the order of the distance between them. It is known for instance that the hardness of polycrystals decreases with the grain size. Such size effects and textures must also be taken into account when comparing the hardnesses of as-implanted films, annealed films or P.V.D. coatings (they can also be used to improve the properties of films).

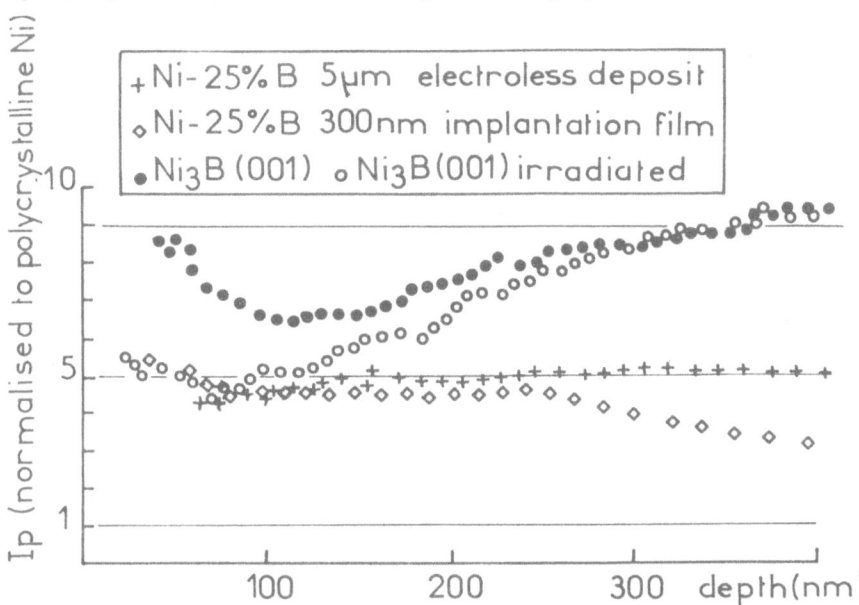

FIGURE 8. Variations with the penetration depth z_p of the plasticity Index I_p (equal to k x H, H being the mean pressure or hardness, and k a geometrical constant about equal to 2.6) of Ni-B films. I_p is normalised with respect to a reference specimen of polycrystalline nickel in order to avoid that any blunting or contamination modifies the value of k.

REFERENCES:
1. "Application of a grazing incidence X-ray diffraction technique to the depth resolved analysis of structural transformations due to surface treatments", F.Pons, S. Megtert, M. Pequignot, D. Mairey, C. Roques-Carmes, J.C. Pivin, Acta Cryst., accepted for publication.
2. H.M. Pollock, D. Maugis, M. Barquins, in " Microindentation Techniques in Materials Science and Engineering", P.J. Blau and B.R. Lawn eds., ASTM Spec. Tech. Publ. 889 (1986) 47.
3a. "Limits to the hardness testing of films thinner than 1um", J.D.J. Ross, H.M. Pollock, J.C. Pivin, J. Takadoum, Thin Solid Films 148 (1987) 171.

3b. "Comparison between fields of plastic deformation in bilayers calculated by means of a kinematic model and experimental data on millimetric or nanometric films", D. Lebouvier, J.C. Pivin, H.M. Pollock, submitted to Thin Solid Films.

4a. Study of the correlation between hardness and structure of N implanted Ti surfaces", J.C. Pivin, F. Pons, J. Takadoum, H.M. Pollock, G. Farges, J. Mat.Sci. $\underline{22}$ (1987) 1087.

4b. "Inhibition of tribo-oxidation preceding wear by single-phased TiN$_x$ films formed by ion implantation into TiAl$_6$V$_4$",F. Pons, J.C. Pivin, G. Farges, J. Mat. Res. (Sept. 1987) and companion paper on wear of Ti and TiAl$_6$V$_4$ not yet published.

5a. "The mechanical properties of B and P implanted Ni discussed in terms of increasing disorder and amorphicity", J. Takadoum, J.C. Pivin, H.M. Pollock, J.D.J. Ross, H. Bernas, Nucl. Instrum. Methods $\underline{B18}$ (1987) 153.

5b. "Surface hardness of trinickel boride: the effect of indent size, disorder and amorphisation", J.C. Pivin, H.M. Pollock, in" Tribology - Fifty Years On", Institution of Mechanical Engineers, London (1987) 179.

6. C. Cohen, A.V. Drigo, H. Bernas, J. Chaumont, K. Krolas, L. Thome, M. Berti, A. Benyagoub, Phys. Rev. Lett. $\underline{48}$ (1982) 1193 and Phys. Rev. $\underline{B31}$ (1985) 5.

7a. " Channeling study of amorphous phase formation in Ni$_3$B by ion implantation", A. Benyagoub, J.C. Pivin, F. Pons, L. Thomé, Phys. Rev. $\underline{B34}$ (1986) 4464.

7b. "Mechanism of ion induced amorphization", A.V. Drigo, M. Berti, A. Benyagoub, H. Bernas, J.C. Pivin, F. Pons, L. Thomé, C. Cohen, Proceed. IBMM 1986, Nucl. Instrum. Methods $\underline{B19-20}$ (1987) 435.

TABLE 2

sample	normalized hardness	sample	normalised hardness
Ni	reference HV=2.20±0.05 GPa	Ti	reference HV=2.50±0.05GPa
Ni-0.5%B	1.6±0.2	Ti-2%N	3.7+0.5
Ni-2.5%B	3.2 "	Ti-10%N	4.2 "
Ni-8%B	3.7 "	Ti-20%N	5.0 "
Ni-16%B	4.5 "	Ti-25%N	5.6 "
Ni-20 to 25%B	4.7 "	Ti-34%N	6.5 "
" relaxed	5.9 "	Ti-44%N	7.4 "
" crystallized	5.3 "	TiN P.V.D.	
Ni-25%B chemical amorphous-submicron	4.9 "	submicron Vickers	21 " 16±2
" -Vickers	3.3 "	TiN (100)	
Ni-25%B chemical crystallized-Vickers	4.9 "	submicron Vickers	13±0.5 8±1
Ni$_3$B (001)-submicron	9±1		
" -Vickers	5±0.5		

LASER PROCESSING

LASER ETCHING AS AN ALTERNATIVE

R. W. DREYFUS and ROGER KELLY

IBM RESEARCH DIVISION, T. J. WATSON RESEARCH CENTER, YORKTOWN HEIGHTS, NY, 10598 U.S.A.

1. INTRODUCTION

Atoms and molecules are removed from surfaces by intense laser beams. This fact has been known almost since the discovery of the laser. Furthermore, laser etching* appears in the literature and technology under many different names (Table 1). While the names in Table 1 often emphasize the desired (or undesired) end product, it is quite evident that similar physical phenomena must underlie these various categories. Within the present overall area of interest, namely understanding ion-beam-induced sputtering, it is equally important both to contrast laser etching to ion sputtering and to understand the underlying physics taking place during laser etching. Beyond some initial broad observations, our specific discussion will be limited to, and aimed at, two areas: (i) short wavelength, UV, laser-pulse effects and (ii) energy fluences sufficiently small that only monolayers (and not microns) of material are removed per pulse. Furthermore, as will be subsequently evident, many of our measurement techniques utilize a second, so-called probe, laser.

2. CONTRASTING LASER ETCHING TO ION SPUTTERING

In very general terms, the comparison of laser etching with ion sputtering centers upon the mechanism of energy deposition and the pathways of energy flow which culminate in surface-atom removal. We shall divide the discussion into the following areas: the density of deposited energy, the mechanism of energy migration, phenomena leading to the freeing of surface atoms, and the perturbing influence of near-surface collisions.

First, note that the densities of locally deposited energy are quite similar regardless of whether one deals with incoming ions or photon pulses. Indeed, one finds that both the deposited energy density (in units of $eV/\text{Å}^3$ per ion or pulse) and the surface energy fluence (in units of J/cm^2 per ion or pulse) have comparable magnitudes, though with the photon-pulse surface energy fluences somewhat higher. Consider, for example, that an incoming ion has an energy of 10 or 100 keV and interacts with a surface area of ~ 10^{-13} cm^2 (Fig. 1) [1,2]. As summarized in Table 2, one then has 0.016 or 0.16 J/cm^2. For photon pulses in the UV range, one requires from ~0.1 J/cm^2 (for polymers) to ~1 J/cm^2 (for inorganic crystals) to produce laser etching.

---

*For present purposes, the term sputtering is defined as ion-induced surface ablation, while laser-induced processes will be referred to as laser etching. We will avoid the term photoablation, which is sometimes, but not always, defined as laser etching by photochemical decomposition.

R. Kelly and M. Fernanda da Silva (eds.), Materials Modification by High-fluence Ion Beams, 559–579.
© 1989 by Kluwer Academic Publishers.

TABLE 1.   AREAS WHICH OVERLAP LASER ETCHING

1. Laser marking
2. Laser damage to optical components
3. Laser welding
4. Fusion pellet implosion
5. (Multiphoton) photochemistry
6. Laser surgery
7. Laser phototherapy for malignancies
8. Desorption induced by electronic processes (DIET)
9. Synchrotron rings for VUV or soft X-rays
10. Laser ablation for mass-spectrometric analysis (LAMA)

TABLE 2.   Comparison of deposited energy for ions and for photon pulses.   The evaluations are made either for individual ions or for individual pulses.

Description of ions or photon pulses	Mean range[a] (Å)	Track area[b] (Å^2)	eV/Å^3	eV/atom[c]	J/cm^2
10 keV Ar$^+$→Si	124	~1000	0.08	1.3	0.016
100 keV Ar$^+$→Si	991	~1000	0.10	1.6	0.16
10 keV Kr$^+$→Ge	62	~1000	0.16	2.5	0.016
100 keV Kr$^+$→Ge	357	~1000	0.28	4.4	0.16
0.1 J/cm^2 photon pulses	~1000	...	0.06	1.0	0.1
1 J/cm^2 photon pulses	~1000	...	0.63	9.8	1

[a] The mean ion ranges are from [36].   The mean "photon ranges" are taken as twice the plasma skin depth (Eq. 2).

[b] See Refs.[1, 2, 37] and Fig. 1.

[c] For an atom spacing of 2.5Å

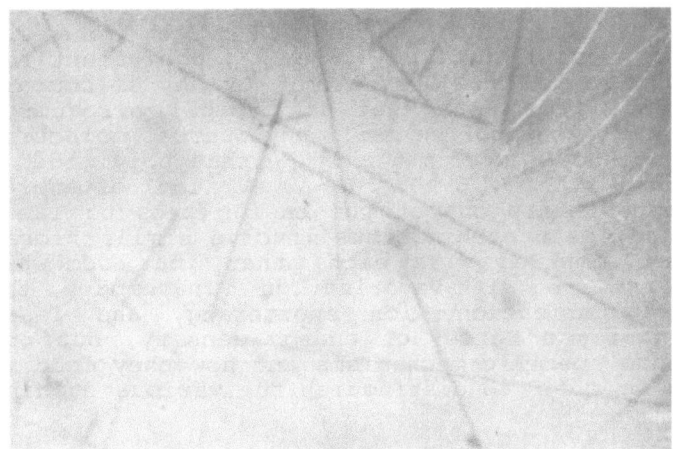

A

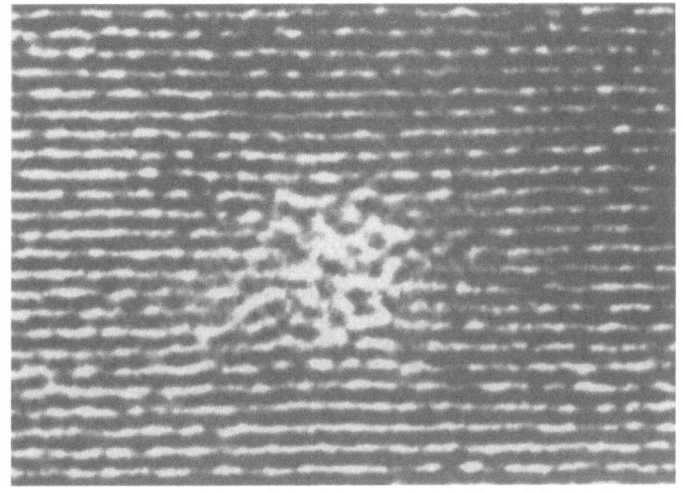

B

Figure 1A. Transmission electron micrograph taken at 100 keV of a mica cleavage which was irradiated with neutrons while in contact with U-coated Al. Fission tracks are resolved which vary in width and length, the length being determined mainly by the thickness of the mica cleavage. Such results serve to define the area of target which is disrupted by a single ion as used, for example, in Table 2. x30,000. Due to Silk and Barnes [1].

Figure 1B. Transmission electron micrograph taken at 80 keV of a (20Ī) platinum phthalocyanine single crystal (lattice spacing 11.94 Å) which was irradiated with neutrons while in contact with a thin U foil. The view is directly along a fission track and the disruption of the atomic planes into an essentially amorphous structure is seen. x1,000,000. Due to Bowden and Chadderton [2].

Similarly, again as summarized in Table 2, both ions and photon pulses deposit roughly 0.06 to 0.6 eV/$\AA^3$, equivalent for an atom spacing of 2.5 $\AA$ to 1 to 10 eV/atom. By way of comparison, it is interesting to note that it typically requires ~ 0.1 to ~ 1eV per entity to vaporize small physisorbed molecules or atoms, while it is known experimentally [3] that polyimide (a polymer) can be etched with ~ 0.6 eV/$\AA^3$ (~10 eV/atom) of energy at 193 nm. This material is one of the easier ones to laser etch. Sputtering and laser etching thus involve similar local energy densities as compared with each other, but somewhat higher densities as compared with vaporization. In summary, the difference between vaporization, ion sputtering, and laser etching is not primarily a matter of energy density, but one must look further into specific mechanisms and how they lead to energy localization in order to distinguish the various etching processes.

In the simplest case, localization of thermal energy, which leads to ordinary vaporization, is simply a statistical process, i.e. a random accumulation of high local energies which may lead to a given atom leaving a surface. Going beyond statistical chance, complete instantaneous vaporization is achieved only at the critical temperature, which may be 60 to 100% greater than the boiling point at 1 atmosphere pressure [4].

Localization of energy in the case of ion sputtering is via one of two processes [5,6]. At low to medium energies, $\leq$ 100 keV, direct momentum transfer through a few "hard" collisions gives a particular surface atom sufficient energy to leave the surface (Fig. 2). Calculations and subtleties related to these lower energy processes are discussed extensively in the adjoining papers [7]. Fig. 2 also reminds us that it is unclear to what extent "hot spots" play a role in atom ejection by ions. At the highest ion energies (~ > 1 MeV), the sputtering is primarily a result of electronic excitation since momentum transfer effects decrease rapidly in going to such energies.

Looking at this electronic energy deposition brings us into the regime of laser-pulse etching effects, as the latter may even be considered as being initially a "pure" form of electronic excitation, though again with an uncertainty as to whether "hot spots" ultimately play a role (Fig. 2). By "pure" we are emphasizing that the laser photons are essentially monochromatic, whereas the electromagnetic fields associated with a moving charge have components from zero energy up to the energy of the particle [8]. The "purest" form of energy-induced desorption results from using monochromatic VUV radiation from a synchrotron [9]. In this case, the photons act singly, usually producing specific inner-shell electronic transitions. This contrasts with laser etching, where a threshold fluence level must be reached before efficient etching ensues, that is to say where the photons act cooperatively [3,10]. In other words, UV laser etching requires several photons within an interactive area in order to produce efficient etching; true multiphoton effects are a special case of this threshold-type behavior. As a result, only order of magnitude figures for the material removal probability are available, as their magnitude is a function of laser energy fluence. The complexities of laser etching

2

LOW-ENERGY IONS
(MOMENTUM)
SECONDARIES
ATOM EJECTION
?
HEAT
?
?
U.V. LASER PULSES
(ELECTRONIC EXCITATION)
BROKEN BONDS
CHEMICAL

3

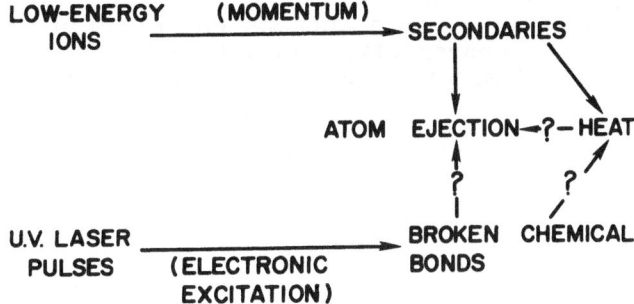

U.V. LASER ETCHING
MORPHOLOGY; PHOTO-SENSITIVE SITES
e⁻ (PLASMA)
LASER PULSE
HEAT
ATOMS AND MOLECULES
PLASMONS; CHEMISTRY
TARGET

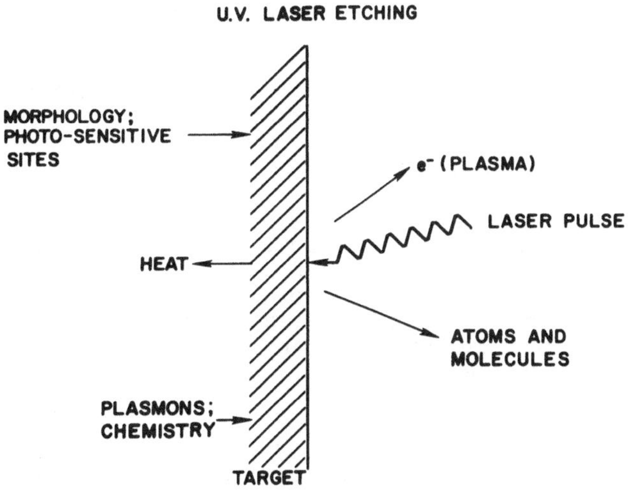

Figure 2. Schematic illustration of the energy flow pathways in ion sputtering and laser etching. Conceptual areas which are currently controversial are indicated by question marks, "?".

Figure 3. Material and photon interactions relevant to laser etching.

extend beyond this specific multiphoton effect, however. There are actually a host of non-linear interactions which influence laser-etching efficiency or, conversely, the interpretation of laser-etching experiments. Some of these interactions are discussed below.

3. BASIC FEATURES AND UNANSWERED QUESTIONS RELATING TO LASER ETCHING

The complicated interactions implicit in laser etching occur at several different locations. Fig. 3 schematically shows a number of these interactions, while examples are also listed in Table 3. In a small number of cases, the interaction can be quantitatively evaluated, as is done below for the photon-plasma interaction or in Ref. [11-13]. In some additional cases, empirical data provide understanding of the physics: see Ref. [13] for observations about plasma-particle interactions. Primarily, however, the understanding of the basic physics is improved by minimizing the etch-product densities and hence minimizing the cross-product interactions. In the present work, we study etching when only ~ 1 monolayer of substrate is removed per laser pulse; hence, the particle-photon and particle-particle interactions are minimized [11,12]. This monolayer etch range may be near or beyond the lower limits of many of the areas listed in Table 1; nevertheless, we are in a regime where the basic physics is discernible. Before discussing how both theoretical evaluations and minimal material removal lead to simplifications of the data, we first discuss typical pulsed-laser characteristics.

While there exist numerous types of commercial lasers, the majority of the etching experiments have utilized one of the five types listed in Table 4. Each laser type has specific features or advantages as noted. For present purposes, the important feature is the wavelength, λ, or, equivalently, the photon energy, $h\nu$. First we show how these are important in the photon-plasma interaction.

The non-resonant interaction of light with an ionized gas is strongest at wavelengths longer than the plasma wavelength, λ_{pe} [13,14]. The plasma frequency, ν_{pe}, is given by

$$\nu_{pe} = (n_e e^2/\pi m)^{1/2} \approx 9 \times 10^3 n_e^{1/2} \text{ (Hz)} \qquad (1)$$

where n_e (cm^{-3}) is the density of free electrons (charge $\underline{e}$, mass $\underline{m}$). If L_{pe} is the plasma skin depth, then for free space we have:

$$2\pi L_{pe} = \lambda_{pe} = 10^7/3n_e^{1/2} \text{ (cm)}. \qquad (2)$$

These formulas are used to construct Table 5. The effect of electrons (and ions) in the etch plume on the laser beam can be seen by supposing that the etch laser removes 0.1 monolayers per nanosecond. Since the plume of etched species departs from the surface with a velocity of ~ 5x10^5 cm/s, the initial plume density may be up to 2x10^{17} atom/cm^3. If a significant fraction of these atoms is ionized [12], then the CO_2 laser beam at 10 μm will be prohibited from striking the target surface. If one

TABLE 3. INTERACTIONS WHICH OCCUR DURING LASER ETCHING.

1. Laser beam attenuated by etch products within the plume
2. Laser beam attenuated by plasma above surface
3. Laser absorption influenced by temporarily or permanently altered surface morphology
4. Laser absorption altered by surface plasmon waves
5. Temperature-dependent thermal conductivity and optical absorption
6. Surface stoichiometry effects on surface chemistry, optical absorption, and thermal conductivity

TABLE 4. LASERS COMMONLY USED FOR LASER ETCHING

TYPE	λ (μm)	$h\nu$ (eV)	Pulse length (ns)	Characteristic
CO_2	10.6	0.12	$\geq$ 600	efficient
Nd:YAG	1.06	1.2	~5	reliable short pulses
XeF* excimer	0.35	3.5	~15 ns to <0.1 ps	intermediate energy photons
ArF* excimer	0.193	6.4	~15	high energy photons
Dye	>0.34	<3.6	~10 ns to <0.1 ps	tuneable

Table 5.   Plasma-photon interactions, including cross
sections

Plasma density, n_e (cm^{-3})	Equivalent pressure	Plasma wave-length, λ_{pe}	Plasma skin depth, L_{pe}	Cross section = $1/L_{pe}n_e$ [a) (cm^2)
3×10^{16}	1 torr	190 μm	30 μm	1×10^{-14}
2×10^{19}	1 atm	7.5 μm	1.2 μm	4×10^{-16}
5×10^{22}	metallic density	1500Å	240Å	8×10^{-18}

[a) By way of comparison, we list three related cross sections: to ionize a free atom (0.1 to 3 x $10^{-17} cm^2$) ; to excite a color center in a solid (1 to 5 x $10^{-16} cm^2$) ; to excite an atomic line which has its Doppler width (0.1 to 1 x $10^{-11} cm^2$) .

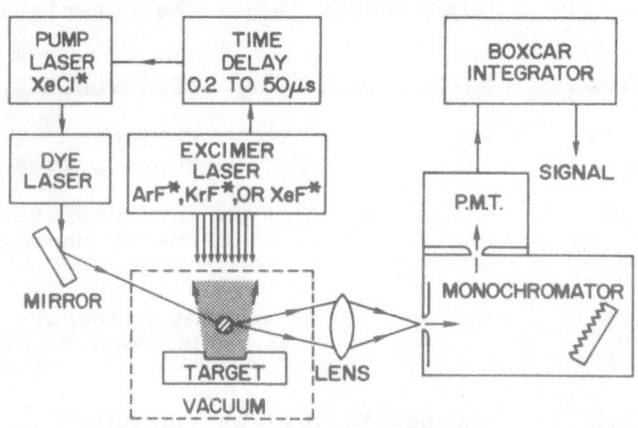

Figure 4.   Schematic illustration of the laser-induced fluorescence technique, LIF, for observing species in the laser etch plume.   Note the time delay which allows the etch products to rise up into the volume probed by the narrow-band dye laser. Due to Dreyfus et al. [18].

increases the etch flux to the level where one micron is etched per pulse and simultaneously ionized, then $n_e \approx 8 \times 10^{20}$ cm^{-3} , and even the YAG laser beam at 1.06 μm will be attenuated. Obviously, one produces a rapidly oscillatory efficiency of etching as the etch product temporarily blocks further etching [13]. Overall, experiments with short-wavelength excimer lasers are preferred, as plasma shielding of the surface is minimized.

Certain other cross sections are included in the footnote to Table 5. The cross section to ionize a free atom is typically 0.1 to 3×10^{-17} cm^2, i.e. much smaller than $\sigma = 4 \times 10^{-16}$ cm^2 for laser-induced plasmas at densities of 2×10^{19} cm^{-3}. Only when dealing with photons in the far UV is it relatively probable that the plasma shielding decreases to the level to permit single-electron absorption. This is one of the reasons for using UV etch lasers. Another major reason is that the photon energy is comparable to typical chemical bond energies and this enhances photochemical effects.

Referring again to the cross sections in Table 5, atom excitation can be expected quite generally.

The observation that hν is comparable to solid-state bond energies leads us into a discussion of one of our primary goals. Specifically, we wish to understand the fundamental mechanisms (pathways in Fig. 2) underlying laser etching. Surface atom removal by laser-induced ion bombardment (similar to ion sputtering) has been considered elsewhere and does not occur at the present near-threshold fluences [15,16]. Macroscopic surface removal mechanisms have also been reported such as hydrodynamical splashing or exfoliation [15] These mechanisms are also not relevant to the present work.

4. THE QUESTION OF PHOTOCHEMICAL ETCHING VS. VAPORIZATION

Probably the most important question in the present context is evaluating the relative importance of electron-excitation-induced photochemistry vs. thermally activated vaporization [3,15-19]. One of the basic approaches to investigating laser etching mechanisms is to ascertain the surface temperature, T_s, during the etching process.

Our experimental approach is to measure the energy (primarily internal energy) distribution of diatomic etch products as they travel away from the surface. This distribution serves as a thermometer to measure surface temperatures, as the distribution [20,21] commonly has Boltzmann form. The distribution is measured by the laser-induced fluorescence (LIF) technique, which is schematically shown in Fig. 4 [18] and has been discussed in detail elsewhere [22]. We recognize that the signal is derived from species in their ground electronic state, and furthermore that the individual rotational/vibrational states are resolved by the narrow wavelength spread of the excitation laser. Validation of this post-ablation (time delay ~0.1 to 50 μs) technique to determine T_s is a primary step. Earlier, LIF was used to measure Na$_2$ temperatures above liquid sodium [23]. While this latter experiment gave temperatures very close to the expected peak surface temperatures, the experimental conditions were not so extreme as occur here, as T_s was only somewhat higher than ambient.

4.1. Graphite

One of our first sets of experiments was aimed at measuring T_s for C_2 molecules above a laser-heated graphite surface. Scanning electron micrographs indicated [18] that the laser fluence was sufficiently low that surface melting did not occur, whence $T < T_m \approx 4600K$ [24]. Nevertheless, the laser etching of graphite would be expected to be a case of thermally activated vaporization. Some of the reasons are: graphite is essentially metallic so the electron-phonon relaxation time is fast, presumably femtoseconds; also, the optical absorption is primarily via the electron plasma in this metal-like solid, so high-energy ($\sim > 1eV$) electrons are not present in the solid.

The expectation that the laser etching of graphite is the result of vaporization is strongly reinforced by the LIF results. In this case, the Mulliken $^1\Sigma_u^+ \leftarrow {}^1\Sigma_g^+$ bands (near 230 nm) are used for the excitation and fluorescence transitions (Fig. 5) [20]. The internal energies of the C_2 molecules indicate a temperature of 3600 ± 100 K (Figs. 6 and 7) [20]. Vapor-pressure data indicate that a temperature of 4000 ± 300 K would produce the observed rates of material removal. One thus sees that both the directly estimated surface temperature and the temperatures deduced from LIF experiments are the same within experimental uncertainty. The conclusion is that the LIF-determined internal energies constitute a valid means of determining surface temperatures with graphite.

4.2. LIF measurements with Al_2O_3

The next step is to consider the results of applying the LIF technique to systems where the etching mechanism cannot be anticipated a priori. The system under consideration is sapphire, Al_2O_3, which is known to etch at a surprisingly low fluence (~ 0.6 J/cm^2) with a 193 nm ArF* laser. Fig. 8 [18] shows the LIF spectrum from AlO diatomics ablated from Al_2O_3. These results show that the diatomic temperature is ~500 K and therefore, since the vaporization of Al_2O_3 at a significant rate requires >4000 K, that the AlO is not produced thermally. The implication is that Al_2O_3 etches due to the direct effects of chemical bond breaking, an electronic process (Fig. 2). While this appears to be a reasonable conclusion, there remain several points to check. First, the mechanism of photon absorption has not yet been identified, since bulk Al_2O_3 is transparent down to ~150 nm. The possibility that a multiphoton ($2h\nu$) transition may be responsible requires investigation. Second, one needs to verify that the few gas-phase collisions which occur after the AlO leaves the surface are not altering the population distribution among the various rotational/vibrational states and thereby giving an incorrect temperature indication. Third, it would be interesting to know something of the surface excited-state density which gives rise to the material expulsion.

5

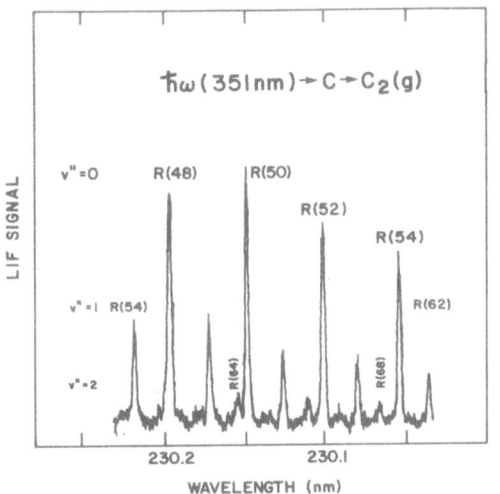

6

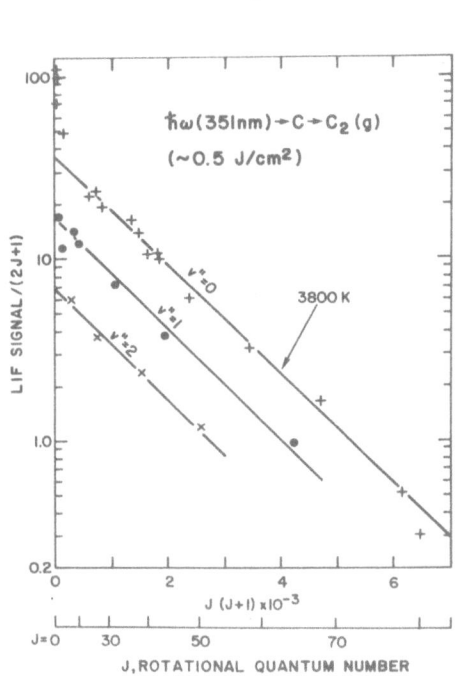

Figure 5. LIF spectrum of C_2 radicals from graphite as a function of probe wave-length using the Mulliken bands. Spectra from the $(v', v'') = (0,0)$, $(1,1)$ and $(2,2)$ vibrational manifolds overlap, as is evident. The rotational lines within each manifold are designated by the standard notation [21]. Due to Dreyfus et al. [20].

Figure 6. LIF signal from C_2 radicals (divided by ground-state degeneracy) plotted as a function of $J(J+1)$ derived from data such as those shown in Fig. 5. The $J(J+1)$ abscissa is proportional to the rotational energy of the J state. The straight-line fit to this semi-log plot indicates that the data follow a Boltzmann-type equation, and yields a temperature of 3800K. Similar to Fig. 4 of Dreyfus et al. [20].

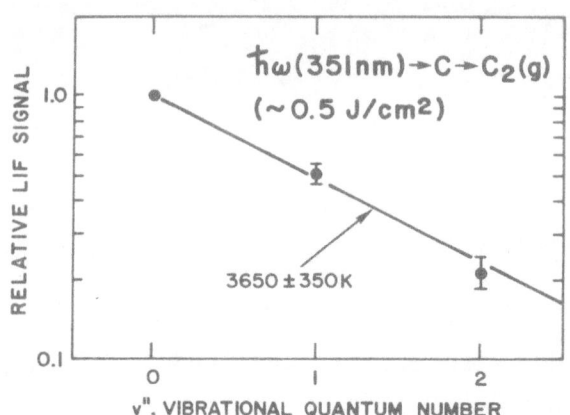

$$\hbar\omega(351nm)\rightarrow C\rightarrow C_2(g)$$
$$(\sim 0.5\ J/cm^2)$$

3650 ± 350K

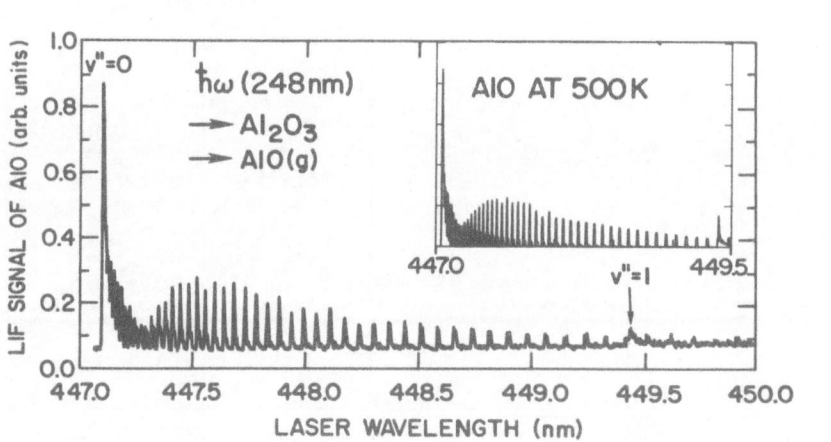

8

Figure 7. Similar to Fig. 6, except that here the signal is plotted vs. v'', which is proportional to the ground-state vibrational energy. The temperature obtained is 3650 ± 350 K. Due to Dreyfus et al. [20].

Figure 8. The LIF spectrum of AlO diatomics from Al_2O_3 etched in the near-threshold region, when each pulse removes ∼ 0.4nm. Bands for both (2←0) and (3←1) excitation are evident, whereas (4←2) was unobservably small. The multiple fine lines are primarily the higher rotational (J) states of the P branch of $v'' = 0$. A calculated AlO spectrum at 500K is shown in the inset and is noted to be a reasonable representation of the measured spectrum. The Al_2O_3 was therefore much colder than the graphite. Due to Dreyfus et al. [18].

4.3. Photothermal measurements with Al₂O₃

Pulsed photothermal deformation (PPTD) measurements were carried out in order to answer the above three questions. PPTD involves measuring the thermal expansion (i.e. deformation) of a surface to determine the energy deposition [25,26]. (Attempts to measure the transmitted and reflected intensities from laser-etched Al_2O_3 surfaces were not successful as only a small fraction of the incident energy is absorbed, and even that is primarily absorbed by the etch plume.) This situation is different from the case of ion sputtering where one has complete and known energy deposition in the near-surface region. The results of PPTD have been already discussed in detail elsewhere [27]. For present purposes we shall restrict our results to the graphical information displayed in Fig. 9 [26]. Since the slope in Fig. 9 is unity, the laser-energy is being absorbed primarily by a one-photon ($h\nu$) process, not a $2h\nu$ process. As detailed in Ref. 27, surface chemical treatment indicates that the absorbing species is not an intrinsic surface state, but occurs as a result of optical polishing. Second, the LIF and PPTD measurements indicate similar temperatures in the vicinity of the etching threshold; hence one concludes that LIF correctly measures surface temperatures. That is to say, the approximately 1 to 20 gas-phase collisions that an AlO diatomic undergoes in the etch plume have not altered the internal-energy distribution and hence the deduced surface temperature. Lastly, the approximately 3% absorption of the Al_2O_3 surfaces can be used to derive a rough estimate of the upper limit for the surface electron-hole density [27]. The result is that this density is ≤ 3 e-h pairs per surface anion. This appears to be a reasonable density range to produce significant photochemical effects leading eventually to material removal. In fact, a density of only $\sim$ 0.3 e-h pairs per surface anion could easily cause a rebonded surface. Furthermore, if we presume that the optical absorption cross section is 4×10^{-16} cm^2 (Table 5), the above 3% absorption corresponds to 10^{14} color centers/cm^2. This number may be taken as an illustration of how electronic energy can be absorbed at one type of site yet disperse into the surrounding area so as to cause extensive decomposition.

5. THE QUESTION OF NORMAL VS. PERTURBED VAPORIZATION

We have discussed data for laser-bombarded graphite and emphasized that certain temperatures lay in the interval 4000 ± 500 K: that inferred from the lack of melting of the graphite, that inferred from the depth of etching interpreted in terms of vaporization, and the internal, i.e. rotational and vibrational, temperatures of C_2 The picture is not that simple, however. As seen in Fig. 10 [20], the time-of-flight (TOF) temperatures of the C_2 can exceed 4000 ± 500 K by a factor of 2 to 3.

A similar discrepancy was noted as long ago as 1978 for the desorption of D_2 from W (Fig. 11) [28], where in addition we see that TOF temperatures measured obliquely to the surface can be too low.

These are relatively subtle effects. Quite unsubtle are the extreme extents of forward peaking found in most laser-etching or equivalent experiments (Figs. 12 and 13) [29,30].

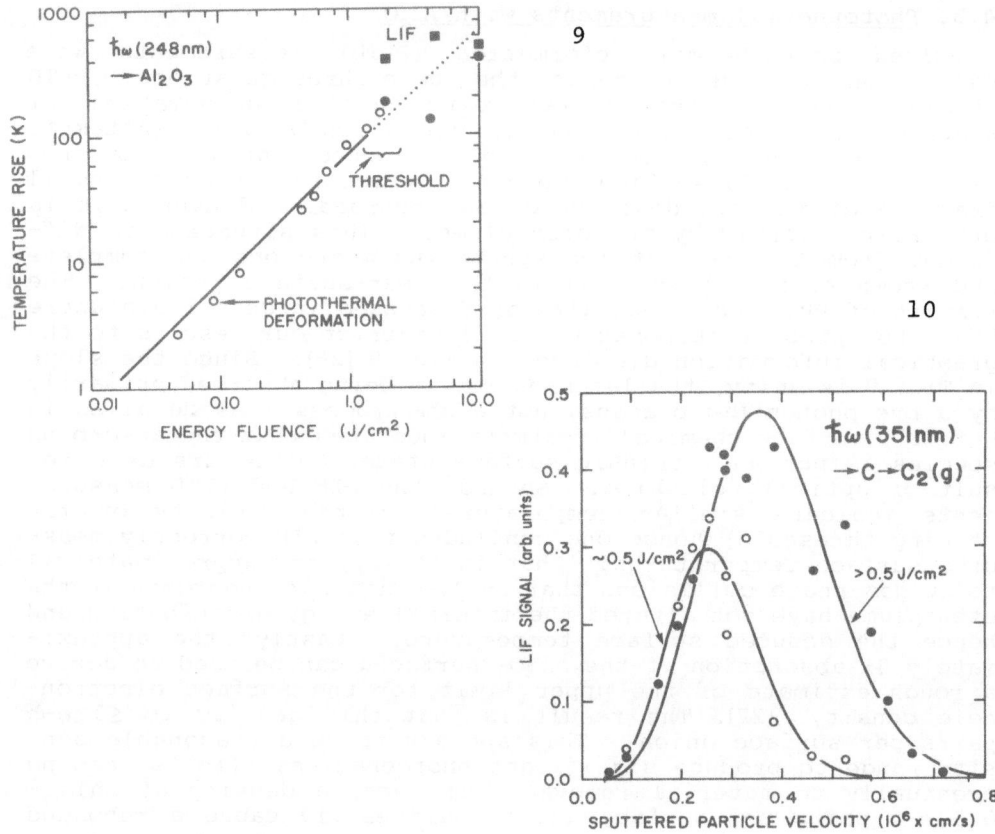

Figure 9.   Surface temperature rise for Al_2O_3 surfaces being etched by 248nm laser pulses.   In the below-threshold region, PPTD ("pulsed photo-thermal deformation") measurements serve as the thermometer (points "O"), while above the threshold, LIF measurements as in Fig. 8 serve the same purpose (points "●" for T_{vib} and "■" for T_{rot}).   Note that the measurements complement each other in that extrapolation through the threshold region does not produce a noticeable discontinuity.   Similar to Fig. 2 of von Gutfeld et al. [26].

Figure 10.   LIF signal from C_2 radicals plotted as a function of velocity for two slightly different fluences as determined by TOF ("time of flight") using the signal of the bandhead of the Mulliken bands.   The solid curves correspond to Boltzmann distributions with, respectively, T = 4600 K for the lower fluence and T = 9100 K for the higher fluence.   These temperatures are derived by the conventional analysis, based on Eq. (3), according to which the maximum LIF signal appears at $\hat{v} = 2(kT/m)^{1/2}$.   They are higher than the surface temperature for reasons which include near-surface collisions, i.e. Knudsen-layer formation.   Similar to Fig. 5 of Dreyfus et al. [20].

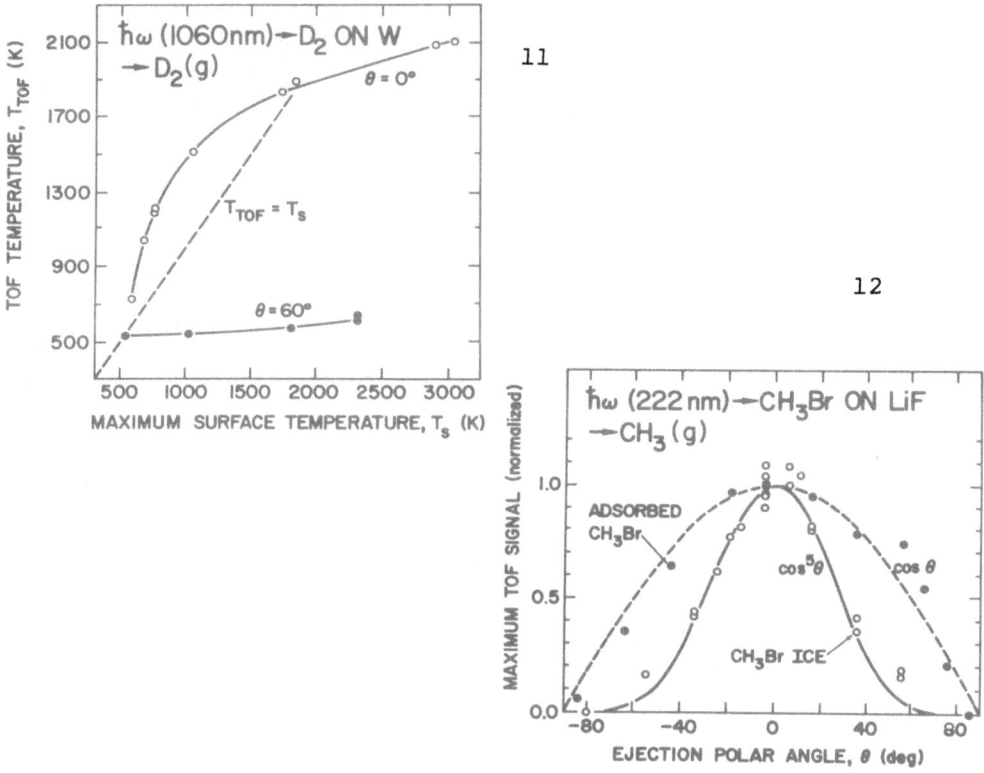

Figure 11. TOF temperature versus the maximum surface temperature for laser-pulse desorption of 0.8 monolayers of D_2 from D_2 adsorbed on W. The adsorption was carried out near room temperature by exposing the W for 20 s to 1.0×10^{-5} Pa of D_2. The pulses had a length of 30 ns. The significant aspects in the present context are that the TOF temperatures are apparently too high (i.e. lie above the dashed line) for normal emission and apparently too low (i.e. lie below the dashed line) for oblique emission. (The fact that the temperatures for normal emission are low for the highest values of T_s is due to source exhaustion and, as such, is not mechanistically significant.) After Cowin et al. [28].

Figure 12. TOF yields versus the ejection polar angle for laser-pulse desorption of CH_3 from CH_3Br adsorbed on LiF. The quantity removed per pulse in the case of the points labeled "adsorbed CH_3Br" can be inferred to be substantially less than a monolayer, while that removed for the points labeled "CH_3Br" ice was probably of order one monolayer. (These numbers are based on the quoted coverages, but the latter are possibly underestimates owing to the use of ion-gauge readings.) The CH_3Br was adsorbed at 110K. The length of the pulses may be assumed to have been 10-20 ns, as is appropriate to an excimer laser. The significant aspect is the pronounced forward peaking which sets in when a quantity similar to one monolayer is removed per pulse. After Bourdon et al. [29]

What is important here is that results such as those of Figs. 10 to 13 have led to postulating the occurrence of deviant forms of thermally activated vaporization. If this scenario were correct it would mean that laser etching, instead of involving a simple decision between photochemistry and vaporization, was a much more complicated process. The deviant forms include:

(i) Vaporization of species which must surmount an <u>activation barrier</u> at the surface, an effect which would explain high TOF temperatures as in Figs. 10 and 11.

(ii) Vaporization of species which do not <u>librate</u> freely (Fig. 12).

(iii) Vaporization as a <u>one-phonon</u> rather than statistical process (Fig. 13). We note here that Fig. 13 relates to the use of <u>phonon</u> not <u>photon</u> pulses but that this distinction is unimportant to the argument.

5.1 Knudsen-layer formation

We have referred several times to the occurrence of near-surface gas-phase collisions, e.g. in posing the question as to whether laser-etched AlO suffers sufficient collisions to give false temperature information (Sect. 4.2). It is now known that as few as 3 collisions per emitted particle cause a major change in the velocity distribution function of etched particles [31,32]. Assuming truly thermal emission, particles leaving a surface with temperature T_S will have a "half-range" Maxwellian:

$$f_S^+ \propto \exp(-m[v_x^2 + v_y^2 + v_z^2]/2kT_S) ; \qquad (3)$$
$$v_x \geq 0 .$$

Eq. (3) clearly implies a center-of-mass motion since we have $v_x \geq 0$. It is physically reasonable that, whenever sufficient collisions occur, a "full-range" Maxwellian will develop but at the same time the system will tend to conserve its initial center-of-mass motion. These two requirements are satisfied if the distribution function evolves from Eq. (3) to a form having a formal center-of-mass velocity, u_K :

$$f_K \propto \exp(-m[(v_x - u_K)^2 + v_y^2 + v_z^2]/2kT_K) ; \qquad (4)$$
$$-\infty \leq v_x \leq \infty .$$

Such a change as that embodied in the evolution of Eq. (3) to (4) is termed "Knudsen-layer formation" [33,34]. Without going into detail [11,35] it leads to the following results:

(i) The temperature of the emitted particles falls from T_S to T_K, with T_K/T_S equal to 0.669 for an atom, 0.782 for a diatomic with 2 internal degrees of freedom ("j = 2"), and 0.837 for a diatomic with j = 4.

(ii) The emitted particles acquire a formal center-of-mass velocity u_K , equal to the velocity of sound at $T = T_K$, namely $u_K = (\gamma k T_K/m)^{1/2}$.

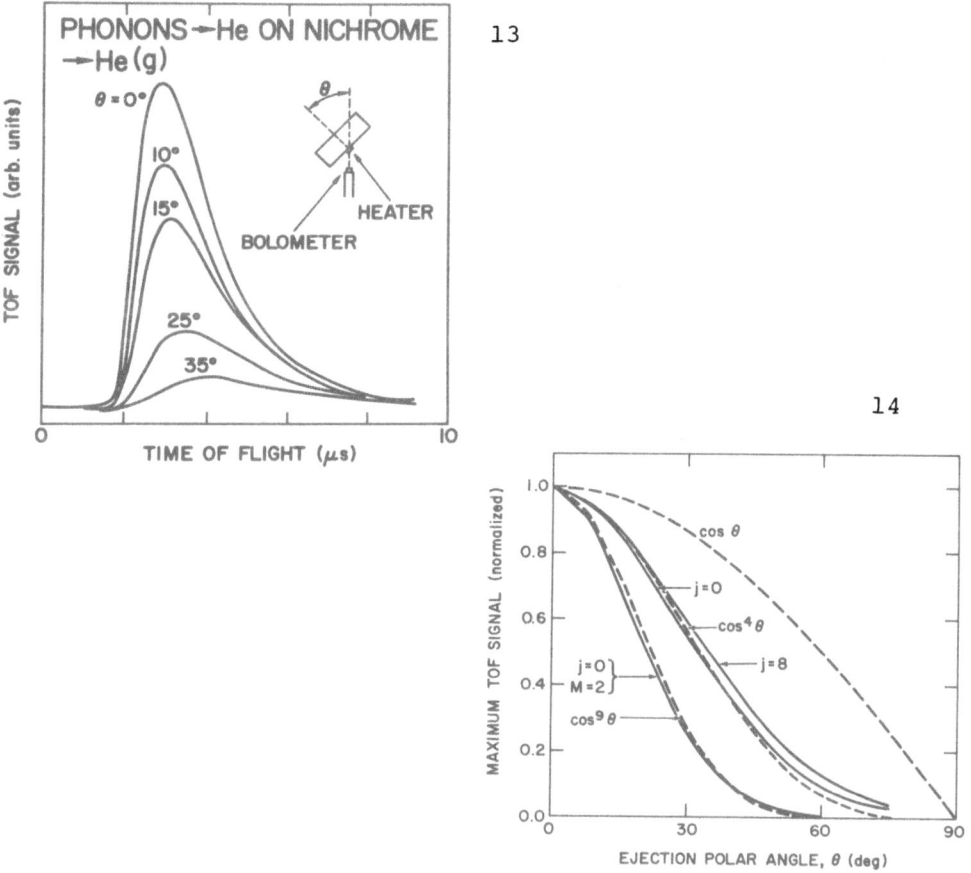

Figure 13. TOF spectra for <u>phonon</u>-pulse desorption of about 2 monolayers [38] of He from a nichrome surface. The adsorption was carried out at 2 K by exposing the nichrome for an unspecified time to 4×10^{-5} Pa of He. The pulses had a length of 100 ns. The significant aspects in the present context are that the TOF temperatures are about a factor-of-two too high at $\theta = 0°$ (namely 17 K) and about correct at $\theta = 35°$ (namely 9 K). After Taborek [30].

Figure 14. Comparison of maximum TOF signals as would be obtained with a small, density-sensitive detector located at various angles θ from a small target. For collisionless emission (when Eq. (3) is valid) the particles should show an angular dependence of the type $\cos\theta$. If sufficient collisions occur per particle for a Knudsen layer to form, then the angular dependence should be approximately of the type $\cos^4\theta$ for any value of j. ("j" is the number of internal degrees of freedom). If the number of collisions is sufficient for the onset of adiabatic expansion beyond the Knudsen layer, then the forward peaking is even more pronounced, for example $\sim \cos^9\theta$ for $u_K = Mu_{sound} = 2u_{sound}$. ("M" is called the Mach number.) After Kelly and Dreyfus [35].

(iii) Recondensation on the surface occurs to an extent 0.18 for atoms, 0.21 for a diatomic with j = 2, and 0.22 for a diatomic with j = 4.

5.2 Reconsidering the examples of apparently deviant vaporization

The formation of a Knudsen layer has important, even severe, implications to TOF experiments as in Figs. 10 to 13. Owing to the different forms of Eqs. (3) and (4), the TOF spectra of normally emitted particles are shifted to shorter time (higher velocity) as if the temperature were excessive $(T >> T_s)$. Corresponding TOF spectra for obliquely emitted particles tend to be shifted to longer time (smaller velocity), although this effect does not become significant unless more collisions occur than are needed for Knudsen-layer formation. Again as a result of the difference between Eqs. (3) and (4), the angular distributions evolve from $\cos\theta$ to $\cos^4\theta$ and to even greater extents of forward peaking if additional collisions occur (Fig. 14) [35]. In principle, significant recondensation will occur, although this effect has never been sought experimentally.

Without going into detail, we find that in all cases the deviant forms of vaporization are only apparent and not real. This is important in that it means that laser etching is a basically simple process described by the ideas of Sect. 4 and does not show the complications suggested in much current work.

6. SUMMARY

The results on Al_2O_3 bring out many of the differences between laser etching and ion sputtering. As a subdivision of sputtering, direct collisional sputtering contrasts very strongly with laser etching. In the former case, the energy is deposited locally and a momentum transfer pathway leads either directly (or indirectly via "hot spots") to ejection of surface atoms (Fig. 2). In laser etching, only a small portion of the impinging photon flux is absorbed and the energy pathway leads via various electronic states to the freeing of surface species (Fig. 2). Whether the etching is by a photochemical (electronic) mechanism or thermal mechanism, high energy densities are required and one has the appearance of an apparent "threshold" for etching. As noted before, a local interaction, e.g. electronic sputtering, can also produce such high energy densities per pulse. Having reached the threshold, the various interaction pathways shown in Fig. 3 may be activated. On the other hand, by etching near threshold, removing ≤ 1 monolayer/pulse and limiting the cross interactions, probing with a second laser yields useful information (Fig. 4). Primary among this information is the fact that vaporization processes (as with graphite) can be distinguished from direct photochemical etching (as with Al_2O_3) due to bond-breaking or bond-weakening effects. These latter effects overlap phenomena observed with very high-energy ions and the laser case actually may be a more easily analyzed than the situation with ions.

We finally note that the mechanisms of laser sputtering are basically simple, normally either thermal or photochemical (electronic), and that the various complications proposed in contemporary work are often trivial results of near-surface collisions. This is because near-surface collisions cause physically incorrect temperatures to be inferred in TOF experiments (Fig. 11) and also lead to strong forward peaking (Figs. 12 and 14).

REFERENCES

[1] E.C.H. Silk and R. S. Barnes, Phil. Mag. 4 (1959) 970.

[2] F. P. Bowden and L. T. Chadderton, Nature 192 (1961) 31.

[3] R. Srinivasan, B. Braren and R. W. Dreyfus, J. Applied Physics 61 (1987) 372.

[4] M. M. Martynyuk, Russ. J. Phys. Chem. 57 (1983) 494.

[5] R. Kelly, Rad. Effects 80 (1984) 273.

[6] R. Kelly, Ch. II of Ion Bombardment Modification of Surfaces, ed. by O. Auciello and R. Kelly (Elsevier, Amsterdam, 1984) p. 27.

[7] This Volume.

[8] W. Panofsky and M. Phillips, Classical Electricity and Magnetism (Addison-Wesley, Cambridge MA, 1955). Chapt. 18 and 19 and particularly Eq. (18-38).

[9] Desorption Induced by Electronic Transitions, ed. by M. L. Knotek and R. H. Stulen (Springer-Verlag, Berlin, 1987).

[10] R. Srinivasan, B. Braren, D. E. Seeger, and R. W. Dreyfus, Macro-molecules 19 (1986) 916.

[11] R. Kelly and R. W. Dreyfus, Surf. Sci. (in press).

[12] R. E. Walkup, J. M. Jasinski, and R. W. Dreyfus, Appl. Phys. Lett. 48 1986) 1690.

[13] M. von Allmen, Laser Beam Interactions with Materials: Physical Principles and Applications (Springer-Verlag, Berlin, 1987).

[14] D. L. Book, Ch. 18 of AIP 50th Anniversary Physics Vade Mecum, ed by H. L. Anderson (Am. Inst. of Phys., New York, NY, 1981) p. 260.

[15] R. Kelly, J. J. Cuomo, P. A. Leary, J. E. Rothenberg, B. E. Braren, and C. F. Aliotta, Nucl. Instr. Meth. B9 (1985) 329.

[16] R. W. Dreyfus, R. E. Walkup, and R. Kelly, Rad. Effects 99 (1986) 199.

[17] R. Srinivasan, B. Braren, R. W. Dreyfus, L. Hadel, and D. E. Seeger, J. Opt. Soc. 3A (1986) 785.

[18] R. W. Dreyfus, R. Kelly, R. E. Walkup and R. Srinivasan, Proc. Soc. Photo-Opt. Instr. Eng. 710 (1986) 46; and R. W. Dreyfus, R. Kelly, and R. E. Walkup, Appl. Phys. Lett. 49 (1986) 1478.

[19] R. W. Dreyfus, J. M. Jasinski, G. S. Selwyn, and R. E. Walkup, Laser Focus 22 (1986) 62.

[20] R. W. Dreyfus, R. Kelly, and R. E. Walkup, Nucl. Instr. Meth. B23 (1987) 557.

[21] G. Herzberg, Molecular Spectra and Molecular Structure I: Spectra of Diatomic Molecules (Van Nostrand, New York,NY, 1950). Useful compilations of spectroscopic data are: K. P. Huber and G. Herzberg, Molecular Spectra and Molecular Structure IV: Constants of Diatomic Molecules (Van Nostrand, New York, NY, 1979); R.W.B. Pearse and A. G. Gaydon, The Identification of Molecular Spectra (Chapman and Hall, New York, NY, 1976).

[22] See, for instance: R. W. Dreyfus, J. M. Jasinski, R. E. Walkup, and G. S. Selwyn, Pure and Appl. Chem. 28 (1987) 349.

[23] G. Miksch and H. Weber, Chem. Phys. Lett. 87 (1982) 544.

[24] A. Cezairliyan, Phys. Rev. Lett. 54 (1985) 1208.

[25] F. A. McDonald, R. J. von Gutfeld, and R. W. Dreyfus, Proc. IEEE Ultrasonics Symp. of 1986, p. 403.

[26] R. J. von Gutfeld, F. A. McDonald, and R. W. Dreyfus, Appl. Phys. Lett. 49 (1986) 1059 and 50 (1987) 1491.

[27] R. W. Dreyfus, F. A. McDonald, and R. J. von Gutfeld, J. Vac. Sci. Tech. B5 (1987) 1521.

[28] J. P. Cowin, D. J. Auerbach, C. Becker, and L. Wharton, Surf. Sci. 78 (1978) 545.

[29] E.B.D. Bourdon, P. Das, I. Harrison, J. C. Polanyi, J. Segner, C. D. Stanners, R. J. Williams, and P. A. Young, Faraday Disc. Chem. Soc. 82 (1986) 1.

[30] P. Taborek, Phys. Rev. Lett 48 (1982) 1737.

[31] I. Noorbatcha, R. R. Lucchese, and Y. Zeiri, J. Chem. Phys. 86 (1987) 5816.

[32] I. Noorbatcha, R. R. Lucchese, and Y. Zeiri, Surf. Sci.
      (in press).

[33] T. Ytrehus, in Rarefied Gas Dynamics, ed. by J. L. Potter
      (AIAA, New York, NY, 1977) p. 1197.

[34] C. Cercignani, in Rarefied Gas Dynamics, ed. by S. S. Fisher
      (AIAA, New York, NY, 1981) p. 305.

[35] R. Kelly and R. W. Dreyfus, Nucl. Instr. Meth. B (in press).

[36] J. F. Gibbons, W. S. Johnson, and S. W. Mylroie, Projected
      Range Statistics: Semiconductors and Related Materials
      (Dowden, Hutchinson, and Ross, Stroudsberg, PA, 1975).

[37] J.Bénit,J-P. Bibring, and F. Rocard, This Volume.

[38] M. Sinvani, P. Taborek, and D. Goodstein, Phys. Lett. 95A
      (1983) 59.

PULSED LASER IRRADIATION OF HEAVILY GE IMPLANTED SILICON

A. AYDINLI*, M. BERTI

University of Padova, Physics Department, Via Marzolo 8, 35131 Padova, Italy

A.V. DRIGO

University of Lecce, Physics Department, Via Arnesano, 73100 Lecce, Italy

P.G. MERLI

Istituto di Chimica e Tecnologia dei Materiali e Componenti per l'Elet-
tronica (La.M.El.) - C.N.R., Via Castagnoli 1, 40126 Bologna, Italy

1.INTRODUCTION

Pulsed laser irradiation of ion implantation-amorphized layers is a complex phenomenon with various mechanisms at work. It is by now well established (1) that the thermal parameters of the amorphized layers are different
from those of crystalline substrates. In particular, both the melting temperature, thermal conductivity and latent heat are lower for amorphized
layers. Furthermore, explosive crystallization feeding on the latent heat
released during recrystallization of very thin molten layers produced at
very low energy densities (2) and resolidification from the surface have
been proposed (3). It must be noted that in all these cases implanted species were not a significant fraction of the amorphous layer composition.

In this paper we report on the first phase of work aimed at characterization of strain and defect formation in mismatched layers produced by high
dose ion implantation and subsequent pulsed laser irradiation.

2.EXPERIMENT

Si(100) wafers were implanted at room temperature by Ge$^+$ ions to a dose
of 2.3E16/cm^2 and to a dose of 5.5E16/cm^2 at 150 keV. Pulsed laser irradiation was performed in air using a Q-switched ruby laser, wavelength 694 nm,
pulse length 20 ns. The laser energy density was varied between 0.5J/cm^2
and 2.0 J/cm^2. A diffuser glass plate was used to obtain a uniform laser
beam over a 4 mm diameter. It is estimated that temporal and spatial variations across the laser spot were about ±5%. A UV-VIS-IR spectrophotometer
was used to meure optical reflectivities at both doses and the results
were found to be 0.35 and 0.39 for the higher and lower doses respectively.
Channeling and TEM were used to assess structural changes in the layers.

* On leave from Hacettepe University, Physics Dept.,Beytepe, Ankara, Turkey.

R. Kelly and M. Fernanda da Silva (eds.), Materials Modification by High-fluence Ion Beams, 581–588.
© *1989 by Kluwer Academic Publishers.*

Impurity profiles were determined by Rutherford Backscattering Spectrometry using a 2 MeV He+ beam. Scattered particles were detected and energy analyzed by a surface barrier detector (energy resolution 14 keV) at a scattering angle of 160°. The depth of the a-c interface as well as the molten layer thickness and layer damage could be inferred by channeling and TEM. The thicknesses of the amorphous layers were found to be 250 nm and 280 nm for lower and higher doses respectively. For 150 keV implant energy, the RBS measured Ge distribution peaked at about a depth of 100 nm with a full width at half-maximum of about 150 nm and 140 nm for higher and lower doses respectively. The depth of the peaks are in agreement with theoretical predictions but widths of the distributions are large due to very heavy doses involved (4).

3.RESULTS
3.1.Damage

Fig. 1 shows the channeling spectra of as-implanted and laser-irradiated layers implanted with 2.3E16 Ge/cm^2. A random spectrum is also shown for comparison. Starting at 0.61J/cm^2 we find that a surface peak followed by a short plateau ends in a damage peak located very close to the a-c interface. While the surface peak and the plateau indicates crystallinity at the surface in the form of large grain polycrystals (5), the damage peak at the a-c interface is a clear sign that the underlying substrate was not molten. The existence of the a-c interface damage with a peak height almost half as that of the amorphous layer suggests that the melt front had reached this point. The two separate regions suggest also the existence of a columnar structure with grain sizes varying from the surface towards the a-c interface. With increasing laser energy densities both the surface peak and the damage peak at the a-c interface decrease in height while the plateau slightly widens. At 0.97J/cm^2 the a-c interface damage still persists although it is much reduced in height. At still higher energy densities shown in Fig. 2 we find that the damage peak at the a-c interface almost disappears at 1.16 J/cm^2 while the interface is evident. Finally, at 1.55 J/cm^2 we obtain the best channeled spectra indicating that not only the amorphous layer but also the underlying substrate was molten and good epitaxial regrowth has been obtained. Further increase of the laser energy density to 2.0J/cm^2 results in laser beam-induced damage. It is however interesting to note at this point that this damage is (i) confined to a thickness equal to the original amorphous layer thickness and (ii) more or less homogeneously distributed throughout this original layer.

Fig. 3 shows channeling spectra of as-implanted and laser-irradiated layers with 5.5E16 Ge/cm^2. A random spectrum is shown for comparison. While the general characteristics of the spectra are the same as before, the situation is drastically different with respect to the energy densities involved as compared with the lower dose.

Here we see that the damage at the a-c interface starts to disappear at as low as 0.71J/cm^2. In fact at 0.82J/cm^2 (not shown) no trace of this damage peak is observed. This value of energy is approximately 40% lower than

that required to obtain the same effect for the lower dose. At energy densities above 0.85J/cm² we find that the channeling yield starts slowly to
increase, Fig. 4. For example, at 1.16J/cm² (not shown) and 1.55J/cm² the
channeling yield is much higher than that at 0.85J/cm². It should be noted,
however, that while the channeling yield is increasing, the a-c interface

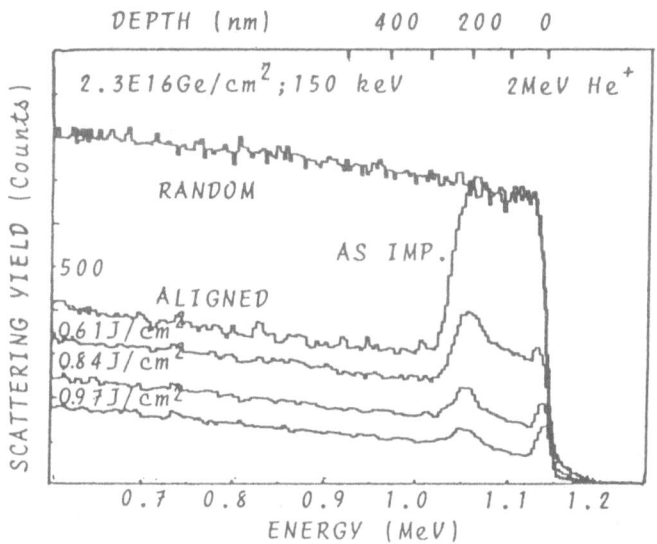

FIGURE 1. Random and <100> channeling spectra of as-implanted as well as
0.61J/cm², 0.84J/cm², 0.97J/cm² laser-irradiated silicon.

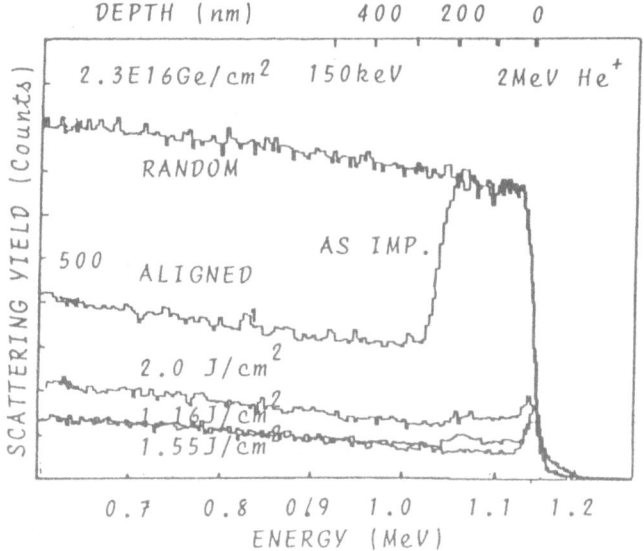

FIGURE 2. Random and <100> channeling spectra of as-implanted as well as
1.16J/cm², 1.55J/cm², 2.0J/cm² laser-irradiated silicon.

584

is not distinguishable. At a higher energy density of 2.0J/cm² laser-induced damage is clearly seen to be confined to the originally amorphous layer again. The dechanneling due to laser beam induced damage at this dose is higher when compared to the lower dose implant.

The ratios of the aligned to random spectra, χ, were calculated by integrating each spectrum over the thickness of the plateau excluding the in-

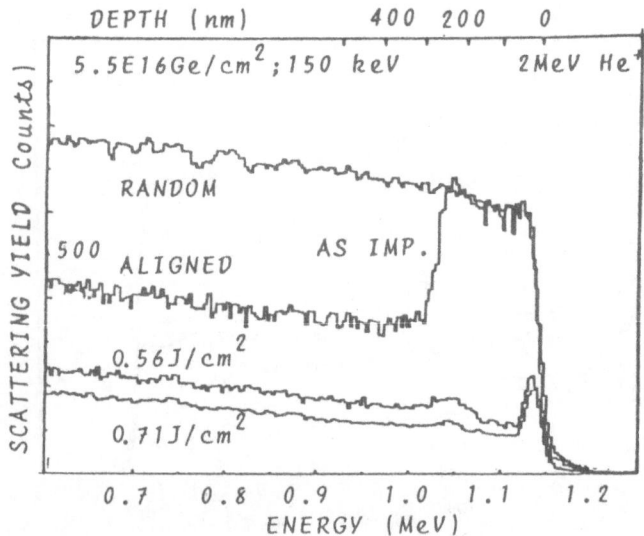

FIGURE 3. Random and <100> channeling spectra of as-implanted as well as 0.56J/cm², 0.71J/cm² laser-irradiated Si.

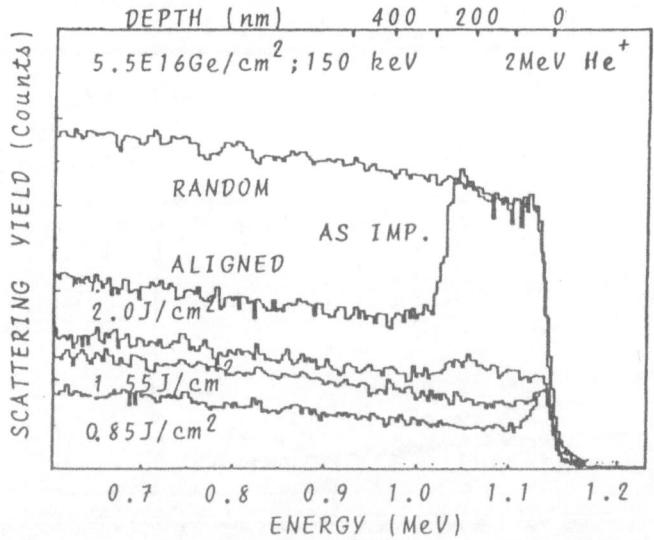

FIGURE 4. Random and <100> channeling spectra of as-implanted as well as 0.85J/cm², 1.55J/cm², 2.0J/cm² laser-irradiated silicon.

terface peak for Si and Ge separately. A plot of χ vs. laser energy density
is given in Fig. 5.

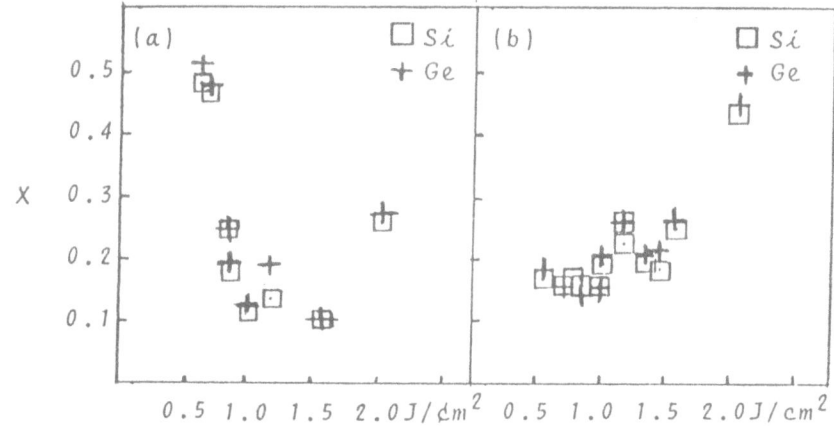

FIGURE 5. χ vs. laser energy density for   (a) 2.3E16 Ge/cm² and (b) 5.5E16
        Ge/cm².

For the lower dose as the laser energy density increases from 0.61J/cm²
to 1.55J/cm² χ drops from 0.47 to about 0.10. A window of energies contered
approximately around 1.35J/cm² (estimated) gives minimum dechanneling. At
lower densities the interface peak dominates and at higher energy densities
laser-induced damage sets in. Ge is seen to be above 95% substitutional. No
loss of Ge was found at any of the energy densities used. For the higher
dose, we find slightly higher χ values but shifted towards lower energy
densities. Ge is again more than 95% substitutional. We do, however, find a
15% loss of Ge at 1.55J/cm² and 25% at 2.0J/cm². This is clear indication
that vigorous evaporation takes place on the surface of higher dose samples
while the lower dose samples have negligible evaporation, if any, at these
energy densities.

3.2.Diffusion Profiles
In order to assess the depth of melting, Ge profiles were obtained for
all as-implanted and laser-irradiated samples. In Fig. 6 we present Ge profiles of lower dose implants irradiated between 0.61 and 1.55J/cm². As-
implanted profiles are also shown for comparison. In all cases the profiles
are wider than in the as-implanted sample, indicating that most of amorphous layer melted   at 0.61J/cm². Note also that no further significant
broadening of the profiles is observed at higher energy densities up to
1.55 J/cm² indicating (i) that the substrate was not molten and (ii) no
substantial change in melt thickness took place in this energy range. Unfortunately, we can't be precise about the exact depth of the molten layer
since unlike some other impurities such as Cu in Si Ge is not a good marker
of molten depth. Only at 1.55J/cm² we find that the Ge tail extends well
beyond the a-c interface. Here the substrate was molten for long durations

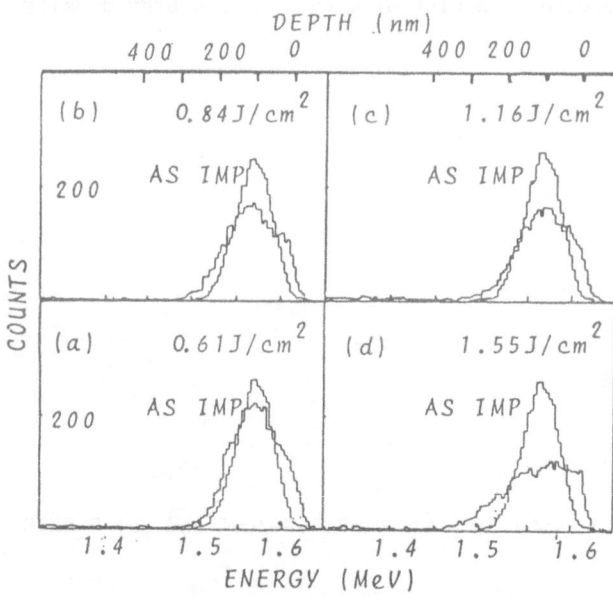

DEPTH (nm)

FIGURE 6. Depth profiles of Ge for as-implanted and laser-irradiated Si
with 2.3E16 Ge/cm².

to allow both a deep diffusion of Ge and flattening of
the profile closer to the
surface. From the extent of
diffusion we estimate that
the threshold for a-c interface penetration is somewhat
lower than 1.55J/cm². As for
the Ge diffusion profiles of
the higher implant dose we
find melting of the amor-
phous layer already at
0.56J/cm² where the Ge profile is broader than in the
as-implanted sample, Fig. 7.
At 0.82J/cm² there is significant diffusion of the Ge
into the substrate indicating that the melt front has
penetrated into the crystalline substrate. Further in
creases in the laser energy
density serve to increase
this diffusion tail and hen-
ce the melt thickness. Note also that at all energy densities shown Ge
reaches the surface.

3.3. TEM Results

Cross sectioned specimens have been observed in Transmission Electron Microscopy with a Philips CM12 operating at 120 keV. A preliminary result
shown in Fig. 8 shows a dark field image of a specimen implanted with a dose of 2.3E16Ge/cm² and laser irradiated at 0.84J/cm². The image was obtained using one of the polycrystalline reflections visible on the diffraction
pattern. The crystalline grains are clearly visible. Their size varies with
depth. Near the surface they have an average size of about 100 nm while in
the region centered at a depth of about 60 nm the dimensions of the grains
are about 10 nm. A band without visible structure extends for 20 nm from
the interface that is at a depth of 250 nm from the surface. At higher
energy densities where complete epitaxial growth was realized, dislocations
were observed to extend from the interface up to the surface. Detailed results will be the subject of a future paper.

4. DISCUSSION

We interpret the data as follows. The differences in otpical reflectivity
are too small to consider different absorbed energies. Preliminary TEM data
suggest that explosive crystallization (5) is taking place for both doses.
In this framework, near surface large grains due to primary melt and under-

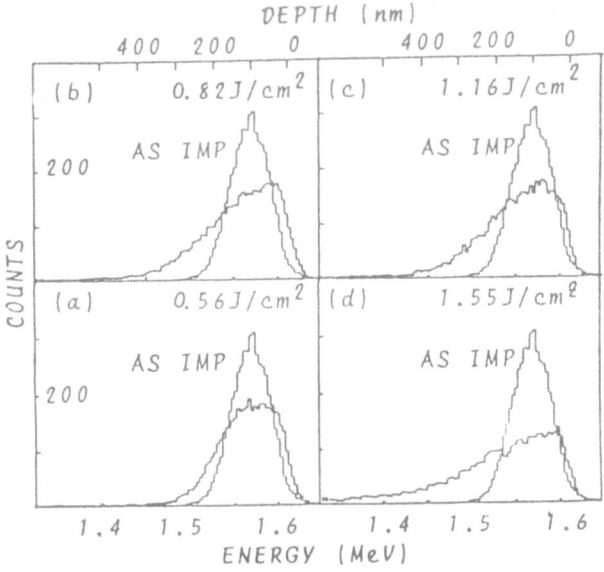

DEPTH (nm)

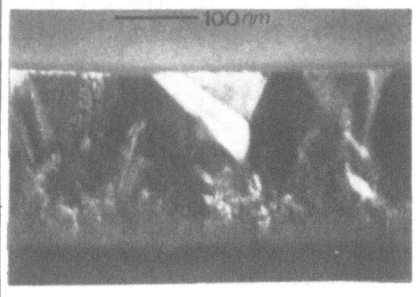

FIGURE 7. Depth profiles of Ge for as-implanted and laser-irradiated Si with
5.5E16 Ge/cm².

FIGURE 8. TEM dark field image of a specimen
implanted with 2.3E
16Ge/cm² and irra-
diated at 0.84J/cm²
Bottom dark band is
crystal substrate.

lying fine grains due to explosive crystallization are expected. The magnitude of dechanneling in the alloy layer of the higher dose is however significantly lower indicating larger grain sizes at equivalent (low) energy
densities. Also, the growth in grain size should be larger for the higher
dose with increasing energy density. This suggests either that the primary
melt depth is larger or the nucleation kinetics is different for the higher
dose. Thicker melt depths would require higher thermal conductivity and
lower melting temperatures whereas changed nucleation kinetics imply lower
nucleation temperatures (1). Rapid evaporation from the surface of the
higher dose sample with optically visible associated surface damage may imply several scenarios: (i) evaporation temperature may be depressed or (ii)
surface temperature of the higher dose reaches the evaporation temperature
(which is the same for both doses) at lower energy densities. The lower
energy threshold for penetration of the a-c interface for the higher dose
also imply larger thermal conductivity for the amorphous layer. Coupled
with lower melting temperatures this would remove the remaining amorphous
layer at the interface at lower energy densities to eliminate the thermal
barrier imposed by it (1).

While the combined effect of several thermal parameters, most of which
probably change even if by small amounts, is difficult to analyze, the idea
of a lower melting temperature is not only suggested by the equilibrium
phase diagram of Si-Ge system where the depression difference for the two
doses is about 30K but also by the recent work of Peercy et al. (6). In

transient conductance measurements during pulsed laser irradiation of heavily As implanted Si they found that increasing concentrations of As reduce the melting temperatures of the amorphous layer by as much as 150 K (at 7 at.%, similar to our higher dose sample). We note that in Si, Ge and As behave very similarly in many ways. Furthermore, experiments on MBE growth of Ge_xSi_{1-x} alloys on Si, where addition of Ge is observed to reduce the minimum epitaxial temperature (7), also lend support in the right direction.

In summary, we have used RBS, channeling and TEM to investigate the dose dependent crystallization of metastable Si-Ge alloys. Evidence suggests significant changes in the thermal parameters of the alloy and especially the lowering of the melting temperature with increasing Ge content. Simulation of the process with numerical solutions of the heat equation should further discriminate between thermal parameters. The use of RBS and channeling correlated with TEM is a powerful way of obtaining information on the distribution and sizes of grains in explosively crystallized systems. More work is needed to clarify the effects of mismatch between the alloy and the substrate.

5. ACKNOWLEDGEMENTS

The authors would like to thank A. Compaan and G. Battaglin for valuable discussions, G. Lotti for implantations, and F. Corticelli for technical assistance. One of us (A.A) would like to thank to International Center for Theoretical Physics (Trieste) and Turkish Scientific and Technical Research Council (TUBITAK) for financial assistance.

REFERENCES

1. Webber H C, Cullis A G, Chew N G: Appl. Phys. Lett. 43 (1983) 669; Lowndes D H, Wood R F, Narayan J: Phys. Rev. Lett. 52 (1984) 561; Baeri P, Campisano S U in "Laser Annealing of Semiconductors", Eds. J. M. Poate and J.W. Mayer, Academic Press, 1982.
2. Sinke W, Saris F W: Proceedings of MRS, Symposium A; Energy Beam-Solid Interactions and Transient Thermal Processing, p. 324, Elsevier, 1984; Sinke W, Saris F W: Phys. Rev. Lett. 53 (1984) 2121.
3. Campisano S U, Jacobson D C, Poate J M, Cullis A G, Chew N G: Appl. Phys. Lett. 46 (1985) 846; Bruines J J P, Van Hal R P M, Boots H M J, Sinke W, Saris F W: Appl. Phys. Lett. 48 (1986) 1252.
4. Gras-Marti A et al.: Nucl. Instr. and Meth. 194 (1982) 449.
5. Narayan J, White C W, Holland O W: Proceedings of MRS, Vol. 23 p. 179, Elsevier, 1984.
6. Peercy P S, Thompson M O,Tsao J Y: Appl. Phys. Lett. 47 (1985) 244.
7. Bean J C, Sheng T T, Feldman L C, Fiory A T: Appl. Phys. Lett. 44 (1984) 102.

Index

596